Communications in Computer and Information Science 1965

Rationale

The CCIS series is devoted to the publication of proceedings of computer science conferences. Its aim is to efficiently disseminate original research results in informatics in printed and electronic form. While the focus is on publication of peer-reviewed full papers presenting mature work, inclusion of reviewed short papers reporting on work in progress is welcome, too. Besides globally relevant meetings with internationally representative program committees guaranteeing a strict peer-reviewing and paper selection process, conferences run by societies or of high regional or national relevance are also considered for publication.

Topics

The topical scope of CCIS spans the entire spectrum of informatics ranging from foundational topics in the theory of computing to information and communications science and technology and a broad variety of interdisciplinary application fields.

Information for Volume Editors and Authors

Publication in CCIS is free of charge. No royalties are paid, however, we offer registered conference participants temporary free access to the online version of the conference proceedings on SpringerLink (http://link.springer.com) by means of an http referrer from the conference website and/or a number of complimentary printed copies, as specified in the official acceptance email of the event.

CCIS proceedings can be published in time for distribution at conferences or as post-proceedings, and delivered in the form of printed books and/or electronically as USBs and/or e-content licenses for accessing proceedings at SpringerLink. Furthermore, CCIS proceedings are included in the CCIS electronic book series hosted in the SpringerLink digital library at http://link.springer.com/bookseries/7899. Conferences publishing in CCIS are allowed to use Online Conference Service (OCS) for managing the whole proceedings lifecycle (from submission and reviewing to preparing for publication) free of charge.

Publication process

The language of publication is exclusively English. Authors publishing in CCIS have to sign the Springer CCIS copyright transfer form, however, they are free to use their material published in CCIS for substantially changed, more elaborate subsequent publications elsewhere. For the preparation of the camera-ready papers/files, authors have to strictly adhere to the Springer CCIS Authors' Instructions and are strongly encouraged to use the CCIS LaTeX style files or templates.

Abstracting/Indexing

CCIS is abstracted/indexed in DBLP, Google Scholar, EI-Compendex, Mathematical Reviews, SCImago, Scopus. CCIS volumes are also submitted for the inclusion in ISI Proceedings.

How to start

To start the evaluation of your proposal for inclusion in the CCIS series, please send an e-mail to ccis@springer.com.

Biao Luo · Long Cheng · Zheng-Guang Wu ·
Hongyi Li · Chaojie Li
Editors

Neural Information Processing

30th International Conference, ICONIP 2023
Changsha, China, November 20–23, 2023
Proceedings, Part XI

Editors
Biao Luo
School of Automation
Central South University
Changsha, China

Long Cheng
Institute of Automation
Chinese Academy of Sciences
Beijing, China

Zheng-Guang Wu
Institute of Cyber-Systems and Control
Zhejiang University
Hangzhou, China

Hongyi Li
School of Automation
Guangdong University of Technology
Guangzhou, China

Chaojie Li
School of Electrical Engineering
and Telecommunications
UNSW Sydney
Sydney, NSW, Australia

ISSN 1865-0929 ISSN 1865-0937 (electronic)
Communications in Computer and Information Science
ISBN 978-981-99-8144-1 ISBN 978-981-99-8145-8 (eBook)
https://doi.org/10.1007/978-981-99-8145-8

This Springer imprint is published by the registered company Springer Nature Singapore Pte Ltd.
The registered company address is: 152 Beach Road, #21-01/04 Gateway East, Singapore 189721, Singapore

Paper in this product is recyclable.

Preface

Welcome to the 30th International Conference on Neural Information Processing (ICONIP2023) of the Asia-Pacific Neural Network Society (APNNS), held in Changsha, China, November 20–23, 2023.

The mission of the Asia-Pacific Neural Network Society is to promote active interactions among researchers, scientists, and industry professionals who are working in neural networks and related fields in the Asia-Pacific region. APNNS has Governing Board Members from 13 countries/regions – Australia, China, Hong Kong, India, Japan, Malaysia, New Zealand, Singapore, South Korea, Qatar, Taiwan, Thailand, and Turkey. The society's flagship annual conference is the International Conference of Neural Information Processing (ICONIP). The ICONIP conference aims to provide a leading international forum for researchers, scientists, and industry professionals who are working in neuroscience, neural networks, deep learning, and related fields to share their new ideas, progress, and achievements.

ICONIP2023 received 1274 papers, of which 394 papers were accepted for publication in Communications in Computer and Information Science (CCIS), representing an acceptance rate of 30.93% and reflecting the increasingly high quality of research in neural networks and related areas. The conference focused on four main areas, i.e., "Theory and Algorithms", "Cognitive Neurosciences", "Human-Centered Computing", and "Applications". All the submissions were rigorously reviewed by the conference Program Committee (PC), comprising 258 PC members, and they ensured that every paper had at least two high-quality single-blind reviews. In fact, 5270 reviews were provided by 2145 reviewers. On average, each paper received 4.14 reviews.

We would like to take this opportunity to thank all the authors for submitting their papers to our conference, and our great appreciation goes to the Program Committee members and the reviewers who devoted their time and effort to our rigorous peer-review process; their insightful reviews and timely feedback ensured the high quality of the papers accepted for publication. We hope you enjoyed the research program at the conference.

October 2023

Biao Luo
Long Cheng
Zheng-Guang Wu
Hongyi Li
Chaojie Li

Organization

Honorary Chair

Weihua Gui	Central South University, China

Advisory Chairs

Jonathan Chan	King Mongkut's University of Technology Thonburi, Thailand
Zeng-Guang Hou	Chinese Academy of Sciences, China
Nikola Kasabov	Auckland University of Technology, New Zealand
Derong Liu	Southern University of Science and Technology, China
Seiichi Ozawa	Kobe University, Japan
Kevin Wong	Murdoch University, Australia

General Chairs

Tingwen Huang	Texas A&M University at Qatar, Qatar
Chunhua Yang	Central South University, China

Program Chairs

Biao Luo	Central South University, China
Long Cheng	Chinese Academy of Sciences, China
Zheng-Guang Wu	Zhejiang University, China
Hongyi Li	Guangdong University of Technology, China
Chaojie Li	University of New South Wales, Australia

Technical Chairs

Xing He	Southwest University, China
Keke Huang	Central South University, China
Huaqing Li	Southwest University, China
Qi Zhou	Guangdong University of Technology, China

Local Arrangement Chairs

Wenfeng Hu	Central South University, China
Bei Sun	Central South University, China

Finance Chairs

Fanbiao Li	Central South University, China
Hayaru Shouno	University of Electro-Communications, Japan
Xiaojun Zhou	Central South University, China

Special Session Chairs

Hongjing Liang	University of Electronic Science and Technology, China
Paul S. Pang	Federation University, Australia
Qiankun Song	Chongqing Jiaotong University, China
Lin Xiao	Hunan Normal University, China

Tutorial Chairs

Min Liu	Hunan University, China
M. Tanveer	Indian Institute of Technology Indore, India
Guanghui Wen	Southeast University, China

Publicity Chairs

Sabri Arik	Istanbul University-Cerrahpaşa, Turkey
Sung-Bae Cho	Yonsei University, South Korea
Maryam Doborjeh	Auckland University of Technology, New Zealand
El-Sayed M. El-Alfy	King Fahd University of Petroleum and Minerals, Saudi Arabia
Ashish Ghosh	Indian Statistical Institute, India
Chuandong Li	Southwest University, China
Weng Kin Lai	Tunku Abdul Rahman University of Management & Technology, Malaysia
Chu Kiong Loo	University of Malaya, Malaysia
Qinmin Yang	Zhejiang University, China
Zhigang Zeng	Huazhong University of Science and Technology, China

Publication Chairs

Zhiwen Chen	Central South University, China
Andrew Chi-Sing Leung	City University of Hong Kong, China
Xin Wang	Southwest University, China
Xiaofeng Yuan	Central South University, China

Secretaries

Yun Feng	Hunan University, China
Bingchuan Wang	Central South University, China

Webmasters

Tianmeng Hu	Central South University, China
Xianzhe Liu	Xiangtan University, China

Program Committee

Rohit Agarwal	UiT The Arctic University of Norway, Norway
Hasin Ahmed	Gauhati University, India
Harith Al-Sahaf	Victoria University of Wellington, New Zealand
Brad Alexander	University of Adelaide, Australia
Mashaan Alshammari	Independent Researcher, Saudi Arabia
Sabri Arik	Istanbul University, Turkey
Ravneet Singh Arora	Block Inc., USA
Zeyar Aung	Khalifa University of Science and Technology, UAE
Monowar Bhuyan	Umeå University, Sweden
Jingguo Bi	Beijing University of Posts and Telecommunications, China
Xu Bin	Northwestern Polytechnical University, China
Marcin Blachnik	Silesian University of Technology, Poland
Paul Black	Federation University, Australia
Anoop C. S.	Govt. Engineering College, India
Ning Cai	Beijing University of Posts and Telecommunications, China
Siripinyo Chantamunee	Walailak University, Thailand
Hangjun Che	City University of Hong Kong, China

Wei-Wei Che	Qingdao University, China
Huabin Chen	Nanchang University, China
Jinpeng Chen	Beijing University of Posts & Telecommunications, China
Ke-Jia Chen	Nanjing University of Posts and Telecommunications, China
Lv Chen	Shandong Normal University, China
Qiuyuan Chen	Tencent Technology, China
Wei-Neng Chen	South China University of Technology, China
Yufei Chen	Tongji University, China
Long Cheng	Institute of Automation, China
Yongli Cheng	Fuzhou University, China
Sung-Bae Cho	Yonsei University, South Korea
Ruikai Cui	Australian National University, Australia
Jianhua Dai	Hunan Normal University, China
Tao Dai	Tsinghua University, China
Yuxin Ding	Harbin Institute of Technology, China
Bo Dong	Xi'an Jiaotong University, China
Shanling Dong	Zhejiang University, China
Sidong Feng	Monash University, Australia
Yuming Feng	Chongqing Three Gorges University, China
Yun Feng	Hunan University, China
Junjie Fu	Southeast University, China
Yanggeng Fu	Fuzhou University, China
Ninnart Fuengfusin	Kyushu Institute of Technology, Japan
Thippa Reddy Gadekallu	VIT University, India
Ruobin Gao	Nanyang Technological University, Singapore
Tom Gedeon	Curtin University, Australia
Kam Meng Goh	Tunku Abdul Rahman University of Management and Technology, Malaysia
Zbigniew Gomolka	University of Rzeszow, Poland
Shengrong Gong	Changshu Institute of Technology, China
Xiaodong Gu	Fudan University, China
Zhihao Gu	Shanghai Jiao Tong University, China
Changlu Guo	Budapest University of Technology and Economics, Hungary
Weixin Han	Northwestern Polytechnical University, China
Xing He	Southwest University, China
Akira Hirose	University of Tokyo, Japan
Yin Hongwei	Huzhou Normal University, China
Md Zakir Hossain	Curtin University, Australia
Zengguang Hou	Chinese Academy of Sciences, China

Lu Hu	Jiangsu University, China
Zeke Zexi Hu	University of Sydney, Australia
He Huang	Soochow University, China
Junjian Huang	Chongqing University of Education, China
Kaizhu Huang	Duke Kunshan University, China
David Iclanzan	Sapientia University, Romania
Radu Tudor Ionescu	University of Bucharest, Romania
Asim Iqbal	Cornell University, USA
Syed Islam	Edith Cowan University, Australia
Kazunori Iwata	Hiroshima City University, Japan
Junkai Ji	Shenzhen University, China
Yi Ji	Soochow University, China
Canghong Jin	Zhejiang University, China
Xiaoyang Kang	Fudan University, China
Mutsumi Kimura	Ryukoku University, Japan
Masahiro Kohjima	NTT, Japan
Damian Kordos	Rzeszow University of Technology, Poland
Marek Kraft	Poznań University of Technology, Poland
Lov Kumar	NIT Kurukshetra, India
Weng Kin Lai	Tunku Abdul Rahman University of Management & Technology, Malaysia
Xinyi Le	Shanghai Jiao Tong University, China
Bin Li	University of Science and Technology of China, China
Hongfei Li	Xinjiang University, China
Houcheng Li	Chinese Academy of Sciences, China
Huaqing Li	Southwest University, China
Jianfeng Li	Southwest University, China
Jun Li	Nanjing Normal University, China
Kan Li	Beijing Institute of Technology, China
Peifeng Li	Soochow University, China
Wenye Li	Chinese University of Hong Kong, China
Xiangyu Li	Beijing Jiaotong University, China
Yantao Li	Chongqing University, China
Yaoman Li	Chinese University of Hong Kong, China
Yinlin Li	Chinese Academy of Sciences, China
Yuan Li	Academy of Military Science, China
Yun Li	Nanjing University of Posts and Telecommunications, China
Zhidong Li	University of Technology Sydney, Australia
Zhixin Li	Guangxi Normal University, China
Zhongyi Li	Beihang University, China

Toshiaki Omori	Kobe University, Japan
Babatunde Onasanya	University of Ibadan, Nigeria
Manisha Padala	Indian Institute of Science, India
Sarbani Palit	Indian Statistical Institute, India
Paul Pang	Federation University, Australia
Rasmita Panigrahi	Giet University, India
Kitsuchart Pasupa	King Mongkut's Institute of Technology Ladkrabang, Thailand
Dipanjyoti Paul	Ohio State University, USA
Hu Peng	Jiujiang University, China
Kebin Peng	University of Texas at San Antonio, USA
Dawid Połap	Silesian University of Technology, Poland
Zhong Qian	Soochow University, China
Sitian Qin	Harbin Institute of Technology at Weihai, China
Toshimichi Saito	Hosei University, Japan
Fumiaki Saitoh	Chiba Institute of Technology, Japan
Naoyuki Sato	Future University Hakodate, Japan
Chandni Saxena	Chinese University of Hong Kong, China
Jiaxing Shang	Chongqing University, China
Lin Shang	Nanjing University, China
Jie Shao	University of Science and Technology of China, China
Yin Sheng	Huazhong University of Science and Technology, China
Liu Sheng-Lan	Dalian University of Technology, China
Hayaru Shouno	University of Electro-Communications, Japan
Gautam Srivastava	Brandon University, Canada
Jianbo Su	Shanghai Jiao Tong University, China
Jianhua Su	Institute of Automation, China
Xiangdong Su	Inner Mongolia University, China
Daiki Suehiro	Kyushu University, Japan
Basem Suleiman	University of New South Wales, Australia
Ning Sun	Shandong Normal University, China
Shiliang Sun	East China Normal University, China
Chunyu Tan	Anhui University, China
Gouhei Tanaka	University of Tokyo, Japan
Maolin Tang	Queensland University of Technology, Australia
Shu Tian	University of Science and Technology Beijing, China
Shikui Tu	Shanghai Jiao Tong University, China
Nancy Victor	Vellore Institute of Technology, India
Petra Vidnerová	Institute of Computer Science, Czech Republic

Shanchuan Wan	University of Tokyo, Japan
Tao Wan	Beihang University, China
Ying Wan	Southeast University, China
Bangjun Wang	Soochow University, China
Hao Wang	Shanghai University, China
Huamin Wang	Southwest University, China
Hui Wang	Nanchang Institute of Technology, China
Huiwei Wang	Southwest University, China
Jianzong Wang	Ping An Technology, China
Lei Wang	National University of Defense Technology, China
Lin Wang	University of Jinan, China
Shi Lin Wang	Shanghai Jiao Tong University, China
Wei Wang	Shenzhen MSU-BIT University, China
Weiqun Wang	Chinese Academy of Sciences, China
Xiaoyu Wang	Tokyo Institute of Technology, Japan
Xin Wang	Southwest University, China
Xin Wang	Southwest University, China
Yan Wang	Chinese Academy of Sciences, China
Yan Wang	Sichuan University, China
Yonghua Wang	Guangdong University of Technology, China
Yongyu Wang	JD Logistics, China
Zhenhua Wang	Northwest A&F University, China
Zi-Peng Wang	Beijing University of Technology, China
Hongxi Wei	Inner Mongolia University, China
Guanghui Wen	Southeast University, China
Guoguang Wen	Beijing Jiaotong University, China
Ka-Chun Wong	City University of Hong Kong, China
Anna Wróblewska	Warsaw University of Technology, Poland
Fengge Wu	Institute of Software, Chinese Academy of Sciences, China
Ji Wu	Tsinghua University, China
Wei Wu	Inner Mongolia University, China
Yue Wu	Shanghai Jiao Tong University, China
Likun Xia	Capital Normal University, China
Lin Xiao	Hunan Normal University, China
Qiang Xiao	Huazhong University of Science and Technology, China
Hao Xiong	Macquarie University, Australia
Dongpo Xu	Northeast Normal University, China
Hua Xu	Tsinghua University, China
Jianhua Xu	Nanjing Normal University, China

Xinyue Xu	Hong Kong University of Science and Technology, China
Yong Xu	Beijing Institute of Technology, China
Ngo Xuan Bach	Posts and Telecommunications Institute of Technology, Vietnam
Hao Xue	University of New South Wales, Australia
Yang Xujun	Chongqing Jiaotong University, China
Haitian Yang	Chinese Academy of Sciences, China
Jie Yang	Shanghai Jiao Tong University, China
Minghao Yang	Chinese Academy of Sciences, China
Peipei Yang	Chinese Academy of Science, China
Zhiyuan Yang	City University of Hong Kong, China
Wangshu Yao	Soochow University, China
Ming Yin	Guangdong University of Technology, China
Qiang Yu	Tianjin University, China
Wenxin Yu	Southwest University of Science and Technology, China
Yun-Hao Yuan	Yangzhou University, China
Xiaodong Yue	Shanghai University, China
Paweł Zawistowski	Warsaw University of Technology, Poland
Hui Zeng	Southwest University of Science and Technology, China
Wang Zengyunwang	Hunan First Normal University, China
Daren Zha	Institute of Information Engineering, China
Zhi-Hui Zhan	South China University of Technology, China
Baojie Zhang	Chongqing Three Gorges University, China
Canlong Zhang	Guangxi Normal University, China
Guixuan Zhang	Chinese Academy of Science, China
Jianming Zhang	Changsha University of Science and Technology, China
Li Zhang	Soochow University, China
Wei Zhang	Southwest University, China
Wenbing Zhang	Yangzhou University, China
Xiang Zhang	National University of Defense Technology, China
Xiaofang Zhang	Soochow University, China
Xiaowang Zhang	Tianjin University, China
Xinglong Zhang	National University of Defense Technology, China
Dongdong Zhao	Wuhan University of Technology, China
Xiang Zhao	National University of Defense Technology, China
Xu Zhao	Shanghai Jiao Tong University, China

Liping Zheng	Hefei University of Technology, China
Yan Zheng	Kyushu University, Japan
Baojiang Zhong	Soochow University, China
Guoqiang Zhong	Ocean University of China, China
Jialing Zhou	Nanjing University of Science and Technology, China
Wenan Zhou	PCN&CAD Center, China
Xiao-Hu Zhou	Institute of Automation, China
Xinyu Zhou	Jiangxi Normal University, China
Quanxin Zhu	Nanjing Normal University, China
Yuanheng Zhu	Chinese Academy of Sciences, China
Xiaotian Zhuang	JD Logistics, China
Dongsheng Zou	Chongqing University, China

Contents – Part XI

Applications

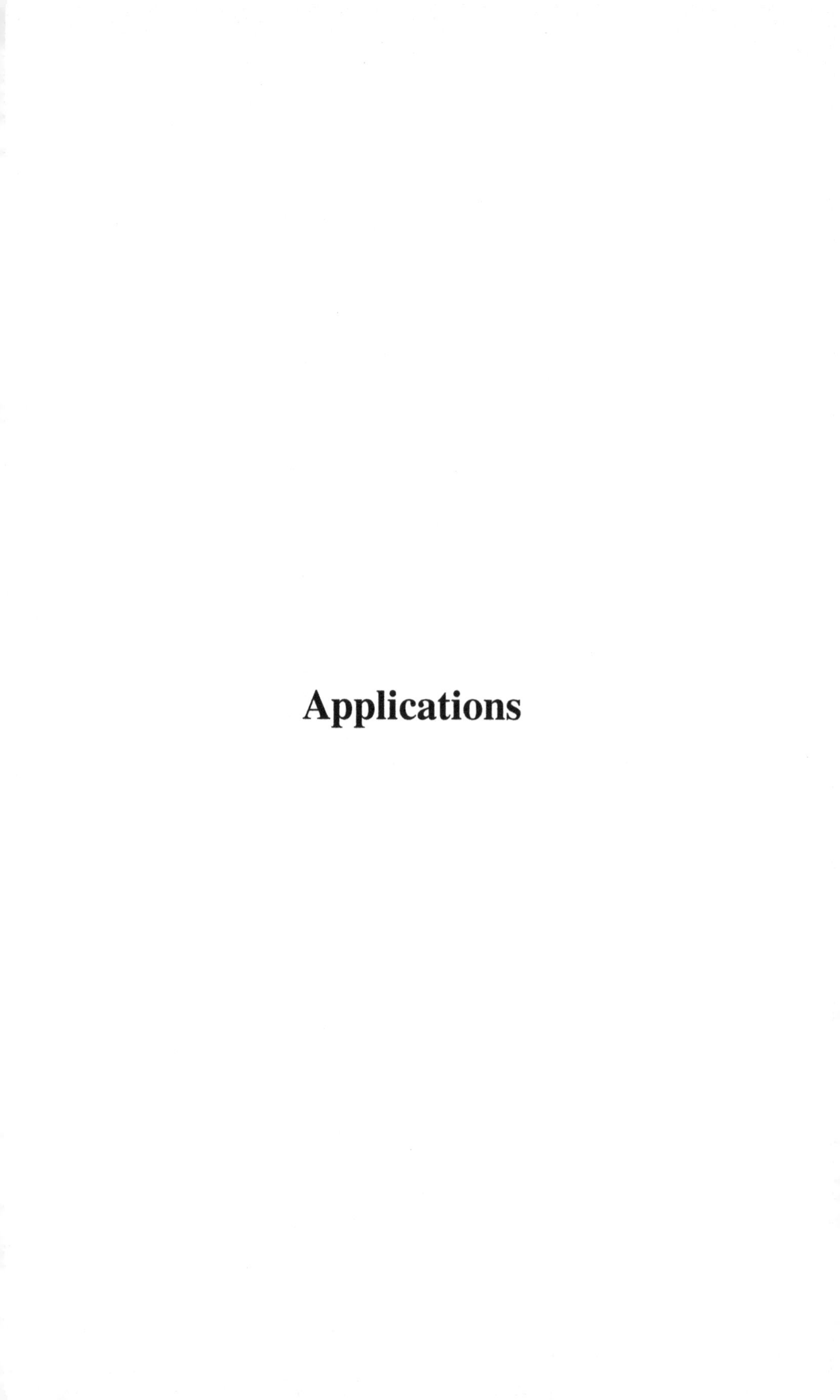

Applications

Multi-intent Description of Keyword Expansion for Code Search

Haize Hu[1](✉), Jianxun Liu[1], and Lin Xiao[2]

[1] Hunan University of Science and Technology, Xiangtan 411201, Hunan, China
hhz@mail.hnust.edu.cn
[2] Hunan Normal University, Changsha 410081, Hunan, China

Abstract. To address the issue of discrepancies between online query data and offline training data in code search research, we propose a novel code search model called multi intent description keyword extension-based code search (MDKE-CS). Our model utilizes offline training data to expand query data, thereby mitigating the impact of insufficient query data and intention differences between training and query data on search results. Furthermore, we construct a multi-intention description keyword vocabulary library based on developers, searchers, and discussants from the StackOverflow Q&A library to further expand the query. To evaluate the effectiveness of MDKE-CS in code search tasks, we conducted comparative experimental analyses using two baseline models, DeepCS and UNIF, as well as WordNet and BM25 extension methods. Our experimental results demonstrate that MDKE-CS outperforms the baseline models in terms of R@1, R@5, R@10, and MRR values.

Keywords: Code search · Multi-intention · Expand query

1 Introduction

With the increasing number of developers who upload and share their code fragments on open source communities, the code resources available in these communities have become increasingly abundant. [1] This continuous enrichment of open source community resources has provided a vital foundation for the development of code search [2]. By searching for existing code fragments in open source communities, software developers can modify and reuse them, thereby improving the utilization of existing code, saving development time, and enhancing software development efficiency [3]. Consequently, the rapid and accurate search for existing code fragments (i.e., code search) has become a crucial area of research in software engineering.

Currently, deep learning-based code search research is mainly divided into offline training and online search [4]. In the offline training phase, a deep learning network model is employed to learn features from a large dataset, and the network model parameters are acquired through learning [5]. During feature learning, the data primarily consists of Code Description pairs, where Code represents the source code fragment and Description corresponds to the statement that describes the function of the source code fragment [6].

B. Luo et al. (Eds.): ICONIP 2023, CCIS 1965, pp. 3–14, 2024.
https://doi.org/10.1007/978-981-99-8145-8_1

The deep learning network model is primarily utilized to learn the syntax and semantic relationships between code language (Code) and natural language (Description), ultimately determining the network model parameters that can map Code and Description. In the online search stage, developers input query content (describing code fragments or representing code functions), and the query content is matched with the code fragments in the dataset to obtain the highest matching results [7]. While Code Description pairs are employed for feature learning in offline training, Query instead of Description is used to match Code in online search. However, there are two differences between Query and Description. Firstly, there is a difference in length, with Query usually being shorter than Description [8]. According to statistics, the average query length entered by search personnel is 2–3, while the average description length for code is 20–30 [9]. Secondly, there are semantic differences between Query and Description. Description refers to the way code developers describe code fragments from their perspective, while Query represents the description of code fragments based on the needs of search personnel, from their own perspective. These differences lead to significant differences in the descriptions of the same code fragment by developers and searchers [10]. Nevertheless, existing research often treats Query and Description equally, ignoring their differences, which can have a significant impact on search results [11].

To address the differences between Query and Description in existing code search research, researchers have proposed query extension studies. Query extension research aims to design extension methods and sources to expand Query, reduce the differences between Query and Description, and improve the accuracy of code search results [12]. Existing research on query extension primarily focuses on two aspects: extension methods and extension sources [13]. Extension methods are primarily divided into keyword extension and source code API extension methods [14]. Keyword extension involves using words in a query statement as keywords, matching them with words in a vocabulary, and extending the query statement with words that have high similarity [15]. The source code API extension method involves using APIs in code fragments as extension sources, matching query statement keywords with API keywords, and extending query statements with API keywords that have high similarity [16]. Existing research on extension sources mainly relies on Q&A pairs (Stack Overflow Q&A) and WordNet as extension sources in question answering libraries [17]. However, existing extension methods and sources for query extension research are still unable to effectively reduce the differences between Query and Description, and cannot effectively improve the query extension effect, resulting in lower code search results. Overall, there are still three main issues with existing research on query extension. (1) The current datasets available for research in code search primarily focus on code search itself, and there is a lack of specialized datasets for query extension research. (2) Disregarding the differences between Query and Description makes it challenging to map Query and Description effectively. (3) The lack of consideration for query intent results in an inability to accurately express search intent.

In order to enhance the effectiveness of query expansion, we propose a multi-intent description of keyword expansion model for code search, abbreviated as MDKE-CS. Firstly, we construct a Query Code Description keyword dataset from the perspectives of code developers, searchers, and reviewers based on the Q&A of Stack Overflow.

Secondly, we utilize a deep learning model to train Query Code data pairs to obtain the best matching Code for Query. Then, we use multiple types of Keywords corresponding to the matched Code and residual information Description to compensate for keyword features and extend the Query, obtaining the first extended Query$_{-1}$. We repeat the best extension times for the extended model to obtain the final Query$_{-n}$. Finally, we match Query$_{-n}$ with the Code in the database one by one to obtain the best code fragment. Our proposed model, MDKE-CS, provides an approach to improve the accuracy of code search results by accounting for the differences between Query and Description while considering multiple intent descriptions for keyword expansion. By leveraging the deep learning model and multi-intent keyword expansion strategy, MDKE-CS enables effective query expansion and improves the accuracy of code search results.

In this research, we make the following contributions:

a. We create a Query-Code-Description-Keyword dataset for query extension research and demonstrate its effectiveness through experimental analysis based on existing code search models.
b. We propose a query extension code search model, MDKE-CS, which utilizes multiple intention description keywords to improve the accuracy of code search results.
c. We conduct a comparative experimental analysis on the code search performance of MDKE-CS based on the Query-Code-Description-Keyword dataset and verify the effectiveness of the proposed model in improving the accuracy of code search results. Our research provides a novel approach to enhancing the accuracy of code search by utilizing a multi-intent keyword expansion strategy and deep learning techniques in a query extension model.

The remaining sections of the paper are organized as follows: Sect. 2 introduces our proposed MDKE-CS model for query extension in code search. Section 3 presents a detailed analysis of the experimental results, including experimental preparation, dataset construction, and analysis of the experimental results. Section 4 provides a summary of our work and outlines the contributions of our research. By organizing the paper in this manner, we aim to provide a clear and structured presentation of our proposed model and experimental results, and to provide readers with a comprehensive understanding of our research.

2 Model Method

In order to make up for the shortcomings of existing query extensions, we propose a query extension code search model MDKE-CS based on multiple intention description keywords. The overall framework of the MDKE-CS model is shown in Fig. 1.

The MDKE-CS model consists of two main parts: offline training and online search. In the offline training section, the primary objective is to learn the Query-Code mapping relationship and obtain the parameters of the feature extraction model. Conventional code search training methods are utilized, and the Query-Code dataset is used for model training. A deep learning model is utilized to extract and learn the feature information of Query and Code, and similarity calculation is employed for matching analysis. The loss function is utilized as the basis for adjusting model parameters. By adjusting the

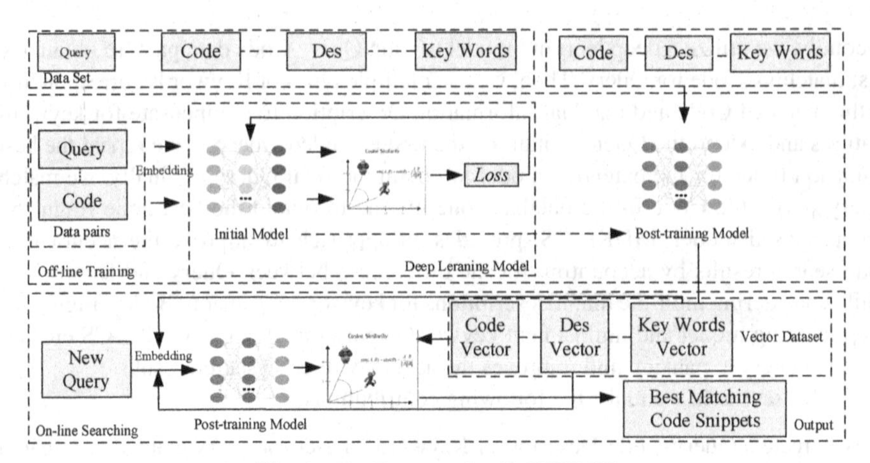

Fig. 1. The Framework of MDKE-CS

parameters of the model through a large amount of Query-Code data, the final model parameters are obtained. Finally, the Code-Description Keyword data pairs in the dataset are embedded using the model to obtain the Code-Description Keyword vector dataset pairs. The online search part primarily involves expanding the query through the expansion method to obtain the expanded query and taking the expanded query statement as the search to obtain the best search results. In the online search, the new query statement is first vectorized using the trained model, and then matched with Code to obtain the best Description vector and Keyword vector. The Description vector and Keyword vector are then extended to the query statement. The search and expansion process is repeated until the best results are achieved.

2.1 Training Model Selection

Although our research focuses on code search, our focus is on query extension methods, without studying the heterogeneous representation models between code language and natural language. Therefore, during the research process, we adopted the Bidirectional Long Short Memory Network (BiLSTM) (as shown in Fig. 2) as a deep learning model for heterogeneous feature extraction. The reason for choosing the Bidirectional Long Short Memory Network (BiLSTM) is because it is based on the manuscript "Deep Code Search" studied by Gu et al. and proposes a DeepCS code search model. The proposal of the DeepCS model represents the beginning of the introduction of deep learning into code search research, aimed at bridging the semantic gap between code language and natural language. Moreover, the DeepCS model has been recognized by a large number of researchers, and research on code search based on "Deep Code Search" and deep learning has developed rapidly.

Figure 2 depicts the structure of the BiLSTM model, which consists of three layers: the Data layer, the LSTM extraction layer, and the feature information hiding layer (h). The Data layer contains n units of data, denoted as $Data_n$. The LSTM extraction layer is composed of a forward LSTM and a reverse LSTM. The forward $LSTM_{1\text{-}n}$ is determined not only by the current input data $Data_n$, but also by the preceding $LSTM_{1\text{-}(n\text{-}1)}$ output. In

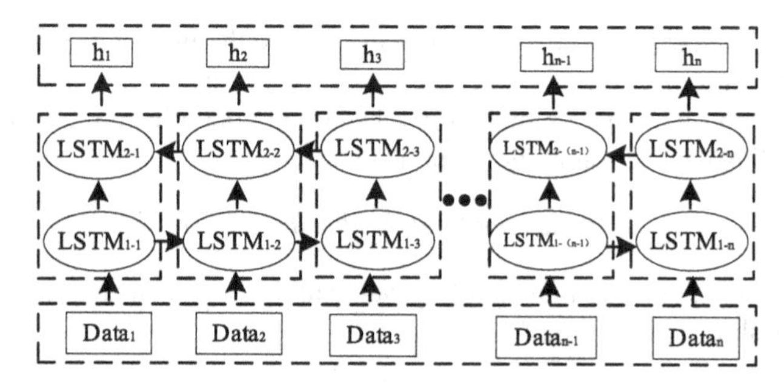

Fig. 2. Structure of the BiLSTM model

contrast, the reverse $LSTM_{2\text{-}(n\text{-}1)}$ output is influenced not only by the preceding output $LSTM_{2\text{-}n}$, but also by the output of the forward $LSTM_{1\text{-}(n\text{-}1)}$. The hidden layer hn is jointly obtained by $LSTM_{1\text{-}n}$ and $LSTM_{2\text{-}n}$ and serves as the output layer, representing the feature information of the input data $Data_n$.

2.2 Joint Embedding

As source code belongs to programming languages and query statements belong to natural languages, there exists a significant semantic gap between the two. Joint embedding is a technique that utilizes a deep learning model to embed source code and query statements into the same vector space. By embedding the two types of data in the same vector space, cosine similarity can be used to calculate the similarity between them, thereby reducing the semantic gap between the two. This technique is commonly employed in code search models to improve the accuracy of search results by accounting for both the programming language and natural language aspects of the query.

As previously mentioned, we constructed a Query-Code-Description-Keyword dataset suitable for code search query extension research. During offline training, our training objectives differ from those of "Deep Code Search." In our training, our goal is to train the mapping relationship between Query and Code in the quad metadata set. Therefore, we replaced the Description in "Deep Code Search" with Query. The Description and Keyword in the quad metadata set serve as query extension data. In our study, sequence preprocessing was performed on the source code to obtain $M = \{m_1, m_2, m_3 \ldots m_M\}$, $A = \{a_1, a_2, a_3 \ldots a_A\}$, $T = \{t_1, t_2, t_3 \ldots t_T\}$, $Q = \{t_1, t_2, t_3 \ldots t_Q\}$, respectively. The Methodname sequence contains M words, the API sequence contains A words, the Token sequence contains T words, and the Query sequence contains Q words. To better illustrate joint embedding, we will use $A = \{a_1, a_2, a_3 \ldots a_A\}$ as an example. Using BiLSTM for feature extraction and learning of the A sequence, the API sequence corresponds to the Data layer in Fig. 3. If the API sequence contains a total of A words, it corresponds to n Data values in Fig. 3. The LSTM layer performs feature extraction on the API sequence to obtain the output layer feature vector values (as shown in formula 1).

$$\{h_1, h_2, h_3 \ldots h_A\} = \text{BiLSTM}(\{a_1, a_2, a_3 \ldots a_A\}) \tag{1}$$

To effectively characterize data features and reduce the impact of noise, we utilize a maximum pooling network to select the extracted features. The maximum pooling network is used to select the optimal hidden layer and obtain feature information. The maximum pooling calculation is shown in formula 2.

$$\mathrm{a} = \max pooling([h_1, h_2, h_3, ...h_A]) \tag{2}$$

After performing the maximum pooling calculation, the final feature information h_{at} is obtained (as shown in formula 3).

$$h_{at} = \tanh(W^M[h_{t-1}; a_{tA}]) \tag{3}$$

where, W^M is the parameter matrix of the API sequence in the BiLSTM network, and a_{tA} is the vector corresponding to the words in the sequence. After feature extraction in the LSTM layer, the output sequence of the hidden layer of the model is $h = \{h_1, h_2, h_3, \ldots, h_A\}$.

Similarly, we utilize BiLSTM to extract features from the other three sequences, obtaining the feature hiding layer vectors of the Methodname sequence vector (*m*), the Tokens sequence vector (*t*), and the Query vector (*q*), respectively. Assuming the number of words in the Methodname sequence is M, the number of words in the Token sequence is T, and the number of words in the Query sequence is Q (as shown in formula 4).

$$\begin{aligned} \mathrm{m} &= \max pooling([h_1, h_2, h_3, ...h_m]) \\ t &= \max pooling([h_1, h_2, h_3, ...h_T]) \\ q &= \max pooling([h_1, h_2, h_3, ...h_Q]) \end{aligned} \tag{4}$$

During the model training process, our goal is to train the mapping relationship between Code and Query. To achieve this, we use the concatenation network "concat" to concatenate vectors *a, m*, and *t* to obtain the source code vector *c*).

2.3 Extended Research

To address the limitations of existing query expansion methods, we propose a multi-intent description keyword expansion model. Our approach integrates the intention description keywords of developers, searchers, and reviewers to improve the accuracy of extension and reduce the semantic gap between queries and descriptions. The multi-intent description keyword expansion method is illustrated in Fig. 3.

Figure 3 illustrates the three-step process of the multi-intent description keyword expansion method. First, a similarity matching is performed with the code database to obtain n sorted best matching codes. Second, from the n best matching codes, the first k keywords corresponding to the code are selected as the extension words for the Query. To compensate for the loss of contextual semantics during the extraction of multi-intent description keywords, we use the Description as residual to supplement the lost semantic information. Finally, the multi-intent description keyword expansion is repeated on the Query until the decision maker meets the set requirements, and outputs the final code sorting result.

The multi-intent description keyword expansion method offers three advantages for code search tasks. First, we utilize the Stack Overflow Q&A dataset, which contains real

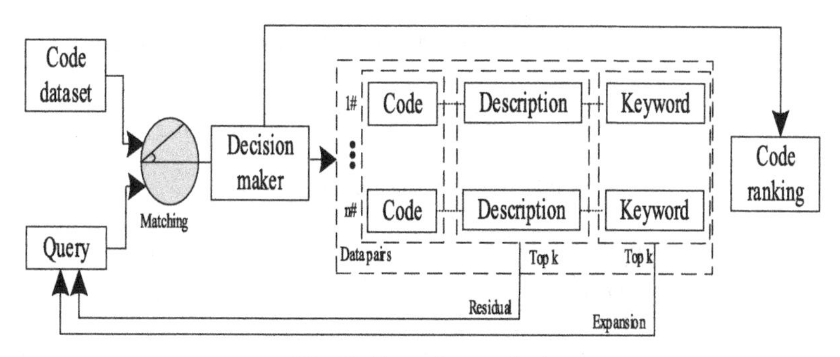

Fig. 3. Extension method

Queries and Descriptions. Queries are proposed by searchers and express their requirements, while Descriptions are provided by code developers and third-party researchers, expressing their descriptions of the requirements. Hence, our data selection is more aligned with real-world code search tasks. Second, we use extended kezywords derived from Descriptions, and use Descriptions as residuals to compensate for any lost information. As the extended information comes from the code, it can match the Query more accurately during the search. Third, we adopt a multi-intent fusion extension method that better considers the intentions of searchers, developers, and other researchers, thereby improving the accuracy of code search.

3 Experimental Analysis

To evaluate the effectiveness of the proposed model in code search tasks, we conducted a comparative experimental analysis on a Linux server. The experimental analysis includes three parts: description and query difference analysis, extension module effect analysis, and extension comparison analysis. Due to the large size of our model parameters and experimental dataset, a separate CPU cannot calculate our model parameters separately. Therefore, we used a Linux server with two Nvidia GTX 2080Ti GPUs, each with 11GB of memory. For the experiments, we implemented the proposed model using Python (version 3.6+) and the PyTorch (version 0.4) experimental simulation platform.

3.1 Experimental Preparation

To address the lack of suitable datasets in existing query extension research, we constructed a four-metadata set [Query-Code-Description-Keyword] based on the Q&A library of the Stack Overflow platform. We defined this dataset as CSExpansion. Query is a question raised by researchers on the Stack Overflow platform regarding the required code. The Description includes not only the explanation of the code developer, but also the explanation of other researchers who participated in the discussion of the code. Keywords are extracted from the Description, using the top 10 words of TFIDF in each Description. To account for the varying length of each Description, we set less than 10 keywords and fill them with 0. If there are more than 10 keywords, we sort the top 10 keywords and delete the remaining ones. Due to the limited data resources and model characteristics of the Stack Overflow platform, our CSExpansion dataset currently only includes Python and Java languages. Table 1 shows the composition of the CSExpansion dataset.

Table 1. CSExpansion Dataset

Data	Python	Java	Total
Number	37234	31297	68531

During the experimental analysis, we used four basic models: DeepCS, UNIF, WordNet, and BM25. DeepCS and UNIF models were used as the basic code search models for feature extraction and data training. WordNet and BM25 models were used as query extension models for comparative analysis. The four basic models used in our experimental analysis are DeepCS [10], UNIF [18], WordNet [19] extension model, and BM25 [20] extended model.

Our research on the proposed model is based on the basic code search models, DeepCS and UNIF. To analyze the effectiveness of the proposed model in code search tasks, we evaluated the search performance of the model using $R@k$ ($k = 1$, $k = 5$, and $k = 10$) and MRR metrics. These metrics are commonly used in information retrieval to evaluate the accuracy of ranking algorithms. $R@k$ measures the percentage of correct results in the top k returned results, while MRR measures the average rank of the first correct result. By using these metrics, we can quantitatively evaluate the effectiveness of the proposed model in improving the accuracy of code search.

3.2 Difference Analysis

The study of query extension focuses on reducing the differences between Query and Description in the code search process, thereby improving the accuracy of code search. To analyze these differences, we used the CSExpansion dataset and conducted a comparative experimental analysis using three metadata pairs (Query, Code, Description). We compared and analyzed the Query-Code and Code-Description data pairs in the experimental study. To evaluate the effectiveness of code search, we used the DeepCS and UNIF basic code search models for comparative analysis. We divided the CSExpansion dataset into training, validation, and testing sets in a 6:3:1 ratio, and conducted the experimental analysis. The comparative experimental results are shown in Table 2.

The experimental results in Table 2 show that using Description as the search object has better search performance than Query for the DeepCS and UNIF models on the CSExpansion dataset. For the Python language, using Description instead of Query in the DeepCS model resulted in an increase of 61.16%, 32.31%, 12.23%, and 12.12% for R@1, R@5, R@10, and MRR, respectively, while in the UNIF model, the increase was 15.07%, 8.21%, 12.67%, and 6.08%, respectively. For the Java language, using Description instead of Query in the DeepCS model resulted in an increase of 81.65%, 15.27%, 17.53%, and 71.89% for R@1, R@5, R@10, and MRR, respectively, while in the UNIF model, the increase was 23.44%, 17.05%, 36.52%, and 60.70%, respectively.

Table 2. Difference Analysis Results

Model	Input	Python				Java			
		R@1	R@5	R@10	MRR	R@1	R@5	R@10	MRR
DeepCS	Description	0.195	0.258	0.367	0.222	0.198	0.234	0.342	0.318
	Query	0.121	0.195	0.327	0.198	0.109	0.203	0.291	0.185
UNIF	Description	0.084	0.224	0.329	0.175	0.237	0.453	0.572	0.413
	Query	0.073	0.207	0.292	0.148	0.192	0.387	0.419	0.257

From the experimental results, we can conclude that there are significant differences between Description and Query in code search tasks, and using Description instead of Query yields better code search results. Moreover, the Java language showed a more significant improvement when using Description, possibly due to its widespread use in engineering and the involvement of more researchers in discussions, leading to more accurate Descriptions.

3.3 Comparative Experimental Analysis

We used the CSExpansion dataset and the widely used WordNet and BM25 query extension models to verify the effectiveness of our proposed MDKE-CS in code search tasks. While there are various existing query extension methods, fair comparative analysis is currently not possible due to the lack of publicly available source code from the authors. The experimental results are shown in Table 3.

Table 3. Comparative Analysis Results

Model	Input	Python				Java			
		R@1	R@5	R@10	MRR	R@1	R@5	R@10	MRR
DeepCS	WordNet	0.081	0.176	0.280	0.127	0.117	0.209	0.384	0.234
	BM25	0.088	0.181	0.288	0.137	0.180	0.339	0.405	0.254
	MDKE-CS	0.351	0.501	0.687	0.301	0.357	0.511	0.698	0.477
UNIF	WordNet	0.120	0.198	0.245	0.148	0.210	0.401	0.511	0.319
	BM25	0.147	0.178	0.225	0.158	0.221	0.429	0.577	0.380
	MDKE-CS	0.178	0.241	0.539	0.297	0.405	0.725	0.843	0.549

The experimental results in Table 3 show that the proposed MDKE-CS has better search performance for the DeepCS and UNIF models compared to the comparative models WordNet and BM25 on the CSExpansion dataset. For the Python language, using MDKE-CS instead of WordNet and BM25 in the DeepCS model resulted in an increase of 333.33% and 298.86%, 184.66% and 176.80%, 145.36% and 138.54%,

137.00% and 119.71% for R@1, R@5, R@10, and MRR, respectively, while in the UNIF model, the increase was 48.33% and 21.09%, 21.72% and 35.39%, 120.00% and 139.56%, 100.68% and 87.97%, respectively. For the Java language, using MDKE-CS instead of WordNet and BM25 in the DeepCS model resulted in an increase of 205.13% and 98.33%, 104.14% and 50.74%, 81.77% and 72.35%, 103.85% and 87.80% for R@1, R@5, R@10, and MRR, respectively, while in the UNIF model, the increase was 92.86% and 83.26%, 80.80% and 69.00%, 64.97% and 46.10%, 72.10% and 44.47%, respectively. The experimental results indicate that MDKE-CS has a more significant effect in the Python language, possibly due to better training and the suitability of the extended model for the language, improving the accuracy of description. Moreover, the proposed MDKE-CS outperforms the comparative models, WordNet and BM25, in code search tasks.

3.4 Analysis of Ablation Experiments

To analyze the structural rationality of the MDKE-CS model, we conducted a separate analysis of the roles of each module in the model. Specifically, we focused on the Keyword extension and Description residual effects in the MDKE-CS model. We compared four models: using Query search alone, using Query+Keyword search, using Query+Description, and using MDKE-CS (Query+Description+Keyword), based on the CSExpansion dataset. The experimental results are shown in Table 4.

The experimental results in Table 4 lead to two conclusions. Firstly, using Description or Keyword for extension results in better search performance than using Query alone, indicating that extending the Query can improve the accuracy of code search. Secondly, using Keyword for extension has a better effect than using Description for extension. Additionally, using both Description and Keyword for extension (MDKE-CS) yields the best results, indicating that the proposed MDKE-CS model can effectively improve the accuracy of code search.

Table 4. Model Structure Analysis

Model	Input	Python				Java			
		R@1	R@5	R@10	MRR	R@1	R@5	R@10	MRR
DeepCS	Query	0.121	0.195	0.327	0.198	0.109	0.203	0.291	0.185
	Query+Keyword	0.317	0.487	0.601	0.287	0.337	0.498	0.684	0.416
	Query+Description	0.297	0.417	0.579	0.259	0.309	0.457	0.611	0.409
	MDKE-CS	0.351	0.501	0.687	0.301	0.357	0.511	0.698	0.477
UNIF	Query	0.073	0.207	0.292	0.148	0.192	0.387	0.419	0.257
	Query+Keyword	0.161	0.379	0.478	0.271	0.350	0.668	0.798	0.495
	Query+Description	0.124	0.324	0.429	0.225	0.347	0.643	0.772	0.483
	MDKE-CS	0.178	0.241	0.539	0.297	0.405	0.725	0.843	0.549

4 Conclusion

In this research, we focused on code search query extension and proposed an MDKE-CS code search model. Through experimental analysis, we have shown that the proposed MDKE-CS model effectively improves the accuracy of code search. Based on our research, we have drawn the following conclusions:

a. There are significant differences between Query and Description during the code search process.
b. The CSExpansion dataset, which we constructed, is suitable for code search research and can improve the accuracy of query expansion.
c. The use of multiple intent keywords and residual descriptions to extend Query can effectively reduce the differences between Description and Query, and improve the accuracy of code search.

References

1. Di Grazia, L., Pradel, M.: Code search: a survey of techniques for finding code. ACM Comput. Surv. **55**(11), 1–31 (2023)
2. Liu, S., Xie, X., Siow, J., et al.: GraphSearchNet: enhancing gnns via capturing global dependencies for semantic code search. IEEE Trans. Software Eng. (2023)
3. Zeng, C., Yu, Y., Li, S., et al.: Degraphcs: embedding variable-based flow graph for neural code search. ACM Trans. Software Eng. Methodol. **32**(2), 1–27 (2023)
4. Zhong, H., Wang, X.: An empirical study on API usages from code search engine and local library. Empir. Softw. Eng. **28**(3), 63 (2023)
5. Hu, F., Wang, Y., Du, L., et al.: Revisiting code search in a two-stage paradigm. In: Proceedings of the Sixteenth ACM International Conference on Web Search and Data Mining, pp. 994–1002 (2023)
6. Li, X., Zhang, Y., Leung, J., et al.: EDAssistant: supporting exploratory data analysis in computational notebooks with in situ code search and recommendation. ACM Trans. Interact. Intell. Syst. **13**(1), 1–27 (2023)
7. Hu, H., Liu, J., Zhang, X., et al.: A mutual embedded self-attention network model for code search. J. Syst. Software 111591 (2023)
8. Martie, L., Hoek, A., Kwak, T.: Understanding the impact of support for iteration on code search. In: Proceedings of the 2017 11th Joint Meeting on Foundations of Software Engineering, pp. 774–785 (2017)
9. Ge, X., Shepherd, D.C., Damevski, K., et al.: Design and evaluation of a multi-recommendation system for local code search. J. Vis. Lang. Comput. **39**, 1–9 (2017)
10. Yang, Y., Huang, Q.: IECS: Intent-enforced code search via extended boolean model. J. Intell. Fuzzy Syst. **33**(4), 2565–2576 (2017)
11. Karnalim, O.: Language-agnostic source code retrieval using keyword & identifier lexical pattern. Int. J. Software Eng. Comput. Syst. **4**(1), 29–47 (2018)
12. Wu, H., Yang, Y.: Code search based on alteration intent. IEEE Access **7**, 56796–56802 (2019)
13. Hu, G., Peng, M., Zhang, Y., et al.: Unsupervised software repositories mining and its application to code search. Software: Pract. Exper. **50**(3), 299–322 (2020)
14. Kim, K., Kim, D., Bissyandé, T.F., et al.: FaCoY: a code-to-code search engine. In: Proceedings of the 40th International Conference on Software Engineering, pp. 946–957 (2018)

15. Sirres, R., Bissyandé, T.F., Kim, D., et al.: Augmenting and structuring user queries to support efficient free-form code search. Empir. Softw. Eng. **23**(5), 2622–2654 (2018)
16. Rahman, M.M.: Supporting Source Code Search with Context-Aware and Semantics-Driven Query Reformulation. University of Saskatchewan (2019)
17. Yan, S., Yu, H., Chen, Y., et al.: Are the code snippets what we are searching for? a benchmark and an empirical study on code search with natural-language queries. In: 2020 IEEE 27th International Conference on Software Analysis, Evolution and Reengineering (SANER), pp. 344–354. IEEE (2020)
18. Cambronero, J., Li, H., Kim, S., et al.: When deep learning met code search. In: Proceedings of the 2019 27th ACM Joint Meeting on European Software Engineering Conference and Symposium on the Foundations of Software Engineering, pp. 964–974 (2019)
19. Azad, H.K., Deepak, A.: A new approach for query expansion using Wikipedia and WordNet. Inf. Sci. **492**, 147–163 (2019)
20. Liu, J., Kim, S., Murali, V., et al.: Neural query expansion for code search. In: Proceedings of the 3rd ACM Sigplan International Workshop on Machine Learning and Programming Languages, pp. 29–37 (2019)

Few-Shot NER in Marine Ecology Using Deep Learning

Jian Wang[1], Ming Liu[1], Danfeng Zhao[1], Shuai Shi[2], and Wei Song[1](✉)

[1] College of Information Technology, Shanghai Ocean University, Shanghai, China
{wangjian,dfzhao,wsong}@shou.edu.cn, m210911550@st.shou.edu.cn
[2] College of Electrical Engineering, Shanghai University of Electric Power, Shanghai, China
shishuai@shiep.edu.cn

Abstract. In the field of marine ecological named entity recognition (NER), challenges arise due to limited domain-specific text, weak semantic representations of input vectors and the neglect of local features. To address these challenges of NER in a low-resource environment, a deep learning-based few-shot NER model was proposed. Firstly, Sequence Generative Adversarial Nets (SeqGAN) was utilized to train on the original text and generated new text, thereby expanding the original corpus. Subsequently, BERT-IDCNN-BiLSTM-CRF was introduced for extracting marine ecological entities. BERT (Bidirectional Encoder Representation from Transformers) was pre-trained on the expanded corpus. The embeddings produced by BERT were then fed into Iterative Dilation Convolutional Networks (IDCNN) and Bidirectional Long Short-Term Memory Networks (BiLSTM) to facilitate feature extraction. Finally, Conditional Random Fields (CRF) was employed to enforce label sequence constraints and yielded the final results. For the proposed few-shot NER method based on deep learning, comparative experiments were conducted horizontally and vertically against BiLSTM-CRF, IDCNN-CRF, BERT-IDCNN-CRF and BERT-BiLSTM-CRF models on both the original and expanded corpora. The results show that BERT-IDCNN-BiLSTM-CRF outperforms BERT-BiLSTM-CRF by 2.48 percentage points in F1-score on the original corpus. On the expanded corpus, BERT-IDCNN-BiLSTM-CRF achieves a F1-score 2.65 percentage points higher than that on the original corpus. This approach effectively enhances entity extraction in the domain of marine ecology, laying a foundation for downstream tasks such as constructing marine ecological knowledge graphs and ecological governance.

Keywords: Few-Shot · Marine Ecology · NER · SeqGAN · IDCNN · BiLSTM

1 Introduction

The deep learning-based NER methods comprise three modules: distributed representations for input, context encoder and tag decoder [1]. The framework of deep learning-based NER is illustrated in Fig. 1. The distributed representations for input are further divided into three types: word-level representation [2], character-level representation [3] and hybrid representation [4]. These representations automatically learn both

B. Luo et al. (Eds.): ICONIP 2023, CCIS 1965, pp. 15–26, 2024.
https://doi.org/10.1007/978-981-99-8145-8_2

semantic and syntactic features of words from the text, enabling representation with low-dimensional real-valued dense vectors. Context encoders use Convolutional Neural Networks (CNN) [5], Recurrent Neural Networks (RNN) [6] or Transformers to capture contextual dependencies. Tag decoders predict sequences derived from the features and generate corresponding tag sequences. Common tag decoders include SoftMax [7], CRF [8] and Capsule Networks.

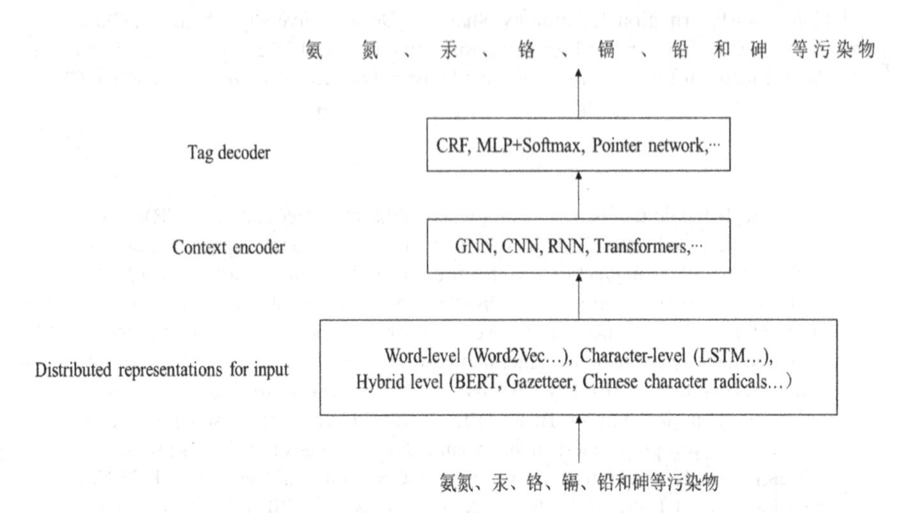

Fig. 1. NER Framework Based on Deep Learning

Deep learning-based NER requires training on a large-scale labeled corpus. In practical applications, due to the scarcity of training data, it's common to employ methods that generate or synthesize data to augment the text. Scale drives machine learning progress [9]. Obtaining a larger corpus can enhance the performance of deep learning, and augmenting the performance can also be achieved by introducing generated or synthesized data. To address the issue of subpar results caused by sparse entity distribution in fishery standard entity recognition tasks, Yang [10] used multivariate combination algorithms involving deletion, insertion and joint replacement to augment the corpus, effectively mitigating the problem of sparse samples for certain entities.

In the field of marine ecology, not only is there a shortage of annotated data but also an insufficiency of overall corpus, thereby presenting challenges for deep learning models to perform at their anticipated level. Furthermore, the marine ecology domain faces challenges such as varying lengths of entity names and unclear entity boundaries leading to polysemy. These challenges make traditional methods inadequate for effectively representing semantic features and ignoring local contextual features [11].

This study constructed a dataset by collecting the Bulletin on the Status of China's Marine Ecology and Environment and conducted research on NER in marine ecology based on this dataset. A deep learning-based few-shot NER model was designed to improve the recognition effect of named entities in marine ecology. The specific work and contributions are as follows:

(1) To address the issue of insufficient marine ecological corpora, SeqGAN [12], utilizing the synergy of reinforcement learning and adversarial concepts, is adopted for dataset augmentation. It can solve the problem of generating discrete sequences, introducing a novel avenue for data augmentation for small-sample texts and effectively alleviating the problems stemming from corpus scarcity and data sparsity.
(2) BERT-IDCNN-BiLSTM-CRF is proposed in response to the issues of varying lengths of entity names and the ambiguity caused by unclear entity boundaries in marine ecology. This model leverage Bert's enhanced semantic representations capabilities to tackle the challenge posed by traditional embeddings struggling to capture marine ecological entities with unclear boundaries. Moreover, IDCNN-BiLSTM is used to address the difficulty of extracting features from marine ecological entities with varying lengths.

The remaining structure of the article is as follows: Sect. 2 introduces the proposed deep learning-based few-shot NER method. Section 3 presents the data preprocessing and evaluation metrics. Section 4 provides a comprehensive analysis of experiments and results. Section 5 summarizes existing work and outlines prospects for future research.

2 Few-Shot NER Based on Deep Learning

The few-shot NER based on deep learning proposed in this study is depicted in Fig. 2. The overall structure of this model comprises two major components. The first part is SeqGAN, addressing the sparsity of marine ecological data through augmentation. This model generates data by training on the original corpus, thus expanding the dataset. The second part is BERT-IDCNN-BiLSTM-CRF for marine ecological NER. It consists of three layers, corresponding to the three modules of deep learning-based NER.

The first layer is BERT Input. BERT is pre-trained on a large-scale corpus to extract word vector representations enriched with semantic features. The second layer is IDCNN-BiLSTM Encoder. This layer combines IDCNN with BiLSTM, effectively fusing feature vectors and extracting contextual features. The third layer is CRF Decoder. CRF uses the Viterbi algorithm to decode the output vectors from the previous layer, yielding the optimal label sequence.

This method uses SeqGAN for data augmentation of marine ecological corpora. It uses BERT to better extract embeddings from the input. The fusion of IDCNN and BiLSTM allows for a more comprehensive information extraction compared to a single model. CRF effectively addresses the problem of predicting illegal label sequences. This method demonstrates favorable performance in few-shot NER.

2.1 Text Generation Using SeqGAN

SeqGAN views the sequence generation problem as a sequence decision problem [13]. It introduces policy gradient and Monte Carlo search from reinforcement learning to address the challenges of using GAN for discrete data generation.

Figure 3 illustrates the training process of SeqGAN. Firstly, the generator is pre-trained using real corpus. Subsequently, the discriminator is pre-trained using both real and generated data. Then, adversarial training follows. The generator is trained with

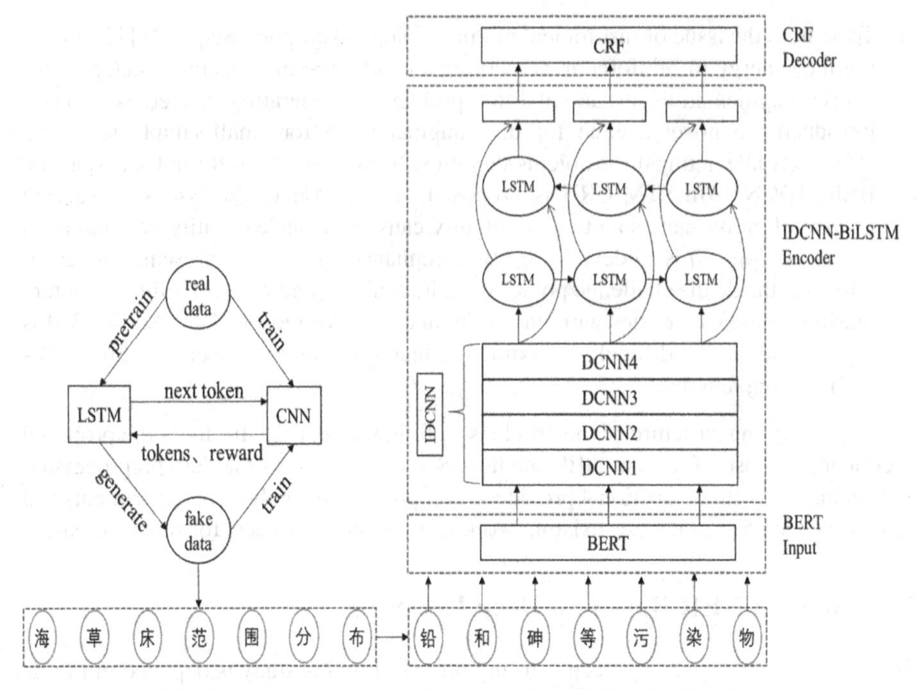

Fig. 2. Structure of SeqGAN and BERT-IDCNN-BILSTM-CRF

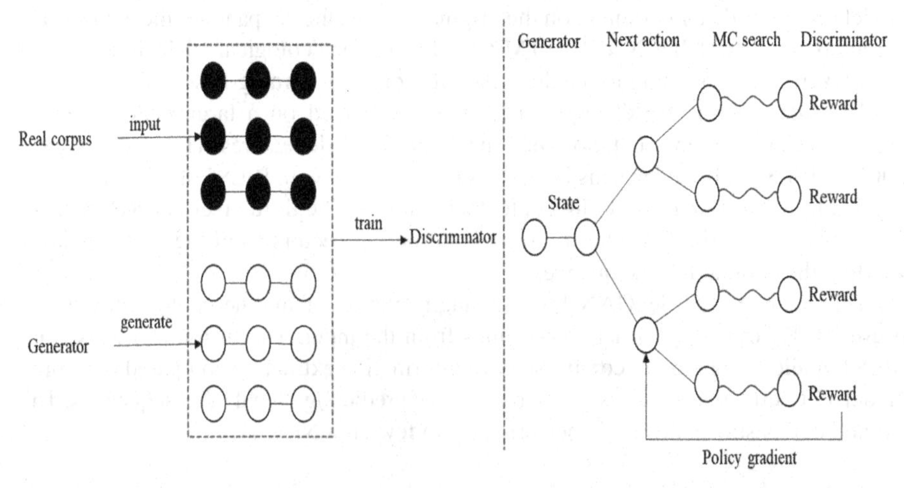

Fig. 3. The training process of SeqGAN. Left: The discriminator D is trained on both real data and data generated by the generator G. Right: G is trained using policy gradient, with the final reward signal coming from D and being retroactively passed to the intermediate action value through Monte Carlo search.

rewards derived from the discriminator's feedback. Simultaneously, the discriminator is retrained using the updated generator.

Taking the example of generating word sequences, assume a word sequence is denoted as $Y_{1:T} = (y_1, \ldots, y_t, \ldots, y_T)$, where $y_t \in \mathcal{Y}$, and Y represents the candidate vocabulary. The current state is the generated word sequence $(y_1, \ldots, y_{t-1})$. The next action involves selecting the next generated word y_t. This choice depends on the t-1 words previously generated. The generator uses the first t-1 words and model parameters to start sampling from the t-th position. Through Monte Carlo search, N alternative word sequences $Y_{t:T}$, are generated to form N complete word sequences $Y_{1:T}$. These sequences are scored by the discriminator and based on these scores, the optimal strategy is selected and policy gradient are adjusted.

2.2 Distributed Representations for Input Using BERT

In distributed representations for input, BERT [14] is used to embed vector representations. BERT, a pre-trained language representation model, is primarily composed of deep bidirectional Transformer encoders. This encoder integrates a self-attention mechanism capable of automatically assigning weights to both long and short sequence entities.

The BERT's embeddings are derived as the summation of token embeddings, segmentation embeddings and position embeddings, as depicted in Fig. 4. Token embeddings represent the vector representation of words, segmentation embeddings differentiate between two sentences in a sentence pair, and position embeddings indicate the sequential order of the input sequence. The combination of these embeddings enables BERT to be trained on a large-scale unlabeled dataset, performing tasks such as masked language modeling and next sentence prediction. The pre-trained BERT is fine-tuned for NER task in the marine ecology domain. The resulting embeddings effectively mitigate the limitations observed in traditional embeddings, as they insufficiently capture the varying length of entity names and unclear boundaries.

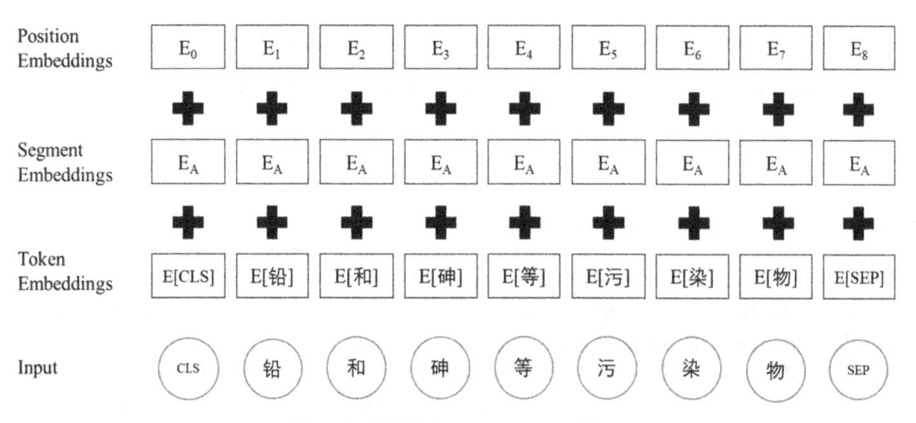

Fig. 4. BERT input representation

2.3 Context Encoder Using IDCNN

In context encoder, the first layer employs IDCNN [15]. In a typical CNN, convolution kernels slide across continuous positions of the input matrix, extracting features by convolution and then utilizing pooling to reduce parameters and prevent overfitting, albeit at the cost of resolution loss. DCNN [16] introduces dilation rates within the network. This means that during feature extraction with convolutional kernels, data between the dilation rates is skipped, expanding the model's receptive field without altering kernel size.

From the comparison shown in Fig. 5, for the same sequence of text, with the same convolution kernel of size 3 and two convolutional layers, traditional convolution captures 5 characters, while dilated convolution extends the contextual scope to 7 characters. The advantage of dilated convolution lies in its ability to expand the receptive field by introducing dilation rates in the convolutional kernels, thereby allowing each convolutional output to encompass more information without resorting to pooling and losing information.

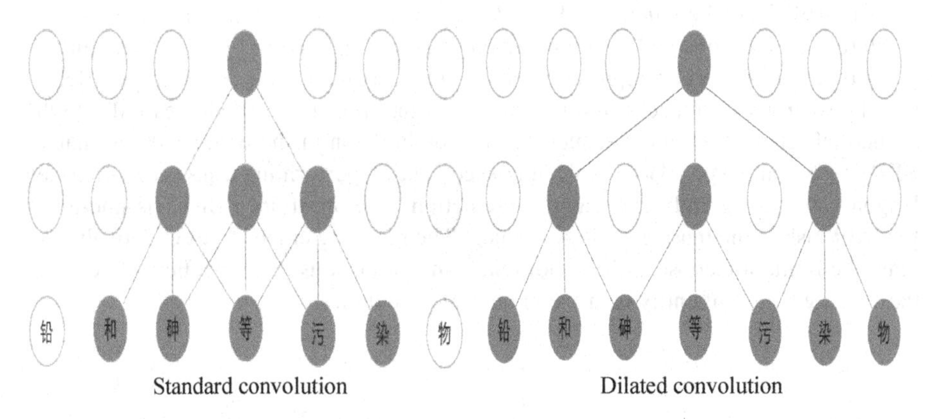

Fig. 5. Difference of text extraction between standard convolution and dilated convolution

2.4 Context Encoder Using BiLSTM

In the second layer of the context encoder, BiLSTM is employed. LSTM [17] control memory cells using input gates, forget gates and output gates to learn and retain long-term dependencies. This enables LSTM to acquire and retain long-term dependencies, addressing the issue of short-term memory. Among them, the input gate is used to update the cell state, the forget gate decides what information to discarded or retain, and the output gate determines the value of the next hidden state. The hidden state contains relevant information from the previous inputs. The operations of the three gates and the memory cell are illustrated in Eqs. (1)–(5):

$$i_t = \sigma\left(W_i \times \left[x_t, h_{t-1}, c_{t-1}\right] + b_i\right) \tag{1}$$

$$f_t = \sigma\left(W_f \times \left[x_t, h_{t-1}, c_{t-1}\right] + b_f\right) \tag{2}$$

$$o_t = \sigma\left(W_o \times \left[x_t, h_{t-1}, c_t\right] + b_o\right) \tag{3}$$

$$c_t = f_t * c_{t-1} + i_t * \tanh\left(W_c \times \left[x_t, h_{t-1}\right] + b_c\right) \tag{4}$$

$$h_t = o_t * \tanh(c_t) \tag{5}$$

where i_t, f_t, o_t, c_t, h_t represent the input gate, forget gate, output gate, cell vectors, and hidden state at time t respectively. The sigmoid activation function is denoted by σ, W represents the weight matrix, and b is the bias term.

In sequence labeling task, BiLSTM can be used to extract both past and future features from sentences. As depicted in Fig. 6, BiLSTM consists of two layers of LSTM with opposite directions, which can effectively utilize past features through the forward state and future features through the backward state.

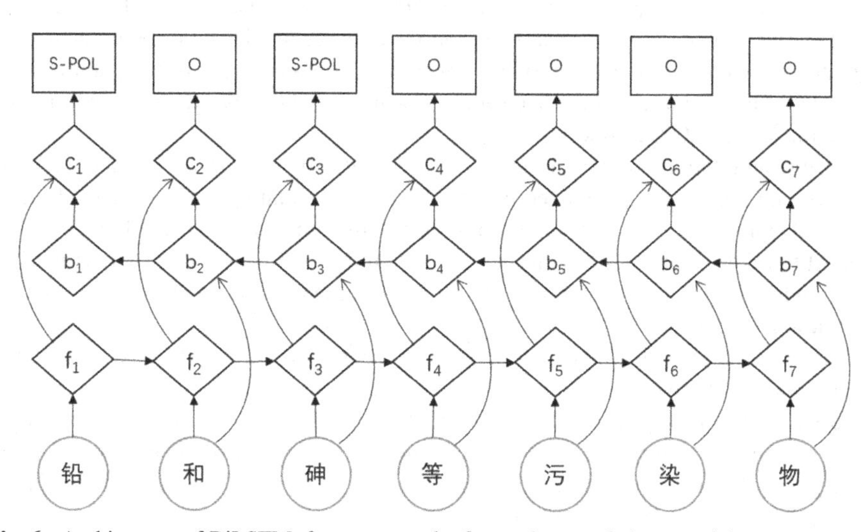

Fig. 6. Architecture of BiLSTM. f_i represents the forward state of character i, b_i represents the backward state of character i. By concatenating these two vectors, the representation c_i of character i in its contextual is obtained.

2.5 Tag Decoder Using CRF

In tag decoder, CRF is utilized. After the previous stage, the output of the context encoder represents the sequence of the highest probabilities at each time step. Treating each label as an independent entity might lead to illegal sequences in the decoded labels. However, CRF is a probabilistic model that outputs a sequence of labels conditioned on a set of input sequences. By introducing custom feature functions, it can capture not only dependencies between observations but also complex dependencies between the current observation and multiple preceding and succeeding states, effectively addressing the issue of label constraints faced by statistical models.

For an input sequence x = $(x_1, x_2, \ldots, x_n)$,P represents the emission matrix output from the context encoder, where $p_{i,j}$ corresponds to the score of the j-th tag for the i-th word in the sentence. The predicted sequence y = $(y_1, y_2, \ldots, y_n)$ has a score as shown in formula (6). A is the transition score matrix for labels, where $A_{i,j}$ represents the score of transitioning from the tag i to tag j.

$$s(X, y) = \sum_{i=1}^{n} p_{i,y_i} + \sum_{i=0}^{n} A_{y_i,y_{i+1}} \tag{6}$$

3 Data Processing and Evaluation Metrics

3.1 Data Preprocessing

Due to the lack of mature NER corpora in the field of marine ecology, this study collected the Bulletin of Marine Ecology and Environment Status of China from 2001 to 2021. An initial marine ecological corpus was formed after applying regular expressions, character formatting normalization and removing unnecessary non-textual and irrelevant textual data related to marine ecology. According to statistics, the cleaned corpus comprises approximately 240,000 characters.

For the annotation process in the initial marine ecological corpus, manual labeling was required due to the customized nature of labels. Annotating entities demands substantial domain-specific knowledge. To address this, the study consulted experts and referred to relevant literature to classify named entities into categories such as marine pollution factors, marine animals and marine plants based on existing marine ecological environmental texts.

Furthermore, to handle mislabeling issues present in the corpus, the most frequently labeled category for each entity in the corpus was chosen as the corrected category. This process aimed to rectify mislabeled entities. Additionally, to address cases of missing labels, text matching techniques were employed to identify and label entities that were not originally labeled. These measures were taken to reduce data noise, optimize training and improve the overall quality of the corpus.

The self-constructed marine ecological corpus was used as the original dataset. According to the characteristics of the corpus, the preprocessed texts were input into SeqGAN for training, resulting in the generation of approximately 200,000 characters of synthetic data. The generated corpus was then combined with the original corpus to form an expanded corpus. The data preprocessing workflow is illustrated in Fig. 7.

3.2 Labeling Guidelines

The dataset uses the BIOES five-label sequence labeling method, as shown in Table 1. The labels are represented as follows: B for Begin (indicating the start position of an entity), I for Inside (indicating the middle positions of an entity), O for Outside (indicating that a character does not belong to any entity), E for End (indicating the end position

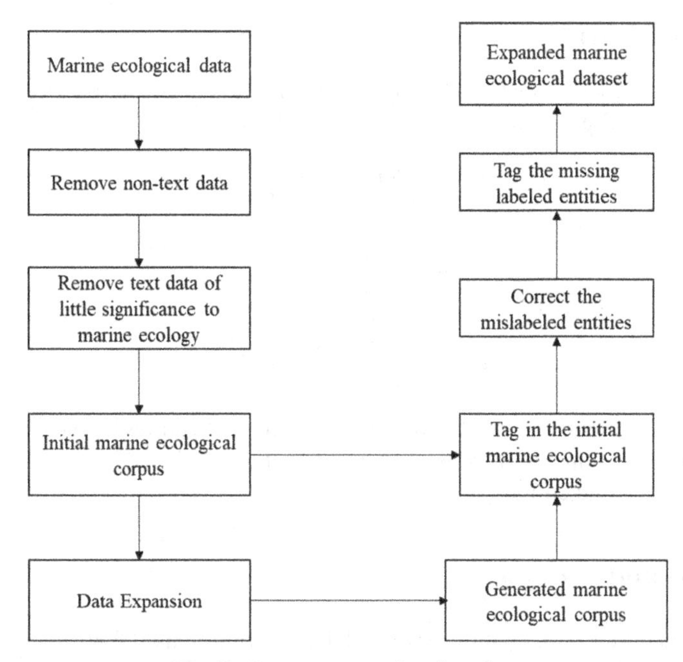

Fig. 7. Data preprocessing flowchart

of an entity), and S for Singleton (indicating a single-character entity). The entity label categories are represented by the following English terms: POL for marine pollution factor, ANI for marine animal and PLA for marine plant.

Table 1. Partial Annotation in Original and Generated Datasets with BIOES Labeling

Original dataset	Label category	Generated dataset	Label category
氨	B-POL	近	O
氮	E-POL	岸	O
、	O	海	O
汞	S-POL	湾	O
、	O	等	O
铬	S-POL	浮	O
、	O	游	O
镉	S-POL	植	O
、	O	物	O
铅	S-POL	海	B-PLA
和	O	草	I-PLA
砷	S-POL	床	E-PLA
等	O	范	O
污	O	围	O
染	O	分	O
物	O	布	O

3.3 Evaluation Metrics

This experiment uses precision (P), recall (R), and the composite metric F1-score to evaluate the model' performance, calculated as follows:

$$P = \frac{Num_T}{Num_p} \times 100\% \tag{7}$$

$$R = \frac{Num_T}{Num_p} \times 100\% \tag{8}$$

$$F1 = \frac{2 \times P \times R}{P + R} \times 100\% \tag{9}$$

where Num_T represents the number of correctly predicted entities by the model; Num_P represents the total number of entities predicted by the model; and Num_R represents the total number of actual entities.

4 Results and Analysis

To validate the effectiveness of the proposed deep learning-based few-shot NER model for marine ecological entities, comparisons were made under the same experimental conditions with the following NER models: word embedding-based BiLSTM-CRF [18], character embedding-based IDCNN-CRF [19], BiLSTM-CRF [20], BERT-IDCNN-CRF, and BERT-BiLSTM-CRF in terms of P, R and F1. The experimental results are shown in Table 2 below:

Table 2. Comparison of NER results across different models and datasetsUnit: %

Model	Original dataset			Expanded dataset		
	P	R	F1	P	R	F1
Word-BiLSTM-CRF	86.64	85.42	86.03	87.62	87.30	87.46
IDCNN-CRF	87.06	85.97	86.51	88.32	88.18	88.25
Character-BiLSTM-CRF	87.08	87.56	87.32	89.25	89.71	89.48
BERT-IDCNN-CRF	89.78	86.16	87.94	90.72	89.85	90.28
BERT-BiLSTM-CRF	88.12	88.23	88.17	90.63	90.87	90.75
BERT-IDCNN-BiLSTM-CRF	**90.35**	**90.96**	**90.65**	**93.16**	**93.44**	**93.30**

The method proposed in this paper achieved optimal results on both the original and expanded dataset. Compared to the best-performing baseline model, BERT-BiLSTM-CRF, BERT-IDCNN-BiLSTM-CRF showed improvements of 2.23% in P, 2.73% in R and 2.48% in F1 on the original dataset. On the expanded dataset, it achieved improvements of 2.53% in P, 2.57% in R and 2.55% in F1.

On the original dataset, IDCNN-CRF and BiLSTM-CRF, both based on character embeddings, outperformed the word embedding-based BiLSTM-CRF by 0.48% and 1.29% in F1 respectively. This is because the marine ecological corpus contains long entities and nesting phenomena, such as "Chinese white dolphin", "seagrass" and "seagrass bed", "coral" and "coral reef", which lead to inaccurate word segmentation. Character-based embeddings can alleviate the impact of inaccurate word segmentation on NER in the marine ecological domain.

Among the character embedding-based models, IDCNN focuses more on the surrounding information and features of entities, enabling better distinction of entity boundaries. BiLSTM learns features of the entire sentence, allowing it to capture more entities from the sentence as a whole. BERT-based embeddings extract context information from both directions (before and after), obtaining entity boundary features and having stronger semantic feature representation capabilities. IDCNN can fully extract local information, while BiLSTM can extract global information. The fusion of IDCNN and BiLSTM improves the limitations of insufficient global feature extraction by IDCNN, resulting in further improvements in recognition performance.

The experiment results show that, compared to the original dataset with limited samples, Character-BiLSTM-CRF showed improvements of 2.17%, 2.15% and 2.16% in P, R and F1 respectively after data augmentation. This improvement is attributed to SeqGAN's ability to generate text data that closely resembles real data, resulting in a more evenly distributed augmented corpus. This experiment underscores the importance of having an ample amount of data for data-driven learning, as models can perform better with sufficient data. Compared to IDCNN, BiLSTM performed better due to its stronger ability to extract global context features. However, IDCNN traded off some accuracy for overall speed improvement.

5 Conclusion

This paper studied NER in the field of marine ecology, proposing a few-shot method. The approach involves several steps. First, SeqGAN was used to expand the marine ecological corpus. Then, BERT was used to obtain input embeddings. Next, these embeddings were input into IDCNN-BiLSTM to extract features. Finally, CRF was used to decode the entities of marine pollution factors, marine animals and marine plants in the marine ecology dataset.

Experimental results show that data augmentation has a positive impact on entity recognition performance. Additionally, the fusion of IDCNN and BiLSTM, compared to a single NER method, also leads to improvements. Unlike traditional NER methods, this approach does not rely on domain-specific rules and feature engineering, making it easily applicable to other domains lacking large-scale corpora.

However, there is still room for improvement in terms of accuracy, particularly regarding nested entities, long entities and the quality of data generated by SeqGAN. Future work will focus on addressing these challenges, expanding the dataset and conducting further research to achieve better recognition results. Additionally, constructing a marine ecological knowledge graph will aid in solving marine ecological management issues more effectively.

References

1. Li, J., Sun, A., Han, J.: A survey on deep learning for named entity recognition. IEEE Trans. Knowl. Data Eng. **34**(1), 50–70 (2022)
2. Liu, B.: Text sentiment analysis based on CBOW model and deep learning in big data environment. J. Ambient. Intell. Humaniz. Comput. **11**(2), 451–458 (2018). https://doi.org/10.1007/s12652-018-1095-6
3. Akbik, A., Blythe, D., Vollgraf, R.: Contextual string embeddings for sequence labeling. In: International Conference on Computational Linguistics (2018)
4. Zhang, X., Guo, R., Huang, D.: Named entity recognition based on dependency. J. Chin. Inform. Process. **35**(6), 63–73 (2021)
5. Li, L., Guo, Y.: Biomedical named entity recognition with CNN-BLSTM-CRF. J. Chin. Inform. Process. **32**(1), 116–122 (2018)
6. Chen, K., Yan, Z., Huo, Q.: A context-sensitive-chunk BPTT approach to training deep LSTM/BLSTM recurrent neural networks for offline handwriting recognition. In: ICDAR (2015)
7. Cui L., Zhang Y.: Hierarchically-refined label attention network for sequence labeling. In: EMNLP-IJCNLP (2019)
8. Lafferty, J.D., McCallum, A., Pereira, F.: Conditional random fields: probabilistic models for segmenting and labeling sequence data. In: International Conference on Machine Learning, pp. 282–289 (2001)
9. Andrew, N.: Machine Learning Yearning. Self-publishing (2018)
10. Yang, H., Yu, H., Liu, J.: Fishery standard named entity recognition based on BERT+BiLSTM+CRF deep learning model and multivariate combination data augmentation. J. Dalian Ocean Univ. **36**(4), 661–669 (2021)
11. Chen, X., Xu, L., Liu, Z., Sun, M., Luan, H.: Joint learning of character and word embeddings. In: International Joint Conference on Artificial Intelligence (2015)
12. Yu, L., Zhang, W., Wang, J., Yu, Y.: SeqGAN: sequence generative adversarial nets with policy gradient. In AAAI (2017)
13. Bachman, P., Precup, D.: Data generation as sequential decision making. In: NIPS (2015)
14. Devlin, J., Chang, M., Lee, K., Toutanova, K: Bert: pre-training of deep bidirectional transformers for language understanding. In: NAACL-HLT, pp. 4171–4186 (2019)
15. Strubell, E., Verga, P., Belanger, D., McCallum, A.: Fast and accurate entity recognition with iterated dilated convolutions. In: Proceedings of the 2017 Conference on Empirical Methods in Natural Language Processing (2017)
16. Yu, F., Koltun, V.: Multi-scale context aggregation by dilated convolutions. In: ICLR (2016)
17. Bengio, Y., Simard, P., Frasconi, P.: Learning long-term dependencies with gradient descent is difficult. IEEE Trans. Neural Networks **5**(2), 157–166 (1994)
18. Sun, J., Yu, H., Feng, Y.: Recognition of nominated fishery domain entity based on deep learning architectures. J. Dalian Ocean Univ. **32**(2), 265–269 (2018)
19. Feng, H., Sun, Y., Wu, T.: Chinese electronic medical record named entity recognition based on multi-features and IDCNN. J. Changzhou Univ. (Natl. Sci. Edn.) **35**(1), 59–67 (2023)
20. Huang, Z., Xu, W., Yu, K.: Bidirectional LSTM-CRF Models for Sequence Tagging (2015)

Knowledge Prompting with Contrastive Learning for Unsupervised CommonsenseQA

Lihui Zhang and Ruifan Li(✉)

Beijing University of Posts and Telecommunications, Beijing 100876, China
rfli@bupt.edu.cn

Abstract. Unsupervised commonsense question answering is an emerging task in NLP domain. In this task knowledge is of vital importance. Most existing works focus on stacking large-scale models or extracting knowledge from external sources. However, these methods suffer from either the unstable quality of knowledge or the deficiency in the model's flexibility. In this paper, we propose a **K**nowledge **P**rompting with **C**ontrastive **L**earning (KPCL) model to address these problems. Specifically, we first consider dropout noise as augmentation for commonsense questions. Then we apply unsupervised contrastive learning in further pre-training to capture the nuances among questions, and thus help the subsequent knowledge generation. After that, we utilize generic prompts to generate question-related knowledge descriptions in a zero-shot manner, facilitating easier transfer to new domains. Moreover, we concatenate knowledge descriptions with the commonsense question, forming integrated question statements. Finally, we reason over them to score the confidence and make predictions. Experimental results on three benchmark datasets demonstrate the effectiveness and robustness of our proposed KPCL, which outperforms baseline methods consistently.

Keywords: Commonsense question answering · Prompt learning · Contrastive learning

1 Introduction

Commonsense question answering (CQA) is a prevalent task in the domain of natural language processing (NLP) with a long-term goal to evaluate the ability of cognitive understanding in machines. As shown in Fig. 1, given a commonsense question, the machine is expected to predict the right answer from multiple choices by combining with external knowledge. Recently, with the popularity of large pre-trained language model (PLM) in many downstream scenarios [8,14], the research on unsupervised CQA gradually comes into view. Specifically, the algorithmic model needs to acquire the commonsense knowledge automatically, and make predictions without any annotated data.

B. Luo et al. (Eds.): ICONIP 2023, CCIS 1965, pp. 27–38, 2024.
https://doi.org/10.1007/978-981-99-8145-8_3

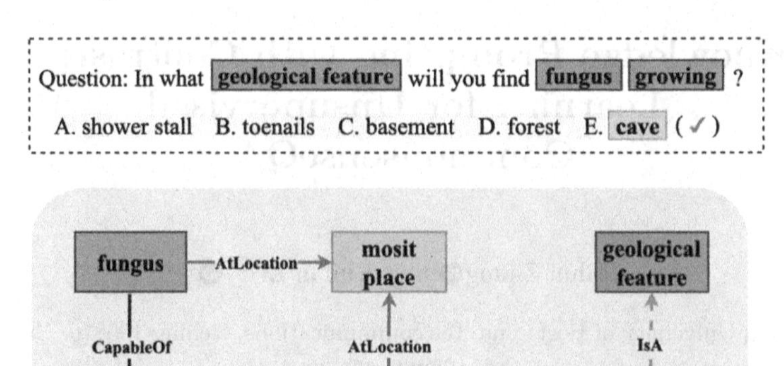

Fig. 1. An example of CQA task using external knowledge. Without using any labeled data, the solution for unsupervised CQA, needs to seek a way of mining relevant knowledge automatically to predict answers.

Existing studies for unsupervised CQA task have mainly developed two types of approaches. The generative approaches [12,17,19,27] devote to eliciting commonsense knowledge from large generative models such as GPT-3 [3] and T5 [22] in a zero-shot way, and do not require any fine-tuning for downstream tasks. Nevertheless, such techniques heavily rely on the prior knowledge that language model learned in the pre-training stage, resulting in unstable quality of generated knowledge in specific fields. Meanwhile, the extraction approaches [1,2,15] utilize external open-source knowledge bases to provide evidences and guide reasoning. However, these methods often use a fixed knowledge source and require pre-defined rules to extract valuable information, which makes it difficult to efficiently adapt to new domains and thus limits the flexibility.

To solve the aforementioned challenges, we propose a novel model, **K**nowledge **P**rompting with **C**ontrastive **L**earning (KPCL) for unsupervised CQA task. **Firstly**, we utilize dropout noise as data augmentation for commonsense questions. Then in further pre-training, unsupervised contrastive learning is applied on the knowledge generative model to perceive semantic nuances between different commonsense questions, thereby improving the quality of generated knowledge in subsequent steps. **Secondly**, we use knowledge prompts with a generic format of the instruction and random instances, having the knowledge generative model elicit question-related knowledge descriptions. Based on the domain feature of different NLP tasks, we only need to make some minor adjustments in prompts to achieve flexible transfer. **Thirdly**, we concatenate each generated knowledge description with the corresponding question, forming a set of integrated question statements. After that, we leverage the knowledge reasoning model to make predictions by scoring the confidence of each choice.

The major contributions can be summarized as follows: **1)** We propose a novel KPCL model with further pre-training for unsupervised CQA task. Unsupervised contrastive learning with dropout augmentation are applied to help the model capture the semantic nuances between various questions. **2)** We present a generic prompt scheme utilizing the instruction with task-specific instances, which is beneficial for generating question-related knowledge descriptions and adapting the model to the downstream task. **3)** We conduct extensive experiments on three benchmark datasets. The experimental results show that our method outperforms the competitive baselines consistently, which fully demonstrates the effectiveness of our KPCL model.

2 Related Works

Prompt Learning. Prompt learning is a rapidly evolving paradigm that enables the adaption of PLMs to various downstream tasks by utilizing textual prompts. The core of prompt learning is to generate desired information without relying on extensive labeled datasets, making it applicable in many zero-shot and few-shot scenarios [3,4,20]. Some works focus on devising prompts manually [25,26]. Typically, the work of PET [25] defines cloze question patterns to reformulate text classification and natural language inference (NLI) problems in the few-shot setting. However, the manual way in prompt learning is labor-intensive and requires handcrafting with rich experience. Recently, many researchers have proposed continuous methods [7,9,10,13] to mitigate the limitations. For instance, P-tuning [13] enables PLMs to learn the optimal prompt automatically by inserting trainable parameters into continuous embedding space. Similarly, Prompt-tuning [9] and Prefix-tuning [10] froze the parameters of the pre-training backbone and only update parameters of prompts in the training phase.

Knowledge-aware Methods for CQA Task. The CQA task commonly focuses on exploring various approaches [31–33] to use massive knowledge graphs and perform structured reasoning over questions. When it comes to unsupervised CQA, some studies are inclined to leverage external knowledge sources to extract evidences or explanations for zero-shot answer prediction [1,2,15]. In particular, Ma et al. [15] devise a neuro-symbolic framework to transform diverse knowledge sources into a form which is effective for PLMs. While another line of research is dedicated to generating relevant knowledge in the specific domain to help reasoning [12,17,19,27]. For example, Self-talk [27] inquires the generative PLMs with information seeking questions to elicit knowledge as clarifications. SEQA [17] uses PLMs to generate a series of plausible answers and then vote for the correct choice by measuring the semantic similarity. To improve the flexibility of sequential models, Liu et al. [12] generate knowledge with an input prompt for answer prediction, reducing the reliance on task-specific adjustments.

3 Methodology

Formally, the unsupervised CQA task can be defined as follows. Given a question q and a candidate answer set $\mathcal{C}$ with M choices, i.e., $\mathcal{C} = \{c_1, c_2, \ldots, c_M\}$, we need

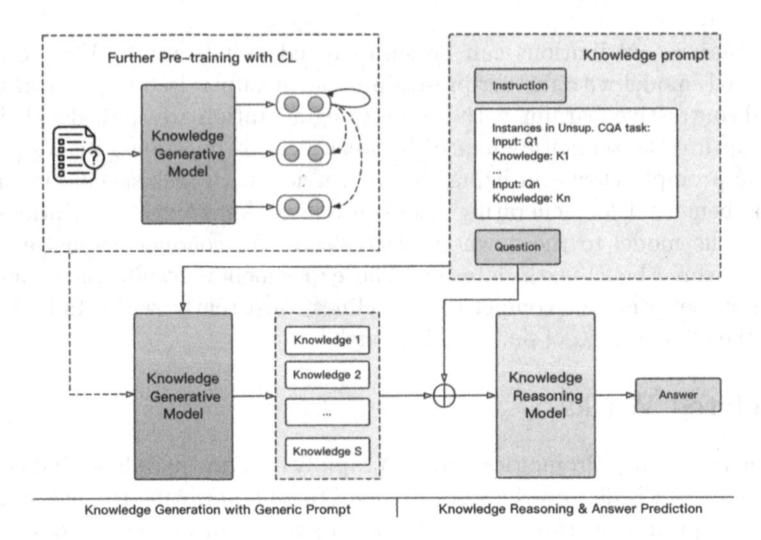

Fig. 2. The architecture of our KPCL model. It consists of three parts: 1) Further Pre-training with Contrastive Learning (Sect. 3.1), 2) Knowledge Generation with Generic Prompt (Sect. 3.2), and 3) Knowledge Reasoning and Answer Prediction (Sect. 3.3).

to identify the best matching answer from $\mathcal{C}$ for the given question q without any labeled data. The architecture of KPCL model is shown in Fig. 2.

3.1 Further Pre-training with Contrastive Learning

We use the commonsense questions in the same domain of the target benchmark in further pre-training, aiming to close the disparity in data distribution. Inspired by SimCSE [5], the dropout in PLMs naturally suits the noise in data augmentation due to the high alignment and uniformity. The neurons in the network are randomly discarded during the forward propagation. Therefore, the dropout noise for augmentation only exists in the neuron-level of the hidden network and cannot hurt the contextual semantics of the input sentence.

As shown in Fig. 3, we first duplicate the input as two copies for each commonsense question q_i. Then we utilize GPT-2 [21] as the knowledge generative model. And the two copies (q_i, q_i^+) are fed into the model with two different dropout masks. The hidden representations (h_i, h_i^+) are obtained as the positive pair. In contrast, the negative pairs are the representations of other instances in the training batch.

We adopt InfoNCE loss [18] as the objective of the contrastive learning in further pre-training. The loss is formally defined as,

$$\mathcal{L}_{CL} = -\sum \log \frac{e^{\mathrm{sim}\left(h_i, h_i^+\right)/\tau}}{\sum_{j=1}^{N}\left(e^{\mathrm{sim}\left(h_i, h_j^+\right)/\tau}\right)} \tag{1}$$

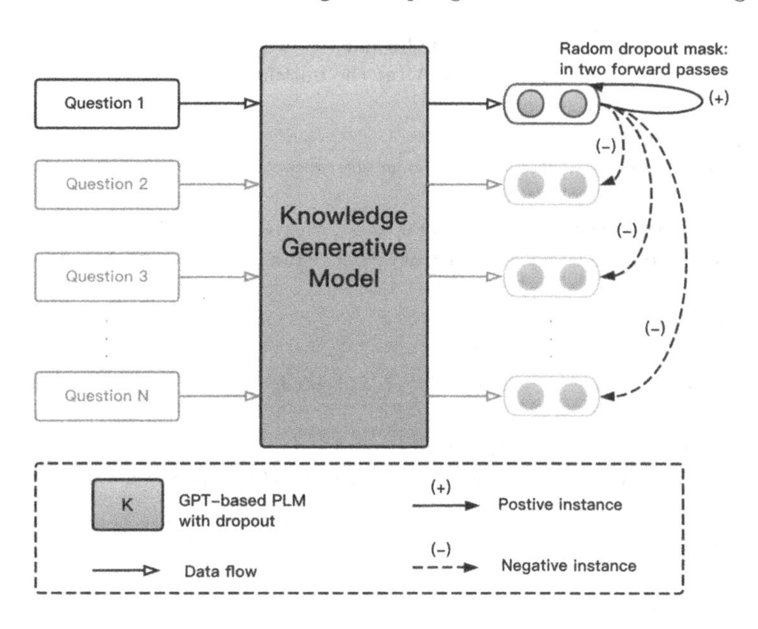

Fig. 3. The strategy of unsupervised contrastive learning in further pre-training.

where N represents the training batch size, $\mathrm{sim}(\cdot, \cdot)$ denotes the cosine similarity between two vectors, and τ is the temperature hyper-parameter to scale and control the attention for hard negative pairs. The idea is that when optimizing the unsupervised comparative loss, the generative model draws closer the semantics between the question itself and its augmented instance, and pulls away the semantics of other different questions. This is an effective way to alleviate the collapse of the pre-trained model and learn better representation of the commonsense questions.

3.2 Knowledge Generation with Generic Prompt

We use generic knowledge prompt to guide the knowledge generative model to produce a series of question-related knowledge descriptions. The main idea is to bridge the formulation gap between different tasks and thus improve the model flexibility. Specifically, the knowledge prompt applies a generic format. It consists of an instruction sentence, several random instances of question-knowledge pairs fixed for each task, and a "<question>" placeholder as the end. For each question q, it will be inserted into the placeholder. Then a series of knowledge descriptions K_q can be obtained as the continuation part of the knowledge prompt by repeated sampling, and it can be formalized as follow,

$$K_q = \{k_s : k_s = f_G(P_q, q), s = 1 \dots S\} \tag{2}$$

where f_G indicates the knowledge generative model, P_q is the knowledge prompt, and k_s are the generated knowledge descriptions with variable lengths.

Table 1. Prompt sketches using generic format for two evaluating datasets, CommonsenseQA (on the top) and OpenbookQA (at the bottom).

Prompt Sketches
Instructions: Generate some knowledge about the concepts in the input
Instances:
Input: The fox walked from the city into the forest, what was it looking for?
Knowledge: Natural habitats are usually away from cities
...
Input: <question>
Knowledge:
Instructions: Generate some elementary science knowledge about the concepts in the input
Instances:
Input: In the wilderness, light pollution is?
Knowledge: As distance to a city decreases, the amount of light pollution will increase
...
Input: <question>
Knowledge:

We showcase the prompt sketches using the generic format in Table 1. It is worth noting that, the knowledge prompt is fixed when generating knowledge descriptions in the same dataset, and the question instances in the prompt are randomly sampled as the representative. In addition, the corresponding knowledge instances in the prompt should not directly make answer prediction, but are expected to give more diverse expressions to imply potential connections between the question and the answer.

3.3 Knowledge Reasoning and Answer Prediction

After obtaining a set of knowledge descriptions, we use knowledge reasoning model to integrate them with the corresponding question and further make prediction by scoring the confidence of each choice. To be specific, we first concatenate each question q with its S generated knowledge descriptions, forming $S+1$ integrated question statements, i.e.,

$$\underbrace{q_0 = q, q_1 = [k_1 \oplus q], \ldots, q_S = [k_S \oplus q]} \tag{3}$$

in which $\oplus$ denotes the concatenation operator. Then we apply the off-the-shelf PLM, in a zero-shot setting, as the knowledge reasoning model. For each choice c_i, the model will compute its confidence under different integrated question statements, and select the most supportive one with the highest score,

$$P(c_i \mid q, K_q) \propto \max_{0 \le s \le S} f_R(c_i \mid q_s) \tag{4}$$

where f_R indicates the knowledge reasoning model. Therefore, we choose the candidate answer with the highest confidence score as the final prediction $\hat{c}$, which can be formulated as follow,

$$\hat{c} = \underset{c_i \in C}{\operatorname{argmax}} \max_{0 \leq s \leq S} f_R\left(c_i \mid q_s\right) \tag{5}$$

The evidence for the final prediction is derived from one certain knowledge description $k_{\hat{s}}$, and its index $\hat{s}$ can be computed as follow,

$$\hat{s} = \underset{0 \leq s \leq S}{\operatorname{argmax}} \max_{c_i \in C} f_R\left(c_i \mid q_s\right) \tag{6}$$

Note that the model is inclined to perform reasoning by screening out those statements that cannot help distinguish choices decisively.

4 Experiments and Results

4.1 Datasets and Metric

We evaluate our model on three benchmark datasets, i.e., CommonsenseQA [29], OpenbookQA [16], and SocialIQA [24]. **1)** The **CommonsenseQA** dataset creates questions from ConceptNet [28] target concepts and semantic relations, and contains 12,102 questions. CommonsenseQA involves a 5-way multiple choice QA task which centers around substantial commonsense knowledge. The official test set of CommonsenseQA is not publicly available, therefore we perform experiments on the in-house data split[1] used in Kagnet [11]. **2)** The **OpenBookQA** dataset comes with an open book of 1,326 science facts, and contains 5,957 questions focusing on elementary science knowledge. It is a 4-way multiple choice QA task. We use the official data split of OpenbookQA[2]. **3)** The **SocialIQA** dataset is designed to evaluate machines' commonsense understanding regarding human behaviors and social interactions. It consists of 37,588 questions and follows a 3-way multiple-choice format. We apply the official split of SocialIQA[3]. Furthermore, to evaluate the performance of different unsupervised CQA methods, we use the **Accuracy** (i.e., Acc) in answer prediction as the metric.

4.2 Baselines and Implementation Details

We compare KPCL with state-of-the-art baselines, including **Self-talk** [27], **SEQA** [17], **DynaGen** [1], **GKP** [12]. In addition, **Vanilla Baseline** is adopted, which developed to assign answers to the corresponding question without utilizing external knowledge.

The experiments are implemented under the PLM backbones using the Huggingface framework[4] [30]. All results are reported on the development sets. Note

[1] https://github.com/INK-USC/KagNet.
[2] https://github.com/allenai/OpenBookQA.
[3] https://leaderboard.allenai.org/socialiqa.
[4] https://github.com/huggingface/transformers.

Table 2. Experimental results (%) on CommonsenseQA, OpenBookQA and SocialIQA. The symbol "†" indicates our reproduced results on these datasets.

Dataset	Method	GPT-2 (S)	GPT-2 (M)	GPT-2 (L)	Published
CommonsenseQA	Vanilla†	25.6	28.2	28.7	-
	Self-Talk	24.8	27.3	31.5	32.4
	SEQA	<u>26.1</u>	<u>30.7</u>	<u>34.6</u>	-
	GKP	-	-	-	47.3
	KPCL	**30.7**	**33.3**	**35.1**	-
OpenbookQA	Vanilla†	15.2	17.3	20.9	-
	Self-Talk	17.4	21.0	23.8	-
	SEQA	<u>27.6</u>	<u>28.6</u>	<u>32.0</u>	-
	KPCL	**35.4**	**37.8**	**38.2**	-
SocialIQA	Vanilla†	37.1	39.7	41.6	-
	Self-Talk	41.2	43.3	45.3	46.2
	SEQA	<u>44.4</u>	<u>44.6</u>	<u>46.6</u>	47.5
	DynaGen	-	-	-	50.1
	KPCL	**45.3**	**46.1**	**47.2**	-

that the labels are invisible in the training stage and only used for the evaluation of model performance. For further pre-training and knowledge generation, we apply three different scale of GPT-2, including GPT-2 (S) with 117M, GPT-2 (M) with 345M, GPT-2 (L) with 762M. The temperature hyper-parameter τ in the unsupervised contrastive learning is empirically set to 0.1. In addition, for each question, we generate $S = 20$ knowledge descriptions utilizing nucleus sampling [6] with the probability $p = 0.5$, and remove duplicates and empty strings. The knowledge generation is terminated either when the token length exceeds more than 64 tokens, or a special token '.' is encountered. For knowledge reasoning, following the configuration of SEQA, we utilize SentenceBERT (L) [23], which is further fine-tuned on NLI task, to measure the semantic relevance between the integrated question statements and each choice.

4.3 Main Results

The experimental results on our evaluated datasets, CommonsenseQA, OpenBookQA, and SocialIQA, are reported in Table 2. It can be seen that our KPCL model consistently outperforms the other baselines on these three datasets. Our method surpasses the previous best method SEQA by an average accuracy of 2.6% on the CommonsenseQA, 7.7% on the OpenBookQA, and 1.0% on the SocialIQA. Besides, the GKP baseline is equipped with larger language model like GPT-3 and T5 to generate commonsense knowledge, and the DynaGen baseline uses external knowledge base to generate inferences for prediction. Therefore, they are well-performed in the published setting. It is noteworthy that, with the

Table 3. Ablation study of our KPCL model on each individual component.

Module	CommonsenseQA	OpenbookQA
Our KPCL - GPT-2 (M)	**33.3**	**37.8**
w/o Unsup. Contrastive Learning	30.7 (2.6↓)	34.1 (3.7↓)
w/o Generative Knowledge Prompting	29.2 (4.1↓)	28.3 (9.5↓)
w/o Knowledge Reasoning	32.4 (0.9↓)	36.1 (1.7↓)

scale of knowledge generative model increases from GPT-2 (S) to GPT-2 (L), the performance improvement of our KPCL model is not as significant as other baselines. It turn out that the KPCL model tends to generate knowledge descriptions with stable quality when using generic prompts. Therefore, the improvement brought by increasing the model scale is relatively limited.

4.4 Ablation Study

To further investigate the effectiveness of each individual component in our KPCL model, we conduct extensive ablation studies on the datasets, CommonsenseQA and OpenbookQA. As shown in Table 3, generative knowledge prompting has the greatest impact on model performance, verifying the critical role of knowledge for answer prediction. Also, we find an average drop of 3.2% in "*w/o* Unsup. Contrastive Learning" setting, which shows that contrastive learning in the further pre-training can bring benefits for subsequent knowledge generation. To sum up, each component in the KPCL model contributes to the entire performance of the unsupervised CQA task.

4.5 Robustness Analysis

To explore the robustness of our KPCL model, we set four different disturbances on prompts. As shown in Table 4, we conduct experiments with GPT-2 (M) in the CommonsenseQA dataset, and the reported results are the average on three runs to avoid the contingency. Under the disturbances of instruction paraphrase[5] and in-domain replacement[6], there are minor drops of 0.4% and 0.3% in model performance, respectively. It demonstrates that the KPCL model is robust to confront the changes in instructions and in-domain instances. While under the disturbances of order exchange[7] and cross-domain replacement[8], the performance suffers significant decline of 1.2% and 2.9%, respectively. It can be

[5] For instruction paraphrase, the keywords in the instruction of prompt are replaced with the synonyms.

[6] For in-domain replacement, two examples are randomly chosen from CommonsenseQA to replace the instances in prompt.

[7] For order exchange, the order of the instruction and instances in prompt is swapped.

[8] For cross-domain replacement, two examples are randomly chosen from OpenbookQA to replace the instances in prompt.

found that the prompt scheme of the instruction's order and instances' domain should be carefully considered, which determines the quality of generated knowledge descriptions and indeed has a crucial impact on the model performance.

Table 4. Robustness analysis of disturbances on prompts.

#	Disturbance	Performance
0	None	**33.3**
1	Instruction Paraphrase	32.9 (0.4↓)
2	Order Exchange	32.1 (1.2↓)
3	In-domain Replacement	33.0 (0.3↓)
4	Cross-domain Replacement	30.4 (2.9↓)

Table 5. Case study for our KPCL and the vanilla baseline. The correct answers are underlined, and the constructive parts of our generated knowledge are marked in blue.

Cases	Vanilla	KPCL
Q1: Riding a bike for a long time can cause what? A.enjoyment B.fatigue C.falling down D.getting lost E.thirst Knowledge: high-intensity outdoor workouts reduce the strength and energy.	C. (✗) Score: 0.69	B. (✓) Score: 0.73
Q2: If you are committing perjury you have done what while under oath? A.crime B.disrespect judge C.embarrassment D.lie E.indictment Knowledge: perjury crime has no reliance to be placed	A. (✗) Score: 0.52	D. (✓) Score: 0.81
Q3: Piece of land in Canada where you can find marmot? A.North America B.United States C.Vancouver Island D.American E.cage Knowledge: the vancouver marmots are rare animals in Canada also in the world.	C. (✓) Score: 0.47	C. (✓) Score: 0.65
Q4: He is house was a mess, he began doing housework to get what? A.boredom B.nice home C.Michigan D.feeling satisfied E.house clean Knowledge: housework involves cleaning and cooking in most families	E. (✓) Score: 0.58	E. (✓) Score: 0.98

4.6 Case Study

We randomly select four cases from the CommonsenseQA dataset and show the results in Table 5. The KPCL model gives the correct predictions in the first two examples while the vanilla baseline fails. Compared with the vanilla baseline, our KPCL model pays more attention on the semantic constraints implied in the question, such as "for a long time" in Q1 and "perjury...under oath" in Q2. Thus, the generated knowledge descriptions can effectively guide the model to perform reasonable predictions towards the correct answers. For the last two examples, both the KPCL model and the vanilla baseline make the right predictions. However, the confidence scores for the correct answers are much higher in our solution. The reason behind is that the knowledge descriptions generated by our KPCL model, such as "the vancouver marmots" in Q3 and "involves cleaning" in Q4, are target-oriented and can help the model reason over commonsense questions in a positive way.

5 Conclusion

In this paper, we propose **K**nowledge **P**rompting with **C**ontrastive **L**earning (KPCL) for unsupervised CQA task. We first use dropout noise in PLM as the augmentation for commonsense questions. To improve the quality of generated knowledge, we then use unsupervised contrastive learning in further pre-training and learn better representations of various questions. Furthermore, to enhance the model flexibility, we leverage prompts with generic format to generate a set of question-related knowledge descriptions without any labeled data. Finally, we integrate generated knowledge descriptions with the question, and predict the answer with the highest confidence. Experiments on three datasets show our proposed KPCL model outperforms baselines. In the future, we will further explore fine-grained ways of knowledge integration to perform effective reasoning.

References

1. Bosselut, A., Le Bras, R., Choi, Y.: Dynamic neuro-symbolic knowledge graph construction for zero-shot commonsense question answering. In: AAAI, vol. 35, pp. 4923–4931 (2021)
2. Bosselut, A., Rashkin, H., Sap, M., Malaviya, C., Celikyilmaz, A., Choi, Y.: Comet: commonsense transformers for automatic knowledge graph construction. In: ACL, pp. 4762–4779 (2019)
3. Brown, T., Mann, B., Ryder, N., Subbiah, M., Kaplan, J.D., Dhariwal, P., Neelakantan, A., Shyam, P., Sastry, G., Askell, A., et al.: Language models are few-shot learners. NeuIPS **33**, 1877–1901 (2020)
4. Cui, L., Wu, Y., Liu, J., Yang, S., Zhang, Y.: Template-based named entity recognition using bart. In: Findings of ACL: ACL-IJCNLP, pp. 1835–1845 (2021)
5. Gao, T., Yao, X., Chen, D.: Simcse: simple contrastive learning of sentence embeddings. In: EMNLP, pp. 6894–6910 (2021)
6. Holtzman, A., Buys, J., Du, L., Forbes, M., Choi, Y.: The curious case of neural text degeneration. In: ICLR (2020)
7. Houlsby, N., Giurgiu, A., Jastrzebski, S., Morrone, B., De Laroussilhe, Q., Gesmundo, A., Attariyan, M., Gelly, S.: Parameter-efficient transfer learning for nlp. In: ICML. pp. 2790–2799 (2019)
8. Khashabi, D., et al.: Unifiedqa: crossing format boundaries with a single qa system. In: Findings of ACL: EMNLP, pp. 1896–1907 (2020)
9. Lester, B., Al-Rfou, R., Constant, N.: The power of scale for parameter-efficient prompt tuning. In: EMNLP, pp. 3045–3059 (2021)
10. Li, X.L., Liang, P.: Prefix-tuning: optimizing continuous prompts for generation. In: ACL-IJCNLP, pp. 4582–4597 (2021)
11. Lin, B.Y., Chen, X., Chen, J., Ren, X.: Kagnet: knowledge-aware graph networks for commonsense reasoning. In: EMNLP-IJCNLP, pp. 2829–2839 (2019)
12. Liu, J., Liu, A., Lu, X., Welleck, S., West, P., Le Bras, R., Choi, Y., Hajishirzi, H.: Generated knowledge prompting for commonsense reasoning. In: ACL, pp. 3154–3169 (2022)
13. Liu, X., Zheng, Y., Du, Z., Ding, M., Qian, Y., Yang, Z., Tang, J.: Gpt understands, too. arXiv preprint arXiv:2103.10385 (2021)

14. Lourie, N., Le Bras, R., Bhagavatula, C., Choi, Y.: Unicorn on rainbow: A universal commonsense reasoning model on a new multitask benchmark. In: AAAI, vol. 35, pp. 13480–13488 (2021)
15. Ma, K., Ilievski, F., Francis, J., Bisk, Y., Nyberg, E., Oltramari, A.: Knowledge-driven data construction for zero-shot evaluation in commonsense question answering. In: AAAI, vol. 35, pp. 13507–13515 (2021)
16. Mihaylov, T., Clark, P., Khot, T., Sabharwal, A.: Can a suit of armor conduct electricity? a new dataset for open book question answering. In: EMNLP, pp. 2381–2391 (2018)
17. Niu, Y., Huang, F., Liang, J., Chen, W., Zhu, X., Huang, M.: A semantic-based method for unsupervised commonsense question answering. In: ACL-IJCNLP, pp. 3037–3049 (2021)
18. Oord, A.v.d., Li, Y., Vinyals, O.: Representation learning with contrastive predictive coding. arXiv preprint arXiv:1807.03748 (2018)
19. Paranjape, B., Michael, J., Ghazvininejad, M., Hajishirzi, H., Zettlemoyer, L.: Prompting contrastive explanations for commonsense reasoning tasks. In: Findings of ACL: ACL-IJCNLP, pp. 4179–4192 (2021)
20. Petroni, F., et al.: Language models as knowledge bases? In: EMNLP-IJCNLP, pp. 2463–2473 (2019)
21. Radford, A., Wu, J., Child, R., Luan, D., Amodei, D., Sutskever, I.: Language models are unsupervised multitask learners. OpenAI Blog **1**(8), 9 (2019)
22. Raffel, C., Shazeer, N., Roberts, A., Lee, K., Narang, S., Matena, M., Zhou, Y., Li, W., Liu, P.J.: Exploring the limits of transfer learning with a unified text-to-text transformer. J. Mach. Learn. Res. **21**(1), 5485–5551 (2020)
23. Reimers, N., Gurevych, I.: Sentence-bert: Sentence embeddings using siamese bert-networks. In: EMNLP-IJCNLP, pp. 3982–3992 (2019)
24. Sap, M., Rashkin, H., Chen, D., Le Bras, R., Choi, Y.: Social iqa: commonsense reasoning about social interactions. In: EMNLP-IJCNLP, pp. 4463–4473 (2019)
25. Schick, T., Schütze, H.: Exploiting cloze-questions for few-shot text classification and natural language inference. In: EACL, pp. 255–269 (2021)
26. Schick, T., Schütze, H.: It's not just size that matters: small language models are also few-shot learners. In: NAACL, pp. 2339–2352 (2021)
27. Shwartz, V., West, P., Le Bras, R., Bhagavatula, C., Choi, Y.: Unsupervised commonsense question answering with self-talk. In: EMNLP, pp. 4615–4629 (2020)
28. Speer, R., Chin, J., Havasi, C.C.: Conceptnet 5.5: An open multilingual graph of general knowledge. In: AAAI, pp. 4444–4451 (2017)
29. Talmor, A., Herzig, J., Lourie, N., Berant, J.: Commonsenseqa: a question answering challenge targeting commonsense knowledge. In: NAACL, pp. 4149–4158 (2019)
30. Wolf, T., et al.: Transformers: State-of-the-art natural language processing. In: EMNLP: system demonstrations, pp. 38–45 (2020)
31. Yasunaga, M., Ren, H., Bosselut, A., Liang, P., Leskovec, J.: Qa-gnn: reasoning with language models and knowledge graphs for question answering. In: NAACL, pp. 535–546 (2021)
32. Zhang, L., Li, R.: Ke-gcl: Knowledge enhanced graph contrastive learning for commonsense question answering. In: Findings of ACL: EMNLP, pp. 76–87 (2022)
33. Zhang, X., Bosselut, A., Yasunaga, M., Ren, H., Liang, P., Manning, C., Leskovec, J.: Greaselm: Graph reasoning enhanced language models for question answering. In: ICLR (2022)

PTCP: Alleviate Layer Collapse in Pruning at Initialization via Parameter Threshold Compensation and Preservation

Xinpeng Hao, Shiyuan Tang, Heqian Qiu, Hefei Mei, Benliu Qiu, Jian Jiao, Chuanyang Gong, and Hongliang Li(✉)

University of Electronic Science and Technology of China, Chengdu, China
{xphao,sytang,hfmei,qbenliu,jij2021,cygong}@std.uestc.edu.cn,
{hqqiu,hlli}@uestc.edu.cn

Abstract. Over-parameterized neural networks have good performance, but training such networks is computationally expensive. Pruning at initialization (PaI) avoids training a full network, which has attracted intense interest. But at high compression ratios, layer collapse severely compromises the performance of PaI. Existing methods introduce operations such as iterative pruning to alleviate layer collapse. However, these operations require additional computing and memory costs. In this paper, we focus on alleviating layer collapse without increasing cost. Therefore, we propose an efficient strategy called parameter threshold compensation. This strategy constrains the lower limit of network layer parameters and uses parameter transfer to compensate for layers with fewer parameters. To promote a more balanced transfer of parameters, we further propose a parameter preservation strategy, using the average number of preserved parameters to more strongly constrain the layers that reduce parameters. We conduct extensive experiments on five pruning methods on Cifar10 and Cifar100 datasets using VGG16 and ResNet18 architectures, verifying the effectiveness of our strategy. Furthermore, we compare the improved performance with two SOTA methods. The comparison results show that our strategy achieves similar performance, challenging the design of increasingly complex pruning strategies.

Keywords: Deep neural network · Model compression · Pruning at initialization · Parameter threshold compensation and preservation

1 Introduction

Deep neural networks have achieved great success in computer vision [1–3], natural language processing [4], autonomous driving [5], and other fields [6]. However, the massive size of modern neural networks hinders model training and deployment process [7]. Therefore, pruning is widely used to compress neural networks. Among many branches of pruning, pruning at initialization (PaI) becomes more and more attractive because it can avoid the training of a full network [8–12].

B. Luo et al. (Eds.): ICONIP 2023, CCIS 1965, pp. 39–51, 2024.
https://doi.org/10.1007/978-981-99-8145-8_4

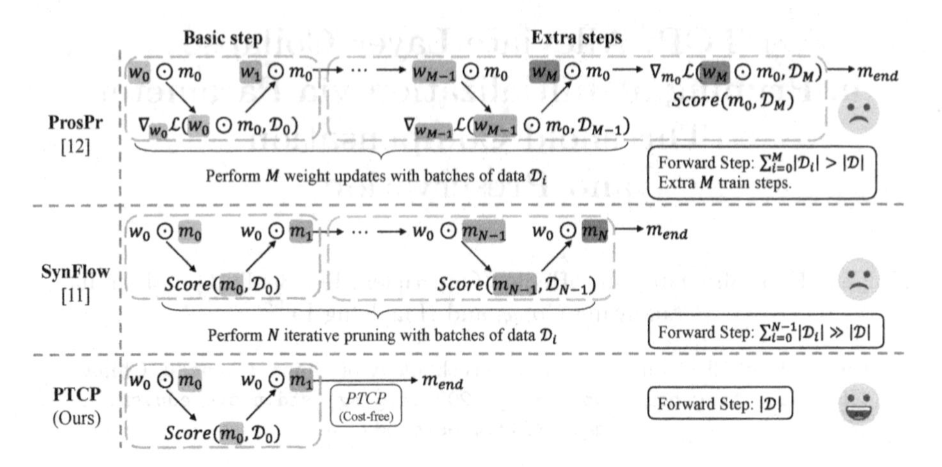

Fig. 1. Comparison of PTCP and SOTA methods. Forward Step means the number of forward propagation steps required for pruning. $|\mathcal{D}_i|$ means the number of graphs in $\mathcal{D}_i$. For example, on Cifar10, the number of forward propagation steps under the optimal settings of ProsPr and SynFlow is $\sum_{i=0}^{M}|\mathcal{D}_i| = 2048$ ($M = 3$, $|\mathcal{D}_i| = 512$) and $\sum_{i=0}^{N-1}|\mathcal{D}_i| = 25600$ ($N = 100$, $|\mathcal{D}_i| = 256$), which is much higher than $|\mathcal{D}| = 100$ required by PTCP.

The earlier PaI methods use criteria such as randomness [8], magnitude, connection sensitivity [9], gradient flow [10], to evaluate the importance of parameters. These methods suffer from layer collapse, which refers to pruning a network layer completely. SynFlow [11] points out that layer collapse severely impairs subnet performance at high sparsity and proposes an iterative pruning strategy to avoid this phenomenon. ProsPr [13] proposes adding several training steps in the pruning process to improve subnet performance, considering that the network will be trained immediately after pruning.

However, existing studies ignore the extra cost of alleviating layer collapse. Therefore, we propose a parameter threshold compensation and preservation strategy (PTCP). The advantages of our strategy compared to other methods are shown in Fig. 1, alleviating layer collapse without extra computational cost.

Specifically, our strategy includes parameter threshold compensation (PTC) and parameter preservation (PP). In PTC, we design a parameter threshold as the lower limit of network layer parameters and compensate layers with fewer parameters by parameter transfer. Furthermore, we propose PP, constraining the layers that reduce parameters strongly by the average number of preserved parameters, to promote a more balanced transfer of parameters.

We test PTCP on twenty combinations of two network architectures (VGG16 [14], ResNet18 [15]), two datasets (Cifar10, Cifar100 [16]), and five pruning methods (Random [8], Magnitude, SNIP [9], GraSP [10], SynFlow [11]). The results show that our strategy can effectively improve the performance of PaI methods in multiple scenarios. Moreover, we compare the performance of our strategy with two SOTA methods (SynFlow [11], ProsPr [13]), showing similar performance at partial sparsity, challenging increasingly complex pruning strategies.

Our contributions can be summarized as follows:

- We analyze that extra computational cost is a key limitation of existing PaI methods for alleviating layer collapse.
- We propose a progressive strategy to alleviate layer collapse without extra computational cost. PTC constrains the lower limit of layer parameters and uses parameter transfer to compensate for layers with fewer parameters. PP uses the average number of preserved parameters as a stronger constraint to promote a more balanced transfer of parameters.
- We show experimentally that our strategy effectively improves the performance of existing PaI methods. And our strategy achieves comparable performance to SOTA methods at partial sparsity.

2 Related Work

Pruning is an important method of network compression. PaI finds sparse subnets from randomly initialized dense networks. According to the different sparsity structures, pruning is divided into structured [17,18] and unstructured [19,20]. Structured pruning usually uses channels or filters as pruning units, while unstructured pruning uses weights. PaI methods usually use unstructured pruning because it is more precise and flexible to adjust the network structure.

The essence of pruning is to find an appropriate mask. According to the basis of mask selection, pruning is divided into the pre-selected mask and the post-selected mask [21]. Pre-selected mask only requires a randomly initialized network, while post-selected mask relies on a pre-trained model.

Pre-selected Mask: SNIP [9] first proposes the concept of pre-selected mask. Specifically, SNIP proposes connection sensitivity based on loss preservation to evaluate parameter importance. Compared with SNIP, GraSP [10] believes that the training dynamics of the neural network are more suitable as pruning indicators. GraSP proposes a pruning method called gradient signal preservation, which improves the pruned subnet's performance. At the same time, AI [22] tries to explain why SNIP works from the perspective of signal propagation. AI proposes a data-free pruning method to break the dynamic isometry between network layers. SynFlow [11] proposes the concept of layer collapse and analyzes its impact on the performance of pruned networks. SynFlow introduces iterative pruning to improve the generalization of existing methods, alleviating the performance gap with pruning after training. ProsPr [13] adds several training steps in the pruning process so that the network can adapt to subsequent training. Part of the follow-up works attempt to improve the pruning strategy [8,12], and the rest attempt to introduce existing methods into more tasks [23,24].

Post-selected Mask: LTH [25] first proposes the concept of post-selected mask. LTH determines the mask from the pre-trained model, then randomly initializes parameters and trains this subnet, which achieves comparable results to a dense network. Part of the follow-up works expand LTH from theoretical and experimental perspectives [26–28], and the rest try to verify its correctness with more ablation studies [29,30].

3 Method

We propose parameter threshold compensation and preservation strategy to alleviate layer collapse without extra computational cost in PaI. First, we formulate the pruning problem described in Sect. 3.1. Then, we propose the parameter threshold compensation strategy in Sect. 3.2. Based on parameter constraints, this strategy limits the network layer's parameter threshold to alleviate layer collapse effectively. To promote a more balanced transfer of parameters, we design the parameter preservation strategy as an improvement in Sect. 3.3. Finally, we present the algorithm implementation in Sect. 3.4. Figure 2 shows the retention ratio of network layer parameters, with or without PTCP.

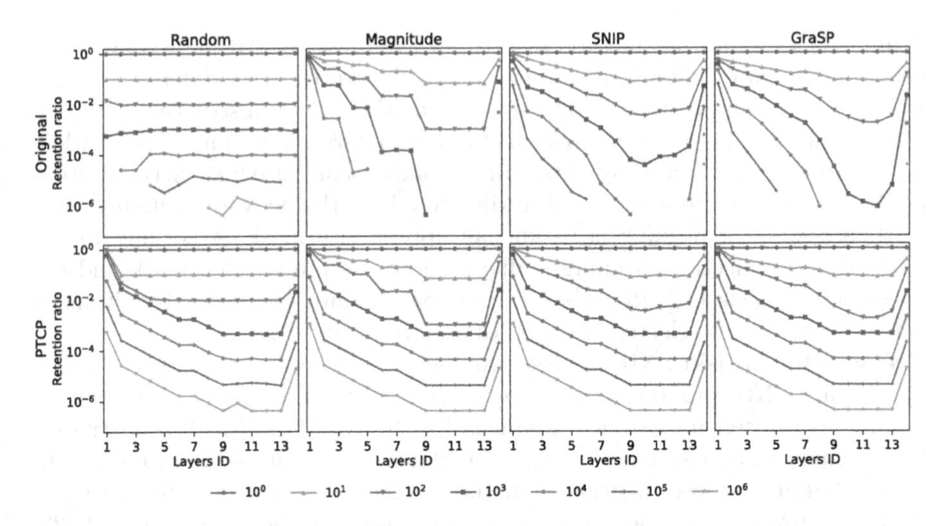

Fig. 2. Layer collapse phenomenon. Retention ratio of parameters at each layer of VGG16 model pruned at initialization with Cifar100 dataset over a range of compression ratios ($10^n, n = 0, 1, \ldots, 6$). The first row represents the original performance, and the second represents the performance processed by PTCP. The presence of missing points in a line indicates that layer collapse occurs.

3.1 Problem Formulation

Over-parameterization of neural networks is a prerequisite for efficient pruning. Typically, neural network pruning is modelled as an optimization problem. Given a training dataset $\mathcal{D} = \{(x_i, y_i)\}_{i=1}^{n}$, a network model f with ω as parameters and

m as masks, and the number of parameters κ expected to be retained, network pruning can be written as a constrained optimization problem as follows:

$$\begin{gathered}\min_{w,m} L\left(f\left(w \odot m; x\right); y\right) = \min_{w,m} \frac{1}{n} \sum_{i=1}^{n} l\left(f\left(w \odot m; x_i\right); y_i\right) \\ s.t.\ w \in \mathcal{R}^d,\ m \in \{0,1\}^d,\ \|m\|_0 \leq \kappa\end{gathered} \tag{1}$$

where $l(\cdot)$ is the standard loss function, d is the total number of neural network parameters, and $\|\cdot\|_0$ is the standard L_0 norm.

3.2 Parameter Threshold Compensation

Layer collapse in PaI severely impairs the performance of subnets, thus attracting researchers' attention. SynFlow [11] points out the hazards of layer collapse and uses iterative pruning to alleviate it. Subsequent studies alleviate this problem indirectly by combining more and more information, such as ProsPr [12], but the effect is limited.

However, existing methods ignore the potential of previous pruning methods, which have the advantages of low cost and easy deployment. To fully exploit the potential of prior methods, we propose the parameter threshold compensation strategy to alleviate layer collapse. Our strategy alleviates layer collapse based on parameter constraints to improve performance without extra cost.

Setting of Parameter Threshold. We first propose three primary constraints on parameter threshold setting.

1) Parameter threshold is less than the number of parameters in any layer.
2) The number of parameters retained needs to satisfy the pruning ratio.
3) Keeping the same number of parameters in each layer is unwise.

We adopt a fixed parameter threshold, that is, set the same threshold for each network layer. We take the minimum number of layer parameters as the baseline to ensure the effectiveness and generalizability of the threshold. The expected network sparsity is denoted as μ, the number of network layers is denoted as N_L, the parameter quantity of each layer is denoted as l_i, $i \in [0, N_L - 1]$, the coefficient of fixed threshold is denoted as α, and the fixed threshold $l_{\min}$ is:

$$l_{\min} = \alpha * \mu * \min_i l_i \tag{2}$$

According to Constraint 1 and 2, α needs to satisfy the following equation:

$$\begin{cases} l_{\min} \leq \min_i l_i \\ N_L l_{\min} \leq \mu \sum_i l_i \end{cases} \Rightarrow \alpha \leq \min\left(\frac{1}{\mu}, \frac{\sum_i l_i}{N_L \min_i l_i}\right) \tag{3}$$

Existing pruning strategies indicate that network layers have different average pruning scores, i.e., different importance [9–12]. Therefore, reserving the same number of parameters for each layer is unreasonable, as stated in Constraint 3. Further, we shrink the upper limit of the fixed threshold coefficient α as follows:

$$\alpha \leq \min\left(\frac{1}{\mu}, FLOOR\left(\frac{\sum_i l_i}{N_L \min_i l_i}\right)\right) \tag{4}$$

where $FLOOR(\cdot)$ is defined as rounding down according to the highest digit.

We set α as its upper limit (UL) to get $l_{\min}$ as shown in Eq. 5. The minimum number of layer parameters bounds the first term, and the pruning ratio bounds the second term.

$$l_{\min} = \min\left(\min_i l_i, \mu * \min_i l_i * FLOOR(\frac{\sum_i l_i}{N_L \min_i l_i})\right) \tag{5}$$

Setting of Parameter Transfer Method. We call the layers that need to retain more parameters as *In-layer*, and the layers that need to be pruned more as *Out-layer*. The parameter transfer method determines which parameters to add in *In-layer* and which to prune in *Out-layer*. Considering that our strategy is to improve existing PaI methods, we fully respect the original scores of parameters. Specifically, we force each *In-layer* to retain the threshold number of parameters according to the ordering of parameter scores. And we treat all *Out-layer* as a new subnet for pruning and threshold compensation again.

3.3 Parameter Preservation

From the perspective of maintaining connectivity, the importance of any parameter is negatively correlated with the parameter retention ratio at its network layer. However, PTC simply considers that *In-layer* and *Out-layer* are complementary, that is, *Out-layer* $=\overline{\textit{In-layer}}$, where $\overline{\textit{In-layer}}$ means all network layers except *In-layer*. The increased pruning ratio may lead to over-pruning of some *Out-layer*, resulting in an unbalanced transfer of parameters.

To avoid over-pruning some *Out-layer* with a few parameters, we impose stricter constraints on selecting *Out-layer*. Specifically, we allow the existence of layers that do not belong to *In-layer* and *Out-layer*, that is, *Out-layer* $\subsetneq \overline{\textit{In-layer}}$, to achieve parameter preservation. The parameters of each layer after pruning are denoted as l_i', $i \in [0, N_L]$. Considering that the average number of parameters reflects the general situation of each layer, we take the average number l_{avg}' of parameters in each layer after pruning as the lower limit of the selection of *Out-layer*, as shown in Eq. 6. Limited by Eq. 4, $l_{\min} \leq l_{\text{avg}}'$ is always satisfied.

$$l_{\text{avg}}' = \frac{1}{N_L} \sum_{i=0}^{N_L-1} l_i' \tag{6}$$

The above constraints are summarized in Eq. 7.

$$\begin{cases} l_i' < l_{\min} & \Rightarrow \textit{In-layers} \\ l_{\min} \leq l_i' \leq l_{\text{avg}}' & \Rightarrow \textit{No-operation} \\ l_i' > l_{\text{avg}}' & \Rightarrow \textit{Out-layers} \end{cases} \tag{7}$$

3.4 Round-by-Round Matching Algorithm

In this section, we design a round-by-round matching algorithm to achieve parameter threshold compensation and preservation strategy. The algorithm performs three steps, including pre-pruning, verification, and matching, until all layers are matched. The pseudocode for the algorithm is shown in Algorithm 1. The whole algorithm does not need additional network training steps, and its cost is negligible compared to expensive training costs.

Algorithm 1: Parameter Threshold Compensation and Preservation

---

Data: Number of network layers N_L, number of reserved parameters N_R, threshold compensation coefficient α, number of parameters pre layer l_i, importance of parameters pre layer $Scores_i$, $i \in [0, N_L]$

Result: Mask of neural network M

1 $l_{\min} \leftarrow$ Calculated by *Eq.* 5;
2 **while** *layers matched incompletely* **do**
3     $M \leftarrow$ Pre-prune unmatched network layers; `// Pre-pruning`
4     $l' \leftarrow$ Number of reserved parameters pre layer;
5     $l'_{\text{avg}} \leftarrow$ Calculated by *Eq.* 6;
6   **for** *unmatched network layers* **do**
7   **if** $l'_i < l_{\min}$ **then** `// Verification`
8             $M_i \leftarrow$ Prune the layer independently;
9   Mark as matched; `// Matching`
10   **else if** $l_{\min} \leq l'_i \leq l'_{\text{avg}}$ **then** `// Verification`
11   Mark as matched; `// Matching`
12   **end**
13   **end**
14 **end**

---

4 Experiment

We conduct various experiments to verify the effectiveness of the PTCP strategy on image classification tasks with twenty combinations of two network structures (VGG16 [14], ResNet18 [15]), two datasets (Cifar10, Cifar100 [16]), and five pruning methods (Random [8], Magnitude, SNIP [9], GraSP [10], SynFlow [11]) over a range of compression ratios (10^n, $n = [0, 0.25, \ldots, 3]$). Furthermore, we compare the performance of classical methods processed by PTCP with two SOTA methods (SynFlow [11], ProsPr [13]) aimed at alleviating layer collapse.

4.1 Experimental Settings

For a fair comparison, we follow the network structure settings of SynFlow [11], namely VGG16 from OpenLTH[1] and ResNet18 from PyTorch Model[2]. All experiments in this paper are performed on Titan Xp GPU. And all results are obtained from three replicate experiments.

---

[1] https://github.com/facebookresearch/open_lth.
[2] https://github.com/pytorch/vision/blob/master/torchvision/models/resnet.py.

Pruning Setting. For random pruning, we sample independently from a standard Gaussian distribution. For magnitude pruning, we take the absolute value of the parameter as the score. For SNIP [9], GraSP [10], SynFlow [11] and ProsPr [13], we follow the methods described in their papers to calculate the scores. SynFlow-Mult represents the original method reported (100 iterations), while SynFlow and SynFlow-PTCP represent the results of only one iteration with or without PTCP. ProsPr uses the 3-round cycle setting corresponding to its optimal performance. We only report the performance of ProsPr on Cifar10 dataset, the same as its reported experiments. For each pruning method, we prune the model with a random subset of the training dataset whose size is ten times the number of classes, i.e. 100 for Cifar10 and 1000 for Cifar100. The batch size is uniformly selected as 256.

Training Setting. On VGG16(ResNet18), the model is optimized by SGD for 160 epochs, the initial learning rate is 0.1(0.01), the batch size is 128, the weight decay is 1e−4(5e−4), the momentum is 0.9, and the learning rate multiple by 0.1(0.2) at 60 and 120 epochs.

4.2 Results and Analysis

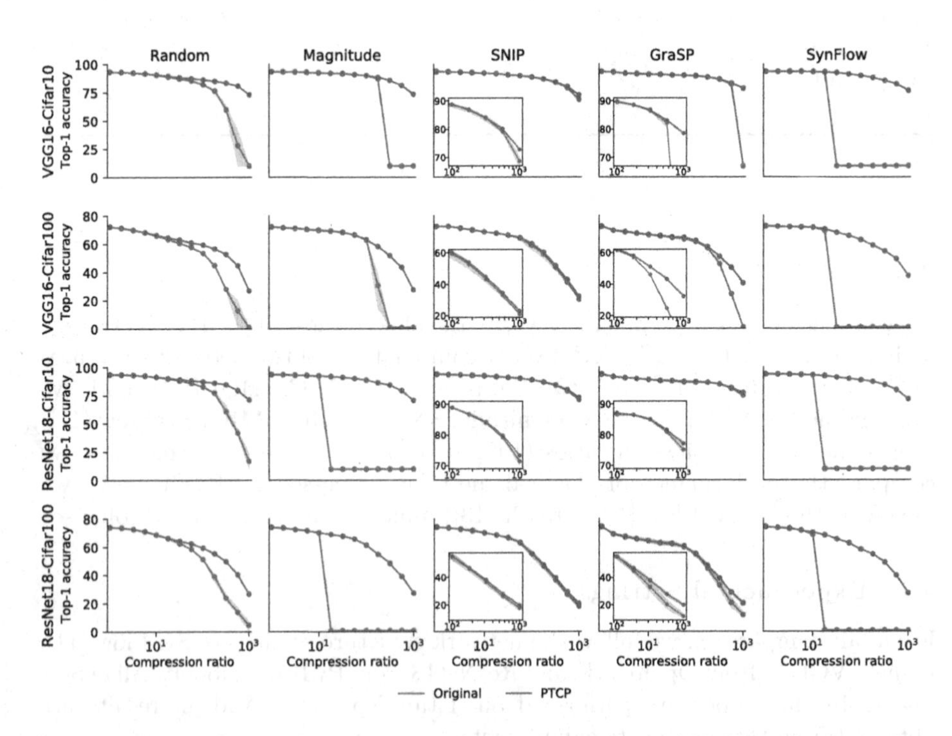

Fig. 3. PTCP effectively improves the performance of PaI methods under any setting. The solid line represents the mean, and the shaded region represents the area between the minimum and maximum performance for the three runs.

Effectiveness of PTCP Strategy. We evaluate the effectiveness of the PTCP strategy on the image classification task with twenty combinations of two networks structures (VGG16 [14], ResNet18 [15]), two datasets (Cifar10, Cifar100 [16]), and five pruning methods (Random [8], Magnitude, SNIP [9], GraSP [10], SynFlow [11]). The results are shown in Fig. 3. The PTCP strategy significantly alleviates layer collapse and improves subnet performance at high compression ratios. Meanwhile, the PTCP strategy demonstrates more stability as its tight interval indicates. Notably, the PTCP strategy is an improvement to existing PaI methods. Therefore, the quality of the original parameter evaluation indicators still affects the performance improved by PTCP.

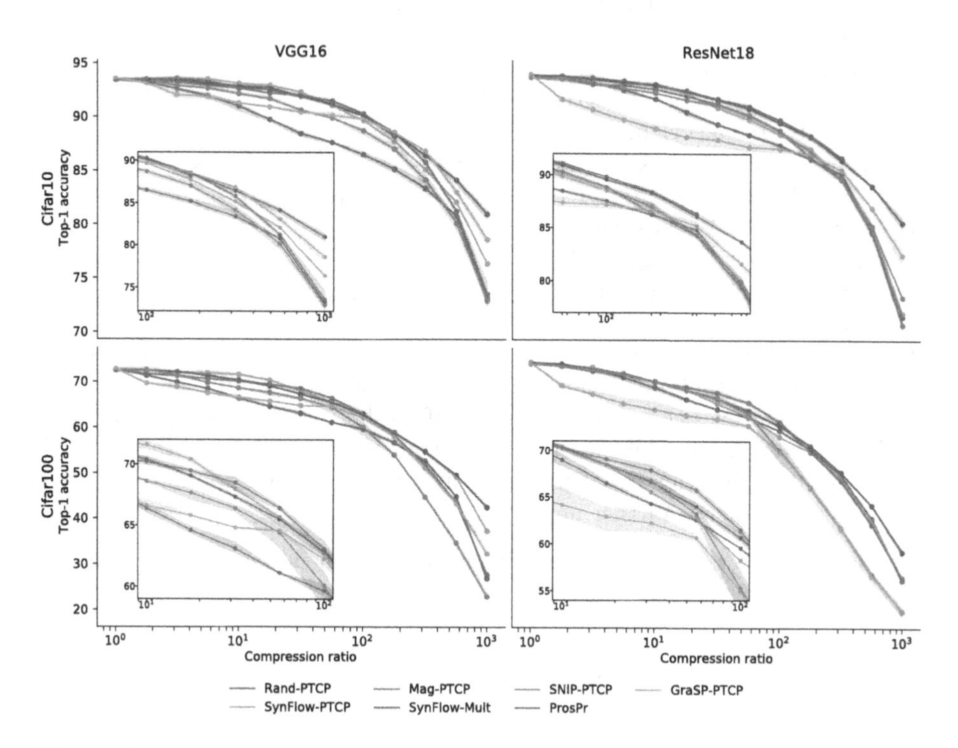

Fig. 4. The performance of the PTCP-improved PaI methods matches the SOTA methods at some sparsity.

Comparing to Expensive SOTA Methods. We compare the performance of the above five PTCP-improved PaI methods with existing SOTA methods that introduce extra costs to alleviate layer collapse. The results are shown in Fig. 4. Some PTCP-improved methods perform the same as SOTA methods and even perform better at medium compression ratios, such as VGG16-Cifar100-SynFlow and ResNet18-Cifar100-Mag. At the same time, it is observed that Mag-PTCP is superior to the SNIP method in most scenarios, indicating that PTCP can exploit the potential of the pruning method based on basic standards such as magnitude, which leads to thinking about the increasingly complex pruning standards.

4.3 Ablation Study

We conduct ablation studies under the setting of VGG16-Cifar100-GraSP.

Parameter Ablation: We ablate the compensation coefficient α in Eq. 2, and the results are shown in Fig. 5. Limited by the constraint shown in Eq. 4, α can only be ablated in the decreasing direction. When α respectively take UL, $UL/2$, and $UL/3$, the average performance growths of PTCP relative to the original method over a range of compression ratios $10^n, n = [2, 2.25, ..., 3]$ are 11.13%, 10.22%, 8.89%. As α gets smaller, the performance of our strategy gradually deteriorates, which verifies the rationality of the parameter setting.

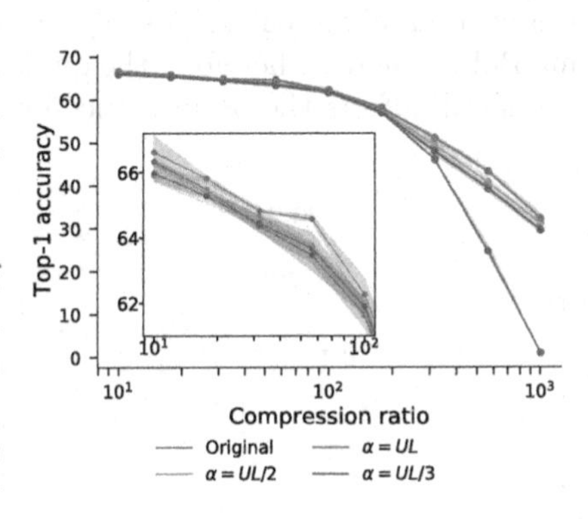

Fig. 5. Ablation results of coefficient α

Modular Ablation: We perform modular ablation on our two modules, PTC and PP, and the results are shown in Table 1. The results indicate that PTC significantly improves the performance of PaI methods by an average growth of 6.21%. However, limited by the unbalanced transfer of parameters, there is still an unavoidable performance degradation in some cases. After adding PP, the average performance is further improved.

Table 1. Modular ablation results. Bold numbers are the highest in the results.

Module		Compression ratio								
PTC	PP	10^1	$10^{1.25}$	$10^{1.5}$	$10^{1.75}$	10^2	$10^{2.25}$	$10^{2.5}$	$10^{2.75}$	10^3
✗	✗	66.34 ± 0.09	65.49 ± 0.08	64.38 ± 0.23	63.46 ± 0.23	61.63 ± 0.38	57.12 ± 0.33	46.14 ± 0.13	24.75 ± 1.10	1.00 ± 0.00
✓	✗	$\mathbf{66.70 \pm 0.42}$	65.62 ± 0.04	$\mathbf{64.93 \pm 0.28}$	64.40 ± 0.23	62.10 ± 0.20	57.58 ± 0.19	50.44 ± 0.14	42.60 ± 0.08	31.82 ± 0.46
✓	✓	66.62 ± 0.37	$\mathbf{65.82 \pm 0.10}$	64.84 ± 0.02	$\mathbf{64.59 \pm 0.08}$	$\mathbf{62.27 \pm 0.42}$	$\mathbf{58.12 \pm 0.14}$	$\mathbf{51.17 \pm 0.59}$	$\mathbf{43.39 \pm 0.50}$	$\mathbf{32.45 \pm 0.76}$

5 Conclusions

In this paper, we show that PaI suffers from severe layer collapse at high compression ratios. We propose the parameter threshold compensation strategy to avoid the extra cost of existing alleviation methods, which improves the performance of PaI methods based on parameter constraints. And to further promote a more balanced transfer of parameters, we propose the parameter preservation strategy to impose more substantial constraints on *Out-layer*. Finally, we conduct experiments on twenty combinations of different network structures, datasets, and pruning methods to verify that our strategy can significantly improve performance. Then we compare the performance improved by our strategy with SOTA methods, showing similar performance in some cases, although we do not require any additional cost. Promising future directions for this work are to (i) design parameter threshold with higher flexibility to compensate network layers of different types and sizes accurately and (ii) look for more improvements besides layer collapse in PaI to improve performance further. Overall, our strategy effectively improves the performance of current PaI methods, challenging the increasingly complex design of pruning strategies.

Acknowledgement. This work was supported in part by National Natural Science Foundation of China (No. 61831005).

References

1. Krizhevsky, A., Sutskever, I., Hinton, G.E.: Imagenet classification with deep convolutional neural networks. Commun. ACM **60**(6), 84–90 (2017)
2. Long, J., Shelhamer, E., Darrell, T.: Fully convolutional networks for semantic segmentation. In: Proceedings of the IEEE Conference on Computer Vision and Pattern Recognition (CVPR), pp. 3431–3440 (2015)
3. Ren, S., He, K., Girshick, R., Sun, J.: Faster R-CNN: towards real-time object detection with region proposal networks. In: Cortes, C., Lawrence, N., Lee, D., Sugiyama, M., Garnett, R. (eds.) Advances in Neural Information Processing Systems, vol. 28. Curran Associates, Inc. (2015)
4. Vaswani, A., et al.: Attention is all you need. In: Guyon, I., Luxburg, U.V., Bengio, S., Wallach, H., Fergus, R., Vishwanathan, S., Garnett, R. (eds.) Advances in Neural Information Processing Systems, vol. 30. Curran Associates, Inc. (2017)
5. Wen, J., Zhao, Z., Cui, J., Chen, B.M.: Model-based reinforcement learning with self-attention mechanism for autonomous driving in dense traffic. In: Tanveer, M., Agarwal, S., Ozawa, S., Ekbal, A., Jatowt, A. (eds.) ICONIP 2022. LNCS, pp. 317–330. Springer, Cham (2023). https://doi.org/10.1007/978-3-031-30108-7_27
6. Graves, A., Mohamed, A.r., Hinton, G.: Speech recognition with deep recurrent neural networks. In: 2013 IEEE International Conference on Acoustics, Speech and Signal Processing, pp. 6645–6649 (2013)
7. Kaplan, J., et al.: Scaling laws for neural language models. arXiv preprint arXiv:2001.08361 (2020)
8. Liu, S., et al.: The unreasonable effectiveness of random pruning: return of the most naive baseline for sparse training. In: International Conference on Learning Representations (2022)
9. Lee, N., Ajanthan, T., Torr, P.: Snip: single-shot network pruning based on connection sensitivity. In: International Conference on Learning Representations (2019)
10. Wang, C., Zhang, G., Grosse, R.: Picking winning tickets before training by preserving gradient flow. In: International Conference on Learning Representations (2020)
11. Tanaka, H., Kunin, D., Yamins, D.L., Ganguli, S.: Pruning neural networks without any data by iteratively conserving synaptic flow. In: Larochelle, H., Ranzato, M., Hadsell, R., Balcan, M., Lin, H. (eds.) Advances in Neural Information Processing Systems, vol. 33, pp. 6377–6389. Curran Associates, Inc. (2020)
12. Alizadeh, M., et al.: Prospect pruning: finding trainable weights at initialization using meta-gradients. In: International Conference on Learning Representations (2022)

13. Frankle, J., Dziugaite, G.K., Roy, D., Carbin, M.: Pruning neural networks at initialization: Why are we missing the mark? In: International Conference on Learning Representations (2021)
14. Simonyan, K., Zisserman, A.: Very deep convolutional networks for large-scale image recognition. In: International Conference on Learning Representations (2015)
15. He, K., Zhang, X., Ren, S., Sun, J.: Deep residual learning for image recognition. In: 2016 IEEE Conference on Computer Vision and Pattern Recognition (CVPR), pp. 770–778 (2016)
16. Krizhevsky, A., Hinton, G.: Learning multiple layers of features from tiny images. Tech. Rep. 0, University of Toronto, Toronto, Ontario (2009)
17. Li, H., Kadav, A., Durdanovic, I., Samet, H., Graf, H.P.: Pruning filters for efficient convnets. In: International Conference on Learning Representations (2017)
18. Liu, Z., Li, J., Shen, Z., Huang, G., Yan, S., Zhang, C.: Learning efficient convolutional networks through network slimming. In: Proceedings of the IEEE International Conference on Computer Vision (ICCV) (2017)
19. Han, S., Pool, J., Tran, J., Dally, W.: Learning both weights and connections for efficient neural network. In: Cortes, C., Lawrence, N., Lee, D., Sugiyama, M., Garnett, R. (eds.) Advances in Neural Information Processing Systems, vol. 28. Curran Associates, Inc. (2015)
20. Le Cun, Y., Denker, J.S., Solla, S.A.: Optimal brain damage. In: Proceedings of the 2nd International Conference on Neural Information Processing Systems, NIPS 1989, pp. 598–605. MIT Press, Cambridge (1989)
21. Wang, H., Qin, C., Bai, Y., Zhang, Y., Fu, Y.: Recent advances on neural network pruning at initialization. In: Raedt, L.D. (ed.) Proceedings of the Thirty-First International Joint Conference on Artificial Intelligence, IJCAI-22, pp. 5638–5645. International Joint Conferences on Artificial Intelligence Organization (2022)
22. Lee, N., Ajanthan, T., Gould, S., Torr, P.H.S.: A signal propagation perspective for pruning neural networks at initialization. In: International Conference on Learning Representations (2020)
23. Mallya, A., Lazebnik, S.: Packnet: adding multiple tasks to a single network by iterative pruning. In: Proceedings of the IEEE Conference on Computer Vision and Pattern Recognition (CVPR) (2018)
24. Wortsman, M., et al.: Supermasks in superposition. In: Larochelle, H., Ranzato, M., Hadsell, R., Balcan, M., Lin, H. (eds.) Advances in Neural Information Processing Systems, vol. 33, pp. 15173–15184. Curran Associates, Inc. (2020)
25. Frankle, J., Carbin, M.: The lottery ticket hypothesis: finding sparse, trainable neural networks. In: International Conference on Learning Representations (2019)
26. Chen, T., et al.: The lottery ticket hypothesis for pre-trained bert networks. In: Larochelle, H., Ranzato, M., Hadsell, R., Balcan, M., Lin, H. (eds.) Advances in Neural Information Processing Systems, vol. 33, pp. 15834–15846. Curran Associates, Inc. (2020)
27. Frankle, J., Schwab, D.J., Morcos, A.S.: The early phase of neural network training. In: International Conference on Learning Representations (2020)
28. Evci, U., Ioannou, Y., Keskin, C., Dauphin, Y.: Gradient flow in sparse neural networks and how lottery tickets win. In: Proceedings of the AAAI Conference on Artificial Intelligence 36(6), 6577–6586 (2022)

29. Su, J., et al.: Sanity-checking pruning methods: Random tickets can win the jackpot. In: Larochelle, H., Ranzato, M., Hadsell, R., Balcan, M., Lin, H. (eds.) Advances in Neural Information Processing Systems, vol. 33, pp. 20390–20401. Curran Associates, Inc. (2020)
30. Liu, N., et al.: Lottery ticket implies accuracy degradation, is it a desirable phenomenon. CoRR (2021)

Hierarchical Attribute-Based Encryption Scheme Supporting Computing Outsourcing and Time-Limited Access in Edge Computing

Ke Li[1(✉)], Changchun Li[2], and Jun Shen[1]

[1] China Telecom Corporation Limited Research Institute, Guangzhou 510000, China
lik8@chinatelecom.cn

[2] Tianyi Cloud Technology Corporation Limited, Guangzhou 510000, China

Abstract. With the rapid increase of user data and traffic, the traditional attribute encryption scheme based on the central cloud will cause the bottleneck of computing performance. And user's access privilege and ciphertext in the existing scheme is not limited by the time duration and the number of attempts, which could be brute force attack. We propose a solution to support computing outsourcing and time-limited access in edge computing. Edge nodes can shorten data transmission distances and eliminate latency issues. To solve the central cloud performance problem during encryption and decryption, massive and complex computing is considered outsource to edge nodes. And set valid time for the ciphertext and the user key, which avoid their permanent validity and significantly improve data security. In addition, all attributes are divided into attribute trees. According to the hierarchical relationship between attributes, we judge the user's access privilege. Finally, we give security proof, performance cost, functional comparison of the scheme.

Keywords: Hierarchical Attribute-Based Encryption · Time-Limited · Computing Outsourcing · Edge Computing

1 Introduction

With the rapid development of information technology and communication technology, data shows an explosive growth. The secure transmission and storage of data have become an important issue that users pay more attention. Cloud computing is unable to manage the increasing amount of user data and huge calculations. In addition, cloud services are not suitable for delay, portability and location-sensitive applications. Edge computing is an important application of cloud computing. Edge cloud is closer to the user or terminal. It can provide more convenient and fast computing power, and solve the problem of limited mobile terminal resources and excessive computing burden of cloud server. However, data is stored in the cloud or edge cloud, which brings privacy and security issues. Such as user access policies and identity tracing still threaten data security. In order to protect the confidentiality of user data, attribute based encryption (ABE) has become one of the important choices for users to encrypt data.

B. Luo et al. (Eds.): ICONIP 2023, CCIS 1965, pp. 52–64, 2024.
https://doi.org/10.1007/978-981-99-8145-8_5

Sahai and waters [1] proposed ABE in 2005. Since then, different scholars have proposed derivative schemes of ABE, such as key-policy ABE (KP-ABE) [2], ciphertext-policy ABE (CP-ABE) [3], hierarchical attribute based encryption (HABE) [4] and multi-authority ABE(MA-ABE) [5]. Huang Q. [6] proposed a HABE scheme with secure and efficient data collaboration in cloud computing. In this scheme, most of the computing cost is entrusted to cloud service providers (CSP), and attribute authorities (AA) are managed in layers. Leng [7] proposed an ABE scheme for encryption outsourcing in cloud. According to the characteristics of the access tree, the scheme [7] constructs the shared access strategy into an equivalent matrix of a general matrix, which greatly reduces the calculation of the user. Qi [8] proposed a multi-authority attribute-based encryption scheme with attribute hierarchy. The encryption and decryption in the above references are based on the cloud server, without considering the performance bottleneck of the cloud server. Peng [9] proposed an ABE scheme in edge computing, which outsources decryption calculation to edge nodes, and uses multi-authority (MA) to meet the performance requirements of users' cross-domain access. K. Huang [10] proposed an ABE scheme that supports online encryption and offline encryption, and supports outsourced decryption. It uses methods such as split encryption and reuse ciphertext to protect data privacy and realize fine-grained access control. Wang [11] combined fog computing technology to propose a micro attribute encryption scheme that supports computing outsourcing. This scheme shortens the length of key and ciphertext, and transfers part of the calculation to the fog node. However, there are not considered the timeliness of key and ciphertext.

Based on the above research, there are problems in the HABE scheme, such as high computational overhead, and inability to support time-sensitive access for users. We propose a hierarchical attribute-based encryption scheme supporting computing outsourcing and time-limited access in edge computing. We use the computing power of edge nodes to handle complex encryption and decryption calculations, which improves computing efficiency and saves resources for users and the central cloud. And we set the time limit of ciphertext and key according to specific scenarios, which enriches the fine-grained access strategy. In order to identify the user, we propose to embed ID identifiers into users' private keys. Different users have different private keys and IDs, which helps to find and lock malicious users and improve the security of users and the system.

Ours main goal is to establish a secure and confidential communication and data storage method in the edge environment, specifically:

Efficiency: Put complex encryption and decryption calculations on edge nodes, and make full use of all system resources to achieve efficiency.

Time Limit: Limit the valid time of users' private keys and ciphertexts to avoid long-term abuse of keys and ciphertexts.

Access control: Use the secret sharing protocol and hierarchical attribute tree, and judge the user's decryption authority according to the attribute path relationship between the private key and the ciphertext.

2 Preliminaries

2.1 Bilinear Groups Pairing

Boneh et al. introduced bilinear groups pairing [12]. Let G_0 and G_1 be two multiplicative cyclic groups of prime order p. Let g be a generator of G_0 and e be a bilinear groups pairing, $e : G_0 \times G_0 \to G_1$ which has the following properties:

Bilinear: All $\mu, v \in G_0$ and all $a, b \in Z_P$ satisfy the equation $e(\mu^a, v^b) = e(\mu, v)^{ab}$.

Non-degeneracy: The pairing does not map all the elements in $G_0 \times G_0$ to the unit of G_1, that is, there exists $g \in G_0$ such that $e(g, g) \neq 1$.

Computable: Randomly select two elements μ, v, there is an effective algorithm to calculate $e(\mu, v)$.

Note that the pairing $e(g, g)$ is symmetric because $\mathrm{e}(g^a, g^b) = e(g, g)^{ab} = \mathrm{e}(g^b, g^a)$.

2.2 Linear Secret Sharing Scheme

The secret sharing scheme is known as the threshold scheme by A. Beimel [13]. A secret is divided into n copies and distributed to n managers (e.g. users or attribute authority). In the (k, n) threshold, the secret shares only meet more than k to restore the original secret. This process can be described by Lagrange interpolation.

Lagrange coefficient:

$$\Delta_{i,S}(x) = \prod_{\vartheta \in S, \vartheta \neq i} \frac{x - \vartheta}{i - \vartheta} \tag{1}$$

Choose any k shares, and restore the secret:

$$F(x) = \sum_{i=1}^{k} \left(y_i \times \Delta_{i,S}(x) \right) \tag{2}$$

Among them, the elements in the set S are composed of Z_P, and $i \in Z_P$, y_i is the secret share and $y_i = F(x_i)$.

2.3 Security Assumption

l-th Billinear Diffie-Hellman Inversion problem($l - wBDHI$): Let G_0 and G_1 be two multiplicative cyclic groups of prime order p. Let g be a generator of G_0 and e be a bilinear groups pairing. Randomly select two elements $\beta, z (\beta, z \in Z_P)$, and let $g, g^\beta, g^{\beta^2}, \ldots\ldots g^{\beta^l} \in G_0, T \in G_1$, then determine whether T is equal to $e(g, g)^{\beta^{l+1}}$, if

$$\begin{aligned} &\left| Pr\left[A\left(g, y_1 = g^\beta, \ldots\ldots y_l = g^{\beta^l}, e(g, g)^{\beta^{l+1}} \right) = 1 \right] \right. \\ &\left. - Pr\left[A(g, y_1 = g^\beta, \ldots\ldots y_l = g^{\beta^l}, e(g, g)^z) = 1 \right] \right| \geq \varepsilon \end{aligned} \tag{3}$$

l-th Billinear Diffie-Hellman Inversion assumption($l - wBDHI$): There is no polynomial time algorithm A that can solve the $l - wBDHI$ problem with the non-negligible advantage of ε.

3 System Structure

3.1 System Model

We design a hierarchical attribute-based encryption scheme that supports computing outsourcing time-limited access in the edge environment. It has timeliness and traceability, and enriches the functional expression of hierarchical attribute-based encryption. Moreover, we use edge nodes as medium of outsourcing encryption and decryption, which is helpful for privacy data protection, improves the calculation efficiency and resource utilization efficiency. This system involves six entities. Data Owner (DO) encrypts data by defining attribute access control policies, uploads and storage the encrypted data to the cloud. Data User (DU) downloads the ciphertext and uses the private key to decrypt the ciphertext. DU is allowed to access the ciphertext when the private key of DU meets the ciphertext access policy. Cloud Server Provider (CSP) is responsible for ciphertext transmission and storage, manages edge nodes, and provides data access and storage services for DU and DO. Edge Node (EN) is also responsible for ciphertext transmission and storage, and provides encryption and decryption operations, reducing the computing burden on the central cloud and users. Central Authority (CA) is a global certificate authority trusted by the system. It accepts the registration of all authorities and all users in the system and issues global unique ID for DU. Attribute Authority (AA) is independent from each other, generates key shares for users, and is responsible for issuing, revoking, and updating user attributes in its management domain. And AA is responsible for generating attribute trees, to help users and edge nodes use hierarchical relationships between attributes for encryption and decryption. The system architecture is shown in Fig. 1.

The system model mainly includes the following algorithms, and the specific description of each algorithm is as follows.

Setup: Input security parameters, output global public key (GPK), master private key (MSK), public key (PK_i) and private key (SK_i) of AA.

EN Encryption: Input GPK, PK_i, and randomly select encryption parameters, output ciphertext of the edge node CT_1.

DO Encryption: Input CT_1, plaintext m, DO sets the ciphertext valid time, and randomly selects parameters, output final ciphertext CT.

DU KeyGeneration: Input MSK and SK_i, select the key valid time, user attribute set and user identity ID, and output the user private key SK_u.

EN KeyGeneration: Input SK_u, output the edge keySK_t.

Time Check: Input the key and ciphertext valid time, check whether their hash values are the same and output 1. If both times are within the validity period, the decryption phase is entered.

EN Decryption: Input SK_t and CT, output the decrypted intermediate ciphertext CT_2.

DU Decryption: Input CT_2 and SK_u, output plaintext m.

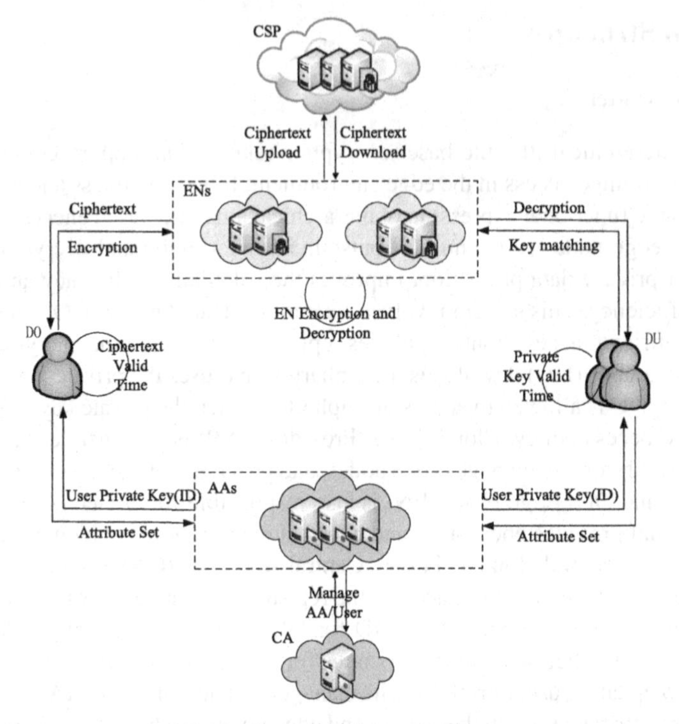

Fig. 1. System Architecture Figure.

3.2 Scheme Description

GlobalSetup. Let G and G_T be two multiplicative cyclic groups of prime order p. Let g be a generator of G and e be a bilinear groups pairing. $e : G \times G \to G_T$. Define two hash functions: $H_1 : \{0, 1\}^* \to Z_P$ and $H_2 : \{0, 1\}^* \to G$. There are N attributes in the system, which are divided into n attribute sets and then constructed into n attribute trees. The root nodes of attribute trees are $U_0 = \{\omega_{10}, \omega_{20}, \ldots\ldots, \omega_{n0}\}$. l_i is the depth of the i-th attribute tree, and the most depth of attribute trees is $l = \max\{l_1, l_2, \ldots\ldots, l_n\}$. Randomly select attribute parameters, denoted as $V = \{v_1, v_2, \ldots\ldots, v_n\}$ and $U = \{u_1, u_2, \ldots\ldots, u_l\}$. Randomly select elements as input parameters. So we can get the system public key is GPK and the system master private key is MSK.

$$GPK = \left\{p, e, g, g^y, V, U\right\} \tag{4}$$

$$MSK = \{y\} \tag{5}$$

AuthoritySetup. There are M attribute authorities in the system, which manage different attribute sets. Each authority manages m attributes. Input the system master key MSK, and each authority randomly selects parameters α_i and β_i to satisfy equation $y = \sum_{i=1}^{M} \alpha_i \cdot \beta_i$. The public key is PK_i and the private key is SK_i of the i-th authority.

$$PK_i = \left\{g^{\alpha_i}, g^{\beta_i}\right\} \tag{6}$$

$$SK_i = \{\alpha_i, \beta_i\} \tag{7}$$

EN Encryption. DO selects the ciphertext attribute set and transfers it to the edge node. There is a hierarchical relationship between attributes. Suppose an attribute is located in the j-th attribute tree, and the depth of the attribute tree is h_1, then the ciphertext attribute can be represented as $t_{j\delta}|_{1 \le j \le n, 1 \le \delta \le h_1}$, where j represents the number of the attribute tree, and δ represents the depth of the attribute. DO sets the ciphertext valid time to prevent the ciphertext from being deciphered for a long time. Using the hash function to represent the valid time as $H_2(time)$. Input PK_i and randomly select an encryption parameter s. Then, the encrypted ciphertext of the edge node is CT_1.

$$CT_1 = \left\{ \begin{array}{c} C_1 = \left(v_j \prod_{\delta=1}^{h_1} u_\delta^{H_1(t_{j\delta})}\right)^s, \\ C_2 = g^{\alpha_i s}, \\ C_3 = g^{\beta_i s}, \\ C_4 = g^{H_2(time) \cdot s}, \\ C_5 = g^s, \\ C_6 = e(g,g)^{ys} \end{array} \right\} \tag{8}$$

DO Encryption. DO randomly selects the encryption parameter t, and obtains the final ciphertext CT based on the intermediate ciphertext encrypted by the edge node.

$$\mathrm{CT} = \left\{ \begin{array}{c} C_0 = me(g,g)^{y(t+s)}, \\ C_1 = \left(v_j \prod_{\delta=1}^{h_1} u_\delta^{H_1(t_{j\delta})}\right)^s, \\ C_2 = g^{\alpha_i s}, \\ C_3 = g^{\beta_i s}, \\ C_4 = g^{H_2(time) \cdot s}, \\ C_5 = g^s, \\ C_6 = e(g,g)^{ys}, \\ C_7 = g^t \end{array} \right\} \tag{9}$$

DU KeyGeneration. The user private key is closely related to the user attributes. The attributes of the user are represented as $t_{i\varphi}|_{1 \le i \le n, 1 \le \varphi \le h_2}$, where i represents the number of the attribute tree and φ is the depth of the attribute. The user private key also has an valid time, and the hash function expresses the valid time of the key as $H_2\left(time'\right)$. In order to trace the user identity, using the hash function $H_1(ID)$ to express identity information. Assuming that the secret segmentation threshold of the authority is d, a $d-1$ degree polynomial $q(x)$ is selected for each attribute authority to satisfy the equation $q(0) = \beta_i$. Randomly select security parameters γ as the input of the user's private key, so we get the user's key is SK_u.

$$SK_u = \left\{ \begin{array}{c} SK_1 = g^{\gamma}, \\ SK_2 = g^{H_2\left(time'\right)\cdot\gamma}, \\ SK_3 = g^{\alpha_i\gamma}, \\ SK_4 = g^{\beta_i\gamma}, \\ SK_5 = g^{\beta_i H_1(ID)}, \\ SK_6 = g^{\gamma+\beta_i H_1(ID)}, \\ SK_j = \left(v_j \prod_{\varphi=1}^{h_2} u_{\varphi}^{H_1\left(t'_{j\varphi}\right)}\right)^{\gamma} \cdot g^{q\left(H_1\left(t'_{j\varphi}\right)\right)}, \\ SK_{j,h+1} = u_{h+1}^{\gamma}, \ldots\ldots, SK_{jl_j} = u_{l_j}^{\gamma} \end{array} \right\} \tag{10}$$

EN KeyGeneration. In the process of outsourcing decryption, the edge node needs to use the user's private key to decrypt the ciphertext according to the needs of different users. So the edge key SK_t serves the user as part of the user's key.

$$SK_t = \left\{ g^{H_2\left(time'\right)\cdot\gamma}, g^{\alpha_i\gamma}, g^{\beta_i\gamma}, \left(v_j \prod_{\varphi=1}^{h_2} u_{\varphi}^{H_1\left(t'_{j\varphi}\right)}\right)^{\gamma} \cdot g^{q\left(H_1\left(t'_{j\varphi}\right)\right)}, u_{h+1}^{\gamma}, \ldots\ldots, u_{l_j}^{\gamma} \right\} \tag{11}$$

Time Check. In order to prevent ciphertext and key from brute force cracking, a time check algorithm is proposed to associate ciphertext and key with time to increase its flexibility and randomness. By calculating the hash value of the ciphertext validity time and the key validity time, it can determine whether the time calibration is reasonable.

$$H_2(time) = H_2\left(time'\right) = 1 \tag{12}$$

EN Decryption. According to the judgment of the time check algorithm, when the Eq. (12) is satisfied, the edge node is preliminarily decrypted. In this stage, using the edge key SK_t to decrypt, and obtaining the intermediate ciphertext CT_2.

$$CT_2 = \prod_{j=1, t\in U} \left[\frac{e(C_2, SK'_j)\cdot e(C_4, SK_4)}{e(C_1, SK_3)\cdot e(C_3, SK_2)} \right]^{\Delta_{H(t),s}(0)} \tag{13}$$

DU Decryption. DU downloads the decrypted intermediate ciphertext CT_2 from the edge node, and uses private key SK_t to further decrypt, and obtains the plaintext m.

$$m = \frac{C_0}{e(C_7, \frac{SK_6}{SK_5})\cdot \prod_{i=1}^{M} CT_2} \tag{14}$$

4 Security Proof

In this scheme, the security specification is based on the probability that the attacker solves the difficult mathematical problem. We use the security game of the attacker and the challenger to prove the security. If the difficulty problem can be solved with a non-negligible probability, it means that the attacker can breach the model with a non-negligible advantage, so the model is unsafe. Conversely, we consider the model safe.

Theorem. If the $l - wBDHI$ assumption is true, the attacker A cannot win the security game in the probability polynomial time. The advantage of security in this paper is the possibility of solve the $l - wBDHI$ problem.

Proof. By using the method of proof by contradiction, it is assumed that an attacker A can attack this scheme with a non-negligible advantage ε, then a simulator B can solve $l - wBDHI$ difficult problems with a non-negligible advantage $\frac{\varepsilon}{2}$.

The challenger C sets two groups as G_1 and G_2 respectively. Both G_1 and G_2 are cyclic groups of order p, and the generator of G_1 is g. Let $y_i = g^{x_i}$. It is assumed that there are N authorization authorities in the system, and n attribute trees for all attributes. The depth of the i-th attribute tree is l_i, and the maximum depth of all attribute trees is l and expressed as $l = max\{l_i\}_{1 \leq i \leq n}$. Define hash functions $H_1 : \{0, 1\}^* \to Z_P$ and $H_2 : \{0, 1\}^* \to G$. The challenger C randomly selects $\mu \in \{0, 1\}$.

If $\mu = 0$, the challenger C selects $T = e(g, g)^{x^{l+1}}$.

If $\mu = 1$, the challenger C selects $T = e(g, g)^{\gamma}$, where $\gamma \in Z_P$.

The challenger C sends $\{g, e, G_1, G_2, y_1, \ldots\ldots, y_l, T\}$ to the simulator B and plays the security game.

Init. The attacker A randomly selects an attribute set U^* to be challenged and sends it to the challenger C. There are v elements in the attribute set U^*, represented by $U^* = \left\{t^*_{i_1}, t^*_{i_2}, \ldots\ldots, t^*_{i_v}\right\}$. The depth of each element in the attribute tree is $\{k_1, k_2, \ldots\ldots, k_v\}$, so the path of the attribute t^* is defined as $\left\{t^*_{i0}, t^*_{i1}, \ldots\ldots, t^*_{i,k_i-1}, t^*\right\}$. The implication is that the attribute t^* is in the attribute tree with root node t^*_{i0} and depth k_i.

Setup. The challenger C is initialized using the system model, and the simulator B generates parameters g and p. Specify $g_1 = y_1$, $g_2 = y_l$, and $u_i = y_{l-i+1}$, where $1 \leq i \leq l$. For $i \notin \{i_1, i_2, \ldots\ldots, i_v\}$, the system randomly selects $a_i \in Z_P$, so there is $x_i = g^{a_i}$. For i $\in \{i_1, i_2, \ldots\ldots, i_v\}$, there is $x_i = \frac{g^{a_i}}{\prod_{\delta=1}^{k_i} y_{l-i+1}{}^{t^*_{i\delta}}}$. The system publishes x_i as a public key. The challenger C gets the system public key, the system master private key, the public and private keys of authorities. The attacker A gets the system public key and the public keys of authorities.

Phase 1. The attacker A arbitrarily constructs the attribute set R so that the elements in the set R do not meet the attribute set U^*. For $\forall\omega \in$ R, suppose the attribute ω is in the d-th attribute tree and the depth is p, then the path is represented as $L_\omega = \left(\omega_{d0}, \omega_{d1}, \ldots\ldots, \omega_{d,p-1}, \omega\right)$. Define a attribute set R^*, where R^* satisfies $R^* \subseteq R$ and $|R^*| = d-1$. According to the characteristics of hierarchical attributes, all attributes in R^* can override attributes in U^* on the path. Let $S = R^* \cup \{0\}$. For $\forall t \in R^*$, $\mu \in Z_P$, there is

$q(H_1(t)) = \mu$. The simulator B constructs a polynomial $q(z)$ for each attribute authority, which is $d - 1$ degree and $q(0) = x$. For $\forall t \notin S$, the system computes $q(H_1(t)) = \sum \Delta_{t,S}(H_1(t))q(H_1(t)) + \Delta_{0,S}(H_1(t))q(0)$. Therefore, the simulator B outputs the final private key and returns it to A.

When $t \in S$,

$$SK_R = \left\{ \begin{array}{c} SK_{j0} = \left(x_j \prod_{\theta=1}^{k_\theta} u_\theta^{H_1(t_{j\theta})}\right)^r g^{q(H_1(t_{j\theta}))}, \\ SK_{j,k_i+1} = u_{k_i+1}^r, \ldots\ldots SK_{jl_j} = u_{l_j}^r, \\ SK_j = g^r \end{array} \right\}, \text{ where } r \in Z_P \tag{15}$$

When $t \notin S$, let $r' = \frac{r \cdot x}{H_1(t_{j\theta}) - H_1\left(t'_{j\theta}\right)}$. Because $t \notin S$ and $t' \in S$, there is an equation $H_1\left(t_{j\theta}\right) \neq H_1\left(t'_{j\theta}\right)$. So

$$SK'_R = \left\{ \begin{array}{c} SK_{j0} = g_2^{\frac{r}{H_1(t_{j\theta}) - H_1\left(t'_{j\theta}\right)}} g_1^{r'} \left(\prod_{\theta=1}^{k_\theta} u_\theta^{H_1(t_{j\theta})}\right)^{r'}, \\ SK_{j,k_i+1} = u_{k_i+1}^{r'}, \ldots\ldots SK_{jl_j} = u_{l_j}^{r'}, \\ SK_j = g_1^{\frac{r}{H_1(t_{j\theta}) - H_1\left(t'_{j\theta}\right)}} \end{array} \right\} \tag{16}$$

Challenge. After the attacker A completes the key query in phase 1, then selects two equal-length plaintexts M_0 and M_1 and sends to the challenger C. C randomly throws a coin to get a bit, then the simulator B also selects the same bit $\mu \in \{0, 1\}$ and encrypts message M_μ which satisfies the attribute set R. B generates a ciphertext CT^* and sends to A. The ciphertext is

$$CT^* = \left\{ M_\mu T, \left(x_j \prod_{\delta=1}^{k_\delta} u_\delta^{t_{j\delta}}\right)^s, g^s, g^{H_2(time)s} \right\} \tag{17}$$

Phase 2. The attacker A performs the second key query to decrypt the challenge ciphertext M_μ. The interaction between A and C is the same as the phase 1.

Guess. The attacker A answers which message is encrypted and outputs μ'. The advantage of winning the game is $Pr\left(\mu = \mu'\right) - \frac{1}{2}$.

If the simulator B outputs $\xi = 0$ and$\mu = \mu'$, it means that the attacker A gets the encrypted ciphertext, which is $T = e(g, g)^{x^{l+1}}$. It is assumed that the advantage of attacking is ε, there is$Pr\left(\mu = \mu' | \xi = 0\right) = \varepsilon + \frac{1}{2}$.

If the simulator B outputs $\xi = 1$, it means that A cannot get the ciphertext. Because of $T = e(g, g)^\gamma$ and γ is a random number. So the attacker A cannot recover plaintext, there is $Pr\left(\mu \neq \mu' | \xi = 1\right) = \frac{1}{2}$.

Based on the above discussion, the advantage that the attacker A wins the game is

$$Pr\left(\mu = \mu'\right) - \frac{1}{2} = Pr\left(\mu = \mu' | \xi = 0\right) \cdot Pr(\xi = 0) + Pr\left(\mu \neq \mu' | \xi = 1\right) \cdot Pr(\xi = 1) - \frac{1}{2} = \left(\varepsilon + \frac{1}{2}\right) \times \frac{1}{2} + \frac{1}{2} \times \frac{1}{2} - \frac{1}{2} = \frac{1}{2}\varepsilon \tag{18}$$

If the attacker A can decrypt the ciphertext with probability ε, it means that B can solve the difficult $l - wBDHI$ problem with probability $\frac{\varepsilon}{2}$. This is inconsistent with the fact and cannot be true, so the assumption that the attacker can decrypt the ciphertext is not valid. Therefore, we confider that the security of this scheme meets IND-sAtr-CPA.

5 Performance Analysis

5.1 Security Analysis

We outsource encryption and decryption, and use the edge node to handle the complex and costly computation. Through the matching of the ciphertext and the key, the first decryption is performed at the edge node to obtain CT_2, and the second decryption is performed at the user side to obtain m.

$$CT_2 = \prod_{j=1, t \in U} \left[\frac{e\left(C_2, SK_j'\right) \cdot e(C_4, SK_4)}{e(C_1, SK_3) \cdot e(C_3, SK_2)} \right]^{\Delta_{H(t),s}(0)}$$

$$= \prod_{j=1, t \in U} \left[\frac{e\left(g^{\alpha_i s}, \left(v_j \prod_{\varphi=1}^{h_2} u_\varphi^{H_1\left(t_{j\varphi}'\right)}\right)^{\gamma}\right) \cdot e\left(g^{\alpha_i s}, g^{q\left(H_1\left(t_{j\varphi}'\right)\right)}\right) \cdot e\left(g^{H_2(time) \cdot s}, g^{\beta_i \gamma}\right)}{e\left(\left(v_j, \prod_{\delta=1}^{h_1} u_\delta^{H_1(t_{j\delta})}\right)^{s}, g^{\alpha_i \gamma}\right) \cdot e\left(g^{\beta_i s}, g^{H_2\left(time'\right) \cdot \gamma}\right)} \right]^{\Delta_{H(t),s}(0)}$$

$$= \prod_{j=1, t \in U} \left[e\left(g^{q\left(H_1\left(t_{j\varphi}'\right)\right)}, g^{\alpha_i s}\right) \right]^{\Delta_{H(t),s}(0)} = e(g, g)^{\alpha_i \beta_i s} \tag{19}$$

We know $y = \sum_{i=1}^{M} \alpha_i \cdot \beta_i$, so we can get

$$\prod_{i=1}^{M} CT_2 = \prod_{i=1}^{M} e(g, g)^{\alpha_i \beta_i s} = e(g, g)^{s \cdot \sum_{i=1}^{M} \alpha_i \beta_i} = e(g, g)^{ys} \tag{20}$$

$$m = \frac{C_0}{e(g^t, g^y) \cdot e(g, g)^{ys}} = \frac{me(g, g)^{y(s+t)}}{e(g^t, g^y) \cdot e(g, g)^{ys}} = m \tag{21}$$

The above reasoning process can prove accuracy and safety of the scheme.

5.2 Performance Analysis

The performance analysis of the scheme mainly includes storage cost, functional comparison, and calculation cost. Reference [14] proposes a representative scheme for time-limited access, and reference [15] improves the hierarchical attribute encryption method. As the closest to our scheme, the following three tables compare the performance parameters of references [14] and [15] and our scheme.

Table 1. Storage Cost.

	GPK		MSK		CT	SK_u
[14]	$(n \cdot l + 13)\|G\| + \|G_T\|$		$2\|G\|$		$(\|U_1\| + 9)\|G\|$	$(\|U_2\| + 7)\|G\|$
[15]	$(n + l + 3)\|G\| + \|G_2\|$		$\|G\|$		$(\|U_1\| + 1)\|G\| + \|G_2\|$	$(\|U_2\| + 1)\|G\|$
Ours	GPK	PK_i	MSK	SK_i	$(\|U_1\| + 4)\|G\| + \|G_2\|$	$(\|U_2\| + 2)\|G\|$
	$(n + l + 3)\|G\|$	$2\|G\| + \|G_2\|$	$\|G\|$	$2\|G\|$		

Table 2. Function Comparison.

	Time access function	Outsourced computing	Hierarchical attribute
[14]	√	×	×
[15]	×	×	√
Ours	√	√	√

Table 3. Calculation Cost.

	Encryption		KeyGeneration	Decryption	
[14]	$14\tau_m + \tau_e + 3\tau_r$		$22\tau_m + 5\tau_r$	Cloud	User
				$(\|U_2\| + 11)\tau_e + 8\tau_m$	$3\tau_e + \tau_m$
[15]	$(\|h_1\| + 3)\tau_m + \tau_e + \tau_r$		$(h_2 + 3)\tau_m + \tau_r$	$4\tau_e + \|U_2\|\tau_m$	
Ours	Edge	User	$(h_2 + 8)\tau_m + \tau_r$	Edge	User
	$(h_1 + 6)\tau_m + \tau_e + \tau_r$	$2\tau_m + \tau_e + \tau_r$		$\|U_2\|\tau_m + 6\tau_e$	$\tau_e + 3\tau_m$

By comparing the performance of Tables 1 and 2, we can get the following conclusions.

- From the perspective of functionality, our scheme meets the timeliness requirements, tracking function, hierarchical attribute matching, etc. The scheme has significant functional advantages compared with other schemes.
- From the perspective of key storage cost, the key of ours includes master private key and user private key. Due to $|U_2| \gg 1$, compared with other schemes, the total cost of master private key and user private key occupies the minimum space.
- From the perspective of ciphertext storage cost, G and G_T are cyclic groups of order p, and p should be a large prime number in application, so we can consider $|G| \approx |G_T|$. Therefore, ours ciphertext storage is small (Table 1).

Calculation cost mainly includes power operation, random number selection and bilinear pair operation. Each calculation cost is represented by τ_m, τ_r and τ_e. We introduce edge nodes to outsource computing. The distributed computing capability of the edge cloud can help users to perform a large number of encryption and decryption operations.

- From encryption, decryption and key generation cost, ours only requires constant level operations on the user side, and other computations are undertaken by edge nodes. Compared with other schemes, this scheme has obvious computational advantages on the user side.
- From the perspective of scalability, with the increase in the number of users and the amount of data, our scheme can be expanded only by operating on edge computing nodes, which is easier to achieve than other central cloud solutions.

6 Conclusion

We propose a hierarchical attribute-based encryption scheme supporting computing outsourcing and time limited access in edge computing, working on the problems such as excessive computing load of cloud server and long validity time of user key and ciphertext. In this scheme, the access permission of users is controlled by time, so that users can access valid ciphertext that meets their access policies within the validity period. In addition, we take edge nodes as outsourcing entities to undertake a large number of encryption and decryption operations, which reduces the computing overhead of users and cloud servers. Based on the selection attribute model and the $l - wBDHI$ difficulty assumption, we prove the security of the scheme. Finally, through the comparison and analysis between the scheme and other schemes, it shows the advantage of ours in storage cost, calculation cost, and function.

References

1. Amit, S., Waters, B.: Fuzzy identity-based encryption. In: Cramer, R. (eds.) Advances in Cryptology – EUROCRYPT 2005. EUROCRYPT 2005. LNCS, vol. 3494. Springer, Berlin, Heidelberg (2005). https://doi.org/10.1007/11426639_27
2. Goyal, V., Pandey, O., Sahai, A., et al.: Attribute-based encryption for fine-grained access control of encrypted data. In: Proceedings of the 13th ACM conference on Computer and communications security. ACM **2006**, 89–98 (2006)

3. Bethencourt, J., Sahai, A., Waters, B.: Ciphertext-policy attribute-based encryption. In: IEEE Symposium on Security and Privacy, pp. 321–334. IEEE Computer Society (2007)
4. Li, J., Wang, Q., Wang, C., et al.: Enhancing attribute-based encryption with attribute hierarchy. Mobile Networks Appl. **16**(5), 553–561 (2011)
5. Chase, M.: Multi-authority attribute based encryption. In: Vadhan, S.P. (ed.) TCC 2007. LNCS, vol. 4392, pp. 515–534. Springer, Heidelberg (2007). https://doi.org/10.1007/978-3-540-70936-7_28
6. Huang, Q., Yang, Y., Shen, M.: Secure and efficient data collaboration with hierarchical attribute-based encryption in cloud computing. Futur. Gener. Comput. Syst. **72**, 239–249 (2017)
7. Leng, Q.S., Luo, W.P.: Attribute-based encryption with outsourced encryption. Commun. Technol. **54**(9), 2242–2246 (2021)
8. Qi, F., Li, K., Tang, Z.: A Multi-authority attribute-based encryption scheme with attribute hierarchy. In: 2017 IEEE International Symposium on Parallel and Distributed Processing with Applications, vol. 2017, pp. 607–613. Guangzhou, China (2017)
9. Peng, H., Ling, J., Qin, S., et al.: Attribute-based encryption scheme for edge computing. Comput. Eng. **47**(1), 37–43 (2021)
10. Huang, K.: Multi-authority attribute-based encryption for resource-constrained users in edge computing. In: 2019 International Conference on Information Technology and Computer Application (ITCA), pp. 323–326, Guangzhou, China (2019)
11. Wang, Z., Sun, X.: A compact attribute-based encryption scheme supporting computi outsourcing in fog computing. Comput. Eng. Sci. **44**(03), 427–435 (2022)
12. Y. Li, Z. Dong, K. Sha, C. Jiang, J. Wan and Y. Wang.: TMO: time domain outsourcing attribute-based encryption scheme for data acquisition in edge computing. IEEE Access **7**, 40240–40257 (2019)
13. Ning, J.T., Huang, X.Y., et al.: Tracing malicious Insider in attribute-based cloud data sharing. Chin. J. Comput. **45**(07), 1431–1445 (2022)
14. Xu, M., Fang, M.: Cloud outsourcing support aging access attributes of anonymous encryption scheme. J. Chin. Comput. Syst. **39**(02), 225–229 (2018)
15. Wang, Z., Wang, J.: Hierarchical ciphertext-policy attribute-based encryption scheme. J. Chin. Comput. Syst. **37**(6), 1263–1267 (2016)

An Ontology for Industrial Intelligent Model Library and Its Distributed Computing Application

Cunnian Gao[1,2], Hao Ren[2(✉)], Wei Cui[1,2], Xiaojun Liang[2], Chunhua Yang[2,3], Weihua Gui[2,3], Bei Sun[3], and KeKe Huang[3]

[1] School of Future Technology, South China University of Technology, Guangzhou, Guangdong, China
[2] Department of Strategic and Advanced Interdisciplinary Research, Peng Cheng Laboratory, Shenzhen, Guangdong, China
renh@pcl.ac.sn
[3] School of Automation, Central South University, Changsha, Hunan, China

Abstract. In the context of Industry 4.0, the paradigm of manufacturing has shifted from autonomous to intelligent by integrating advanced communication technologies. However, to enable manufacturers to respond quickly and accurately to the complex environment of manufacturing, knowledge of manufacturing required suitable representation. Ontology is a proper solution for knowledge representation, which is used to describe concepts and attributes in a specified domain. This paper proposes an ontology-based industrial model and significantly improves the interoperability of the models. Firstly, we conceptualize the attribute of the industrial models by providing concept and their properties in the schema layer of the ontology. Then, according to the data collected from the manufacturing system, several instances are created and stored in the data layer. In addition, we present a prototype distributed computing application. The result suggests that the ontology can optimize the management of industrial models and achieve interoperability between models.

Keywords: Ontology · Knowledge representation · Industry 4.0 · Industrial model

1 Introduction

The integration of advanced communication technologies, e.g., cyber-physical systems (CPS), the Internet of Things (IoT), edge computing, and artificial intelligence, is guiding traditional manufacturing to next-generation intelligent manufacturing which is generally known as Industrial 4.0. The intelligent manufacturing system enables collecting data from sensors and making smart decisions based on real-time communication and collaboration with machines, thus, can improve the performance of the manufacturing system significantly. Meanwhile,

B. Luo et al. (Eds.): ICONIP 2023, CCIS 1965, pp. 65–76, 2024.
https://doi.org/10.1007/978-981-99-8145-8_6

the large amount of data collected from manufacturing systems and knowledge of manufacturing processes are the cornerstone of intelligent manufacturing.

The most crucial aspect of intelligent manufacturing is the self-learning and self-decision-making capabilities enabled by manufacturing knowledge. In general, knowledge plays a crucial role in providing necessary information and controlling the manufacturing system automatically. On the one hand, knowledge of data analysis generated by machine learning and data visualization tool can assist manufacturing systems in identifying the status and trends of manufacturing processes. On the other hand, based on the knowledge of artificial intelligent, manufacturing systems can predict equipment failures and fix them timely, thus, production lines can avoid unnecessary work and reduce production costs. In this context, the system can ensure that the products meet the required quality standards and improve production efficiency. Nowadays, knowledge modeling has attracted much attention from academia and industry. H.L. Wei et al. [1] focused on mechanistic models of addictive manufacturing for predicting experimental observations correctly. To provide an advanced solution for the smart factory, M. Ghahramani et al. [2] deployed machine learning and data analysis technologies in semiconductor manufacturing processes. However, due to different development standards among engineers, e.g., the meaning of symbols and programming language, most models are not interoperable. Moreover, these models are scattered across different domains, thus, it is a time-consuming and high-cost task that searching for the specific model from different model libraries and deploying it in the manufacturing system. In addition, intelligent manufacturing is a complex domain that is comprised of different concepts, and it is a challenging task that represents manufacturing knowledge in a generic way. Hence, there is an urgent demand for designing an appropriate knowledge representation mechanism for intelligent manufacturing.

Ontology is a proper solution to achieve interoperability and knowledge representation, which maintains the machine-interoperable concepts in a specific domain and the relation among them by providing a standard platform that supports data process and information exchange. [3] Since ontology represents concepts and relationships in a semantic way, the ontology-based model can be applied in various domains and enables semantic interoperability. Nowadays, various ontology-based models have been developed to represent and share manufacturing knowledge. Järvenpää et al. [4] developed an OWL-based manufacturing resource capability ontology (MaRCO), which describes the functionality and constraint of manufacturing resource and assists the system in adjusting to the change in manufacturing system. Dinar et al. [5] developed an addictive manufacturing ontology, which stores the concept of domain and experiential knowledge of addictive manufacturing and provides guidelines for designers. A new manufacturing ontology was developed by Saha et al. [6] to model manufacturing operations and sequencing knowledge and the result of the experiment suggests that the model can greatly facilitate manufacturing engineers in making decisions. In order to achieve the goal of reasoning automatically and assisting decision-making, Sanfilippo [7] and his colleagues reused the Descriptive Ontology for

Linguistic and Cognitive Engineering (DOLCE) ontology and extended several branches by adding concepts related to the addictive manufacturing process. However, most previous works focus on conceptualizing manufacturing processes, equipment, and operations, while ignoring the knowledge of reaction mechanisms during the manufacturing processes. Moreover, most models developed in previous work are only applicable to specific domains. To the best of our knowledge, there is no production mechanism model involved in manufacturing processes.

This paper is aimed at developing an ontology-based industrial intelligent manufacturing model to fill this gap. In this paper, a new industrial intelligent manufacturing ontology is developed to model knowledge of manufacturing processes and make it easily sharable. Due to the explicit concept in the proposed ontology, the model is suitable for various applications. Firstly, our ontology is comprised of two layers: the schema layer and the data layer. We conceptualize basic information of the model and provide the properties of every concept in the schema layer. Then, according to the data of the industrial model, a number of instances are created and stored in the data layer. Furthermore, to verify the applicability and interoperability of the proposed ontology, we experiment by applying the ontology to distributed computing. The result suggests that our proposal is feasible and the ontology-based model is easy to manage. The knowledge of manufacturing processes is represented semantically, thus, users without expertise in a specific domain can quickly obtain useful information about the model. The contribution of this paper is listed as follows:

- Develop a new ontology-based intelligent manufacturing model by extending the model into two layers, the former layer is the schema layer which is comprised of the concepts of the industrial model and relationships between different entities, and the latter layer is the data layer, which contains the instances of the ontology and stored data collected from manufacturing system.
- To the best of our knowledge, we are the first to develop an ontology-based industrial model library for wet-process zinc smelting. The number of models in our library is more than 100 and will continue to grow in the future. Moreover, we map the ontology into the neo4j database and visualize the relationship between models explicitly, a further benefit is that the model can be classified by machine learning algorithms.
- Conduct a prototype experiment in distributed computing application and validate the interoperability characteristic of the proposed ontology. Due to the explicit concept of the model, the edge node can quickly obtain the main information of the model, e.g., the functionality of the model, and the usage of the model, thus, the time consumption of distributed computing is decreased significantly.

The rest of the paper is organized as follows: Sect. 2 reviews the state-of-the-art ontology development and summarizes the shortcomings of previous works and the challenges we are facing. Section 3 not only demonstrates the development of a new ontology-based industrial intelligent model in detail but also presents the construction and visualization of the industrial intelligent model library. Section 4 presents the application of our proposal and the experiment suggests that the proposed ontology is feasible and interoperable. The conclusion of this paper and a roadmap for research directions for future works are outlined in Sect. 5.

2 Related Work

The Semantic Web, coined by Tim Berners-Lee, is an extension of the World Wide Web, which aims to encode the data and make it machine-readable [13]. To encode the semantic data and represent knowledge from heterogeneous resources, the WWW Consortium (W3C) develops the Resource Description Framework and Web Ontology Language(OWL) technologies. Indeed, ontology plays an essential role in representing knowledge on the Semantic Web and making it understandable to electronic agents. In this section, we review the existing ontology developed for intelligent manufacturing. Then we conclude the challenge we are facing.

2.1 Ontology for Intelligent Manufacturing

Ontology is an intelligent solution to capture knowledge in a specific domain and enable machines to reason the change in complex environments. In particular, integrating ontology in manufacturing can improve the intelligence level of the manufacturing system, and a great number of previous works have developed ontologies for different purposes. For instance, Lemaignan et al. [8] developed a manufacturing semantics ontology(MASON) for capturing the semantics of manufacturing operations. López et al. [11] created a chemical ontology to represent chemical substances and elements. Farazi et al. [12] encapsulated the chemical kinetic reaction mechanism into the ontology and created instances to construct a knowledge base. To validate the accuracy and correctness of the ontology, they applied the ontology for chemical kinetic reaction mechanisms (OntoKin) to three use cases, which include querying across mechanisms, modeling atmospheric pollution dispersion, and a mechanism browser tool. Although the result indicates that chemical ontology can help chemical knowledge exchange, they focus on combustion chemistry particularly.

In another study, to achieve flexibility and adaptability in Cyber-Physical Production Systems (CPPS), Günter Bitsch et al. [14] created an Intelligent Manufacturing Knowledge Ontology Repository (IMKOR), which addresses the interoperability issue by connecting different domain ontologies. According to the manufacturing process planning document(STEP-NC), Zhao et al. [15] developed a self-learning ontology for representing the capability of the machine in cloud manufacturing, the core method is the combination of algorithm and analysis result. However, while the common purpose of previous works is to overcome the interoperability issue, most of them focus on containing concepts in a specific domain as much as possible and ignore the practicality of the ontologies. In this context, we develop an ontology-based approach for the standard representation of the industrial models to enable interoperability and flexibility. Meanwhile, by maintaining the industrial model library, manufacturers can fetch the models that they are searching for.

2.2 Open Challenges

The review of previous works exposes three main challenges concerning intelligent manufacturing ontology. From the application perspective, most existing ontologies are designed for a specific domain, thus, it is difficult to reuse the existing ontology to model knowledge of another domain. In addition, most of the previous works focus on conceptualizing manufacturing operations, materials, and equipment, while lacking knowledge of reaction mechanisms. Finally, the existing ontologies are difficult to manage, and it is time-consumption work to find the relevant ontologies and reuse them to reduce development costs and avoid ontology redundancy. In this context, we aim to address these issues by developing a new ontology-based industrial intelligent model for manufacturing knowledge representation and model management.

3 The Development of the Ontology

Ontology is useful for knowledge representation and sharing by enabling interoperability. There are several methodologies for developing ontology, such as Integrated Definition for Ontology Description Capture Method(IDEF5) [10], CyC [9]. In this paper, we choose the IDEF5 methodology for developing the proposed ontology. The reason we choose this methodology is that it can capture the real-world object in an intuitive and natural form. In this section, we introduce ontology development, which includes two layers. Section 3.1 presents the schema layer of the ontology, and the data layer is introduced in Sect. 3.2.

3.1 The Schema Layer of the Ontology

In the schema layer, we conceptualize the attribute of the industrial models by provide concept and their properties. To guarantee the semantics of ontology, we choose Web Ontology Language(OWL) as the ontology encoding language, because OWL can recognize various formats of data, e.g., XML, RDF, JSON-LD, and turtle syntax. Meanwhile, we take Protégé 5.5.0 as the ontology editor and visualize the structure of the ontology by using the OntoGraf plug-in. Protégé 5.5.0 is a useful open-source ontology editor tool, which is developed by Standford University. The key steps required in creating an ontology as follows.

1. identify the key concepts of the industrial model and design the hierarchical relationships between classes by using Protégé 5.5.0.
2. identify the attribute of the concepts, including data properties and object properties. The object property represents the relationship between concepts, while the data property connects different formats of attribute data with the concepts (e.g. the value of the input variable.).

To ensure the interoperability of the ontology, designing the schema layer is the first step in developing the ontology. The structure of the schema layer is shown in Fig. 1.

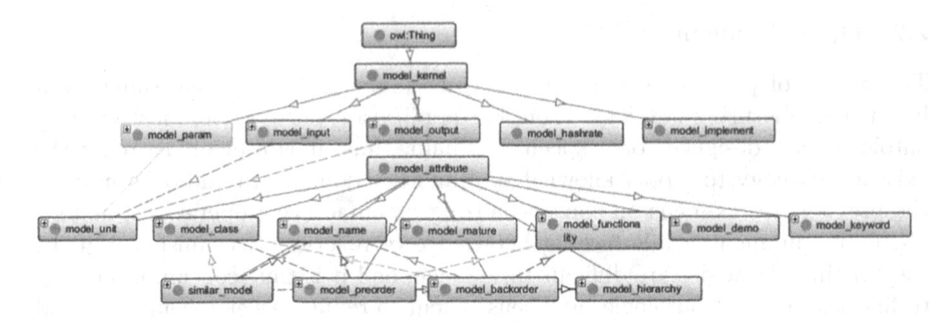

Fig. 1. The structure of the schema layer of the industrial models. The solid arrows represent the inheritance between the classes, and the dotted arrows represent the object properties.

The core concepts and description of the industrial model are given in Table 1 and Table 2 reports the properties used in the ontology. For instance, **model_kernel** is the main class of the model. Furthermore, **model_input**, **model_output**, and **model_param** are the subclasses of **model_kernel**. For instance, as for a manufacturing process status prediction model based on Long Short-Term Memory (LSTM), some features of the system represents the input of the model and the status prediction represents the output. Meanwhile, some hyperparameters (e.g. learning rate, epoch, batch size) are denoted by **model_param**. In order to enhance the machine-readability of the ontology, some basic attributes information of the industrial models is classified into different classes, which inherit their superclass **model_attribute**. The class **model_class** declares the class of industrial model, e.g., material type, industrial process, or industrial equipment. The maturity level of the model indicates the quality of the model, the machine can judge the maturity level according to the instance of the class **model_mature**. Similarly, the class **model_functionality** tells the machine the main functionality and usage of the model and the class **model_keyword** indicates which fields the model is related to. In the process of industrial manufacturing, many processes are interrelated, and the output of the previous process will be the input of the next process. Therefore, the class **model_preorder**, **model_backorder**, **similar_model** and **model_hierarchy** is designed to show the model id to get the input variable from the previous process or similar process.

The models can be implemented in a variety of ways. One of the important features is the inconsistency of the programming language (e.g. Python, Golang, and Java), which significantly hinders the interoperability of the models. The class **model_implement** is responsible for maintaining the interoperability of the models. The value of **data_property** *code_language* is the programming

Table 1. Description of main classes and their subclasses in industrial ontology

Class	Description
model_attribute	"A series of attributes including the basic information of the models."
model_class	"According to the application scenario, divide the models into different classes."
model_name	"The unique identity of the model."
model_functionality	"The functionality of the model."
model_keyword	"The labels of the model, which facilitate retrieval of models from model libraries."
model_unit	"The dimensional unit of a variable."
model_demo	"A test version of the model implement."
model_mature	"The development level of the model."
model_preorder	"The model that is inherited by this model."
model_backorder	"The model that inherits this model."
similar_model	"The model equipped similar functionality."
model_hierarchy	"There is a cascading inheritance relationship between models, the hierarchical dimension of the model is beneficial for indexing the interconnected models."
model_input	"Input variable of the model."
model_output	"Result of computation."
model_param	"The parameters of the model."
model_implement	"Detail information of the implement, such as the coding, programming language, the contributors, etc."
model_hashrate	"Detail information of the computation, such as the IP or the API of cloud computing."

language, and the machine can pull a related docker to run the code, which can get from the property *code_description*.

Finally, accelerating the computational speed of the models is another purpose of ontology. Distribute computing is an excellent solution, which has received considerable attention in the past ten years. The **model_hashrate** is a general class for recording the running results of the model.

3.2 The Data Layer of the Ontology

In the data layer, according to different mechanisms, we create the individual instances of the classes defined in the schema layer. All of the data and basic information are stored in this layer. For instance, we can create an individual instance *Implement1* to represent the implement of the model, which belong to the class **model_implement**. This instance has the properties value as follows.

Table 2. The properties used in the ontology

Property	Domain	Range	Description
has_id	model_input, model_output, model_param, model_implement, model_hashrate, model_class	int	"The unique identity of the instance."
has_name	model_input, model_output, model_param, model_implement, model_hashrate, model_class, model_mature, model_functionality, model_keyword, model_hierarchy	string	"The symbol of the instance."
has_value	model_input, model_output, model_param	int, float64, string	"The value of the instance."
has_description	model_input, model_output, model_param, model_hashrate, model_class, model_mature, model_functionality, model_keyword, model_hierarchy	string	"Description of the instance."
has_timestamp	model_input, model_output	string	"The timestamp of the instance creation."
code_language	model_implement	string	"The kind of programming language."
code_description	model_implement	string	"The code of the model."
hashrate_result	model_hashrate	string, float, int	"The result after running the code."
hashrate_api	model_hashrate	string	"The API of the computing service."
preorder_id	model_preorder	int	"The unique identity of model, which is inherited by this model."
backorder_id	model_backorder	int	"The unique identity of model, which inherits this model."
similar_id	similar_model	int	"The unique identity of model, whose functionality is similar to this model."

has_id: 20221106
has_name: "zinc ion leaching rate"
code_language: "Python"
code_description: "import owlready2..."

Industrial manufacturing, especially metallurgy, is composed of multiple processes. Taking the calculation of zinc ion leaching rate in wet zinc smelting as an example, it is necessary to collect the value of the input variables first. And the next step is to add a model description and the concrete implementation code to the ontology. Finally, we save the ontology by using the standard RDF/XML format.

The owlready2 package, a third-party extension package for Python, can be used to parse and develop the ontology. In this paper, based on the attribute of the industrial model and the data we collect from the manufacturing system, we write scripts to automatically generate instances by using the owlready2 package.

4 Application of the Ontology

In the Internet of Things environment, due to the limitations of size, IOT devices are often equipped with slow processors and limited memory storage. Therefore, most devices can't execute local computing. Computing technology has accelerated the process of the industrial revolution, effectively separating data collection from data computing tasks, and breaking through the limitation of the computing power of industrial equipment. Ontology technology makes industrial models machine-readable and universal. Therefore, integrating distributed computing and ontology technology can provide a platform for data process and exchange between multi-structure systems. Figure 2 shows the architecture of the ontology-based distributed computing application. Firstly, we define the actors involved in the architecture we proposed. Secondly, we demonstrate the working flow of distributed computing. Finally, the experiment result is shown to prove that the ontology we design is feasible and effective.

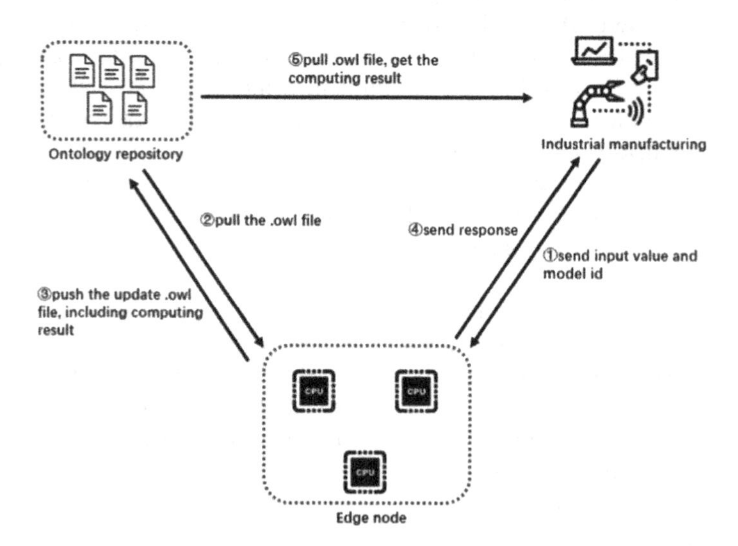

Fig. 2. The architecture of the distributed computing application.

4.1 Actors

The distributed computing system involves three entities. Some detailed descriptions of these entities are listed as followed:
1. Industrial manufacturing: Industrial manufacturing includes the equipment in the factory, e.g. temperature sensor, control center, and assembly line. These devices can collect data generated in the production environment and make requests to nodes in the edge network.
2. Ontology repository: The ontology repository manages the ontologies created by the users. Based on the model id, every owl file is named as a standard format, which can improve the retrieval efficiency significantly.
3.Edge node: Due to the low storage capacity and computing power of industrial devices, edge computing technology has received considerable attention from a large number of researchers. The edge nodes in this system represent the servers close to industrial equipment, which can provide a large number of computing services.

4.2 Distribute Computing Working Flow

The distributed computing process consists of the following steps:
1. Send request: In the first step, the industrial devices collect data, e.g. the reaction temperature, ion concentration, etc. All of these data will be encoded in JSON format, which is a common format for data propagation in networks. Based on the HTTP protocol, the IOT devices broadcast their request package and wait for the response package.
2. Pull the *.owl file: If the edge nodes capture the request package successfully, they will pull the *.owl file from the ontology repository and get the implemented code. According to the implementing language, they run corresponding dockers for supporting multiple programming language environments.
3. Push the update *.owl file: In this step, based on the PBFT (Practical Byzantine Fault Tolerance) algorithm, these edge nodes reach a consensus before they update the computing result. The result will be stored in the data layer and the leader node is responsible for pushing the new *.owl file to the ontology repository.
4. Send response: After the leader, node update the *.owl file, it packages the filename and sends the response package to clients.
5. Get the computing result: The client waits for the response package until it catches the file name, and pulls the *.owl file from the ontology repository. Parsing the *.owl file by the owlready2 python package, the client can fetch the computing result from the data layer.

5 Conclusion

The paper presents the development of computational ontology with a particular focus on encoding machine-readable knowledge of industrial models. Meanwhile, we constitute a prototype distributed computing application that is used to facilitate the data computing of the manufacturing process. Reviewing existing ontologies, it is clear that all of them are suffering problems such as developing with different levels of granularity, lacking interoperability, and redundant definitions of the same entities. The strategy we proposed provides a standardized development of ontology-based industrial models in an attempt to optimize interoperability between industrial models.

The data collected from heterogeneous resources may involve with the privacy of industrial manufacturing. If the data transmission is intercepted by hackers, the industrial production system has to face large-scale network paralysis. Therefore, enhancing the security of data and adding access control is the future work we will focus on.

References

1. Wei, H.L., Mukherjee, T., Zhang, W., Zuback, J.S., Knapp, G.L., De, A., DebRoy, T.: Mechanistic models for additive manufacturing of metallic components. Progress Mater. Sci. **116**, 100703(2021)
2. Ghahramani, M., Qiao, Y., Zhou, M.C., d O'Hagan, A., Sweeney, J.: AI-based modeling and data-driven evaluation for smart manufacturing processes. IEEE/CAA J. Automatica Sinica **7**(4), 1026–1037(2020)
3. Noy, F.N., McGuinness, D.L.: Ontology development 101: a guide to creating your first ontology. Stanford knowledge systems laboratory technical report KSL-01-05 (2001)
4. Järvenpää, E., Siltala, N., Hylli, O., Lanz, M.: The development of an ontology for describing the capabilities of manufacturing resources. J. Intell. Manuf. **30**(2), 959–978 (2019)
5. Dinar, M., Rosen, D.W.: A design for additive manufacturing ontology. J. Comput. Inf. Sci. Eng. **17**(2) (2017)
6. Saha, S., Li, W.D., Usman, Z., Shah, N.: Core manufacturing ontology to model manufacturing operations and sequencing knowledge. Service Oriented Computing and Applications, 1–13 (2023)
7. Sanfilippo, E.M., Belkadi, F., Bernard, A.: Ontology-based knowledge representation for additive manufacturing. Comput. Ind. **109**, 182–194 (2019)
8. Lemaignan, S., Siadat, A., Dantan, J.Y., Semenenko, A.: MASON: a proposal for an ontology of manufacturing domain. In: IEEE Workshop on Distributed Intelligent Systems: Collective Intelligence and Its Applications (DIS'06), pp. 195–200 (2006)
9. Elkan, C., Greiner, R.: Building large knowledge-based systems: representation and inference in the cyc project: DB Lenat and RV Guha. Artificial Intelligence (1993)
10. Mayer, R.J.: Information integration for concurrent engineering (IICE). In: IDEF3 Process description capture method report (1995)
11. López, M.F., Gómez, P.A., Sierra, J.P., Sierra, A.P.: Building a chemical ontology using methontology and the ontology design environment. IEEE Intell. Syst. Appl. **14**, 37–46 (1999)
12. Farazi, F.: OntoKin: an ontology for chemical kinetic reaction mechanisms. J. Chem. Inf. Model. **60**(1), 108–120 (2019)

13. Tim, B.L., Hendler, J., Lassila, O.: The semantic web. Sci. Am. **284**(5), 34–43 (2001)
14. Bitsch, G., Senjic, P., Askin, J.: Dynamic adaption in cyber-physical production systems based on ontologies. Procedia Comput. Sci. **200**, 577–584 (2022)
15. Zhao, Y.Y., Liu, Q., Xu, W.J., Yuan, H.Q., Lou, P.: An ontology self-learning approach for CNC machine capability information integration and representation in cloud manufacturing. J. Ind. Inf. Integr. 25, 100300 (2022)

Efficient Prompt Tuning for Vision and Language Models

Bing Li(✉), Feng Li, Shaokun Gao, Qile Fan, Yuchen Lu, Reyu Hu, and Zhiyuan Zhao

Nanjing University of Posts and Telecommunications, Nanjing, China
b20150120@njupt.edu.cn

Abstract. Recently, large-scale pre-trained visual language models have demonstrated excellent performance in many downstream tasks. A more efficient adaptation method for different downstream tasks is prompt tuning, which fixes the parameters of the visual language model and adjusts only prompt parameters in the process of adapting the downstream tasks, using the knowledge learned by the visual language model during pre-training to solve the problems in the downstream tasks. However, the loss of the downstream task and the original loss of the visual language model are not exactly same during model training. For example, CLIP uses contrast learning loss to train the model, while the downstream image classification task uses the cross-entropy loss commonly used in classification problems. Different loss has different guiding effects on the task. The trend of the accuracy of the visual language model task during training is also different from that with the downstream task. The choice of an appropriate loss function and a reasonable prompt tuning method have a great impact on the performance of the model. Therefore, we pro-pose a more efficient method of prompt tuning for CLIP, experiments on 11 datasets demonstrate that our method achieves better performance and faster convergence in the downstream task.

Keywords: Deep Learning · Visual Language Models · CLIP · Prompt tuning · Few-shot learning

1 Introduction

The visual language pre-training model performs well in many downstream tasks, such as CLIP [1], ALIGN [2]. An important feature of the visual language pre-training model is to map text and images into a common vector space. For example, image encoder and text encoder of CLIP model are used to extract features of images and text respectively. CLIP model utilizes the idea of contrast learning to maximize the cosine similarity between matched image text pairs and minimize the cosine similarity of unmatched image text pairs. In contrast, there are usually two methods for adapting visual language pre-training models to downstream tasks, fine tuning and prompt tuning [3]. Fine tuning

B. Li, F. Li and Q. Fan — These authors contributed equally to this article and should be considered as co-first authors.

B. Luo et al. (Eds.): ICONIP 2023, CCIS 1965, pp. 77–89, 2024.
https://doi.org/10.1007/978-981-99-8145-8_7

pre-training models need to consume a lot of storage and computational resources to adjust the parameters of the whole model, while prompt tuning adapts downstream tasks by fixing the pre-training model parameters and adding additional trainable parameters. So prompt tuning only needs to save the parameters of the pre-trained model and add a few additional parameters for different downstream tasks [3].

The visual language model usually consists of an image encoder and a text encoder to extract image features and text features, respectively. Therefore, there are three prompt tuning methods, namely visual prompt tuning, text prompt tuning, visual and text prompt tuning. For visual prompt tuning, such as VPT [4], a small number of learnable parameters will be added to the vision transformer. For text prompt tuning, such as CoOp [5], trainable parameters are added instead of manual fixed text prompts to find the optimal solution matching the current task in a continuous parameter space. For visual and text prompt tuning, such as UPT [6], unified prompt is input to the transformer for processing and then shunted to serve as separate prompt for the image and text encoders, respectively. The above approach has made significant progress and achievements in many downstream tasks, such as few-shot learning.

However, the above approaches to prompt tuning exploration for visual language model lack consideration for downstream tasks. For example, during our replication, we found that the trend of the accuracy of the visual language model task during training was different from that with the downstream task. We illustrate this problem with an example of CLIP model adapted to downstream task of few-shot learning. In Fig. 1 (left), we showed the accuracy of training CLIP with contrast learning loss. We found that when the contrast learning accuracy leveled off, the classification accuracy with few-shot learning did not fully converge, which may be due to the different training difficulty of different tasks. In Fig. 1 (right), we showed the change in accuracy when training the CLIP model with classification loss of few-shot learning task. We found that the classification accuracy was high, but the contrast learning accuracy was low. However, a model with good performance should perform well in both.

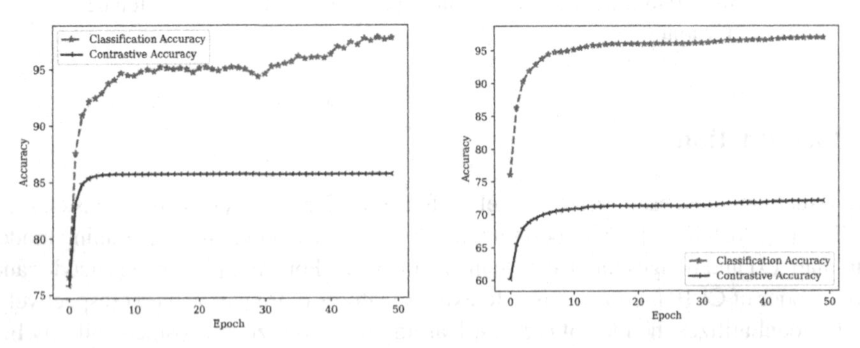

Fig. 1. The trend of accuracy during training CLIP with comparative learning loss (left). The trend of accuracy during training CLIP with classification loss (right).

Therefore, to solve the above problem, we propose a more efficient method of prompt tuning called **E**fficient **P**rompt **T**uning (**EPT**). With this approach, EPT can be better adapted to different downstream tasks and improve the performance on these tasks.

The main contributions of our paper are as follows: 1) We propose a new prompt tuning method, called Efficient Prompt Tuning (EPT), for downstream task adaptation of visual language models; 2) We firstly propose incorporating downstream task loss into the prompt tuning process of visual language model; 3) We perform EPT method on 11 datasets extensive experiments to demonstrate that it outperforms all other existing prompt tuning methods. We hope that our work will stimulate more in-depth research in the field of multimodal prompt tuning.

2 Related Work

At present, prompt tuning methods for visual language models are still a major challenge. In general, deep learning-based approaches can be divided into two categories: **2.1. Single-modal prompt tuning** and **2.2. Muti-modal prompt tuning**. In this section, related work from both perspectives is presented in detail.

2.1 Single-modal Prompt Tuning

Large-scale pre-trained models can be adapted to downstream tasks by prompt tuning. For different downstream tasks, only different prompts need to be designed [3]. Compared with fine-tuning pre-trained models, prompt engineering has a higher ac-curacy with less data and does not need to adjust the parameters of the whole model, saving computational resources. While the setting of prompts can greatly affect the model performance and it is a time and effort consuming task to design the prompt templates manually [5]. The current unimodal prompt tuning methods can be broadly classified into two categories: text prompt tuning and visual prompt tuning.

Prompt tuning originated from natural language processing techniques [3]. Excellent prompt tuning methods allow large-scale pre-trained models to effectively adapt to downstream tasks, such as text classification. To this end, Shin et al. proposed auto prompt based on gradient descent to find the prompt that adapts to the downstream task in a discrete space [7]. Soft prompt method proposed by Qin et al. used continuous optimizable vector space instead of the traditional hard prompt which were always fixed manual templates with single structure, circumventing the problem of poor performance on a particular corpus [8].

A common approach to image recognition problems in computer vision is to use pre-trained convolutional models to fine-tune a subset of parameters, such as classifier heads or bias terms, in order to achieve an improvement in the accuracy of the model for downstream tasks [4]. However, fine-tuning pre-trained model suffered from the problem of low accuracy. Moreover, fine-tuning the whole pre-trained model required a lot of storage resources and computational resources. There also exist some researchers in computer vision who draw inspiration from prompt tuning in NLP. For example, visual prompt tuning (VPT) proposed by Jia et al. introduced a small number of task-specific learnable parameters into the input space and froze the entire pre-trained Transformer backbone during training in the downstream tasks [4]. This approach reduced the utilization of computational resources because only a few prompt parameters need to be tuned. The current experiments demonstrated that VPT performed well in the field of few-shot learning.

2.2 Multi-modal Prompt Tuning

The multimodal prompt tuning technique originated from the popularity of large-scale pre-trained multimodal models. The current mainstream visual language models usually contain dual-stream and single-stream structured Transformer models, such as LXMERT [9], Oscar [10], ViLBERT [11], etc. Oscar proposed by Li et al. improved the performance of cross-modal models by increasing the recognition of picture objects and text connection between them. However, cross-modal models based on contrast learning also performed well in many tasks, such as CLIP [1] and ALIGN [2], which extracted features of different modalities by image encoder and text encoder respectively and mapped them to the same vector space and then computed the cosine similarity of different texts and images, showing excellent performance in downstream tasks, such as few-shot learning.

In the cross-modal domain, Zhou et al. proposed to use contextual optimization (CoOp) on text modalities to achieve prompt tuning of CLIP, and obtained excellent performance in the field of few-shot learning [5]. Other methods such as CoCoOp [12], DualCoOp [13] and ProGrad [14] emerged subsequently after this. However, this method does not use prompt parameters on image modality. Unified prompt tuning (UPT) [6] proposed by Zang et al. adapted the unified prompt parameters to multimodal features using Transformer. And then shunted them and embedded them into text encoder and image encoder of CLIP model respectively. The problem with this approach is that Transformer structure is huge compared to the prompt parameters. On the other hand, the initial aim of prompt tuning was to efficiently adapt pre-trained models to downstream tasks using a small number of prompt parameters.

However, a common problem with the above methods is that the design of the prompt tuning method does not adequately consider the impact on downstream tasks. Different loss guides the task differently. During training, accuracy trend for the visual language model task also differs from downstream tasks. For example, the trend of accuracy when the CLIP model is trained under contrast learning loss is not the same as that in a few-shot learning task. To solve the above problem, we propose an efficient method of prompt tuning called EPT, and we will present our work in detail in Sect. 3.

3 Approach

After extensive experimental and reproduction work, we propose an efficient prompt tuning method for visual language models, called Efficient Prompt Tuning (EPT). Our prompt tuning method is based on the CLIP model, so we first introduce the CLIP visual language model in Sect. 3.1 Visual and language pre-training. We will then introduce prompt tuning method on image encoders in Sect. 3.2 Visual prompt tuning and prompt tuning method on text encoders in Sect. 3.3 Text prompt tuning. Finally, to solve the problem mentioned above, we will introduce loss fusion methods specific to downstream tasks in 3.4 Downstream task-related loss fusion.

3.1 Visual and Language Pre-training

CLIP [1] consists of an image encoder and a text encoder. The image encoder is usually built with ResNet50 [15] or ViT [16] as the backbone, while the text encoder is usually built on Transformer [17]. A pair of image-text data (image, text) is input to the image encoder and text encoder respectively to extract the corresponding features. For the encoded image features and text features, CLIP is applied to maximize the cosine similarity of matched image-text data pairs and minimize the cosine similarity of other mismatched image-text data pairs.

To construct the text description, the label of the image is introduced into the manual template "a photo of [class]" and then, the encoded features are extracted by the text encoder. For the extracted visual features and text features, the final predicted class probabilities are expressed as follows:

$$p(y = i|\boldsymbol{x}) = \frac{exp(cos(\omega_i, z)/\tau)}{\sum_{j=1}^{N} exp(cos(\omega_j, z)/\tau)} \tag{1}$$

For a given image x and a text set y consisting of N image categories, ω_i denote the text features extracted by the text encoder, z denote the visual features of the image extracted by the image encoder. $cos(\cdot, \cdot)$ is used to calculate the cosine similarity between the text features and the visual features. τ refers to a fixed temperature coefficient.

3.2 Visual Prompt Tuning

VPT [4] was the first means to introduce prompt engineering as a large-scale pre-trained model, such as ViT [16] for image processing. Simple trainable prompt parameters that were simply added were difficult to adapt to complex image information and realize the potential of pre-trained visual models. To expand the space of input prompt parameters, we apply a fully connected neural network [18] to encode high-dimensional prompt parameters, which are subsequently combined with image features as the input to the image encoder. After extensive experiments, we found that simply adding parameters may cause the model to overfit the training data, so we added the dropout layer to the fully connected neural network. The architecture is shown in Fig. 2. Thus, original prompt parameter of dimension d_1 (d_1 can be a large value) is first encoded and downscaled by the fully connected neural network to output a prompt parameter of dimension d_2 (d_2 can be a value that matches the image encoder). Our approach takes ViT as the reference model. The prompt tuning method for the visual part is represented as follows, where the green color ■indicates the parameters that can be tuned during the training of the model. The rest of the parameters in ViT are fixed.

$$\begin{aligned} \boldsymbol{P}_1 &= FCN(\boldsymbol{P}_0) \\ [x_1, Z_1, E_1] &= L_1([x_0, \boldsymbol{P}_1, E_0]) \\ [x_i, Z_i, E_i] &= L_i([x_{i-1}, Z_{i-1}, E_{i-1}]) \\ y &= Head(x_k) \end{aligned} \tag{2}$$

In Eq. (2), $\boldsymbol{P}_0$represents the initial soft prompt parameters, $\boldsymbol{P}_1$denotes the soft prompt parameters encoded by the fully connected neural network (FCN). Z_i represents the feature characteristics computed by the ith transformer layer. In the context of ViT, these parameters are integrated prior to the position encoding. Thus, the relative localization of x_k to the prompt is preserved.

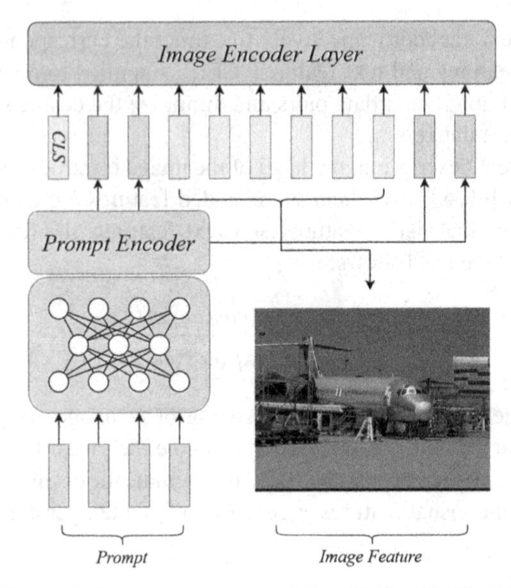

Fig. 2. The trend of accuracy during training CLIP with classification loss.

3.3 Text Prompt Tuning

In this section, we introduce text prompt tuning part of EPT. Since the text prompt tuning method proposed by CoOp [5] has made great progress, we still use the method in CoOp, which use trainable continuous parameters instead of discrete words as "prompts". Prompt parameters and image labels are stitched together and fed into the text encoder, so that the corresponding text is described as "[soft] [soft] [soft] [soft] [soft] [class]". Figure 3 shows the detailed architecture of the text prompt tuning, where the soft tokens represent the optimizable prompt parameters.

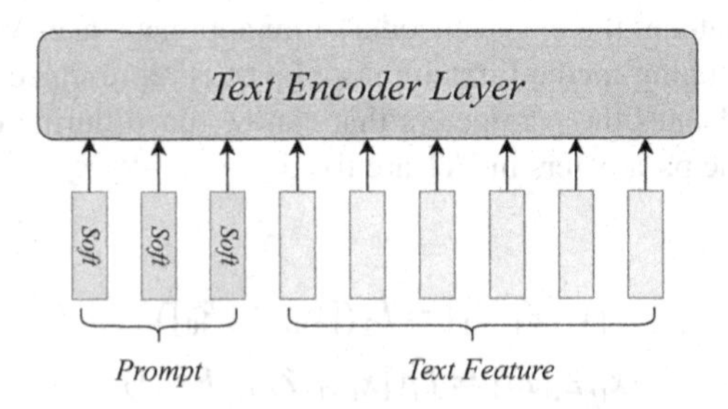

Fig. 3. The architecture of text prompt tuning method practiced in CLIP image encoder.

Thus, a given text description is fed into the text encoder to generate the probability of a visual feature falling into a category i, as shown in Eq. (3). The ***[class]*** token in the prompt t_i is replaced with the corresponding category of the image i^{th}, such as "airplane" and "dog". $g(t_i)$ denotes the features extracted by the text encoder from text description consisting of optimizable prompt parameters and the label of ith image.

$$p(y = i|\boldsymbol{x}) = \frac{exp(cos(\boldsymbol{g}(\boldsymbol{t_i}), z)/\tau}{\sum_{j=1}^{N} exp(cos(\boldsymbol{g}(\boldsymbol{t_j}), z)/\tau} \tag{3}$$

3.4 Downstream Task-Related Loss Fusion

In this section, we will detail the implementation of loss specific to downstream tasks. In CLIP model, the image and text pairs are trained with the goal of contrast learning, which is to maximize the cosine similarity of N matched image text pairs at diagonal positions and minimize the cosine similarity of $N^2 - N$ mismatched image text pairs at other positions in image text pairs of batch size N. InfoNCE loss is used in CLIP [1]. The loss for image encoder is as follows:

$$L_I = -\frac{1}{N}\sum_{i=1}^{N} log \frac{exp(cos(\boldsymbol{\omega_i}, z)/\tau)}{\sum_{j=1}^{N} exp(cos(\boldsymbol{\omega_j}, z)/\tau)} \tag{4}$$

The loss of the text encoder L_T and the loss of the image encoder L_I are symmetric [19]. Loss of CLIP model L_{CLIP} is the arithmetic average of the loss of the text encoder and the loss of the image encoder, so L_{CLIP} can be expressed as:

$$L_{CLIP} = Average(L_T + L_I) \tag{5}$$

As far as we know, the downstream task loss and the original training loss have different effects on the results when adapting the visual language to the downstream task. Therefore, in our approach, we integrate the loss of the downstream task into the training task of the visual language model. We choose the common classification task in few-shot learning as the reference downstream task. The cross entropy loss [20] of the classification task L_{down} with few shot learning is shown as follows:

$$L_{down} = -\frac{1}{N}\sum_{i=1}^{N}\sum_{j=1}^{M} \boldsymbol{f_{i,j}} log(\boldsymbol{p_{i,j}}) \tag{6}$$

In an image classification problem with batch size N and number of classes M. For image i,$\boldsymbol{f_{i,j}}$ denotes the binary indicator (0 or 1) if class label j is the correct classification for image i. log denotes the natural logarithm. $\boldsymbol{p_{i,j}}$ denotes the probability that image i is predicted to be class j. Therefore, in order to integrate the loss of downstream task into prompt tuning of the visual language model, we define the loss function of EPT as follows. The parameter α in the formula is preset to 0.5.

$$L_{EPT} = (1 - \alpha)L_{CLIP} + \alpha L_{down} \tag{7}$$

4 Experiments and Discussions

In this section, we first test few-shot learning performance of our method in Sect. 4.1 Few-shot learning. To verify the improvement of the model performance by the fused loss function, we test the performance of different loss functions on the model performance in Sect. 4.2 Performance of the model with different loss functions.

4.1 Few-Shot Learning

Baselines. We compare our approach with **1) Zero-shot CLIP**, which utilizes manually constructed prompts and does not use new training data. **2) The single-modal prompt tuning approach.** This approach used prompt tuning on the text or image modality of CLIP model for text and image, respectively. For visual prompt tuning, we chose VPT-deep [4] as the comparison model. For text prompt tuning, we chose CoOp [5] as the comparison model. **3) Multimodal prompt tuning approach.** This approach applied the prompt parameter on both image and text modalities of the visual language model at the same time. We choose UPT [6] as the comparison model.

Datasets. We follow Zhou et al. [5] to test the model's few-shot learning performance using 11 datasets (ImageNet [21], Caltech101 [22], OxfordPets [23], StanfordCars [24], Flowers102 [25], Food101 [26],FGVC-Aircraft [27], SUN397 [28], UCF101 [29], DTD [30], EuroSAT [31]) as our benchmarks. For image feature extraction, we used ViT-B/16 as part of visual prompt tuning. Following Zhou et al. we samely used 1/2/4/8/16 samples as training data and test data from the entire dataset as evaluation data. We recorded the average results of different random seeds as the final results. The results of all experiments are shown in Fig. 4. All the details of the training follow Zhou et al.

EPT vs Single-modal Prompt Tuning Approach. From the average results, our method beats VPT by 0.59%, 2.03%, 2.43%, and 3.82% at 2/4/8/16 training shots, respectively. Our method outperforms CoOp 1.31%, 2.53%, and 4.84% at 4/8/16 training shots, respectively. In general, our method has more obvious advantages over CoOp, VPT and other unimodal prompt tuning methods. In particular, on the datasets of Food101, FGV-CAircraft, DTD, EuroSAT, and UCF101, our method has made great progress compared with the unimodal prompt tuning method. However, we observe that the performance of our method decreases compared to the previous method when the sample size is 1. This may be due to the loss of the downstream task addition that causes overfitting to some of the data. Also, on some datasets, such as OxfordPet, Flowsers102, StanfordCars, the improvement of EPT is less, which may be caused by the excessive noise of the data.

EPT vs Multimodal Prompt Tuning Approach. From Fig. 4, we observe that EPT achieves approximately the same excellent performance as UPT in most cases, such as Caltech101, OxfordPets, EuroSAT. It is worth noting that EPT outperforms UPT on a few datasets, such as Food101, FGVCAircraft, DTD, UCF101. From the average results, EPT performs essentially the same as UPT at 1/2/4/8 training samples. At 16 training shots, EPT outperformed UPT by 1.55% on average on 11 data sets. In addition, EPT only needs to adjust the image encoder and text encoder prompting parameters during prompt tuning. In contrast, UPT needs to adjust the whole Transformer parameters in addition to the image and text modalities in order to achieve consistent performance. In general, EPT performs well in the adaptation of few-shot learning due to the addition of downstream tasks to guide the visual language model.

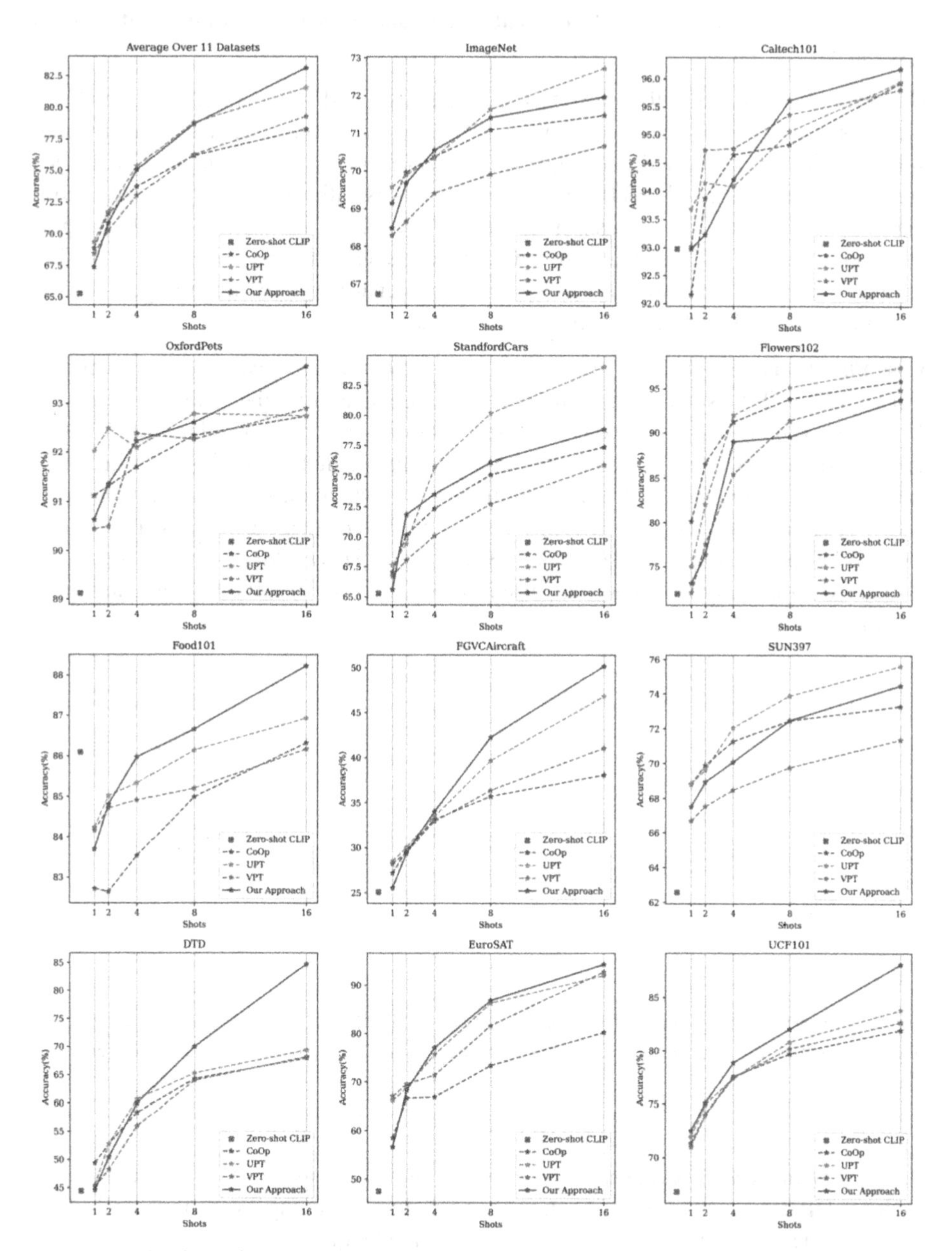

Fig. 4. Main results over 11 datasets under the few-shot learning setting.

4.2 Performance of the Model with Different Loss Functions

In this section, to test the performance of the new loss in downstream tasks and exclude the effect of the number of dataset categories on the experimental results, we set up two different scenarios (**datasets with few categories** and **datasets with large categories**) separately.

Datasets. For datasets with few categories, we choose RAF-DB dataset [32], which is a face expression recognition dataset contains 7 basic expressions (i.e., neutral, happy, sad, surprised, fearful, disgusted, and angry). The training data consisted of 12,271 images and the test data consisted of 3,068 images. For datasets with large categories, we choose the Food101 dataset. This dataset includes 101 food categories with 101,000 images. For each category, 250 manually reviewed test images are presented along with 750 training images. In the two different experimental scenarios, all training images are used as training data. All test images are used as test data.

Loss Functions. For the loss function, we choose ***1) Contrast learning loss***, which is usually InfoNCE loss in the CLIP model, which is to maximize the cosine similarity of N image text pairs at diagonal positions and minimize the cosine similarity of $N^2 - N$ image text pairs at other positions. The specific implementation is shown in Eq. 5. ***2) Loss of downstream task***, in CLIP adaptation to downstream task of few-shot learning, which is to calculate the cross entropy loss of predicted image labels and real image labels. The specific implementation is shown in Eq. 6. ***3) Loss fusion***, which is the loss associated with the downstream task used in the EPT. The specific implementation is shown in Eq. 7.

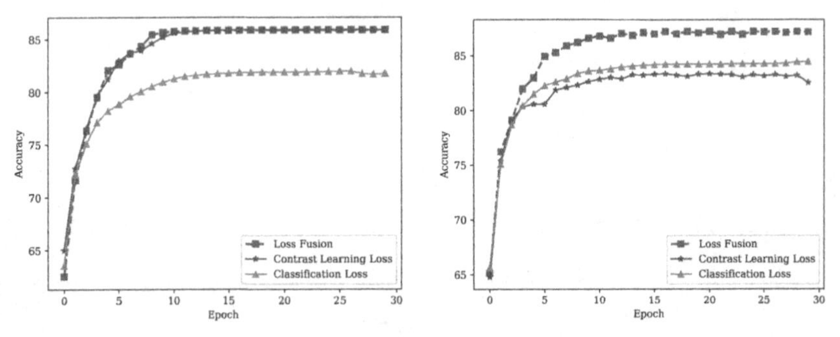

Fig. 5. Classification accuracy of the model trained on the RAF-DB dataset with different loss (left). Classification accuracy of the model trained on the Food101 dataset with different loss (right).

Datasets with Few Categories. Figure 5 (left) shows the influence of different loss in the RAF-DB dataset on the classification accuracy during training. We found that the loss of contrast learning and the loss of fusion performed similarly. The classification loss for the downstream task is slightly worse than the former. It is worth acknowledging that the fusion loss of the downstream task used in EPT outperforms. While classification loss is significantly weaker than contrast learning loss and loss fusion. This suggests that loss fusion can combine the properties of CLIP itself and improve the weakness of classification loss.

Datasets with Large Categories. Figure 5 (right) shows the effect of different loss in the Food101 dataset on the classification accuracy during training. In the face of dataset containing 101 categories, we found that the loss after fusion outperformed the loss from comparison learning and the classification loss from downstream task. At the 5th epoch, both the contrast learning loss and the classification loss converged, however the fused loss did not converge. This may be the main reason why EPT outperformed UPT, CoOp and VPT. Thus, the fusion loss of the downstream task utilized in EPT still performed well in the face of datasets with large number of categories.

5 Conclusion

With the rapid expansion and growth of the number of visual language model parameters, efficient and computationally efficient adaptation methods are critical for pre-trained models for downstream tasks. Our paper provides a novel solution to the problem of adapting large visual language models like CLIP from the perspective of model structure and design of loss functions. Our study sheds light on the problem of loss in downstream tasks that has been overlooked in previous studies and gives a solution called EPT. Performance comparable to the effect of previous studies can be achieved in EPT by simply adjusting the loss function and adding simple prompt parameters. The results show that the fused loss achieve excellent performance in both the CLIP model itself and in downstream tasks. Overall, we believe that multimodal prompt learning is a promising area of research. We hope that our study will stimulate more lively discussions and deeper research.

References

1. Radford, A., et al.: Learning transferable visual models from natural language supervision. In: International Conference on Machine Learning. PMLR (2021)
2. Jia, C., et al.: Scaling up visual and vision-language representation learning with noisy text supervision. In: International Conference on Machine Learning. PMLR (2021)
3. Liu, P., et al.: Pre-train, prompt, and predict: a survey of prompting methods in NLP. ACM Comput. Surv. **55**(9), 1–35 (2023)
4. Jia, M., et al.: Visual prompt tuning. In: Computer Vision–ECCV 2022: 17th European Conference, Tel Aviv, 23–27 October 2022, Proceedings, Part XXXIII, pp. 709–727. Springer, Cham (2022)
5. Zhou, K., et al.: Learning to prompt for vision-language models. Int. J. Comput. Vision **130**(9), 2337–2348 (2022)
6. Zang, Y., et al.: Unified vision and language prompt learning. arXiv preprint arXiv:2210.07225 (2022)
7. Shin, T., et al.: Autoprompt: eliciting knowledge from language models. arXiv preprint arXiv:2010.15980 (2020)
8. Qin, G., Eisner, J.: Learning how to ask: Querying LMs with soft prompts. arXiv preprint arXiv:2104.06599 (2021)

9. Tan, H., Bansal, M.: Lxmert: learning cross-modality encoder representations from transformers. arXiv preprint arXiv:1908.07490 (2019)
10. Li, X., et al.: Oscar: object-semantics aligned pre-training for vision-language tasks. In: Vedaldi, A., Bischof, H., Brox, T., Frahm, J.-M. (eds.) ECCV 2020. LNCS, vol. 12375, pp. 121–137. Springer, Cham (2020). https://doi.org/10.1007/978-3-030-58577-8_8
11. Lu, J., et al.: Vilbert: pretraining visiolinguistic representations for vision-language tasks. Adv. Neural Inf. Process. Syst. **32** (2019)
12. Zhou, K., et al.: Conditional prompt learning for vision-language models. In: Proceedings of the IEEE/CVF Conference on Computer Vision and Pattern Recognition, pp. 16816–16825 (2022)
13. Sun, X., et al.: Dualcoop: fast adaptation to multi-label recognition with limited annotations. arXiv preprint arXiv:2206.09541 (2022)
14. Xing, Y., et al.: Class-aware visual prompt tuning for vision-language pre-trained model. arXiv preprint arXiv:2208.08340 (2022)
15. He, K., et al.: Deep residual learning for image recognition. In: Proceedings of the IEEE Conference on Computer Vision and Pattern Recognition, pp. 770–778. IEEE (2016)
16. Dosovitskiy, A., et al.: An image is worth 16x16 words: transformers for image recognition at scale. arXiv preprint arXiv:2010.11929 (2020)
17. Vaswani, A., et al.: Attention is all you need. Adv. Neural Inf. Process. Syst. **30** (2017)
18. Farhat, N.: Optoelectronic neural networks and learning machines. IEEE Circuits Devices Mag. **5**(5), 32–41 (1989)
19. Li, Y., et al.: Supervision exists everywhere: a data efficient contrastive language-image pre-training paradigm. arXiv preprint arXiv:2110.05208 (2021)
20. Shannon, C.E.: A mathematical theory of communication. ACM SIGMOBILE Mob. Comput. Commun. Rev. **5**(1), 3–55 (2001)
21. Deng, J., et al.: Imagenet: a large-scale hierarchical image database. In: 2009 IEEE Conference on Computer Vision and Pattern Recognition, pp. 248–255. IEEE (2009)
22. Fei-Fei, L., et al.: Learning generative visual models from few training examples: an incremental Bayesian approach tested on 101 object categories. In: 2004 Conference on Computer Vision and Pattern Recognition Workshop, p. 178. IEEE (2004)
23. Parkhi, O.M., et al.: Cats and dogs. In: 2012 IEEE Conference on Computer Vision and Pattern Recognition, pp. 3498–3505. IEEE (2012)
24. Krause, J., et al.: 3D object representations for fine-grained categorization. In: Proceedings of the IEEE International Conference on Computer Vision Workshops, pp. 554–561 (2013)
25. Nilsback, M.E., Zisserman, A.: Automated flower classification over a large number of classes. In: 2008 Sixth Indian Conference on Computer Vision, Graphics & Image Processing, pp. 722–729. IEEE (2008)
26. Bossard, L., Guillaumin, M., Van Gool, L.: Food-101 – mining discriminative components with random forests. In: Fleet, D., Pajdla, T., Schiele, B., Tuytelaars, T. (eds.) ECCV 2014. LNCS, vol. 8694, pp. 446–461. Springer, Cham (2014). https://doi.org/10.1007/978-3-319-10599-4_29
27. Maji, S., et al.: Fine-grained visual classification of aircraft. arXiv preprint arXiv:1306.5151 (2013)
28. Xiao, J., et al.: Sun database: large-scale scene recognition from abbey to zoo. In: Proceedings of the IEEE Conference on Computer Vision and Pattern Recognition, pp. 3485–3492. IEEE (2010)
29. Soomro, K., et al.: UCF101: a dataset of 101 human actions classes from videos in the wild. arXiv preprint arXiv:1212.0402 (2012)
30. Cimpoi, M., et al.: Describing textures in the wild. In: Proceedings of the IEEE Conference on Computer Vision and Pattern Recognition, pp. 3606–3613 (2014)

31. Helber, P., et al.: Eurosat: a novel dataset and deep learning benchmark for land use and land cover classification. IEEE J. Sel. Top. Appl. Earth Observ. Remote Sens. **12**(7), 2217–2226 (2019)
32. Li, S., Deng, W.: Blended emotion in-the-wild: Multi-label facial expression recognition using crowdsourced annotations and deep locality feature learning. Int. J. Computer Vision **127**(6–7), 884–906 (2019)

Spatiotemporal Particulate Matter Pollution Prediction Using Cloud-Edge Intelligence

Satheesh Abimannan[1], El-Sayed M. El-Alfy[2(✉)], Saurabh Shukla[3], and Dhivyadharsini Satheesh[4]

[1] Amity School of Engineering and Technology (ASET), Amity University, Maharashtra, Mumbai, India
[2] Fellow SDAIA-KFUPM Joint Research Center for Artificial Intelligence, Interdisciplinary Research Center of Intelligent Secure Systems, Information and Computer Science Department, King Fahd University of Petroleum and Minerals, Dhahran, Saudi Arabia
alfy@kfupm.edu.sa
[3] Department of Information Technology, Indian Institute of Information Technology, Lucknow (IIIT-L), India
[4] School of Computer Science and Engineering, Vellore Institute of Technology, Vellore, India

Abstract. This study introduces a novel spatiotemporal method to predict fine dust (or $PM_{2.5}$) concentration levels in the air, a significant environmental and health challenge, particularly in urban and industrial locales. We capitalize on the power of AI-powered Edge Computing and Federated Learning, applying historical data spanning from 2018 to 2022 collected from four strategic sites in Mumbai: Kurla, Bandra-Kurla, Nerul, and Sector-19a-Nerul. These locations are known for high industrial activity and heavy traffic, contributing to increased pollution exposure. Our spatiotemporal model integrates the strengths of Convolutional Neural Networks (CNNs) and Long Short-Term Memory (LSTM) networks, with the goal to predict $PM_{2.5}$ concentrations 24 h into the future. Other machine learning algorithms, namely Support Vector Regression (SVR), Gated Recurrent Units (GRU), and Bidirectional LSTM (BiLSTM), were evaluated within the Federated Learning framework. Performance was assessed using Mean Absolute Error (MAE), Root Mean Squared Error (RMSE), and R^2. The preliminary findings suggest that our CNN-LSTM model outperforms the alternatives, with a MAE of 0.466, RMSE of 0.522, and R^2 of 0.9877.

Keywords: Air Quality · $PM_{2.5}$ Pollution · CNN-LSTM Networks · Edge Intelligence · Federated Learning

1 Introduction

The escalating concern of air pollution, specifically $PM_{2.5}$, has intensified health risks globally [3,5,11,12], necessitating an urgent requirement for effective air

B. Luo et al. (Eds.): ICONIP 2023, CCIS 1965, pp. 90–100, 2024.
https://doi.org/10.1007/978-981-99-8145-8_8

quality monitoring and prediction systems. $PM_{2.5}$ denotes fine particulate matter with a diameter of fewer than 2.5 μm, associated with serious environmental and health concerns, including cardiovascular and respiratory diseases. As technologies advance, supporting the emergence of smart cities, the efficacy of these systems becomes increasingly essential. Traditional air quality monitoring systems rely on centralized monitoring stations. Despite their widespread use, these systems bear inherent limitations, including high costs, sluggish response times, and limited coverage, demanding an innovative solution [13].

In response to these challenges, our paper presents a novel spatiotemporal approach leveraging Cloud-Edge Intelligence and Federated Learning, revolutionizing the air quality monitoring, and forecasting landscape. This approach is a distributed computing paradigm that allows integration of both cloud and edge capabilities. By putting data storage and artificial intelligence functionalities nearer to the data sources, it reduces the communication overhead and augments prediction accuracy. Furthermore, Federated Learning enables multiple decentralized edge devices to collaboratively train a global machine learning model while keeping the data locally on the devices, thereby bolstering data privacy [2,4,6–10,14]. It also allows real-time predictions of $PM_{2.5}$ while delivering a more precise and cost-effective solution.

This study also incorporates an evaluation of multiple machine learning algorithms, namely SVR, GRU, and BiLSTM, and contrast them against our proposed CNN-LSTM model. Edge servers facilitate this process, utilizing a secure Message Queuing Telemetry Transport (MQTT) protocol over a Transport Layer Security (TLS) connection, ensuring data integrity and security during transmissions. The necessity of using the CNN-LSTM model stems from its ability to capture both spatial and temporal dependencies in the data effectively, a feature that sets it apart from other models. We employ different performance metrics to assess its prediction accuracy in terms of Mean Absolute Error (MAE) and Root Mean Squared Error (RMSE), and to assess the proportion of variance using R-squared (R^2). In short, our key contributions include the development of a novel prediction model using CNN-LSTM, an in-depth comparison of the model against other conventional methods, and an exploration into the practical deployment of the model.

The paper is structured as follows. Section 2 elaborates on the proposed method. Section 3 documents the experimental work and discusses the results of various approaches. Section 4 concludes the paper, highlights the limitations of our work, offers suggestions for scalability and practical deployment of our model as well as directions for future work.

2 Proposed $PM_{2.5}$ Prediction Model

As depicted in Fig. 1, our spatiotemporal framework integrates Cloud-Edge Intelligence with Federated Learning for real-time prediction of $PM_{2.5}$. The main motivation behind this approach is to combine the advantages both technologies to offer a solution that addresses the limitations of traditional air quality monitoring systems, ensuring robustness and keeping data and decision-making near the data source. At the bottom layer, edge devices, equipped with air quality sensors deployed across multiple locations, collect and preprocess real-time data to train local models. The central server, equipped with a robust processor and ample memory, is tasked with aggregating a global model. The training is performed using federated learning algorithms that prioritize keeping data on local devices while only communicating model parameters to tackle problems of band-limited and unreliable communications such as wireless settings. The edge server acts as an intermediary level between the edge devices and the central server. In our experiments, the central server is hosted on a GPU-enabled AWS EC2 instance and communicates with the edge server using a secure Message Queuing Telemetry Transport (MQTT) protocol over a Transport Layer Security (TLS) connection. More details are explained in the following subsections.

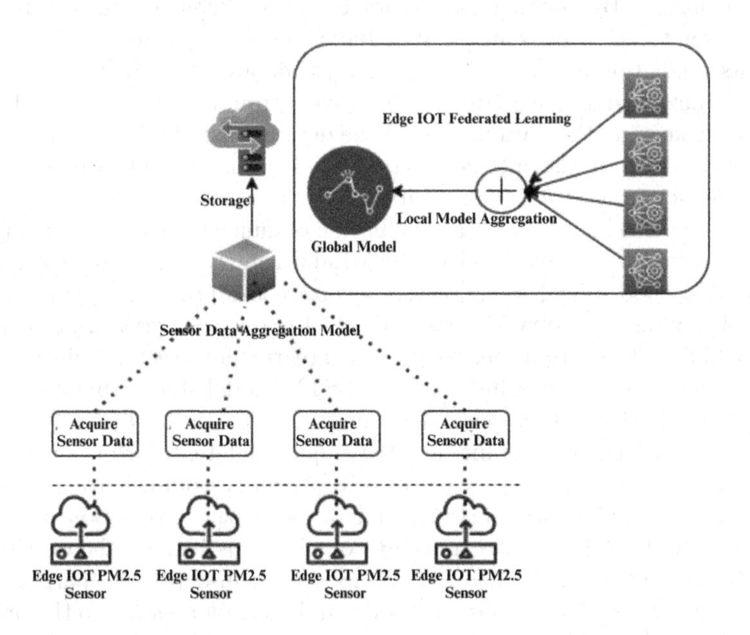

Fig. 1. Federated Edge IoT Framework for $PM_{2.5}$ Prediction.

2.1 Data Collection and Preprocessing

The data collection process is accomplished through a network of edge devices that are equipped with a variety of sensors, including ones for measuring $PM_{2.5}$ levels, temperature, humidity, Carbon Monoxide (CO), and Nitrogen Oxides (NOx). These devices sample the environmental parameters at regular intervals (e.g., every minute), and each data point is time-stamped and stored locally. Prior to utilizing the data for model training, a significant step involves pre-processing this data to handle specific conditions like data standardization and missing value imputation.

Data Standardization: This is critical to bring different variables to a common scale without distorting the differences in the range of values or losing information. The standardization process not only ensures that data from different sensors are compatible, but also that the data feeding into the predictive models are consistent in scale, enhancing the accuracy of the predictions. It is essential to remove unit limits in the various data fields, thereby transforming the data into a pure dimensionless quantity or value. This standardization enables comparison and weighting across different units or magnitudes. In this research, we employ the zero-mean standardization method, a common approach for raw data fields, utilizing the mean (μ) and standard deviation (σ). This method standardizes the data to comply with the standard normal distribution where the average value is zero, and the standard deviation is one. The standardization process is defined mathematically by:

$$z_t = (y_t - \mu_y)/\sigma_y \tag{1}$$

where z_t represents the standardized data, y_t is the original data value, and μ_y and σ_y are the mean and standard deviation of the data, respectively.

Missing Value Imputation: Inevitably, some values in the dataset might be missing due to various reasons such as instrument failure, maintenance, manual check-in, or invalid values. Handling missing values is crucial in time series forecasting, as an incomplete dataset can introduce bias or inaccuracies in the model predictions. Filling these missing values is crucial to maintain the continuity and quality of the dataset. As our data predominantly contains time series of $PM_{2.5}$ and other gas data, we opt for Akima [1] interpolation, a well-suited method for time-series data, to impute the missing values. Akima interpolation method requires the adjacent four observation points in addition to the two observations at which the internal difference is being calculated, making it necessary to have six observation points when performing Akima interpolation. This method ensures that the interpolated values are more reflective of the actual trends in the data, preserving the quality and integrity of our dataset.

2.2 CNN-LSTM Model Architecture

Our proposed model architecture is a Convolutional Neural Network - Long Short-Term Memory (CNN-LSTM) hybrid model, which combines the pattern recognition capabilities of CNN with LSTM's ability to handle sequential data. The main motivation for using the CNN-LSTM model is its proven efficiency in both spatial and temporal data recognition. The input data is fed through a series of convolutional layers for feature extraction, which identify patterns and correlations in the data by applying multiple filters. This abstract representation is then passed to LSTM layers for time-series prediction. The LSTM layers leverage memory cells that can retain information from previous time steps, allowing for predictions based on historical trends.

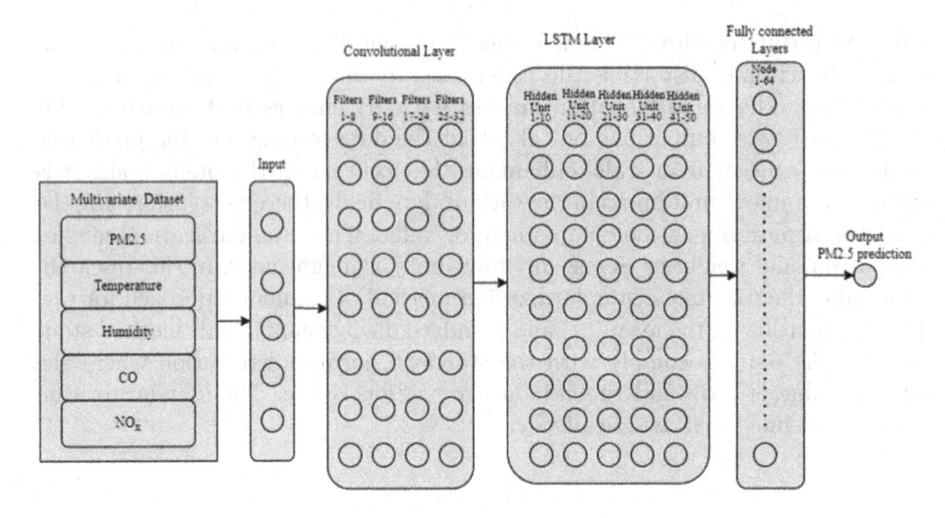

Fig. 2. Architecture of CNN-LSTM Model

The architecture depicted in Fig. 2 showcases the connectivity of the CNN-LSTM model, which has been specifically designed to efficiently predict $PM_{2.5}$ concentrations. When compared to other models like GRU, SVR, and BiLSTM, CNN-LSTM offers a blend of the best features-making it ideal for our specific problem statement. The model begins with the input layer, where individual nodes represent distinct environmental factors such as $PM_{2.5}$ levels, temperature, humidity, CO, and NOx. These nodes establish connections with subsequent convolutional layers, consisting of 32 filters, enabling the capture of specific features and patterns from the input data. The output of each filter undergoes processing through an activation function, introducing non-linearity to the model.

Max-pooling layers follow the convolutional layers, facilitating down-sampling and retaining the most salient features. The output from the max-pooling layers is then fed into LSTM layers, which possess recurrent connections capable of capturing long-term dependencies within sequential data. The LSTM layers, consisting of 50 hidden units, effectively identify temporal dependencies within the sequence of features, providing valuable insights into the evolving patterns of $PM_{2.5}$ levels.

Following the LSTM layers, fully connected layers are introduced, comprising 64 nodes. Each node establishes connections with all nodes in the preceding layer, enabling comprehensive feature transformation and learning processes. The interconnectedness of these nodes facilitates the integration of information from different channels, allowing the model to capture higher-level representations of the data.

Finally, the output layer receives the transformed features and generates the predicted $PM_{2.5}$ concentration. By comparing this architecture to traditional statistical or physical models, we found that the CNN-LSTM approach delivers superior prediction accuracy, especially for datasets with complex interdependencies. This methodology was chosen for its ability to provide real-time predictions, scalability, and a decentralized approach that respects user privacy. The interconnected layers within the CNN-LSTM model establish a powerful flow of information, facilitating efficient processing, feature extraction, and the capturing of temporal dependencies. This comprehensive architecture enables the model to make accurate predictions of $PM_{2.5}$ concentrations, contributing to improved air quality monitoring and forecasting.

2.3 Cloud-Edge Model Training and Inference

In the Federated Learning approach, each edge device trains a local model using its collected data. This methodology was chosen for its ability to provide real-time predictions, scalability, and a decentralized approach that respects user privacy. This model training is performed using the CNN-LSTM model architecture with an ADAM optimizer and mean-squared error loss function. It's essential to note that the entire training process, including forward propagation (calculating predicted $PM_{2.5}$ concentrations), loss computation, and backpropagation (computing gradients and updating weights), occurs locally on each edge device. After training, each edge device shares only its model parameters or gradients (not the actual data) with the Central Server. This federated approach, while complex, offers benefits of data privacy as raw data remains on local devices and is not shared with the central server. The Central Server then aggregates the received model parameters from all edge devices, computes a global update, and shares this updated global model with all the edge devices. The edge devices then use this updated global model to make inferences. By employing Federated Learning, our approach ensures that the data and decision-making processes remain close to the data sources, maximizing robustness, scalability, and privacy. The Cloud-Edge training and inference process provides an optimal blend of centralized intelligence and distributed intelligence, achieving real-time predictions while ensuring data privacy.

3 Experimental Work

3.1 Setup and Dataset

The proposed system for air quality monitoring and forecasting was implemented using the Python programming language. The implementation leveraged TensorFlow's Keras API for developing the CNN and LSTM components of the algorithm. To ensure efficient training of the CNN-LSTM algorithm and accurate air quality predictions, an Amazon Web Services (AWS) EC2 instance with a GPU-enabled instance type was utilized.

For data collection, a Raspberry Pi 4 Model B with 4 GB of RAM and a 1.5 GHz quad-core ARM Cortex-A72 CPU was employed. This hardware configuration provided the necessary computational power and memory capacity for real-time data acquisition and processing. The following sensors were utilized to collect specific environmental parameters:

(a) MQ135 Sensor: This sensor was used to measure Carbon Monoxide (CO) levels in the air.
(b) MQ7 Sensor: This sensor was employed for detecting Nitrogen Oxides (NOx) concentrations.
(c) DHT11 Sensor: This sensor was utilized to measure temperature and humidity levels.
(d) PMS5003 Sensor: This sensor was used for collecting Particulate Matter (PM) data, specifically $PM_{2.5}$ concentrations.

These sensors were strategically placed in four locations (Mumbai: Kurla, Bandra-Kurla, Nerul, and Sector-19a-Nerul) to capture the relevant environmental parameters for air quality monitoring.

The experimental setup involved training the CNN-LSTM algorithm on the collected dataset using the GPU-enabled EC2 instance on AWS. The training process utilized the historical air quality data[1] and the corresponding sensor readings to optimize the model's parameters and enable accurate predictions of $PM_{2.5}$ levels.

The dataset utilized in our study comprises air quality measurements for $PM_{2.5}$ obtained from multiple monitoring stations in Mumbai over an extended period. This dataset offers a comprehensive and long-term perspective on the air quality conditions in various locations within Mumbai, including Kurla, Bandra-Kurla, Nerul, and Sector-19a-Nerul. It includes daily average $PM_{2.5}$ values as well as other relevant environmental factors such as Carbon Monoxide (CO), nitrogen oxide (NOx), temperature, and humidity.

3.2 Results and Discussions

For our evaluation metrics, we utilized Mean Absolute Error (MAE), Root Mean Squared Error (RMSE), and Coefficient of Determination (R^2) to assess the model's performance. It's vital to underline the significance of these metrics in model evaluation, as they provide a comprehensive overview of the model's accuracy, prediction consistency, and generalization capabilities. MAE measures the difference between predicted and actual values without considering the direction of the errors, thus providing an average magnitude of the errors. RMSE amplifies and harshly penalizes large errors, making it sensitive to outliers. On the other hand, the R^2 metric determines how well observed outcomes are replicated by the model, as the proportion of total variation of outcomes is explained by the model. By comparing our results with those of state-of-the-art models, it's evident that our CNN-LSTM model offers superior performance in predicting $PM_{2.5}$ concentrations. This side-by-side analysis reinforces our choice of the model and showcases its robustness in handling real-world data complexities. These metrics are defined as follows:

$$MAE = \frac{1}{n}\sum_{t=0}^{n-1} |y_t - \hat{y}_t|, RSME = \sqrt{\frac{1}{n}\sum_{t=0}^{n-1} (y_t - \hat{y}_t)^2}, R^2 = 1 - \frac{SSR}{SST} \quad (2)$$

[1] https://aqicn.org/forecast/mumbai/.

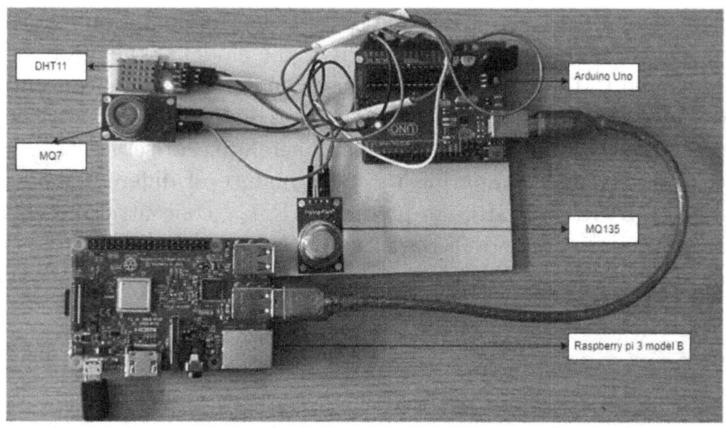

(a) CO, NO, temperature and humidity

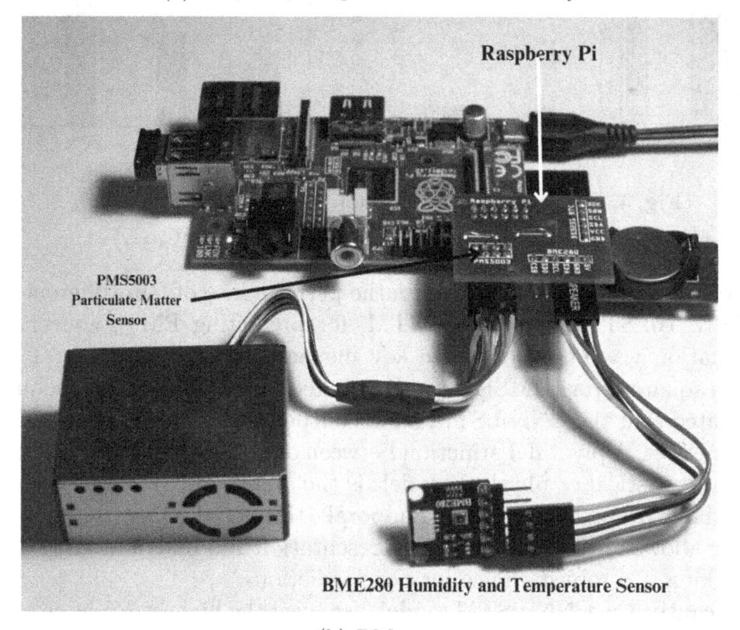

(b) $PM_{2.5}$

Fig. 3. Hardware implementation for the data collection

where y_t represents the actual concentration, $\hat{y}_t$ represents the predicted concentration, SSR (Sum of Squared Residuals) represents the sum of the squared differences between the predicted values and the actual values, and SST (Total Sum of Squares) represents the sum of the squared differences between the actual values and the mean of the dependent variable.

A lower MAE or RMSE value indicates a closer alignment between the predicted and actual concentrations. However, it is important to note that RMSE assigns more weight to larger errors due to the squaring operation. On the other hand, a higher R^2 value signifies a better fit of the model to the data, indicating that the predicted $PM_{2.5}$ concentrations closely match the actual concentrations. This metric provides insights into the models' predictive power and their ability to capture the variability in $PM_{2.5}$ levels. Therefore, R^2 serves as a valuable metric for assessing and comparing the performance of different models, such as CNN-LSTM, SVR, and GRU, in predicting $PM_{2.5}$ concentrations. It offers a quantitative measure of how well these models capture the underlying patterns and trends in the data, enabling researchers to evaluate their predictive accuracy.

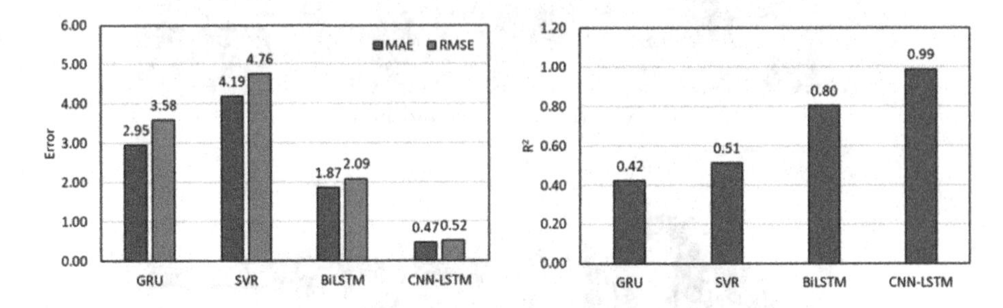

Fig. 4. Performance Comparison: MAE, $RMSE$, and R^2

The research paper aimed to assess the performance of four different models: GRU, SVR, BiLSTM, and CNN-LSTM, for predicting $PM_{2.5}$ concentrations. The evaluation was based on three key metrics: mean absolute error (MAE), root mean square error (RMSE), and R-squared (R^2). The results unequivocally demonstrated that the CNN-LSTM model outperformed the other models across all three metrics. A pivotal distinction between our CNN-LSTM model and other traditional statistical or physical models is the ability of CNN-LSTM to process both spatial (through CNN) and temporal (through LSTM) data. This dual processing allows for a more holistic representation and understanding of PM2.5 levels, making it a robust choice for our predictions.

Significantly, the CNN-LSTM model displayed the highest R^2 value of 0.9877, which signifies its ability to explain a substantial portion of the variance in the $PM_{2.5}$ concentrations. This high R^2 value underscores the strong relationship between the predicted and actual values, illustrating the model's proficiency in capturing underlying patterns and trends within the data. In comparison, the GRU model demonstrated a moderate performance with an MAE of 2.954, RMSE of 3.5818, and R^2 of 0.422. Similarly, the SVR model exhibited slightly better performance with an MAE of 4.193, RMSE of 4.7561, and R^2 of 0.512. On the other hand, the BiLSTM model showcased better performance with an MAE of 1.866, RMSE of 2.0912, and R^2 of 0.8032.

In summary, the evaluation metrics consistently demonstrated that the CNN-LSTM model outperformed the other models, positioning it as the most accurate and reliable for predicting $PM_{2.5}$ concentrations. The findings highlight the

CNN-LSTM model as an excellent choice for accurate $PM_{2.5}$ predictions, while the other models may benefit from further refinement to enhance their performance.

4 Conclusion and Future Work

To address the growing problem of $PM_{2.5}$ pollution, especially in urban and industrial areas, we've developed a new technique. We use Edge Intelligence and Federated Learning to predict $PM_{2.5}$ levels in real-time accurately. This innovative method combines the use of Convolutional Neural Networks (CNN) and Long Short-Term Memory (LSTM) networks. The choice of using the CNN-LSTM model was due to its proven ability to handle time-series data efficiently, especially when combined with the spatial features processed by the CNN. We've found it works remarkably well compared to standard machine learning models like Support Vector Regression (SVR), Gated Recurrent Units (GRU), and Bidirectional LSTM (BiLSTM) and also when contrasted with traditional statistical or physical models commonly used for fine dust prediction.

We focused on three key metrics to evaluate performance: Mean Absolute Error (MAE), Root Mean Squared Error (RMSE), and R-squared (R^2). The reason behind selecting these metrics lies in their capability to provide a comprehensive view of the model's accuracy, deviation from the true values, and its overall explanatory power. Furthermore, the comparison of our results with existing state-of-the-art models showcases the advancements we achieved. The CNN-LSTM model stood out, showing superior ability in predicting $PM_{2.5}$ levels. It was closest to the actual values, with MAE and RMSE values of 0.466 and 0.522 respectively, and a high R^2 score of 0.9877. This high score shows that our model can account for most of the changes we see in $PM_{2.5}$ levels, highlighting its strong predictive ability and reliability. While the other models showed some predictive ability, they didn't perform as well as the CNN-LSTM model. This suggests that there are opportunities to improve these models in the future. Our study's results show the powerful potential of using AI-based Edge Computing and Federated Learning for predicting $PM_{2.5}$ levels. This paves the way for creating efficient, accurate, and privacy-focused air quality prediction systems. Furthermore, the deployment of our model on an edge server using a secure MQTT protocol over a TLS connection ensures data security and real-time predictions.

Looking ahead, we see several interesting areas for future research. First, we could look at how increasing the number of edge devices and having more diverse data could improve the performance of the CNN-LSTM model. This will also help assess the scalability of our proposed model. Second, exploring different Federated Learning algorithms could help speed up the learning process. Lastly, we could look into using our model to predict other types of air pollutants. As we continue to learn more about Federated Learning, we expect to see more advancements that will make real-time air quality prediction systems even better. Additionally, incorporating a comprehensive introduction to Cloud Edge intelligence in the earlier sections will further solidify our methodology and its innovative approach.

Acknowledgment. The second author would like to acknowledge the support received from King Fahd University of Petroleum and Minerals (KFUPM) and the fellowship support from Saudi Data and AI Authority (SDAIA) and KFUPM under SDAIA-KFUPM Joint Research Center for Artificial Intelligence Fellowship Program Grant no. JRC-AI-RFP-04.

References

1. Akima, H.: A new method of interpolation and smooth curve fitting based on local procedures. J. ACM (JACM) **17**(4), 589–602 (1970)
2. Alam, F., Alam, T., Ofli, F., Imran, M.: Robust training of social media image classification models. IEEE Trans. Comput. Soc. Syst. (2022)
3. Biondi, K., et al.: Air pollution detection system using edge computing. In: IEEE International Conference in Engineering Applications (ICEA), pp. 1–6 (2019)
4. Chang, Y.S., Chiao, H.T., Abimannan, S., Huang, Y.P., Tsai, Y.T., Lin, K.M.: An LSTM-based aggregated model for air pollution forecasting. Atmos. Pollut. Res. **11**(8), 1451–1463 (2020)
5. Gandotra, P., Lall, B.: Evolving air pollution monitoring systems for green 5G: from cloud to edge. In: 8th International Conference on Reliability, Infocom Technologies and Optimization (Trends and Future Directions)(ICRITO), pp. 1231–1235 (2020)
6. Le, D.D., et al.: Insights into multi-model federated learning: an advanced approach for air quality index forecasting. Algorithms **15**(11), 434 (2022)
7. Lin, C.Y., Chang, Y.S., Chiao, H.T., Abimannan, S.: Design a hybrid framework for air pollution forecasting. In: IEEE International Conference on Systems, Man and Cybernetics (SMC), pp. 2472–2477 (2019)
8. Liu, L., Zhang, J., Song, S., Letaief, K.B.: Client-edge-cloud hierarchical federated learning. In: IEEE International Conference on Communications (ICC), pp. 1–6 (2020)
9. Liu, Y., Nie, J., Li, X., Ahmed, S.H., Lim, W.Y.B., Miao, C.: Federated learning in the sky: aerial-ground air quality sensing framework with UAV swarms. IEEE Internet of Things J. **8**(12) (2021)
10. Nguyen, D.V., Zettsu, K.: Spatially-distributed federated learning of convolutional recurrent neural networks for air pollution prediction. In: IEEE International Conference on Big Data (Big Data), pp. 3601–3608 (2021)
11. Putra, K.T., Chen, H.C., Ogiela, M.R., Chou, C.L., Weng, C.E., Shae, Z.Y.: Federated compressed learning edge computing framework with ensuring data privacy for PM2. 5 prediction in smart city sensing applications. Sensors **21**(13), 4586 (2021)
12. Ramu, S.P., et al.: Federated learning enabled digital twins for smart cities: concepts, recent advances, and future directions. Sustain. Urban Areas **79**, 103663 (2022)
13. Zheng, Z., Zhou, Y., Sun, Y., Wang, Z., Liu, B., Li, K.: Applications of federated learning in smart cities: recent advances, taxonomy, and open challenges. Connect. Sci. **34**(1), 1–28 (2022)
14. Zhu, Z., Wan, S., Fan, P., Letaief, K.B.: Federated multiagent actor-critic learning for age sensitive mobile-edge computing. IEEE Internet Things J. **9**(2), 1053–1067 (2021)

From Incompleteness to Unity: A Framework for Multi-view Clustering with Missing Values

Fangchen Yu[1], Zhan Shi[1], Yuqi Ma[1], Jianfeng Mao[1,2], and Wenye Li[1,2(✉)]

[1] The Chinese University of Hong Kong, Shenzhen, China
{fangchenyu,zhanshi1,yuqima1}@link.cuhk.edu.cn, {jfmao,wyli}@cuhk.edu.cn
[2] Shenzhen Research Institute of Big Data, Shenzhen, China

Abstract. The assumption of data completeness plays a significant role in the effectiveness of current Multi-view Clustering (MVC) methods. However, data collection and transmission would unavoidably breach this assumption, resulting in the Partially Data-missing Problem (PDP). A common remedy is to first impute missing values and then conduct MVC methods, which may cause performance degeneration due to inaccurate imputation. To address these issues in PDP, we introduce an imputation-free framework that utilizes a matrix correction technique, employing a novel two-stage strategy termed 'correction-clustering'. In the first stage, we correct distance matrices derived from incomplete data and compute affinity matrices. Following this, we integrate them with affinity-based MVC methods. This approach circumvents the uncertainties associated with inaccurate imputations, enhancing clustering performance. Comprehensive experiments show that our method outperforms traditional imputation-based techniques, yielding superior clustering results across various levels of missing data.

Keywords: Multi-view Clustering · Incomplete Data · Matrix Correction

1 Introduction

Multi-view data, stemming from diverse sources, encompasses multiple representations ubiquitous in real-world scenarios, like videos with audible and visual facets or images with raw RGB space data paired with descriptive text. These diverse views offer both consistent and supplementary information. The aim of Multi-view Clustering (MVC) is to assimilate this multi-faceted information into a unified structure, facilitating the unsupervised clustering of data samples with similar structures. However, a predominant assumption in most MVC methodologies [7,13,27,31,32] is the complete observation of sample information across all views. In reality, data collection and transmission can breach this assumption, resulting in Incomplete Multi-view Clustering (IMVC) challenges [15,22,28].

B. Luo et al. (Eds.): ICONIP 2023, CCIS 1965, pp. 101–112, 2024.
https://doi.org/10.1007/978-981-99-8145-8_9

In the realm of IMVC, inconsistencies in views or incomplete data contribute to data gaps. These voids, illustrated in Fig. 1, stem either from *Partially View-missing Problem (PVP)* [15,22,28] due to inconsistent views (*i.e.*, one sample has empty vectors in some views, as shown in Fig. 1 (a)), or from the more pervasive, yet less explored, ***Partially Data-missing Problem (PDP)*** caused by data incompleteness (*i.e.*, one sample has representations in all views but with some missing values, as shown in Fig. 1 (b)). Traditional strategies aimed at remedying PDP often involve padding missing values via imputation techniques [5,6,9] before applying MVC on the imputed data. However, this imputation-based approach can falter, especially when applied without domain knowledge on data structures, risking damage to intrinsic structures. Moreover, inaccurately imputed views can distort the fusion process in existing MVC techniques, potentially undermining clustering outcomes.

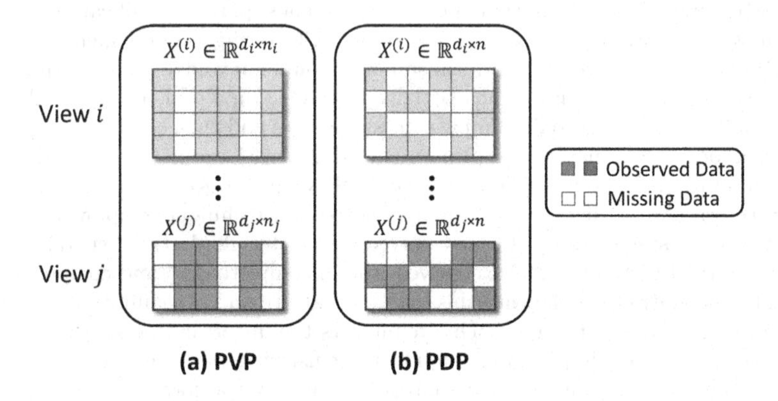

Fig. 1. Interpretation of PVP and PDP in IMVC problems.

To address the above issues, our contributions are threefold:

- We propose an imputation-free framework with the matrix correction method to deal with partially data-missing problems in the IMVC community, which can naturally avoid the uncertainty error caused by inaccurate imputation and directly obtain high-quality affinity matrices through matrix correction.
- We introduce a matrix correction algorithm to effectively and efficiently estimate distance matrices on incomplete data with a theoretical guarantee. Specifically, it starts with initial distance matrices estimated from incomplete data and then corrects these estimates to satisfy specific properties via a convex optimization approach.
- We design a two-stage strategy, *i.e.*, correction-clustering (as shown in Fig. 2), to combine with all affinity-oriented MVC methods, which makes existing MVC methods great again on IMVC problems. Extensive experiments demonstrate that our strategy achieves superior and robust clustering performance under a wide range of missing ratios, compared to the imputation-based approaches.

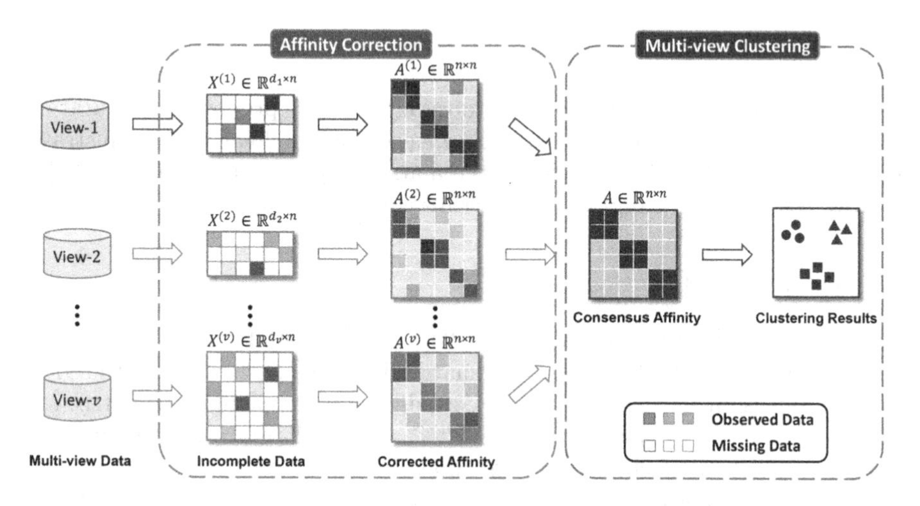

Fig. 2. The framework of our imputation-free approach for PDP problems. It adopts a novel two-stage strategy that first corrects affinity (distance) matrices in all views, and then combines with MVC methods to achieve clustering.

2 Related Work

Multi-view Clustering. MVC assumes that all samples are fully observed in all views. One popular roadmap of MVC is to separate the data based on affinity matrices constructed in all views. Those affinity-based clustering methods include spectral clustering [13,14,32,42], kernel-based clustering [7,21,31,41], and graph clustering [27,29,36,37].

Incomplete-view Multi-view Clustering. Incomplete-view MVC methods focus on scenarios where partial samples have missing views or there is no view containing all samples, and they use the observed-view information to recover the missing views. Traditional IMVC models generate the consensus representation or affinity of view-missing data via matrix factorization [15,22,28], kernel-based [10,23,24], or graph-based [30,38,40] methods.

Incomplete-value Multi-view Clustering. In contrast to view-level missing, incomplete-value MVC aims at value-level (data-level) missing, where each sample in any view may contain some missing values. A feasible solution is to first impute missing values and then perform multi-view clustering methods. In practice, statistical imputation techniques [12,20], such as zero, mean imputation, and k-nearest neighbor (kNN) imputation [39], have been widely used. Besides, matrix completion [5,6,9] is a representative machine learning-based technique that solves a matrix factorization problem. Unfortunately, it is difficult to accurately estimate missing values based on the observed data, especially for a large missing ratio, and there is no guarantee on the quality of imputation. This motivates us to design an imputation-free approach.

3 Methodology

To seek a reliable solution to incomplete-value MVC, we propose an imputation-free framework with a two-stage strategy, *i.e.*, correction-clustering, as illustrated in Fig. 2. Our work mainly resides in the matrix correction technique [18,34] to improve the distance/affinity matrix estimation for incomplete data.

3.1 Distance Estimation

Consider a multi-view dataset with v views and n samples. Denote $X^{(i)} = \{x_1^{(i)}, \cdots, x_n^{(i)}\} \in \mathbb{R}^{d_i \times n}$ as the data matrix in view i, where d_i is the view-specific dimension. For simplicity, we consider an incomplete data matrix in a single-view, *i.e.*, $X^o = \{x_1^o, \cdots, x_n^o\} \in \mathbb{R}^{d \times n}$, where x_i^o represents the observed i-th sample that may contain missing values.

If samples are not fully observed, their pairwise distance has to be estimated. For two incompletely observed samples $x_i^o, x_j^o \in \mathbb{R}^d$, denote $I_{i,j} \subseteq \{1, \cdots, d\}$ as the index set recording the positions of features that are observed in both samples. Assuming $I_{i,j}$ is not empty, denote $x_i^o(I_{i,j}) \in \mathbb{R}^{|I_{i,j}|}$ as a vector of selected values in x_i^o on $I_{i,j}$. Then, the pairwise Euclidean distance $d_{i,j}^o$ can be estimated from their commonly observed features by [18]

$$d_{i,j}^o = \|x_i^o(I_{i,j}) - x_j^o(I_{i,j})\|_2 \cdot \sqrt{\frac{d}{|I_{i,j}|}} \in [0, +\infty). \tag{1}$$

The estimated Euclidean distance matrix are obtained by $D^o = \{d_{i,j}^o\} \in \mathbb{R}^{n \times n}$. Moreover, all distance-based kernel matrices can be calculated from D^o accordingly, such as the widely used Gaussian kernel $K^o = \{\exp(-(d_{i,j}^0)^2/\sigma^2)\} \in \mathbb{R}^{n \times n}$.

3.2 Distance and Affinity Correction

To correct an initial distance matrix D^o calculated via Eq. (1), we introduce the distance correction method [18,34] and resort to a Laplacian kernel matrix, *i.e.*, $K^o = \exp(-\gamma D^o)$. Considering the PSD property of the Laplacian kernel [25], we correct the initial distance matrix D^o by solving the following problem:

$$\hat{D} = \underset{D \in \mathbb{R}^{n \times n}}{\arg\min} \; \|D - D^o\|_F^2 \tag{2}$$

subject to

$$\begin{cases} d_{i,i} = 0, \; \forall \; 1 \leq i \leq n \\ d_{i,j} = d_{j,i} \geq 0, \; \forall \; 1 \leq i \neq j \leq n \\ \exp(-\gamma D) \succeq 0 \end{cases}$$

where $\succeq 0$ denotes the positive semi-definiteness (PSD). However, the problem defined above is hard to solve due to the PSD constraint in the exponential form.

Thus, we change the decision variable from D to $K = \exp(-\gamma D)$ and reformulate the problem under an efficient approximation:

$$\hat{K} = \underset{K \in \mathbb{R}^{n \times n}}{\arg\min} \ \|K - K^o\|_F^2 \tag{3}$$

subject to

$$\begin{cases} k_{i,i} = 1, \ \forall\, 1 \leq i \leq n \\ k_{i,j} = k_{j,i} \in [0,1], \ \forall\, 1 \leq i \neq j \leq n \\ K \succeq 0 \end{cases}$$

which can be solved by the Dykstra's projection method [2,8,16–19,34,35].

Algorithm 1. Correction-clustering Strategy Based on Distance Correction

---

Input: $\{X^{(1)}, \cdots, X^{(v)}\}$: an incomplete multi-view dataset with $X^{(i)} \in \mathbb{R}^{d_i \times n}$, $\forall\, 1 \leq i \leq v$.

Output: $L \in \mathbb{R}^n$: clustering labels.

1: $\triangleright$ *Stage-I. Affinity (Distance) Correction*
2: **for** $i = 1, 2, \cdots, v$ **do**
3: Calculate $D_o^{(i)}$ from incomplete $X^{(i)}$ via Eq. (1).
4: Initialize $K_o^{(i)} = \exp(-\gamma D_o^{(i)})$ with $\gamma = \frac{0.02}{\max\{D_o^{(i)}\}}$.
5: Obtain $\hat{K}^{(i)}$ by solving Eq. (3) with the Dykstra's projection.
6: Obtain $\hat{D}^{(i)} = -\frac{1}{\gamma} \log(\hat{K}^{(i)})$.
7: Calculate a distance-based affinity matrix $\hat{A}^{(i)}$ from $\hat{D}^{(i)}$.
8: **end for**
9: $\triangleright$ *Stage-II. Multi-view Clustering*
10: Obtain a consensus affinity matrix $\hat{A}$ from $\{\hat{A}^{(1)}, \cdots, \hat{A}^{(v)}\}$ by MVC.
11: Obtain the final clustering labels L based on $\hat{A}$.

3.3 Theoretical Analysis

Theoretical Guarantee. Despite the convergence guarantee [4], the proposed method also have a nice theoretical guarantee [18] on the correction performance that we provide an improved estimate of the unknown ground-truth.

Theorem 1. $\|D^* - \hat{D}\|_F^2 \leq \|D^* - D^o\|_F^2$. *The equality holds if and only if* $K^o \succeq 0$, *i.e.*, $\hat{K} = K^o, \hat{D} = D^o$.

Complexity Analysis. The time complexity of distance correction is $O(n^3)$ per iteration, which mainly comes from the eigenvalue decomposition (EVD) in the projection operation onto the PSD space. Nevertheless, we can apply highly efficient algorithms for EVD operation to accelerate the algorithms, such as parallel algorithm [3] and the randomized singular value decomposition algorithm [11]. The storage complexity is $O(n^2)$ to store the whole matrix in memory.

4 Experiments

4.1 Experimental Setup

Datasets. The experiments are carried out on two benchmark datasets as shown in Table 1: 1) **MSRC-v1**[1] [33]: a scene recognition dataset containing 210 images with 6 views; 2) **ORL**[2]: a face-image dataset containing 400 images with 3 views. All the experiments are carried out for 10 random seeds on a ThinkStation with 2.1 GHz Intel i7-12700 Core and 32GB RAM.

Table 1. Statistic of two benchmark multi-view datasets.

Datasets	# of Samples	# of Views	# of Clusters	# of Dimensions
MSRC-v1	210	6	7	1302/48/512/100/256/210
ORL	400	3	40	4096/3304/6750

Implementation. All data is normalized to $[-1, 1]$. In each view, the values of each sample vector are missing completely at random (MCAR) with a given missing ratio r, *e.g.*, 70%. The incomplete clustering task is to first obtain multiple distance/affinity matrices through imputation or correction methods and then conduct multi-view clustering algorithms to get clustering results.

Baselines. The proposed distance correction method is compared with several representative imputation methods from two categories: 1) statistical methods: **ZERO**, **MEAN**, ***k*NN** [39] impute the missing value by zero, mean value or an average value of k-nearest neighbors ($k = 10$), respectively; 2) machine learning methods: **SVT** [5] makes low-rank matrix completion with singular value thresholding, **GROUSE** [1] conducts low-rank matrix completion via Grassmanian rank-one update subspace estimation, **FNNM** [6] performs low-rank matrix completion by factor nuclear norm minimization, and **KFMC** [9] utilizes a kernelized-factorization matrix completion.

Multi-view Clustering Algorithms. To verify the quality of affinity matrices obtained by imputation or correction methods, we choose popular affinity-based multi-view clustering algorithms to perform clustering, including: 1) spectral clustering: **CRSC** [14] and **AASC** [13] employ the co-regularization or affinity aggregation to optimize spectral clustering, respectively; 2) graph-based clustering: **AWP** [27] and **CGD** [29] generate a weighted graph by fusing multiple graph matrices; 3) kernel-based clustering: **RMKKM** [7] and **MVCLFA** [31] improve the robustness via multiple kernel k-means clustering or late fusion alignment maximization, respectively.

Evaluation Metrics. The clustering performance is evaluated by three commonly used metrics, *i.e.*, clustering accuracy (**ACC**), normalized mutual information (**NMI**), and purity (**PUR**), which range between 0 and 1. The higher the better. The average results of 10 random seeds are reported for all experiments.

[1] https://mldta.com/dataset/msrc-v1/.
[2] https://cam-orl.co.uk/facedatabase.html.

4.2 Incomplete Clustering on Single-View Data

We select the first view in the multi-view dataset as a newly single-view dataset, where the values of samples are missing completely at random with a given missing ratio r varying from 20% to 80%. To deal with it, we first obtain an estimated Euclidean distance matrix $\hat{D}$ from incomplete data X^o. Then we calculate a corresponding Gaussian kernel $\hat{K} = \exp(-\hat{D}^2/\sigma^2)$ with $\sigma = \text{median}\{\hat{D}\}$ as the input of the standard spectral clustering algorithm [26], which applies the normalized cut and is a popular package[3] in the MATLAB library. As shown in Fig. 3, the proposed method shows significant improvement in clustering metrics (*i.e.*, ACC, NMI, PUR) in almost all experiments with the least performance degeneration. Even for a large missing ratio (*e.g.*, 80%), our method still maintains reliable performance and shows its robustness.

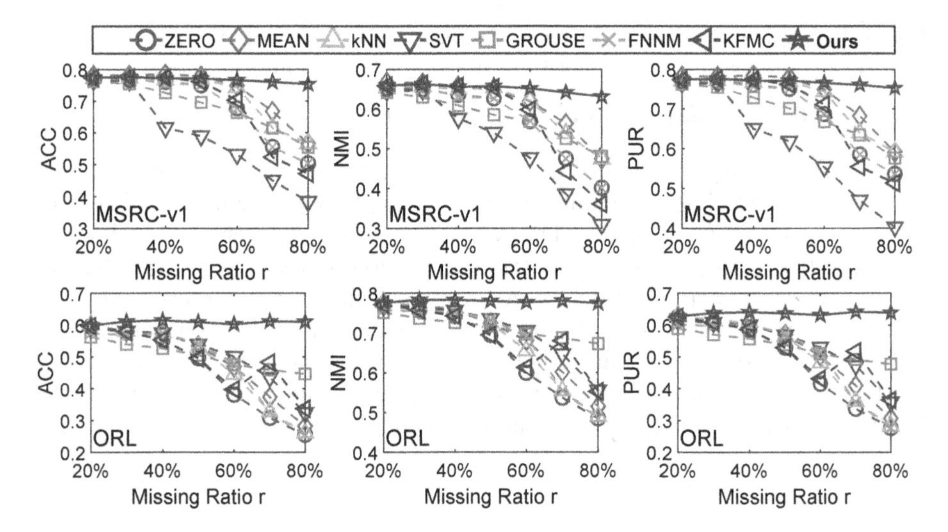

Fig. 3. Incomplete single-view clustering results on MSRC-v1 and ORL datasets using the standard spectral clustering algorithm.

4.3 Incomplete Clustering on Multi-view Data

Now, we further investigate the multi-view clustering performance. All samples in each view randomly replace their values with the NA values under a missing ratio r. Same as the setting in Sect. 4.2, the results on distance-based Gaussian kernels are shown in Fig. 4 with a fixed 70% missing ratio. The experimental results show that our approach consistently achieves better performance, in terms of ACC, NMI and PUR, against all compared imputation methods. Thus, the proposed framework shows effectiveness and robustness and therefore more reliable on incomplete multi-view clustering tasks.

[3] https://ww2.mathworks.cn/help/stats/spectralcluster.html.

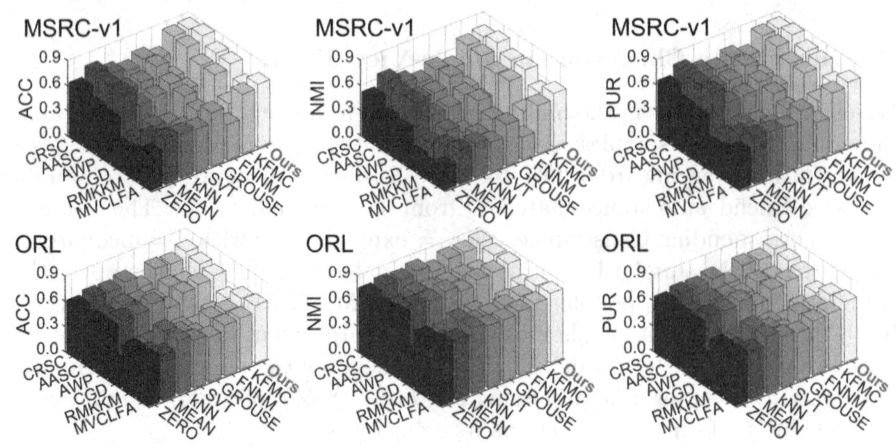

Fig. 4. Incomplete multi-view clustering results on MSRC-v1 and ORL datasets using different multi-view clustering algorithms under a fixed 70% missing ratio.

4.4 Quality Visualization of Affinity Matrices

To assess the quality of affinity matrices, we visualize consensus affinity matrices. We obtain Euclidean distance matrices through imputation methods or our correction method for all views. The AASC multi-view clustering algorithm [13] is used to fuse multiple distance matrices to a consensus affinity matrix. A high-quality affinity matrix should have a clear block diagonal structure. Our consensus affinity matrices, illustrated in Fig. 5, demonstrate a remarkable ability to capture a strong clustering structure that is nearly identical to the ground-truth. This, in turn, leads to improved clustering performance as compared to the ZERO and MEAN methods whose affinity matrices lack clear block structures and are plagued with noise.

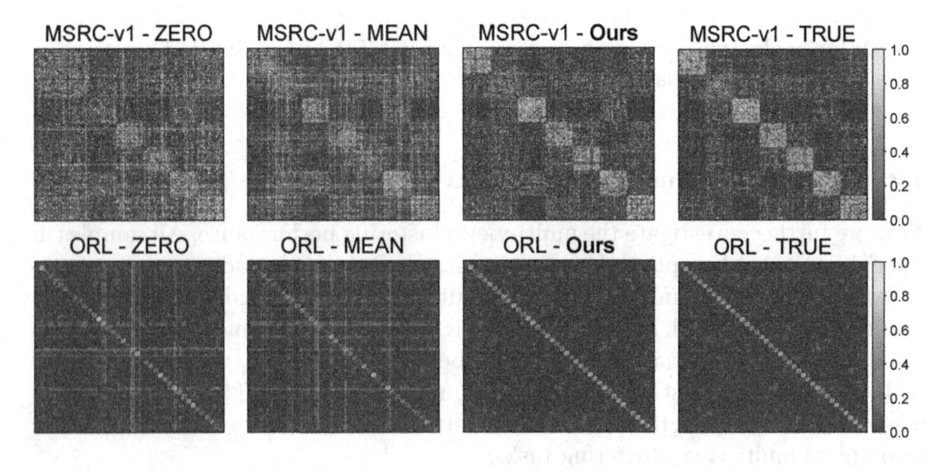

Fig. 5. Consensus affinity matrices obtained by the AASC multi-view clustering algorithm on MSRC-v1 and ORL datasets with a 70% missing ratio.

4.5 Motivation and Results Summary

When dealing with incomplete data, common imputation methods rely on domain knowledge of data structures and lack theoretical guarantees for the imputed data. To tackle this issue, we introduce a matrix correction technique that utilizes convex optimization to correct an estimated distance matrix and ensure certain properties are satisfied. Our approach leverages the positive semi-definite (PSD) property of the Laplacian kernel to improve the estimated distance with a theoretical guarantee of accuracy. As a result, our correction-clustering strategy outperforms traditional imputation-based strategies on incomplete clustering tasks in both single-view and multi-view datasets.

5 Conclusion and Future Work

Partially missing data is a significant issue in incomplete multi-view clustering, yet it has received relatively little attention in the research community. Traditional imputation methods can lead to inaccurate results and degrade performance. To address this challenge, we propose a novel imputation-free and unified framework for incomplete-value multi-view clustering. Our framework includes a distance correction method, combined with a two-stage correction-clustering strategy that integrates with existing multi-view clustering algorithms.

Our proposed framework outperforms existing imputation-based strategies, as demonstrated by extensive experiments. Our matrix correction algorithm provides high-quality Euclidean distance matrices that are closely aligned with the unknown ground-truth, resulting in improved performance in single-view spectral clustering. Additionally, our algorithm achieves better multi-view clustering performance by improving consensus affinity matrices. Overall, our framework provides a valuable tool for various data mining applications, particularly those involving incomplete clustering.

In future work, we plan to study missing data imputation in incomplete multi-view clustering and extend our framework to handle other types of missing data, such as missing views or modalities. We also aim to apply our framework to other real-world datasets and practical applications to further validate its effectiveness.

Acknowledgements. We appreciate the anonymous reviewers for their helpful feedback that greatly improved this paper. The work of Fangchen Yu and Wenye Li was supported in part by Guangdong Basic and Applied Basic Research Foundation (2021A1515011825), Guangdong Introducing Innovative and Entrepreneurial Teams Fund (2017ZT07X152), Shenzhen Science and Technology Program (CUHKSZWDZC0004), and Shenzhen Research Institute of Big Data Scholarship Program. The work of Yuqi Ma was supported in part by CUHKSZ-SRIBD Joint PhD Program. The work of Jianfeng Mao was supported in part by National Natural Science Foundation of China under grant U1733102, in part by the Guangdong Provincial Key Laboratory of Big Data Computing, The Chinese University of Hong Kong, Shenzhen under grant B10120210117, and in part by CUHK-Shenzhen under grant PF.01.000404.

References

1. Balzano, L., Nowak, R., Recht, B.: Online identification and tracking of subspaces from highly incomplete information. In: 2010 48th Annual Allerton conference on Communication, Control, and Computing (Allerton), pp. 704–711. IEEE (2010)
2. Bauschke, H.H., Borwein, J.M.: Dykstra's alternating projection algorithm for two sets. J. Approx. Theory **79**(3), 418–443 (1994)
3. Berry, M.W., Mezher, D., Philippe, B., Sameh, A.: Parallel algorithms for the singular value decomposition. In: Handbook of Parallel Computing and Statistics, pp. 133–180. Chapman and Hall/CRC (2005)
4. Boyle, J.P., Dykstra, R.L.: A method for finding projections onto the intersection of convex sets in Hilbert spaces. In: Advances in Order Restricted Statistical Inference, pp. 28–47. Springer, New York (1986). https://doi.org/10.1007/978-1-4613-9940-7_3
5. Cai, J.F., Candès, E.J., Shen, Z.: A singular value thresholding algorithm for matrix completion. SIAM J. Optim. **20**(4), 1956–1982 (2010)
6. Candes, E., Recht, B.: Exact matrix completion via convex optimization. Commun. ACM **55**(6), 111–119 (2012)
7. Du, L., et al.: Robust multiple kernel K-means using L21-Norm. In: 24th International Joint Conference on Artificial Intelligence (2015)
8. Dykstra, R.L.: An algorithm for restricted least squares regression. J. Am. Stat. Assoc. **78**(384), 837–842 (1983)
9. Fan, J., Udell, M.: Online high rank matrix completion. In: Proceedings of the IEEE/CVF Conference on Computer Vision and Pattern Recognition, pp. 8690–8698 (2019)
10. Guo, J., Ye, J.: Anchors bring ease: an embarrassingly simple approach to partial multi-view clustering. In: Proceedings of the AAAI Conference on Artificial Intelligence, vol. 33, pp. 118–125 (2019)
11. Halko, N., Martinsson, P.G., Tropp, J.A.: Finding structure with randomness: probabilistic algorithms for constructing approximate matrix decompositions. SIAM Rev. **53**(2), 217–288 (2011)
12. Hasan, M.K., Alam, M.A., Roy, S., Dutta, A., Jawad, M.T., Das, S.: Missing value imputation affects the performance of machine learning: a review and analysis of the literature (2010–2021). Inf. Med. Unlocked **27**, 100799 (2021)
13. Huang, H.C., Chuang, Y.Y., Chen, C.S.: Affinity aggregation for spectral clustering. In: 2012 IEEE Conference on Computer Vision and Pattern Recognition, pp. 773–780. IEEE (2012)
14. Kumar, A., Rai, P., Daume, H.: Co-regularized multi-view spectral clustering. In: Advances in Neural Information Processing Systems 24 (2011)
15. Li, S.Y., Jiang, Y., Zhou, Z.H.: Partial multi-view clustering. In: Proceedings of the AAAI Conference on Artificial Intelligence, vol. 28 (2014)
16. Li, W.: Estimating jaccard index with missing observations: a matrix calibration approach. In: Advances in Neural Information Processing Systems, vol. 28, pp. 2620–2628. Canada (2015)
17. Li, W.: Scalable calibration of affinity matrices from incomplete observations. In: Asian Conference on Machine Learning, pp. 753–768. PMLR, Bangkok, Thailand (2020)
18. Li, W., Yu, F.: Calibrating distance metrics under uncertainty. In: Joint European Conference on Machine Learning and Knowledge Discovery in Databases, pp. 219–234. Springer (2022). https://doi.org/10.1007/978-3-031-26409-2_14

19. Li, W., Yu, F., Ma, Z.: Metric nearness made practical. In: Proceedings of the AAAI Conference on Artificial Intelligence, vol. 37, pp. 8648–8656 (2023)
20. Lin, W.-C., Tsai, C.-F.: Missing value imputation: a review and analysis of the literature (2006–2017). Artif. Intell. Rev. **53**(2), 1487–1509 (2020). https://doi.org/10.1007/s10462-019-09709-4
21. Liu, J., et al.: Optimal neighborhood multiple kernel clustering with adaptive local kernels. IEEE Trans. Knowl. Data Eng. (2020)
22. Liu, J., et al.: Self-representation subspace clustering for incomplete multi-view data. In: Proceedings of the 29th ACM International Conference on Multimedia, pp. 2726–2734 (2021)
23. Liu, X., et al.: Efficient and effective regularized incomplete multi-view clustering. IEEE Trans. Pattern Anal. Mach. Intell. **43**(8), 2634–2646 (2020)
24. Liu, X.: Multiple kernel k k-means with incomplete kernels. IEEE Trans. Pattern Anal. Mach. Intell. **42**(5), 1191–1204 (2019)
25. Nader, R., Bretto, A., Mourad, B., Abbas, H.: On the positive semi-definite property of similarity matrices. Theoret. Comput. Sci. **755**, 13–28 (2019)
26. Ng, A., Jordan, M., Weiss, Y.: On spectral clustering: analysis and an algorithm. In: Advances in Neural Information Processing Systems 14 (2001)
27. Nie, F., Tian, L., Li, X.: Multiview clustering via adaptively weighted procrustes. In: Proceedings of the 24th ACM SIGKDD International Conference on Knowledge Discovery & Data Mining, pp. 2022–2030 (2018)
28. Shao, W., He, L., Yu, P.S.: Multiple incomplete views clustering via weighted nonnegative matrix factorization with $L_{2,1}$ regularization. In: Appice, A., Rodrigues, P.P., Santos Costa, V., Soares, C., Gama, J., Jorge, A. (eds.) ECML PKDD 2015. LNCS (LNAI), vol. 9284, pp. 318–334. Springer, Cham (2015). https://doi.org/10.1007/978-3-319-23528-8_20
29. Tang, C., et al.: CGD: multi-view clustering via cross-view graph diffusion. In: Proceedings of the AAAI Conference on Artificial Intelligence, vol. 34, pp. 5924–5931 (2020)
30. Wang, S., et al.: Highly-efficient incomplete large-scale multi-view clustering with consensus bipartite graph. In: Proceedings of the IEEE/CVF Conference on Computer Vision and Pattern Recognition, pp. 9776–9785 (2022)
31. Wang, S., et al.: Multi-view clustering via late fusion alignment maximization. In: 28th International Joint Conference on Artificial Intelligence, pp. 3778–3784 (2019)
32. Xia, R., Pan, Y., Du, L., Yin, J.: Robust multi-view spectral clustering via low-rank and sparse decomposition. In: Proceedings of the AAAI Conference on Artificial Intelligence, vol. 28 (2014)
33. Xu, N., Guo, Y., Wang, J., Luo, X., Kong, X.: Multi-view clustering via simultaneously learning shared subspace and affinity matrix. Int. J. Adv. Rob. Syst. **14**(6), 1729881417745677 (2017)
34. Yu, F., Bao, R., Mao, J., Li, W.: Highly-efficient Robinson-Foulds distance estimation with matrix correction. In: (to appear) 26th European Conference on Artificial Intelligence (2023)
35. Yu, F., Zeng, Y., Mao, J., Li, W.: Online estimation of similarity matrices with incomplete data. In: Uncertainty in Artificial Intelligence, pp. 2454–2464. PMLR (2023)
36. Zhan, K., Nie, F., Wang, J., Yang, Y.: Multiview consensus graph clustering. IEEE Trans. Image Process. **28**(3), 1261–1270 (2018)
37. Zhan, K., Zhang, C., Guan, J., Wang, J.: Graph learning for multiview clustering. IEEE Trans. Cybern. **48**(10), 2887–2895 (2017)

38. Zhang, P., et al.: Adaptive weighted graph fusion incomplete multi-view subspace clustering. Sensors **20**(20), 5755 (2020)
39. Zhang, S.: Nearest neighbor selection for iteratively KNN imputation. J. Syst. Softw. **85**(11), 2541–2552 (2012)
40. Zhao, H., Liu, H., Fu, Y.: Incomplete multi-modal visual data grouping. In: 25th International Joint Conference on Artificial Intelligence, pp. 2392–2398 (2016)
41. Zhou, S., et al.: Multiple kernel clustering with neighbor-kernel subspace segmentation. IEEE Trans. Neural Netw. Learn. Syst. **31**(4), 1351–1362 (2019)
42. Zong, L., Zhang, X., Liu, X., Yu, H.: Weighted multi-view spectral clustering based on spectral perturbation. In: Proceedings of the AAAI Conference on Artificial Intelligence, vol. 32 (2018)

PLKA-MVSNet: Parallel Multi-view Stereo with Large Kernel Convolution Attention

Bingsen Huang, Jinzheng Lu(✉), Qiang Li, Qiyuan Liu, Maosong Lin, and Yongqiang Cheng

Southwest University of Science and Technology, Mianyang, China
lujinzheng@163.com

Abstract. In this paper, we propose PLKA-MVSNet to address the remaining challenges in the in-depth estimation of learning-based multi-view stereo (MVS) methods, particularly the inaccurate depth estimation in challenging areas such as low-texture regions, weak lighting conditions, and non-Lambertian surfaces. We ascribe this problem to the insufficient performance of the feature extractor and the information loss caused by the MVS pipeline transmission, and give our optimization scheme. Specifically, we introduce parallel large kernel attention (PLKA) by using multiple small convolutions instead of a single large convolution, to enhance the perception of texture and structural information, which enables us to capture a larger receptive field and long-range information. In order to adapt to the coarse-to-fine MVS pipeline, we employ PLKA to construct a multi-stage feature extractor. Furthermore, we propose the parallel cost volume aggregation (PCVA) to enhance the robustness of the aggregated cost volume. It introduces two decision-making attentions in the 2D dimension to make up for information loss and pixel omission in the 3D convolution compression. Particularly, our method shows the best overall performance beyond the transformer-based method on the DTU dataset and achieves the best results on the challenging Tanks and Temples advanced dataset.

Keywords: Multi-view Stereo · 3D Reconstruction · Deep Learning

1 Introduction

Nowadays, learning-based MVS methods [1,2] has made great progress. Generally, MVS networks utilize a set of 2D convolutions to extract features and compute paired cost volumes based on a set of assumed depths through plane sweeping. These paired cost volumes are fused into a unified cost volume and regularized to obtain the final depth map. However, there are still two unresolved issues in the above MVS pipeline. (A) The feature extractor lacks the capability to capture robust features, which leads to difficulties in handling weak texture structures or non-Lambertian surfaces. Furthermore, features captured

B. Luo et al. (Eds.): ICONIP 2023, CCIS 1965, pp. 113–125, 2024.
https://doi.org/10.1007/978-981-99-8145-8_10

by the low-performance extractor seriously affect the subsequent operation of the MVS pipeline and deteriorate the 3D reconstruction. (B) The paired cost aggregation strategies usually depend on variance, average, or 3D convolution. For the variance-based and average-based, these methods often make all pixels visible by default, which results in the introduction of wrong pixels. For the 3D convolution-based, it is prone to loss of information, and the aggregation weight is overly dependent on local information and 3D convolution performance.

In our work, we propose a novel deep neural network called PLKA-MVSNet for MVS. Firstly, we introduce parallel large kernel attention (PLKA). It uses LKA to capture and integrate long-range information, local information and global information of features, while using gated convolution to learn the correlation and importance between different channels, which improves the perception of rich features. By constructing a multi-stage feature extractor with the PLKA, it provides more robust feature encoding for different stages of the MVS pipeline. Secondly, In order to obtain more effective cost volume weights, we propose the parallel cost volume aggregation (PCVA) for aggregating paired cost volumes. It contains a parallel 2D channel for global information capture, a parallel 3D channel for spatial information matching, and then a reliable cost weight is obtained by pointwise multiplication. Lastly, we integrate several practical components from recent MVS networks, such as the group-wise [3] correlation, binary search [4], and coarse-to-fine strategy [5], to further improve the reconstruction quality. In summary, our contributions are as follows:

1. We propose a new MVS feature extractor. It captures both global context information and local details through PLKA, which can provide richer features for subsequent MVS pipelines.
2. To leverage the global information in the paired views, we propose a novel strategy called PCVA to aggregate paired cost volumes. It suppresses the interference of invisible information and produces more robust cost volume weights.
3. Our method shows the best overall performance beyond the transformer-based method on the DTU dataset and achieves the best results on the challenging Tank and Temple advanced dataset.

2 Related Work

Recently, learning-based methods have dominated the research of MVS. The most common method is based on the depth map [1,5,6], which usually runs in a similar way and exceeds most traditional methods. MVSNet [1] takes a reference image and multiple source images as input to extract depth map features. It encodes the camera parameters in the network and constructs a 3D cost volume using differentiable homography. Finally, the depth map is obtained through 3D convolution regression. Although it has high memory requirements, MVSNet has laid the foundation for subsequent MVS research.

Some methods based on recursion [7,8] propose the recursive regularization of the 3D cost volume and utilize a Recurrent Neural Network (RNN) to propagate features between different depth hypotheses. These methods trade time for space, which allows them to process high-resolution images, but their inference speed is slow. Some studies devoted to cost volume aggregation [2,9,10] attempt to regress 2D visibility maps from paired cost volumes. They use multiple sets of 3D convolutions to learn the visibility relationship between paired views and use it as weight information to suppress the influence of mismatched pixels. However, their probability data distribution severely relies on 3D convolution, which can lead to loss of information and extreme values.

Gu et al. [5] integrate a coarse-to-fine strategy into the learning-based MVS reconstruction, which successfully reduces memory consumption to support deeper backbone networks and higher-resolution outputs. This idea is widely used in subsequent methods. To further explore the potential of this pipeline, many variants have emerged. TransMVSNet [11] attempts to introduce the self-attention mechanism of the transformer and achieves excellent results, which validate the importance of a large receptive field in extracting useful features.

Although the above learning-based MVS methods have achieved promising results, they have not focused on improving the feature-capturing capability of convolutions themselves and have overlooked the information loss that occurs in the MVS pipeline. This leads to inaccurate estimations in challenging regions and severely occluded areas.

3 Proposed Method

3.1 Network Overview

In this section, we introduce the detailed structure of PLKA-MVSNet. The overall framework is illustrated in Fig. 1. It consists of two main components: the image feature extractor and the MVS pipeline network for iterative depth estimation. To adapt the requirements of the MVS pipeline from coarse to fine strategy, we construct the feature extractor similar to the pyramid structure. At each stage of the feature pyramid, we enhance the capturing capability of global and long-range information by incorporating our proposed PLKA mechanism (Sect. 3.2). Subsequently, following the MVS pipeline, we construct paired cost volumes by applying homographic transformations to the features at different resolutions (Sect. 3.3). These paired cost volumes are then fed into the PCVA to obtain visibility maps by weight fusion. Using the visibility maps, we aggregate the paired cost volumes into a unified cost volume (Sect. 3.4), which is further regularized by 3D CNNs to generate the predicted depth map. Finally, we treat the MVS task as a classification problem and employ cross-entropy loss to optimize the maximum likelihood estimator (Sect. 3.5).

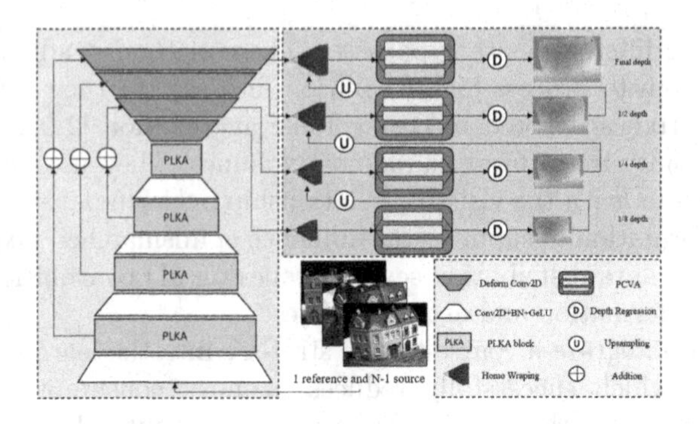

Fig. 1. Network architecture of the proposed PLKA-MVSNet.

3.2 Feature Extractor

Parallel Large Kernel Attention (PLKA): As shown in Fig. 2a, refer to [12], we decompose a large kernel $K \times K$ convolution into 3 components: a $\lceil \frac{K}{d} \times \frac{K}{d} \rceil$ depth-wise dilated convolution ($Conv_{DWD}$) with dilation d, a 1×1 point-wise convolution ($Conv_{PW}$) along the channel dimension, and a $(2d-1) \times (2d-1)$ depth-wise convolution ($Conv_{DW}$). The $Conv_{PW}$ is used for channel-wise adaptation, and learning long-range dependencies. The $Conv_{DWD}$ is used to increase the overall receptive field, learning global correlations. The $Conv_{DW}$ is used to integrate local contextual information. We can obtain the importance of each point in the overall context by the LKA. We then perform a Hadamard product with the input feature (f_{in}), which can be represented as follows:

$$LKA = f_{in} \otimes Conv_{PW}(Conv_{DWD}(Conv_{DW}(f_{in}))) \tag{1}$$

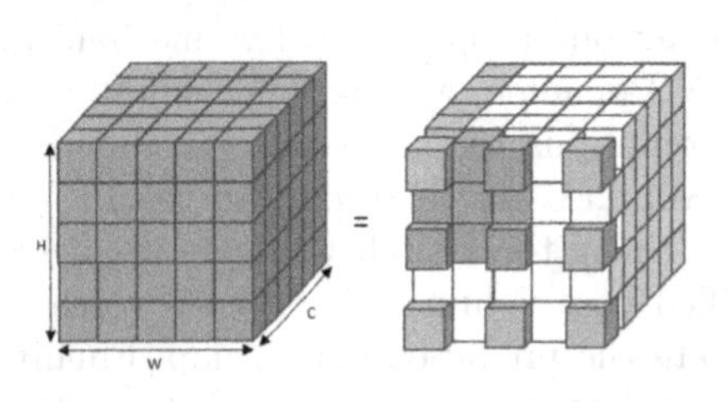

(a) The decomposition model of LKA.

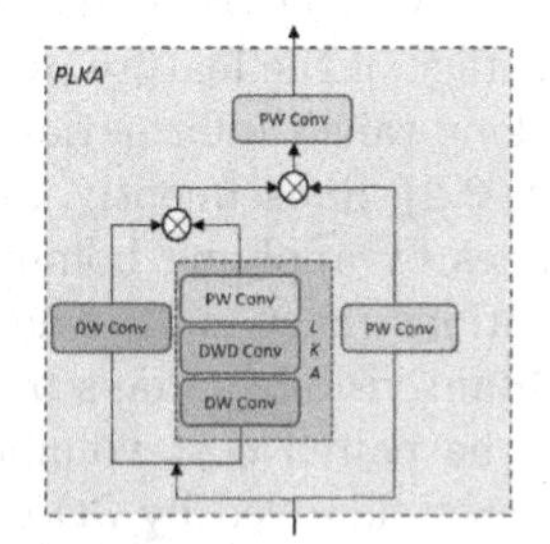

(b) The structure of PLKA.

Fig. 2. The construction process of PLKA.

We set $K = 21$, $d = 3$ as the basis for the next step. To enhance the robustness of feature capture, we propose parallel large kernel attention (PLKA), which consists of three components: a LKA for capturing large receptive fields, a parallel $Conv_{DW}$ for capturing local features, and a gated aggregation for adaptive weight adjustment. The structure is illustrated in Fig. 2b. A specific PLKA can be represented as follows:

$$PLKA = Conv_{PW}(Conv_{DW}(f_{in}) \otimes LKA(f_{in})) \otimes Conv_{PW}(f_{in}) \quad (2)$$

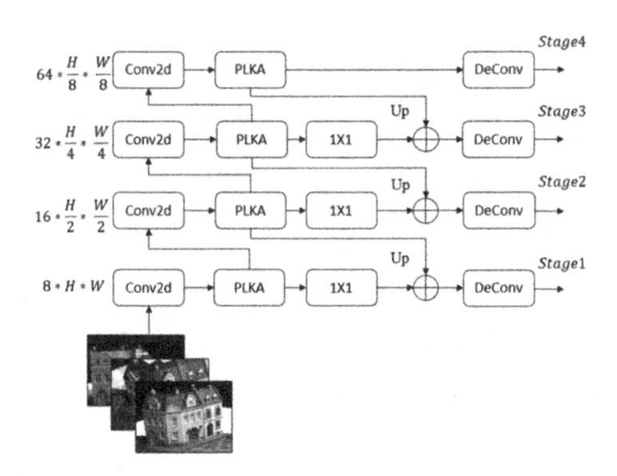

Fig. 3. The concrete structure of the multi-stage feature extractor.

Multi-stage Feature Extractor: To align with the coarse-to-fine MVS strategy, we construct a multi-stage feature extractor. The structure is illustrated in Fig. 3. Firstly, we perform downsampling with 2D convolutions to obtain preliminary features in 4 stages. In each stage, we employ a proposed PLKA block to expand the receptive field and capture long-range information. Subsequently, we fuse features from different stages through interpolation and dimension reduction operations. Additionally, to make the fused features more flexible, we incorporate deformable convolutions to adaptively capture informative patterns in the scene. Besides, we experiment with different positions for the PLKA blocks and feature fusion methods, which are further described in the ablation experiments.

3.3 Cost Volume Construction

According to [1], the features of $N-1$ source views are mapped to the hypothetical depth plane of the reference view. Under the depth hypothesis d, the pixel p in the source view can be warped to the pixel $\hat{p}$ in the reference view according to the following equation:

$$\hat{p} = K\left[R\left(\mathrm{K}_0^{-1} pd\right) + t\right] \quad (3)$$

where R and t are the rotation and translation between the two views. K_0 and K are the intrinsic matrices of the reference camera and source camera respectively. All pixels can be mapped to different depth planes according to the current assumed depth value d.

Through the relationship of reference view and source view, we can calculate N-1 paired cost volumes V_i:

$$V_i^{(d)}(p) = \langle F_0(p), \hat{F}_i^{(d)}(p) \rangle \tag{4}$$

where $\hat{F}_i^{(d)}(p)$ is the pixel p of the $i_{th} \in [1, N-1]$ source feature map at depth d, and $F_0(p)$ is the pixel p of the reference feature map.

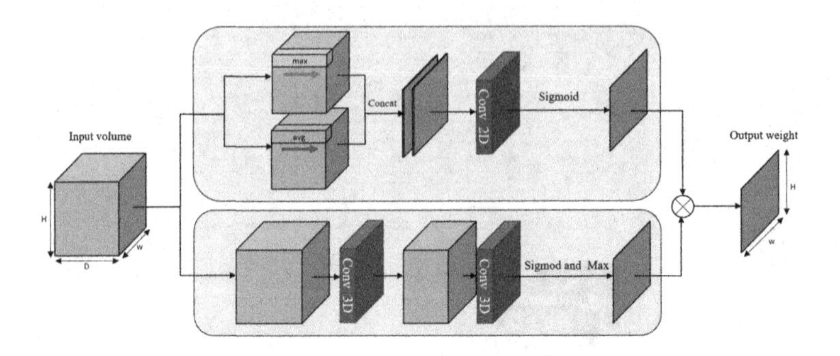

Fig. 4. The structure of proposed PCVA. The input is a paired cost volume, and the output is the weight of it to aggregate.

3.4 Parallel Cost Volume Aggregation (PCVA)

To address the potential information loss and pixel omission in learning the visibility distribution of the cost volume through local 3D convolutions, we introduce the 2D channel weight to enhance the capture of visibility information.

As shown in Fig. 4, we construct 2 parallel channels specifically designed for paired cost volumes. These channels aim to capture different types of information (both global information in the 2D channel and adaptive pixel weights in the 3D channel).

For the 2D channel, we first perform max-pooling and avg-pooling along the d dimension to reduce the dimensionality of V_i. Then, we concatenate the outputs with the *Concat* operation and apply a depth-wise convolution ($Conv_{DW}$) to compress them into a single feature F_i with a size of $1 \times H \times W$. After this, it is normalized to facilitate the weight fusion:

$$F_i = Conv_{DW}(Concat(max(V_i), avg(V_i))) \tag{5}$$

$$Att_{2d}(i) = \frac{1}{1 + \exp(-F_i)} \tag{6}$$

For the 3D channel, we apply several 3D convolutions ($Conv_{3D}$) to progressively reduce the dimension of the cost volume, resulting in feature matrices of size $D \times H \times W$. Finally, we obtain attention maps of the same size as the 2D channel by applying a Sigmoid layer and MAX operation:

$$Att_{3d}(i) = MAX(\frac{1}{1+\exp(-Conv_{3D}(V_i))}) \tag{7}$$

We perform element-wise multiplication between the two attention maps to obtain the final attention weights $Att_{pair}(i) = Att_{2d}(i) \otimes Att_{3d}(i)$. Subsequently, we aggregate N-1 paired cost volumes into a unified cost volume V with $Att_{pair}(i)$:

$$V = \frac{\sum_{i=1}^{N-1} V_i \cdot Att_{pair}(i)}{\sum_{i=1}^{N-1} Att_{pair}(i)} \tag{8}$$

The cost volume V after fusion is regularized by 3D CNNs, which gradually reduces the channel size of V to 1 and outputs a volume of size $(D \times H \times W)$. Finally, we perform the Softmax function along the D dimension to generate a probability volume P for calculating training loss and label.

3.5 Loss Function

We perform one-hot encoding on the true depth value with the mask to obtain the ground truth volume $G(j,q)$. Afterward, we compute the standard cross-entropy loss for each pixel and take the average to calculate the final loss value:

$$Loss = \sum_{q \in Q_v} \sum_{i=1}^{d} -\mathbf{G}(j,q) \log \mathbf{P}(j,q) \tag{9}$$

where q is the set of valid pixels, d is the D dimension in probability volume P $(D \times H \times W)$, j is each predicted depth plane of d, and Q_v is the set of pixels with valid truth values.

4 Experiments

4.1 Datasets

The **DTU** dataset [13] is an indoor dataset captured with high-precision cameras. It consists of 27,097 training samples, which we use for training and evaluation. The **BlendedMVS** [14] dataset is a large-scale dataset that includes both indoor and outdoor scenes. It contains a total of 16,904 training samples. The **Tanks and Temples** [15] benchmark is a large-scale dataset with various outdoor scenes. It includes 8 intermediate subsets and 6 advanced subsets. We use the fine-tuned model on the BlendedMVS to evaluate on the Tanks and Temples benchmark.

4.2 Implementation

Training: For training PLKA-MVSNet on the DTU dataset, we first crop the original images provided by the official website from the size of 1200 × 1600 to 1024 × 1280. Then, we randomly crop them to a size of 512 × 640. This is done to facilitate learning larger image features from smaller images. During all training stages, we set the number of input images N = 5. The model is trained using the PyTorch Adam optimizer for 16 epochs with an initial learning rate of 0.0001. The learning rate is halved after the 10, 12, and 14 epochs. The total training batch size is 2, distributed across 2 NVIDIA RTX 3090 GPUs. Following [11], we fine-tune PLKA-MVSNet on the BlendedMVS dataset using the original image resolution of 576 × 768 with 10 epochs.

Testing: For the test of DTU dataset, we set N = 5, and the resolution of each image is 1152 × 1600. For the Tanks and Temples benchmark, we use the fine-tuned model on the BlendedMVS dataset and set N = 7 with each image resolution of 1024 × 1920. Follow the dynamic inspection strategy proposed in [16], We filter and fuse depth maps of a scene into one point cloud.

4.3 Experimental Results

Evaluation on DTU Dataset: We evaluate the PLKA-MVSNet on the DTU evaluation sets with the official evaluation metrics. As shown in Fig. 5, PLKA-MVSNet is capable of generating more complete point clouds and preserving more details compared with other methods. This is due to the large receptive field characteristics of PLKA and the information compensation of PCVA for cost volumes aggregation. The quantitative comparison is presented in Table 1, where **Accuracy** and **Completeness** are the evaluation metrics proposed by the official benchmark, and the **Overall** represents the average of both metrics. PLKA-MVSNet outperforms in terms of accuracy (best in a learning-based MVS) and exhibits comparable overall performance to state-of-the-art methods. In particular, we achieve higher overall performance than the transformer-based method.

Fig. 5. Comparison of reconstructed results with state-of-the-art methods on the DTU evaluation set [13]. The red box shows the contrast between the reconstructed results of the challenging areas in the image. (Color figure online)

Table 1. Quantitative evaluation of DTU [13] (lower is better). **Bold** figures indicate the best and ***italic bold*** figures indicate the second best.

	Acc. (mm)	Comp. (mm)	Overall (mm)
Camp [17]	0.835	0.554	0.695
Furu [18]	0.613	0.941	0.777
Gipuma [19]	**0.283**	0.873	0.578
COLMAP [20]	0.400	0.664	0.532
MVSNet [1]	0.456	0.646	0.551
R-MVSNet [8]	0.383	0.452	0.417
P-MVSNet [2]	0.406	0.434	0.420
D2HC-RMVSNet [16]	0.395	0.378	0.386
Point-MVSNet [6]	0.342	0.411	0.376
Vis-MVSNet [9]	0.369	0.361	0.365
CasMVSNet [5]	0.325	0.385	0.355
UCS-Net [21]	0.338	0.349	0.344
AA-RMVSNet [7]	0.376	0.339	0.357
IterMVS [22]	0.373	0.354	0.363
PatchmatchNet [23]	0.427	**0.277**	0.352
TransMVSNet [11]	0.321	***0.289***	***0.305***
PLKA-MVSNet	***0.312***	0.291	**0.301**

Evaluation on Tanks and Temples: To demonstrate the generalization capability of our method in outdoor scenes, we test the PLKA-MVSNet on the Tanks and Temples benchmark. We upload the reconstructed point clouds to the benchmark website, and the quantitative evaluation result is shown in Table 2. Our method achieves highly comparable results to state-of-the-art techniques on the intermediate subset. It obtains the best average score on the more challenging advanced subset and performs the best in 3 out of the 6 scenes. The visualization results are shown in Fig. 6.

4.4 Ablation Study

In this section, we conduct ablation experiments on the combination forms of PLKA in the feature extractor and the effectiveness of our proposed PCVA. Baseline is largely based on [5], which applies the group-wise correlation [3] and binary search strategy [4].

Table 2. Quantitative evaluation on the Intermediate and Advanced subsets of the Tanks and Temples benchmark [15]. The evaluation metric is the F-score (higher is better). The Mean is the average score of all scenes. **Bold** figures indicate the best and underlined figures indicate the second best.

	Advanced							Intermediate								
	Mean	Aud.	Bal.	Cou.	Mus.	Pal.	Tem.	Mean	Fam.	Fra.	Hor.	Lig.	M60	Pan.	Pla.	Tra
COLMAP [20]	27.24	16.02	25.23	34.70	41.51	18.05	27.94	42.14	50.41	22.25	25.63	56.43	44.83	46.97	48.53	42.04
MVSNet [1]	–	–	–	–	–	–	–	43.48	55.99	28.55	25.07	50.79	53.96	50.86	47.9	34.69
Point-MVSNet [6]	–	–	–	–	–	–	–	48.27	61.79	41.15	34.20	50.79	51.97	50.85	52.38	43.06
R-MVSNet [16]	29.55	19.49	31.45	29.99	42.31	22.94	31.10	50.55	73.01	54.46	43.42	43.88	46.80	46.69	50.87	45.25
UCSNet [21]	–	–	–	–	–	–	–	54.83	76.09	53.16	43.03	54	55.6	51.49	57.38	47.89
CasMVSNet [5]	31.12	19.81	38.46	29.10	43.87	27.36	28.11	56.42	76.36	58.45	46.2	55.53	56.11	54.02	58.17	46.56
Vis-MVSNet [9]	33.78	20.79	38.77	32.45	44.20	28.73	**37.70**	60.03	77.4	60.23	47.07	63.44	62.21	57.28	60.54	52.07
PatchmatchNet [23]	32.31	23.69	37.73	30.04	41.80	28.31	32.29	53.15	66.99	52.64	43.24	54.87	52.87	49.54	54.21	50.81
AA-RMVSNet [7]	33.53	20.96	40.15	32.05	46.01	29.28	32.71	61.51	77.77	59.53	51.53	**64.02**	**64.05**	59.47	60.85	54.90
TransMVSNet [11]	37.00	24.84	**44.59**	34.77	46.49	**34.69**	36.62	**63.52**	**80.92**	65.83	**56.94**	62.54	63.06	**60.00**	60.20	**58.67**
PLKA-MVSNet (ours)	**37.11**	**30.30**	41.60	**35.51**	**47.87**	30.70	36.66	61.49	79.31	**66.56**	53.45	61.16	60.13	56.53	**60.99**	53.99

Fig. 6. Point clouds of Tanks and Temples [15] reconstructed by PLKA-MVSNet.

The Combination forms of PLKA in the Feature Extractor: To unleash the potential of PLKA, we design various feature extractor structures (The 4 most representative ones shown in Fig. 7). We perform experiments on the DTU dataset to compare their performance, and the results are presented in Table 3. From the results, we observe two points:

(1) The application of PLKA in the upsampling process has better performance than the application of PLKA after upsampling (Refer to Fig. 7(a) and (c)). We think the reason is that PLKA can grasp richer global information in shallow features than in deep features.
(2) In the process of feature downsampling fusion, using independent features yields better results than using mixed features. For example, when fusing high-stage features in stage 2, using the features of stage 3 is better than using the mixed features of stage 3 and stage 4 (Refer to Fig. 7(a) and (b)).

Component Analysis of PCVA: We use the best-performing architecture Table 3(d) as the baseline to test the performance of our proposed PCVA. We visualize the weights obtained from PCVA, as shown in Fig. 8. PCVA can capture the distribution of visibility information from different source images to reference images, and then obtain high-quality weights for cost volume aggregation.

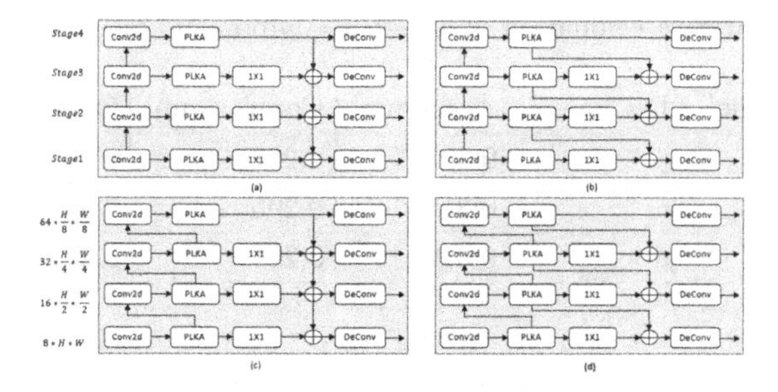

Fig. 7. Different variations of PLKA integrated into the feature extractor.

Table 3. Test results for 4 variations with PLKA and baseline without PLKA.

method	Acc. (mm)	Comp. (mm)	Overall (mm)
baseline	0.327	0.312	0.320
a	0.315	0.309	0.312
b	0.317	0.301	0.309
c	0.309	0.300	0.304
d	**0.307**	**0.297**	**0.302**

We conduct ablation experiments on different components of PCVA to test their effectiveness. The result is shown in Table 4, where the 3D channel refers to the original method in the baseline. From the results, it is evident that using only the 2D channel leads to poorer performance. Although 2D decision-making can capture long-range and global information, it lacks the ability to capture the fine details inside the cost volume. When combining 2D and 3D channel weights, we observe a significant improvement in performance. We think the reason is that the 2D channel compensates for the information loss in the 3D channel and enhances the global representation capability of the cost volume after aggregation.

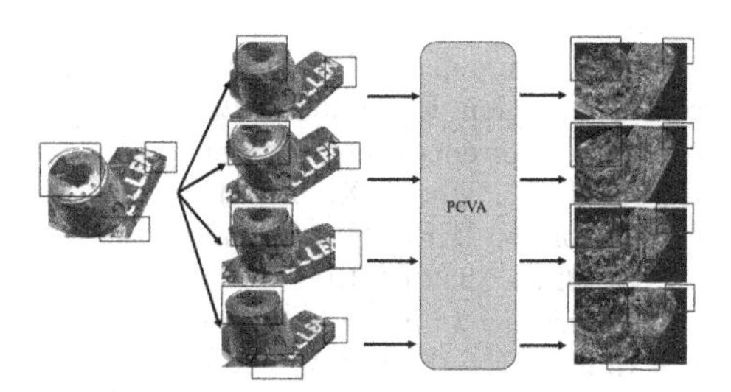

Fig. 8. The output weights of PCVA. The sample input is 1 reference and 4 source images of Scan 1 in DTU dataset [13].

Table 4. Quantitative evaluation of different components in PCVA on the DTU.

3D Channel	2D Channel	Acc. (mm)	Comp. (mm)	Overall (mm)
√		**0.307**	0.297	0.302
	√	0.319	0.295	0.307
√	√	0.312	**0.291**	**0.301**

5 Conclusion

In this paper, we introduce a novel learning-based MVS network called PLKA-MVSNet. Specifically, we design the PLKA, which consists of a LKA for capturing large receptive fields, parallel depth-wise convolution for capturing local features, and gated aggregation for adaptive weight adjustment. Additionally, we propose the PCVA to generate more reliable cost volume weights for aggregating paired cost volumes. Through extensive experiments, we observe that PLKA-MVSNet achieves outstanding performance in both indoor and outdoor scenes. However, due to the memory limitation of MVS pipeline network, we can only use $K = 21$, $d = 3$ combination to realize large receptive field for the LKA setting. It still leads to large memory consumption, and our next work will be to design a lighter-weight LKA alternative for MVS. Finally, we hope that our exploration can provide some insights and encourage further research on the potential of large kernel convolution in the MVS framework.

Acknowledgements. This work was supported in part by the Heilongjiang Provincial Science and Technology Program under Grant 2022ZX01A16, and in part by the Sichuan Provincial Science and Technology Program under Grant 2022ZHCG0001.

References

1. Yao, Y., Luo, Z., Li, S., Fang, T., Quan, L.: MVSNet: depth inference for unstructured multi-view stereo. In: Ferrari, V., Hebert, M., Sminchisescu, C., Weiss, Y. (eds.) ECCV 2018. LNCS, vol. 11212, pp. 785–801. Springer, Cham (2018). https://doi.org/10.1007/978-3-030-01237-3_47
2. Luo, K., Guan, T., Ju, L., Huang, H., Luo, Y.: P-MVSNet: learning patch-wise matching confidence aggregation for multi-view stereo. In: Proceedings of the IEEE/CVF International Conference on Computer Vision, pp. 10452–10461 (2019)
3. Guo, X., Yang, K., Yang, W., Wang, X., Li, H.: Group-wise correlation stereo network. In: Proceedings of the IEEE/CVF Conference on Computer Vision and Pattern Recognition, pp. 3273–3282 (2019)
4. Mi, Z., Di, C., Xu, D.: Generalized binary search network for highly-efficient multi-view stereo. In: Proceedings of the IEEE/CVF Conference on Computer Vision and Pattern Recognition, pp. 12991–13000 (2022)
5. Gu, X., Fan, Z., Zhu, S., Dai, Z., Tan, F., Tan, P.: Cascade cost volume for high-resolution multi-view stereo and stereo matching. In: Proceedings of the IEEE/CVF Conference on Computer Vision and Pattern Recognition, pp. 2495–2504 (2020)

6. Chen, R., Han, S., Xu, J., Su, H.: Point-based multi-view stereo network. In: Proceedings of the IEEE/CVF International Conference on Computer Vision, pp. 1538–1547 (2019)
7. Wei, Z., Zhu, Q., Min, C., Chen, Y., Wang, G.: AA-RMVSNet: adaptive aggregation recurrent multi-view stereo network. In: Proceedings of the IEEE/CVF International Conference on Computer Vision, pp. 6187–6196 (2021)
8. Yao, Y., Luo, Z., Li, S., Shen, T., Fang, T., Quan, L.: Recurrent MVSNet for high-resolution multi-view stereo depth inference. In: Proceedings of the IEEE/CVF Conference on Computer Vision and Pattern Recognition, pp. 5525–5534 (2019)
9. Zhang, J., Yao, Y., Li, S., Luo, Z., Fang, T.: Visibility-aware multi-view stereo network. arXiv preprint arXiv:2008.07928 (2020)
10. Xu, Q., Tao, W.: PVSNet: pixelwise visibility-aware multi-view stereo network. arXiv preprint arXiv:2007.07714 (2020)
11. Ding, Y., et al.: TransMVSNet: global context-aware multi-view stereo network with transformers. In: Proceedings of the IEEE/CVF Conference on Computer Vision and Pattern Recognition, pp. 8585–8594 (2022)
12. Guo, M.-H., Lu, C.-Z., Liu, Z.-N., Cheng, M.-M., Hu, S.-M.: Visual attention network. arXiv preprint arXiv:2202.09741 (2022)
13. Aanæs, H., Jensen, R.R., Vogiatzis, G., Tola, E., Dahl, A.B.: Large-scale data for multiple-view stereopsis. Int. J. Comput. Vision **120**(2), 153–168 (2016)
14. Yao, Y., et al.: BlendedMVS: a large-scale dataset for generalized multi-view stereo networks. In: Proceedings of the IEEE/CVF Conference on Computer Vision and Pattern Recognition, pp. 1790–1799 (2020)
15. Knapitsch, A., Park, J., Zhou, Q.Y., Koltun, V.: Tanks and temples: benchmarking large-scale scene reconstruction. ACM Trans. Graph. **36**(4CD), 78.1–78.13 (2017)
16. Yan, J., et al.: Dense hybrid recurrent multi-view stereo net with dynamic consistency checking. In: Vedaldi, A., Bischof, H., Brox, T., Frahm, J.-M. (eds.) ECCV 2020. LNCS, vol. 12349, pp. 674–689. Springer, Cham (2020). https://doi.org/10.1007/978-3-030-58548-8_39
17. Campbell, N.D.F., Vogiatzis, G., Hernández, C., Cipolla, R.: Using multiple hypotheses to improve depth-maps for multi-view stereo. In: Forsyth, D., Torr, P., Zisserman, A. (eds.) ECCV 2008. LNCS, vol. 5302, pp. 766–779. Springer, Heidelberg (2008). https://doi.org/10.1007/978-3-540-88682-2_58
18. Furukawa, Y., Ponce, J.: Accurate, dense, and robust multiview stereopsis. IEEE Trans. Pattern Anal. Mach. Intell. **32**(8), 1362–1376 (2009)
19. Galliani, S., Lasinger, K., Schindler, K.: Massively parallel multiview stereopsis by surface normal diffusion. In: Proceedings of the IEEE International Conference on Computer Vision, pp. 873–881 (2015)
20. Xu, Q., Tao, W.: Multi-scale geometric consistency guided multi-view stereo. In: Proceedings of the IEEE/CVF Conference on Computer Vision and Pattern Recognition, pp. 5483–5492 (2019)
21. Cheng, S., et al.: Deep stereo using adaptive thin volume representation with uncertainty awareness. In: Proceedings of the IEEE/CVF Conference on Computer Vision and Pattern Recognition, pp. 2524–2534 (2020)
22. Wang, F., Galliani, S., Vogel, C., Pollefeys, M.: IterMVS: iterative probability estimation for efficient multi-view stereo. In: The IEEE/CVF Conference on Computer Vision and Pattern Recognition, pp. 8606–8615 (2022)
23. Wang, F., Galliani, S., Vogel, C., Speciale, P., Pollefeys, M.: PatchmatchNet: learned multi-view patchmatch stereo. In: Proceedings of the IEEE/CVF Conference on Computer Vision and Pattern Recognition, pp. 14194–14203 (2021)

Enhancement of Masked Expression Recognition Inference via Fusion Segmentation and Classifier

Ruixue Chai, Wei Wu(✉), and Yuxing Lee

Department of Computer Science, Inner Mongolia University, Hohhot, China
{32109026,32109019}@mail.imu.edu.cn, cswuwei@imu.edu.cn

Abstract. Despite remarkable advancements in facial expression recognition, recognizing facial expressions from occluded facial images in real-world environments remains a challenging task. Various types of occlusions randomly occur on the face, obstructing relevant information and introducing unwanted interference. Moreover, occlusions can alter facial structures and expression patterns, causing variations in key facial landmarks. To address these challenges, we propose a novel approach called the Occlusion Removal and Information Capture (ORIC) network, which fuses segmentation and classification. Our method consists of three modules: the Occlusion Awareness (OA) module learns about occlusions, the Occlusion Purification (OP) module generates robust and multi-scale occlusion masks to purify the occluded facial expression information, and the Expression Information Capture (EIC) module extracts comprehensive and robust expression features using the purified facial information. ORIC can eliminate the interference caused by occlusion and utilize both local region information and global semantic information to achieve facial expression recognition under occluded conditions. Through experimental evaluations on synthetic occluded RAF-DB and AffectNet datasets, as well as a real occluded dataset FED-RO, our method demonstrates significant advantages and effectiveness. Our research provides an innovative solution for recognizing facial expressions from occluded facial images in wild environments.

Keywords: Expression Recognition · Fusion Model · Occlusion Handle

1 Introduction

Facial expressions are a primary means of conveying emotions, playing a pivotal role in human communication alongside language [1]. In recent years, the field of facial expression recognition (FER) has witnessed extensive applications across diverse domains, including human-computer interaction [2], psychological health assessment [3], driver safety [4], and computer animation [5]. With the developments in deep learning, researchers have made remarkable progress in facial expression recognition [6–8]. However, the recognition of facial expressions in real-world scenarios presents significant challenges, primarily due to the

B. Luo et al. (Eds.): ICONIP 2023, CCIS 1965, pp. 126–138, 2024.
https://doi.org/10.1007/978-981-99-8145-8_11

ubiquitous and diverse nature of occlusions. These occlusions can come from various sources, such as objects, food items, toys, gloves, or any other entities capable of occluding the eyes, mouth, or other crucial facial regions associated with expressions. The recognition of occluded facial expressions involves two main challenges. Firstly, occlusions reduce the availability of original facial region features, especially when occlusions occur in the most discriminative areas for different expressions, masking the underlying expression cues and significantly impacting overall expression recognition. Secondly, occlusions introduce interfering features. The location, shape, color, texture, and other characteristics of occlusions are unknown, which leads to error recognition during the extraction of expression features.

To address the issue of reduced discriminative regions in occluded facial expression features, a straightforward approach is to restore the occluded region information and then perform expression classification. However, if the generated parts do not accurately reflect the intended expressions, they may introduce additional interference. To mitigate the interference caused by occlusions, some work [9,10] have employed attention mechanisms to emphasize unoccluded regions and disregard occluded regions. While this approach considers the interference introduced by occluded regions, improper region partitioning may mistakenly assign low weights to important areas surrounding the occlusions. Another solution [11,12] involves learning the occlusion positions within the network and removing the occlusion information at a specific layer. However, relying on a single-scale mask may lead to either excessive or incomplete removal of the occluded regions, limiting its adaptability to multi-scale occlusions. And, occlusion can distort relevant facial features, making it difficult to establish their connection with expressions based solely on extracted local features.

In this paper, we propose an end-to-end network that fuses segmentation and classification, called the Occlusion Removal and Information Capture (ORIC) network for occluded FER. Our network combines occlusion discarding and feature extraction to achieve effective feature representation. Specifically, we introduce an Occlusion Awareness (OA) module using an encoder-decoder structure to determine the presence of occluded pixels and utilize this information in the Occlusion Purification (OP) module. The OP module generates robust occlusion masks and applies them to remove occlusion information from facial expressions. Furthermore, we incorporate an Expression Information Capture (EIC) module that extracts the most relevant components from the purified information for accurate expression recognition. To overcome the limitations of local features, we introduce a Global Information Capture (GIC) module that extracts comprehensive features. The main contributions of our work are as follows:

- We have proposed an Occlusion Removal and Information Capture (ORIC) network, which can effectively eliminate occlusion information from facial features at varying scales while capturing comprehensive facial expression information.

Fig. 1. Examples of synthetic occlusion in the dataset.

- The OA module and OP module are introduced to effectively learn occlusion information at various scales, enabling the generation of accurate occlusion masks to purify expression features.
- The EIC module is designed to comprehensively integrate global and local information, facilitating feature fusion for improved facial expression recognition.

2 Related Work

With the advancement of facial expression recognition methods, there has been a growing interest among researchers in occluded facial expression recognition. These approaches can generally be classified into three categories: subregion-based methods, reconstruction-based methods, and discard-based methods.

2.1 Subregion-Based Methods

In subregion-based methods, the face is divided into multiple regions, and different weights are assigned to these regions during training. Li et al. [10] proposed PG-CNN, which decomposes the facial image into 24 regions and utilizes PG-Unit to determine the importance of each region and extract local features. Building upon PG-CNN, Li et al. [13] introduced the ACNN framework, where GG-Unit is added to extract global information, balancing the local representations and the global representation. Wang et al. [9] presented the RAN framework, which consists of self-attention and relational attention modules. The self-attention module generates corresponding scores for different regions. The relational attention module considers local and global features to improve the scores. Gera et al. [14] proposed a novel spatial channel attention network that obtains local and global attention for each channel at every spatial position. Additionally, they introduced the CCI branch to complement the extracted features.

2.2 Reconstruction-Based Methods

Reconstruction-based methods aim to restore the features of occluded facial regions. Lu et al. [15] proposed a method based on Wasserstein Generative Adversarial Networks (GANs) for the restoration of occluded facial images. In a different approach, Pan et al. [16] utilized unoccluded images to guide the training of an occlusion-aware network at both the feature and label levels, with the aim of minimizing the discrepancy between occluded and unoccluded image features. Dong et al. [17] proposed a method where they first utilized a GAN to remove occlusions and then used the result as an additional input to a second GAN to synthesize the final de-occluded surface.

2.3 Discard-Based Methods

The occlusion-discarding approach involves removing the occluded parts of facial regions. Yang et al. [12] utilize facial landmarks to generate masks for locating relevant facial regions and combining attention mechanisms while employing two branches to extract comprehensive expressive features. Xing et al. [11] proposed the integration of occlusion discard and feature completion to mitigate the influence of occlusion on facial expression recognition. To reduce the dependency of feature completion on occlusion discard, guidance from discriminative regions was introduced for joint feature completion.

3 Proposed Method

Inspired by MSML [18], We propose a network called Occlusion Removal and Information Capture (ORIC) for addressing the challenges of occlusion in facial expression recognition. The overall architecture of ORIC, as illustrated in Fig. 2, consists of three key modules: Occlusion Awareness (OA), Occlusion Purification (OP), and Expression Information Capture (EIC). The OA module employs an encoder-decoder structure to generate multi-scale occlusion segmentation predictions. These predictions are represented as $Y_o^j = \{Y_o^1, Y_o^2, ..., Y_o^m\}$, where j = 1, ..., m. Here, m denotes the number of features generated by the decoder, and Y_p^j represents the output of the j-th decoder. Similarly, the EIC module generates multi-scale expression features, denoted as $Y_e^i = \{Y_e^1, Y_e^2, ..., Y_e^m\}$. In this representation, m indicates the number of features. The initial feature, Y_e^1, represents the raw feature before passing through the EIC block, while Y_e^m corresponds to the feature obtained after the final EIC block. Both the OA and EIC modules generate an equal number of feature maps, hence the use of m as the representation. The OP blocks receive the occlusion segmentation predictions at different scales and process them to produce occlusion masks, which are used to purify the occlusion information in the input image. Subsequently, the resulting clean expression images are fed into the EIC module to extract comprehensive and robust facial expression information for the expression recognition stage. By integrating the OA, OP, and EIC modules, our ORIC network aims to effectively address occlusion challenges, enabling improved facial expression recognition performance.

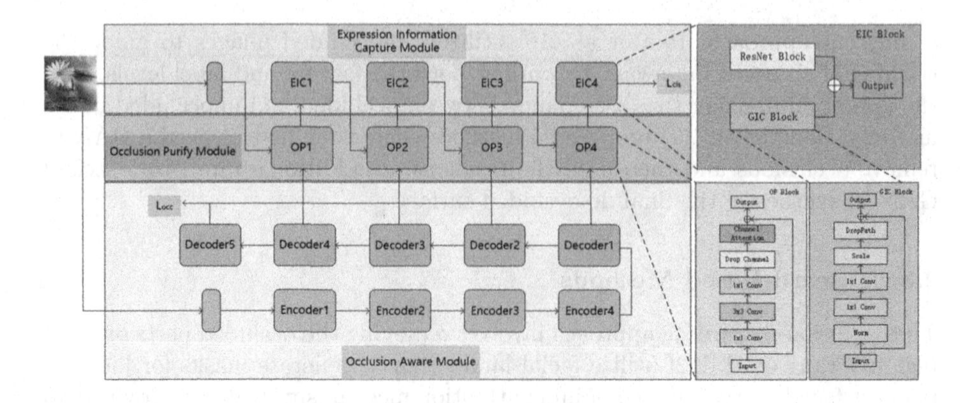

Fig. 2. Overall of our proposed method

3.1 Occlusion Awareness Module

The purpose of the OA module is to identify the position and shape of occluding objects. We generate potential masks at different scales to capture occlusion information across various levels of image detail. The OA module adopts the encoder-decoder structure of U-Net. The decoder part of the OA module produces multi-scale features denoted as $Y_o^j = \{Y_o^1, Y_o^2, ..., Y_o^m\}$. The deeper layers of the decoder output, represented by Y_o^j, contain more accurate occlusion position information as well as higher-level semantic details. On the other hand, the shallower layers of the EIC module, denoted as Y_e^i, have smaller receptive fields and lack semantic information. Therefore, we establish a bridge between the two modules using the loss function shown in Eq. (1), accurately representing the information about the occlusion positions.

$$L_{occ} = \sum_{j=1}^{m} w_{side}^j l_{side}^j + w_{fuse} l_{fuse} \tag{1}$$

where l_{side}^j represents the loss between the occlusion segmentation map obtained by the j-th decoder and the target segmentation map. Similarly, l_{fuse} represents the loss computed between the segmentation result obtained by fusing the outputs of all stage decoders and the target segmentation map. Finally, w_{side}^j and w_{fuse} indicate the respective weights associated with each loss.

3.2 Occlusion Purification Module

The OP module transforms the multi-scale occlusion segmentation predictions obtained from the OA module into multi-scale masks. These masks are generated to remove occlusions from the extracted facial expression information in the EIC module, resulting in clean facial expression features. The OP module first concatenates the features from both the OA module and the EIC module and subsequently performs mask conversion. As shown in Eq. (2):

$$M^i = OP([Y_e^i, Y_o^j]) = F(Conv1([Y_e^i, Y_o^j], W_i) + [Y_e^i, Y_o^j]) \tag{2}$$

where Conv1(·) represents a combination operation of 1×1 convolution, 3×3 convolution, and 1×1 convolution, W_i denotes the parameter matrix for the residual operation, where i takes values from 1 to m-1, and j equals m-i. Y_e^i and Y_o^j represent the i-th expression feature and the corresponding predicted occlusion position feature generated by the j-th decoder, respectively. [·] denotes concatenation, F(·) represents the activation function, and M^i indicates the generated occlusion mask.

To enhance the model's robustness, we introduce the DropChannel [19] and channel attention mechanisms. These enable the model to better adapt to different occlusion scenarios and handle uncertainties for occlusion positions, shapes, and degrees. When occlusions occur, the information in dropped channels can be compensated by other channels, thereby mitigating the impact of occlusions on model performance. The channel attention mechanism dynamically adjusts the channel weights based on the location and severity of occlusions to capture relevant features of the occluded areas more effectively. The final structure of the OP block is illustrated in Fig. 2. By employing this approach, Eq. (3) can be reformulated as follows:

$$M^i = OP([Y_e^i, Y_o^j]) = F(CA(DC(Conv1([Y_e^i, Y_o^j], W_i))) + [Y_e^i, Y_o^j]) \quad (3)$$

where DC(·) represents channel drop, and CA(·) represents channel attention. Then, multiple levels of occlusion masks are utilized to purify facial expressions. The purified facial expression feature representation is denoted as Eq. (4):

$$Y_p^i = M^i \circ Y_e^i \quad (4)$$

where $\circ$ denotes the Hadamard product.

3.3 Expression Information Capture Module

The EIC module aims to extract the most relevant and advantageous information for facial expression recognition. Typically, ResNet structures designed for local feature extraction are used. However, in the process of facial expression recognition, relying solely on local information may fail to accurately extract the key features relevant to specific expressions. Due to the presence of occlusions, facial structures, and expression patterns can vary, causing variations in specific key positional features. Consequently, recognition methods that depend on local features alone may lack accuracy. To address this issue, we propose a Global Information Capture (GIC) module that extracts global expression information, aiding the model in identifying key positional features and their relationship with expressions. The structure of each block in the GIC module is illustrated in Fig. 2.

The GIC module extracts global information from multi-scale purified facial expression features. Specifically, the output Y_p^i from the OP module serves as the input to the EIC module. It is simultaneously fed into both the residual block for original local feature extraction and the GIC block for global feature extraction.

Within the GIC block, the data undergoes group normalization, which is equivalent to grouping the feature maps along the channel dimension. Subsequently, two 1×1 convolutional layers are applied. The first convolutional layer reduces the number of input feature channels, integrating and compressing the original feature information to facilitate the learning of representative features. Then, the second convolution is used to restore the feature dimension, allowing the compressed and integrated features to be fused with the original features, preserving fine-grained details while incorporating global channel information for comprehensive modeling. Channel scaling and DropPath operations are employed to adjust the weights and importance of the channel features, introducing randomness and regularization to enhance the model's expressiveness, flexibility, and generalization ability. This process is represented as Eq. (5):

$$R_{GIC}^{i} = DP(Scale(Conv2(Norm(Y_{p}^{i})))) \tag{5}$$

where Norm(·) is group normalization, Conv2(·) is the combination of two 1×1 convolutions, Scale(·) is channel scaling, and DP(·) is droppath. The obtained global feature information is combined with the local features extracted by the ResNet block to obtain comprehensive and robust feature representations. This is mathematically represented as Eq. (6):

$$Y_{e}^{i} = \theta(Y_{p}^{i}) + R_{GIC}^{i} \tag{6}$$

where θ represents the residual operation and Y_{e}^{i} represents the obtained global representation. It is important to note that the previous m-1 expression features Y_{e}^{i} obtained will be fed into the OP module before the next block. Only the last global feature Y_{e}^{m}, produced by the final convolutional block of the network, will not undergo expression purification but instead proceed to the classification process.

The final loss function is a combination of classification loss and segmentation loss represented as Eq. (7):

$$L_{total} = L_{cls} + \alpha L_{occ} \tag{7}$$

where L_{cls} is the cross-entropy loss function and α is a weighting factor.

4 Experiments

4.1 Datasets

AffectNet [20] has collected over 1M facial images from the internet. In our experiment, only images labeled with basic facial expressions are utilized, including 280,000 images in the training set and 3,500 images in the testing set. RAF-DB [21] contains 29672 facial images tagged with basic or compound expressions by 40 independent taggers. In this work, only images with 7 categories of basic emotions are used, including 12,271 images in the training set and 3,068 images

Table 1. Comparison with state-of-the-art methods.*represents models trained on RAF-DB and AffectNet.

Methods	RAF-DB	AffectNet	*FED-RO
PG-CNN [10]	78.05	52.47	64.25
gACNN [13]	80.54	54.84	66.50
Pan et al. [16]	81.97	56.42	69.75
Xia and Wang [23]	82.74	57.46	70.50
Co-Completion [11]	82.82	57.69	73.00
ORIC (Ours)	**83.57**	**59.59**	**73.68**

in the testing set. FED-RO [13] is the first facial expression database with realistic occlusion in the wild. It consists of 400 facial images with various occlusions and includes annotations for seven emotions. Considering the limited capacity of FED-RO, we simply utilize it to conduct cross-database validation.

In order to simulate occlusion scenarios in real-world settings, we artificially synthesized occluded images by randomly placing occluding objects on datasets other than FED-RO. The occluding objects used were collected by searching for more than 20 keywords, including hands, beverages, food, toys, computers, cups, and others, in search engines. As Benitez-Quiroz et al. [22] have demonstrated that small localized occlusions have no impact on current FER algorithms, we set the size of the occluding objects to be between 25% and 60% of the size of the facial expression images. The occlusion example is illustrated in Fig. 1.

4.2 Implementation Details

The backbone network of the proposed architecture is a pre-trained ResNet-18, trained on MS-Celeb-1M [24], which also be the baseline for this research paper. All images are aligned using MTCNN [25] and then resized to 224×224. The Adam optimizer is used for the OA module, while the OP and EIC modules utilize stochastic gradient descent (SGD) for optimizing the model over 110 epochs. The initial learning rate for the OA module is set to the default value of 0.001 on the RAF-DB, while for the AffectNet, it is set to 0.0005. The initial learning rates for the EIC and OP modules are set to 0.01 on the RAF-DB, and to 0.005 on the AffectNet, and the batch size is set to 64. The accuracy serves as the performance metric in this paper.

4.3 Results Comparison

Comparison on Synthesised Occlusion: On datasets with synthesized occlusions, we make a comparison with state-of-the-art methods. According to the results shown in Table 1, we observed the following findings: PG-CNN and gACNN adopt facial key points and attention mechanisms to learn key facial expression information, but they lack an effective recognition of occluded regions,

Table 2. Verify the importance of each proposed module.

Method	ResNet18	+OA+OP	+GIC
RAF-DB	79.07	81.55	**83.57**

resulting in relatively lower accuracy. In contrast, Pan et al.'s method and Xia and Wang's method leverage information from both occluded and non-occluded images, achieving relatively better performance. Building upon this, Co-Completion further considers the removal of occlusion information. However, as mentioned earlier, single-layer occlusions may be challenging to completely remove. To address this, we employ a multi-scale masking approach to eliminate occluded regions and focus on the global context during feature extraction, which leads to the best performance on the RAF-DB and AffectNet datasets. We employ a similar occlusion handling approach to Co-Completion, but the generated occlusion datasets may still differ.

Comparison on Realistic Occlusion: To evaluate the performance of our model in real occluded scenes and assess its generalization ability, we conducted cross-dataset validation on FED-RO. We merged images from RAF-DB and AffectNet for training purposes. The experimental results are shown in Table 1. It can be observed that our method outperforms previous approaches in terms of generalizing well to real occluded scenes. Our method effectively addresses occlusion interference and utilizes valuable facial expression information.

4.4 Ablation Study

To evaluate the effectiveness of our proposed method, we compared the accuracy exhibited by each stage of the network on the occluded RAF-DB dataset. It is important to note that the OA module and the OP module are interdependent. Therefore, in our ablation study, both components were added simultaneously. As shown in Table 2, we observed the following results: Firstly, our method achieved a 4.5% higher accuracy on RAF-DB compared to the baseline (ResNet18), proving the effectiveness of our proposed network in suppressing interference caused by occluded information. Secondly, the experiments conducted on the dataset revealed that both the OP module and the GIC module played crucial roles in recognizing occluded expressions. This finding provides evidence that the OP module effectively removes interference caused by occluded information, while the GIC module captures a broader range of semantic information, thereby enhancing the expressive capability of features. To assess the significance of multi-scale occlusion removal, we conducted an exploration of the optimal number of OP blocks. Considering our task of using extracted high-level features for expression classification, we systematically incorporated OP blocks in a stepwise manner, following the sequence OP4, OP3, OP2, and OP1. As shown in Table 3, the result reveals an incremental improvement in model

Table 3. Effect of multi-scale OP blocks on occlusion FER: incremental addition (+) on the previous blocks.

Method	w/o OP	+OP4	+OP3	+OP2	+OP1
RAF-DB	80.96	81.42	82.01	82.73	**83.57**

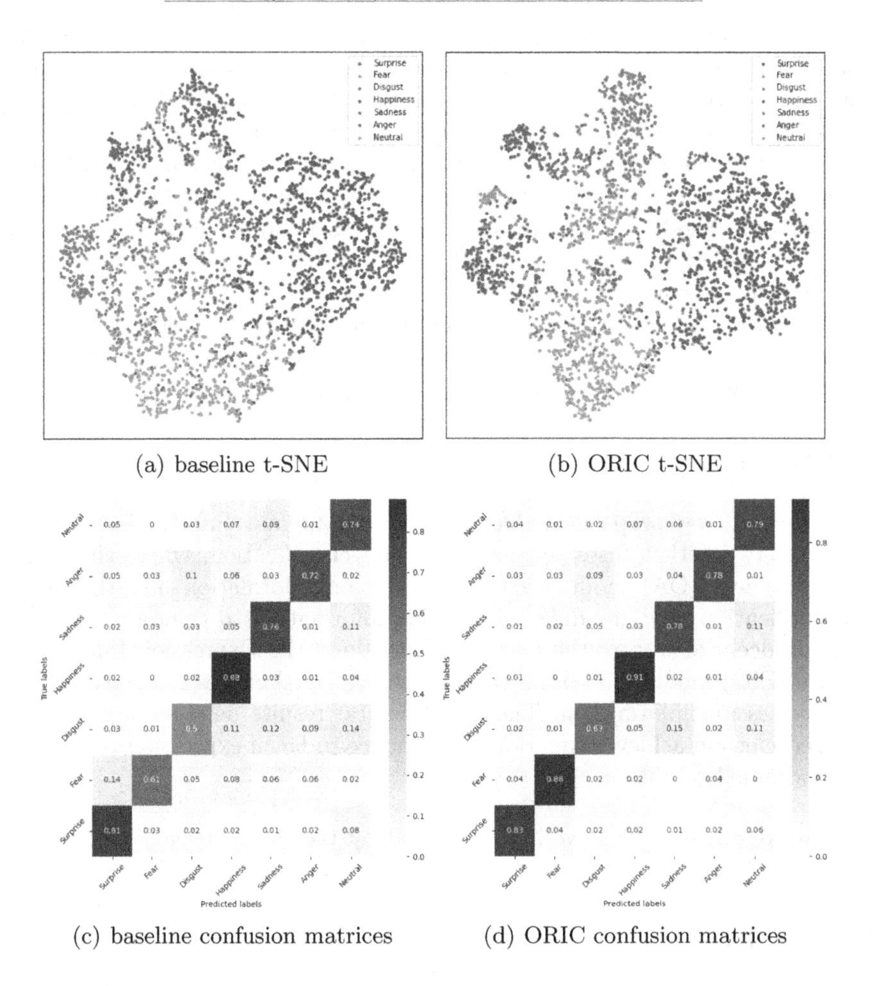

(a) baseline t-SNE (b) ORIC t-SNE

(c) baseline confusion matrices (d) ORIC confusion matrices

Fig. 3. Results of t-SNE and confusion matrix with baseline and ORIC.

performance with the successive addition of OP blocks. This provides evidence supporting the efficacy of the multi-scale occlusion removal technique in achieving a more comprehensive comprehension and localization of occluded areas.

4.5 Visualization

To show the efficacy of our method, we present the t-SNE and confusion matrices plots in Fig. 3, for the outputs of RAF-DB after processing with the baseline and ORIC. Analyzing the baseline's confusion matrix reveals a relatively low recognition rate for fear, and it is inclined to be mistaken for a surprise. This could be attributed to the facial expression similarities between fear and surprise, such as widened eyes and raised eyebrows. These shared facial features contribute to the visual resemblance between the two expressions, increasing the likelihood of confusion. However, it is noteworthy that the ORIC's confusion matrix shows a significant improvement in fear recognition accuracy. This indicates that our method effectively captures crucial fear-related features. Additionally, our method demonstrates varying degrees of improvement in recognition accuracy for other categories compared to the baseline, highlighting its superiority in reducing occlusion interference and extracting comprehensive features. In the t-SNE plot, the ORIC shows clearer classification boundaries compared to the baseline, with similar categories being more clustered, demonstrating that the distribution distortion is alleviated with our method.

5 Conclusion

In this paper, we propose an Occlusion Removal and Information Capture (ORIC) network that fuses segmentation and classification, where the Occlusion Awareness (OA) module extracts occlusion information and the Occlusion Purification (OP) module utilizes this information to generate masks for removing occlusions from the feature. This allows the Expression Information Capture (EIC) module to obtain comprehensive facial expression features from clean expression information. The experimental results demonstrate that the proposed method achieves superior performance in facial expression recognition under occluded conditions.

Acknowledgements. This work is supported by the Inner Mongolia Science and Technology Project No. 2021GG0166 and NSFC project No. 61763035.

References

1. Ekman, P.: Universals and cultural differences in facial expressions of emotion. In: Nebraska Symposium on Motivation. University of Nebraska Press (1971)
2. Abdat, F., Maaoui, C., Pruski, A.: Human-computer interaction using emotion recognition from facial expression. In: 2011 UKSim 5th European Symposium on Computer Modeling and Simulation, pp. 196–201. IEEE (2011)
3. Fei, Z., et al.: Deep convolution network based emotion analysis towards mental health care. Neurocomputing **388**, 212–227 (2020)
4. Hachisuka, S., Ishida, K., Enya, T., Kamijo, M.: Facial expression measurement for detecting driver drowsiness. In: Harris, D. (ed.) EPCE 2011. LNCS (LNAI), vol. 6781, pp. 135–144. Springer, Heidelberg (2011). https://doi.org/10.1007/978-3-642-21741-8_16

5. Lo, H.-C., Chung, R.: Facial expression recognition approach for performance animation. In: Proceedings Second International Workshop on Digital and Computational Vide, pp. 132–139. IEEE (2001)
6. Gan, Y., Chen, J., Yang, Z., Xu, L.: Multiple attention network for facial expression recognition. IEEE Access **8**, 7383–7393 (2020)
7. Li, H., Wang, N., Ding, X., Yang, X., Gao, X.: Adaptively learning facial expression representation via CF labels and distillation. IEEE Trans. Image Process. **30**, 2016–2028 (2021)
8. Savchenko, A.V., Savchenko, L.V., Makarov, I.: Classifying emotions and engagement in online learning based on a single facial expression recognition neural network. IEEE Trans. Affect. Comput. **13**(4), 2132–2143 (2022)
9. Wang, K., Peng, X., Yang, J., Meng, D., Qiao, Y.: Region attention networks for pose and occlusion robust facial expression recognition. IEEE Trans. Image Process. **29**, 4057–4069 (2020)
10. Li, Y., Zeng, J., Shan, S., Chen, X.: Patch-gated CNN for occlusion-aware facial expression recognition. In: 2018 24th International Conference on Pattern Recognition (ICPR), pp. 2209–2214. IEEE (2018)
11. Xing, Z., Tan, W., He, R., Lin, Y., Yan, B.: Co-completion for occluded facial expression recognition. In: Proceedings of the 30th ACM International Conference on Multimedia, pp. 130–140 (2022)
12. Yang, B., Jianming, W., Hattori, G.: Face mask aware robust facial expression recognition during the COVID-19 pandemic. In: 2021 IEEE International conference on image processing (ICIP), pp. 240–244. IEEE (2021)
13. Li, Y., Zeng, J., Shan, S., Chen, X.: Occlusion aware facial expression recognition using CNN with attention mechanism. IEEE Trans. Image Process. **28**(5), 2439–2450 (2018)
14. Gera, D., Balasubramanian, S.: Landmark guidance independent spatio-channel attention and complementary context information based facial expression recognition. Pattern Recogn. Lett. **145**, 58–66 (2021)
15. Lu, Y., Wang, S., Zhao, W., Zhao, Y.: WGAN-based robust occluded facial expression recognition. IEEE Access **7**, 93 594–93 610 (2019)
16. Pan, B., Wang, S., Xia, B.: Occluded facial expression recognition enhanced through privileged information. In: Proceedings of the 27th ACM International Conference on Multimedia, pp. 566–573 (2019)
17. Dong, J., Zhang, L., Zhang, H., Liu, W.: Occlusion-aware GAN for face de-occlusion in the wild. In: 2020 IEEE International Conference on Multimedia and Expo (ICME), pp. 1–6. IEEE (2020)
18. Yuan, G., Zheng, H., Dong, J.: MSML: enhancing occlusion-robustness by multi-scale segmentation-based mask learning for face recognition. In: Proceedings of the AAAI Conference on Artificial Intelligence, vol. 36, no. 3, pp. 3197–3205 (2022)
19. Hinton, G.E., Srivastava, N., Krizhevsky, A., Sutskever, I., Salakhutdinov, R.R.: Improving neural networks by preventing co-adaptation of feature detectors, arXiv preprint arXiv:1207.0580 (2012)
20. Mollahosseini, A., Hasani, B., Mahoor, M.H.: AffectNet: a database for facial expression, valence, and arousal computing in the wild. IEEE Trans. Affect. Comput. **10**(1), 18–31 (2017)
21. Li, S., Deng, W., Du, J.: Reliable crowdsourcing and deep locality-preserving learning for expression recognition in the wild. In: Proceedings of the IEEE Conference on Computer Vision and Pattern Recognition, pp. 2852–2861 (2017)

22. Benitez-Quiroz, C.F., Srinivasan, R., Feng, Q., Wang, Y., Martinez, A.M.: EmotioNet challenge: recognition of facial expressions of emotion in the wild, arXiv preprint arXiv:1703.01210 (2017)
23. Xia, B., Wang, S.: Occluded facial expression recognition with step-wise assistance from unpaired non-occluded images. In: Proceedings of the 28th ACM International Conference on Multimedia, pp. 2927–2935 (2020)
24. Guo, Y., Zhang, L., Hu, Y., He, X., Gao, J.: MS-Celeb-1M: a dataset and benchmark for large-scale face recognition. In: Leibe, B., Matas, J., Sebe, N., Welling, M. (eds.) ECCV 2016, Part III. LNCS, vol. 9907, pp. 87–102. Springer, Cham (2016). https://doi.org/10.1007/978-3-319-46487-9_6
25. Zhang, K., Zhang, Z., Li, Z., Qiao, Y.: Joint face detection and alignment using multitask cascaded convolutional networks. IEEE Signal Process. Lett. **23**(10), 1499–1503 (2016)

Semantic Line Detection Using Deep-Hough with Attention Mechanism and Strip Convolution

Hang Zhang[1], Heping Li[2], and Liang Wang[1,3](✉)

[1] Faculty of Information Technology, Beijing University of Technology, Beijing, China
wangliang@bjut.edu.cn
[2] Chinese Institute of Coal Science, Beijing, China
[3] Engineering Research Center of Digital Community, Ministry of Education, Beijing, China

Abstract. Semantic lines are some particular dominant lines in an image, which divide the image into several semantic regions and outline its conceptual structure. They play a vital role in image analysis, scene understanding, and other downstream tasks due to their semantic representation ability for the image layout. However, the accuracy and efficiency of existing semantic line detection methods still couldn't meet the need of real applications. So, a new semantic line detection method based on the deep Hough transform network with attention mechanism and strip convolution is proposed. Firstly, the detection performance is improved by combining the channel attention mechanism with the feature pyramid network to alleviate the influence of redundant information. Then, the strip convolution and mixed pooling layer are introduced to effectively collect the remote information and capture the long-range dependencies between pixel backgrounds. Finally, the strategy of GhostNet is adopted to reduce the computational cost. Results of experiments on open datasets validate the proposed method, which is comparable to and even outperforms the state-of-the-art methods in accuracy and efficiency. Our code and pretrained models are available at: https://github.com/zhizhz/sml.

Keywords: Deep learning · semantic line detection · attention mechanism · strip convolution · deep Hough network

1 Introduction

Semantic lines [1] are a particular type of straight lines that can outline the conceptual structure of an image by dividing it into several semantic regions.

Supported by NSFC under Grant No. 61772050 and the Key Project of Science and Technology Innovation and Entrepreneurship of TDTEC (No. 2022-TD-ZD004).

B. Luo et al. (Eds.): ICONIP 2023, CCIS 1965, pp. 139–152, 2024.
https://doi.org/10.1007/978-981-99-8145-8_12

Fig. 1. Results of line detection for an image randomly selected from NKL [7] datasets. (a) Semantic lines detected by the proposed method. (b) Semantic lines detected by the DHT [7]. (c) Lines detected by the classical Hough transform.

As shown in Fig. 1(a), semantic lines detected by the proposed method divide the image into grassland, water, and sky semantic regions. Semantic lines can provide high-level priors of images by characterizing the image layout and are essential components in high-level image understanding. So, semantic line detection is vital for computer vision applications, such as lane detection [2], horizon detection [1], sea level detection [3], and artistic creation [4], and further provides cues to artificial intelligence generated content (AIGC).

Most existing methods perform line detection by exploiting handcrafted [5] or deep [6] features. However, only redundant short line segments or apparent line structures rather than semantic lines are detected. Although recently emerging SLNet [1] and Deep Hough Transform (DHT) [7] can successively detect semantic lines by exploiting the deep convolutional neural network (CNN), the accuracy and efficiency still need further improvement. For example, as shown in Fig. 1(b), the line detected by [7] on the top has redundant false semantic lines.

So, a new semantic line detection method based on DHT is proposed. Firstly, the attention mechanism is introduced into the feature pyramid network (FPN) using the feature selection module (FSM) [8] to remove the interference of redundant feature information. Then, the ResNet of the feature extraction part is improved by introducing the strip convolution layer and mixed strip pooling layer. Finally, the strategy of GhostNet [9] is applied to replace the convolution module in the original convolution network to reduce the computational cost. Extensive experiments on open datasets validate that the proposed method improves the accuracy and efficiency of semantic line detection, whose performance is comparable or even superior to that of the state-of-the-art methods.

The main contributions can be summarized as follows.

- A new semantic line detection method based on the deep Hough network with the attention mechanism and strip convolution is proposed.
- The attention mechanism is incorporated into the FPN to emphasize the essential features and suppress redundant information.
- The strip convolution and mixed pooling strategy are proposed to capture remote dependency of semantic contexts and aggregate global and local information.

– The strategy of GhostNet is adopted to reduce the computational cost.

The rest of this paper is organized as follows. Section 2 introduces the related work. Section 3 elaborates the proposed method. Experiments are reported in Sect. 4. Section 5 concludes the paper.

2 Related Work

Automatic detection of straight lines in images [1–3,10,11] is essential in computer vision. Currently, line detection can be fulfilled by the Hough transform and its variants and the deep CNN. Different from ordinary straight lines, semantic lines are a particular type of straight lines that can outline the conceptual structure of an image. So, this section introduces standard straight line detection methods based on Hough transform and CNN first. Then, the definition of semantic line is presented. Finally, semantic line detection methods are reviewed.

2.1 Hough Transform

Hough transform [12] is a reliable and robust scheme for line detection. Due to its simplicity and effectiveness, it is widely used. However, it exploits slope-offset parameters to represent a line, which results in an infinite parameter space. So, Ballard [13] proposed the angle-radius representation to generalize the Hough transform (GHT). Then, Fernandes et al. [14] improved the voting scheme to enhance its efficiency and return final line detection results. Kiryati et al. [15] proposed the probabilistic Hough transform to improve the computational efficiency further. However, the line detector based on the classical Hough transform usually ignores the semantic information of lines. Moreover, it is sensitive to brightness changes and occlusion, as shown in Fig. 1(c).

2.2 CNN-Based Line Detection

With the progress of deep learning, some CNN-based line detection methods have emerged. Law et al. [16] first proposed that the object detector CornerNet can be used for line detection. Dai et al. [17] proposed masked convolution to extract CNN features to detect line segments. Zhuo et al. [18] proposed a single end-to-end CNN to directly output vectorized box lines. However, the receptive field of the convolutional layer of the traditional CNN is square, and there is redundant information to obtain the context information of the line. Although significant progress has been made in CNN-based line detection, the efficiency and accuracy still need improvements to meet the need of real applications.

2.3 Semantic Lines

Semantic lines are particular straight lines that can outline the conceptual structure of an image. Similar to semantic segmentation, semantic lines [1] can also divide an image into several semantic regions and outline its conceptual structure. Unlike semantic segmentation, semantic lines are a particular type of straight lines. On the contrary, contours of semantic segmentation regions generally are curves in any shape. In addition, semantic lines can provide cues for AIGC to produce visually balanced images, while contours of semantic segmentation regions are only related to the specific areas.

2.4 Semantic Lines Detection

The concept of semantic line first proposed by Lee et al. [1] refers to salient lines that outline the layout of images. Semantic line detection is considered an uncommon instance of object detection in [1], where a network similar to Faster-RCNN [19] is proposed to detect and localize semantic lines. An open dataset for semantic line detection, SEL dataset, is also provided in [1]. Zhao et al. [7] incorporated CNN with Hough transform to aggregate more content information to detect semantic lines and built a larger open dataset, NKL dataset. Jin et al. [20] proposed a network consisting of the selection network (S-Net) and the coordination network (H-Net) to compute semantic lines' probability and offset and finally exploited the maximal weight clique selection to determine final semantic lines. Similar to the case of CNN-based line detection, the efficiency and accuracy of semantic line detection still don't meet the needs of real applications.

Most related to this work is DHT [7], which incorporates CNN and Hough transform to detect semantic lines. Different from DHT, the proposed method exploits the attention mechanism and strip convolution to improve the robustness and efficiency of semantic line detection. The attention mechanism is incorporated into the FPN to emphasize essential features and suppress redundancy. The strip convolution and the pooling strategy combining strip and spatial pooling are proposed to improve the DHT via capturing the remote dependency of semantic contexts and aggregating the global and local information. In addition, the GhostNet is adopted in the FPN of DHT to reduce computational costs.

3 Proposed Method

This section firstly presents the network structure of proposed method. Then, the main contributions of the proposed method are elaborated: 1. the attention mechanism is introduced into the feature extraction part via incorporating the feature selection module (FSM); 2. the strip convolution is introduced to substitute for the traditional convolution for line extraction; 3. a new pooling strategy combining the strip pooling and average pooling is proposed; 4. the GhostNet is applied to the network backbone to meet the demand of lightweight.

3.1 Network Structure

The network structure of proposed method is shown in Fig. 2. It consists of three parts: feature extraction, deep Hough, and regression prediction. Finally, it outputs detected semantic lines. Details of each part are as follows.

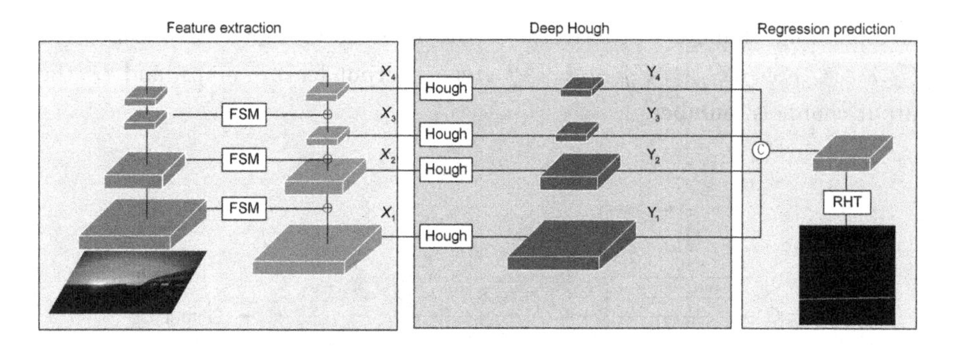

Fig. 2. The network structure of proposed method. It mainly consists of feature extraction, deep Hough, and regression prediction.

Feature extraction part has a reverse U-Net structure like that of original DHT, which consists of bottom-up and top-down sub-networks. Unlike DHT, the strip convolution and new pooling strategy combining the strip pooling and average pooling are incorporated into the bottom-up sub-network. The deep features of the input image are first extracted by the bottom-up sub-network, then transferred into the top-down sub-network. Besides general connections, there are skip connections in the reverse U-Net [21], where FSM explicitly models the weight value of each connection. The output of feature extraction part is intermediate X_i, which contains both semantic and detailed information.

Deep Hough part is the same as that of DHT. The outputs of the top-down sub-network of feature extraction part, X_1, X_2, X_3, and X_4, are fed into the Hough transform, whose size is 1/4, 1/8, 1/16, and 1/16 of that of the input image, respectively. Then, feature maps X_i in different scales extracted in the spatial domain are independently transformed into Y_i in the parameter domain via Hough transform. Finally, the bilinear interpolation is applied to adjust the size of each output to the size of Y_1 for aggregation operation.

Prediction part aggregates deep features of the parameter domain, Y_i, along the channel dimension, then performs 1×1 convolution to output the prediction result. To visualize the prediction results, the reverse Hough transform is performed finally. The cross-entropy loss function is taken to train the proposed network via back-propagation using the Adan [22] optimizer.

3.2 Feature Selection Module

The semantic line detection method should emphasize critical feature maps that contain much more spatial information and suppress redundant feature maps to balance the accuracy and computational cost. Feature selection module shown in Fig. 3 can fulfill this task by assigning weights to input feature maps along the channel dimension and explicitly modeling the importance of feature maps.

The input feature map $\mathcal{C}_i$ has the form of $\mathcal{C}_i = [C_1, \cdots, C_d, \cdots, C_D]$, where $C_d \in \mathbb{R}^{H_i \times W_i}$, H_i and W_i refers to the height and width of the corresponding channel respectively, D the number of input channels. $\hat{C}_i = \left[\hat{C}_1, \cdots, \hat{C}_{d'}, \cdots, \hat{C}_{D'}\right]$, $\hat{C}_{d'} \in \mathbb{R}^{H_i \times W_i}$ refers output feature maps, and D' the output channels' number.

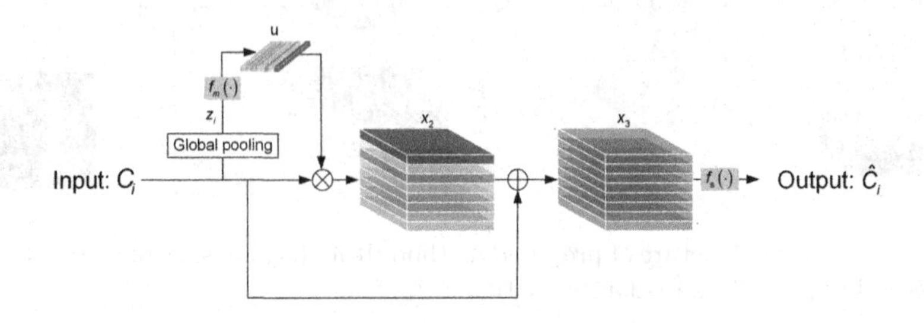

Fig. 3. Feature selection module, i.e., FSM.

The global information $\mathbf{z}_i = [z_1, \cdots, , z_d, \cdots, z_D]$ of feature map C_i is extracted by the global pooling operation.

$$\mathbf{z}_i = \frac{1}{H_i \times W_i} \sum_{h=1}^{H_i} \sum_{w=1}^{W_i} C_d(h, w) \tag{1}$$

where $C_d(h, w)$ represents pixel value in the coordinate (h, w) of the dth channel.

Then, z_i is fed into the feature importance modeling layer $f_m(\cdot)$,

$$f_m(\cdot) = \begin{cases} 1 \times 1 \text{ conv} \\ \text{sigmoid}(\cdot) \end{cases} \tag{2}$$

$f_m(\cdot)$ can model the channel weights of feature maps, whose output is a weight vector $\mathbf{u} = [u_1, \cdots, u_d, \cdots, u_D]$,

$$\mathbf{u} = f_m(\mathbf{z}), \tag{3}$$

where u_d is the importance weight of the dth channel input feature map. Then, the weight vector $\mathbf{u}$ is used to scale the input feature map. After that, the scaled feature map is added to the input feature map pixel-by-pixel. The rescaled feature map $x_3 = C_i + u_d \cdot C_i$ is obtained.

The feature selection layer, i.e., $f_s(\cdot) = 1 \times 1$ conv layer, is introduced to the rescaled feature map. The effects of $f_s(\cdot)$ layer are two folds: 1. selectively maintain essential feature maps. 2. discard unnecessary feature maps to reduce information channels.

$$\hat{C}_i = f_s(C_i + u_d * C_i) \tag{4}$$

FSM is incorporated into the FPN of feature extraction part, which replaces the original 1×1 conv layer between the bottom-up and top-down in the original DHT as shown in Fig. 2.

3.3 Strip Convolution

To extract deep features of semantic lines, it is necessary to capture the remote dependency relationship between semantic backgrounds. The strip convolution of $1 \times n$ and $n \times 1$ is proposed to replace the square convolution of the original DHT. The strip convolution is conducive to aggregate context and local context information of straight lines.

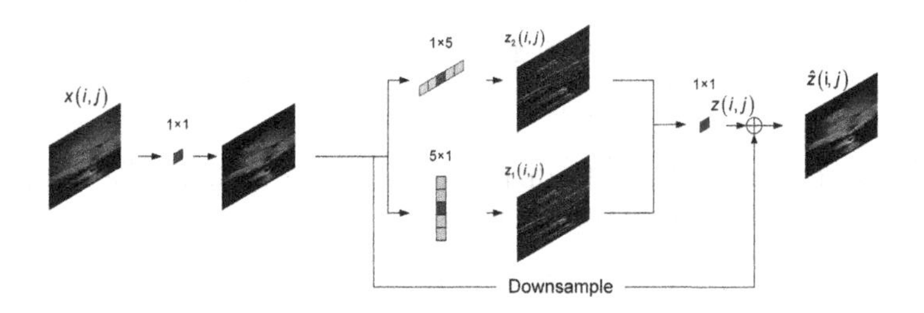

Fig. 4. The improved strip convolutional network is divided into three layers: 1×1, 5×1, and 1×5.

As shown in Fig. 4, the 3×3 convolution layers in the bottleneck module of ResNet of original DHT are substituted by the parallel strip convolution layer of 5×1 and 1×5. The 1×1 convolution layer before and after the parallel strip convolution reduces and increases the dimensions, respectively. The output combining vertical and horizontal strip convolution, $z(i,j)$, satisfies

$$z(i,j) = f_s(z_1(i,j) + z_2(i,j)) \tag{5}$$

$$z_1(i,j) = \sum_{j-2 \leq j \leq j+2} \hat{w} \cdot f_s(x(i,j)) \tag{6}$$

$$z_2(i,j) = \sum_{i-2 \leq i \leq i+2} \hat{w} \cdot f_s(x(i,j)) \tag{7}$$

where $z_1(i,j)$ and $z_2(i,j)$ is the output of vertical strip convolution and the output of horizontal strip convolution respectively, $\hat{w}$ is the corresponding weight of the stripe convolution, $x(i,j)$ is the input of the position (i,j), $f_s(\cdot)$ is the 1×1 convolution layer.

The final output of the strip convolution, $\hat{z}(i,j)$, is the pixel-by-pixel sum of the residual downsampling of input and $z(i,j)$,

$$\hat{z}(i,j) = z(i,j) \oplus f_d(x(i,j)) \tag{8}$$

where $f_d(\cdot)$ is the downsampling function.

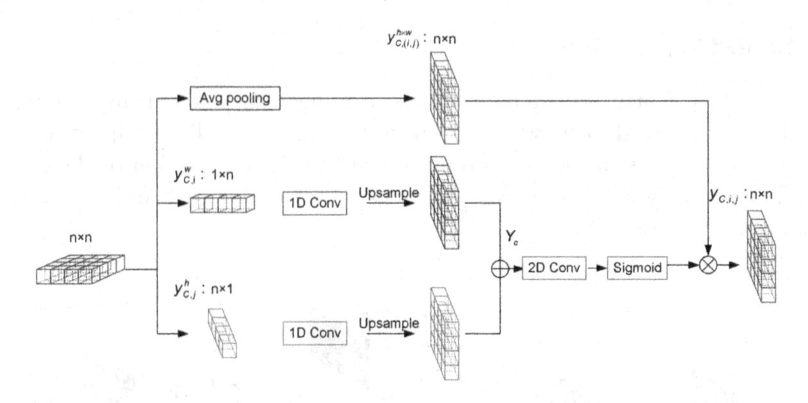

Fig. 5. The structure of the mixed strip pooling layers. It combines the average pooling and the horizontal and vertical strip pooling.

3.4 Mixed Strip Pooling Layer

Considering that the strip convolution can combine remote information, a mixed pooling layer combining the strip pooling and average pooling is proposed to substitute the global pooling of the original DHT, which can better gather global and local context information.

The mixed pooling layer shown in Fig. 5 consists of two modules that capture short- and long-range dependencies between different locations. Its output is

$$y_{C,i,j} = w_1 y_{C,i}^{h} + w_2 y_{C,j}^{w} + w_3 y_{C,(i,j)}^{h \times w}. \tag{9}$$

$$y_{C,i}^{h} = \frac{1}{W} \sum_{0 \leq i < W} x_{i,j} \tag{10}$$

$$y_{C,j}^{w} = \frac{1}{H} \sum_{0 \leq j < H} x_{i,j}, \tag{11}$$

where $y_{C,j}^{w}$ and $y_{C,i}^{h}$ is the output of horizontal and vertical strip pooling, respectively, $y_{C,(i,j)}^{h \times w}$ is the output of average pooling corresponding to the short-range information collection, $H \times W$ is the height and width of the input, h, w is the range of average pooling. Then, $y_{C,j}^{w}$ and $y_{C,i}^{h}$ are up-sampled to $n \times n$ by one-dimensional convolution. After that, $y_{C,j}^{w}$ and $y_{C,i}^{h}$ are aggregated pixel-by-pixel according to (9). The resulting $y_{C,i,j}$ reduces the channel dimension via the 1×1 convolution layer. The outputs of average and strip pooling layers are summed and fused with the weight w_i pixel-by-pixel along each channel.

3.5 Lightweight Network Using the Strategy of GhostNet

To reduce the amount of floating point computation and speed up the network inference, some convolutional layers in feature extraction part are modified using the strategy of GhostNet [9]. The core strategy of GhostNet is to upgrade features extracted by the classical convolution into two parts: intrinsic features generated by classical convolution and ghost features generated by cheap operations. Intrinsic features generated by classical convolution are fused with ghost features generated by cheap operations along the channel dimension. A new feature map with equivalent dimension of the original classical convolution's output is obtained. In fact, the number of parameters in the backbone of original DHT is 2.753×10^7, while that of proposed network is 2.636×10^7. It is reduced by 4.4%, while some extra manipulations, such as attention mechanism and strip convolution, are incorporated.

4 Experiments

Experiments are conducted with an RTX2080ti GPU and Pytorch framework on two open datasets: SEL [1] and NKL [7] dataset. SEL [1] contains 1,715 images and 2,791 semantic lines, where the training set includes 1,541 images and 2,493 lines, and the remaining forms the evaluation set. NKL [7] consists of 6,500 images and 13,148 semantic lines, where the training set contains 5,200 images and 10,498 lines, and the remaining forms its evaluation set. The qualitative and quantitative results of the proposed method are reported in comparison with existing methods on NKL. In addition, results of generalization ability on SEL and ablation study are also presented.

4.1 Comparison of Experimental Results

Experimental details are as follows. The input images are unified and resized to 400×400. The batch size is 8. Adan optimizer is adopted. The loss function is cross entropy. The learning rate is 0.0002, the momentum is 0.9, and the attenuation is 0.1. The quantization parameters of θ and R [7] are 100, and the threshold parameter is 0.01. The training epoch is 30. The proposed network is initialized with the model using ResNet pre-trained on ImageNet.

Evaluation Metrics. Two evaluation metrics, the Chamfer distance (CD) [24] and EA-score [7], are used to evaluate the similarity between the estimated line and the ground truth. The CD satisfies $d_{CD}(l_i, l_j) = \frac{1}{n}\sum_{p_i \in l_i} \min_{p_j \in l_j} \|p_i - p_j\|$, while p is the point on line l and n is the number of points. The EA-score [7], $S = (S_d \cdot S_\theta)^2$, considers both the Euclidean distance $S_d = 1 - D(l_i, l_j)$ and the angular distance $S_\theta = 1 - \frac{\theta(l_i,l_j)}{\pi/2}$ between the estimated and ground truth semantic lines pair (l_i, l_j), where $D(l_i, l_j)$ is the Euclidean distance between midpoints of two lines l_i and l_j and $\theta(l_i, l_j)$ is the angle between l_i and l_j. Both of them indicate the matching degree between the prediction line and the truth line. In addition, a series of precision, recall, and F-measure scores are obtained based on CD and EA-score to evaluate the proposed method.

Avg. P, Avg. R, and Avg. F are used to evaluate the overall performance after obtaining the accuracy of semantic line detection on 1300 images in the validation

Table 1. Quantitative results on the NKL dataset

Method	CD			EA		
	Avg.P	Avg.R	Avg.F	Avg.P	Avg.R	Avg.F
DHT(ResNet50) [7]	<u>0.766</u>	0.864	<u>0.812</u>	<u>0.679</u>	<u>0.766</u>	<u>0.719</u>
DHT(VGG16) [7]	0.750	0.864	0.803	0.659	0.759	0.706
HED [23]+HT [12]	0.301	**0.878**	0.448	0.213	0.612	0.318
Ours	**0.773**	<u>0.872</u>	**0.819**	**0.697**	**0.783**	**0.733**

Table 2. Quantitative comparison of the efficiency of DHT and Ours.

Method	Parameters(M)	FLOPs(G)	FPS	Forward time(s)
DHT(ResNet50)	27.53	140.065	14.308	61.17
Ours	**26.36**	**119.71**	**17.057**	**51.82**

stage and then taking the average value. Avg. P represents the average accuracy, Avg. R represents the average recall, and Avg. F is the overall performance indicator combining precision and recall. Precision represents the proportion of correctly predicted semantic lines to the number of predicted semantic lines. The recall rate represents the proportion of correctly predicted semantic lines to the number of ground-truth lines.

FPS (Frames Per Second) indicates the speed at which the model processes images for semantic lines detection.

Parameters indicate the size of the learnable parameters of the model.

FLOPs refer to the number of floating point operations used to measure the complexity of the model.

Forward time refers to the time for forward propagation of all valid images. The total time-consuming is calculated from inputting all images into the model to output results. It evaluates the model operation efficiency.

Comparison of Quantitative Results. Experimental results of the proposed method and the state-of-the-art methods, i.e., evaluation metrics EA [7] and CD [24], are shown in Table 1, where the best are in bold and the second best are marked by underlines. It can be seen that our model outperforms other methods except that the average recall of CD is slightly lower than that of the HED [23]+HT [12]. The baseline network, DHT(ResNet50), achieves the second best. Then, the quantitative efficiency comparison is performed between the baseline method and the proposed method, as shown in Table 2. It can be seen that the processing time of the proposed network is reduced by 15.2% compared with the baseline network, DHT(ResNet50), and the speed of image processing is increased by 19.2%.

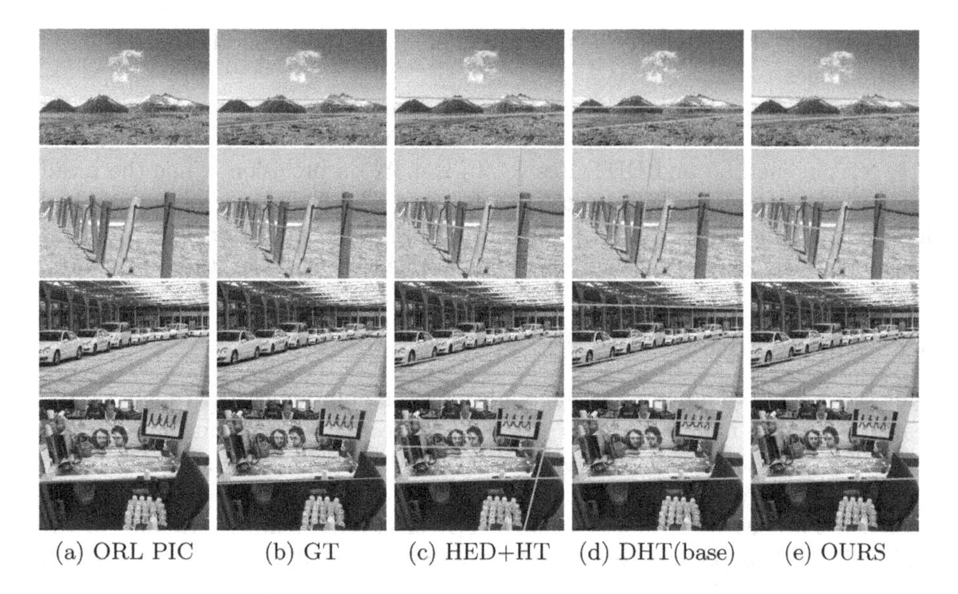

(a) ORL PIC (b) GT (c) HED+HT (d) DHT(base) (e) OURS

Fig. 6. Compared visualization results on the NKL dataset.

Table 3. Results of generalization ability on the SEL dataset.

Method	EA		
	Avg.P	Avg.R	Avg.F
HED [23]+HT [12]	0.356	0.42	0.385
SLNet-iter1 [1]	0.654	**0.803**	0.721
SLNet-iter10 [1]	0.762	0.729	0.745
DHT(ResNet50) [7]	**0.819**	0.755	**0.786**
DHT(VGG16) [7]	0.756	0.774	0.765
Ours	<u>0.776</u>	<u>0.779</u>	<u>0.777</u>

Comparison of Qualitative Results. Figure 6 shows the visual results of the proposed method for semantic line detection on several randomly selected images.

In comparison with the ground truth, it can be seen that the proposed method can accurately predict all semantic lines except the 2nd image, where one semantic line is omitted, while other methods have more errors. It is consistent with the results shown in Table 1 that the recall rate of the DHT(ResNet50) [7] and HED [23]+HT [12] is lower that there are more irrelevant lines in their results. On the whole, our method outperforms the other methods.

Generalization Ability. The generalization ability of our method is qualitatively compared with the state-of-the-art methods on the SEL dataset. Results are shown in Table 3, where the best are in bold and the second best are marked by underlines.

It can be seen in Table 3 that all metrics of the proposed method rank the 2nd in compare with the state-of-the-art methods on the SEL dataset. The difference between our method and DHT(ResNet50) is 4.3% on precision, while the recall rate is better than DHT(ResNet50) with 2.4%. The F-measure of our method has a slight difference of 0.9% from that of DHT(ResNet50). So, the proposed method has the generalization ability comparable to that of the state-of-the-arts.

4.2 Ablation Study

This section tests four improved modules via the ablation study, which are the feature selection module (FSM) presented in Sect. 3.2, strip convolution (SC) shown in Sect. 3.3, mixed strip pooling layer (MSP) presented in Sect. 3.4, and lightweight network module with the strategy of GhostNet (GHB) shown in Sect. 3.5. The baseline model is the DHT(ResNet50). Then, the validity of the FSM, SC, MSP, and GHB are verified. Experimental results are shown in Table 4.

Table 4. Results of ablation study.

FSM	SC	MSP	GHB	F-measure	FPS
				0.684	14.308
✓				0.703	15.981
✓	✓			0.724	15.954
✓		✓		0.729	14.596
✓			✓	0.657	18.373
✓	✓	✓	✓	0.733	17.057

Table 4 shows the quantitative results of the ablation study. The first row shows the performance of the baseline, i.e., DHT(ResNet50). It can be seen in Table 4 that different components of the proposed method can improve the performance of semantic line detection. Compared with the baseline, they all improved the F-measure values by 2–4%. The GHB can reduce 1.91% of the computation number. Therefore, each improved component of the proposed method has a positive effect on the performance of semantic line detection.

5 Conclusion

This paper proposes a semantic line detection method using a deep Hough network with the strip convolution and attention mechanism. By combining the channel attention into the FPN, the influence of redundant information is alleviated. The strip convolution is incorporated into the deep Hough network to collect the remote information and reduce the computational cost. The mixed pooling strategy is proposed to gather global and local content information better. Besides, the GhostNet module is adopted to replace the convolutional module in the FPN to reduce the computational cost further. Extensive experimental results on open datasets validate the proposed method. Its performance is comparable with and even superior to some of the performance of the state-of-the-art methods in accuracy and efficiency on the NKL and SEL datasets.

References

1. Lee, J.T., Kim, H.U., Lee, C., Kim, C.S.: Semantic line detection and its applications. In: 2017 IEEE International Conference on Computer Vision (ICCV), pp. 3229–3237 (2017)
2. Fan, R., Wang, X., et al.: SpinNet: spinning convolutional network for lane boundary detection. Comput. Vis. Media **5**(4), 417–428 (2019)
3. Zardoua, Y., Astito, A., Boulaala, M.: A survey on horizon detection algorithms for maritime video surveillance: advances and future techniques. Visual Comput. **39**(1), 197–217 (2021)
4. Bhattacharya, S., Sukthankar, R., Shah, M.: A framework for photo-quality assessment and enhancement based on visual aesthetics. In: Proceedings of the 18th ACM International Conference on Multimedia (2010)
5. Akinlar, C., Topal, C.: EDLines: a real-time line segment detector with a false detection control. Pattern Recogn. Lett. **32**(13), 1633–1642 (2011)
6. Lin, Y., Pintea, S.L., van Gemert, J.C.: Deep Hough-transform line priors. In: Vedaldi, A., Bischof, H., Brox, T., Frahm, J.-M. (eds.) ECCV 2020. LNCS, vol. 12367, pp. 323–340. Springer, Cham (2020). https://doi.org/10.1007/978-3-030-58542-6_20
7. Zhao, K., Han, Q., Zhang, Xu, J., Cheng, M.M.: Deep Hough transform for semantic line detection, IEEE Trans. Pattern Anal. Mach. Intell. **44**(9): 4793–4806 (2022)
8. Huang, S., Lu, Z., Cheng, R., He, C.: FAPN: feature-aligned pyramid network for dense image prediction. In: 2021 IEEE International Conference on Computer Vision (ICCV), pp. 864–873 (2021)
9. Han, K., Wang, Y., Xu, C.J.: GhostNets on heterogeneous devices via cheap operations. Int. J. Comput. Vis. **130**(4), 1050–1069 (2022)
10. Duan, F., Wang, L.: Calibrating central catadioptric cameras based on spatial line projection constraint. In: IEEE International Conference on Systems, Man and Cybernetics, pp. 2088–2093 (2010)
11. Wang, L., Zhao, D.: Robust orientation estimation from only a plane and a line in Manhattan world. In: 34th Chinese Control and Decision Conference, pp. 5060–5065 (2022)
12. Duda, R.O., Hart, P.E.: Use of the Hough transformation to detect lines and curves in pictures. Commun. ACM **15**(1), 11–15 (1972)
13. Ballard, D.H.: Generalizing the Hough transform to detect arbitrary shapes. Pattern Recogn. **13**(2), 111–122 (1981)
14. Fernandes, L.A.F., Oliveira, M.M.: Real-time line detection through an improved Hough transform voting scheme. Pattern Recogn. **41**(1), 299–314 (2008)
15. Kiryati, N., Eldar, Y., Bruckstein, A.M.: A probabilistic Hough transform. Pattern Recogn. **24**(4), 303–316 (1991)

16. Law, H., Deng, J.: CornerNet: detecting objects as paired keypoints. In: Ferrari, V., Hebert, M., Sminchisescu, C., Weiss, Y. (eds.) Computer Vision – ECCV 2018. LNCS, vol. 11218, pp. 765–781. Springer, Cham (2018). https://doi.org/10.1007/978-3-030-01264-9_45
17. Dai, J., He, K., Sun, J.: Convolutional feature masking for joint object and stuff segmentation. In: 2015 IEEE Conference on Computer Vision and Pattern Recognition (CVPR), pp. 3992–4000 (2015)
18. Zhou, Y., Qi, H., Ma, Y.: End-to-end wireframe parsing. In: 2019 IEEE International Conference on Computer Vision (ICCV), pp. 962–971 (2019)
19. Ren, S., Girshick, R., Sun, J.: Faster R-CNN: towards real-time object detection with region proposal networks. In: Advances in Neural Information Processing Systems (NIPS) (2015)
20. Jin, D., Park, W., Jeong, S.G., Kim, C.S.: Harmonious semantic line detection via maximal weight clique selection. In: 2021 IEEE/CVF Conference on Computer Vision and Pattern Recognition (CVPR) (2021)
21. Lin, T.Y., Dollar, P., et al.: Feature pyramid networks for object detection. In: 2017 IEEE Conference on Computer Vision and Pattern Recognition (CVPR), pp. 936–944 (2017)
22. Xie, X.Y., Zhou, P., Li, et al.: Adan: adaptive Nesterov momentum algorithm for faster optimizing deep models, arXiv: 2208.06677 (2022)
23. Xie, S., Tu, Z.: Holistically-nested edge detection. In: 2015 IEEE International Conference on Computer Vision (ICCV), pp. 1395–1403 (2015)
24. Borgefors, G.: Distance transformations in digital images. Comput. Vis. Graph. Image Process. **34**(3), 344–371 (1986)

Adaptive Multi-hop Neighbor Selection for Few-Shot Knowledge Graph Completion

Xing Gong[1], Jianyang Qin[1], Ye Ding[2], Yan Jia[1,3], and Qing Liao[1,3](✉)

[1] School of Computer Science and Technology,
Harbin Institute of Technology, Shenzhen, China
{gongxing,22b351005}@stu.hit.edu.cn, {jiayan2020,liaoqing}@hit.edu.cn

[2] School of Cyberspace Security,
Dongguan University of Technology, Dongguan, China
dingye@dgut.edu.cn

[3] Peng Cheng Laboratory, Shenzhen, China

Abstract. Few-shot Knowledge Graph Completion (FKGC) is a special task proposed for the relations with only a few triples. However, existing FKGC models face the following two issues: 1) these models cannot fully exploit the dynamic relation and entity properties of neighbors to generate discriminative representations; 2) these models cannot filter out noise in high-order neighbors to obtain reliable entity representations. In this paper, we propose an adaptive multi-hop neighbor selection model, namely AMBLE, to mitigate these two issues. Specifically, AMBLE first introduces a query-aware graph attention network (QAGAT) to obtain entity representations by dynamically aggregating one-hop neighbors based on relations and entities. Then, AMBLE aggregates high-order neighbors by iterating QAGAT and LSTM, which can efficiently extract useful and filter noisy information. Moreover, a Transformer encoder is used to learn the representations of subject and object entity pairs. Finally, we build an attentional matching network to map the query to few support triples. Experiments show that AMBLE outperforms state-of-the-art baselines on two public datasets.

Keywords: Knowledge graph completion · Few-shot learning · Link prediction · Graph attention network · Multi-hop aggregation

1 Introduction

Knowledge graphs (KGs) as a kind of structured data can assist many artificial intelligence downstream applications, such as question answering systems [23], recommendation systems [18], etc. KGs usually represent every fact with a triple (s, r, o), where s, o are the subject entity and object entity, and r is the relation between s and o. Due to the incompleteness of KGs, knowledge graph completion (KGC) has become one of the most important research tasks in the field of

B. Luo et al. (Eds.): ICONIP 2023, CCIS 1965, pp. 153–164, 2024.
https://doi.org/10.1007/978-981-99-8145-8_13

knowledge graphs. The task is to infer missing facts based on the existing entity and relation by answering queries such as $(s, r, ?)$.

Existing large-scale knowledge graphs [2,17] often suffer from the long-tail distribution problem, i.e., a large number of relations contain only a few triples. However, traditional KGC models require a large number of triples for each relation for training to obtain discriminative representations. As a result, these models have poor completion ability for the relation with only a small number of triples. To alleviate this problem, few-shot knowledge graph completion (FKGC) has been proposed recently. These methods only need a small number of triples as references for queries of each relation to achieve the completion task in few-shot scenarios.

Existing FKGC models [12,13,21,22] obtain entity representations by aggregating neighbor information, but these methods have two major limitations. **1) Dynamic neighbor properties:** Dynamic neighbor properties mean that the influence of neighbors on entity varies with the relation of different completion tasks. Dynamic neighbor properties are determined by both the entity and relation information. However, GMatching [21], FSRL [22] and GANA [12] ignore the dynamic properties of entities, these methods cannot dynamically assign neighbor weights based on the completion task, resulting in inaccurate encoding of entities. Although FAAN [13] considers the problem of dynamic neighbor properties, it only considers the effect of relations and ignores the effect of entities. **2) High-order neighbor noise:** In real-world knowledge graph datasets [2,17], there are a large number of entities that contain only a very small number of one-hop neighbors. Existing FKGC approaches only aggregate one-hop neighbors, resulting in their inability to obtain reliable representations of entities. Although traditional KGC model [11] aggregates high-order neighbors to obtain supplementary neighbor information, it ignores the noise problem in high-order neighbors. Thus, how to efficiently filter out these noisy high-order neighbors remains a challenging problem.

To address the above problems, we propose an **A**daptive **M**ulti-hop neigh**B**or se**LE**ction for few-shot knowledge graph completion (AMBLE). Specifically, we firstly propose a query-aware graph attention network (QAGAT) to obtain entity representations by aggregating one-hop neighbors. QAGAT can fully make use of both entity and relation information to dynamically assign weights to the neighbors. Secondly, we iterate QAGAT and LSTM to aggregate high-order neighbors which can effectively extract useful information from high-order neighbors and filter out the noisy information. Thirdly, we use a Transformer to learn the representations of entity pairs. Finally, an attentional matching network is applied to calculate the score of each query. Main contributions of this paper are summarized as follows:

- We propose an adaptive multi-hop neighbor selection model, namely AMBLE, to solve dynamic neighbor properties and high-order neighbor noise problems in FKGC.

- We devise a novel query-aware graph attention network (QAGAT), which can take advantage of both entity and relation information to adaptively assign neighbor weights based on different tasks.
- We design a new high-order neighbor aggregation and selection structure by iterating QAGAT and LSTM, which can efficiently extract useful and filter noisy information from high-order neighbors.
- We demonstrate the superiority of the AMBLE over state-of-the-art baselines by conducting extensive experiments on two public datasets.

2 Related Work

Due to the long-tail phenomenon in real-world KGs, FKGC has become a popular research area. Existing FKGC models can be grouped into two categories:

Metric-Based Models. GMatching [21] learns entity representations by aggregating neighbors, and then introduces a matching processor to evaluate the similarity between queries and support triples. FSRL [22] introduces a neighbor aggregator based on attention mechanism to aggregate neighbor information. FAAN [13] considers the problem of dynamic properties in neighbors and proposes a relation-aware attentional neighbor aggregator to learn entity representations. Thus, it can dynamically aggregate neighbors with the change of the completion task. YANA [8] aims to mitigate the issue of generating reliable embeddings for solitary entities in FKGC tasks, and introduces four novel abstract relations to represent inner- and cross- pair entity correlations and constructs a local pattern graph from the entities. MFEN [20] aims to capture the heterogeneous influence of neighbor characteristics by devising a single-layer CNN with differently sized filters to capture multi-scale characteristics while controlling model complexity.

Optimization-Based Models. MetaR [3] designs a fast gradient descent update procedure based on the idea of MAML [4] to achieve the completion task by transferring relational meta-information from support triples to queries. Based on MetaR, GANA [12] proposes a gated and attentive neighbor aggregator to filter noise in one-hop neighbors. In addition, benefiting from TransH [19], GANA designs a MTransH to deal with the complex relations.

However, the above models cannot utilize the information of entities and relations in dynamic properties problem on the one hand, and do not consider the noise problem in high-order neighbors on the other hand.

3 Preliminaries

In this section, we give formal definitions of the knowledge graph, the few-shot knowledge graph completion task, and the corresponding few-shot learning setting. Specific notations and their descriptions are listed in Table 1.

Table 1. Notations and descriptions.

Notation	Description
$\mathcal{G}, \mathcal{E}, \mathcal{R}, \mathcal{F}$	Knowledge Graph, entity set, relation set and fact set
(s, r, o)	A triple of subject entity, relation and object entity
$\mathcal{T}_i, \mathcal{S}_i, \mathcal{Q}_i$	Task i and its support set and query set
$\mathcal{Q}_i^+, \mathcal{Q}_i^-$	The positive and negative query set of task i
$\mathcal{G}^{'}, \mathcal{C}$	Background knowledge graph and candidate entity set
$\mathcal{N}_e$	Neighbor set of entity e
$\mathbf{e}^{(t)}, \mathbf{e}_i^{(t)}$	t-layer entity e and its neighbors' representations
$\mathbf{r}, \mathbf{r}_i$	Query relation and neighbor's relation representations
$\mathbf{q}, \mathbf{s}_i$	Query and support entity pairs representations
$\mathbf{W}_1, \mathbf{W}_2, \mathbf{W}_3$	$d \times d$ dimensional parameters of liner
b_1, b_2, b_3	d dimensional bias of liner
$\mathbf{W}_f^{(t)}, \mathbf{W}_i^{(t)}, \mathbf{W}_o^{(t)}, \mathbf{W}_C^{(t)}$	$d \times d$ dimensional parameters of t-layer LSTM
$b_f^{(t)}, b_i^{(t)}, b_o^{(t)}, b_C^{(t)}$	d dimensional bias of t-layer LSTM
σ	Activation function *sigmoid*
λ, γ	Hyperparameters

Knowledge Graph. A knowledge graph $\mathcal{G}$ is represented by a collection of triples: $\mathcal{G} = \{(s, r, o) | s, o \in \mathcal{E}, r \in \mathcal{R}\}$. For each triple (s, r, o), s, o denote the subject entity and object entity, and r is the relation between s and o. $\mathcal{E}$, $\mathcal{R}$ denote the entity set and relation set of $\mathcal{G}$, respectively.

Few-Shot Knowledge Graph Completion. Few-shot knowledge graph completion is a specialized task proposed for the relations with only a few triples, which are called few-shot relations. Each few-shot relation r corresponds to one knowledge graph completion task $\mathcal{T}_r$. Each task have a support set and a query set, i.e., $\mathcal{T}_r = \{\mathcal{S}_r, \mathcal{Q}_r\}$. Support set $\mathcal{S}_r = \{(s_i, r, o_i) | (s_i, r, o_i) \in \mathcal{G}\}$ contains support triples of task $\mathcal{T}_r$, and $|\mathcal{S}_r| = K$ suggests a K-shot knowledge graph completion task. Besides, query set $\mathcal{Q}_r$ contains all query triples of task $\mathcal{T}_r$, including positive query triples $\mathcal{Q}_r^+ = \{(s_i, r, o_i^+) | (s_i, r, o_i^+) \in \mathcal{G}, o_i^+ \in \mathcal{C}\}$ and corresponding negative query triples $\mathcal{Q}_r^- = \{(s_i, r, o_i^-) | (s_i, r, o_i^-) \notin \mathcal{G}, o_i^- \in \mathcal{C}\}$. $\mathcal{C}$ is the candidate entity set. A few-shot knowledge graph completion task is to find the best completion entity for each query from the candidate entity set using the support set as a reference.

Few-Shot Learning Setting. We follow the same few-shot settings proposed by GMatching [21]. We divide all few-shot relations into three disjoint subsets $\mathcal{R}_{train}$, $\mathcal{R}_{valid}$ and $\mathcal{R}_{test}$ for model training, validation and testing. Therefore, the training, validation and testing phases of our model correspond to a series of

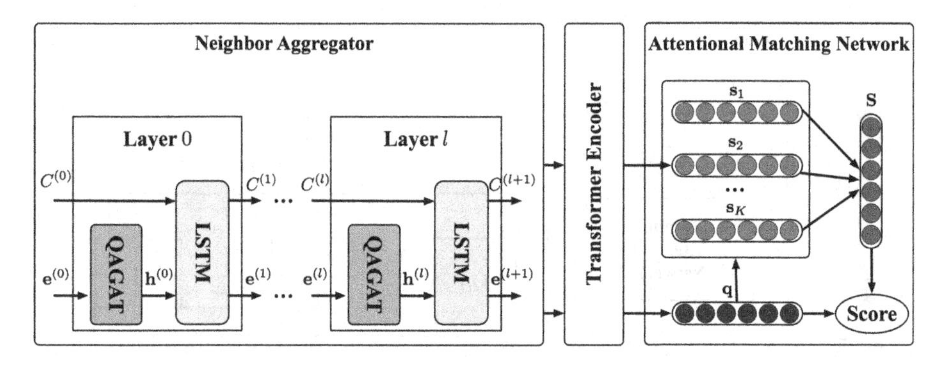

Fig. 1. Overall framework of our model AMBLE.

few-shot knowledge graph completion tasks. The training, validation and testing phases are defined as $\mathcal{T}_{train} = \{\mathcal{T}_i | i \in \mathcal{R}_{train}\}$, $\mathcal{T}_{valid} = \{\mathcal{T}_i | i \in \mathcal{R}_{valid}\}$ and $\mathcal{T}_{test} = \{\mathcal{T}_i | i \in \mathcal{R}_{test}\}$ respectively. In addition to few-shot relations, the other relations in knowledge graph $\mathcal{G}$ have sufficient triples, called high-frequency relations, and the triples containing high-frequency relations constitutes the background knowledge graph $\mathcal{G}'$.

4 Methodology

Given a task $\mathcal{T}_i = \{\mathcal{S}_i, \mathcal{Q}_i\}$, the purpose of AMBLE is to find the best candidate entity by matching the input query $q \in \mathcal{Q}_i$ to the given support set $\mathcal{S}_i$. To achieve this goal, as shown in Fig. 1, AMBLE consists of three major parts: (1) Neighbor aggregator to learn entity representations by aggregating neighbor information; (2) Transformer encoder to learn relational representations for entity pairs; (3) Attentional matching network to match the query to the given support set. Finally, we present the loss function and training details of our model.

4.1 Neighbor Aggregator

Neighbor aggregator is proposed to learn entity representations, which aggregate high-order neighbors through multiple layers of iterative aggregation. Each aggregation layer of the neighbor aggregator is shown in Fig. 2, and each layer consists of QAGAT and LSTM.

Query-Aware Graph Attention Network. The influence of neighbors on one entity keeps changing with the relation of current task, i.e., when completing different queries, neighbors have different weights of influence on the target entity. This dynamic property depends on the relevance between the target entity and neighbor entity on the one hand, and on the relevance between neighbor relation and query relation on the other hand. However, existing FKGC methods

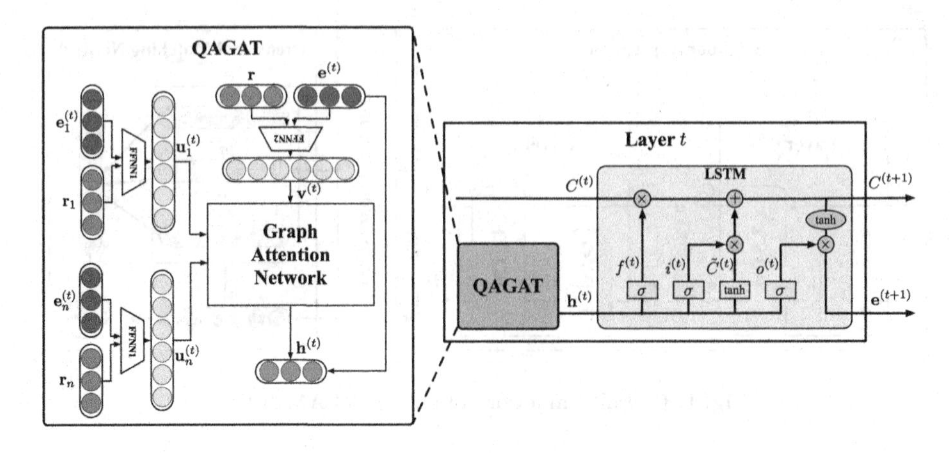

Fig. 2. Details of the t-layer of the neighbor aggregator

[12,13,21,22] cannot simultaneously consider these two influencing factors to obtain discriminative entity representations.

To tackle the above issue, we design a query-aware graph attention network (QAGAT) to dynamically aggregate one-hop neighbors. For each entity e, we constructs the neighbors of e, i.e., $\mathcal{N}_e = \{(e_i, r_i)|, (e, r_i, e_i) \in \mathcal{G}^{'}\}$, by searching for the triples in background knowledge graph $\mathcal{G}^{'}$ whose subject entity is e. e_i is the object entity considered as entity e's neighbor, r_i is the relation between e and e_i. At the t-layer, we first use two different FFNNs [14] to learn the integration representations of the neighbors and target entity as follows:

$$\mathbf{u}_i^{(t)} = \mathbf{W}_1(\mathbf{e}_i^{(t)} \| \mathbf{r}_i) + b_1 \tag{1}$$

$$\mathbf{v}^{(t)} = \mathbf{W}_2(\mathbf{e}^{(t)} \| \mathbf{r}) + b_2 \tag{2}$$

where $\|$ denotes concatenation operation. $\mathbf{u}_i^{(t)}$ denotes neighbor e_i's entity and relation integration representation, $\mathbf{v}^{(t)}$ denotes target entity e and query relation r integration representation. We obtain $\mathbf{r}$ given its support set $\mathcal{S}_r = \{(s_i, r, o_i)|s_i, o_i \in \mathcal{E}\}$ by TrasnE [1]: $\mathbf{r} = \mathbf{Mean}(\mathbf{e}_{o_i} - \mathbf{e}_{s_i})$.

Having obtained the integration representations of the neighbors and target entity, the weight $\alpha_i^{(t)}$ of the neighbor e_i for target entity e can be calculated as follows:

$$\alpha_i^{(t)} = \frac{\exp\ f(\mathbf{v}^{(t)}, \mathbf{u}_i^{(t)})}{\sum_{(e_j, r_j) \in \mathcal{N}_e} \exp\ f(\mathbf{v}^{(t)}, \mathbf{u}_j^{(t)})} \tag{3}$$

We use *softmax* function to apply over $f(x, y)$, and we want to take benefit of the relevance of both entities and relations, so the $f(x, y)$ is defined as follows:

$$f(x, y) = LeakyReLU(x^\top \mathbf{W}_3 y + b_3) \tag{4}$$

$\mathbf{W}_3$ is the similarity matrix to calculate the relevance between x and y. Following GAT [15], we use activation function *LeakyReLU* [10] here. Thus, we make full use of entity and relation information to dynamically assign neighbor weights.

Then, we can learn entity e's representation by adaptively aggregating neighbor information and its own information as follows:

$$\mathbf{h}^{(t)} = Relu(\lambda \sum_{(e_i, r_i) \in \mathcal{N}_e} \alpha_i^{(t)} \mathbf{e}_i^{(t)} + (1 - \lambda)\mathbf{e}^{(t)}) \tag{5}$$

where $\mathbf{h}^{(t)}$ is the output representation of entity e at t-layer QAGAT. λ is a trade-off hyperparameter, and $Relu$ denotes activation function. Thus, QAGAT adaptively aggregates one-hop neighbors using both entity and relation information.

Adaptive Multi-hop Neighbor Selection. We expand from aggregating one-hop neighbors to aggregating multi-hop neighbors, aiming to find more complementary information from high-order neighbors. However, as the distance increases, more noise information is trapped in high-order neighbors [9]. To solve the high-order neighbor noise problem, we add a LSTM [6] after QAGAT of each layer aggregator for information filtering, because LSTM has excellent memory and forgetting functions. As shown in Fig. 2, the detailed calculation process for each step is as follows:

$$f^{(t)} = \sigma(\mathbf{W}_f^{(t)} \mathbf{h}^{(t)} + b_f^{(t)}),\ i^{(t)} = \sigma(\mathbf{W}_i^{(t)} \mathbf{h}^{(t)} + b_i^{(t)}),\ o^{(t)} = \sigma(\mathbf{W}_o^{(t)} \mathbf{h}^{(t)} + b_o^{(t)}) \tag{6}$$

$$\tilde{C}^{(t)} = tanh(\mathbf{W}_C^{(t)} \mathbf{h}^{(t)} + b_C^{(t)}),\ C^{(t+1)} = f^{(t)} \cdot C^{(t)} + i^{(t)} \cdot \tilde{C}^{(t)} \tag{7}$$

$$\mathbf{e}^{(t+1)} = o^{(t)} \cdot tanh(C^{(t+1)}) \tag{8}$$

where $\tilde{C}^{(t)}$ is the newly added neighbor information in the t-layer aggregation. The gated $i^{(t)}$ is used to extract useful information from newly added neighbors, and the extracted information is added to the memory by $i^{(t)} \cdot \tilde{C}^{(t)}$. The gated $f^{(t)}$ is used to filter noisy information from the old memory by $f^{(t)} \cdot C^{(t)}$. As such, we are able to filter the memory for entity e as $C^{(t+1)}$. The gated unit $o^{(t)}$ is used to select the output information from $C^{(t+1)}$. Then, we obtain representation $\mathbf{e}^{(t+1)}$ of the t-layer aggregation of entity e. After l-layer aggregation, we aggregate l-hop neighbor information and obtain entity e's final representation $\mathbf{e}^{(l)}$.

Through the above, we effectively extract useful information from the t-hop neighbors and filter out the noisy information. Therefore, the neighbor aggregator of our model can efficiently achieve the information aggregation and selection of high-order neighbors by iterating QAGAT and LSTM.

4.2 Transformer Encoder

With the neighbor aggregator, we have obtained the entity representation. Inspired by FAAN [13], which uses a Transformer module to learn the representation of entity pairs. We use a Transformer encoder to learn representations of entity pairs. We use the encoder to interact information between subject and object entities to learn more reliable representations of entity pairs. For each triple (s, r, o) in support set or query set, we input them into Transformer as follows:

$$\mathbf{z}_1^0 = \mathbf{e}_s^{(l)} + \mathbf{x}_1^{pos}, \mathbf{z}_2^0 = \mathbf{r}_{mask} + \mathbf{x}_2^{pos}, \mathbf{z}_3^0 = \mathbf{e}_o^{(l)} + \mathbf{x}_3^{pos} \tag{9}$$

where $\mathbf{e}_s^{(l)}$ and $\mathbf{e}_o^{(l)}$ are the representations of entity s and o obtained by neighbor aggregator. $\mathbf{r}_{mask}$ is a randomly initialized mask. $\mathbf{x}_1^{pos}$, $\mathbf{x}_2^{pos}$ and $\mathbf{x}_3^{pos}$ are the position embeddings. Later, we feed them into a stack of P Transformer blocks as follows:

$$\mathbf{z}_i^p = Transformer(\mathbf{z}_i^{p-1}), i = 1, 2, 3. \tag{10}$$

where $\mathbf{z}_i^p$ is the hidden state of the p-th layer Transformer. After P layer Transformer, the final hidden state $\mathbf{z}_2^P$ is the representation of entity pair (s, o). By this way, we can obtain the representations $(\mathbf{s_1}, \mathbf{s_2}, ..., \mathbf{s_K})$ of entity pairs in support set $\mathcal{S}_r$ and the query entity pairs representation $\mathbf{q}$ of task $\mathcal{T}_r$.

4.3 Attentional Matching Network

Having obtained the representations of entity pairs in support set and query set by Transformer encoder, we adopt the idea of matching network [16] to calculate the similarity between query and support set to achieve FKGC task. Due to the semantic divergence in support set, different support triples have different weights for a query [5]. To enable our model to dynamically aggregate support triples when matching different queries, we adopt the attentional matching network in FAAN [13]. The similarity score of query $\mathbf{q}$ is calculated as follows:

$$\beta_i = \frac{\exp(\mathbf{q}^\top \mathbf{s}_i)}{\sum_{j=1}^{K} \exp(\mathbf{q}^\top \mathbf{s}_j)}, \quad \mathbf{S} = \sum_{i=1}^{K} \beta_i \mathbf{s}_i \tag{11}$$

$$\text{Score}(\mathbf{q}, \mathcal{S}_r) = \mathbf{S}^\top \mathbf{q} \tag{12}$$

where β_i denotes the attention weight of support triple (s_i, r, o_i), and $\mathbf{S}$ is support set representation. Thus, we can obtain adaptive support set representation for different queries by Eq. 11. We take the inner product of the representations of query and support set as their similarity score.

4.4 Loss Function

Our model is trained on a training task set $\mathcal{T}_{train}$ with the goal of high similarity scores for positive queries and low similarity scores for negative queries. The objective function is a hinge loss defined as follows:

$$\mathcal{L} = \sum_{r}^{\mathcal{R}} \sum_{q^+ \in \mathcal{Q}_r^+, q^- \in \mathcal{Q}_r^-} \left[\gamma + Score(\mathbf{q}^-, \mathcal{S}_r) - Score(\mathbf{q}^+, \mathcal{S}_r)\right]_+ \tag{13}$$

where γ is a hyperparameter represents safety margin distance, and $[x]_+ = max(0, x)$ is the standard hinge loss.

Table 2. Statistics of the experimental datasets.

Dataset	#Relation	#Entity	#Triples	#Task-Train	#Task-Valid	#Task-Test
NELL-One	358	68545	181109	51	5	11
Wiki-One	822	4838244	5859240	133	16	34

5 Experiments

5.1 Datasets and Baselines

We conduct experiments on two public datasets NELL-One and Wiki-One. Following GMatching [21], we regard relations containing more than 50 but less than 500 triples as few-shot relations, and others as high-frequency relations. The task relation ratios for training/validation/testing on NELL-One and Wiki-One are 51/5/11 and 133/16/34, respectively. The statistics of the two datasets are shown in Table 2.

The existing FKGC models: including GMatching [21], MetaR [3], FSRL [22], FAAN [13], GANA [12], YANA [8] and MFEM [20]. To evaluate the performance of our model and the baselines for FKGC task, we utilize two traditional metrics MRR and Hits@1/5/10 on both datasets. The results of GMatching and FSRL are derived from the paper of FAAN, and the results of the other FKGC models are obtained from their corresponding original papers.

5.2 Implementation

In our model, all entities and relations representations are initialized randomly with dimension of 100 and 50 for NELL-One and Wiki-One. The few-shot size K is set to 5 for the following experiments. The two hyperparameters of this model, margin γ and trade-off λ, are set to 10 and 0.6 respectively. For Neighbor Aggregator, our model aggregates 2-hop neighbors on both datasets to achieve optimal performance, i.e., $l = 2$. The number of Transformer layers P is set to 3 and 4 for NELL-One and Wiki-One respectively, and the number of attention heads is set to 4 and 8 respectively. We implement all experiments with PyTorch and use Adam optimizer [7] to optimize model parameters with a learning rate of 0.0001.

5.3 Experimental Comparison with Baselines

We compare AMBLE with baselines on NELL-One and Wiki-One datasets to evaluate the effectiveness of AMBLE. The performances of all models are reported in Table 3, where the best results are highlighted in bold, and the best performance of baselines is underlined. AMBLE achieves general improvements compared to the baselines. To be concrete, 1) For NELL-One dataset, AMBLE achieves an improvement of 6.1/4.9/11.4/8.3% in MRR/Hits@1/5/10 compared to the best performing baseline GANA. These results illustrate that it

Table 3. Experimental results for all methods. The best results are marked in **bold**, and the best results of the baseline are underline.

Models (5-shot)	NELL-One				Wiki-One			
	MRR	Hits@1	Hits@5	Hits@10	MRR	Hits@1	Hits@5	Hits@10
GMatching (MaxP)	0.176	0.113	0.233	0.294	0.263	0.197	0.337	0.387
GMatching (MeanP)	0.141	0.080	0.201	0.272	0.254	0.193	0.314	0.374
GMatching (Max)	0.147	0.090	0.197	0.244	0.245	0.185	0.295	0.372
MetaR (Pre-train)	0.209	0.141	0.280	0.355	0.323	0.270	0.385	0.418
MetaR (In-train)	0.261	0.168	0.350	0.437	0.221	0.178	0.264	0.302
FSRL	0.153	0.073	0.212	0.319	0.158	0.097	0.206	0.287
FAAN	0.279	0.200	0.364	0.428	0.341	0.281	0.395	0.463
GANA	0.344	0.246	0.437	0.517	0.351	0.299	0.407	0.446
YANA	0.294	0.230	0.364	0.421	0.380	0.327	0.442	0.523
MFEN	0.310	0.236	0.369	0.443	0.331	0.253	0.398	0.470
AMBLE (Ours)	**0.365**	**0.258**	**0.487**	**0.560**	**0.392**	**0.335**	**0.463**	**0.546**

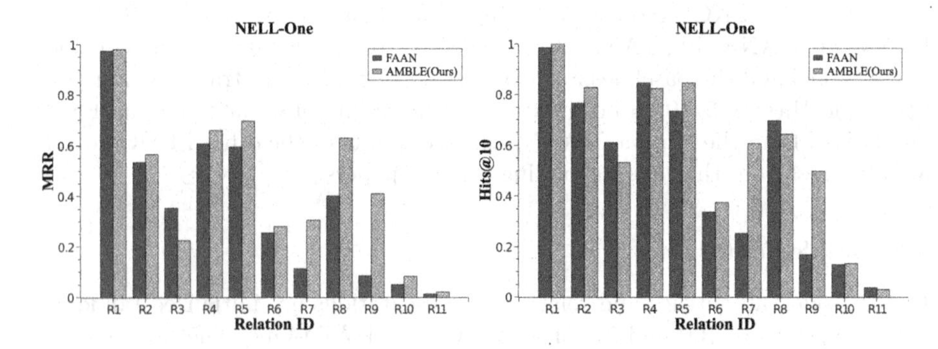

Fig. 3. The MMR and Hits@10 of AMBLE and FAAN for each relation on NELL-One.

is more effective to adaptively aggregate multi-hop neighbors based on relation and entity information. 2) For Wiki-One dataset, AMBLE achieves an improvement of 3.2/2.4/4.8/4.4% in MRR/Hits@1/5/10 compared to the best performing baseline YANA. Although YANA utilizes the information from the subgraphs, our model achieves a better performance. This indicates that our model can effectively extract useful and filter noisy information from high-order neighbors by iterating QAGAT and LSTM.

5.4 Comparison over Different Relations

To demonstrate the superiority of our model in more detail, we set up comparative experiments with FAAN [13] on NELL-One over different relations. The experimental results are shown in Fig. 3, where Relation ID represents a class of relation. AMBLE outperforms FAAN in MRR metric with 10 out of 11 relations,

Table 4. The results of ablation experiment.

Variants	NELL-One				Wiki-One			
	MRR	Hits@1	Hits@5	Hits@10	MRR	Hits@1	Hits@5	Hits@10
w/o QAGAT	0.265	0.187	0.334	0.408	0.342	0.254	0.388	0.463
w/o Neighbor selection	0.337	0.239	0.425	0.476	0.364	0.265	0.408	0.495
AMBLE (Ours)	**0.365**	**0.258**	**0.487**	**0.560**	**0.392**	**0.335**	**0.463**	**0.546**

and in Hits@10 metric with 7 out of 11 relations. Experimental results indicate that our model is robust for different task relations.

5.5 Ablation Study

We perform experiments on all the datasets with several variants of AMBLE to provide a better understanding of the contribution of each module to AMBLE. The ablative results are shown in Table 4. The experimental results demonstrate the effectiveness of each module in AMBLE. 1) **w/o QAGAT** means we replace QAGAT with the heterogeneous neighbor encoder in FSRL [22]. Experimental results show that QAGAT can obtain discriminative entity and relation representations by dynamically aggregating neighbors based on relation and entity information. 2) **w/o Neighbor selection** means that we remove the LSTM from the neighbor aggregator. Experimental results prove that using high-order neighbor information can improve the performance of our model. In addition, neighbor selection by LSTM can effectively extract useful and filter noisy information from high-order neighbors.

6 Conclusion

In this paper, we propose a novel model AMBLE to address the dynamic neighbor properties and high-order neighbor noise issues in few-shot knowledge graph completion. We propose a query-aware graph attention network (QAGAT) to dynamically aggregate neighbors based on relation and entity information, so as to capture the dynamic neighbor properties in completion task. In addition, we iterate QAGAT and LSTM to aggregate multi-hop neighbors, which can efficiently extract useful and filter noisy information from high-order neighbors. The experimental results on the datasets NELL-One and Wiki-One show the superiority of our model and the effectiveness of each component of our model.

Acknowledgements. This work was partially supported by the National Natural Science Foundation of China (Grant No. U19A2067, 61976051) and Major Key Project of PCL (Grant No. PCL2021A09, PCL2021A02, PCL2022A03).

References

1. Bordes, A., Usunier, N., Garcia-Duran, A., Weston, J., Yakhnenko, O.: Translating embeddings for modeling multi-relational data. In: NIPS (2013)
2. Carlson, A., Betteridge, J., Kisiel, B., Settles, B., Hruschka, E., Mitchell, T.: Toward an architecture for never-ending language learning. In: AAAI, pp. 1306–1313 (2010)
3. Chen, M., Zhang, W., Zhang, W., Chen, Q., Chen, H.: Meta relational learning for few-shot link prediction in knowledge graphs. In: EMNLP-IJCNLP, pp. 4217–4226 (2019)
4. Finn, C., Abbeel, P., Levine, S.: Model-agnostic meta-learning for fast adaptation of deep networks. In: ICML, pp. 1126–1135 (2017)
5. Gong, X., Qin, J., Chai, H., Ding, Y., Jia, Y., Liao, Q.: Temporal-relational matching network for few-shot temporal knowledge graph completion. In: DASFAA, pp. 768–783 (2023)
6. Hochreiter, S., Schmidhuber, J.: Long short-term memory. In: Neural Computation, pp. 1735–1780 (1997)
7. Kingma, D.P., Ba, J.: Adam: a method for stochastic optimization. arXiv preprint arXiv:1412.6980 (2014)
8. Liang, Y., Zhao, S., Cheng, B., Yin, Y., Yang, H.: Tackling solitary entities for few-shot knowledge graph completion. In: KSEM, pp. 227–239 (2022)
9. Liu, Z., Chen, C., Li, L., Zhou, J., Li, X., Song, L., Qi, Y.: GeniePath: graph neural networks with adaptive receptive paths. In: AAAI, pp. 4424–4431 (2019)
10. Maas, A.L., Hannun, A.Y., Ng, A.Y., et al.: Rectifier nonlinearities improve neural network acoustic models. In: Proceedings of the ICML, p. 3 (2013)
11. Nathani, D., Chauhan, J., Sharma, C., Kaul, M.: Learning attention-based embeddings for relation prediction in knowledge graphs. In: ACL, pp. 4710–4723 (2019)
12. Niu, G., et al.: Relational learning with gated and attentive neighbor aggregator for few-shot knowledge graph completion. In: SIGIR, pp. 213–222 (2021)
13. Sheng, J., et al.: Adaptive attentional network for few-shot knowledge graph completion. In: EMNLP, pp. 1681–1691 (2020)
14. Svozil, D., Kvasnicka, V., Pospichal, J.: Introduction to multi-layer feed-forward neural networks. In: Chemometrics and Intelligent Laboratory Systems, pp. 43–62 (1997)
15. Veličković, P., Cucurull, G., Casanova, A., Romero, A., Lio, P., Bengio, Y.: Graph attention networks. arXiv preprint arXiv:1710.10903 (2017)
16. Vinyals, O., Blundell, C., Lillicrap, T., Wierstra, D., et al.: Matching networks for one shot learning. In: NIPS (2016)
17. Vrandečić, D., Krötzsch, M.: Wikidata: a free collaborative knowledgebase. Commun. ACM **57**, 78–85 (2014)
18. Wang, X., Wang, D., Xu, C., He, X., Cao, Y., Chua, T.S.: Explainable reasoning over knowledge graphs for recommendation. In: AAAI, pp. 5329–5336 (2019)
19. Wang, Z., Zhang, J., Feng, J., Chen, Z.: Knowledge graph embedding by translating on hyperplanes. In: AAAI (2014)
20. Wu, T., Ma, H., Wang, C., Qiao, S., Zhang, L., Yu, S.: Heterogeneous representation learning and matching for few-shot relation prediction. Pattern Recogn. **131**, 108830 (2022)
21. Xiong, W., Yu, M., Chang, S., Guo, X., Wang, W.Y.: One-shot relational learning for knowledge graphs. In: EMNLP, pp. 1980–1990 (2018)
22. Zhang, C., Yao, H., Huang, C., Jiang, M., Li, Z.J., Chawla, N.: Few-shot knowledge graph completion. In: AAAI, pp. 3041–3048 (2020)
23. Zhang, Y., Dai, H., Kozareva, Z., Smola, A.J., Song, L.: Variational reasoning for question answering with knowledge graph. In: AAAI, pp. 1–8 (2018)

Applications of Quantum Embedding in Computer Vision

Juntao Zhang[1], Jun Zhou[1], Hailong Wang[1], Yang Lei[1(✉)], Peng Cheng[2], Zehan Li[3], Hao Wu[1], Kun Yu[1,3], and Wenbo An[1]

[1] Institute of System Engineering, AMS, Beijing, China
leiyang1983@163.com
[2] Coolanyp L.L.C., Wuxi, China
[3] University of Electronic Science and Technology of China, Chengdu, China

Abstract. Nowadays, Deep Neural Networks (DNNs) are fundamental to many vision tasks, including large-scale visual recognition. As the primary goal of the DNNs is to characterize complex boundaries of thousands of classes in a high-dimensional space, it is critical to learn higher-order representations for enhancing nonlinear modeling capability. Recently, a novel method called Quantum-State-based Mapping (QSM) has been proposed to improve the feature calibration ability of the existing attention modules in transfer learning tasks. QSM uses the wave function describing the state of microscopic particles to map the feature vector into the probability space. In essence, QSM introduces a novel higher-order representation to improve the nonlinear capability of the network. In this paper, we extend QSM to Quantum Embedding (QE) for designing new attention modules and Self-Organizing Maps, a class of unsupervised learning methods. We also conducted experiments to validate the effectiveness of QE.

Keywords: Quantum mechanics · Embedding · Attention mechanism · SOM · Image classification

1 Introduction

Researchers have recognized the connection between quantum theory and machine learning in the past two decades and published many high-quality works. Quantum computation uses the mathematic rules of quantum physics to redefine how computers create and manipulate data. These properties imply a radically new way of computing, using qubits instead of bits, and give the possibility of obtaining quantum algorithms that could be substantially faster than classical algorithms. There have been proposals for quantum machine learning algorithms that have the potential to offer considerable speedups over the corresponding

J. Zhou—Co-first author.

B. Luo et al. (Eds.): ICONIP 2023, CCIS 1965, pp. 165–177, 2024.
https://doi.org/10.1007/978-981-99-8145-8_14

classical algorithms, such as q-means [11], QCNN [12], QRNN [1]. Another concept in literature is quantum logic, which refers to the non-classical logical structure and logical system originating from the mathematical structure of quantum mechanics. For example, Garg et al. [7] proposed a Knowledge-Base embedding inspired by quantum logic, which allows answering membership-based complex logical reasoning queries. Unlike the above works, Zhang et al. [23] proposed a novel method called Quantum-State-based Mapping (QSM) to improve the feature calibration ability of the attention module in transfer learning tasks. QSM uses the wave function describing the state of microscopic particles to embed the feature vector into the high dimensional space.

In this paper, we extend QSM to Quantum Embedding (QE). QE uses wave functions and quantum effects between particles as mappings, which helps to represent the complex interactions between features. We propose a new attention method called Quantum-State-based Attention Networks (QSAN) based on QE. We evaluate the proposed method on large-scale image classification using ImageNet and compare it with state-of-the-art counterparts. Convolutional neural networks (CNNs) have been the mainstream architectures in computer vision for a long time until recently when new challengers such as Vision Transformers [5] (ViT) emerged. We train ResNet-50 with QSAN and adopt the advanced training procedure to achieve 79.6% top-1 accuracy on ImageNet-1K at 224×224 resolution in 300 epochs without extra data and distillation. This result is comparable to Vision Transformers under the same conditions. In addition, we construct a SOM-based image classifier using the QE method. Experiments show that the proposed QE method can improve the classification performance by modifying the similarity measure.

2 Related Works

2.1 Revisiting QSM

In order to highlight the plug-and-play ability of QSM, [23] use QSM only once at the appropriate layer in the module and do not change the data dimension. Specifically, assuming that the feature vector after pooling is $\mathbf{X} \in \mathbb{R}^{1\times d}$, and then the probability density function of one-dimensional particles under a coordinate representation in an infinite square well is used as a mapping:

$$\begin{cases} |\Psi(\mathbf{X})|^2 = QSM(\mathbf{X}) = \frac{2}{a}\sum_{n=1}^{N} c_n sin^2(\frac{n\pi}{a}\mathbf{X}) \\ a = max(|\mathbf{X}|) \end{cases} \tag{1}$$

This mapping does not change the dimension of $\mathbf{X}$. As analyzed in [23], training $|\Psi(\mathbf{X})|^2$ allows the neural network to exploit the probability distribution of the global information, which is beneficial for generating more effective attention. Clearly, this method is straightforward but preliminary. Instead of simply using the wave function as the mapping, QSAN in this paper uses the quantum effect between two identical particles as the gating.

2.2 Attention Mechanism in CNNs

SENet [10] proposed a channel attention mechanism in which SE blocks comprise a squeeze module and an excitation module. Global spatial information is collected in the squeeze module by global average pooling. The excitation module captures channel-wise relationships and outputs an attention vector using fully connected (FC) layers and non-linear layers (ReLU and Sigmoid). Then, each channel of the input feature is scaled by multiplying the corresponding element in the attention vector. Some later work attempts to improve the outputs of the squeeze module (e.g., GSoP-Net [6]), improve both the squeeze module and the excitation module (e.g., SRM [14]), or integrate with other attention mechanisms (e.g., CBAM [20]). Besides these methods, ECANet [18] aims to reduce the complexity of the model by improving the excitation module. ECANet uses a convolution layer with adaptive kernel size to replace FC layers in the excitation module. Qin et al. [17] rethought global information captured from the compression viewpoint and analyzed global average pooling in the frequency domain. They proved that global average pooling is a special case of the discrete cosine transform (DCT) and used this observation to propose a novel multi-spectral channel attention called FcaNet. Similarly, QSAN uses a two-particle quantum mechanical model to enhance global spatial information.

2.3 Self-organizing Map

Self-organizing map (SOM) is a classical artificial neural network introduced by Teuvo Kohonen in the 1980s [13], which can generate a discretized low-dimensional representation of the input space of unsupervised training samples. As introduced in [4], the SOM algorithm was designed for data modeled as numerical vectors and belonging to a subset $\mathbb{X}$ of Euclidean space, for instance, $\mathbb{R}^d$. For the discrete setting considered in this work, the input space $\mathbb{X}$ comprises N data points $\mathbf{x}_1, ..., \mathbf{x}_N$ stored beforehand or generated online. The learning map is a lattice of M neurons, e.g., string-like for one dimension or grid-like for two dimensions, with each neuron having a weight vector $\mathbf{w}_i \in \mathbb{R}^d$, where i is the index of each neuron. The learning process for the SOM can be described as follows. At each time instant, a random sample $\mathbf{v}$ from the input space $\mathbb{X}$ is selected, and the best matching unit (BMU) corresponding to sample $\mathbf{v}$ is determined by

$$r = \arg\min_{s} \sum_{i} h_{si} D(\mathbf{v}, \mathbf{w}_i) \tag{2}$$

where r is the index of the *winning node*. Here function $D(\cdot, \cdot)$ represents a similarity measure used to compare the closeness between two vectors. And $h_{s\cdot}$ is the neighborhood function, a non-increasing function of the distance among node s and all the other nodes in the lattice. Then the weight update of each neuron follows

$$\mathbf{w}_i(n+1) = \mathbf{w}_i(n) + \epsilon h_{ri} D(\mathbf{v}, \mathbf{w}_i) \tag{3}$$

where ϵ is the learning rate. The closer a node is to the BMU, the more its weights get altered, and the farther away the neighbor is from the BMU, the less it learns. And throughout the recursive calculation of BMU and weight update, the map tends to approximate the distribution of given input samples.

Although the SOM algorithm was originally designed for unsupervised data, it is also available for supervised and semi-supervised learning. Kohonen's algorithm is a neighborhood-preserving vector quantization tool working on the winner-take-all principle, where the winner is the most similar node to the input in an instant.

3 Proposed Method

In this section, we introduce two specific applications of Quantum Embedding (QE). One is to modify the SOM similarity measure by using the wave function of one-dimensional particles in an infinite square well. The other is to use the relative position distribution probability of two identical free particles in multidimensional space to generate more efficient channel attention.

3.1 QE-SOM

We know that BMU selection is the most critical step since the following procedures are all closely related to the winning node. In addition, the design of similarity measure $D(\cdot,\cdot)$ influences the BMU computation and thus affects the performance of SOM. Traditional SOMs use Euclidean distance as the similarity measure:

$$D(\mathbf{v}, \mathbf{w}_i) = \|\mathbf{v} - \mathbf{w}_i\|^2 \tag{4}$$

where, $\mathbf{v} = (v^{(1)}, v^{(2)}, ..., v^{(d)}) \in \mathbb{R}^d$ is the input sample, and $\mathbf{w}_i \in \mathbb{R}^d$ represents the weight of the ith neuron.

The suboptimality of those original SOMs is due to the global nature of MSE cost functions. The outliers inevitably affect the updating of each neuron's weight, i.e., the data with low probabilities. It leads to oversampling in the low-probability subspaces and undersampling in high-probability subspaces during the learning process. For the output feature vector extracted by previous layers, we choose the same stationary state wave function of a particle in the one-dimensional infinite well as the mapping function with the highest energy level as 3. Then the similarity measure of the proposed QE-SOM is modified as follows:

$$\begin{cases} p^{(k)} = \sum_{n=1}^{3} \frac{2}{a} sin^2(\frac{n\pi}{a} v^{(k)}), k = 1, 2, ..., d \\ D^p(\mathbf{v}, \mathbf{w}_j) = \|\mathbf{p} - \mathbf{w}_j\|^2 \end{cases} \tag{5}$$

where a is the width of potential well, and vector $\mathbf{p} = (p^{(1)}, p^{(2)}, ..., p^{(d)})$ is composed of transformed feature elements. Since we can normalize $\mathbf{v}$ into a unit vector, then $a = 1$.

Note that the computation cost of this modification is constant, which can hardly impact the efficiency of training and prediction of a neural network. An immediate impact is that we change the self-organizing mode of the learning map, i.e., how those neurons cluster and update. Experiments indicate that measuring the similarity in probability space can improve the classification accuracy on the test set.

3.2 Quantum-State-Based Attention Networks

In this section, we present a specific form of Quantum-State-based Attention Networks (QSAN), which is a channel attention. In order to highlight the role of quantum embedding, we do not use a complex network structure. As shown in Fig. 1, the only difference between QSAN and classical SE blocks [10] is that the feature vector after the max-pooling operation is used to generate quantum effect gating, which is used to adjust spatial information intensity.

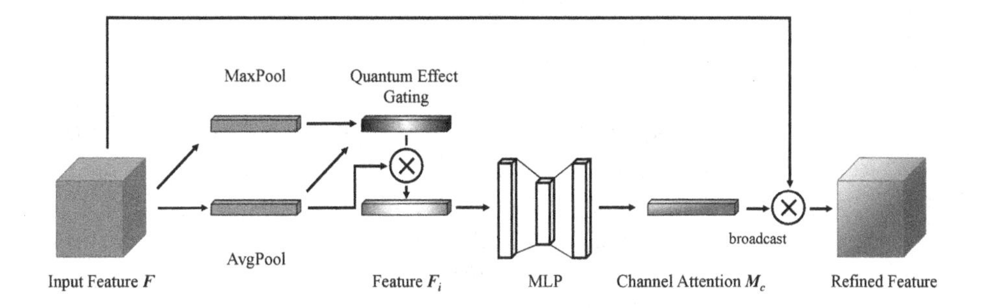

Fig. 1. Illustration of the QSAN.

Given an intermediate feature map $\mathbf{F} \in \mathbb{R}^{C \times H \times W}$ as input, the feature map $\mathbf{F}_i \in \mathbb{R}^{C \times 1 \times 1}$ is computed as:

$$\mathbf{F}_i = AvgPool(\mathbf{F}) \otimes QEG\big(AvgPool(\mathbf{F}), MaxPool(\mathbf{F})\big) \tag{6}$$

where $\otimes$ denotes element-wise multiplication, $AvgPool$ and $MaxPool$ represent the average-pooling operation and the max-pooling operation, respectively.

Quantum Effect Gating (QEG). In the attention mechanism, average-pooling is usually used to aggregate spatial information. But [20] argues that max-pooling gathers another important clue about distinctive object features to infer finer channel-wise attention. However, CBAM [20] concatenates the features after average-pooling and max-pooling operation, which ignores the intrinsic relationship between the two kinds of pooling. Therefore, Quantum Effect Gating (QEG) is proposed to solve this problem. Let the feature $\mathbf{F}$ after average-pooling operation be $\mathbf{F}_{Avg}$ and the feature $\mathbf{F}$ after max-pooling operation be $\mathbf{F}_{Max}$. Obviously, $\mathbf{F}_{Avg}$ and $\mathbf{F}_{Max}$ express the same spatial global information from two different perspectives. We consider $\mathbf{F}_{Avg}$ and $\mathbf{F}_{Max}$ as position vectors of two identical free particles, both of which are in momentum eigenstates (eigenvalues are $\hbar \boldsymbol{k}_\alpha$, $\hbar \boldsymbol{k}_\beta$). Without the exchange symmetry, the wave function of two free particles can be expressed as:

$$\psi \boldsymbol{k}_\alpha \boldsymbol{k}_\beta(\boldsymbol{r}_1, \boldsymbol{r}_2) = \frac{1}{(2\pi\hbar)^3} e^{i(\boldsymbol{k}_\alpha \cdot \boldsymbol{r}_1 + \boldsymbol{k}_\beta \cdot \boldsymbol{r}_2)} \tag{7}$$

where $\boldsymbol{r}_1$ and $\boldsymbol{r}_2$ represent the spatial coordinates of the two particles respectively, and $\hbar$ is reduced Planck constant. In order to facilitate the study of the distribution probability of relative positions, let:

$$\begin{cases} \boldsymbol{r} = \boldsymbol{r}_1 - \boldsymbol{r}_2 \\ \boldsymbol{k} = (\boldsymbol{k}_\alpha - \boldsymbol{k}_\beta)/2 \\ \boldsymbol{R} = \dfrac{1}{2}(\boldsymbol{r}_1 + \boldsymbol{r}_2) \\ \boldsymbol{K} = \boldsymbol{k}_\alpha + \boldsymbol{k}_\beta \end{cases} \tag{8}$$

so Eq. 7 can be reduced to:

$$\psi \boldsymbol{k}_\alpha \boldsymbol{k}_\beta(\boldsymbol{r}_1, \boldsymbol{r}_2) = \frac{1}{(2\pi\hbar)^{\frac{3}{2}}} e^{i(\boldsymbol{K} \cdot \boldsymbol{R})} \phi_{\boldsymbol{k}}(\boldsymbol{r}) \tag{9}$$

where

$$\phi_{\boldsymbol{k}}(\boldsymbol{r}) = \frac{1}{(2\pi\hbar)^{\frac{3}{2}}} e^{i\boldsymbol{k} \cdot \boldsymbol{r}} \tag{10}$$

we only discuss the probability distribution of relative positions. The probability of finding another particle around a particle in the spherical shell with radius $(r, r + dr)$ is:

$$4\pi r^2 P(r) dr \equiv r^2 dr \int |\phi_{\boldsymbol{k}}(\boldsymbol{r})|^2 d\Omega = r^2 dr \frac{4\pi}{(2\pi\hbar)^3} \tag{11}$$

so the probability density $P(r)$ is constant, independent of r. However, when we consider the exchange symmetry, the wave function is:

$$\psi^S \boldsymbol{k}_\alpha \boldsymbol{k}_\beta(\boldsymbol{r}_1, \boldsymbol{r}_2) = \frac{1}{\sqrt{2}} (1 + P_{12}) \psi \boldsymbol{k}_\alpha \boldsymbol{k}_\beta(\boldsymbol{r}_1, \boldsymbol{r}_2) \tag{12}$$

where P_{12} is the exchange operator. We directly give the probability density of relative positions under the condition that the exchange symmetry is satisfied:

$$P^S(r) = \frac{1}{(2\pi\hbar)^3} (1 + \frac{sin2kr}{2kr}) \tag{13}$$

let x represent the relative distance between two particles and omit the constant factor $1/(2\pi\hbar)^3$, we get:

$$P(x) \propto 1 + \frac{sinx}{x} \tag{14}$$

more elaborated derivation of the above equations can refer to most textbooks of quantum mechanics such as [9]. According to Eq. 14, QEG can be expressed as:

$$QEG(\mathbf{F}_{Avg}, \mathbf{F}_{Max}) = \mathbf{1C} + \frac{sin((\mathbf{F}_{Avg} - \mathbf{F}_{Max}) \otimes (\mathbf{F}_{Avg} - \mathbf{F}_{Max}))}{(\mathbf{F}_{Avg} - \mathbf{F}_{Max}) \otimes (\mathbf{F}_{Avg} - \mathbf{F}_{Max})} \tag{15}$$

where $\mathbf{1C}$ represents a C-dimensional vector whose elements are all 1, $\otimes$ denotes element-wise multiplication. As shown in Eq. 15, QEG has no parameters to train, and the computation is negligible.

4 Experiments

In this section, the experiments verify the effectiveness of QE-SOM and QSAN, respectively.

4.1 Experiments on MNIST

We trained a CNN model from scratch. The CNN Model consists of two custom CNN layers followed by two fully connected (dense) layers.

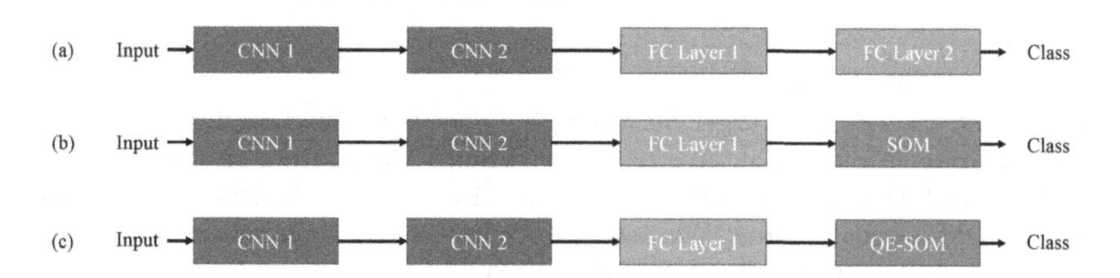

Fig. 2. Full architecture of training pipeline including the CNN model, SOM and QE-SOM.

Fig. 2(a) illustrates the full architecture of the basic pipeline: CNN 1 (10 filters, kernel size 5×5, stride 1×1, padding 2×2, max pooling 2×2), CNN 2 (20 filters, kernel size 3×3, stride 1×1, padding 1×1), fully connected layer 1 (FC layer 1 with 500 neurons) and fully connected layer 2 (FC layer 2 with 10 neurons). In Fig. 2(b), we described the SOM training pipeline, in which we removed FC layer 2 and extracted features to feed SOM with 500 input dimensions. More specifically, we extract the features before the classifier layer and use them as input to SOM. Figure 2(c) illustrates the QE-SOM training pipeline, i.e., the previous FC layer 2 in Fig. 2(a) is removed, and the features extracted from FC layer 1 are then fed to the proposed QE-SOM. Note that CNN 1, CNN 2, and FC layer 1 are determined by the CNN model and kept fixed during both the training and test stages of SOM and QE-SOM. The practice is a typical operation in transfer learning, i.e., training a base network and then copying its first n layers

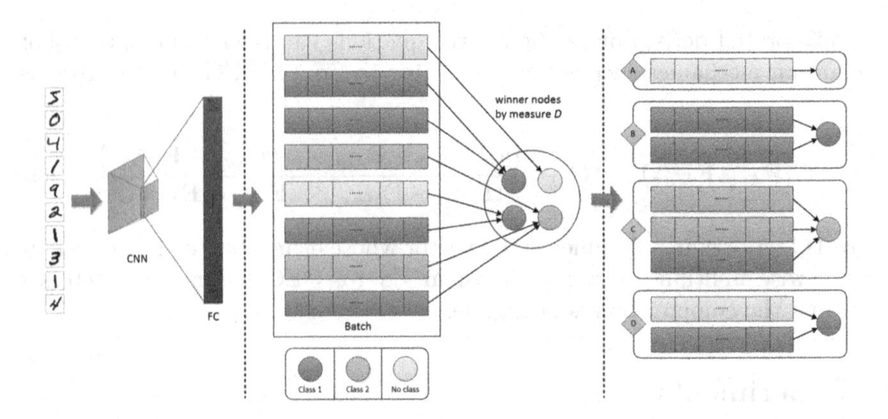

Fig. 3. The basic operation performed by SOM or QE-SOM when a batch is given, and its resulting cases.

Table 1. Performance comparison under different batch sizes.

	Number of samples			
	600	6000	30000	48000
QE-SOM	93.51%	93.69%	94.16%	93.88%
SOM	93.30%	93.60%	93.83%	93.31%

to the first n layers of a target network [22]. That is to say, transfer learning is the improvement of learning in a new task through transferring knowledge from a related task that has already been learned [16]. Figure 3 depicts more details of the SOM-based pipeline. We randomly select a certain number of samples from the training set as a batch for training SOM for one single iteration in this experiment. We report the average accuracy of 100 experiments to reduce the impact of randomness.

We start with the SOM neurons arranged in a 25×25 rectangle grid. Table 1 shows the classification accuracy of original SOM and QE-SOM under different sample sizes with the same other configurations in only one iteration.

Table 2 shows the same comparison under different iterations with the same number of neurons and batch size. At this time, the batch size is set as 600. Experiment results show that the performance gain is limited as the batch size and the iteration number are increased separately, and the effect of QE cannot be ignored.

Finally, with a batch size of 600 and the iteration number 100, we set the topology of SOM neurons as 2×5, 5×5, 15×15 and 30×30, respectively. Table 3 shows that QE can still achieve consistent performance improvements, especially when the number of neurons is small.

Considering that the SOM can be used for both supervised and unsupervised learning, we deleted labels of some samples from the training set for semi-

Table 2. Performance comparison under different iterations.

	Iterations			
	5	10	20	100
QE-SOM	97.59%	98.05%	98.06%	98.13%
SOM	97.47%	97.75%	97.85%	97.96%

Table 3. Performance comparison under different numbers of neurons.

	Number of neurons			
	2×5	5×5	15×15	30×30
QE-SOM	93.47%	96.27%	98.03%	98.27%
SOM	92.69%	95.24%	97.94%	98.16%

supervised learning with a fixed batch size of 600 and iteration number of 100 at last. Furthermore, empirically, the SOM architecture is again set as a 25×25 rectangle grid. Table 4 shows that when the energy level in Eq. 5 is 3, as the proportion of unlabeled samples increases, the performance advantage of QE-SOM gradually disappears or is even surpassed by the original SOM. However, with a higher energy level, such as 10 in Table 4, the proposed QE-SOM outperforms the original SOM in all proportion cases.

4.2 Experiments on ImageNet

We evaluate the QSAN on large-scale image classification using ImageNet and compare it with state-of-the-art counterparts.

Implementation Details. To evaluate our QSAN on ImageNet-1K classification, we employ three widely used CNNs as backbone models, including ResNet-18, ResNet-34, and ResNet-50. To train ResNets with QSAN, we adopt the same data augmentation and hyperparameter settings. Specifically, the input images are randomly cropped to 224×224 with random horizontal flipping. The parameters of networks are optimized by stochastic gradient descent (SGD)

Table 4. Performance comparison between QE-SOM and SOM under different proportions of unlabeled samples.

	Energy level	Proportion of unlabeled samples				
		0%	10%	25%	50%	90%
QE-SOM	3	98.13%	98.34%	98.16%	98.03%	97.04%
	10	98.32%	98.30%	98.28%	98.16%	97.26%
SOM	–	97.96%	98.25%	98.15%	97.79%	97.13%

with weight decay of 1e-4, a momentum of 0.9, and a mini-batch size of 256. All models are trained within 100 epochs by setting the initial learning rate to 0.1, which is decreased by a factor of 10 per 30 epochs. All programs run on a server equipped with two RTX A6000 GPUs and two Xeon Gold 6230R CPUs.

Comparisons Using Different Deep CNNs. We compare different attention methods using ResNet-18 and ResNet-34 on ImageNet-1K. The evaluation metrics include both efficiency (i.e., network parameters, floating point operations per second (FLOPs)) and effectiveness (i.e., Top-1/Top-5 accuracy). The results are listed in Table 5, where the results of ResNet, SENet, CBAM, and ECA-Net are duplicated from [18]. Table 5 shows that QSAN improves the original ResNet-18 and ResNet-34 over 0.46% and 0.96% in Top-1 accuracy, respectively. Compared with SENet, CBAM, and ECA-Net, our QSAN achieves better performance.

Table 5. Comparison of different methods using ResNet-18 (R-18) and ResNet-34 (R-34) on ImageNet-1K in terms of network parameters (Param.), floating point operations per second (FLOPs), and Top-1/Top-5 accuracy (in %).

Method	CNNs	Years	Param.	GFLOPs	Top-1	Top-5
ResNet	R-18	CVPR 2016	11.148M	1.699	70.40	89.45
SENet [10]		CVPR 2018	11.231M	1.700	70.59	89.78
CBAM [20]		ECCV 2018	11.234M	1.700	70.73	89.91
ECA-Net [18]		CVPR 2020	11.148M	1.700	70.78	89.92
QSAN (ours)			11.231M	1.700	**70.86**	**89.92**
ResNet	R-34	CVPR 2016	20.778M	3.427	73.31	91.40
SENet [10]		CVPR 2018	20.938M	3.428	73.87	91.65
CBAM [20]		ECCV 2018	20.943M	3.428	74.01	91.76
ECA-Net [18]		CVPR 2020	20.778M	3.428	74.21	91.83
QSAN (ours)			20.938M	3.428	**74.27**	**92.02**

ResNet-50. ResNet-50 is one of the most widely adopted backbones, and many attention modules report their performance on ResNet-50. We compare our QSAN with several state-of-the-art attention methods using ResNet-50 on ImageNet-1K, including SENet [10], SRM [14], CBAM [20], A^2-Nets [3], ECA-Net [18], GCT [21] and FcaNet [17]. For comparison, we report the results of other compared methods in their original papers, except FcaNet. As shown in Table 6, compared with state-of-the-art counterparts, QSAN obtains better or more competitive results.

Recently, most of the state-of-the-art vision models have adopted the Transformer architecture. To boost the performance, these works resort to large-scale pre-training and complex training settings, resulting in excessive demands of

Table 6. Comparison of different methods using ResNet-50 (R-50) on ImageNet-1K in terms of network parameters (Param.), floating point operations per second (FLOPs), and Top-1/Top-5 accuracy (in %). *: FcaNet uses more advanced training schedules, such as cosine learning rate decay and label smoothing. We reimplement it with the same settings as the other methods for a fair comparison.

Method	CNNs	Years	Param.	GFLOPs	Top-1	Top-5
ResNet	R-50	CVPR 2016	24.37M	3.86	75.20	92.52
SENet [10]		CVPR 2018	26.77M	3.87	76.71	93.39
A^2-Nets [3]		NeurIPS 2018	25.80M	6.50	77.00	93.69
CBAM [20]		ECCV 2018	26.77M	3.87	77.34	93.69
SRM [14]		ICCV 2019	25.62M	3.88	77.13	93.51
ECA-Net [18]		CVPR 2020	24.37M	3.86	77.48	93.68
GCT [21]		CVPR 2020	24.37M	3.86	77.30	**93.70**
FcaNet-LF* [17]		ICCV 2021	26.77M	3.87	77.32	93.55
QSAN (ours)			26.77M	3.87	**77.57**	93.56

data, computing, and sophisticated tuning of many hyperparameters. Therefore, it is unfair to judge an architecture by its performance alone. We employ rsb-A2 [19], a strong training procedure, to train QSAN-ResNet50 and ResNet50. Due to hardware limitations, we adjusted the minibatch to 512 according to the Linear Scaling Rule [8], which inevitably led to performance reduction. All models were trained for 300 epochs without extra data. As shown in Table 7, the strong training procedure reduces the performance gap between QSAN-ResNet50 and ResNet50, but QSAN-ResNet50 still outperforms ResNet50. In addition, the performance of QSAN-ResNet50 is competitive with ViT-SAM, especially when the number of parameters is close.

Table 7. Comparison of different architectures on ImageNet-1K in terms of network parameters (Param.), and Top-1 accuracy (in %). *: The result was reimplemented using the code provided by the open source toolbox MMClassification [15].

Method	Years	Resolution	Param.	Top-1
ViT-S/16-SAM [2]	ICLR 2022	224×224	22M	78.1
ViT-B/16-SAM [2]	ICLR 2022	224×224	87M	79.9
ResNet50 (rsb-A2)* [19]	NeurIPS 2021 Workshop	224×224	26M	79.2
QSAN-ResNet50 (ours)		224×224	27M	79.6

5 Conclusion

In this paper, we extend QSM to QE and introduce its two specific applications. One is to modify the SOM similarity measure by using the wave function of one-dimensional particles in an infinite square well. The other is to use the relative position distribution probability of two identical free particles in multidimensional space to generate more efficient channel attention. Future work will focus on exploring the application of time-dependent wave functions and quantum effects to QE.

References

1. Bausch, J.: Recurrent quantum neural networks. In: NeurIPS, pp. 1368–1379 (2020)
2. Chen, X., Hsieh, C.J., Gong, B.: When vision transformers outperform resnets without pre-training or strong data augmentations. In: ICLR, pp. 869 (2022)
3. Chen, Y., Kalantidis, Y., Li, J., Yan, S., Feng, J.: A^2-nets: Double attention networks. In: NeurIPS, pp. 350–359 (2018)
4. Cottrell, M., Olteanu, M., Rossi, F., Villa-Vialaneix, N.: Theoretical and applied aspects of the self-organizing maps. In: Merényi, E., Mendenhall, M.J., O'Driscoll, P. (eds.) Advances in Self-Organizing Maps and Learning Vector Quantization. AISC, vol. 428, pp. 3–26. Springer, Cham (2016). https://doi.org/10.1007/978-3-319-28518-4_1
5. Dosovitskiy, A., et al.: An image is worth 16 × 16 words: transformers for image recognition at scale. In: ICLR (2021)
6. Gao, Z., Xie, J., Wang, Q., Li, P.: Global second-order pooling convolutional networks. In: CVPR, pp. 3024–3033 (2019)
7. Garg, D., Ikbal, S., Srivastava, S.K., Vishwakarma, H., Karanam, H.P., Subramaniam, L.V.: Quantum embedding of knowledge for reasoning. In: NeurIPS, pp. 5595–5605 (2019)
8. Goyal, P., et al.: Accurate, large minibatch SGD: training imagenet in 1 hour. CoRR abs/1706.02677 (2017). http://arxiv.org/abs/1706.02677
9. Griffiths, D.J.: Introduction to quantum mechanics. Am. J. Phys. **63**(8), 1–12 (2005)
10. Hu, J., Shen, L., Sun, G.: Squeeze-and-excitation networks. In: CVPR, pp. 7132–7141 (2018)
11. Kerenidis, I., Landman, J., Luongo, A., Prakash, A.: q-means: a quantum algorithm for unsupervised machine learning. In: NeurIPS, pp. 4136–4146 (2019)
12. Kerenidis, I., Landman, J., Prakash, A.: Quantum algorithms for deep convolutional neural networks. In: ICLR (2020)
13. Kohonen, T.: Self-organized formation of topologically correct feature maps. Biol. Cybern. **43**(1), 59–69 (1982)
14. Lee, H., Kim, H., Nam, H.: SRM: a style-based recalibration module for convolutional neural networks. In: ICCV, pp. 1854–1862 (2019)
15. MMClassification Contributors: Openmmlab's image classification toolbox and benchmark. https://github.com/open-mmlab/mmclassification (2020)
16. Olivas, E.S., Guerrero, J.D.M., Martinez-Sober, M., Magdalena-Benedito, J.R., Serrano, L., et al.: Handbook of Research on Machine Learning Applications and Trends: Algorithms, Methods, and Techniques. IGI Global (2009)
17. Qin, Z., Zhang, P., Wu, F., Li, X.: FCANet: frequency channel attention networks. In: ICCV, pp. 763–772 (2021)
18. Wang, Q., Wu, B., Zhu, P., Li, P., Zuo, W., Hu, Q.: ECA-Net: efficient channel attention for deep convolutional neural networks. In: CVPR, pp. 11531–11539 (2020)

19. Wightman, R., Touvron, H., Jegou, H.: Resnet strikes back: an improved training procedure in TIMM. In: NeurIPS 2021 Workshop on ImageNet: Past, Present, and Future (2021)
20. Woo, S., Park, J., Lee, J.-Y., Kweon, I.S.: CBAM: convolutional block attention module. In: Ferrari, V., Hebert, M., Sminchisescu, C., Weiss, Y. (eds.) ECCV 2018. LNCS, vol. 11211, pp. 3–19. Springer, Cham (2018). https://doi.org/10.1007/978-3-030-01234-2_1
21. Yang, Z., Zhu, L., Wu, Y., Yang, Y.: Gated channel transformation for visual recognition. In: CVPR (2020)
22. Yosinski, J., Clune, J., Bengio, Y., Lipson, H.: How transferable are features in deep neural networks? In: Proceedings of the 27th International Conference on Neural Information Processing Systems - Volume 2, pp. 3320–3328 (2014)
23. Zhang, J., et al.: An application of quantum mechanics to attention methods in computer vision. In: ICASSP (2023)

Traffic Accident Forecasting Based on a GrDBN-GPR Model with Integrated Road Features

Guangyuan Pan[1,2], Xiuqiang Wu[1], Liping Fu[2], Ancai Zhang[1], and Qingguo Xiao[1](✉)

[1] Linyi University, Linyi 276000, Shandong, China
xiaoqingguo@lyu.edu.cn
[2] University of Waterloo, Waterloo, ON N2L3G1, Canada

Abstract. Traffic accidents pose a significant challenge in modern society, leading to substantial human loss and economic damage. Therefore, accurate forecasting of such accidents holds a paramount importance in road safety status evaluation. However, models in many studies often prioritize individual factors like accuracy, stability, or anti-interference ability, rather than considering them comprehensively. Toward this end, this study presents a novel traffic accident forecasting model, known as the Gaussian radial Deep Belief Net - Gaussian Process Regression (GrDBN-GPR). This model integrates feature engineering and predictive algorithms to capture the intricate relationships among various traffic factors. This model comprises two key components: firstly, the GrDBN uses the Gaussian-Bernoulli Restricted Boltzmann Machine (GBRBM) and the Gaussian activation functions to extract valuable features more effectively and stably. This feature extraction mechanism enhances the ability to uncover meaningful patterns within the data. Secondly, the GPR can achieve stable predictions based on the extracted informative features achieved by GrDBN. Finally, this model is applied to evaluate the road safety status of Highway 401 in Ontario, Canada, using a set of collision data collected for over eight years. In comparison to six commonly used benchmark models, the predictive accuracy, stability, and resistance to interference of the proposed model are evaluated.

Keywords: Machine learning · Neural network · Road safety · Forecasting · GPR · DBN

1 Introduction

Forecasting traffic accidents plays an important role in analyzing accident trends and related factors in specific traffic situations. Moreover, it serves as a valuable method to conduct road safety research, evaluate the effectiveness of safety measures, and estimate the expected safety level in specific locations under given conditions. Therefore, by effectively predicting road accidents, it is possible to reduce their occurrence.

Furthermore, the road Safety Performance Function (SPF), a representative example, was introduced in the first edition of the Highway Safety Manual (HSM) by the National

B. Luo et al. (Eds.): ICONIP 2023, CCIS 1965, pp. 178–190, 2024.
https://doi.org/10.1007/978-981-99-8145-8_15

Highway and Transportation Association of the United States in 2010. Traditionally, various statistical methods have been employed as road accident prediction models. Among them, Generalized Linear Models (GLMs), such as negative binomial regression models and Poisson regression models, have been widely used [1–7]. In addition, other statistical models have been also used for collision prediction, including Bayesian models [8], autoregressive models, moving average models, Autoregressive Integrated Moving Average (ARIMA) models, and hidden Markov models [9, 10]. For instance, in Tennessee USA, a model was developed and implemented by the state to predict daily accident risk [11]. By forecasting accidents for the upcoming week, the model provides valuable insights for managers to take preventive measures, resulting in a significant reduction in traffic accident fatalities.

Along with the rapid development of intelligent transportation systems, the amount of data obtained is constantly increasing, and Machine Learning (ML) models are getting more attention than before. Therefore, several scholars have researched accident prediction issues based on ML models, such as the Support Vector Machine (SVM) [12, 13], the Random Forests (RF) [14], the gradient boosting decision trees [15], and the Deep Neural Networks (DNNs) [16–18].

In recent years, Deep Learning (DL), a subset of ML, has gained widespread application in traffic accident prediction, emerging as the dominant trend in ML across various application domains. For instance, Deep Belief Networks (DBNs), a classic DL model, have also shown excellent performance in traffic accident prediction due to the combination of both the unsupervised and the supervised learning processes [18]. Moreover, Pan et al. introduced a visual feature importance method to identify and remove noise features, leading to significant improvements in model performance [19]. Moreover, Franco et al. employed a CNN architecture and Deep Convolutional Generative Adversarial Networks (DCGAN) technology with random under-sampling, achieving better performance in accident prediction [17].

However, in real cases, traffic accident data is often in tabular format, representing heterogeneous data. Despite the remarkable performance of DNNs in various domains, recent studies have shown that they are not the optimal choice for tabular datasets, unlike homogeneous datasets such as images and speech [20–22].

To sum up, traditional modeling methods were initially used for traffic accident prediction due to the challenges in collecting and obtaining data. While these methods have a solid theoretical foundation and are easy to implement, they have limitations when dealing with big data problems, missing values, and outliers as well as transferring knowledge to other datasets [12]. As modern transportation has evolved, data is available in a bigger quantity, leading to the widespread adoption of data-driven methods. Compared to traditional theory-based and experience-based approaches, data-driven methods focus on patterns and regularities within the data itself. Moreover, they can uncover information and knowledge that traditional methods may overlook [17]. For example, data-driven methods encompass both traditional ML and DL techniques. Traditional ML models are relatively straightforward and can handle big data to some extent. Their performance is acceptable in small-sample learning scenarios. However, for complex tasks, ML methods often require extensive feature engineering and hyperparameter tuning, which can be time-consuming. Additionally, their performance evolution with the increasing dataset volume is not as significant as those of DL. On the other hand, DL models can automatically extract and select features, reducing therefore the need for extensive feature engineering. They excel in handling large-scale, high-dimensional data and tackling complex tasks. However, DL methods rely heavily on a large amount of training data and may not perform effectively when dealing with limited data. Their performance is also highly influenced by parameters' initialization, such as the initial weights. Moreover, since traffic accident data is typically tabular and heterogeneous, it may contain noise or irrelevant features. As a result, the performance of DL methods on such data is not particularly promising.

Considering the aforementioned studies and the need for accurate and stable prediction method for traffic accidents, this paper proposes a new model called the Gaussian radial Deep Belief Net - Gaussian Process Regression (GrDBN-GPR). This model combines an improved version of DBN with GPR to overcome the limitations of existing methods. Therefore, the main contributions of this work are as follows:

(1) On the theoretical side, the model controls the continuously restricted Boltzmann machines within DBN to minimize data loss during the unsupervised learning process and employs some Gaussian activation functions to enhance the model's stability performance during the supervised learning process. Then, the integrated features are incorporated with GPR technology for regression prediction.
(2) In engineering applications, the study collects real-world traffic accident data spanning over eight years. Through a series of experiments, the model was rigorously tested for prediction accuracy, stability, and sensitivity-to-noise features analysis.

2 Methodology

2.1 Gaussian Radial Deep Belief Network

The DBN is a hierarchical model composed of multiple layers of Restricted Boltzmann Machines (RBMs) [23]. In traditional RBMs, the visible and hidden layers consist of binary units, i.e., their values are either 0 or 1. However, some input variables are continuous; therefore, binary unit models are not suitable for handling continuous data. To address this issue, the traditional RBM model is extended to incorporate these continuous values by introducing independent Gaussian distributions. This modification restricts the output values of the RBM's structure units to continuously vary between 0 and 1. This variation of RBM is known as the Gaussian-Bernoulli Restricted Boltzmann Machine (GBRBM), consisting of a visible layer with real-valued nodes and another hidden layer with Boolean nodes. Furthermore, Fig. 1(a) depicts the schematic diagram of the GBRBM.

The learning process of GBRBM is facilitated through an energy function that is mathematically defined as follows:

$$E(v, h; \theta) = \sum_{i=1}^{D} \frac{(v_i - b_i)^2}{2\sigma_i^2} - \sum_{i=1}^{D}\sum_{j=1}^{F} \frac{v_i}{\sigma_i} W_{ij} h_j - \sum_{j=1}^{F} a_j h_j \quad (1)$$

where v represents the visible units, h indicates the hidden units, and $\theta = \{a, b, W, \sigma\}$ presents the complete set of model parameters in RBM. Moreover, parameter a represents the bias terms associated with the hidden units, parameter b indicates the bias terms associated with the visible units, W represents the weight matrix that connects the visible and hidden units, and, finally, σ shows the standard deviation of the Gaussian activation function used in this model.

Furthermore, the probability density functions of the visible and hidden layers are given by:

$$P(v;\theta) = \frac{1}{Z(\theta)} \sum_h \exp(-E(v,h;\theta)) \tag{2}$$

$$Z(\theta) = \int_v \sum_h \exp(E(v,h;\theta))dv \tag{3}$$

where $P(v;\theta)$ represents the probability density of the visible layer with respect to the model parameters θ. The normalization constant $Z(\theta)$ ensures that the probability distribution sums up to the unit. It is calculated by integrating all possible values of the visible layer while summing all possible configurations of the hidden layer.

Therefore, the conditional distribution can be given by the following equations:

$$P(h|v;\theta) = \prod_{j=1}^{F} p(h_j v) \tag{4}$$

$$p(h_j = 1|v) = sigmoid\left(\sum_{i=1}^{D} W_{ij}\frac{v_i}{\sigma_i} + a_j\right) \tag{5}$$

$$P(v|h;\theta) = \prod_{i=1}^{D} p(v_i h) \tag{6}$$

$$v_i|h \sim \mathcal{N}\left(b_i + \sigma_i \sum_{j=1}^{F} W_{ij}h_j, \sigma_i^2\right) \tag{7}$$

where $N(\mu, \sigma^2)$ is a Gaussian distribution having a mean μ and a variance σ^2. With predefined training samples, the model parameters can be optimized using pre-training methods.

Figure 1(b) shows the structure of the Gr-DBN, where the bottom and top layers represent, respectively, the input and the output layers. The training process of GrDBN starts by inserting the training data into the first layer; then, the unsupervised pre-training calculates the parameters of the hidden layer of the first GBRBM. The output of this hidden layer serves as an input for the subsequent RBM's hidden layer. This iterative process continues until reaching the final hidden layer. Finally, a regression layer is added at the top to enable the regression prediction. Once the unsupervised training phase is completed, the entire network parameters undergo supervised fine-tuning using the Back Propagation (BP) and the gradient descent techniques. The specific process can be outlined as shown in Fig. 1.

In this study, the GrDBN method that incorporates GBRBM, is used to extract hierarchical features from the data. Furthermore, the use of a Gaussian activation function

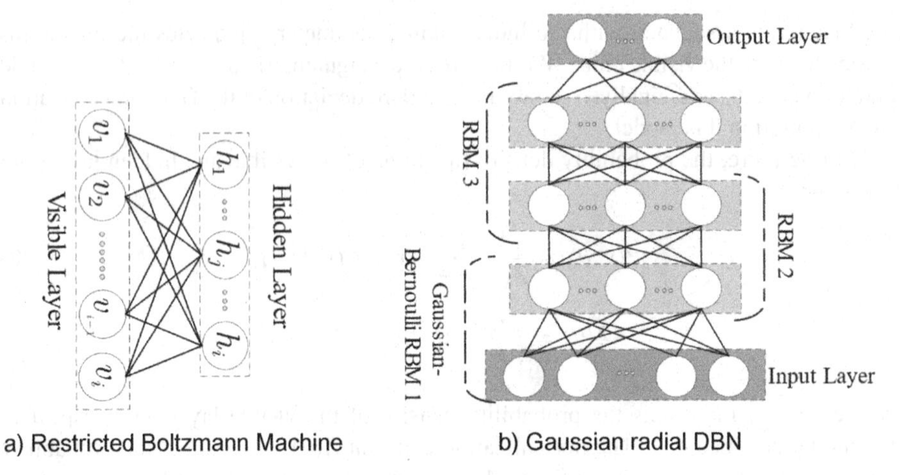

Fig. 1. Diagram of the improved DBN

(refer to Eq. (8)) in the BP process provides additional benefits in terms of robustness and accuracy. The relation expression is expressed as follows:

$$f(x) = e^{-x^2} \tag{8}$$

2.2 Gaussian Process Regression

GPR is a supervised learning algorithm that can determine the relationship between the input and the output variables based on empirical data. It, not only predicts the expected values of unknown variables, but also provides information about their distribution.

In GPR, assuming that the observed target variable y is contaminated with noise ε_i, which differs from the true function output $f(x_i)$, the following relationship can be determined:

$$y = f(x_i) + \varepsilon_i \tag{9}$$

where ε_i is an independent random variable with a Gaussian distribution, having a zero mean and variance σ_n^2.

The matrix form of the covariance function is:

$$C(X, X) = K(X, X) + \sigma_n^2 I_n \tag{10}$$

where I_n is an n-dimensional identity matrix, $K(X, X)$ is an n × n kernel matrix where $K_{ij} = k(x_i, x_j)$ representing the correlation between x_i and x_j measured by the kernel function, and σ_n^2 is the variance of the n-dimensional data sample.

The joint Gaussian distribution between the observed values y of the training samples and the output vector f_* of the testing data is expressed as follows:

$$\begin{bmatrix} y \\ f_* \end{bmatrix} \sim N\left(0, \begin{bmatrix} K(X, X) + \sigma_n^2 I_n & K(X, x_*) \\ K(x_*, X) & K(x_*, x_*) \end{bmatrix}\right) \tag{11}$$

where $K(X, x_*) = K(x_*, X)^T$ as it measures the n × 1 covariance matrix between the test point x_* and the training set X; moreover, $k(x_*, x_*)$ is the covariance matrix of x_* itself.

The squared exponential kernel is the most common kernel function [24], given by Eq. (12), where σ_f and l are two adjustable parameters that must be greater than zero; therefore, one can have:

$$\kappa_{ij} = \sigma_f^2 \cdot exp\left(-\frac{(x_i - x_j)^2}{2l}\right) \tag{12}$$

Using the above equations, the posterior distribution of the predicted values y can be calculated as follows:

$$f_*|y \sim N\left(\overline{f_*}, \sigma_{f_*}^2\right) \tag{13}$$

$$\overline{f_*} = K(x_*, X)_*[K(X, X) + \sigma_n^2 I_n]^{-1} y \tag{14}$$

$$\sigma_{f_*}^2 = K(x_*, x_*) - K(x_*, X)[K(X, X) + \sigma_n^2 I_n]^{-1} K(X, x_*) \tag{15}$$

The point prediction result of the GPR is the mean of the posterior distribution, denoted as $\overline{f_*}$.

2.3 Proposed Method

The proposed method combines GrDBN with GPR to achieve reliable and probabilistic predictions. Firstly, the GrDBN utilizes the Gaussian function as an activation function, inspired by the Radial Basis Function Network and overcoming the limitations of the traditional RBMs when dealing with continuous input data. During the pre-training phase, a GBRBM is employed whereas, during the fine-tuning phase, the Gaussian function is used as the activation function. This combination enables the GrDBN to effectively capture and analyze the features from the data distribution without the need for manual feature selection and extraction. The hierarchical structure of the GrDBN allows it to extract the features at different levels, enhancing its ability to represent the data complex structure.

Furthermore, the integration of GrDBN with GPR improves the prediction performance of the model. Compared to neural networks that can be sensitive to parameter initialization, GPR offers a more stable and accurate prediction. The core idea of this approach is to first train a DNN on the training set and use the GrDBN to extract features from both the training and test sets. The features extracted from the last hidden layer of the GrDBN are then used as inputs for GPR, which is trained using original labels. Finally, the trained GPR model is used to predict the output using the features extracted from the test set.

Therefore, the specific process, presented in Fig. 2, is as follows:

(1) Train a GrDBN: The GrDBN model is trained using the training data in a layered process. The first layer network, the RBM, extracts the low-level features from the

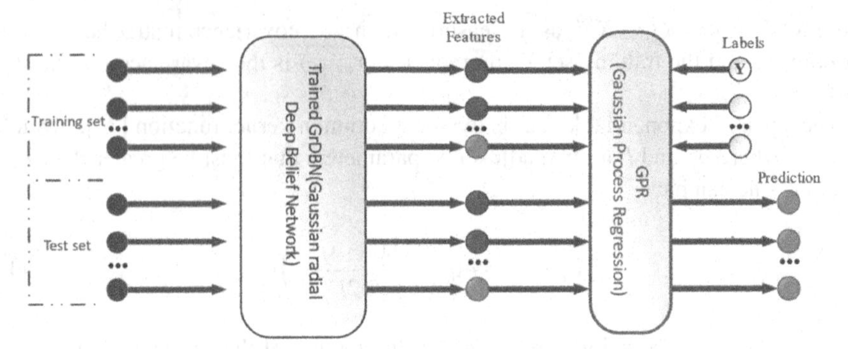

Fig. 2. Flowchart of GrDBN-GPR

input data. The output of the first layer is then used as input to train the second layer network, which captures the higher-level abstract features. This process continues for subsequent layers until the entire DBN network is trained. Finally, fine-tuning is performed using BP to further optimize the network;

(2) Feature extraction: Once the GrDBN is trained, it can be used to extract features from both the training and the test datasets. The output of each layer network represents the feature representation of the input data. Typically, the output of the last layer contains the highest-level abstract features, which are used as input features for the regressor. In this study, the features extracted from the last layer are used for further analysis.

The extracted feature F' can be obtained using the following formula:

$$F' = g\left(w_2^T * g\left(w_1^T x + b_1\right) + b_2\right) \tag{16}$$

where w and b stand for the weight and the bias parameters in the DBN, respectively. The subscripts indicate the layer number, and g(x) denotes the activation function used.

(3) Train the GPR: In the final step, GPR is trained using the features extracted from the DBN and the original labels.

3 Case Study

3.1 Data Description and Preprocessing

To assess the effectiveness of the GrDBN-GPR model, an empirical study was conducted using data obtained from Highway 401 provided by the Ministry of Transportation Ontario (MTO) in Canada.

To facilitate the analysis, homogeneous road segments, with similar road features, are classified into 418 sets. The collision and traffic data of eight years is combined to create a unified dataset, comprising 3,762 entries of collision, road geometry, and traffic data from Ontario, Canada. Moreover, Table 1 offers a concise overview of the variables present in the dataset. The target variable of interest is the "accident". In addition, Fig. 3 illustrates the histogram of all the used variables, consisting of six

Table 1. Description of Variables

Feature No	Feature Name (variables)	Description
1	Exposure	Degree of exposure of the road
2	AADT_Comm	Annual average daily commercial traffic volume (vehicles per day)
3	ShldLeft	Shoulder width on the left side of the highway (in meter)
4	MedianWidth	Median width (in meter)
5	ShldRight	Shoulder width on the right side of the highway (in meter)
6	CurveDeflection	Horizontal curve deflection (per km)
7	Accident	Number of accidents per year

feature variables and one target variable. The horizontal axis represents the variable values, while the vertical axis denotes the frequency of their occurrences. This figure shows that the variables are not uniformly distributed, with data points being concentrated within specific intervals. As a preprocessing step, all variables were normalized (ranging between 0 and 1). Following the training phase, the performance of each model was evaluated using two metrics: Mean Absolute Error (MAE) and Root Mean Squared Error (RMSE), as defined in Eqs. (17) and (18), respectively. They are represented as follows:

$$\mathrm{MAE} = \frac{1}{n}\sum_{i=1}^{n}|y_i - Y_i| \tag{17}$$

$$\mathrm{RMSE} = \sqrt{\frac{1}{\mathrm{n}}\sum_{\mathrm{i}=1}^{\mathrm{n}}(\mathrm{y_i} - \mathrm{Y_i})^2} \tag{18}$$

where y_i is the true label of the ith input and Y_i represents the prediction or estimated value of the i th input. Moreover, these metrics are commonly used to assess the accuracy of the prediction models and provide valuable insights into their performance.

3.2 Experiment 1

In this experiment, seven models are compared: GrDBN-GPR, Back Propagation Neural Network (BPNN), DBN, GPR, RF, Gradient-Boosted Decision Trees (GBDT), and SVM. The main objective is to evaluate the accuracy and the stability of GrDBN-GPR and to examine the impact of the data volume on the model performance.

Consequently, Table 2 and Table 3 provides a comparison of performance metrics between the proposed model and other methods. Furthermore, Fig. 4(a, b) depict the performance index obtained by each model regarding the test set as a function of the percentage of the training set used. The performance index includes metrics, such as the

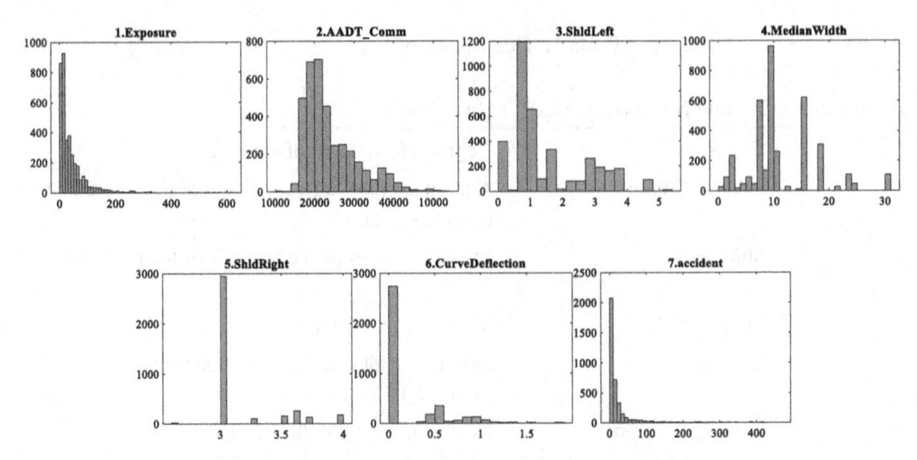

Fig. 3. Frequency distribution of the seven variables

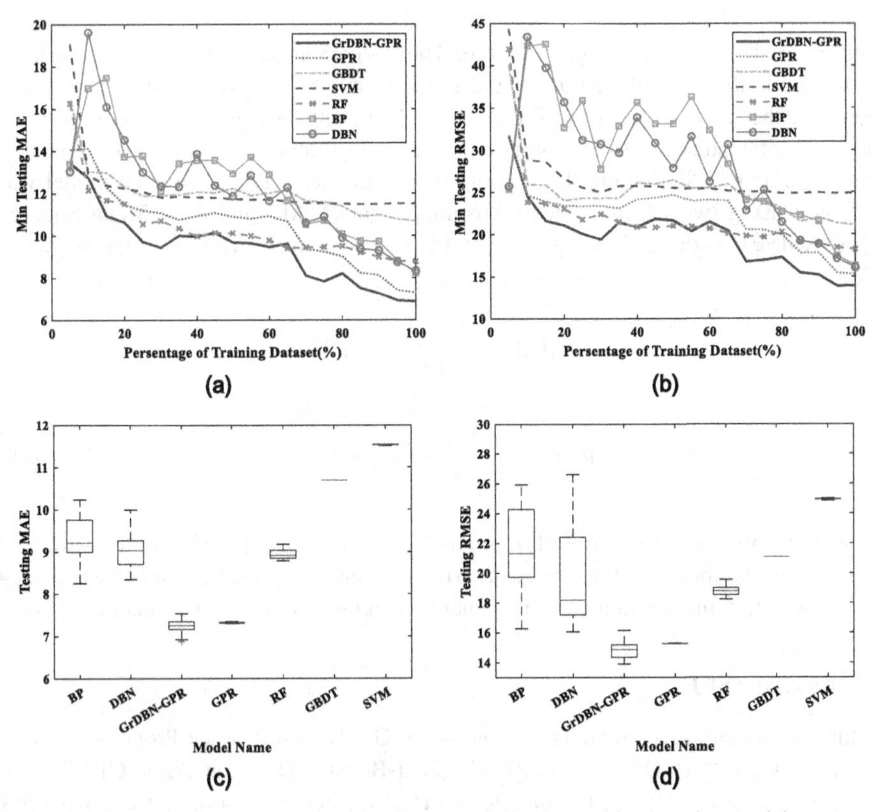

Fig. 4. Performance comparison between different machine learning models

minimum MAE, the minimum RMSE, the average MAE, and the average RMSE, derived from 20 repeated experiments. Moreover, Fig. 4(c, d) present the MAE and RMSE values of all seven models on the test set when all training sets are used. The box plot displays the distribution of the experimental results from the 20 repeated experiments. In general, a flatter box plot suggests a better stability model, indicating consistent performance or prediction results across repetitive experiments.

According to Fig. 4, the below conclusions can be proposed: (1) The GrDBN-GPR model achieved high accuracy, as indicated by the minimum MAE and RMSE, mean MAE, and RMSE and consistently outperformed other models in terms of average MAE and RMSE across multiple experiments; (2) The proposed model exhibited high stability, as evidenced by the box plots. Although it showed higher stability compared to BPNN and DBN, there is still room for improvement compared to some ML models; (3) **The GrDBN-GPR model, not only demonstrated strong performance with large amounts of training data, but it also exhibited good performance even when trained with limited data.**

3.3 Experiment 2

To test the model's resistance to interference, five sets of random numbers are generated at this level to retrieve the interference features. These features are uniformly distributed between 1 and 10, and they are denoted as Extra_F1, Extra_F2, ..., and Extra_F5. To facilitate replication, random number seeds 1–5 are used to generate these interference features. These latter are combined with the original dataset D_0 to generate five new datasets, denoted as D_1, D_2, D_3, D_4 and D_5. RF are trained on all six datasets to obtain the feature importance. The same comparison model, used for Experiment 1, is trained and tested on these datasets, and the results are shown in Tables 2 and 3.

Furthermore, Tables 2 and 3 provide the performance metrics for the proposed model and other models when being applied to the datasets with interference features. The proposed GrDBN-GPR consistently exhibits the best performance among all models, both before and after adding interference features. For example, even though GPR initially performs well, its performance significantly deteriorates after adding the interference features. Conversely, GrDBN-GPR consistently achieves an optimal performance in

Table 2. Min. MAE of seven models

	GrDBN_GPR	BP	DBN	GPR	SVM	GBDT	RF
D_0	**6.87**	8.25	8.33	7.30	11.51	10.69	8.78
D_1	**6.79**	9.03	8.84	9.80	11.51	10.69	8.61
D_2	**7.64**	9.01	9.08	10.52	11.54	10.63	8.67
D_3	**7.04**	9.04	8.85	10.07	11.51	10.67	8.64
D_4	**6.88**	9.28	8.54	10.29	11.41	10.60	8.74
D_5	**7.19**	8.77	9.06	10.52	11.48	10.71	8.74

terms of minimum MAE and minimum RMSE, regardless of the presence of interference features. This may be due to GrDBN's feature extraction phase, which filters out the interference features and learns useful features.

Table 3. Min. RMSE of seven models

	GrDBN_GPR	BP	DBN	GPR	SVM	GBDT	RF
D_0	**13.90**	16.27	16.05	15.24	24.85	21.10	18.22
D_1	**13.23**	18.45	18.29	20.36	24.93	21.10	17.92
D_2	**15.52**	17.98	17.59	22.78	24.99	20.93	17.95
D_3	**14.76**	19.12	18.32	22.27	24.80	21.12	18.01
D_4	**13.50**	18.63	16.98	22.58	24.59	21.09	18.15
D_5	**14.55**	16.56	18.10	23.50	24.67	21.30	18.05

4 Conclusions

The proposed method, GrDBN-GPR, demonstrates significant potential in accurately and reliably predicting traffic accidents, which is crucial for traffic planning and decision-making. By incorporating the GBRBM and Gaussian activation function, GrDBN enhances feature extraction capability and ensures the extraction of effective features in a stable environment. Moreover, the GPR is employed to predict these extracted features, resulting in stable and highly accurate predictions.

The application of GrDBN-GPR to predict the number of accidents on Ontario's Highway 401 in Canada involved comparing it to six other models and evaluating its performance using the RMSE and MAE validation indicators. The experimental findings show that GrDBN-GPR consistently delivers accurate predictions while being robust against noise. Moreover, the proposed method showcases promising performance even with limited data, making it suitable for scenarios with small dataset volumes.

It is important to acknowledge that traffic accident datasets frequently suffer from imbalanced data, *e.g.*, there is a significant disparity in the number of instances for different classes or outcomes. Moreover, these datasets are typically structured in a tabular form, showing heterogeneity in the data, which can be challenging for accurate modeling. It is worth noting that the current study did not explicitly address these issues; however, future research can concentrate on enhancing the modeling approach to specifically tackle these challenges.

Acknowledgement. This research is supported by National Natural Science Foundation of China under grant no. 62103177, Shandong Provincial Natural Science Foundation for Youth of China under grant no. ZR2023QF097 and the National Science and Engineering Research Council of Canada (NSERC) under grant no. 651247. The authors would also like to thank the Ministry of Transportation Ontario Canada for technical support.

References

1. Tang, H., Gayah, V.V., Donnell, E.T.: Evaluating the predictive power of an SPF for two-lane rural roads with random parameters on out-of-sample observations. Accid. Anal. Prev. **132**, 105275 (2019)
2. Gao, L., Lu, P., Ren, Y.: A deep learning approach for imbalanced crash data in predicting highway-rail grade crossings accidents. Reliab. Eng. Syst. Saf. **216**, 108019 (2021)
3. Gayah, V.V., Donnell, E.T.: Estimating safety performance functions for two-lane rural roads using an alternative functional form for traffic volume. Accid. Anal. Prev. **157**, 106173 (2021)
4. Part D: Highway Safety Manual. American Association of State Highway and Transportation Officials: Washington (2010)
5. Lim, K.K.: Analysis of railroad accident prediction using zero-truncated negative binomial regression and artificial neural network model: a case study of national railroad in South Korea. KSCE J. Civ. Eng. **27**(1), 333–344 (2023)
6. Champahom, T., et al.: Spatial zero-inflated negative binomial regression models: application for estimating frequencies of rear-end crashes on Thai highways. J. Transp. Safety Secur. **14**(3), 523–540 (2022)
7. Tang, H., Gayah, V.V., Donnell, E.T.: Evaluating the predictive power of an SPF for two-lane rural roads with random parameters on out-of-sample observations. Accident Anal. Prevent. **132**, 105275 (2019)
8. Fawcett, L., Thorpe, N., Matthews, J., et al.: A novel Bayesian hierarchical model for road safety hotspot prediction. Accid. Anal. Prev. **99**, 262–271 (2017)
9. Ihueze, C.C., Onwurah, U.O.: Road traffic accidents prediction modeling: an analysis of Anambra State, Nigeria. Accident Anal. Prevent. **112**, 21–29 (2018)
10. Li, J., He, Q., Zhou, H., Guan, Y., Dai, W.: Modeling driver behavior near intersections in hidden Markov model. Int. J. Environ. Res. Public Health **13**(12), 1265 (2016)
11. Dedon, L., Gander, S., Hill, K., Kambour, A., Locke, J., McLeod, J.: State Strategies to Reduce Highway and Traffic Fatalities and Injuries: A Road Map for States (2018)
12. Mokhtarimousavi, S., Anderson, J.C., Azizinamini, A., Hadi, M.: Improved support vector machine models for work zone crash injury severity prediction and analysis. Transp. Res. Rec. **2673**(11), 680–692 (2019)
13. Dong, C., Xie, K., Sun, X., Lyu, M., Yue, H.: Roadway traffic crash prediction using a state-space model based support vector regression approach. PLoS ONE **14**(4), e0214866 (2019)
14. Zhou, X., Lu, P., Zheng, Z., Tolliver, D., Keramati, A.: Accident prediction accuracy assessment for highway-rail grade crossings using random forest algorithm compared with decision tree. Reliab. Eng. Syst. Saf. **200**, 106931 (2020)
15. Lu, P., et al.: A gradient boosting crash prediction approach for highway-rail grade crossing crash analysis. J. Adv. Transp. **2020**, 1–10 (2020)
16. Devan, P., Khare, N.: An efficient XGBoost–DNN-based classification model for network intrusion detection system. Neural Comput. Appl. **32**, 12499–12514 (2020)
17. Basso, F., Pezoa, R., Varas, M., Villalobos, M.: A deep learning approach for real-time crash prediction using vehicle-by-vehicle data. Accid. Anal. Prev. **162**, 106409 (2021)
18. Pan, G., Fu, L., Thakali, L.: Development of a global road safety performance function using deep neural networks. Int. J. Transp. Sci. Technol. **6**(3), 159–173 (2017)
19. Pan, G., Fu, L., Chen, Q., Yu, M., Muresan, M.: Road safety performance function analysis with visual feature importance of deep neural nets. IEEE/CAA J. Autom. Sinica **7**(3), 735–744 (2020)
20. Borisov, V., Leemann, T., Seßler, K., Haug, J., Pawelczyk, M., Kasneci, G.: Deep neural networks and tabular data: a survey. IEEE Trans. Neural Netw. Learn. Syst. (2022)

21. Shwartz-Ziv, R., Armon, A.: Tabular data: deep learning is not all you need. Information Fusion **81**, 84–90 (2022)
22. Grinsztajn, L., Oyallon, E., Varoquaux, G.: Why do tree-based models still outperform deep learning on tabular data? arXiv preprint arXiv:2207.08815 (2022)
23. Hinton, G.E.: Deep belief networks. Scholarpedia **4**(5), 5947 (2009)
24. Wang, J.: An intuitive tutorial to Gaussian processes regression. arXiv preprint arXiv:2009.10862 (2020)

Phishing Scam Detection for Ethereum Based on Community Enhanced Graph Convolutional Networks

Keting Yin(✉) and Binglong Ye

School of Software Technology, Zhejiang University, Ningbo, China
22251160@zju.edu.cn

Abstract. Blockchain technology has garnered a lot of interest recently, but it has also become a breeding ground for various network crimes. Cryptocurrency, for example, has suffered losses due to network phishing scams, posing a serious threat to the security of blockchain ecosystem transactions. To create a favorable investment environment, we propose a community-enhanced phishing scam detection model in this paper. We approach network phishing detection as a graph classification task and introduce a network phishing detection graph neural network framework. Firstly, we construct an Ethereum transaction network and extract transaction subgraphs, and corresponding content features from it. Based on this, we propose a community-enhanced graph convolutional network (GCN)-based detection model. It extracts more reasonable node representations in the GCN neighborhoods and explores the advanced semantics of the graph by defining community structure and measuring the similarity of nodes in the community. This distinguishes normal accounts from phishing accounts. Experiments on different large-scale real-data sets of Ethereum consistently demonstrate that our proposed model performs better than related methods.

Keywords: Phishing Scam Detection · Graph convolutional networks · Community analysis · Ethereum · Graph embedding

1 Introduction

The emergence of Bitcoin has ushered in a new era of cryptographic currencies. According to coinmarketcap.com, there are currently over 5,000 cryptographic currencies (or tokens) with a total market capitalization of over $200 billion [10]. The key technology behind these cryptographic currencies is blockchain technology.

Blockchain [24] is a distributed database that maintains a continually growing list of data records that are distributed, jointly replicated, and linked in a chain, making them tamper-proof. By leveraging blockchain technology, data is distributed globally, and people no longer need to rely on traditional third

B. Luo et al. (Eds.): ICONIP 2023, CCIS 1965, pp. 191–206, 2024.
https://doi.org/10.1007/978-981-99-8145-8_16

parties, providing a more reliable form of trust while also reducing costs associated with intermediaries. Ethereum is currently the largest blockchain platform supporting smart contracts, and its corresponding cryptocurrency, Ethereum, is the second largest cryptocurrency [30]. However, with the rapid development of Ethereum, it has also become a breeding ground for various network crimes [13]. Therefore, identifying these scams has become a pressing and critical issue in the blockchain ecosystem, and has attracted great attention from researchers. Network phishing scams have emerged with the rise of network business and have now been found in the blockchain ecosystem [17]. According to a report by Chainalysis[1], more than 50% of network crime revenue comes from phishing scams since 2017. In the first half of 2017, 30,287 victims lost $225 million, indicating that financial security has become a key issue in the blockchain ecosystem. A well-known example is the Bee Token ICO phishing scam[2], in which the phisher ultimately collected about $1 million from investors in just 25 h. These examples demonstrate that detecting and preventing phishing fraud is a pressing issue in the blockchain ecosystem.

Due to the open and transparent nature of blockchain technology, extracting information from transaction records is an intuitive way of detecting phishing fraud on the Ethereum platform. The Ethereum transaction history can be modeled as a directed transaction network, where nodes are unique addresses. When utilizing the Ethereum transaction records for fraud detection, we may face three challenges that impede the performance of fraud detection: (i) Lack of node features. The data we collect can be considered as weighted directed edges, thus there is no longitudinal information on the nodes themselves. To effectively implement a phishing fraud detection scheme, we need to extract features that can accurately differentiate between phishing and non-phishing addresses. (ii) Extreme data imbalance. One of the biggest obstacles to phishing detection on Ethereum is the data imbalance. According to a report from etherscan.io[3], a well-known Ethereum block explorer and analytics platform, there are over 500 million addresses and 3.8 billion transactions on Ethereum. In contrast, the total number of labeled phishing network addresses on etherscan is only 2,041. (iii) Massive node data. If we study the graph from a topological perspective, we will inevitably face time and space constraints.

Some researchers are dedicated to developing effective methods for detecting network phishing fraud in the Ethereum ecosystem. These methods typically involve constructing a graph structure that takes into account addresses in Ethereum and forms a graph based on the transactional connections between them. Graph embedding methods are then used to learn the network representations of Ethereum addresses [8,29,32], and further phishing detection is achieved through downstream machine learning classifiers. However, traditional graph embedding methods such as graph neural networks [21], may embed some irrelevant neighbor information on the current node, leading to a lower accu-

[1] https://blog.chainalysis.com/the-rise-of-cybercrime-on-ethereum/.
[2] https://therippleсryptocurrency.com/bee-token-scam/.
[3] https://etherscan.io/charts.

racy of the node representation, and therefore, the classification results are not satisfactory.

We are presenting the idea of community structure [22] to overcome the drawbacks of graph representation in our work. We propose a community-enhanced phishing detection model (CEGCN-PD) for detecting phishing scams on Ethereum. In our research, we first gather edge information on each node as a feature of the node. Then, we use graph convolutional networks [15] to learn the node representation of addresses in the graph. In this process, we aggregate neighbor features of the node and embed the adjacency relations in the graph structure into the node. To make the node representation of the addresses more comprehensive, we incorporate a modularity measure that can quantify the strength of community structure into the GCN to maintain community information. In this way, we merge the modularity and node representation in the loss function, making it possible to update the node representation while considering community structure information. Finally, we use the graph representation of addresses as the input of lightGBM to obtain the classification results and optimize the parameters through a new loss function. In summary, the contributions of this work are as follows: (i) Our model can well aggregate the node features and spatial structure of the network, and can easily be extended to general feature engineering. (ii) We propose a new loss function to consider community structure information in graph node representation, which will produce a more comprehensive node representation and thus better for classification. (iii) We conduct extensive experimental studies using a publicly crawled dataset from Ethereum to fully evaluate the performance of the proposed model. By comparing with a series of baseline methods, the effectiveness of the method is demonstrated.

2 Related Works

2.1 Phishing Detection Methods

In order to create a favorable investment environment in the Ethereum ecosystem, many researchers have paid close attention to the effective detection methods of network phishing frauds. Wu et al. [14] proposed a method called trans2vec, which detects network phishing frauds by mining Ethereum's transaction records. By taking into account the transaction amount and timestamp, the method uses random walks to perform network embedding and finally uses an SVM classifier to distinguish normal accounts from network phishing accounts. Chen et al. [8] proposed a method (PSDECN) based on graph convolutional networks and autoencoders for phishing detection. This method aggregates account features and network topology and uses classifiers to recognize phishing accounts in Ethereum. Li et al. [16] introduced a graph neural network-based self-supervised incremental depth graph learning model for Ethereum's phishing fraud detection problem. The aforementioned methods mainly construct phishing account detection as a node classification task, unable to capture more potential global structural features of phishing accounts. Wang et al. [29] introduced a subgraph network (SGN) mechanism into the Ethereum transaction network to expand the feature space of accounts. They constructed the transaction subgraph network (TSGN) by extracting the first-order subgraph of the accounts and used GCN and Diffpool and other graph neural networks to identify phishing accounts. However, these methods still suffer from the issue of insufficient graph embedding learning.

2.2 Network Representation Learning

Network representation learning aims to learn low-dimensional latent representations of nodes in a network while preserving network topology, node content, and other auxiliary information [34]. The learned feature representations can be used as features for various graph-based tasks such as node classification [1], graph classification [33], and link prediction [7]. One class of methods is based on random walks. Perozzi et al. [23] proposed generating node representations by using the co-occurrence probability of node sequences generated by random walks. Grover et al. [12] proposed a biased random walk to balance the trade-off between breadth-first and depth-first sampling, achieving a balance between homophily and structural equivalence. These methods use the Skip-gram model to learn node embeddings and generate node sequences by random walks on the graph. Another class of methods is based on graph structural information, such as LINE [27] and GraRep [5]. LINE uses a complex loss function to maintain first-order and second-order proximity of nodes in the embedding space, while GraRep represents each node by aggregating the embeddings of its neighbors based on the graph structure. There are also some graph generative model-based methods, such as GCN [15], and SDNE [28]. SDNE uses an autoencoder structure to optimize first-order and second-order similarity, learns to preserve vector representations of both local and global structures, and is robust to sparse networks. GCN learns node embeddings by performing multi-layer graph convolution on node features. Kipf et al. [15] proposed a network topology-based graph convolutional network that iteratively aggregates and updates node features to achieve higher representation performance. Narayanan et al. [21] proposed graph2vec which can learn graph-level representation of networks for graph classification tasks. However, these network representation learning methods all have the problem of incomplete representation of graphs, which affects the performance of network phishing detection and classification.

3 Proposed Methods

In this section, we will provide a detailed overview of some initial data preprocessing, including content features generated from feature engineering and context features obtained from subgraph extraction. Next, we propose an embedding layer based on GCN to learn graph representations of addresses. Finally, we use LightGBM for classification to obtain the results.

Table 1. Notations and explanations

Notation	Explanation
G, G_s	The original Ethereum transaction graph and subgraphs
V, V_s	The original set of Ethereum address nodes and subsets
E, E_s	The original Ethereum address transaction set and subset
X	Content features of the subgraph
A	The adjacency matrix of the subgraph
n, N	Number of address nodes of the subgraph
d	Dimension of content features
h^l	l-th layer of node representation
W^l	weights of l-th GCN layer
e	the representation of nodes
B	the modularity matrix
α	The hyperparameter that controls the impact of community structure
y_i	True classification of target addresses
$\hat{y}_i$	Predicted classification of target addresses

3.1 Problem Statement

Given a destination address and the corresponding transaction data in the Ethereum network, our goal is to determine whether that destination address is a phishing scam address. More formally, given a destination address x and its transaction data, a prediction function f is learned to determine whether the address is a phishing scam address such that $f : x \rightarrow y = \{y|y \in 1, 0\}$. The notations used in this paper are shown in Table 1.

3.2 Feature Extraction

Context Feature. The Ethereum transaction network is a multi-graph, which possesses rich context information, for example, the network structure allows for multiple edges between nodes, and each edge is directed and contains a weight. Specifically, the direction of remittance is the key information of the transaction described by the direction of edges and the weight of edges, such as the transaction amount (in units of Ether) and the timestamp of execution (here represented as an integer). The scale of the original graph is very large, so we use Algorithm 1 for sampling. It is worth noting that transaction relationships (i.e., edges in the graph) are directed, but the neighbor relationships are undirected. Our goal is to obtain a subgraph with the same distribution through random walks of neighbor relationships. Therefore, during sampling, the graph is treated as an undirected graph. Here, during the sampling phase, we directly convert the directed edges into undirected edges.

The natural choice of the sampling idea is to conduct random walks. During the walking phase, we randomly choose a node and start from it to obtain a

Algorithm 1. Node Sampling

Input:Undirected Graph G.
Sampled Size L of nodes.
Start Node V_{src}
Output: Sampled node's set S.

1: $S = set(V_{src});$
2: $U = V_{src};$
3: **while** $len(S) < L$ **do**
4: $V_{cur} = U;$
5: V_{next} = a random neighbor ofV_{cur} ;
6: $S.add(V_{next});$
7: $U = V_{next};$
8: **end while**
9: **return** S.

fixed-size subgraph. The method of walking is clear, that is, to choose one of the current node's neighbors to move forward until a fixed number of nodes have been successfully collected. Then we extract a subgraph based on these nodes.

For ease of description, we provide the definition of basic variables and the sampling algorithm used in this stage. We define the graph G(V, E) that is collected for analysis, where $V = \{v_1, v_2, \ldots, v_n\}$ represents the set of nodes, $|V| = n$, and $E \in V \times V$ represents the set of edges, $|E| = m$. Then, we further assume $G_s = (V_s, E_s)$, a subgraph of G, where each subgraph $V_s \subseteq V$ and retains all edges Es among these nodes. Using Algorithm 1, we obtain Gs and extract the node's feature content $X \in R^{n \times d}$ and the adjacency matrix $A \in R^{n \times n}$. Where n and d represent the number of nodes and the dimension of features respectively.

Content Feature. In this study, we set the dimension of node features, d, to be 8, indicating that we have extracted 8 features for each address node. The specific features and their description are as follows: (1) `FT1`. In-degree: the number of incoming edges [3], which refers to the number of ether transactions sent to the current node. On average, non-phishing node samples have a larger number of in-degree transactions, while phishing node samples have a smaller number. (2) `FT2`. Out-degree: the number of outgoing edges [3], which refers to the number of ether transactions sent by the current node to other nodes. According to statistics, the number of phishing nodes is generally greater than the number of non-phishing nodes. This is somewhat counterintuitive, and we believe it is due to the bait transactions sent by the phishing account's auxiliary accounts. (3) `FT3`. Degree: the total number of transactions sent and received by the current node [3]. (4) `FT4`. In-strength: the total amount of ETH received by the current node [2]. In other words, FT4 is the sum of the weights of the incoming edges of the node. (5) `FT5`. Out-strength: the total amount of ETH transferred from the current node to other nodes is [2]. In other words, FT5 is the sum of the weights of outgoing edges. (6) `FT6`. Strength: the total amount of ETH transferred and received by the current node, or the total transaction amount of the node is

[2]. By observing the data, we know that the values of FT4, FT5, and FT6 for phishing accounts are generally higher than those for non-phishing accounts. (7) FT7. The number of neighbors: as the data is based on a multigraph, the number of neighbors and degree (FT3) are not equal. For example, an account may have only one neighbor but the degree will be greater than the number of neighbors due to many transactions. We can see that the number of neighbors for phishing accounts is generally higher than that for normal users. This is consistent with our previous conclusion that the process of network phishing is usually to cast a wide net in order to increase the number of successful scams. (8) FT8. Reciprocal of transaction frequency: FT8 is defined as the time interval between the first and last transactions of the account divided by FT3. The smaller the FT8, the more frequent the transactions. Network phishing nodes attempt to scam more money by trying multiple times, plus transferring stolen money, resulting in a higher frequency.

After the above operations, we obtained contextual features $A \in R^{n \times n}$ and content features $X \in R^{n \times d}$.

3.3 Community Enhanced Phishing Scam Detection

After obtaining the content features X and the adjacency matrix A, we input them into a graph convolutional neural network to learn the underlying graph representation. The overall system architecture is depicted in Fig. 1. GCN is a semi-supervised graph embedding method that utilizes the Laplacian transformation to aggregate high-order neighborhood features. GCN performs convolution operations in the spectral domain, with each operation aggregating additional layers of features. We use the following GCN layers in our approach:

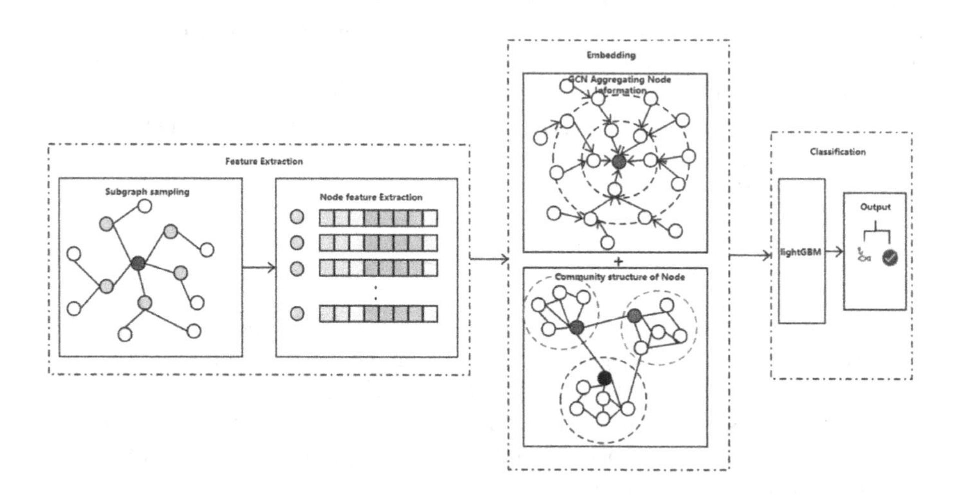

Fig. 1. The overall framework of CEGCN-PD

$$h^{l+1} = \sigma\left(\left(Ph^l\right)W^l\right) \tag{1}$$

The node representation h^{l+1} is obtained. Specifically, h^0 is the input provided at the beginning, in addition to the adjacency matrix $A.P = \tilde{D}^{-1/2}\tilde{A}\tilde{D}^{-1/2}$, the matrix $\tilde{A} = A + I$ represents the adjacency matrix A plus its identity matrix I,$\tilde{D} = diag\left(\tilde{A}\mathbf{1}\right)$, where $\mathbf{1}$ is the all-ones vector,$\tilde{D}_{i,i} = \sum_j \tilde{A}_{i,j}$, and W^l is the shared trainable weight matrix for all nodes in layer l. The activation function $\sigma(\cdot)$ is applied, where $ReLU(\cdot) = max\{0,\cdot\}$. The variable l represents the l-th layer of graph convolution. The information from adjacent nodes is aggregated through a graph convolution layer, and after passing through three layers of graph convolution, the graph representation of Ethereum address nodes is obtained, which can be represented as.

$$h^1 = ReLU\left(\left(PX\right)W^0\right) \tag{2}$$

$$h^2 = ReLU\left(\left(Ph^1\right)W^1\right) \tag{3}$$

$$h^3 = ReLU\left(\left(Ph^2\right)W^2\right) \tag{4}$$

$$e = h^3 \tag{5}$$

However, the above process ignores the measurement of community structure and fails to consider the relationships between communities. Community structure is an advanced semantic of the graph. The connections between nodes in a community are dense, while the connections between nodes in different communities are sparse. Therefore, the representation of nodes within the same community should be more similar than that of nodes belonging to different communities [18]. For example, in social networks, users from the same organization have closer contacts, while users from different organizations have farther contacts; papers within the same field or theme have more similarities in citation networks. Therefore, both the neighbor information and community structure information of the graph should be considered within a unified framework. Therefore, to make the graph representation of entities more complete, we introduce modularity, which can measure the strength of community structure, and merge it into a unified loss function for optimization. Modularity is a measure of community structure strength [22], by defining the modularity matrix $B \in R^{n\times n}$,

$$B_{ij} = A_{ij} - \frac{k_i k_j}{2r} \tag{6}$$

where A_{ij} represents the weight of the edge between node i and node j, k_i and k_j respectively represent the degrees of these two nodes, and r represents the total number of edges in this network. Modularity measures the difference between the number of edges within a community and the expected number of edges in an equivalent random network. Finally, to obtain community information, the following loss function is defined:

$$Loss_{com} = \frac{1}{2r}tr(e^T Be) \tag{7}$$

Algorithm 2. Generating the representations of nodes

Input: the given adjacency matrix $A \in R^{n \times n}$.
the content feature matrix $X \in R^{n \times m}$ of nodes.
n is number of nodes, m is number of features;
Output: The final node representations of address $e \in R^{n \times d}$.
d is the representation dimension.

1: Initialize e;
2: Calculate $\tilde{A}, \tilde{D}, P$ according to Formula 1;
3: Let $h^0 \Leftarrow X$;
4: Calculate h^1 by Formula 2;
5: Calculate h^2 by Formula 3;
6: Calculate h^3 by Formula 4;
7: Calculate modularity matrix B and $Loss_{com}$ by Formula 6 and Formula 7 ;
8: $e \Leftarrow h^3$;
9: **return** e.

The larger the value of $Loss_{com}$ the stronger the community structure strength, indicating that the quality of community partitioning is better. The algorithm framework is shown in Algorithm 2.

In the final step, the generated nodes are embedded in e as the input for the tree-based model, LightGBM. A stratified 5-fold cross-validator is used to obtain classification results. LightGBM [14] is an efficient classifier, especially suitable for handling large datasets in classification problems. It employs tree-based algorithms and leverages the idea of gradient boosting, resulting in a significant acceleration in classification speed. Additionally, LightGBM possesses excellent computational efficiency and can complete the calculation in a relatively short time, making large-scale dataset classification tasks more feasible. Furthermore, LightGBM has a high level of accuracy, performing well in many practical applications. It has an adaptive learning rate algorithm that automatically adjusts the learning rate in each step, thereby enhancing the model's generalization capability. Moreover, LightGBM also supports automatic feature selection, which can automatically identify useful features during the training process and ignore irrelevant ones, thus improving the model's efficiency. The binary cross-entropy loss function is adopted as the target loss function for node classification.

$$Loss_{ce} = \frac{1}{N}\sum_{i} -[y_i log(\hat{y}_i) + (1 - y_i) log(1 - \hat{y}_i)] \tag{8}$$

The overall loss function is as follows:

$$Loss = Loss_{ce} - \alpha Loss_{com} \tag{9}$$

where α is the hyperparameter controlling the impact of community structure. The weight matrix $W = W^0, W^1, W^2$ is trained by minimizing the final loss function. The gradients are computed through backpropagation and the parameters are optimized using Adam.

4 Experiment

4.1 Datasets

In this study, we collected labeled data on phishing scams from two authoritative websites reporting various illegal activities on the Ethereum network. The first website is EtherScamDB[4], which collects information on network fraud to guide Ethereum investors away from potential scams. The second website is Etherscan[5], which serves as an Ethereum block explorer. Reports of various scams on these two websites not only show the content of the scams but also display the addresses suspected of being involved in the scams. From the various scams reported on these two websites, we extracted addresses related to network phishing scams. We obtained over 600 million addresses and 4.5 billion transaction records. However, only over a thousand addresses were labeled as phishing addresses. Through further screening, we obtained the largest connected graph, which includes 2,973,382 nodes and 13,551,214 edges, with 1262 labeled phishing nodes. We commence by sampling subgraphs of sizes 30000, 40000, and 50000, respectively, from the initial node by utilizing random walk, to assess the performance of our proposed method under different data scales.

4.2 Evaluation Metrics

To measure effectiveness, we adopt four widely used evaluation protocols [19,25]: AUC (Area Under the Curve), Recall, Precision, and F1 score.

The AUC score evaluates the overall effectiveness of the model, while F1 is a comprehensive evaluation of the Precision and Recall scores. Recall, in this context, refers to the proportion of positive samples correctly predicted in the sample. As the number of suspicious nodes is small, we tend to prioritize the recall of these nodes. In everyday life, we are more tolerant of risk warnings than being deceived. Therefore, our goal is to identify as many network fishing suspects as possible within a reasonable range. At the same time, we supplement Precision, which shows the proportion of samples that are judged as positive and truly positive samples.

4.3 Comparison with Other Methods

For each model, we treat them as inputs for LightGBM. These models are designed to explore the effective combination of features and structural information as much as possible, ultimately improving the performance of classification.

To evaluate the performance of the proposed approach, we consider various baseline methods as follows: **Features Only**: The eight features (FT1 to FT8) of the nodes were directly input into the LightGBM classifier. **DeepWalk** [23]: A network representation learning method. The inspiration for DeepWalk comes

[4] https://etherscamdb.info/scams.
[5] https://etherscan.io/.

from Word2Vec. This method aims to obtain a sequence distribution similar to that of words in natural language through short random walks and to follow a power law. **Node2Vec** [12]: An extension of DeepWalk, which differs from its random walk strategy. Node2Vec is a dynamic combination of depth-first and breadth-first search. **LINE** [27]: A technique for transforming graph-structured data into a natural language processing task using the random walk method to generate node sequences. **Trans2vec** [14]: A method that detects phishing frauds on Ethereum's network. It uses transaction records, considers amount and timestamp, employs random walks for network embedding, and an SVM classifier to differentiate regular and phishing accounts.

We also compare our model against graph convolutional network methods: **ChebNet** [11]: A variant of graph convolutional network that uses Chebyshev polynomials for graph convolution. **PSDECN** [8]: A method using graph convolutional networks and autoencoders to detect phishing in Ethereum. The method uses classifiers to identify phishing accounts accurately.

To implement these network embedding methods, we need to set the following parameters. Firstly, the dimension of the embedding vectors is set to 64. For DeepWalk and Node2Vec, we set the number of random walks per node to 20 and the length of the random walks to 20. The p and q of Node2Vec are set to 0.25 and 0.75 according to [12]. The number of layers for GCN and ChebNet is set to 3. Incorporating the eight features suggested in the paper, we fed them into the baseline methods, except for Trans2vec and PSDECN. Afterward, we utilized LightGBM to derive the classification results.

Table 2. Overall performance comparison for 30,000, 40,000 and 50,000 sampled nodes

Method	GraphSize = 30000				GraphSize = 40000				GraphSize = 50000			
	AUC	Recall	Precision	F1	AUC	Recall	Precision	F1	AUC	Recall	Precision	F1
Features	0.5631	0.1120	0.6154	0.1895	0.5637	0.1120	0.5829	0.1879	0.5783	0.1589	0.6022	0.2515
DeepWalk	0.5741	0.1263	0.7146	0.2147	0.5789	0.1371	0.6634	0.2272	0.5801	0.1575	0.6729	0.2553
Node2Vec	0.5535	0.1083	0.6841	0.1870	0.5733	0.1436	0.6721	0.2366	0.5759	0.1549	0.6733	0.2511
LINE	0.5780	0.1421	0.5338	0.2245	0.5721	0.1346	0.5567	0.2168	0.5823	0.1739	0.5233	0.2610
ChebNet	0.5821	0.1431	0.7210	0.2388	0.5833	0.1502	0.7049	0.2476	0.5921	0.1664	0.7042	0.2692
Trans2vec	0.5902	0.1534	0.7321	0.2546	0.5951	0.1572	0.7178	0.2598	0.6014	0.1825	0.7089	0.2891
PSDECN	0.5830	0.1492	0.7258	0.2475	0.5887	0.1531	0.7122	0.2520	0.5986	0.1781	0.7051	0.2844
CEGCN-PD	**0.6325**	**0.1625**	**0.7523**	**0.2673**	**0.6386**	**0.1702**	**0.7428**	**0.2769**	**0.6418**	**0.1923**	**0.7329**	**0.3046**

As can be seen from Table 2, models based on graph convolutional networks typically produce the best performance across all datasets. Among all metrics, we are most concerned with recall, as in such a complex and practical transaction network, the accounts of potential network phishing victims can only be lightly reminded of risks. Such risk reminders are common in daily life, so we may be more inclined to recall suspected phishing accounts. Since in blockchain networks, transaction accounts are anonymous and there is no user profile, the differences between nodes are mainly reflected in the structure and attributes of the edges. Models based on graph convolutional networks take into account node features (i.e., information extracted from edge information) and integrate them well into the spatial structure. We believe that the adjacency matrix formed by the transaction network is usually sparse, and graph attention networks have better handling of sparsity in graph data compared to traditional graph convolutional networks.

The method proposed in this paper takes into consideration the significant graph information of community structure, resulting in a more comprehensive representation of the graph nodes and a more effective capture of the relationships between nodes in the graph. By considering the community structure as a factor, the CEGCN-PD model demonstrates a significantly better classification performance compared to the GCN model (PSDECN) which does not utilize community structure. As shown in Table 2, the CEGCN-PD model improved the AUC by 7.2% and the Recall, Precision, and F1 scores by 7.9%, 3.9%, and 7.1%, respectively. It is believed that this is because phishing fraud addresses often transfer funds to other accounts after acquiring them, and these accounts, forming a community with the phishing fraud address, have a higher degree of similarity and may also be fraud addresses. Thus, incorporating community structure information is highly effective in enhancing the classification performance of graph embeddings. In comparison to other baseline methods, CEGCN-PD also exhibits superior performance in various evaluation metrics.

Table 3. Comparing Performance Across Various Classifiers.

Method	AUC	Recall	Precision	F1
XGBoost	0.6152	0.1324	0.7145	0.2253
Random Forest	0.6018	0.1189	0.6952	0.2012
Naive Bayes	0.5123	0.0927	0.6421	0.1619
SVM	0.5915	0.1046	0.7013	0.1823
LightGBM	**0.6325**	**0.1625**	**0.7523**	**0.2673**

To ensure optimal system performance, the choice of classification model is crucial. We conducted a comprehensive analysis comparing LightGBM to several established machine learning models including Naive Bayes [26], XGBoost [9], Random Forest [4], and Support Vector Machines (SVM) [6]. The results, presented in Table 3, clearly demonstrate that LightGBM outperforms all other models in all evaluation metrics. This superior performance is due to its high-performance gradient boosting framework, which effectively manages large-scale datasets with exceptional scalability. Additionally, LightGBM's ability to handle high-dimensional and sparse features, as well as its robustness to noise and missing values, makes it an ideal choice for our phishing detection framework.

4.4 Evaluation of Model Parameters

In this study, we investigated the impact of the number of GCN layers on model performance. As shown in Fig. 2, we conducted experiments on datasets of different graph sizes. It is clear that the performance of the proposed model varies with the number of layers (i.e. 1, 2, 3, 4, and 5), and it can be seen that as the depth of GCN increases, the performance of the model is enhanced and reaches its peak at 3. However, as the number of convolutional layers continues to increase, the overall performance no longer increases, but slightly decreases. Currently, GCN is mostly applied to shallow layers, and simply stacking more layers can result in over-smoothing obstacles [20]. We originally believed that as more information is obtained, the efficiency of the model would also increase correspondingly, but in fact, aggregating multi-order neighbors results in the loss of individual information [31], thereby weakening performance.

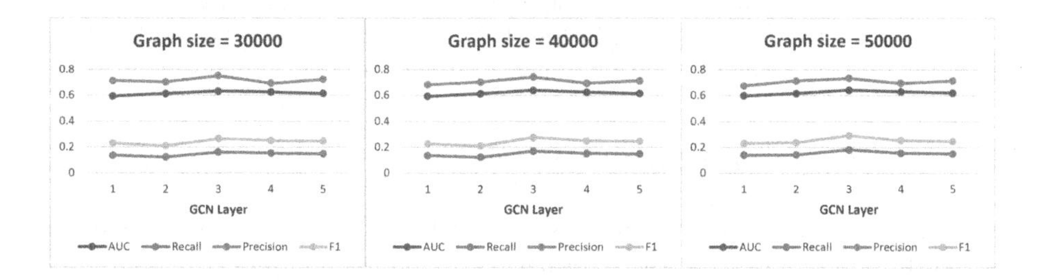

Fig. 2. Performance of GCN layers at different Graph sizes

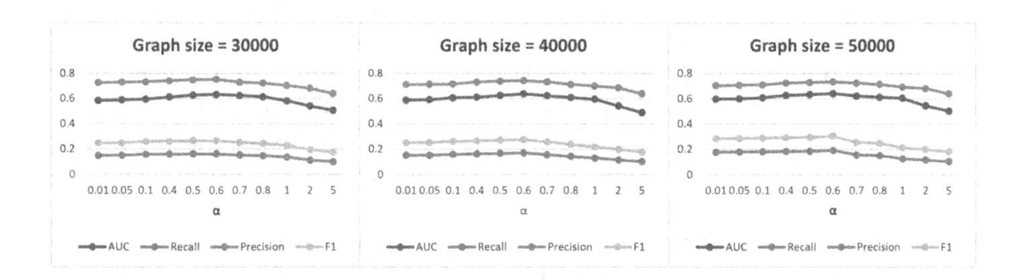

Fig. 3. Performance of α at different Graph sizes

We also studied the hyperparameter α, which controls the impact of community structure on the performance of the model, as shown in Fig. 3. We found that the best performance was achieved when α was around 0.6, and $\alpha \in [0, 1]$ worked well. When α is too large, it ignores the information from the neighbors, which affects the decision results.

In this study, we also investigate the impact of embedding dimensions on performance. Specifically, we explore the effect of varying the dimensionality of

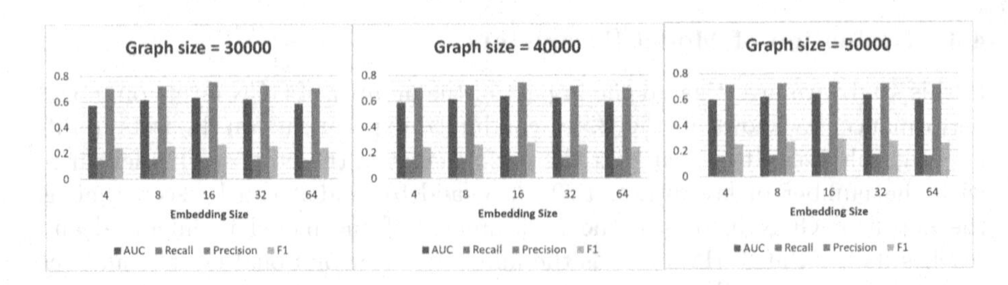

Fig. 4. Performance of d at different Graph sizes

embeddings, denoted as d, on the performance of a classifier. High-dimensional embeddings can lead to overfitting and increased computational complexity, while low-dimensional embeddings can degrade representation capabilities. To study how dimensionality affects classification performance, we varied the dimensionality from 4 to 64. As shown in Fig. 4, we observe a general trend that suggests the best performance is usually achieved with an embedding size of 16.

5 Conclusion and Future Work

This paper presents a community-structure-based phishing scam detection model. In order to capture the information of address nodes on Ethereum, we construct a transaction graph among address nodes. Guided by GCN to capture the information between multiple entities, while taking into account the community structure information between nodes, this leads to a more complete graph representation of nodes. In the transaction network, due to the anonymity of accounts, feature engineering capability is limited. The idea of combining structural features from the topological graph seems like an extension of this. In order to validate the effectiveness of the proposed method, experiments were conducted on real-world data, and satisfactory results were obtained.

However, it is necessary to recognize some limitations and flaws in our research. Firstly, the evaluation of our model focused exclusively on real-world data from the Ethereum network, thereby necessitating further evaluation of diverse blockchain networks to gain comprehensive insights. Secondly, the anonymity of accounts in transaction networks poses a challenge, leading to limited feature engineering capabilities. For future work, we will extend and improve our model to fully exploit Ethereum transaction information, making it adaptable to various blockchain data mining tasks, including Ponzi scheme detection and transaction tracking. These advancements are aimed at creating a safer transaction environment and improving the security of the blockchain system.

Acknowledgements. This research was supported by the National Key R&D Program of China No. 2022YFB2702504. It is also supported by the Fundamental Research Funds for the Central Universities (226-2022-00064).

References

1. Abu-El-Haija, S., Kapoor, A., Perozzi, B., Lee, J.: N-GCN: multi-scale graph convolution for semi-supervised node classification. In: Uncertainty in Artificial Intelligence, pp. 841–851. PMLR (2020)
2. Bellingeri, M., Bevacqua, D., Scotognella, F., Cassi, D.: The heterogeneity in link weights may decrease the robustness of real-world complex weighted networks. Sci. Rep. **9**(1), 10692 (2019)
3. Boccaletti, S., Latora, V., Moreno, Y., Chavez, M., Hwang, D.U.: Complex networks: structure and dynamics. Phys. Rep. **424**(4–5), 175–308 (2006)
4. Breiman, L.: Random forests. Mach. Learn. **45**, 5–32 (2001)
5. Cao, S., Lu, W., Xu, Q.: GraRep: learning graph representations with global structural information. In: Proceedings of the 24th ACM International on Conference on Information And Knowledge Management, pp. 891–900 (2015)
6. Chang, C.C., Lin, C.J.: LIBSVM: a library for support vector machines. ACM Trans. Intell. Syst. Technol. (TIST) **2**(3), 1–27 (2011)
7. Chen, J., Zhang, J., Chen, Z., Du, M., Xuan, Q.: Time-aware gradient attack on dynamic network link prediction. IEEE Transactions on Knowledge and Data Engineering (2021)
8. Chen, L., Peng, J., Liu, Y., Li, J., Xie, F., Zheng, Z.: Phishing scams detection in Ethereum transaction network. ACM Trans. Internet Technol. (TOIT) **21**(1), 1–16 (2020)
9. Chen, T., Guestrin, C.: XGBoost: a scalable tree boosting system. In: Proceedings of the 22nd ACM SIGKDD International Conference on Knowledge Discovery and Data Mining, pp. 785–794 (2016)
10. Chen, W., Zhang, T., Chen, Z., Zheng, Z., Lu, Y.: Traveling the token world: a graph analysis of Ethereum ERC20 token ecosystem. In: Proceedings of The Web Conference 2020, pp. 1411–1421 (2020)
11. Defferrard, M., Bresson, X., Vandergheynst, P.: Convolutional neural networks on graphs with fast localized spectral filtering. In: Advances in Neural Information Processing Systems, vol. 29 (2016)
12. Grover, A., Leskovec, J.: node2vec: scalable feature learning for networks. In: Proceedings of the 22nd ACM SIGKDD International Conference on Knowledge Discovery and Data Mining, pp. 855–864 (2016)
13. Holub, A., O'Connor, J.: Coinhoarder: Tracking a Ukrainian bitcoin phishing ring DNS style. In: 2018 APWG Symposium on Electronic Crime Research (eCrime), pp. 1–5. IEEE (2018)
14. Ke, G., et al.: LightGBM: a highly efficient gradient boosting decision tree. In: Advances in Neural Information Processing Systems, vol. 30 (2017)
15. Kipf, T.N., Welling, M.: Semi-supervised classification with graph convolutional networks. arXiv preprint arXiv:1609.02907 (2016)
16. Li, S., Xu, F., Wang, R., Zhong, S.: Self-supervised incremental deep graph learning for Ethereum phishing scam detection. arXiv preprint arXiv:2106.10176 (2021)
17. Liu, J.: E-commerce Agents: Marketplace Solutions, Security Issues, and Supply And Demand, vol. 2033, 1st edn. Springer Science & Business Media, Heidelberg (2001). https://doi.org/10.1007/3-540-45370-9
18. Liu, Y., Wang, Q., Wang, X., Zhang, F., Geng, L., Wu, J., Xiao, Z.: Community enhanced graph convolutional networks. Pattern Recogn. Lett. **138**, 462–468 (2020)

19. Liu, Z., Chen, C., Yang, X., Zhou, J., Li, X., Song, L.: Heterogeneous graph neural networks for malicious account detection. In: Proceedings of the 27th ACM International Conference on Information and Knowledge Management, pp. 2077–2085 (2018)
20. Mikolov, T., Chen, K., Corrado, G., Dean, J.: Efficient estimation of word representations in vector space. arXiv preprint arXiv:1301.3781 (2013)
21. Narayanan, A., Chandramohan, M., Venkatesan, R., Chen, L., Liu, Y., Jaiswal, S.: graph2vec: Learning distributed representations of graphs. arXiv preprint arXiv:1707.05005 (2017)
22. Newman, M.E.: Modularity and community structure in networks. Proc. Natl. Acad. Sci. **103**(23), 8577–8582 (2006)
23. Perozzi, B., Al-Rfou, R., Skiena, S.: Deepwalk: online learning of social representations. In: Proceedings of the 20th ACM SIGKDD International Conference on Knowledge Discovery and Data Mining, pp. 701–710 (2014)
24. Pilkington, M.: Blockchain technology: principles and applications. In: Research Handbook on Digital Transformations, pp. 225–253. Edward Elgar Publishing (2016)
25. Qiu, J., Tang, J., Ma, H., Dong, Y., Wang, K., Tang, J.: DeepInf: social influence prediction with deep learning. In: Proceedings of the 24th ACM SIGKDD International Conference on Knowledge Discovery & Data Mining, pp. 2110–2119 (2018)
26. Rish, I., et al.: An empirical study of the Naive Bayes classifier. In: IJCAI 2001 Workshop on Empirical Methods in Artificial Intelligence, vol. 3, pp. 41–46 (2001)
27. Tang, J., Qu, M., Wang, M., Zhang, M., Yan, J., Mei, Q.: Line: large-scale information network embedding. In: Proceedings of the 24th International Conference on World Wide Web, pp. 1067–1077 (2015)
28. Wang, D., Cui, P., Zhu, W.: Structural deep network embedding. In: Proceedings of the 22nd ACM SIGKDD International Conference on Knowledge Discovery and Data Mining, pp. 1225–1234 (2016)
29. Wang, J., Chen, P., Yu, S., Xuan, Q.: TSGN: transaction subgraph networks for identifying Ethereum phishing accounts. In: Dai, H.-N., Liu, X., Luo, D.X., Xiao, J., Chen, X. (eds.) BlockSys 2021. CCIS, vol. 1490, pp. 187–200. Springer, Singapore (2021). https://doi.org/10.1007/978-981-16-7993-3_15
30. Wang, S., Ouyang, L., Yuan, Y., Ni, X., Han, X., Wang, F.Y.: Blockchain-enabled smart contracts: architecture, applications, and future trends. IEEE Trans. Syst. Man Cybern. Syst. **49**(11), 2266–2277 (2019)
31. Watts, D.J., Strogatz, S.H.: Collective dynamics of 'small-world' networks. Nature **393**(6684), 440–442 (1998)
32. Wu, J., et al.: Who are the phishers? Phishing scam detection on Ethereum via network embedding. IEEE Trans. Syst. Man Cybern. Syst. **52**(2), 1156–1166 (2020)
33. Xuan, Q., et al.: Subgraph networks with application to structural feature space expansion. IEEE Trans. Knowl. Data Eng. **33**(6), 2776–2789 (2019)
34. Zhang, D., Yin, J., Zhu, X., Zhang, C.: Network representation learning: a survey. IEEE Trans. Big Data **6**(1), 3–28 (2018)

DTP: An Open-Domain Text Relation Extraction Method

Yuanzhe Qiu[1], Chenkai Hu[2], and Kai Gao[1(✉)]

[1] School of Information Science and Engineering, Hebei University of Science and Technology, Shijiazhuang 050081, Hebei, China
gaokai@hebust.edu.cn

[2] The 54th Research Institute of China Electronics Technology Group Corporation, Shijiazhuang 050081, Hebei, China

Abstract. Open-domain text relation extraction (OpenTRE) is a subfield of information extraction that focuses on extracting relational facts from open-domain corpora. Recent OpenTRE researches have shown that clustering unlabeled instances leveraging knowledge from labeled data is effective, but most of them are based on the assumption that the testing set only contains open relations, which is inconsistent with real-world scenarios where known and open relations are mixed. Therefore, a novel OpenTRE method based on Dynamic Thresholds and Pair-based Self-weighting Loss (DTP) is proposed. It performs text data processing by categorizing instances and predicting unknown relations, which can handle a more diverse range of data. Specifically, we break down OpenTRE into two stages: detecting and discovering, which makes the OpenTRE process more understandable. Wherein, sample-based dynamic threshold strategy is employed to clarify the relation boundaries. Additionally, pair-based self-weighted loss improves the capture of semantic knowledge in labeled data and guides clustering. Experimental results indicate that this method outperforms strong baseline models on two datasets and has significant improvements.

Keywords: Open Relation Extraction · Dynamic Thresholds · Pair-based Self-weighting Loss

1 Introduction

Open-domain text relation extraction (OpenTRE) is a critical information extraction task that involves detecting and discovering relational triples containing new relation types from open-domain text corpora. The extracted triples are subsequently utilized for various downstream applications, including but not limited to question and answer, information retrieval and dialogue [1].

Previous OpenTRE methods have been widely regarded as a completely unsupervised task. [2] rely on syntax or syntactic patterns to extract salient relational facts, however, such patterns are difficult to cover the diverse range

B. Luo et al. (Eds.): ICONIP 2023, CCIS 1965, pp. 207–219, 2024.
https://doi.org/10.1007/978-981-99-8145-8_17

of relation expressions. Unsupervised clustering methods [3], linguistic feature-based clustering [4], and relational siamese networks [5] have been suggested to address these limitations, but they are susceptible to irrelevant information or can only transfer knowledge to corpora in the same domain. A self-supervised learning framework [6] has also been raised, but it increases learning instability and cannot leverage the knowledge inherent in existing known relation instances. Deep metric learning, a technique widely applied in computer vision, can be used to learn a sample-to-feature mapping $|f(\cdot)|$ through a specific loss function. A recent OpenTRE study employed a metric learning framework [7] to obtain rich supervision signals from known relation instances directly. Nevertheless, this framework only uses known relations for training, making it difficult to discover new relations with large differences in distribution. To sum up, current OpenTRE methods have not been able to fully exploit the relational semantic knowledge contained in known relations, and their ability to transfer knowledge is often limited to the same domain. Additionally, in real-world scenarios, both labeled and unlabeled data can exist that contain both known and unknown relations. It is important to note that existing methods assume that the testing phase only involves new relation instances.

In view of the above problems, this paper proposes a novel semi-supervised framework for open-domain text relation extraction called **DTP**, which includes the use of **D**ynamic **T**hresholds and a **P**air-based self-weighted loss. Specifically, DTP first extracts relation representations from the BERT model [8] and embeds known relations into dense vectors. Then, it joints each dense vector and relation representation to obtain logits for open relation detection. Simultaneously, the relation representation and a zero vector are concatenated to obtain the sample-dependent dynamic threshold. The proposed method leverages the dynamic threshold to detect unknown relations and integrates them with negative samples to learn thresholds adaptable for open-domain scenarios. DTP method creates dynamic thresholds without additional parameters and network design, which enables the thresholds to automatically adapt to different samples. Additionally, the unsupervised learning by clustering is similar to the DeepCluster approach [7], but DTP utilizes a deep metric loss to train the model without a classifier. Moreover, we only require to know whether the instance pairs belong to the same class, which helps to avoid the mismatch of pseudo-labels during each training epoch. Our contributions are summarized below:

- A novel approach is formulated for open-domain text relation extraction that can handle more diverse data and enhance the interpretability of the relation extraction process.
- A sample-based dynamic threshold is constructed without prior knowledge of the open class, which is combined with two negative sample generation techniques to learn thresholds that are more suitable for open scenarios.
- A pairwise self-weighting loss is designed to fully exploit the semantic knowledge of relations in labeled data and guide the clustering process.

2 Methodology

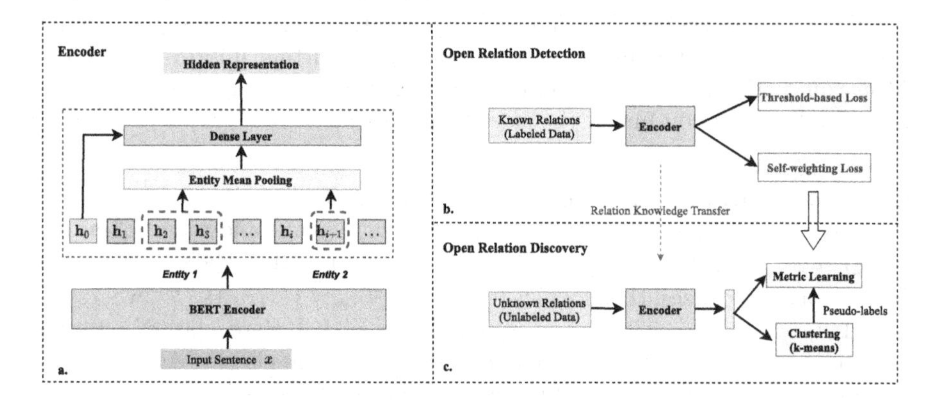

Fig. 1. Overall architecture of open relation extraction.

This paper presents a novel method for extracting both known and unknown relations in text. In particular, we decomposes open text relation extraction into two stages: open relation detection and open relation discovery. The first stage aims to distinguishing instances with N known relations and one unknown relation. While it can classify known relations and detect the new one, it cannot discover specific open relations. To tackle this issue, the second stage employs clustering techniques to group analogous open relation instances into more specific relation clusters. The comprehensive framework of our proposed approach is depicted in Fig. 1.

2.1 Encoder

Given an input sentence x, all token embeddings $[\mathbf{h}_0, \mathbf{h}_1, \ldots, \mathbf{h}_l] \in \mathbb{R}^{(l+1)\times d}$ are get from the last hidden layer of the pre-trained language BERT [8].

To obtain the relation semantic representation between two entities, the entity representations are obtained by using the mean-pooling method first:

$$\mathbf{e}_* = \text{mean-pooling}\left(\left[\mathbf{h}_{i_{e_*}}, \ldots, \mathbf{h}_{j_{e_*}}\right]\right) \tag{1}$$

Then, we concatenate the output corresponding special token $\mathbf{h}_0$ and two entity representations for extract hidden representation $\mathbf{h} \in \mathbb{R}^d$:

$$\mathbf{h} = \text{ReLU}(\mathbf{W}_h \left[\mathbf{h}_0; \mathbf{e}_1; \mathbf{e}_2\right] + \mathbf{b}_h), \tag{2}$$

In addition, known relation labels are embedded as a high-dimensional vector:

$$[\mathbf{c}_1, \mathbf{c}_2, \ldots, \mathbf{c}_N] = \mathbf{E}\left([c_1, c_2, \ldots, c_N]\right), \tag{3}$$

2.2 Open Relation Detection

In this section, we present the formal definition of the open relation detection task, followed by the introduction of dynamic thresholds based on samples. Additionally, we introduce two techniques for generating negative samples, namely manifold mixup [9] and entity boundary sliding, to further improve the threshold learning. Meanwhile, utilizing deep metric learning with a pair-based self-weighting loss function makes the class boundaries clearer.

Formal Definition. The model is trained using $\mathcal{D}_{tr} = \{(\mathrm{x}_i, y_i)\}_{i=1}^{|\mathcal{D}_{tr}|}$, where each instance x_i consists of a token sequence $\{w_1, w_2, \ldots\}$ and a pair of entities (e_1, e_2), and $y_i \in Y = \{1, 2, \ldots, N\}$ is the corresponding class label. In the presence of open classes, the model is tested with $\mathcal{D}_{te} = \{(\mathrm{x}_i, y_i)\}_{i=1}^{|\mathcal{D}_{te}|}$, where $y_i \in Y = \{1, 2, \ldots, N, N+1\}$ is the corresponding class label. Here, class $N+1$ refers to a set of novel categories that may consist of multiple classes. As there is no prior information available about these novel classes, they cannot be decomposed into subcategories.

Dynamic Thresholds. For the training instances with known relations, we concatenate $\mathbf{c}_i$ with relation representation and output logits by:

$$\begin{aligned} \mathbf{z}_i &= \mathrm{ReLU}(\mathbf{W}_z [\mathbf{h}; \mathbf{c}_i] + \mathbf{b}_z), \\ logit_i &= \mathrm{Proj}(\mathbf{z}_i), \end{aligned} \tag{4}$$

where $\mathbf{W}_z \in \mathbb{R}^{2d \times 2d}$ and $\mathbf{b}_z \in \mathbb{R}^d$ are learnable parameters, $\mathrm{Proj} : \mathbb{R}^{2d \to 1}$ is a linear projection. The logits $\{logit_i\}_{i=1}^N$ are used to calculate the confidence of relations in subsequent computations.

In closed-set classification, open class samples are typically assigned to the class with the highest predicted probability. In order to detect the open class, an intuitive method [10] is to treat the highest output probability as confidence scores and set a threshold to extend the classifier:

$$\hat{y} = \begin{cases} \mathrm{argmax}_{k \in Y} \hat{p}_k & conf > th \\ N+1 & \text{otherwise} \end{cases} \tag{5}$$

where th is a threshold, $\hat{p}_k$ is the probability that the model prediction belongs to class k, $Y = \{1, \cdots, N\}$ denote the known classes, and $conf = \max_{k=1,\cdots,N} \hat{p}_k$. To enable adaptive detection of open samples, a sample-dependent threshold is proposed for reducing the need for manual tuning. The threshold is defined as:

$$\begin{aligned} \mathbf{z}_0 &= \mathrm{ReLU}(\mathbf{W}_z [\mathbf{h}; \mathbf{0}] + \mathbf{b}_z), \\ th &= \mathrm{Proj}(\mathbf{z}_0), \end{aligned} \tag{6}$$

The loss function of dynamic threshold is defined as follows:

$$\mathcal{L}_{th} = -\log \frac{\exp(logit_y)}{\exp(th) + \exp(logit_y)} - \log \frac{\exp(th)}{\exp(th) + \sum_{j=1, j \neq y}^{N} \exp(logit_j)} \tag{7}$$

where y is the index corresponding to the ground-truth label. Minimizing the loss function $\mathcal{L}_{th}$ enables the model to learn a threshold that adapts to each individual sample, leading to more precise detection of open classes. Since the designed threshold depends on the sample itself, it can perform well even when there are multiple sub-classes within the open class during testing.

Generative Negative Samples. Although the dynamic threshold detects open classes, it is worth mentioning that the model has never seen open class samples before. Therefore, we generate negative samples as open samples for threshold learning.

Manifold Mixup. Inspired by [11], we generate the negative relation representation instead of origin input. The Manifold Mixup method [9] is used to generate negative samples:

$$\tilde{\mathbf{r}} = \lambda \mathbf{r}_i + (1-\lambda)\mathbf{r}_j, y_i \neq y_j, \tag{8}$$

where $\lambda \in [0,1]$ is sampled from Beta distribution, $\tilde{\mathbf{r}} \in \mathbb{R}^d$ is a negative sample obtained by mix different classes of relation representations.

The negative sample $\tilde{\mathbf{r}}$ is used to calculate the output $[logit_1^m, \ldots, logit_N^m, th^m]$. We treat $\tilde{\mathbf{r}}$ as open samples and minimize the following loss:

$$\mathcal{L}_{mix} = -\log \frac{\exp(th^m)}{\exp(th^m) + \sum_{j=1}^{N} \exp(logit_j^m)}. \tag{9}$$

Manifold mixup generates $\tilde{\mathbf{r}}$ to enhance the compactness of the boundary of known classes and adapt the dynamic threshold for open scenarios.

Entity Boundary Sliding. Entity boundary sliding is proposed to handle complex data distributions and account for instances that may not express any relations during testing, in contrast to manifold mixup which may not be effective for open classes with different high-dimensional space distributions than all known classes. To generate new entity representations, the method randomly slides the boundary of the entity pair, denoted as $i_e = i+\delta$ and $j_e = j+\delta$, where i and j are the start and end indices of the single entity. The scalar δ is randomly sampled from the interval $[-a,-1] \cup [1,b]$, where a and b are the distances from the start and end indices of the entity to the start and end of the sequence, respectively. The open entity representation $\mathbf{e}*$ is obtained using Eq. 1, and the negative sample $\hat{\mathbf{r}}$ is generated based on $\mathbf{e}*$ using Eq. 2. The output $[logit_1^s, \ldots, logit_N^s, th^s]$ is computed using Eq. 4 and Eq. 6, and we minimize the following loss:

$$\mathcal{L}_{slid} = -\log \frac{\exp(th^s)}{\exp(th^s) + \sum_{j=1}^{N} \exp(logit_j^s)}. \tag{10}$$

The dynamic threshold loss based on generative negative samples is:

$$\mathcal{L}_{neg} = \mathcal{L}_{mix} + \mathcal{L}_{slid}. \tag{11}$$

Metric Learning with Pair-Based Self-weighting Loss. To optimize the semantic space and capture the semantic knowledge present in the data, we build upon prior work on metric learning [7,12], and employ it to pre-train the encoder and create semantic embeddings directly. Within each batch, we consider sample pairs belonging to the same class as positive pairs and those of different classes as negative pairs for each batch. Our goal is to push the negative pairs beyond the negative boundary α_n while simultaneously pulling the positive pairs closer than the positive boundary α_p. In order to leverage informative examples [12,13] for fast convergence and good performance, we adopt the practice of mining non-trivial instances. For a given instance x_i, the non-trivial positive set after mining is defined as $\mathbf{P}_i = \{x_j | i \neq j, y_i = y_j, s_{ij} > \alpha_p\}$ and the negative set is defined as $\mathbf{N}_i = \{x_j | y_i \neq y_j, s_{ij} < \alpha_n\}$. To make the most of these non-trivial samples, we need to consider their distance from the boundaries as well as their relative distance from other non-trivial samples in the set when designing the loss function. To achieve this, the pair-based self-weighting (PSW) loss is:

$$\begin{aligned} \mathcal{L}_{PSW} &= \mathcal{L}_{\mathrm{P}} + \mathcal{L}_{\mathrm{N}} \\ \mathcal{L}_{\mathrm{P}} &= \log\left[\sum_{i=1}^{B_1} \exp\left(\gamma \left[s_i^p - \alpha_p\right]_+^2\right)\right] \\ \mathcal{L}_{\mathrm{N}} &= \log\left[\sum_{i=1}^{B_2} \exp\left(\gamma \left[\alpha_n - s_i^n\right]_+^2\right)\right] \end{aligned} \tag{12}$$

where $s_i^n \in \mathbf{N}_i$ and $s_i^p \in \mathbf{P}_i$ are positive sample pairs and negative sample pairs, respectively. B_1 and B_2 are the sizes of non-trivial sample sets. $\gamma > 0$ is a scalar temperature parameter.

The derivative concerning model parameters θ can be calculated as:

$$\frac{\partial \mathcal{L}_s}{\partial \theta} = \sum_{i=1}^{m} \frac{\partial \mathcal{L}_s}{\partial s_i} \frac{\partial s_i}{\partial \theta}, \tag{13}$$

where m is the size of a batch. Note that $\frac{\partial \mathcal{L}_s}{\partial s_i}$ is regarded as a constant scalar that not in the gradient θ. We regard $\frac{\partial \mathcal{L}_s}{\partial s_i}$ as the weight of $\frac{\partial s_i}{\partial \theta}$ and rewritten as:

$$\mathcal{L}_{PSW} = \sum_{i=1}^{m} w_i s_i \tag{14}$$

where $w_i = \frac{\partial \mathcal{L}_s}{\partial s_i}$. From Eq. 12, $s_i \in \mathbf{N}_i \cup \mathbf{P}_i$ and we further analyze for a non-trivial positive sample pair:

$$\frac{\partial \mathcal{L}_s}{\partial s_i^p} = \frac{\partial \mathcal{L}_P}{\partial s_i^p} = \frac{2\gamma[s_i^p - \alpha_p]_+}{\sum_{j=1}^{m} \exp(\gamma([s_j^p - \alpha_p]_+^2 - [s_i^p - \alpha_p]_+^2))} \tag{15}$$

From Eq. 15, we know that the weight of a positive sample pair is not only determined by $[s_i - \alpha_p]_+$ but also affected by other non-trivial samples. When

s_i is close to boundary α_p, the weight of the sample pair will decrease, and it will increase otherwise. Moreover, when other non-trivial samples are closer than the point, $([s_j^p - \alpha_p]_+^2 - [s_i^p - \alpha_p]_+^2) < 0$ will increase the overall impact after exp scaling, making it focus on optimizing such samples. The analysis presented above demonstrates that our proposed loss function assigns different weights to different sample pairs, allowing us to flexibly mine non-trivial samples.

Finally, our optimized training loss is:

$$\mathcal{L} = \mathcal{L}_{th} + \mathcal{L}_{neg} + \mathcal{L}_{PSW}. \tag{16}$$

In the test phase, the model outputs a dynamic threshold th for each sample and then detect open classes through Eq. 5.

2.3 Open Relation Discovery

In order to enable the model to better discover new relations, the model undergoes pre-training of the Encoder with labeled data in Sect. 2.2, and then further transfers the learned semantic knowledge to discover new relations from unlabeled data. This section presents the complete framework for discovering new relations in unlabeled data through clustering-based methods.

Formal Definition. Given a set of unlabeled relation instances X^u, the goal is to group the underlying semantic relations of the entity pairs in X^u. $\mathrm{X}^u = \{(x_i, h_i, t_i)\}_{i=1}^N$, where x_i is a sentence and (h_i, t_i) is a named entity pair corresponding to x_i. In semi-supervised scenarios, high-quality labeled data $\mathrm{X}^l = \{(x_i, h_i, t_i)\}_{i=1}^M$ is available to improve the model performance. Note that the supervised data X^l is separate from the label space and may be different from the corpus domain of X^u.

Unsupervised Clustering. We employ a clustering-based approach to extract previously unknown relations from unlabeled data. Specifically, we use an online deep clustering framework that combines label assignment and feature learning, inspired by the ODC method [14]. The framework involves of two memories: the sentence memory, which stores the features and pseudo-labels of the unlabeled training set, and the centroids memory, which stores the features of class centroids. In this context, a class represents a temporary relational cluster that evolves continuously during training. With the use of two memories, the clustering process becomes more stable and obviates the need for additional feature extraction steps.

The complete iteration consists of four steps. Firstly, we extract relational representations $\tilde{Z} = [\tilde{\mathbf{z}}_1, \tilde{\mathbf{z}}_2, \ldots, \tilde{\mathbf{z}}_B]$ in a batch of input, which are described in Sect. 2.2. Secondly, read the pseudo-labels from the sentence memory as a supervision signal to train the entire network, which minimizes $\mathcal{L}$. Thirdly, $\mathbf{z}$ after L2 normalization is reused to update the sentence memory:

$$\mathrm{S} \longleftarrow m\frac{\tilde{Z}}{||\tilde{Z}||_2} + (1 - m)\mathrm{S} \tag{17}$$

where S is the relational memory, $m \in (0, 1]$ is a momentum coefficient. Simultaneously, each involved instance is assigned with a new label by finding the nearest centroid following:

$$\min_{y \in 1,\ldots,M} < \tilde{\mathbf{z}}, \mathbf{c}_y > \quad (18)$$

where $\mathbf{c}_y$ denotes the centroid feature of class y. Finally, every k iterations, the centroid memory is updated by averaging all the features belonging to the corresponding centroid in the sentences memory.

3 Experiments

To demonstrate the effectiveness of DTP, we conduct three types of validation experiments: main experiments, ablation experiments, and visualization experiments on two publicly accessible datasets and report the comparison results. The statistics of two datasets are shown in Table 1.

Table 1. The statistics of the relation datasets (# indicates the total number of instances.).

Dataset	#Training	#Validation	#Test	Relations
SemEval [15]	6500	1500	2717	19
FewRel [16]	46400	4000	5600	80

3.1 Baselines and Evaluation Metrics

Baselines. We consider 3 advanced baselines: RSN [5], a supervised open-domain text relation extraction framework, uses the siamese network structure to learn the similarity of the relation from the label data. DeepAligned [17] aids the discovery of new classes by labeled data and proposes an alignment strategy to tackle the label inconsistency problem during clustering assignments. MORE [7] is a metric learning based framework, which uses deep metric learning to obtain rich supervision signals in labeled data and drive the model to learn semantic relational representation directly.

Evaluation Metrics. We use B^3 metric [18], Normalized Mutual Information (NMI), Adjusted Rand Index (ARI), and Clustering Accuracy (ACC) for comprehensive evaluation. We use the Hungarian algorithm [19] to calculate clustering accuracy by aligning predicted and ground-truth labels.

3.2 Parameters Setting and Training Details

DTP is built on top of the pre-trained BERT model (base-uncased, with 12-layer transformer) implemented in PyTorch [20] with default hyperparameter settings. To speed up the training procedure, we freeze all but the last transformer layer parameters of BERT. The training batch size is 64, the d is 768, and the learning rate is 1e−4. The scaled temperature γ is set to 50. The positive margin α_p is 0.7, and the negative margin α_n is 1.4.

We vary known classes number in the training set by using proportions of 25%, 50%, and 75%, and use all available classes for testing. For each ratio, we report the average performance over 5 runs of experiments. Instances from open classes will not be used during training or validation for open relation detection stage, and unlabeled open class training instances will be used during training for open relation discovery stage.

3.3 Result and Discussion

Table 2. Results (%) of relation extraction with different known relation proportions (KRP) on two dataset.

KRP	Method	SemEval				FewRel			
		ACC	ARI	NMI	B^3	ACC	ARI	NMI	B^3
25%	RSN	35.69	22.29	39.41	29.78	43.75	31.89	63.19	35.31
	DeepAligned	33.05	21.08	33.33	26.65	45.47	32.55	62.89	36.26
	MORE	35.44	22.61	35.80	28.36	49.79	37.31	66.29	40.67
	DTP	**41.66**	**28.87**	**43.33**	**35.77**	**51.83**	**38.75**	**71.31**	**47.27**
50%	RSN	48.61	34.01	50.89	41.04	44.94	32.80	63.76	36.03
	DeepAligned	52.68	39.39	51.29	44.32	63.19	51.86	76.03	54.70
	MORE	53.15	39.41	52.68	45.17	62.71	51.68	76.10	54.91
	DTP	**57.96**	**43.46**	**57.53**	**50.82**	**65.97**	**54.31**	**78.87**	**59.66**
75%	RSN	50.99	34.65	54.56	43.39	44.18	32.58	63.18	35.19
	DeepAligned	68.83	53.63	66.35	60.14	72.77	62.41	81.60	64.61
	MORE	68.04	52.19	66.23	59.31	66.40	56.15	78.93	59.79
	DTP	**69.94**	**54.03**	**68.33**	**62.16**	**74.88**	**64.40**	**82.86**	**67.18**

Open Relation Extraction Result. Table 2 shows the results of the proposed methods and baselines compared on SemEval and FewRel datasets. Firstly, DTP outperforms baselines on ACC, ARI, NMI, and B^3-score and significantly improves B^3-score compared with the results of best baseline up to 6.6%. It shows that DTP can effectively learn the relation semantic knowledge and discover new relations. Secondly, in the setting of 25% known classes, DTP significantly outperforms all baselines on all datasets, which indicates that it can learn latent relation semantic knowledge with only a few known relations instances. What's more, "Other" class in SemEval may cause poor clustering performance, but DTP excels over current state-of-the-art methods, showing effectiveness in complex, open scenarios. Finally, compared with the MORE, which also uses metric learning, the B3 scores of DTP are improved by 5.65% and 4.75% in the 50% setting on the SemEval and FewRel datasets, respectively.

Ablation Study. Ablation studies were conducted to evaluate the effectiveness of generative negative samples-based dynamic thresholds and pair-based self-weighting loss in the proposed method for open-domain text relation extraction. The results of the ablation study are shown in Table 3. Results shows that both techniques made significant contributions to the model, and negative sample training is essential for optimal performance. Removing self-weighting loss also affected the model's performance, but to a lesser extent than removing negative samples. The proposed method was found to be practical and effective for handling various types of data.

Table 3. Ablation study of DTP at detection stage on SemEval and FewRel dataset.

Ratio	Method	SemEval				FewRel			
		Known	Open	Overall	Acc	Seen	Unseen	Overall	Acc
25%	$\mathcal{L}_t + \mathcal{L}_P$	41.85	20.71	37.62	32.97	52.97	31.97	51.97	37.52
	$\mathcal{L}_t + \mathcal{L}_n$	62.71	80.08	66.18	74.66	63.48	69.54	63.77	63.47
	$\mathcal{L}_t + \mathcal{L}_n + \mathcal{L}_P$	**67.98**	**86.56**	**71.69**	**81.72**	**67.30**	**78.66**	**67.84**	**72.04**
50%	$\mathcal{L}_t + \mathcal{L}_P$	64.02	35.21	60.82	55.30	69.23	34.01	68.37	55.02
	$\mathcal{L}_t + \mathcal{L}_n$	75.25	78.56	75.62	78.54	73.77	63.63	73.53	67.83
	$\mathcal{L}_t + \mathcal{L}_n + \mathcal{L}_P$	**75.96**	**80.29**	**76.44**	**79.95**	**75.31**	**68.74**	**75.15**	**71.01**
75%	$\mathcal{L}_t + \mathcal{L}_P$	75.39	38.61	72.76	68.98	78.97	34.13	78.23	70.16
	$\mathcal{L}_t + \mathcal{L}_n$	**81.88**	75.60	**81.44**	**81.89**	78.85	49.63	78.37	72.22
	$\mathcal{L}_t + \mathcal{L}_n + \mathcal{L}_P$	78.91	**76.79**	78.76	81.41	**80.05**	**56.05**	**79.66**	**74.13**

Visualize Relation Representation. The effectiveness of DTP for relation representation extraction was evaluated using t-SNE [21] to visualize the high-dimensional relation representation (see Fig. 2). Four randomly selected known and four unknown relation types were chosen from the FewRel test set, and the sample points were colored according to their ground-truth labels. Results showed that DTP at detection stage outperformed the other two baselines in distinguishing between known and unknown relations. Our model was also better than ADB at handling classes with complex boundaries. DTP and ADB were effective at extracting representations of unknown relation instances. In addition, DTP at discovery stage learned more compact and separable representations between classes than MORE and DeepAligned, indicating that the proposed approach learns cluster-friendly representations.

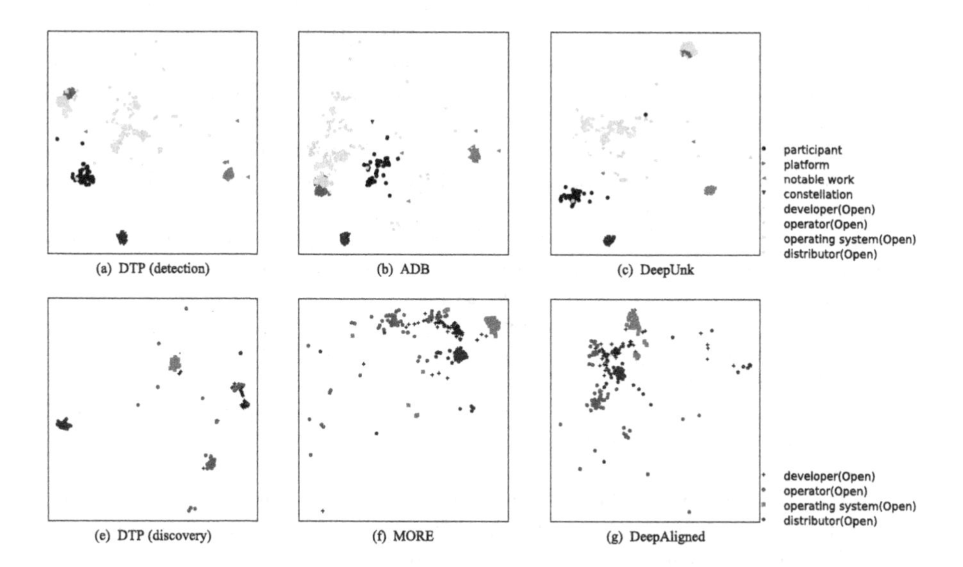

Fig. 2. Visualization of relation representation.

4 Conclusions and Future Works

This paper proposes an open-domain text relation extraction method to handle diverse relation data. Specifically, we propose dynamic thresholds to detect unknown relations and generate negative samples to learn thresholds that are more suitable for open scenarios. To better capture the semantic knowledge in the labeled data, pair-based self-weighted loss is used to learn feature representation effectively. In addition, a well-designed learning framework can better transfer relational semantic knowledge in labeled data to guide clustering. In the future, we will continue to study the combination of cross-domain open relation extraction and continuous learning.

Acknowledgements. This paper is funded by Natural Science Foundation of Hebei Province, China (No. F2022208006). We appreciate the assistance provided by the THUIAR Laboratory of Tsinghua University.

References

1. Fu, D., Zhang, C., Yu, J., et al.: Improving dialogue generation with commonsense knowledge fusion and selection. In: Memmi, G., Yang, B., Kong, L., Zhang, T., Qiu, M. (eds.) Proceedings of the 15th Conference on KSEM. LNCS, vol. 13368, pp. 93–108. Springer, Heidelberg (2022). https://doi.org/10.1007/978-3-031-10983-6_8
2. Xu, Y., Kim, M.Y., Quinn, K., et al.: Open information extraction with tree kernels. In: Proceedings of the 2013 Conference on NAACL-HLT, pp. 868–877 (2013)
3. Yuan, C., Eldardiry, H.: Unsupervised relation extraction: a variational autoencoder approach. In: Proceedings of the 2021 Conference on EMNLP, pp. 1929–1938. Dominican Republic (2021)
4. Yao, L., Haghighi, A., Riedel, S., et al.: Structured relation discovery using generative models. In: Proceedings of the 2011 Conference on EMNLP, Edinburgh, Scotland, UK, pp. 1456–1466 (2011)
5. Wu, R., Yao, Y., Han, X., et al.: Open relational knowledge transfer from supervised data to unsupervised data. In: Proceedings of the 2019 Conference on EMNLP-IJCNLP, Hong Kong, China, pp. 219–228 (2019)
6. Hu, X., Wen, L., Xu, Y., et al.: SelfORE: self-supervised relational feature learning for open relation extraction. In: Proceedings of the 2020 Conference on EMNLP, pp. 3673–3682 (2020)
7. Wang, Y., Lou, R., Zhang, K., et al.: MORE: a metric learning based framework for open-domain relation extraction. In: Proceedings of the 2021 Conference on ICASSP, pp. 7698–7702 (2021)
8. Devlin, J., Chang, M.W., Lee, K., et al.: BERT: pre-training of deep bidirectional transformers for language understanding. In: Proceedings of the 2019 Conference on NAACL-HLT, Minneapolis, Minnesota, pp. 4171–4186 (2019)
9. Verma, V., Lamb, A., Beckham, C., et al.: Manifold Mixup: better representations by interpolating hidden states (2019)
10. Hendrycks, D., Gimpel, K.: A baseline for detecting misclassified and out-of-distribution examples in neural networks. In: International Conference on Learning Representations (2017)
11. Zhou, D.W., Ye, H.J., Zhan, D.C.: Learning placeholders for open-set recognition. In: Proceedings of the 2021 Conference on CVPR, pp. 4399–4408 (2021)
12. Sun, Y., Cheng, C., Zhang, Y., et al.: Circle loss: a unified perspective of pair similarity optimization. In: Proceedings of the 2020 Conference on CVPR, pp. 6397–6406 (2020)
13. Wang, X., Han, X., Huang, W., et al.: Multi-similarity loss with general pair weighting for deep metric learning. In: Proceedings of the 2019 Conference on CVPR, Long Beach, CA, pp. 5017–5025 (2019)
14. Zhan, X., Xie, J., Liu, Z., et al.: Online deep clustering for unsupervised representation learning. CoRR abs/2006.10645 (2020)
15. Hendrickx, I., Kim, S.N., Kozareva, Z., et al.: SemEval-2010 task 8: multi-way classification of semantic relations between pairs of nominals. In: Proceedings of the 5th Conference on SEMEVAL, Uppsala, Sweden, pp. 33–38 (2010)
16. Han, X., Zhu, H., Yu, P., et al.: FewRel: a large-scale supervised few-shot relation classification dataset with state-of-the-art evaluation. In: Proceedings of the 2018 Conference on EMNLP, Brussels, Belgium, pp. 4803–4809 (2018)
17. Zhang, H., Xu, H., Lin, T.E., et al.: Discovering new intents with deep aligned clustering. In: Proceedings of the AAAI Conference on Artificial Intelligence, vol. 35, pp. 14365–14373 (2021)

18. Bagga, A., Baldwin, B.: Algorithms for scoring coreference chains. In: Proceedings of the 1st Conference on LREC, vol. 1, pp. 563–566 (1998)
19. Priya, D., Ramesh, G.: The Hungarian method for the assignment problem, with generalized interval arithmetic and its applications. J. Phys: Conf. Ser. **1377**(1), 012046 (2019)
20. Wolf, T., Debut, L., Sanh, V., et al.: Transformers: state-of-the-art natural language processing. In: Proceedings of the 2020 Conference on EMNLP: System Demonstrations, pp. 38–45 (2020)
21. Maaten, L.V.D., Hinton, G.E.: Visualizing data using t-SNE. J. Mach. Learn. Res. **9**, 2579–2605 (2008)

Exploring the Capability of ChatGPT for Cross-Linguistic Agricultural Document Classification: Investigation and Evaluation

Weiqiang Jin(✉), Biao Zhao, and Guizhong Liu

School of Information and Communications Engineering, Xi'an Jiaotong University, Innovation Harbour, Xi'an, Shaanxi, China
weiqiangjin@stu.xjtu.edu.cn, {biaozhao,liugz}@xjtu.edu.cn

Abstract. In the sustainable smart agriculture era, a vast amount of agricultural knowledge is available on the internet, making it necessary to explore effective document classification techniques for enhanced accessibility and efficiency. Over the past few years, fine-tuning strategies based on pre-trained language models (PLMs) have gained popularity as mainstream deep learning approaches, showcasing impressive performance. However, these approaches face several challenges, including a limited availability of training data, poor domain transferability, lack of model interpretability, and the challenges in deploying large models. Inspired by ChatGPT's significant success, we investigate its capability and utilization in the field of agricultural information processing. We explore various attempts to maximize ChatGPT's potential, including various prompting construction strategies, ChatGPT question-answering (Q&A) inference, and intermediate answer alignment technique. Our preliminary comparative study demonstrates that ChatGPT effectively addresses research challenges and bottlenecks, positioning it as an ideal solution for agricultural document classification. This findings encourage the development of a general-purpose agricultural document processing paradigm. Our preliminary study also indicates the trend towards achieving Artificial General Intelligence (AGI) for sustainable smart agriculture in the future. Code is available on Github (https://github.com/albert-jin/agricultural_textual_classification_ChatGPT).

Keywords: ChatGPT · Natural language processing · Agricultural document classification · Very large language model

1 Introduction

With the rapid development of the sustainable smart agriculture ecosystem, the quantity of various Internet news related to agricultural fields has seen an explosive increase. Artificial intelligence agricultural document classification enables the automatic management of this massive agricultural news and facilitates the

B. Luo et al. (Eds.): ICONIP 2023, CCIS 1965, pp. 220–237, 2024.
https://doi.org/10.1007/978-981-99-8145-8_18

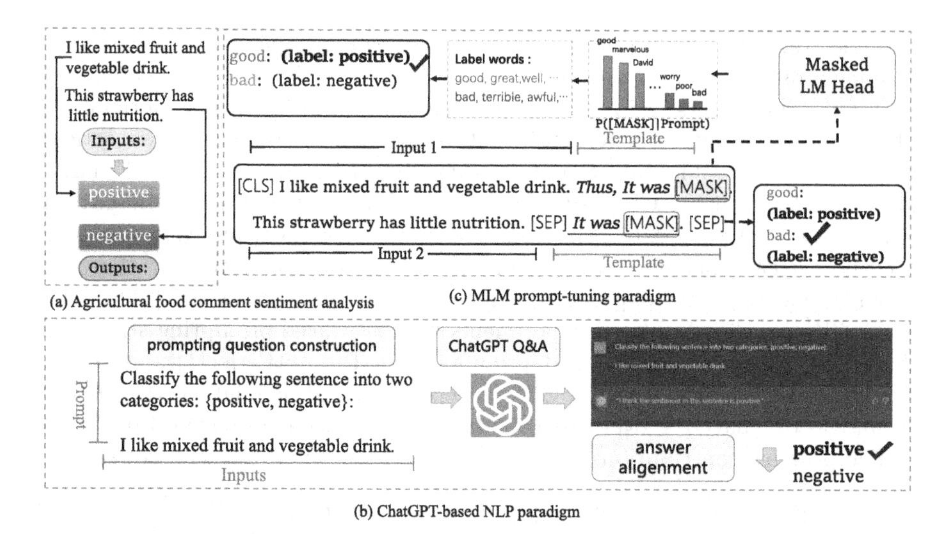

Fig. 1. Comparisons of ChatGPT-based NLP solutions with existing prompt learning paradigms by using an agricultural sentiment analysis example. Part (a) represents the task prototype of the agricultural sentiment analysis, while Part (b) and Part (c) illustrate the standard workflows of ChatGPT-based approaches and Masked LM prompt-tuning methods, respectively.

indexing of unstructured data, which is a crucial step towards agricultural digitization [22,25] and the agricultural Internet of Things development [9,26].

As a powerful very large-scale pre-trained language model (VLLM) for dialogue [19], ChatGPT has rapidly exhibited remarkable language comprehension abilities, attracting increasing attention in many cross-disciplinary researches [11,17,23]. Considering ChatGPT's impressive capabilities, it is natural to explore its potential in optimizing sustainable agricultural applications, including agricultural document classification.

In this paper, we investigate the potential of ChatGPT by focusing on the concise classification of agricultural text. This work introduces a novel NLP paradigm based on ChatGPT, which distinguishes it from existing methods. Figure 1 provides a clear illustration of the major similarities and differences between ChatGPT-based NLP paradigm (a) and MLM prompt-tuning paradigm (b). The prompt-tuning paradigm based on pre-trained language models (PLMs) can be divided into three primary procedures: template engineering, reasoning using pre-trained language models, and answer mapping engineering. Similarly, as depicted in part (b) of Fig. 1, natural understanding tasks based on ChatGPT can be organized into several phases [12,28]: 1) prompting question construction engineering; 2) ChatGPT question-answering (Q&A) inference; and 3) answer alignment (or normalization) engineering. To optimize the process, we considered several key factors: 1) designing task-specific inquiries to intuitively trigger the understanding capability of ChatGPT; and 2) developing an accurate label mapping strategy from the outputs of ChatGPT to the final classified categories.

In our experiments, we initially obtained datasets from various agricultural sub-fields. These datasets were primarily sourced from Internet news articles that covered topics like insect pests, natural hazards, and agricultural market comments. By utilizing ChatGPT for tasks such as pest and disease identification, agricultural news categorization, and market comment analysis, we showcased how it can contribute to agricultural information management. Through a comprehensive set of experiments, including comparisons with state-of-the-art (SOTA) baselines, ablated experiments with advanced prompting strategies, ChatGPT triggered prompts and prompts from PromptPerfect[1], and several additional case studies, we thoroughly investigated and highlighted the distinct advantages of ChatGPT over other methods in advanced agricultural practices.

The primary contributions of this work can be summarized as follows:

- Motivated by the significant application success of ChatGPT, we conducted a preliminary study to explore the potentials of ChatGPT in agricultural document classification tasks and proposed a ChatGPT-based solution for agricultural document classification;
- ChatGPT-based document classification method outperforms existing PLM-based approaches when evaluated on multi-linguistic agricultural datasets, showcasing its impressive cross-lingual understanding capability;
- Our ChatGPT-based method that equipped with advanced prompts shows impressive performance gains, which demonstrates and highlights the importance and significance of the strategies about how to design prompts.
- The reliance of ChatGPT-based method subverts complex and power-intensive PLM-based methods, promising the low-cost and general artificial intelligence (AGI) techniques for smart agriculture. Code is released on Github[2].

2 Related Work

2.1 Agricultural Document Classification

Over the past decade, traditional machine learning approaches, such as SVM, CNNs, and RNNs, have dominated document classification research [24]. Azeez *et al.* utilized support vector machines (SVM) for regional agricultural land texture classification [1]. Dunnmon et al. demonstrated the superiority of CNNs over RNNs in sentiment classification of agriculturally-relevant tweets [6,13]. The introduced pre-trained language models like BERT [5] have significantly improved agricultural document processing tasks, gradually replacing traditional methods. Shi et al. used BERT to extract representative information from unlabeled sources, improving the efficiency of constructing corpora for agricultural

[1] PromptPerfect service: https://promptperfect.jinaai.cn/prompts.

[2] https://github.com/albert-jin/agricultural_textual_classification_ChatGPT.

news [18]. Jiang et al. classified French plant health bulletins using fine-tuned BERT, making the data easily searchable [9]. Cao *et al.* introduced an improved BERT-based sentiment analysis model for evaluating agricultural products using internet reviews [4]. The integration of large pretrained language models like BERT into agricultural document classification has shown promising results, advancing the field and improving efficiency and accuracy.

However, the current scale of existing PLMs is still not capable enough to meet the requirements for artificial general intelligence (AGI). There is an urgent need to accelerate the exploration of the capability of very large language models (VLLMs). Therefore, the tremendous interest in ChatGPT has sparked exploration into its potential and possibilities in the field of agriculture.

2.2 ChatGPT

ChatGPT [3,15], developed by OpenAI, is a leading conversational language model that provides expert knowledge in various fields. It is a disruptive revolution in research domains, extending beyond NLP and has made significant contributions to various application scenarios. As more and more users engage with ChatGPT, a multitude of novel applications emerge, showcasing its immense potential. ChatGPT has shown remarkable proficiency in multilingual translations, especially in high-resource languages such as various European and American languages [20]. ChatGPT's capabilities extend beyond translations to coding tasks, and it can assist in code debugging and generation. Recent empirical studies [2,7,25] have shown that ChatGPT can provide code snippets adhering to syntax and semantics in languages like Python, Java, and JavaScript. Additionally, ChatGPT has been explored for event extraction and information extraction tasks [7,8,21].

Overall, ChatGPT is a remarkable language model with a wide range of applications. It revolutionizes the accessibility and capabilities of large language models, opening up new possibilities for research and development. The groundbreaking nature of ChatGPT has sparked curiosity and exploration, including its potential applications in the agricultural field. While specific agricultural use cases are limited, researchers continue to investigate and uncover new possibilities for the sustainable agricultural development.

3 ChatGPT-Based Agricultural Document Classification

This paper focuses on the feasibility of applying ChatGPT to agricultural document classification. We present a series of exploratory experiments as a preliminary study on ChatGPT-based agricultural applications. As we know, there were no existing works that systematically utilized ChatGPT for document classification. To fill this gap, we discuss the general workflow for ChatGPT-based agricultural document classification. Based on extensive literature [21,28], we identify three phases that most ChatGPT-assisted applications can be divided into, as illustrated in Fig. 2:

i. Prompting Question Construction: This initial stage focuses on providing appropriate prompting strategies to input into ChatGPT;
ii. ChatGPT Q&A Inference: The second stage involves the ChatGPT question-answering reasoning procedure, which is transparent to us that treated as a black box;
iii. Answer Alignment: In the final stage, the natural language intermediate response is transformed into the target label within predefined categories.

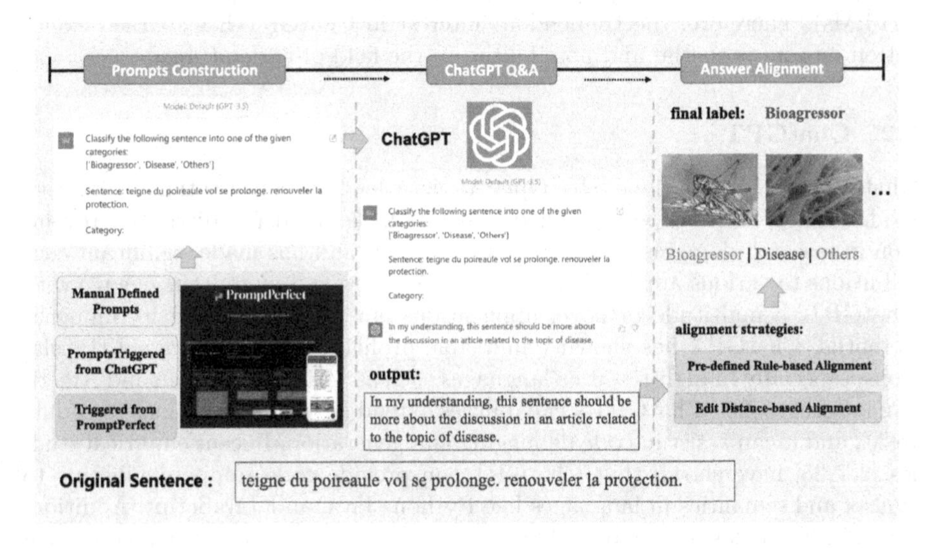

Fig. 2. The ChatGPT-based document classification framework is exemplified using the French Plant Health Bulletin dataset. It involves designing various prompts generation strategies (left), obtaining a response from ChatGPT (center), and aligning the intermediate answer to pre-defined categories (right).

Among these steps, while the question-answering inference conducted by ChatGPT remains static and beyond our modification, we can optimize the prompting construction and answer alignment engineering during our experiments. In the followings, we will introduce several novel strategies that we utilized to fully leverage the enormous potential and superiority of ChatGPT.

3.1 Prompt Question Construction

Prompts serve as a crucial component in determining the quality and relevance of generated outputs from language models. As inputs and guides for the model's output, the quality of prompts has a significant impact on the performance of the model. By leveraging prompt optimization engineering, users can unlock the full potential of large-scale language models and achieve better results with less effort and time. The engineering for constructing prompting question is widely recognized as a complex task that involves extensive historical experience and manual trial-and-error [14,28]. As illustrated on the bottom-left of Fig. 2, we adopt several prompt generation strategies in our experiments, including: 1) manually defined prompts, 2) prompts triggered from ChatGPT, and 3) prompts triggered from the PromptPerfect platform.

Next, we discuss these prompt generation strategies in detail.

Table 1. The partial manually devised prompts. `[Res]` denotes the response provided by ChatGPT.

No.	prompting template
1	Classify the following sentence into one of the given categories: `[CATE]` \n Sentence: `[SENT]` \n Category: \t `[Res]`
2	Which categories do you think sentence: \n `[SENT]` \n belongs to, out of `[CATE]`? \n `[Res]`
3	......

Manually Designed Prompts: Following the communication habits, we manually create several prompting templates. Table 1 showcases part of the designed templates. There are two essential elements that be provided to ChatGPT, denoting the sentence context and the pre-defined categories, respectively. For simplicity, we include two extra slots in the prompts to combine the respective mentions: slot `[SENT]` as the sentence, and slot `[CATE]` as the categories.

Note that not all manually designed templates are always the most suitable. It is necessary to select representative templates from the candidate set to be used for subsequent experiments. For this purpose, we utilize a data down-sampling evaluation as described in [2]. Specifically, we randomly select a predetermined number (100 by default) of samples from the specified dataset, Twitter Natural Hazards. Subsequently, we evaluate the performance of each template on this subset and choose the best-performed template as the most suitable manually defined prompt. Through a series of thorough down-sample comparison experiments, we selected the prompt for the subsequent experiments, which is depicted as follows.

```
> Categorize the given sentence into one of the provided
categories. Please provide a clear and concise response
that identifies the category of the sentence to allow for
categorizations.
The sentence is: [SENT].
The categories are specified in the [CATE].
The sentence to be classified is: {ChatGPT Response}.
```

Moreover, it should be noted that we have additionally included the following command in the prompts.

```
> ......, Please only answer the category.
```

This additional request is aimed at reducing unnecessary explanations generated by ChatGPT regarding its reply. Such explanations have the potential to disrupt the decision-making process for labeling subsequent text. We have also considered this factor when devising the following prompting methods.

ChatGPT Triggered Prompts: Taking cues from the extensive research in the field [10,27], we hypothesize that exploring ChatGPT directly might offer valuable insights into the creation of top-notch templates. With this in mind, we turn to ChatGPT itself, seeking its guidance on the recommended approaches for generating templates. Please note that the following human inquiries can serve as triggers for task-specific prompts:

```
> Provide five concise prompts or templates that can make
you deal with the [x] task.
```

, where the slot `[x]` represents task-specific types in the context.

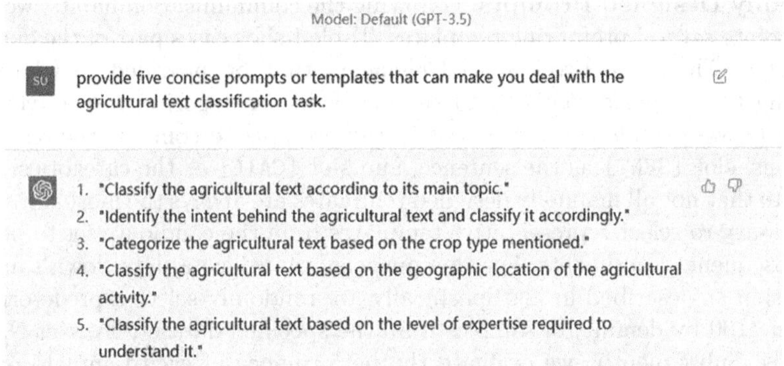

Model: Default (GPT-3.5)
SU
provide five concise prompts or templates that can make you deal with the agricultural text classification task.
1. "Classify the agricultural text according to its main topic."
2. "Identify the intent behind the agricultural text and classify it accordingly."
3. "Categorize the agricultural text based on the crop type mentioned."
4. "Classify the agricultural text based on the geographic location of the agricultural activity."
5. "Classify the agricultural text based on the level of expertise required to understand it."

Fig. 3. Candidate prompt templates triggered by requests to ChatGPT (Model: GPT-3.5, Query Date: 2023.4.02).

As depicted in Fig. 3, we have intuitively formulated our request as follows:

```
> Provide five concise prompts or templates that can make
you deal with the agricultural document classification task.
```

The generated prompts demonstrate sensible and coherent semantic content, with discernible variations in their respective formats. Employing the sampling-based evaluation method described earlier, we have chosen the best-performing prompt to represent the ChatGPT-triggered prompts for subsequent comparative experiments. The selected prompt is as follows:

```
> Classify the agricultural text: [SENT] according to its
main topic [CATE].
```

Experiments prove that this like ChatGPT triggered prompting strategy can enhance the quality and accuracy of generated outputs, making them more useful and effective.

Prompts Triggered from PromptsPerfect: PromptPerfect[3] is a cutting-edge intelligent prompt optimizer for large-scale language models, which helps users craft prompts that yield exceptional results. PromptPerfect has the ability to fine-tune prompts and generate high-quality results for a wide range of applications. PromptPerfect can revolutionize users' experience with models like ChatGPT, MidJourney, and StableDiffusion.

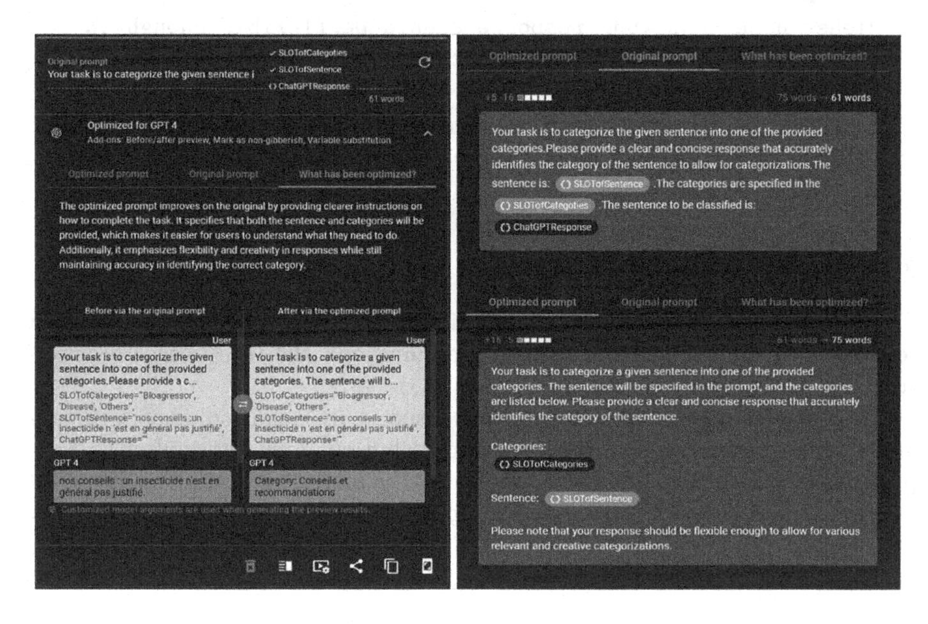

Fig. 4. The prompt optimization procedure of the PromptPerfect interface, exemplified by optimizing the manually designed prompt.

Here, we explore the impact of prompts optimized by PromptPerfect on the output quality of ChatGPT, taking advantage of PromptPerfect's advanced capabilities. Specifically, in the experiment, we simultaneously optimized the manually designed prompt and the ChatGPT-triggered prompt selected in the previous two steps. Figure 4 illustrates the optimization effects of the PromptsPerfect website service on the manually designed prompt.

The optimization result for the manually designed prompt is as follows:

[3] PromptPerfect service is available at https://promptperfect.jinaai.cn/prompts.

```
> Please classify the following agricultural document
according to its main topic:
Sentence: [CATE]
Categories: [SENT].
```

And the optimization result for the ChatGPT triggered prompt is as follows:

```
> Your task is to categorize a given sentence into one of
the provided categories. The sentence will be specified in
the prompt, and the categories are listed below. Please
provide a clear and concise response that accurately
identifies the category of the sentence.
Categories: [SENT]
Sentence: [CATE]
Please note that your response should be flexible enough to
allow for various relevant and creative categorizations.
```

Experimental results demonstrate that the overall performance of utilizing PromptPerfect for optimization are superior, surpassing the original prompts in a slight range of improvement.

3.2 ChatGPT Q&A Inference

The conversational prower of ChatGPT lies in its capacity to generate coherent sentence through sequence-to-sequence learning and the transformer architecture.

During inference, ChatGPT generates response by conditioning on a prompt and sampling words from a probability distribution. This distribution is computed using the softmax operation applied to the model's output. The output at each step is influenced by the previously generated tokens, enabling a generative process that produces coherent text.

Formally, the token generation process of ChatGPT can be expressed as follows:

$$p(y|x) = \prod_{i=1}^{L} p(y_i|y_1, ..., y_{i-1}, x) \tag{1}$$

where $\prod$ denotes the probability multiplication operator. The probability distribution $p(y_i|y_1, ..., y_{i-1}, x)$ represents the likelihood of generating token y_i at the t-th time step, given the preceding tokens $y_1, ..., y_{i-1}$ and the input prompt x. The length of the generated sequence is denoted as L.

To maintain consistency and enable independent thinking, we adopted a unique conversation thread for each prompt during the ChatGPT interaction process. This approach ensures that ChatGPT's responses are unaffected by the previous conversation history. By leveraging the user-provided information, ChatGPT consistently delivers optimal responses.

3.3 Answer Alignment

Different from traditional PLM-based classification models, the answers generated by ChatGPT do not directly correspond to predefined labels, posing challenges for subsequent output analysis and evaluation. Therefore, additional alignment strategies are required to convert these intermediate answers into final labels that can be used to compute various performance metrics. In our experiments, we designed and implemented two distinct alignment strategies: rule-based matching and similarity-based matching.

- **Rule-based matching:** This method utilizes predefined patterns or rules based on token attributes (e.g., part-of-speech tags) to match token sequences in unstructured textual data. During our experiments, we use the Matcher[4] in spaCy v3 to find the matched tokens in context.
- **Similarity-based matching:** This method computes the similarity between the generated answer and each label, selecting the label with the highest similarity score. First, we aggregate and synthesize commonly used expressions by ChatGPT for each category and establish a reference answer repository for each category. Then, we employ the Levenshtein distance algorithm to calculate the minimum edit distance between each reference answer and the input answer to be classified. The answer with the minimum distance is selected as the final category label.

To address the challenge of answer mapping, we adopt a pipeline approach that combines both rule-based and similarity-based matching strategies. We first use the rule-based strategy to parse the intermediate answer. If the category remains uncertain, we then employ the similarity-based strategy to calculate the similarity between the intermediate answer and each category's answer examples, selecting the most similar category as the final label.

4 Experiments

4.1 Setups

Datasets. We collected three datasets for agricultural document classification: **PestObserver-France**[5], **Natural-Hazards-Type**[6], and **Agri-News-Chinese**[7], ranging from different types of categories (e.g. plant diseases, insect pests, and twitter natural hazards) and numbers of categories to various languages, including French, English, and Chinese. Note that Natural-Hazards-Type is a disaster categories classification dataset re-organized from the original sentiment classification dataset, Natural-Hazards-Twitter[8]. The details are depicted in Table 2.

[4] Spacy can be accessed on https://spacy.io.
[5] PestObserver-France download: https://github.com/sufianj/fast-camembert.
[6] https://github.com/albert-jin/agricultural-_textual_classification_ChatGPT.
[7] More details can be accessed from http://zjzx.cnki.net/.
[8] https://github.com/Dong-UTIL/Natural-Hazards-Twitter-Dataset.

Table 2. The statistical meta-information of these datasets. N-train, n-test, n-label denote the count of training samples, test samples, and categories, respectively.

Datasets	n-train	n-test	language	labels	n-label
PestObserver-France	322	80	French	Bioagressor, Disease, Others	3
Natural-Hazards-Type	5000	1000	English	Hurricane, Wildfires, Blizzard, Floods, Tornado	5
Agri-News-Chinese	52000	6500	Chinese	农业经济, 农业工程, 水产渔业, 养殖技术, 林业, 园艺, 农作物	7

Baselines. To conduct a comprehensive comparison with previous machine learning methods, such as feature engineering-based method, representation learning method, PLM-based fine-tuning and prompt-tuning. We selected a series of representative approaches as baselines. These baselines include SVM [1], TextRNN [13], BERT-based fine-tuning [5] and T5-based prompt-tuning methods [16], each representing a different learning paradigm mentioned above. For TextRNN, we utilized the pre-trained word vectors from GloVe[9] as embeddings. The PLM models used in our study are the "bert-base-uncased" and "t5-base" versions obtained from facebook/huggingface.io[10].

Metrics. In agricultural document classification tasks with multiple labels, accuracy and F1-score are commonly used metrics. Accuracy measures the proportion of correctly predicted samples among all predicted samples, while F1-score considers precision and recall.

Experimental Environment. The meta experimental settings can be summarized as follows: The experimental hardware environment consists of an Intel Core i9-9900k CPU and a single Nvidia GPU, specifically the GTX 1080Ti. The code implementation is done in Python 3.7 and PyTorch 1.9.0.

4.2 Main Results

Table. 3 details the comparison results of our method and these state-of-the-art baselines on the three cross-linguistic datasets. For comparison simplicity, we regarded the approach which adopts the manually defined prompts as the vanilla ChatGPT-based classification method and took it for comparison.

Through the comparative experimental results, our approach exhibits competitive performance across all datasets. ChatAgri demonstrates a significant performance advantage of 10% to 20% compared to traditional SVM and TextRNN methods. Furthermore, our approach shows impressive performance compared to the latest methods based on large-scale pre-trained models. Compared to PLM BERT fine-tuning and PLM T5 prompt-tuning methods, our approach improves accuracy or weighted F1 scores. For example, compared to BERT fine-tuning, our approach achieves an accuracy improvement of approximately

[9] GloVe embedding: https://nlp.stanford.edu/projects/glove/.

[10] Huggingface transformers: https://huggingface.co/docs/transformers/index.

Table 3. Performance statistics of the baselines and ChatAgri on the adopted datasets. The best and second-best scores across all models are respectively highlighted in bold and underlined. [ChatGPT Query Date: March 16, 2023].

Learning Paradigms	Baselines	PestObserver -France		Natural-Hazards -Type-English		Agri-News -Chinese	
		Acc	Weighted -F1	Acc	Weighted -F1	Acc	Weighted -F1
Feature Engineering	SVM	0.672	0.645	0.811	0.811	0.623	0.622
Word Representation	TextRNN	0.707	0.697	0.931	0.931	0.812	0.801
PLM-based Fine-tuning	BERT-based fine-tuning	0.736	0.714	0.945	0.945	0.826	0.819
PLM-based Prompt-tuning	T5-based prompt-tuning	<u>0.764</u>	<u>0.753</u>	<u>0.966</u>	<u>0.966</u>	<u>0.859</u>	<u>0.854</u>
ChatGPT-based Prompt QA	(Our Method)	**0.794**	**0.789**	**0.978**	**0.978**	**0.863**	**0.856**

5.8% and 3.7% on the PestObserver-France and Agri-News-Chinese datasets, respectively. Compared to the T5 prompt-tuning method, our approach achieves slightly higher accuracy on the Agri-News-Chinese and Natural-Hazards-Type datasets, approximately 3.0% and 1.1% respectively.

In addition, ChatGPT performs excellently in cross-language understanding. Unlike traditional PLM-based methods, ChatGPT utilizes comprehensive and high-quality training corpora covering most languages used in various countries. Moreover, the enormous parameter size of ChatGPT enables it to retain and master more language knowledge, not limited to English. Thus, ChatGPT exhibits significant superiority over traditional PLM models (such as BERT, RoBERTa, and BART) in terms of cross-language understanding.

4.3 Improved with ChatGPT Triggered Prompts

Here, we discuss the performance influence of ChatGPT triggered prompting strategy. The overall results of ChatGPT equipped with ChatGPT triggered prompts are presented in Table 4. Specifically, from the third and fourth columns of Table 4, our method that equipped with ChatGPT triggered prompts yielded an approximately 2.1% accuracy and 1.4% weighted-F1 improvements on PestOb-server-France over that using manually constructed prompts, and outperforms the basic counterpart by a performance margin about 1.2% both of accuracy and weighted-F1 on Agri-News-Chinese.

However, there was no significant improvement in performance for the Natural-Hazards-Type. We speculate that this may be attributed to the characteristics of the dataset, where the semantics related to classifying natural disaster categories are prominently evident, often relying solely on a few fixed trigger words and phrases. Consider the following sentence found in the dataset: "Matthew charges toward Jamaica, Haiti, Cuba: Islanders stock up on food, water and other supplies." In this sentence, the referenced word "Matthew" is intrinsically associated with the document topic of an American hurricane disaster.

Table 4. The performance comparison of our method using the manually constructed prompts and the prompts generated by ChatGPT. The colored numbers shown in percentage indicated the increased performances.

Prompting Strategies		Manually Defined - Prompts (Vanilla)	ChatGPT Triggered - Prompts
PestObserver -France	Accuracy	0.794	**0.815** ↑ 2.1%
	Weighted-F1	0.789	**0.812** ↑ 1.4%
Natural-Hazards-Type	Accuracy	**0.978**	**0.978** ↑ 0.0%
	Weighted-F1	**0.978**	**0.978** ↑ 0.0%
Agri-News -Chinese	Accuracy	0.863	**0.874** ↑ 1.1%
	Weighted-F1	0.856	**0.867** ↑ 1.1%

One additional notable experimental phenomenon is that prompts obtained through ChatGPT tend to be concise and succinct, while manually crafted prompts often become excessively lengthy due to the inclusion of a series of constraining conditions. This finding further substantiates the lack of a direct relationship, in theory, between the length of the provided prompt and the quality of feedback and classification performance achieved by ChatGPT. Accordingly, appropriately providing brief prompts can still effectively stimulate ChatGPT's exceptional language comprehension capabilities.

In summary, with the help of the advanced prompts triggered by ChatGPT, the proposed classification method achieves impressive performance improvements and shows its significance and effectiveness. Furthermore, designing appropriate templates to stimulate the potential generative capacity of ChatGPT is an valuable research area that warrants thorough exploration and discussion.

4.4 Improved with PromptPerfect Prompt Optimization

Here, we provided a detailed analysis towards the PromptPerfect's prompt optimization effect on our ChatGPT-based classification method. Table 5 statistics the comparative experimental results between the use of PromptPerfect to optimize prompts before and after.

In this table, the first and second groups show the increased performance of manually constructed prompts and ChatGPT triggered prompts equipped with

PromptPerfect, respectively. It can be clearly observed that prompts optimized through PromptPerfect facilitate ChatGPT in providing more accurate feedback responses and achieving higher metrics. Although there was only a slight improvement in the Natural-Hazards-Type dataset, it can be mainly attributed to the fact that the dataset already had high prediction performance, approaching saturation. For instance, after being optimized with PromptPerfect, our methods, based on the two prompting designs, showed an averaged improvements of over 1% on the PestObserver-France dataset. Furthermore, our method utilizing manually constructed prompts that enhanced by PromptPerfect exhibited an increase of 2.6% in accuracy and a increase of 2.3% in weighted-F1 on the Agri-News-Chinese dataset.

Table 5. The performance statistics of our method enhanced by the PromptPerfect tool. The first group and the second group respectively represent the two designed prompts, manually defined prompts and ChatGPT triggered prompts.

Prompting Strategies	PestObserver -France		Natural-Hazards -Type		Agri-News -Chinese	
	Acc	Weighted -F1	Acc	Weighted -F1	Acc	Weighted -F1
Manually Defined - Prompts (Vanilla)	0.794	0.789	0.978	0.978	0.863	0.856
Manually Defined - Prompts + PromptPerfect	0.808 ↑1.4%	0.802 ↑1.3%	0.982 ↑0.4%	0.982 ↑0.4%	0.889 ↑2.6%	0.879 ↑2.3%
ChatGPT Triggered - Prompts	0.815	0.812	0.978	0.978	0.874	0.867
ChatGPT Triggered - Prompts + PromptPerfect	0.832 ↑1.7%	0.827 ↑1.5%	0.980 ↑0.2%	0.980 ↑0.2%	0.886 ↑1.2%	0.876 ↑0.9%

The experiments conducted have provided compelling evidence of the significant and profound influence of different designed prompts on the performance of agricultural document classification. This finding can be extrapolated to various other language understanding tasks as well. We believe that meticulously crafted prompts possess the capability to guide AI models in generating outputs of higher quality and greater value, thereby facilitating more intelligent information processing. Recently, Baidu's founder, Robin Li, stated that in ten years, prompt engineering would encompass half of the world's jobs, highlighting the immense employment opportunities in this field[11]. As we venture into the future of artificial intelligence, the proficiency in prompt design will determine the boundaries

[11] Baidu Ernie Bot: https://wenxin.baidu.com/wenxin/nlp.

of advancement for large-scale language models. Good prompts have the potential to effectively unlock greater potential in large-scale language models like the ChatGPT series, thus enabling their wider application and impact in real-world scenarios.

5 Prominent Cross-Linguistic Supports

To evaluate the level of support that ChatGPT and traditional PLMs provide for different languages, especially low-resource languages, we selected three typical French document fragments from the PestObserver-France dataset. We tokenized these fragments using the tokenizer provided by the official transformers library[12]. The intermediate tokenization results are shown in Table 5.

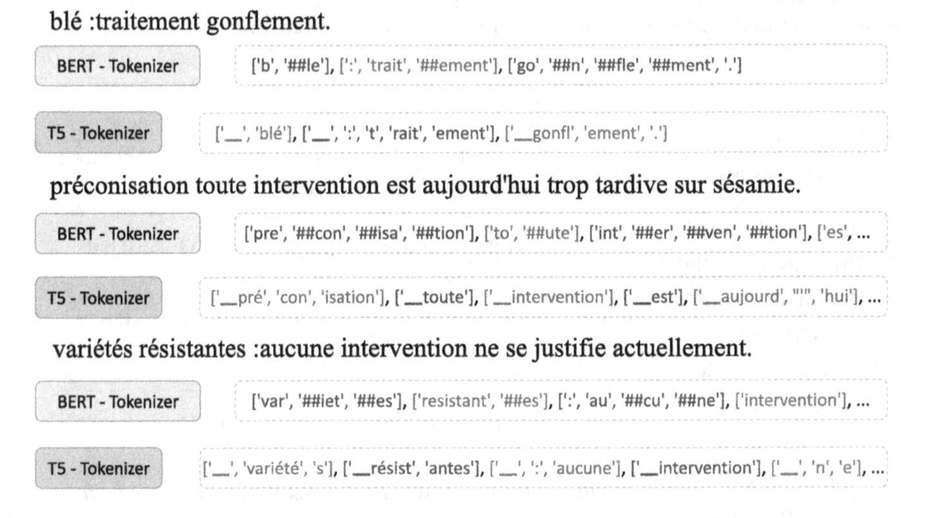

Fig. 5. Three tokenized cases generated by the tokenizers of BERT and T5 on the PestObserver-France dataset. Each group of the same color represents a independent word segmentation result.

From the table, it is evident that BERT and T5 have insufficient vocabulary coverage and support for small languages like French. Specifically, due to the limited vocabulary size during the pre-training phase, most of the French words in these examples are incorrectly segmented. For instance, in the first example, the French word `gonflement` is incorrectly segmented into four subword units with semantic ambiguities: `'go'`, `'##n'`, `'##fle'`, and `'##ment'`. In the second example, `intervention` is segmented into four unrelated individual words: `'int'`, `'##er'`, `'##ven'`, and `'##tion'`. Similarly, in the final example, `variétés` is segmented into `'var'`, `'##iet'`, and `'##es'`. Such inappropriate

[12] https://huggingface.co/docs/transformers/v4.29.1/en/index.

segmentations greatly affect the performance of subsequent language models in semantic understanding during fine-tuning. The primary reason is that the current mainstream pre-trained language models are still limited by the training data, which prioritizes scale and thus restricts their effective and comprehensive coverage of various small languages.

In contrast, one significant advantage of ChatGPT lies in its enormous parameter size. ChatGPT is based on a corpus of 800 billion words (or 45 terabytes of textual data) and contains 175 billion parameters. The 800 billion refers to the training data used for ChatGPT, while the 175 billion model parameters can be considered as the knowledge and representation distilled from this training data. Previously, the largest models only had parameters in the billions, not in the hundreds of billions. Therefore, this larger parameter size enables ChatGPT to capture more complex language patterns and relationships, thereby improving the accuracy of complex natural language processing tasks.

6 Conclusion

Agricultural document classification is a crucial step in managing the vast and ever-increasing amount of agricultural information for sustainable agricultural application. However, mainstream PLM-based classification approaches face several challenges, such as dependency on well-annotated corpora, cross-linguistic transferability. The emergence of ChatGPT, a leading OpenAI conversational technology, has brought a turning point to this dilemma. In this study, we propose a novel ChatGPT-based agricultural document classification framework, to explore the feasibility and potential of using ChatGPT for agricultural document classification. We conducted extensive experiments on datasets that included various languages, comparing our ChatGPT-based classification approach with existing baselines from different learning paradigms. We also developed several effective prompting strategies to further stimulate the generative capability of ChatGPT. The experimental results demonstrate the superiority of ChatGPT in agricultural document classification. Our empirical exploration has opened up new milestones for the development of techniques of ChatGPT-based agricultural information management, without exaggeration. We look forward to proposing more ChatGPT-based applications in sustainable agricultural development, which will help promote future agricultural digital transformation and sustainable agricultural development simultaneously.

Acknowledgements. This work was carried out during the first author's (Wei-qiang Jin) and the second author's (Biao Zhao) research time at Xi'an Jiaotong University. Weiqiang Jin is the corresponding author. The authors declare that they have no known competing financial interests or personal relationships that could have appeared to influence this work. We would like to thank Guizhong Liu for providing helpful discussions and recommendations. Thanks are also due to the anonymous reviewers and action editors for improving the paper with their comments, and recommendations.

References

1. Azeez, N., Al-Taie, I., Yahya, W., Basbrain, A., Clark, A.: Regional agricultural land texture classification based on GLCMS, SVM and decision tree induction techniques. In: 2018 10th Computer Science and Electronic Engineering (CEEC), pp. 131–135 (2018)
2. Bang, Y., et al.: A multitask, multilingual, multimodal evaluation of ChatGPT on reasoning, hallucination, and interactivity (2023)
3. Brown, T.B., et al.: Language models are few-shot learners. In: Proceedings of the 34th International Conference on Neural Information Processing Systems. NIPS 2020, Red Hook, NY, USA, pp. 182–207. Curran Associates Inc. (2020)
4. Cao, Y., Sun, Z., Li, L., Mo, W.: A study of sentiment analysis algorithms for agricultural product reviews based on improved Bert model. Symmetry **14**(8), 1604 (2022)
5. Devlin, J., Chang, M.W., Lee, K., Toutanova, K.: BERT: pre-training of deep bidirectional transformers for language understanding. In: Proceedings of the 2019 Conference of the North American Chapter of the Association for Computational Linguistics: Human Language Technologies, Volume 1 (Long and Short Papers), Minneapolis, Minnesota, pp. 4171–4186. Association for Computational Linguistics (2019)
6. Dunnmon, J., Ganguli, S., Hau, D., Husic, B.: Predicting us state-level agricultural sentiment as a measure of food security with tweets from farming communities (2019)
7. Gao, J., Yu, H., Zhang, S.: Joint event causality extraction using dual-channel enhanced neural network. Knowl.-Based Syst. **258**, 109935 (2022)
8. Gao, J., Zhao, H., Yu, C., Xu, R.: Exploring the feasibility of ChatGPT for event extraction (2023)
9. Jiang, S., Angarita, R., Cormier, S., Rousseaux, F.: Fine-tuning BERT-based models for plant health bulletin classification (2021)
10. Jiao, W., Wang, W., tse Huang, J., Wang, X., Tu, Z.: Is ChatGPT a good translator? yes with GPT-4 as the engine (2023)
11. Jin, W., Zhao, B., Liu, C.: Fintech key-phrase: a new Chinese financial high-tech dataset accelerating expression-level information retrieval. In: Wang, X., et al. (eds.) Database Systems for Advanced Applications, pp. 425–440. Springer, Cham (2023). https://doi.org/10.1007/978-3-031-30675-4_31
12. Jin, W., Zhao, B., Yu, H., Tao, X., Yin, R., Liu, G.: Improving embedded knowledge graph multi-hop question answering by introducing relational chain reasoning. Data Mining and Knowledge Discovery (2022)
13. Liu, P., Qiu, X., Huang, X.: Recurrent neural network for text classification with multi-task learning. In: Proceedings of the Twenty-Fifth International Joint Conference on Artificial Intelligence, pp. 2873–2879. IJCAI 2016, AAAI Press (2016)
14. Liu, P., Yuan, W., Fu, J., Jiang, Z., Hayashi, H., Neubig, G.: Pre-train, prompt, and predict: a systematic survey of prompting methods in natural language processing. ACM Comput. Surv. **55**(9), 1–35 (2023)
15. OpenAI: Gpt-4 technical report (2023)
16. Raffel, C., et al.: Exploring the limits of transfer learning with a unified text-to-text transformer. J. Mach. Learn. Res. **21**(1), 5485–5551 (2020)
17. Shen, Y., et al.: Parallel instance query network for named entity recognition. In: Proceedings of the 60th Annual Meeting of the Association for Computational Linguistics (Volume 1: Long Papers), Dublin, Ireland, pp. 947–961. Association for Computational Linguistics (2022)

18. Yunlai, S., Yunpeng, C., Zhigang, D.: A classification method of agricultural news text based on Bert and deep active learning. J. Lib. Inf. Sci. Agricult. **34**(8), 19 (2022)
19. Vaswani, A., et al.: Attention is all you need. In: Proceedings of the 31st International Conference on Neural Information Processing Systems, Red Hook, NY, USA, pp. 6000–6010. NIPS 2017. Curran Associates Inc. (2017)
20. Wang, J., et al.: Is Chatgpt a good NLG evaluator? A preliminary study (2023)
21. Wei, X., et al.: Zero-shot information extraction via chatting with ChatGPT (2023)
22. Xia, N., Yu, H., Wang, Y., Xuan, J., Luo, X.: DAFS: a domain aware few shot generative model for event detection. Mach. Learn. **112**(3), 1011–1031 (2023)
23. Xiao, Y., Du, Q.: Statistical age-of-information optimization for status update over multi-state fading channels (2023)
24. Xu, J.L., Hsu, Y.L.: Analysis of agricultural exports based on deep learning and text mining. J. Supercomput. **78**(8), 10876–10892 (2022)
25. Zhao, B., Jin, W., Chen, Z., Guo, Y.: A semi-independent policies training method with shared representation for heterogeneous multi-agents reinforcement learning. Front. Neurosci. **17** (2023)
26. Zhao, B., Jin, W., Ser, J.D., Yang, G.: ChataGRI: exploring potentials of ChatGPT on cross-linguistic agricultural text classification (2023)
27. Zhong, Q., Ding, L., Liu, J., Du, B., Tao, D.: Can ChatGPT understand too? a comparative study on ChatGPT and fine-tuned BERT (2023)
28. Zhou, C., et al.: A comprehensive survey on pretrained foundation models: a history from BERT to ChatGPT (2023)

Multi-Task Feature Self-Distillation for Semi-Supervised Machine Translation

Yuxian Wan, Wenlin Zhang(✉), and Zhen Li

School of Information System Engineering, University of Information Engineering, Zhengzhou 450000, China
wenlinzzz@163.com

Abstract. The performance of the model suffers dramatically when the scale of supervision data is constrained because the most sophisticated neural machine translation models are all data-driven techniques. Adding monolingual data to the model training process and using the back-translation method to create pseudo-parallel sentences for data augmentation is the current accepted approach to address data shortage. However, this method's training procedures are laborious, it takes a lot of time, and some of the created pseudo-parallel sentences are of poor quality and detrimental to the model. In this paper, we propose a semi-supervised training method—Multi-Task Feature Self-Distillation (MFSD), which can train models online on a mixed dataset consisting of bilingual and monolingual data jointly. Based on the supervised machine translation task, we propose a self-distillation task for the encoder and decoder of the source and target languages to train two kinds of monolingual data online. In the self-distillation task, we build a teacher model by integrating the student models of the previous rounds, and then use the feature soft labels output by the teacher model with more stable performance to guide the student model training online, and realize the online mining single High-level knowledge of language data by comparing the consistency of the output features of the two models. We conduct experiments on the standard data set and the multi-domain translation data set respectively. The results show that MFSD performed better than mainstream semi-supervised and data augmentation methods approaches. Compared with supervised training, our method improves the BLEU score by an average of +2.27, and effectively improves the model's domain adaptability and generalization ability.

Keywords: Back-translation · Multi-Task · Machine Translation

1 Introduction

In the research field of Natural Language Processing (NLP), Machine Translation (MT) is a topic with high research value and an important task that is relatively difficult to achieve. Machine translation technology needs to comprehend the source language sentences and establish a corresponding relationship

B. Luo et al. (Eds.): ICONIP 2023, CCIS 1965, pp. 238–254, 2024.
https://doi.org/10.1007/978-981-99-8145-8_19

from the source language to the target language, to construct target-side sentences that follow the source and are relatively fluent. Therefore, the quality of machine translation depends on the machine's understanding of natural language and the quality of its generation. If some achievements can be made in machine translation research, it will also be of great help to the promotion of research on other tasks of natural language processing. Both the traditional Statistical Machine Translation [2,23,39] and the current particularly popular Neural Machine Translation (NMT) [5,18,19,37] are data-driven machine translation methods. Apparently, massively parallel corpora is one of the driving factors behind the leap in translation quality. However, in many cases, domain-specific parallel corpora are very scarce, in which case only using domain-specific supervised or unsupervised data is not an optimal solution. In addition, for most languages, large-scale parallel corpus data is still difficult to obtain, and it is even more difficult to improve the performance of NMT in languages with relatively scarce resources. As a result, how to enhance model performance when dealing with a limited parallel corpus is one of the main research areas in the field of MT [1,20,29,36].

Semi-supervised learning approaches [42] based on monolingual data and small-scale bilingual data have grown to be a very popular strategy for obtaining a powerful model in the situation of limited supervised data. Among them, the team at the University of Edinburgh proposed for the first time to use back-translation (BT) technology for NMT [33], using the monolingual data on the target side to effectively improve translation performance. Specifically, BT first generates translation sentences from the target language to the source language through the trained MT model, and mixes the constructed pseudo-parallel sentences with natural parallel sentences to train the translation model, so as to realize the use of monolingual data in training. On the basis of BT, [11] further proposes iterative BT, that is, the process of BT is repeated continuously until a better translation effect is obtained. [13,14] construct higher-quality pseudo-bilingual data through multiple translations to screen target sentences to improve translation quality. [44] proposes a bilingual data extraction method called "Extract-Edit" to replace the widely used BT method to produce high-quality bilingual data. [40] proposes a constrained stochastic BT decoding method based on an automatic post-editing data augmentation architecture. [16] enhances sampling technique by choosing the most informative single-sentence phrases to supplement bilingual data to improve translation quality.

These methods have achieved the desired results, which greatly aid in improving model quality, particularly in low-resource translation scenarios, and have greatly promoted the progress of semi-supervised MT. Their internal logic is to indirectly enhance the model by constructing pseudo-parallel sentences. However, this approach faces three general problems. First of all, the produced pseudo-parallel corpus data has poor quality, which is not always beneficial to the model and will affect the translation performance to a certain extent. Secondly, the BT method's phases are laborious and expensive in terms of training time because they repeatedly need offline reconstruction of the pseudo-parallel corpus. Finally, BT generally only uses one of the source-end sentences and the target-end sentences, and the data usage rate is not high.

Recalling the rationale for the semi-supervised learning approach, the model will be better able to acquire the underlying logic of the language by being trained on a monolingual population. Is it possible to train the model online to extract the characteristics of monolingual data? Along this line, inspired by recent progress in contrastive learning [25,35,43,47], we propose a semi-supervised machine translation method based on Multi-Task Feature Self-Distillation (MFSD). Our semi-supervised task consists of a supervised machine translation task for bilingual data and two self-distillation tasks for monolingual data, which enables online training with mixed data consisting of bilingual data and two kinds of monolingual data through multi-task learning, The step of constructing a pseudo-parallel corpus is omitted, which effectively alleviates the problems existing in the above methods. Specifically, for source and target languages, we combine contrastive learning and knowledge distillation to propose a self-supervised learning method for encoder and decoder feature self-distillation, which extends knowledge distillation to the case without labels by comparing the consistency of model output features. During training, we build the teacher model dynamically using the momentum encoder concept [9] and use its output features as soft labels [45] to guide the training of the student model online. In order to directly mine the underlying high-level knowledge of the monolingual data itself online, we minimize the difference between the output characteristics of the two models. Our approach is flexible enough to work with any other modified strongly supervised task, without modifying the model architecture or adjusting internal norms. In addition, our method and BT method are complementary and can be used instead of or combined with BT in scenarios such as domain adaptation and fine-tuning.

We undertake experiments on typical datasets to test the efficacy of our strategy, and the findings demonstrate that it can efficiently harvest valuable knowledge from monolingual data, enhance model performance, and have additive effects with other approaches. Compared to the base Transformer model, MFSD improves the scores on the IWSLT'14 German-English and IWSLT'15 English-Vietnamese by +2.5 BLEU and +2.04 BLEU, respectively. When combined with other methods, model performance is further improved. We also performed tests on tiny sample data sets to better highlight the function of MFSD in low-resource circumstances. The experimental findings revealed that the model trained on monolingual data consistently improved. Additionally, we test our method's efficacy in the domain adaption scenario. In the domain adaptation scenario, the distinction between the source domain and the target domain is the variation in terminology and vocabulary weight even though both domains are the same language and have the same grammatical structure. In our approach, we train the source-language-to-target-language mapping with bilingual data from the source domain, which is the same across domains, and train the encoder and decoder with two monolingual data from the target domain to learn Linguistic features within domain data. We run trials on multi-domain translation datasets, and the findings demonstrate that our method performs well in different domains and performs better than BT methods in domain adaptation scenarios.

2 Method

This section presents the concrete implementation of our method. Our approach's main goal is to realize online training of bilingual data and monolingual data through multi-task learning. We divide the semi-supervised task into a supervised machine translation task for bilingual data and two self-distillation tasks for monolingual data, where the two self-distillation tasks for monolingual data are encoder self-distillation task for the source language and decoder self-distillation task for the target language. The model structure is shown in the Fig. 1. The overall framework of our suggested architecture for training on monolingual data is the same as that of self-supervised techniques [31]. However, we introduce our method from this viewpoint since it shares certain parallels with knowledge distillation [4,7,10] in terms of its operating concept. We introduce the teacher model structure in Sect. 2.1 and the working specification of our method in Sect. 2.2.

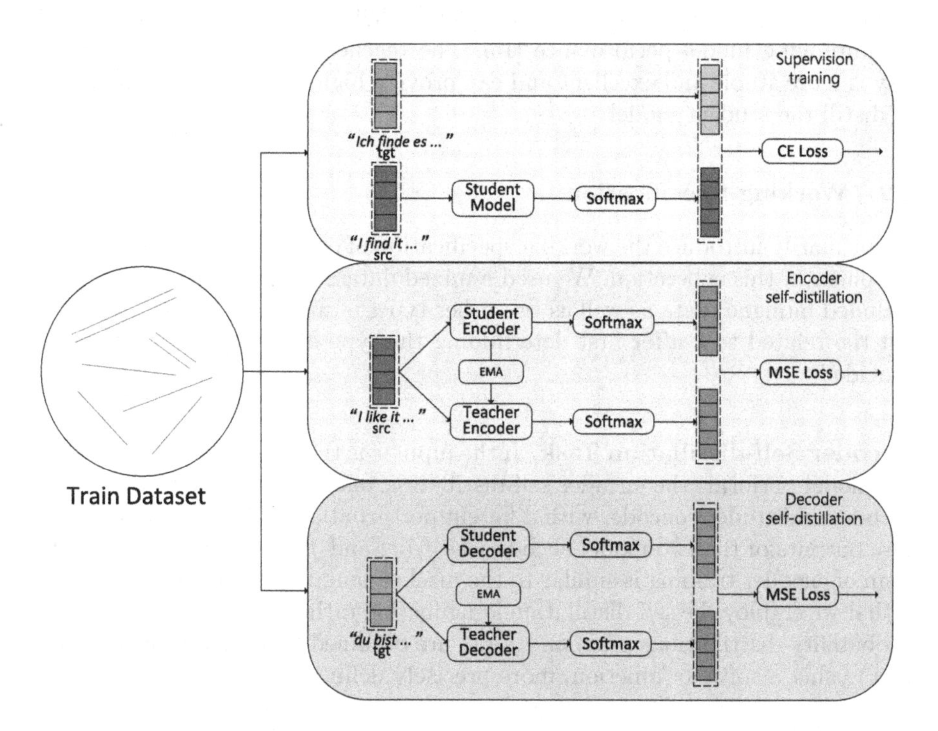

Fig. 1. Multi-Task Feature Self-Distillation for Semi-Supervised Machine Translation. From top to bottom in the figure are supervised machine translation task, encoder self-distillation task, and decoder self-distillation task. The training data set is a mixed data set consisting of labeled data and unlabeled data.

2.1 Teacher Model Structure

Our framework requires training a student model and a teacher model. Different from traditional knowledge distillation using pre-trained teacher models, we follow the idea of momentum encoder to dynamically build teacher models during the training phase. The architectural design of the teacher model and the student model are identical. Using Exponential Moving Average (EMA) on the parameters of the student model, we determined the parameter update rule for the instructor model as follows:

$$\theta_t \leftarrow \lambda\theta_t + (1 - \lambda)\theta_s \tag{1}$$

where θ_s is the parameter of the student model, which is updated through back-propagation. θ_t is the parameter of the teacher model. During training, λ adheres to a cosine schedule from 0.996 to 1 to increase the stability of the mean model update [8]. Momentum encoders were originally proposed to improve the problems of insufficient negative samples and lack of consistency between samples in contrastive learning [9]. In our method, the momentum encoder's function for self-training is more akin to that of the mean teacher [38]. During training, the teacher model performs a form of model ensemble similar to Polyak-Ruppert averaging with exponential decay [27,32], which is a standard practice for improving model performance [15]. The teacher model constructed in this way is more stable and excellent and can provide high-quality feature knowledge to distill the student model.

2.2 Working Specification

We primarily introduce the working specification of our method during the training phase in this subsection. We used a mixed dataset to train our model, which included bilingual data as well as two other types of monolingual data. We carry out the related task after first determining the type of the input that has been provided.

Encoder Self-distillation Task. If the input x is the source monolingual data, the model performs the encoder self-distillation task. We input x to the teacher encoder and student encoder with different perturbations [22] and denote the feature outputs of the two model encoders as $f_t(x)$ and $f_s(x)$. The internal mechanism of encoder training is similar to the masked language model, the distinction is that we employ the self-distillation loss function rather than cross-entropy loss. Probability distributions $p_t(x)$ and $p_s(x)$ are obtained by normalizing $f_t(x)$ and $f_s(x)$ using a softmax function, more precisely defined as:

$$p_{t,i}(x) = \frac{\exp\left(f_{t,i}(x)/\tau_t\right)}{\sum_{k=1}^{K} \exp\left(f_{t,k}(x)/\tau_t\right)} \tag{2}$$

$$p_{s,i}(x) = \frac{\exp\left(f_{s,i}(x)/\tau_s\right)}{\sum_{k=1}^{K} \exp\left(f_{s,k}(x)/\tau_s\right)} \tag{3}$$

where K is the dimension of $p_t(x)$ and $p_s(x)$. τ is the temperature coefficient of the softening probability distribution.

We use the output of the teacher encoder as the target feature to guide the training of the student encoder. Referring to the work of [3, 38, 43], we investigate the performance of the model when using KL Divergence, JS divergence, MAE loss, and MSE loss as the self-distillation task training objective function in Sect. 5.1. The formula is specifically expressed as follows:

$$\mathcal{L}_{KL} = \sum_{i} D_{KL}\left(p_{t,i}(x) \| p_{s,i}(x)\right) \tag{4}$$

$$\mathcal{L}_{MAE} = \frac{1}{m} \sum_{i}^{m} \left\| p_{t,i}(x) - p_{s,i}(x) \right\|_2 \tag{5}$$

$$\mathcal{L}_{MSE} = \frac{1}{m} \sum_{i}^{m} \left\| p_{t,i}(x) - p_{s,i}(x) \right\|_2^2 \tag{6}$$

where m is the number of samples in each batch. Research shows that updating the parameters of the student encoder works best in our architecture by minimizing the MSE loss between the two outputs:

$$\mathcal{L}_{en} = \frac{1}{m} \sum_{i}^{m} \left\| p_{t,i}(x) - p_{s,i}(x) \right\|_2^2 \tag{7}$$

Decoder Self-distillation Task. If the input x is target-side monolingual data, the model performs the decoder self-distillation task. Similar to the encoder self-distillation task, we give the input x to the teacher decoder and student decoder respectively, denoting the feature outputs of the two model decoders as $g_t(x)$ and $g_s(x)$. The internal mechanism of decoder training is similar to the causal language model, we use the self-distillation objective function. The predicted probability distributions $p_t\left(y_i \mid \mathbf{x}, \mathbf{y}_{<i}\right)$ and $p_s\left(y_i \mid \mathbf{x}, \mathbf{y}_{<i}\right)$ are obtained by normalizing $g_t(x)$ and $g_s(x)$ using the softmax function. Likewise, we define the training objective for the decoder self-distillation task by minimizing the MSE loss between two outputs:

$$\mathcal{L}_{de} = \frac{1}{m} \sum_{i}^{m} \left\| p_{t,i}\left(y_i \mid \mathbf{x}, \mathbf{y}_{<i}\right) - p_{s,i}\left(y_i \mid \mathbf{x}, \mathbf{y}_{<i}\right) \right\|_2^2 \tag{8}$$

Table 1. Statistics for different datasets.

Dataset	Standard		Low-Resource		Domains		
	De-En	En-Vi	De-En (Low)	En-Vi (Low)	Medical	Koran	Law
Train	160k	133k	40k	40k	17k	248k	467k
Vaild	7283	1553	2000	1553	2000	2000	2000
Test	6750	1268	2000	1268	2000	2000	2000

Supervised Machine Translation Task. If the input x is labeled bilingual data, the model performs supervised machine translation tasks. The task of machine translation supervision for bilingual data is not the focus of this paper, and will not be introduced here. We follow a standard end-to-end machine translation approach [41].

3 Experiments

3.1 Datasets

To test the effectiveness of our approach generally and it is capacity to mine monolingual knowledge mining, we conduct experiments on two small-sample standard datasets. We conduct experiments on IWSLT'14 German-English (De-En) and IWSLT'15 English-Vietnamese (En-Vi). We follow the steps of [46] to deal with De-En[1] and En-Vi[2]. We sampled standard data sets to simulate small sample data sets in low-resource scenarios. 40k training samples were extracted for each pair of languages. The monolingual data in the experiment all come from News Crawl[3] and the scale is twice that of the bilingual data. Additionally, we do experiments on the multi-domain translation dataset to confirm that our technique works just as well in the domain adaption situation. Our experiments consider domains including Medical, Koran, and Law. We use the pre-processed dataset[4] provided by [48]. The construction of the unaligned corpus is processed according to [12] and the scale is twice that of the bilingual data. The dataset's sentence statistics are displayed in Table 1. We applied Byte Pair Encoding to extract shared subword units [34].

3.2 Baselines

MFSD is also a data augmentation approach in another sense, so our main baseline is powerful and widely used data augmentation and semi-supervised techniques:

---

[1] https://github.com/pytorch/fairseq/blob/main/examples/translation/prepare-iwslt14.sh.

[2] https://nlp.stanford.edu/projects/nmt/.

[3] http://data.statmt.org/news-crawl/.

[4] https://github.com/zhengxxn/adaptive-knn-mt.

Transformer. We conduct supervised training on the most basic Transformer model and only use parallel corpus in the training process.

Back [17]. This is the basic BT baseline. BT first generates translation statements from the target language to the source language through the trained reverse MT model and then mixes the constructed pseudo-parallel statements with natural parallel statements to train the translation model.

UNCSAMP [16]. This is an improved BT algorithm. In the stage of extracting monolingual data, the most informative monolingual lines are selected to supplement the bilingual lines instead of random sampling subsets to construct the composite data.

CMLM [6]. This is a data enhancement approach based on the conditional masking language model and soft word substitution. Using deep, bidirectional CMLM can enhance semantic consistency between generated and raw data during data enhancement by conditioning the source and target.

3.3 Evaluation

As is customary, we use multi-bleu.perl[5] to measure case-sensitive BLEU [26] for small sample standard datasets. We closely follow [18] to assess the Sacre-BLEU [28] results for multi-domain translation datasets in order to provide a fair comparison.

3.4 Implementation Details

We adopted the open-source fairseq toolkit [24] to implement our algorithms. We use transformer_iwslt_de_en configuration. In the transformer model, the encoder and decoder each have 6 layers, 4 heads, and 512 dimensions. During training, the data of each batch is composed of three kinds of data mixed in proportion. We train for 500 epochs and then perform 30 epochs of supervised machine translation training at the breakpoint with the highest score to achieve the purpose of the fitting. When performing the decoder self-distillation task, we remove the Encoder-Decoder Attention layer, and the model training is similar to causal prediction. With a beam size of 5, we employ the beam search. The final model was chosen based on having the best perplexity on the validation set for all tests, which were carried out on a machine with 4 NVIDIA V100 GPUs.

[5] https://github.com/moses-smt/mosesdecoder/blob/master/scripts/genric/multi-bleu.perl.

Table 2. Experimental results of Standard Datasets, Low Resources Datasets, and Domains Datasets.

Models	Standard		Low-Resource		Domains		
	De-En	En-Vi	De-En (Low)	En-Vi (Low)	Medical	Koran	Law
Transformer(base)	34.01	30.65	22.43	20.83	27.46	13.74	60.71
Back	35.50	31.87	24.51	23.74	31.65	17.19	60.82
UNCSAMP	36.31	32.55	25.89	25.01	34.35	19.75	61.54
CMLM	35.93	32.01	25.23	24.12	33.16	20.61	61.11
MFSD	**36.51**	**32.69**	**26.76**	**25.75**	**38.74**	**21.47**	**61.50**
+Back	36.76	32.73	27.65	27.03	39.45	23.43	61.59
+UNCSAMP	**37.02**	**33.26**	**28.68**	**28.09**	**40.26**	24.27	**61.82**
+CMLM	36.94	32.88	28.15	27.46	40.12	**24.86**	61.61

4 Experimental Results

We first validate the MFSD framework used in this study on standard and small sample datasets, and compare the model's performance against mainstream semi-supervised and data-enhanced techniques. We then conduct experiments on the multi-domain translation dataset and investigate the effectiveness of MFSD in domain adaptation scenarios. We'll think about using MFSD to run tests in the fine-tuning scenario to more thoroughly confirm how well the model performs when our method is used in conjunction with pre-training methods in the future.

4.1 Results of Standard Datasets

Table 2 displays the findings from our experiments using common datasets. We contrasted our approach with baseline algorithms and supervised training to demonstrate its efficacy. Our method introduces a self-supervised model as a feature extractor for online training on mixed data and achieves excellent performance on two datasets. Specifically, on the basic Transformer model, MFSD has achieved very good performance, improving +2.5 BLEU score and +2.04 BLEU score on De-En and En-Vi respectively, indicating that our method can effectively learn the internal features of monolingual data to enhance the model performance. MFSD demonstrated better performance than BT and data enhancement technology. We also showed how our strategy works in conjunction with other strategies. The results indicate that the self-distillation method is complementary to the already well-liked semi-supervised methods and can be used in conjunction with them to further improve model performance. The model performs better when MFSD is paired with other baseline methods.

4.2 Results of Low Resource Datasets

Table 2 shows our experimental results on a small sample data set. When using only Transformer models without other related technologies, too little training data will result in difficult model training to fit and poor predictive effect. When the data is enhanced by other techniques such as BT, the model performance is improved to some extent. However, due to the poor quality of pseudo-parallel statements produced by BT, the performance improvement is not obvious. MFSD directly trains monolingual data by mining high-level knowledge in the language and makes more effective use of monolingual data for data enhancement. The results demonstrate that using MFSD significantly improves the model's performance, effectively improving the issue that it is challenging to train the model when resources are limited.

4.3 Results of Domains Datasets

We further tested the effectiveness of MFSD on the multi-domain translation dataset in addition to the standard dataset, and the experimental results are displayed in Table 2. MFSD again outperforms all baselines in all domains. Our proposed method achieves +6.6 BLEU score on average compared to the base Transformer in three domains. This further demonstrates that our strategy successfully enhances the performance of the model and has a great ability to exploit monolingual data features.

4.4 Domain Adaptation

To confirm the generalizability of our strategy, we also run experiments in the semi-supervised domain adaption scenario. The Table 3 displays the findings from the experiment. We further develop our approach to apply to instances when there are only bilingual data in the source domain and monolingual data in the destination domain in the semi-supervised domain adaptation scenario. We define multi-task as a supervised machine translation task on source domain bilingual data and two self-distillation tasks on target domain monolingual data. We train the encoder and decoder on the two monolingual datasets in the target domain to acquire the high-level knowledge included in the domain data, and we train the model on bilingual data in the source domain to learn the mapping connection between the source language and the target language. The outcomes demonstrate that our approach dramatically enhances the model in the semi-supervised domain adaption scenario, and our MFSD performs noticeably better than the basic Transformer and the model utilizing BT technology. This further demonstrates the generalizability of our method, confirming our conclusions in Sect. 1.

5 Analysis

We outline our analytical MFSD experimentation in this section. We look into how various components affect a model's performance. The whole study is conducted on the De-En dataset.

Table 3. Experimental results of Domain Adaptation. "Law-Medical" means that we train the model on the Law domain and directly apply it to the Medical domain, and vice versa.

Models	Law-Medical	Medical-Law
Transformer (base)	18.76	3.05
Back	24.64	17.57
MFSD	29.46	24.48

5.1 Effect of Language Modeling Task Training Objectives

We show the effect of using other self-supervised learning training objectives on the training of our framework. Referring to the work of [3,38,43], we further tested the performance of the model when using KL divergence, JS divergence, MAE loss, and MSE loss as training objectives for self-distillation tasks. The Table 4 displays the experimental outcomes. When other variables stay the same, it is clear that the MSE loss with the best performance is better suited for our framework.

Table 4. BLEU scores of model when using different losses as training objectives for self-distillation tasks.

Training Objectives	KL divergence	JS divergence	MAE loss	MSE loss
BLEU	36.47	36.42	36.36	36.51

5.2 Effect on Different Part

In the Table 5, we show the model's performance when only one of the encoder self-distillation task and the decoder self-distillation task is added to the supervised machine translation task. In the experiment, the monolingual data is twice that of the bilingual data. It can be seen that when only one self-distillation task is added, the model's performance has greatly increased, which shows that both self-distillation tasks have effectively enhanced the model's ability to mine monolingual data knowledge. The results also show that these two tasks are complementary. They work best when combined and help the model more accurately represent both the source and target languages.

Table 5. BLEU scores of model when adding different modeling tasks.

Tasks	Part			
	None	Encoder	Decoder	All
BLEU	34.01	35.89	35.63	36.51

5.3 Monolingual Data Size

We investigate how our models perform in relation to the volume of monolingual data. The Fig. 2 displays the experimental findings. As can be observed, self-distillation utilizing 160k monolingual data with the same size as bilingual data can greatly enhance NMT performance, and adding more monolingual data of a larger size will not result in further model improvements. When using 480k monolingual data which is three times the size of bilingual data, the final performance of the model only reaches 36.42 BLEU score. This indicates that merely increasing the amount of monolingual data is not a viable way to enhance self-training and that more advanced techniques are required to utilize monolingual data.

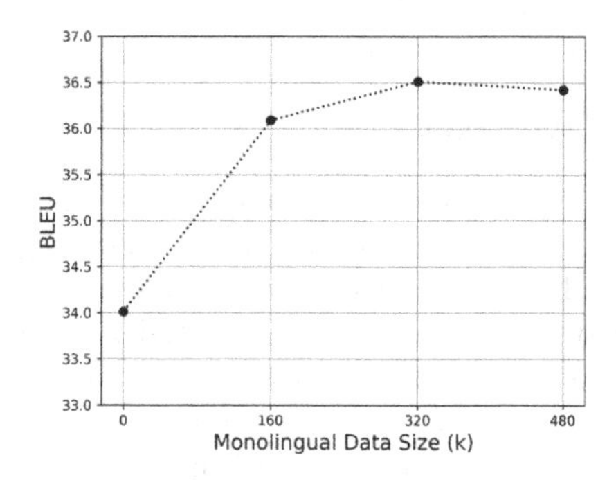

Fig. 2. BLEU scores of model with the increased size of monolingual data.

5.4 Hallucinations

The model's attention mechanism might not accurately reflect the model's actual attention. [21] proposed the concept of hallucinations to further understand the NMT model. If modest input changes cause rapid changes in the output, the model is hallucinating and is not really paying attention to the input. In order to verify that the model is more robust, we followed the algorithm of [30] and used 50 and 100 most common subwords as perturbations respectively, and tested the model's performance under input perturbations.

Table 6 shows the number of hallucinations of the model on the IWSLT De-En test set in the baseline and MFSD. The baseline is a supervised MT model.

In MFSD training, tests were performed using both the student and teacher models. The number of hallucinations dropped on average by 30% in the student model and 40% in the instructor model as compared to the supervised MT. This indicates that the model in our approach is more robust to interference and more focused on input content. The results show that there are fewer hallucinations in the teacher model than in the student model, proving that it is a more stable model overall. This further confirms the implementability of the internal logic of our method.

Table 6. Number of distinct sentences which cause hallucinations in the baseline and MFSD models.

Models	Hallucinations	
	50 subwords	100 subwords
Transformer	24	47
Student Model	16	33
Teacher Model	13	29

5.5 Computation Overhead

In our method, monolingual data is added to the training process for data enhancement, which increases the complexity of training to a certain extent. We compare the model calculation overhead on the IWSLT De-En standard data set. We performed the experiment on 4 NVIDIA V100 GPUs. According to our experiment, it takes a total of 20 h to train the model using only supervised training. BT model increases the training time by about 85%, from 20 h to 38 h. Our approach increases the training time compared to supervised training by about 50%, from 20 h to 30 h. Our approach is computatively expensive compared to supervised training because MFSD uses two models and requires two model generations. However, MFSD is much less complex and faster than BT training because our approach does not require textual reasoning every round, which is serial and time-consuming. Our method alleviates the problem of high training complexity in BT.

6 Conclusion

In this paper, we introduce a semi-supervised machine translation method based on Multi-Task Feature Self-Distillation, which divides the semi-supervised task into one supervised machine translation task for bilingual data and two self-distillation tasks for monolingual data. For monolingual data, we propose a self-distillation training form to enable the teacher model to guide the student model to train and complete the online mining of language features. Our method effectively solves many problems existing in the classical method of BT. We conducted a large number of experiments to demonstrate the effectiveness of our strategy and to further demonstrate the generalizability of our model on semi-supervised domain adaptation tasks. However, because our method uses two models, it will increase the computational overhead compared with supervised training. In the future, we will consider further improving our method in other scenarios such as pre-training and fine-tuning, and apply MFSD to other tasks.

Acknowledgements. This work was supported by the National Natural Science Foundation of China under Grant No. 62171470, Natural Science Foundation of Henan Province of China (No. 232300421240), Zhongyuan Science and Technology Innovation Leading Talent Project of Henan Province of China (No. 234200510019). We appreciate the anonymous reviewers for their helpful comments.

References

1. Bengio, Y., Courville, A.C., Vincent, P.: Representation learning: a review and new perspectives. IEEE Trans. Pattern Anal. Mach. Intell. **35**, 1798–1828 (2012)
2. Brown, P.F., Pietra, S.D., Pietra, V.J.D., Mercer, R.L.: The mathematics of statistical machine translation: parameter estimation. Comput. Linguist. **19**, 263–311 (1993)
3. Caron, M., et al.: Emerging properties in self-supervised vision transformers. In: 2021 IEEE/CVF International Conference on Computer Vision, ICCV 2021, Montreal, QC, Canada, 10–17 October 2021, pp. 9630–9640 (2021)
4. Chen, T., Kornblith, S., Swersky, K., Norouzi, M., Hinton, G.E.: Big self-supervised models are strong semi-supervised learners. arXiv arXiv:2006.10029 (2020)
5. Chen, Y., Gan, Z., Cheng, Y., Liu, J., Liu, J.: Distilling knowledge learned in BERT for text generation. In: Proceedings of the 58th Annual Meeting of the Association for Computational Linguistics, ACL 2020, Online, 5–10 July 2020, pp. 7893–7905 (2020)
6. Cheng, Q., Huang, J., Duan, Y.: Semantically consistent data augmentation for neural machine translation via conditional masked language model. In: International Conference on Computational Linguistics (2022)
7. Fang, Z., Wang, J., Wang, L., Zhang, L., Yang, Y., Liu, Z.: SEED: self-supervised distillation for visual representation. arXiv arXiv:2101.04731 (2021)
8. Grill, J.B., et al.: Bootstrap your own latent: a new approach to self-supervised learning. arXiv arXiv:2006.07733 (2020)
9. He, K., Fan, H., Wu, Y., Xie, S., Girshick, R.B.: Momentum contrast for unsupervised visual representation learning. In: 2020 IEEE/CVF Conference on Computer Vision and Pattern Recognition, CVPR 2020, Seattle, WA, USA, 13–19 June 2020, pp. 9726–9735 (2020)
10. Hinton, G.E., Vinyals, O., Dean, J.: Distilling the knowledge in a neural network. CoRR abs/1503.02531 (2015)
11. Hoang, C.D.V., Koehn, P., Haffari, G., Cohn, T.: Iterative back-translation for neural machine translation. In: NMT@ACL (2018)
12. Hu, J., Xia, M., Neubig, G., Carbonell, J.G.: Domain adaptation of neural machine translation by lexicon induction. In: Annual Meeting of the Association for Computational Linguistics (2019)

13. Imankulova, A., Dabre, R., Fujita, A., Imamura, K.: Exploiting out-of-domain parallel data through multilingual transfer learning for low-resource neural machine translation. arXiv arXiv:1907.03060 (2019)
14. Imankulova, A., Sato, T., Komachi, M.: Improving low-resource neural machine translation with filtered pseudo-parallel corpus. In: WAT@IJCNLP (2017)
15. Jean, S., Cho, K., Memisevic, R., Bengio, Y.: On using very large target vocabulary for neural machine translation. arXiv arXiv:1412.2007 (2014)
16. Jiao, W., Wang, X., Tu, Z., Shi, S., Lyu, M.R., King, I.: Self-training sampling with monolingual data uncertainty for neural machine translation. In: Annual Meeting of the Association for Computational Linguistics (2021)
17. Jin, D., Jin, Z., Zhou, J.T., Szolovits, P.: A simple baseline to semi-supervised domain adaptation for machine translation. arXiv arXiv:2001.08140 (2020)
18. Khandelwal, U., Fan, A., Jurafsky, D., Zettlemoyer, L., Lewis, M.: Nearest neighbor machine translation. CoRR abs/2010.00710 (2020)
19. Lample, G., Conneau, A.: Cross-lingual language model pretraining. In: Neural Information Processing Systems (2019)
20. Lample, G., Conneau, A., Denoyer, L., Ranzato, M.: Unsupervised machine translation using monolingual corpora only. In: 6th International Conference on Learning Representations, ICLR 2018, Conference Track Proceedings, Vancouver, BC, Canada, 30 April–3 May 2018 (2018)
21. Lee, K., Firat, O., Agarwal, A., Fannjiang, C., Sussillo, D.: Hallucinations in neural machine translation (2018)
22. Lewis, M., et al.: BART: denoising sequence-to-sequence pre-training for natural language generation, translation, and comprehension. In: Annual Meeting of the Association for Computational Linguistics (2019)
23. Och, F.J., Ney, H.: The alignment template approach to statistical machine translation. Comput. Linguist. **30**, 417–449 (2004)
24. Ott, M., et al.: fairseq: a fast, extensible toolkit for sequence modeling. In: Ammar, W., Louis, A., Mostafazadeh, N. (eds.) Proceedings of the 2019 Conference of the North American Chapter of the Association for Computational Linguistics: Human Language Technologies, NAACL-HLT 2019, Demonstrations, Minneapolis, MN, USA, 2–7 June 2019, pp. 48–53 (2019)
25. Pan, X., Wang, M., Wu, L., Li, L.: Contrastive learning for many-to-many multilingual neural machine translation. arXiv arXiv:2105.09501 (2021)
26. Papineni, K., Roukos, S., Ward, T., Zhu, W.: BLEU: a method for automatic evaluation of machine translation. In: Proceedings of the 40th Annual Meeting of the Association for Computational Linguistics, 6–12 July 2002, Philadelphia, PA, USA, pp. 311–318 (2002)
27. Polyak, B., Juditsky, A.B.: Acceleration of stochastic approximation by averaging. SIAM J. Control. Optim. **30**, 838–855 (1992)
28. Post, M.: A call for clarity in reporting BLEU scores. In: Bojar, O., et al. (eds.) Proceedings of the Third Conference on Machine Translation: Research Papers, WMT 2018, Belgium, Brussels, 31 October–1 November 2018, pp. 186–191 (2018)
29. Raina, R., Battle, A.J., Lee, H., Packer, B., Ng, A.Y.: Self-taught learning: transfer learning from unlabeled data. In: Ghahramani, Z. (ed.) Proceedings of the Twenty-Fourth International Conference on Machine Learning, ICML 2007, Corvallis, Oregon, USA, 20–24 June 2007, vol. 227, pp. 759–766 (2007)
30. Raunak, V., Menezes, A., Junczys-Dowmunt, M.: The curious case of hallucinations in neural machine translation. arXiv arXiv:2104.06683 (2021)

31. Ruiter, D., Klakow, D., van Genabith, J., España-Bonet, C.: Integrating unsupervised data generation into self-supervised neural machine translation for low-resource languages. arXiv arXiv:2107.08772 (2021)
32. Ruppert, D.: Efficient estimations from a slowly convergent Robbins-Monro process (1988)
33. Sennrich, R., Haddow, B., Birch, A.: Improving neural machine translation models with monolingual data. arXiv arXiv:1511.06709 (2015)
34. Sennrich, R., Haddow, B., Birch, A.: Neural machine translation of rare words with subword units. arXiv arXiv:1508.07909 (2015)
35. Siddhant, A., et al.: Leveraging monolingual data with self-supervision for multilingual neural machine translation. In: Annual Meeting of the Association for Computational Linguistics (2020)
36. Siddhant, A., et al.: Leveraging monolingual data with self-supervision for multilingual neural machine translation. In: Proceedings of the 58th Annual Meeting of the Association for Computational Linguistics, ACL 2020, Online, 5–10 July 2020, pp. 2827–2835 (2020)
37. Sutskever, I., Vinyals, O., Le, Q.V.: Sequence to sequence learning with neural networks. In: Advances in Neural Information Processing Systems 27: Annual Conference on Neural Information Processing Systems 2014, 8–13 December 2014, Montreal, Quebec, Canada, pp. 3104–3112 (2014)
38. Tarvainen, A., Valpola, H.: Mean teachers are better role models: weight-averaged consistency targets improve semi-supervised deep learning results. In: 5th International Conference on Learning Representations, ICLR 2017, Workshop Track Proceedings, Toulon, France, 24–26 April 2017 (2017)
39. Taskar, B., Lacoste-Julien, S., Klein, D.: A discriminative matching approach to word alignment. In: Proceedings of the Conference on Human Language Technology Conference and Conference on Empirical Methods in Natural Language Processing, HLT/EMNLP 2005, 6–8 October 2005, Vancouver, British Columbia, Canada, pp. 73–80 (2005)
40. Tong, Y., Chen, Y., Zhang, G., Zheng, J., Zhu, H., Shi, X.: Generating diverse back-translations via constraint random decoding. In: CCMT (2021)
41. Vaswani, A., et al.: Attention is all you need. In: Advances in Neural Information Processing Systems 30: Annual Conference on Neural Information Processing Systems 2017, 4–9 December 2017, Long Beach, CA, USA, pp. 5998–6008 (2017)
42. Wang, R., Tan, X., Luo, R., Qin, T., Liu, T.Y.: A survey on low-resource neural machine translation. In: International Joint Conference on Artificial Intelligence (2021)
43. Wei, Y., et al.: Contrastive learning rivals masked image modeling in fine-tuning via feature distillation. arXiv arXiv:2205.14141 (2022)
44. Wu, J., Wang, X.E., Wang, W.Y.: Extract and edit: an alternative to back-translation for unsupervised neural machine translation. In: North American Chapter of the Association for Computational Linguistics (2019)
45. Xie, Q., Hovy, E.H., Luong, M.T., Le, Q.V.: Self-training with noisy student improves ImageNet classification. In: 2020 IEEE/CVF Conference on Computer Vision and Pattern Recognition (CVPR), pp. 10684–10695 (2019)
46. Yang, Z., Sun, R., Wan, X.: Nearest neighbor knowledge distillation for neural machine translation. In: Carpuat, M., de Marneffe, M., Ruíz, I.V.M. (eds.) NAACL, pp. 5546–5556 (2022)
47. Zhang, T., et al.: Frequency-aware contrastive learning for neural machine translation. In: AAAI Conference on Artificial Intelligence (2021)

48. Zheng, X., et al.: Adaptive nearest neighbor machine translation. In: Zong, C., Xia, F., Li, W., Navigli, R. (eds.) Proceedings of the 59th Annual Meeting of the Association for Computational Linguistics and the 11th International Joint Conference on Natural Language Processing, ACL/IJCNLP 2021, (Volume 2: Short Papers), Virtual Event, 1–6 August 2021, pp. 368–374 (2021)

ADGCN: A Weakly Supervised Framework for Anomaly Detection in Social Networks

Zhixiang Shen, Tianle Zhang(✉), and Haolan He

School of Computer Science and Engineering, University of Electronic Science and Technology of China, Chengdu, China
tianle_zhang_@outlook.com

Abstract. Detecting abnormal users in social networks is crucial for protecting user privacy and preventing criminal activities. However, existing graph learning methods have limitations. Unsupervised methods focus on topological anomalies and may overlook user characteristics, while supervised methods require costly data annotations. To address these challenges, we propose a weakly supervised framework called Anomaly Detection Graph Convolutional Network (ADGCN). Our model includes three modules: information-preserving compression of user features, collaborative mining of global and local graph information, and multi-view weakly supervised classification. We demonstrate that ADGCN generates high-quality user representations using minimal labeled data and achieves state-of-the-art performance on two real-world social network datasets. Ablation experiments and performance analyses show the feasibility and effectiveness of our approach in practical scenarios.

Keywords: Anomaly Detection · Social Networks · Weak Supervision · Graph Neural Network · Graph Autoencoder

1 Introduction

Social networks have gained immense importance in the era of information explosion due to their vast user privacy information and commercial value. Unfortunately, they are often the primary targets of criminal activities, with abnormal users being a common means of attack [17, 23]. Identifying abnormal users accurately and effectively is crucial for safeguarding user privacy and preventing criminal activities, as highlighted by recent statistics [23]. Recent research has attempted to integrate advances in graph learning to detect anomalies in social networks. For instance, one study utilized graph unsupervised learning for anomaly data mining, which helped reduce the processing cost of data labeling [13]. Additionally, graph-based supervised learning has been employed by designing an appropriate spatial message passing mechanism to capture abnormal information [7].

Z. Shen, T. Zhang and H. He—Contributed equally to this research.

B. Luo et al. (Eds.): ICONIP 2023, CCIS 1965, pp. 255–266, 2024.
https://doi.org/10.1007/978-981-99-8145-8_20

However, the above methods have several drawbacks. With unsupervised graph learning, existing works on anomaly detection primarily focus on using auto-encoders to reconstruct the adjacency matrix of the graph, which lacks effective discrimination of user anomaly attributes [8]. Conversely, supervised methods employ classification loss functions coupled with well-designed message passing mechanisms to learn abnormal features or attributes. However, a significant drawback of supervised tasks is the high cost of data label collection and annotation, making it challenging to deploy and update in actual scenarios [9].

To address the challenges of capturing anomalous features and high costs associated with data labeling, we propose a weakly supervised anomaly detection framework called Anomaly Detection Graph Convolutional Network (ADGCN). Our model comprises three parts: information-preserving compression of user features, collaborative mining of global and local graph information, and a multi-view weakly supervised classifier. The most significant technological innovation is that our ADGCN model exploratively combines generative and contrastive self-supervised mechanisms to enhance model representation in weakly supervised scenarios and we conducted extensive experiments on two real-world social network datasets to demonstrate that our model can produce high-quality and easily classifiable representations for user nodes while using only about 1% of the labeled data and achieve state-of-the-art performance in this task. We also performed ablation experiments and sensitivity analysis of the model hyperparameters, demonstrating the essentiality of the three modules and the feasibility and superiority of our weakly supervised learning framework in detecting anomalies in real social networks. In the Methodology section, we will show the entire streaming workflows of our model in detail as well. And in the Experiment and Discussion sections, specific details of experimental settings and results will be shown. The main contributions are summarized as follows:

- We identify the limitations of previous graph-based anomaly detection methods and propose a novel GCN-based encoder that jointly mines user content features and network topology information to overcome these issues.
- We introduce a weakly supervised framework combining generative and contrastive self-supervised mechanisms (ADGCN) for anomaly detection in social networks that generates high-quality user embeddings using only a small amount of labeled data (1%).
- We conduct extensive experiments on two real-world benchmark datasets, demonstrating the effectiveness and superiority of our approach over state-of-the-art methods.[1]

2 Related Work

2.1 Anomaly Detection in Social Network

Traditional social network anomaly detection algorithms primarily use content features of users' statements under different posts and the complex network relationships between different users and posts. These content features are often processed into high-dimensional vectors through the Linguistic Inquiry and Word Count (LIWC) method

[1] The source code is available at https://github.com/zxlearningdeep/ADGCN-project.

[16]. For network information processing, matrix factorization is employed to reduce the dimension of the adjacency matrix [18].

2.2 Graph Unsupervised Learning

Graph Auto-Encoder (GAE) and Variational Graph Auto-Encoder (VGAE) are two popular graph embedding methods that leverage the topological information of graphs to reconstruct the adjacency matrix but have limitations in learning node features. Recently, contrastive learning has been extended to the graph learning framework (GCL), which constructs different views of data using data augmentation techniques and maximizes the mutual information of the same node representation from two views while reducing the similarity between different node representations. Some researchers have proposed adaptive data augmentation methods based on topological centrality to improve the heuristic data augmentation approach (GCA) [11, 22, 24].

2.3 Message Passing Neural Networks

The Message Passing Neural Network consists of two parts: message passing and readout [4], with the former being the core of graph neural network and explored extensively in previous works [2, 20]. The Graph Convolutional Networks (GCN) introduce the convolution from computer vision into graph neural networks, performing feature aggregation between neighboring nodes [10, 21]. Graph Attention Networks (GAT) enhance representation by introducing an attention mechanism to assign different weights to different nodes in a neighborhood [19]. GraphSAGE proposes an inductive learning framework for graph convolution by sampling from multi-hop neighbors and trying various aggregation methods [5]. Recent works have combined spectral and spatial convolutions to balance global and local feature modeling [1].

2.4 Weakly Supervised Learning

Some weakly supervised methods mainly focus on semi-supervised learning using the pseudo-labels generated by the model, while this method requires the original distribution of the dataset to present significant clustering distribution and requires high learning time cost [12, 14]. Another academic path is to use data augmentation to expand the training data [6].

3 Methodology

In this section, we describe our proposed ADGCN framework in detail, as shown in Fig. 1. ADGCN firstly uses the Auto-Encoder to extract the high-quality low-dimensional representation of the original node features, then mines the global and local information through the asymmetric centrality graph augmentation and node feature aggregation, and finally collaboratively learns the high-quality representation of nodes with the multi-view weakly supervised classifier.

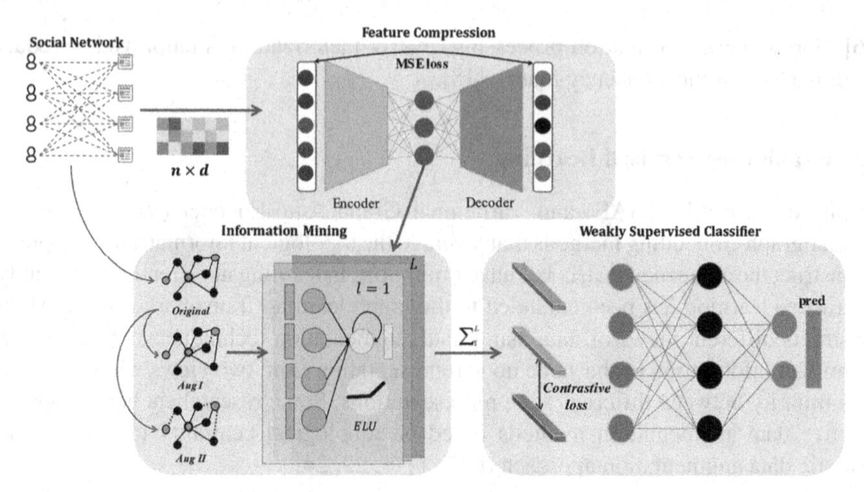

Fig. 1. ADGCN contains three modules: information-preserving compression of user features, GCN-based mining of global-local information, and multi-view weakly supervised classifier.

Problem Definition: Anomaly detection in social networks can be abstracted as a classification problem. The input is the adjacency matrix and the feature matrix of the network, where each node is a user or a post, each edge represents an information release by the user on the post, and the feature matrix of the node is obtained by LIWC method from the content published by the user. The final task is to train a model with very little annotated data to embed each user node in order to facilitate classification.

3.1 Information-Preserving Feature Compression

The initial features of nodes are usually high-dimensional vectors. In order to obtain a low-dimensional embedding that can be operated efficiently, we use the Auto-Encoder for information-preserving feature compression. Auto-Encoder contains encoder and decoder, both implemented using MLP. Each layer of the multilayer perceptron is calculated as follows:

$$\mathbf{X}^{(\ell+1)} = \sigma(\mathbf{X}^{(\ell)}\mathbf{W}^{(\ell)}_{\mathrm{MLP}}) \tag{1}$$

where $\mathbf{X}^{(\ell)}$ and $\mathbf{W}^{(\ell)}_{\mathrm{MLP}}$ denote the input matrix and parameter matrix at the layer ℓ of the MLP respectively. $\sigma(\cdot)$ is an active function such as "ELU" [3], "Sigmoid" et al. In order to make the extracted embeddings retain the information of node features, we take the Mean-Squared Loss as the loss function. The specific process is as follows:

$$\mathbf{H}^{(0)} = \mathrm{MLP}_{\mathrm{E}}(\mathbf{X}), \hat{\mathbf{X}} = \mathrm{MLP}_{\mathrm{D}}(\mathbf{H}^{(0)}) \tag{2}$$

$$\mathcal{L}_{\mathbf{e}} = ||\hat{\mathbf{X}} - \mathbf{X}||^2_{\mathrm{F}} \tag{3}$$

where $\mathbf{X}$ and $\hat{\mathbf{X}} \in \mathbb{R}^{n\times q}$ denote the Node Original Feature Matrix and the Reconstruction Feature Matrix respectively. $\mathrm{MLP}_{\mathrm{E}}(\cdot)$ and $\mathrm{MLP}_{\mathrm{D}}(\cdot)$ are Encoder and Decoder of the

Auto-Encoder respectively. $\mathbf{H}^{(0)} \in \mathbb{R}^{n\times d}$ represents the low-dimensional embedding after dimensionality reduction, which is also the input matrix of the first layer of GCN.

3.2 Collaborative Mining of Global and Local Information

Previous work also pointed out that GCN is equivalent to a low-pass filter, which can effectively capture the low-pass component of graph signal and mine the local information [15]. The aggregation process of each layer can be expressed as follows:

$$\mathbf{H}^{(\ell+1)} = \sigma(\tilde{\mathbf{D}}^{-\frac{1}{2}}\tilde{\mathbf{A}}\tilde{\mathbf{D}}^{-\frac{1}{2}}\mathbf{H}^{(\ell)}\mathbf{W}^{(\ell)}_{\mathrm{GCN}}), \mathbf{Z}_{\mathbf{i}} = \sum_{\ell=0}^{\mathrm{K}} \mathbf{H}^{(\ell)}_{\mathbf{i}} \tag{4}$$

where $\tilde{\mathbf{D}}^{-\frac{1}{2}}\tilde{\mathbf{A}}\tilde{\mathbf{D}}^{-\frac{1}{2}}$ is the symmetrically normalized Laplacian matrix of the graph with self-loop; $\ell = 0, 1, 2, \ldots\ldots, K-1$ denotes the GCN layer; $\mathbf{H}^{(\ell)}/\mathbf{H}^{(\ell+1)}$ stand for the embedding matrix in the ℓ-th $/(\ell+1)$-th layer of GCN and $\mathbf{W}^{(\ell)}_{\mathrm{GCN}}$ denotes the parameter matrix at the layer ℓ of GCN. To overcome the over-smoothing problem, we employ skip-connection to form the final node representation. $\mathbf{Z}_{\mathbf{i}}$ is the final representation of node u_i and $\mathbf{H}^{(\ell)}_{\mathbf{i}}$ indicates the i-th row of the embedding matrix $\mathbf{H}^{(\ell)}$, which is also the representation vector of the node u_i at ℓ-th layer.

In a social network, user nodes that post messages frequently have a greater influence on the community by contributing more to the global information. Therefore, we selected the degree of nodes as a measure of importance and employed asymmetric data augmentation techniques to ensure the model to focus on globally important information. The process is described as follows:

$$s^c_{u_i} = degree(u_i) = \sum_{j=1}^{N} \tilde{\mathbf{A}}_{ij}, s^e_{uv} = log(\frac{s^c_u + s^c_v}{2}) \tag{5}$$

$$p_{uv} = min(\frac{s^e_{max} - s^e_{uv}}{s^e_{max} - \mu^e_s} \cdot p_e, p_\tau) \tag{6}$$

where N is the number of the node; $s^c_{u_i}$ and s^e_{uv} refer to the importance coefficients of node u_i and edge e_{uv}, respectively. The probability p_{uv} that an edge e_{uv} is discarded is calculated from the above equation, where p_e controls the total number of edges that are removed and p_τ is the truncation probability preventing the probability of deleting edges from being too large to damage the topological information of the graph [24].

Note that the data augmentation using this method will retain the more important edges, which can extract the significant topological information in the original graph. InfoNCE loss is used as the loss function with the following formula:

$$\mathbf{Z}^q = \mathrm{GCN}(f_1(\tilde{\mathbf{A}}), \varsigma_1(\mathbf{H}^{(0)})), \mathbf{Z}^k = \mathrm{GCN}(f_2(\tilde{\mathbf{A}}), \varsigma_2(\mathbf{H}^{(0)})) \tag{7}$$

$$\mathcal{L}_c = -\frac{1}{2N}\sum_{i=1}^{N}[\ell(\mathbf{Z}^q_i, \mathbf{Z}^k_i) + \ell(\mathbf{Z}^k_i, \mathbf{Z}^q_i)] \tag{8}$$

$$\ell(\mathbf{Z}_i^q,\mathbf{Z}_i^k)=log\frac{exp(s(\mathbf{Z}_i^q\mathbf{W},\mathbf{Z}_i^k\mathbf{W})/\tau)}{exp(s(\mathbf{Z}_i^q\mathbf{W},\mathbf{Z}_i^k\mathbf{W})/\tau)+\sum_{j\neq i}exp(s(\mathbf{Z}_i^q\mathbf{W},\mathbf{Z}_j^k\mathbf{W})/\tau)+\sum_{j\neq i}exp(s(\mathbf{Z}_j^q\mathbf{W},\mathbf{Z}_i^k\mathbf{W})/\tau)} \tag{9}$$

where $f_1(\cdot)$ and $f_2(\cdot)$ are two topology augmentation functions defined above with different probabilities of edge deletion; $\varsigma_1(\cdot)$ and $\varsigma_2(\cdot)$ are two feature mask functions with different probabilities according. As usual, $s(\cdot)$ is the cosine similarity of two vectors, $\mathbf{W}$ is a linear layer playing the role of feature augmentation and τ is the temperature coefficient in the Contrastive Loss. The combination of GCN and asymmetric centrality graph data augmentation can collaboratively mine the global and local information in the spatial domain.

3.3 Multi-view Weakly Supervised Classifier

In practical scenarios, the acquisition of a large number of labels is costly, so we choose to use data augmentation methods to expand the training data, thereby increasing the amount of labeled data, and finally generating high-quality node embeddings. Here, we implement multi-view weakly supervised learning by exploiting the augmented graphs constructed during the graph contrastive learning phase. We choose a two-layer MLP as a weakly supervised classifier, and jointly perform supervised training on the original graph embedding and the augmented graph embedding as follows:

$$\mathbf{P}_i=Softmax(\mathrm{MLP}_W(\mathbf{Z}_i)),\ \mathcal{L}_r=-\frac{1}{N}\sum_{i=1}^{N}\sum_{c=1}^{M}y_{ic}log(\mathbf{P}_{ic}) \tag{10}$$

$$\mathcal{L}_w=-[\frac{1}{N}\sum_{i=1}^{N}\sum_{c=1}^{M}y_{ic}log(\mathbf{P}_{ic}^q)+\frac{1}{N}\sum_{i=1}^{N}\sum_{c=1}^{M}y_{ic}log(\mathbf{P}_{ic}^k)] \tag{11}$$

where $\mathrm{MLP}_W(\cdot)$ is the Multi-view Weakly Supervised Classifiers; $\mathbf{P}_i$, $\mathbf{P}_i^q$ and $\mathbf{P}_i^k$ refer to the probability distribution vector of the predicted class for node u_i on the original graph and the two augmented graphs, respectively. y denotes the ground truth.

To prevent overfitting, we add L2 Regularization to all parameters of the model. The unsupervised loss can be jointly optimized with the multi-view weakly supervised loss, so the total loss function can be expressed as:

$$\mathcal{L}=\mathcal{L}_r+\lambda_1\cdot\mathcal{L}_w+\lambda_2\cdot\mathcal{L}_e+\lambda_3\cdot\mathcal{L}_c+\lambda_4\cdot||\Theta||_2^2 \tag{12}$$

where $\lambda_1,\lambda_2,\lambda_3,\lambda_4$ are the hyperparameters used to balance out the loss function.

4 Experiment

To verify the superiority and effectiveness of the proposed ADGCN model, we perform abundant experiments on two real-world datasets to answer the following research questions:

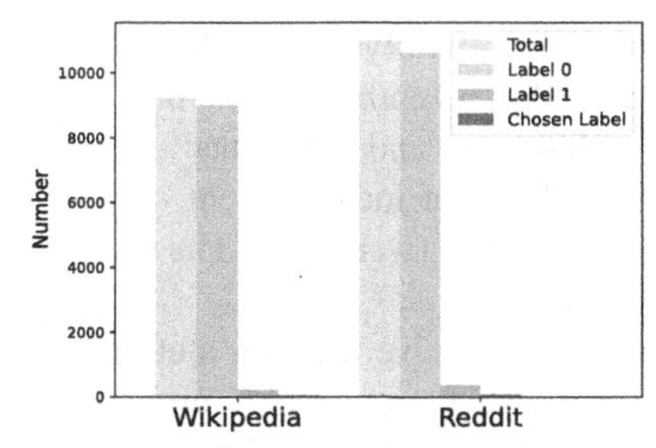

Fig. 2. Label distribution

Table 1. ADGCN performance

	Reddit	Wikipedia
AUC	69.69 (0.2)	84.13 (0.4)
Recall	73.24 (0.3)	82.96 (0.3)
Precision	94.62 (0.1)	97.27 (0.2)
F1score	81.60 (0.4)	88.92 (0.5)

- RQ1: How does ADGCN perform on different datasets compared to various baselines?
- RQ2: How does the local-global integrated contrastive learning method contribute to the performance of ADGCN?
- RQ3: How does node representation generated by ADGCN improve the classification performance?

4.1 Dataset

We evaluate our model and the baselines on two real-world datasets focused on anomaly detection task. These datasets are bipartite graphs, divided into two sets: users and items. Edges only exist between nodes of different set. We choose the following dataset which contain all scenarios.

Wikipedia Dataset.[2] This dataset collects top edited pages and active users as graph nodes, while edges indicate users' editing action on Wiki pages. Node labels indicate if users are banned from editing.

Reddit Dataset.[3] This dataset collects active users and their posts under subreddits on social news and discussion website Reddit. Users and subreddits are regarded as nodes while the posting action of users under subreddits are regarded as edges. Node labels indicate whether users are banned from posting.

Taking a deeper look at these two datasets, we observe that the distribution of the node label is exceedingly imbalanced. Nodes with label 1 account for merely 2%–3% of the entire dataset, which is perfectly aligned with our anomaly detection task. The label distribution is provided in Fig. 2.

4.2 Baselines (RQ1)

We compare our model ADGCN with the following four baseline models [5, 11, 19] using the same experimental settings. And we summarize the results in Table 2, with the following observations and conclusions:

[2] http://snap.stanford.edu/jodie/reddit.csv.
[3] http://snap.stanford.edu/jodie/wikipedia.csv.

In Table 2, all the results are converted to percentage by multiplying by 100, and the standard deviations are computed over 10 runs (displayed in parenthesis). The best and second-best results of each dataset are respectively in **bold** font and underlined.

We then evaluate our model's performance using different measurement on two datasets: AUC, recall (weighted), precision (weighted), F1score (weighted). The results are shown in Table 1.

The proposed model utilizes local-global integrated contrastive learning method to improving its performance. And it outperforms the second-best model by 8% on Reddit and 2% on Wikipedia. We observe that supervised models (GAT and GraphSAGE) generally outperform unsupervised models based on Autoencoder (GAE and VGAE). And graph contrastive method (GCA) achieves competitive results.

The results in Table 1 and Table 2 demonstrate the state-of-the-art performance on anomaly detection task due to the local-global collaborative data mining. Interestingly, our model even outperforms full supervised models on this task. We infer that the separately trained classifier and the feature augmentation method may result in this better performance.

Table 2. Performance comparison with baselines

	Reddit	Wikipedia
GAE	58.40 (0.5)	74.85 (0.6)
VGAE	57.98 (0.6)	73.66 (0.7)
GAT	64.52 (0.5)	82.34 (0.8)
GraphSAGE	61.24 (0.6)	82.40 (0.7)
GCA	65.98(0.2)	81.78 (0.6)
ADGCN	**69.69 (0.2)**	**84.13 (0.4)**

4.3 Ablation Experiment (RQ2)

To investigate the effectiveness of our feature augmentation method and information-preserving feature compression scheme (with Auto-Encoder) without losing the important information of the graph, we perform the ablation experiment on Reddit to answer the question whether we can provide improvement to anomaly detection task with our local-global approach. To this end, we implement four different versions of our model with AUC as our measurement index: the complete ADGCN, ADGCN without feature augmentation, ADGCN without information-preserving feature compression (whose loss function is MSE), and the raw model only uses supervised method on 1%–1.6% labels.

As shown in Table 3, with the feature augmentation and undamaged feature compression, the model is able to reach satisfactory results, and it can outperform the raw model by around 10%. The results indicate that both feature augmentation by contrastive learning and information-preserving feature compression can bring improvement to the raw model.

Table 3. Ablation results

Settings	$\mathcal{L}_r$	$\mathcal{L}_w$	$\mathcal{L}_e$	AUC (%)	Recall (%)	Precision (%)
i	√	√	√	69.69(0.2)	73.24(0.3)	94.62(0.1)
ii	√		√	67.52(0.1)	61.05(0.4)	94.23(0.2)
iii	√	√		66.89(0.3)	58.55(0.6)	94.25(0.1)
iv	√			65.98(0.2)	57.25(0.4)	94.38(0.3)

4.4 Evaluation of Node Representation (RQ3)

To verify the high quality of the node representations generated by the ADGCN framework, we freeze the ADGCN model parameters and measure the embedding by training a two-layer MLP. In the category processing, we set the anomaly label as Class 1 and the normal label as Class 0, so that the anomaly detection problem can be converted into a binary classification problem. In terms of dataset partition, we divided the training dataset, validation dataset and test dataset by stratified sampling according to the ratio of 50%–25%–25%.

Figure 3 is the running result of the classifier on the node representations generated by ADGCN. It's worth noting that the classifier converges quickly on the test dataset after about 35 epochs and achieves an excellent AUC of close to 70%. The rapid convergence and high results of AUC on the test dataset can well reflect the high quality and separability of the node representations generated by ADGCN, which is reflected in the high-dimensional space that node representations with abnormal attributes of features will form spatial clusters, which is due to the collaborative mining of global a global information by ADGCN.

5 Discussion

5.1 Hyperparameter Analysis

In this section, we investigate the sensitivity of our model in relation to several key hyperparameters. In order to analyze the mutual influence and interaction among the three modules in our model on the final results, we focus on weight for augmentation loss λ_1 and weight for feature compression loss (the MSE loss) λ_2. We use AUC, Recall (weighted) and Precision (weighted) on dataset Reddit as the measurement.

In each graph, the ordinate of each node indicates the average value of performance over 10 runs, while the radius of the colored fog around each node indicates the standard deviation over 10 runs.

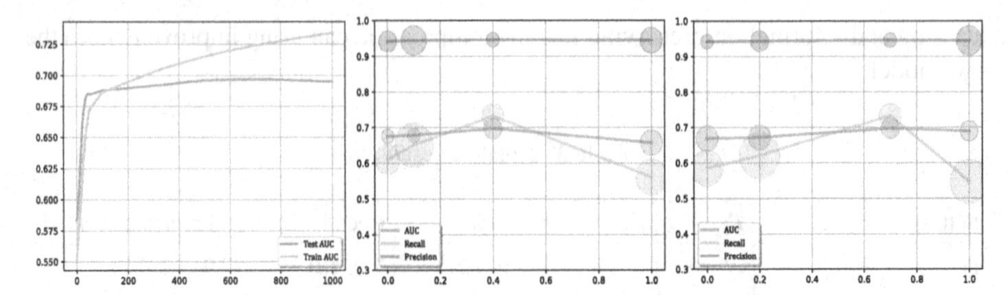

Fig. 3. AUC on Reddit

Fig. 4. Impact of parameter λ_1 (left) and λ_2 (right)

- ·The impact of λ_1. As illustrated in Fig. 4, the performance reaches the peak when augmentation loss weight $\lambda_1 = 0.4$, and presents a more concentrated distribution. It can be noted that λ_1 with the range of [0.0, 1.0] can hardly perturb model performance on Precision.
- ·The impact of λ_2. As illustrated in Fig. 4, the performance reached the peak when the weight of MSE loss $\lambda_2 = 0.7$. We observe that λ_2 with the range of [0.0, 1.0] cause a more significant impact on Recall.

5.2 Loss Convergence

We further notice that our model present fast convergence during the experiment. As illustrated in Fig. 5, the MSE loss can converge within 250 epochs both on Wikipedia and Reddit. Model loss can converge within 200 epochs on Reddit and 300 epochs on Wikipedia. And the model loss presents a sharp decline at the beginning. We attribute the result to two reasons. Firstly, we use exponential-decayed learning rate with the decaying rate of 0.993 to train the classifier. At the very beginning, the learning rate is relatively high, and exponentially decreases during the run. Secondly, ADGCN generates high quality node representations (stated in the preceding paragraph), so the classifier can capture node information more efficiently and quickly converge. As the model loss function include the supervised loss on augmented graph, the randomness of graph augmentation can lead to tiny fluctuations of the total loss.

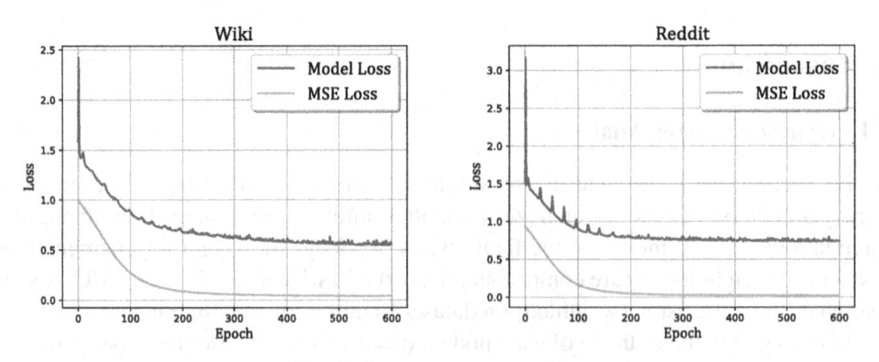

Fig. 5. Loss converges rapidly.

6 Conclusion

In conclusion, the ADGCN framework proposed in this paper addresses the problem of weakly supervised anomaly user identification in social networks. By exploratively combining generative and contrastive self-supervised mechanisms to enhance model representation, our model generates high-quality user representations with very few labeled data. Experimental results show that ADGCN achieves state-of-the-art performance using only 1% of the labeled data and outperforms previous unsupervised and supervised models. However, as social network data is continually evolving, there is a need for further research to explore how to leverage temporal information effectively when incorporating limited labeled data into our model.

References

1. Bo, D., Wang, X., Shi, C., Shen, H.: Beyond low-frequency information in graph convolutional networks. In: Proceedings of the AAAI Conference on Artificial Intelligence, vol. 35, pp. 3950–3957 (2021)
2. Cai, S., et al.: Rethinking graph neural architecture search from message-passing. In: Proceedings of the IEEE/CVF Conference on Computer Vision and Pattern Recognition, pp. 6657–6666 (2021)
3. Clevert, D.A., Unterthiner, T., Hochreiter, S.: Fast and accurate deep network learning by exponential linear units (ELUS). arXiv preprint arXiv:1511.07289 (2015)
4. Gilmer, J., Schoenholz, S.S., Riley, P.F., Vinyals, O., Dahl, G.E.: Neural message passing for quantum chemistry. In: International Conference on Machine Learning, pp. 1263–1272. PMLR (2017)
5. Hamilton, W., Ying, Z., Leskovec, J.: Inductive representation learning on large graphs. In: Advances in Neural Information Processing Systems **30** (2017)
6. Hu, T., Qi, H., Huang, Q., Lu, Y.: See better before looking closer: weakly supervised data augmentation network for fine-grained visual classification. arXiv preprint arXiv:1901.09891 (2019)
7. Jiang, J., et al.: Anomaly detection with graph convolutional networks for insider threat and fraud detection. In: MILCOM 2019–2019 IEEE Military Communications Conference (MILCOM), pp. 109–114. IEEE (2019)
8. Khan, W., Haroon, M.: An efficient framework for anomaly detection in attributed social networks. Int. J. Inf. Technol. **14**(6), 3069–3076 (2022)
9. Khan, W., Haroon, M.: An unsupervised deep learning ensemble model for anomaly detection in static attributed social networks. Int. J. Cognit. Comput. Eng. **3**, 153–160 (2022)
10. Kipf, T.N., Welling, M.: Semi-supervised classification with graph convolutional networks. arXiv preprint arXiv:1609.02907 (2016)
11. Kipf, T.N., Welling, M.: Variational graph auto-encoders. arXiv preprint arXiv:1611.07308 (2016)
12. Lee, D.H., et al.: Pseudo-label: the simple and efficient semi-supervised learning method for deep neural networks. In: Workshop on Challenges in Representation Learning, ICML, vol. 3, p. 896 (2013)
13. Li, Y., Huang, X., Li, J., Du, M., Zou, N.: Specae: spectral autoencoder for anomaly detection in attributed networks. In: Proceedings of the 28th ACM International Conference on Information and Knowledge Management, pp. 2233–2236 (2019)

14. Liu, F., Tian, Y., Chen, Y., Liu, Y., Belagiannis, V., Carneiro, G.: ACPL: anti-curriculum pseudo-labelling for semi-supervised medical image classification. In: Proceedings of the IEEE/CVF Conference on Computer Vision and Pattern Recognition, pp. 20697–20706 (2022)
15. Nt, H., Maehara, T.: Revisiting graph neural networks: all we have is low-pass filters. arXiv preprint arXiv:1905.09550 (2019)
16. Pennebaker, J.W., Francis, M.E., Booth, R.J.: Linguistic Inquiry and Word Count: LIWC 2001, vol. 71. Lawrence Erlbaum Associates, Mahway (2001)
17. Savage, D., Zhang, X., Yu, X., Chou, P., Wang, Q.: Anomaly detection in online social networks. Soc. Netw. **39**, 62–70 (2014)
18. Tosyali, A., Kim, J., Choi, J., Kang, Y., Jeong, M.K.: New node anomaly detection algorithm based on nonnegative matrix factorization for directed citation networks. Ann. Oper. Res. **288**, 457–474 (2020)
19. Veličković, P., Cucurull, G., Casanova, A., Romero, A., Lio, P., Bengio, Y.: Graph attention networks. arXiv preprint arXiv:1710.10903 (2017)
20. Vignac, C., Loukas, A., Frossard, P.: Building powerful and equivariant graphneural networks with structural message-passing. In: Advances in Neural Information Processing Systems, vol. 33, pp. 14143–14155 (2020)
21. Wu, F., Souza, A., Zhang, T., Fifty, C., Yu, T., Weinberger, K.: Simplifying graph convolutional networks. In: International Conference on Machine Learning, pp. 6861–6871. PMLR (2019)
22. You, Y., Chen, T., Sui, Y., Chen, T., Wang, Z., Shen, Y.: Graph contrastivelearning with augmentations. In: Advances in Neural Information Processing Systems, vol. 33, pp. 5812–5823 (2020)
23. Yu, R., Qiu, H., Wen, Z., Lin, C., Liu, Y.: A survey on social media anomaly detection. ACM SIGKDD Explor. Newsl **18**(1), 1–14 (2016)
24. Zhu, Y., Xu, Y., Yu, F., Liu, Q., Wu, S., Wang, L.: Graph contrastive learning with adaptive augmentation. In: Proceedings of the Web Conference 2021, pp. 2069–2080 (2021)

Light Field Image Super-Resolution via Global-View Information Adaption and Angular Attention Fusion

Wei Zhang[1,3](✉), Wei Ke[1], and Hao Sheng[2]

[1] Faculty of Applied Sciences, Macao Polytechnic University, Macao 999078, SAR, People's Republic of China
{wei.zhang,wke}@mpu.edu.mo

[2] State Key Laboratory of Virtual Reality Technology and Systems, School of Computer Science and Engineering, Beihang University, Beijing 100191, People's Republic of China
shenghao@buaa.edu.cn

[3] Technology Department (Chief Engineer Office), China Mobile Group Design Institute Co., Ltd., Beijing 100080, People's Republic of China

Abstract. Light field (LF) is a emerging technology, which can be used in many fields. Furthermore, LF cameras can capture spatial and angular information of 3D real-world scenes. This information is beneficial for image super-resolution (SR). However, most existing LF approaches have the limitation of utilizing the global-view information, which contains the correlation information among all LF. Moreover, to exploit the complementary information from different views of an LF image, we propose a novel SR method that adapts each view to a global domain with the guidance of global-view information. Our method, called LF-GAGNet, uses a dual-branch network to align features across views with deformable convolutions and fuse them with an attention mechanism. The upper branch extracts global-view information and generates adaptive guidance factors for each view through a global-view adaptation-guided module (GAGM). The lower branch uses these factors as offsets for deformable convolutions to achieve feature alignment in the global domain. Furthermore, we design an angular attention fusion module (AAFM) to enhance the angular features of each view according to their importance. We evaluate our method on various real-world scenarios and show that it surpasses other state-of-the-art methods in terms of SR quality and performance. We also demonstrate that our method can handle complex realistic LF scenarios effectively.

Keywords: Light field · Super-resolution · Deformable convolution

1 Introduction

Light field (LF) images, captured by commercial LF cameras, enable various applications such as 3D reconstruction [1] and salience detection [2]. However,

Supported by Macao Polytechnic University.

B. Luo et al. (Eds.): ICONIP 2023, CCIS 1965, pp. 267–279, 2024.
https://doi.org/10.1007/978-981-99-8145-8_21

LF cameras have a fixed sensor resolution, which requires a trade-off between spatial and angular resolutions. This results in low spatial resolution for each sub-aperture image (SAI), which limits the LF applications. Therefore, a key challenge is to recover high-resolution (HR) LF images from low-resolution (LR) LF images using LF image super-resolution (SR) methods.

LF images capture the parallax structure and angular dimension of 3D scenes, which creates high correlations among SAIs. Recently, some learning-based methods [3–8] have been proposed for LF image SR, which leverage both spatial and angular information to enhance the resolution of SAIs. Jin et al. [4] developed an all-to-one model named LF-ATO that reconstructs a reference view by combining its features with those of other views. Subsequently, several works [5,6] designed multiple extractors based on the LF characteristics to achieve the high-quality reconstruction of each SAI. However, existing learning-based methods still face two challenges. First, they have difficulty in adaptively supplementing complementary information for different views, especially when there are occlusion and non-Lambertian reflections in the captured LF scenes. Second, they do not fully exploit the global-view correlations among all SAIs, but only rely on 2D CNNs to model stacked SAIs or perform local-view feature alignment.

We propose a novel method, LF-GAGNet, to enhance the resolution of sub-aperture images (SAIs) in light field (LF) image super-resolution (SR), addressing the two challenges of existing learning-based methods. Our method has two dual branches that integrate complementary information among different views and exploit the global-view information from all LF views. The upper branch has a global-view adaptation-guided module (GAGM), which dynamically generates adaptive guidance factors for each SAI, containing global-view correlation information. These factors are used as offsets for deformable convolutions in the lower branch, which align the features of each view with the global-domain feature, supplementing angular information for each view. After a shared convolution, the features are further fused by an angular attention fusion module (AAFM), which preserves the geometric structure of LF by considering the complementary visual informativeness between each view and the reference view. We conduct comparative experiments on various real-world scenarios and show that our LF-GAGNet outperforms other state-of-the-art LF image and single image SR methods in terms of performance and generalization. The main contributions of this paper are summarized as follows:

1. To achieve state-of-the-art performance, we propose an LF-GAGNet that dynamically incorporates angular information among all LF views. This method adaptively considers the distinctive global-view information for each view.
2. Our method is based on a GAGM, which has an extractor and an adjuster that generate distinctive offsets for each view. By using deformable convolution, our method can effectively exploit spatial and angular information among all LF views for high-quality reconstruction.
3. We introduce an AAFM based on the centre view to further enhance each view with complementary information. Our network can not only generate

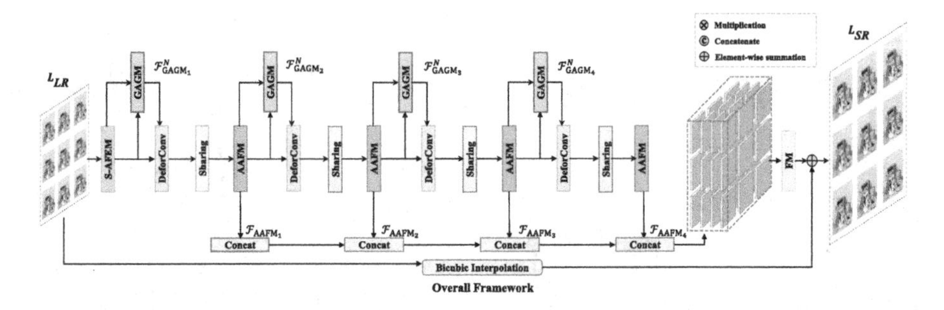

Fig. 1. Network architecture of the proposed LF-GAGNet. The overall network is composed of four modules (S-AFEM, GAGM, AAFM and FM). Note that, the input of our network consists of SAIs with dimensions $N \times C \times H \times W$.

high-quality reconstruction results but also preserve LF parallax structure by using these two modules.

The rest of this paper is structured as follows. We present the architecture and implementation details of our LF-GAGNet in Sect. 2. We report comparative experiments and ablation studies using real-world datasets in Sect. 3. Finally, We conclude this paper in Sect. 4.

2 Architecture and Pipeline

Two planes can represent the 4D LF, which includes spatial and angular information. As mentioned by [9], the LR LF images can be formulated as a 4D function L_{LR} ($\mathbb{R}^{U\times V\times H\times W}$) and the results of SR LF images are L_{SR} ($\mathbb{R}^{U\times V\times \alpha H\times \alpha W}$), where U and V represent angular dimensions, H and W represent spatial dimensions, and α presents the up-scaling factor. We convert the RGB colour space of LF image to the YCbCr colour space and apply our method only on the Y channel in this paper. We use bicubic interpolation to generate the LR SAIs. We describe the proposed network in detail in this section.

2.1 Overview

Figure 1 shows an overview framework of our LF-GAGNet. First, a spatial-angular feature extraction module (S-AFEM) extracts the initial features of all SAIs. Then our network splits into two branches. The top branch has four GAGMs to generate distinctive guidance factors. The bottom branch has four DeforConvs (deformable convolutions), four Sharing (sharing convolutions), four AAFMs and an FM. DeforConv connects the top branch and the bottom branch.

2.2 Spatial-Angular Feature Extraction Module (S-AFEM)

Figure 2 (a) shows that our LF-GAGNet first performs an S-AFEM on the LR SAI ($L_{\mathrm{LR}} \in \mathbb{R}^{U\times V\times H\times W}$) input to extract shallow features with spatial-angular information ($\mathcal{F}_{\mathrm{SAI}}^{N} \in \mathbb{R}^{C\times H\times W}$, $N = U \times V$). The S-AFEM consists

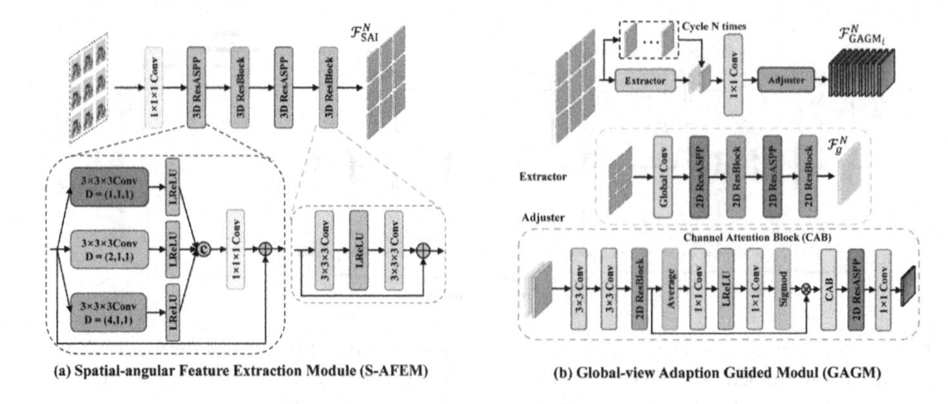

Fig. 2. The architectures of the S-AFEM and GAGM.

of a 3D ResASPP block and a 3D residual block (Resblock). This module can preserve correlation among different SAIs and extract rich spatial information. Our S-AFEM extracts deep and hierarchical features ($\mathcal{F}^N_{\text{SAI}} \in \mathbb{R}^{C\times H\times W}, N = 1, 2, \cdots, u \times v$) of each SAIs. Section 3.4 demonstrates the effectiveness of our S-AFEM.

2.3 Global-View Adaptation-Guided Module (GAGM)

The structure of GAGM is depicted in Fig. 2 (b), and the performance of our extractor and adjustor is evaluated in Sect. 3.4. The extractor of GAGM aims to extract the correlation features in the global angular domain, while the adjustor generates adaptive factors for different views, which can help DeforConv align the features in the global angular domain. Moreover, the proposed module can enrich the angular information for each view. Specifically, the extractor applies a 1×1 convolution, two 2D ResASPP blocks and two 2D Resblocks to the shallow features ($\mathcal{F}^N_{\text{SAI}}$). Each 2D ResASPP block consists of three dilated convolutions with different dilation rates of 1, 2 and 4. The extractor produces a deep feature representation ($\mathcal{F}_g \in \mathbb{R}^{C\times H\times W}$) as the output. The adjustor of GAGM generates the offsets of DeforConv for each view adaptively, which can enrich the angular information. Specifically, we first concatenate $\mathcal{F}_g$ and each $\mathcal{F}^N_{\text{SAI}}$ along the channel dimension and use a 1×1 convolution to fuse the global-domain feature and each-view feature. This process can be expressed as,

$$\mathcal{F}^N_{fus} = \left(f_{\text{Conv}_1}\left[\mathcal{F}_g, \mathcal{F}^N_{\text{SAI}}\right]\right), \tag{1}$$

where $[\cdot]$ represents concatenation the operation, f_{Conv_1} denotes the 1×1 convolution and $\mathcal{F}^N_{fus}$ is the input of the adjuster.

This adjuster is inspired by [10], whose channel attention block (CAB) is the core block. This block contains a 2D Resblock, a global average pooling (*Average*), two 1×1 convolutions and a sigmoid function. The input features ($\mathcal{F}^N_{fus}$) of our adjuster are first fed into a 2D Resblock to obtain deep-level

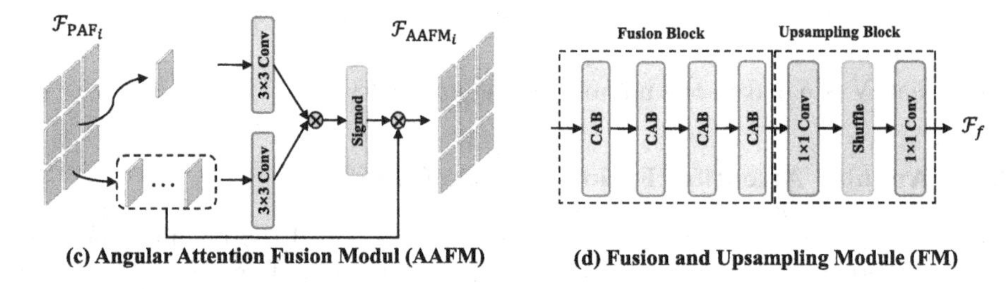

Fig. 3. The architectures of the AAFM and FM.

Table 1. PSNR/SSIM/LPIPS values achieved by different methods for 2× and 4 × SR, the best results are in bold and the second best results are underlined.

Methods	scale	Datasets				Average
		EPFL	INRIA	STFgantry	STFlytro	
Bicubic	2×	29.74/0.9376/0.198	31.33/0.9577/0.200	31.06/0.9498/0.156	33.32/0.9528/0.201	31.36/0.9495/0.189
VDSR [11]		32.63/0.9606/0.092	34.65/0.9750/0.092	35.93/0.9808/0.030	36.31/0.9717/0.102	34.88/0.9720/0.079
EDSR [12]		33.10/0.9632/0.080	35.02/0.9768/0.082	36.36/0.9823/0.023	36.79/0.9737/0.102	35.32/0.9740/0.069
LFSSR [13]		33.83/0.9746/0.041	35.51/0.9833/0.056	37.04/0.9874/0.026	38.46/0.9818/0.053	36.21/0.9818/0.044
resLF [3]		33.63/0.9702/0.044	35.45/0.9803/0.056	37.38/0.9881/0.026	37.85/0.9790/0.053	36.08/0.9794/0.044
LF-InterNet [5]		34.00/0.9752/0.040	35.72/0.9839/0.054	37.27/0.9880/0.024	38.60/0.9824/0.048	36.40/0.9824/0.042
LF-DFNet [14]		34.50/0.9760/0.039	**36.43**/0.9846/0.052	38.51/0.9909/0.018	38.81/0.9830/0.048	37.06/0.9836/0.039
MEG-Net [15]		33.64/0.9725/0.046	35.50/0.9821/0.057	36.16/0.9847/0.034	38.07/0.9806/0.053	35.84/0.9800/0.047
DPT [9]		34.30/0.9760/0.040	36.13/0.9845/0.053	38.75/0.9913/0.015	38.82/0.9829/0.049	37.00/0.9837/0.039
LF-IINet [16]		34.47/0.9764/0.037	36.26/0.9847/0.053	38.10/0.9900/0.021	38.89/0.9832/0.048	36.93/0.9836/0.040
DistgSSR [6]		34.42/0.9770/0.036	36.27/0.9851/**0.049**	**38.96**/**0.9917**/0.014	39.02/0.9837/0.044	37.17/0.9844/0.036
Ours		**34.51**/**0.9777**/**0.033**	36.37/**0.9854**/0.050	38.80/0.9915/**0.012**	**39.16**/**0.9841**/**0.042**	**37.21**/**0.9847**/**0.034**
Bicubic	4×	25.26/0.8324/0.435	26.95/0.8867/0.412	26.09/0.8452/0.432	27.97/0.8564/0.451	26.57/0.8552/0.433
VDSR [11]		27.18/0.8773/0.288	29.17/0.9207/0.278	28.54/0.9012/0.190	29.55/0.8896/0.314	28.61/0.8972/0.268
EDSR [12]		27.84/0.8852/0.269	29.76/0.9262/0.264	28.72/0.9095/0.161	30.04/0.8976/0.298	29.09/0.9046/0.248
LFSSR [13]		28.26/0.9071/0.260	30.32/0.9436/0.231	29.78/0.9317/0.160	31.13/0.9178/0.257	29.87/0.9250/0.220
resLF [3]		27.41/0.8838/0.265	29.45/0.9276/0.260	28.64/0.9067/0.215	29.96/0.8970/0.283	28.86/0.9038/0.256
LF-InterNet [5]		28.16/0.9041/0.253	30.33/0.9434/0.248	29.26/0.9218/0.197	31.05/0.9171/0.267	29.70/0.9216/0.241
LF-DFNet [14]		28.63/0.9097/0.236	30.83/0.9481/0.227	30.28/0.9378/0.148	31.41/0.9215/0.255	30.29/0.9293/0.217
MEG-Net [15]		27.00/0.8791/0.287	29.02/0.9243/0.285	28.04/0.8971/0.259	29.68/0.8931/0.309	28.43/0.8984/0.285
DPT [9]		27.64/0.8925/0.264	29.84/0.9344/0.258	28.83/0.9115/0.199	30.34/0.9044/0.287	29.16/0.9107/0.252
LF-IINet [16]		28.57/0.9109/0.233	30.63/0.9471/0.231	29.90/0.9322/0.165	31.42/0.9217/0.258	30.13/0.9280/0.222
DistgSSR [6]		28.70/0.9130/0.223	30.84/0.9482/0.225	30.30/0.9379/0.140	31.45/0.9221/0.255	30.32/0.9303/0.211
Ours		**28.88**/**0.9160**/**0.221**	**30.88**/**0.9503**/**0.217**	**30.47**/**0.9399**/**0.138**	**31.78**/**0.9265**/**0.243**	**30.50**/**0.9332**/**0.205**

representation information of each view. After being processed by another CAB and a 2D ResASPP block, the output features are fed into a 1×1 convolution to compress the channel numbers from C to C_1. After this, the distinctive guidance factors of each view are generated. Each factor is different, which can guide the deformable convolution to supplement angular information from other views. This process can be specifically expressed as,

$$\mathcal{F}^{N}_{\text{GAGM}_i} = f_{\text{Conv}_2} \circ f_{aspp} \circ f_{\text{CAB}_2} \circ f_{\text{CAB}_1} \circ \mathcal{F}^{N}_{fus}, \tag{2}$$

where $\mathcal{F}_{\mathrm{GAGM}_i}^{N}$ is the output of our GAGM, f_{Conv_2} is the 1×1 convolution, f_{aspp} is a 2D ResASPP block, and f_{CAB_1} and f_{CAB_2} denote the block of our CAB. Note that N is equal to the number of angular views.

2.4 Angular Attention Fusion Module (AAFM)

Due to the parallax structure of LF, each view provided the angular information is different, and the associated pixel information in different views contains some shifts. Our network is composed of several cyclic structures. The output of the later structure may be seriously affected by previous parts, when the outputs of before modules have some problems such as feature misalignment and inadequate fusion. Therefore, an attention mechanism should be adopted to dynamically achieve further interaction with the angular information. Inspired by the method [17], we propose an AAFM to assign different weights for different views. This module can further improve the situation of angular information misalignment. The effectiveness of our AAFM is demonstrated in Sect. 3.4.

The structure of AAFM is shown in Fig. 3 (c). We first selected the central position feature as the reference view. That is because each view has a certain offset compared with the reference view under the parallax structure of LF. Note that the further views have less correlation with the centre view, whose views have less weight through our attention mechanism. Specifically, the output ($\mathcal{F}_{\mathrm{GAGM}_i}^{N}$) through a DeforConv and a 1×1 convolution generates preliminary aligned features ($\mathcal{F}_{\mathrm{PAF}\,i}$). We selected a centre-view feature ($\mathcal{F}_{ref}$) from $\mathcal{F}_{\mathrm{PAF}\,i}$ as our reference feature. $\mathcal{F}_{ref}$ and $\mathcal{F}_{\mathrm{PAF}\,i}$ features are fed into a 3×3 convolution, respectively. Through a sigmoid operation, the attention maps between the reference view and each view can be expressed as,

$$\mathcal{F}_{map_i} = f_{sigmoid}\left[f_{\mathrm{Conv}_3} \circ \mathcal{F}_{ref}, f_{\mathrm{Conv}_4} \circ \mathcal{F}_{\mathrm{PAF}\,i}\right], \tag{3}$$

where $\mathcal{F}_{map_i}$ is the angular attention map.

Finally, the output features ($\mathcal{F}_{\mathrm{AAFM}\,i}$) of AAFM can be generated by multiplying $\mathcal{F}_{map_i}$ and $\mathcal{F}_{\mathrm{PAF}\,i}$ in a pixel-wise manner. This process can be specifically expressed as,

$$\mathcal{F}_{\mathrm{AAFM}\,i} = \mathcal{F}_{\mathrm{PAF}\,i} \odot \mathcal{F}_{map_i}, \tag{4}$$

where $\odot$ denotes the element-wise multiplication.

2.5 Fusion and Upsampling Module (FM)

Our FM aims to fuse hierarchical features and construct an HR residual map for LF, which contains a fusion block and an upsampling block. The output of $\mathcal{F}_{\mathrm{AAFM}\,i}$ indicates different deep representations of L_{LR}. This structure of FM following [16] is shown in Fig. 3 (d). Through these four CABs, these hierarchical features achieve a more compact representation. The first stage can be expressed as,

$$M_{fus}^{M} = f_{\mathrm{CAB}_M} \circ \ldots f_{\mathrm{CAB}_1} \circ \left[\mathcal{F}_{\mathrm{AAFM}\,1}, \ldots, \mathcal{F}_{\mathrm{AAFM}\,M}\right], \tag{5}$$

where the number of M is 4, and f_{CAB_M} is the block of our CAB. The effectiveness of the CAB is demonstrated in Sect. 3.

The goal of the second stage is to produce an HR residual map ($\mathcal{F}_f \in \mathbb{R}^{N\times 1\times H\times W}$). The upsampling block consists of a 1×1 convolution, a pixel-shuffle layer and another 1×1 convolution, which can transform the feature depth into the feature spatial dimension. The final output (L_{SR}) has only one channel, and its spatial size of L_{SR} is $\alpha H \times \alpha W$. The L_{SR} can be expressed as follows,

$$
\begin{aligned}
\mathcal{F}_f &= f_{UP} \circ M_{fus}^{M}, \\
L_{SR} &= \mathcal{F}_f + L_{\mathrm{LR}},
\end{aligned}
\tag{6}
$$

where f_{UP} is an upsampling block.

3 Experiments

In this section, we first introduce the LF public datasets of real scenarios and metrics in Sect. 3.1, which are used to evaluate the performance of our LF-GAGNet in the real world. Then in Sect. 3.2, we provide the settings and implementation details. After that, we compare our LF-GAGNet with state-of-the-art methods from two aspects (quantitative results and qualitative results) in Sect. 3.3. Finally, we perform a series of ablation studies and analyses in Sect. 3.4.

3.1 LF Public Datasets and Evaluation Metrics

There are a total of 4 real-world scene datasets, which are used for training and testing in our experiment. These datasets are EPFL [18], INRIA [19], STF-gantry [20] and STFlytro [21]. We divide all LF datasets into two parts (training datasets and testing datasets) in a nearly 6:1 ratio. To solve the problem of LF image SR in real scenes, we only use the real-world dataset to train our LF-GAGNet, which is more targeted for application to LF cameras. Because these cameras are always used to capture real-world scenes. In total, there are in total 364 and 67 images for training and test datasets, respectively. To evaluate the performance of LF-GAGNet, the metrics of PSNR and SSIM are used on the Y channel. Moreover, the perceptual metric named LPIPS [22] is also used to evaluate the performance of different methods on RGB images. This metric can reflect the difference between the SR image and the ground-truth image.

3.2 Settings and Implementation Details

Following [23], the angular resolution of SAI is set to 5×5. We crop and down-sample each SAI with α ($\alpha = 2, 4$) to generate LR patches, whose sizes are 32×32. Due to limited datasets, the flipping and rotating operations are used to augment training data. The initial learning rate is set to $2e^{-4}$, multiplying an attenuation rate by 0.5 for every 15 epochs. The channel depth C and C_1 of

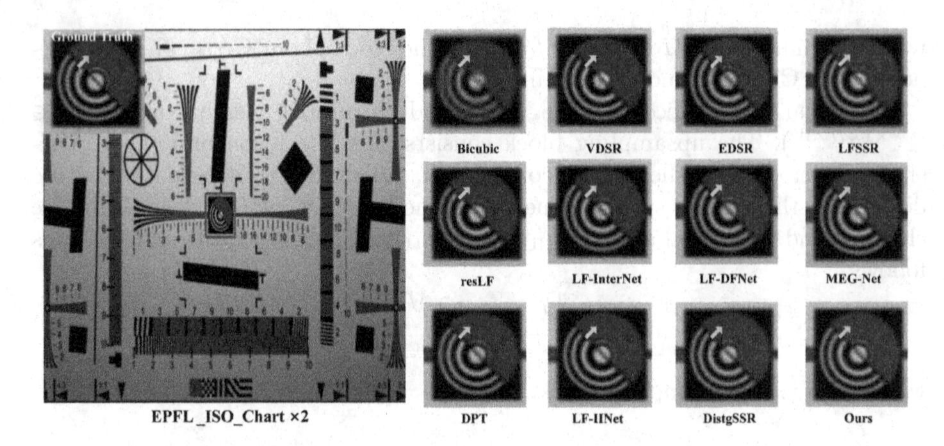

Fig. 4. Visual results for 2× SR. The enlarged patches highlighted using red boxes are generated by VDSR [11], EDSR [12], LFSSR [13], resLF [3], LF-InterNet [5], LF-DFNet [14], MEG-Net [15], DPT [9], LF-IINet [16], DistgSSR [6] and ground truth, respectively. Zoom in for better observation.(Color figure online)

convolution are set to 64 and 18, respectively. N represents the amount of angular resolution. In comparative experiments, N is set to 25. Our network adopts the $L1$ Loss function following previous works [5,14,16]. An Adam optimizer is used with $\beta_1 = 0.9$ and $\beta_2 = 0.999$. The training epoch is set to 100 to get better performance.

3.3 Comparison to State-of-the-Art Methods

We compared our LF-GAGNet with the state-of-the-art SISR and LF image SR methods on a total of five datasets. The SISR methods contain two algorithms, which are VDSR [11] and EDSR [12]. The remaining methods belong to LF image SR, which are LFSSR [13], resLF [3], LF-InterNet [5], LF-DFNet [14], MEG-Net [15], DPT [9], LF-IINet [16] and DistgSSR [6]. For fair comparisons, we retrain the learning-based methods (e.g., LFSSR [13], resLF [3] and DistgSSR [6]) with the same training datasets to produce HR LF images. The bicubic interpolation is used as a baseline in this part.

Quantitative Results. The quantitative results including three metrics are listed in Table 1. All methods are tested on four public LF datasets, and we also provide the average result to evaluate the generalization on complex real-world scenes. We use the bold and underlined types to show the best results and second-best results, respectively. In Table 1, it can be observed that our LF-GAGNet has the best performance of generalization for 2× and 4× SR. Although our method is slightly inferior to some state-of-the-art methods on EPFL, INRIA and STF-gantry datasets, our method achieves the best performance for more challenging 4× SR. Compared with the learning-based SISR methods (VDSR and EDSR), we achieve 2.33 dB (0.0127 and −0.045) higher for VDSR and 1.89 dB (0.0107

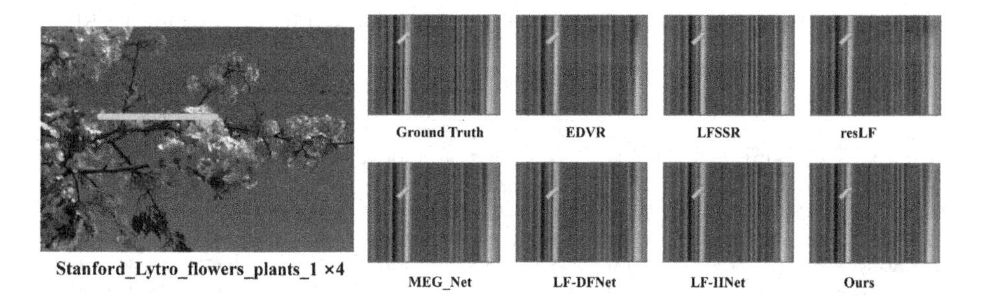

Fig. 5. The visual results of EPIs, which indicate the parallax structure for 4× SR.

and −0.035) higher for EDSR in terms of the average PSNR (SSIM and LPIPS) for 2× SR. That is because directly using the SISR method to SR in each view ignores the correlations among all SAIs, which hinders the performance in SR. Due to the representation capability of CNN, many methods based on CNN are proposed. In contrast to these methods, our LF-GAGNet achieves a state-of-the-art method with the best value of PSNR, SSIM and LPIPS. Compared with LFSSR [13], our method achieves an average gain of 1.00 dB (0.0029 and −0.010) and 0.63 dB (0.0082 and −0.015) in terms of average PSNR (SSIM and LPIPS) for 2× and 4× SR, respectively. This is mainly attributed to the guide of dynamical global-guided factors in our GAGM. This module can capture the global-view information and supplement complementary information.

Qualitative Results. The qualitative results for 2× SR testing on *ISO_ Chart* are shown in Fig. 4. As we can see, our LF-GAGNet can accurately preserve the fine details of the letter, which is pointed out by the green arrow. Specifically, the results of VDSR and EDSR cannot clearly distinguish each letter. The results of LF-IINet and DPT produce more faithful details than SISR methods, but their performance is a little worse than our method. The EPIs of different state-of-the-art methods are shown in Fig. 5, which can reflect the ability to keep the parallax consistency. The results of SISR methods (EDSR) present blurry EPIs. That is because they cannot preserve the parallax structure by using SISR methods directly. Compared with other LF image SR methods, our LF-GAGNet can present straight lines in EPIs, which are more fine-grained. That is mainly benefited by designed modules (GAGM and AAFM). Our GAGM can integrate angular information and AAFM can effectively utilize the correlation among different SAIs by using 3D convolution.

3.4 Ablation Study

In this subsection, we conduct several ablation experiments to investigate the effectiveness of different components in our LF-GAGNet, which mainly contain the S-AFEM, GAGM, AAFM and FM. We provide two types of comparison results containing numerical and visual experimental results. The numerical

Table 2. PSNR/SSIM values achieved by LF-FANet and its variants for 4× SR

Models	#Params.	Datasets				Average
		EPFL	INRIA	STFgantry	STFlytro	
Bicubic	–	25.26/0.8324	26.95/0.8867	26.09/0.8452	27.97/0.8564	26.57/0.8552
LF-GAGNet w/o S-AFEM	5.34M	28.78/0.9137	30.85/0.9494	30.32/0.9387	31.63/0.9254	30.40/0.9318
LF-GAGNet w/o GAGM_only	5.59M	28.72/0.9146	30.72/0.9498	30.43/0.9392	31.74/0.9260	30.40/0.9324
LF-GAGNet w/o GAGM (Extractor)	5.40M	28.76/0.9148	30.84/0.9495	30.34/0.9388	31.74/0.9259	30.42/0.9323
LF-GAGNet w/o GAGM (Adjuster)	4.94M	28.84/0.9147	30.88/0.9496	30.36/0.9389	31.65/0.9257	30.24/0.9322
LF-GAGNet w/o AAFM_only	5.76M	28.49/0.9093	30.57/0.9472	30.24/0.9341	31.67/0.9260	30.24/0.9292
LF-GAGNet w/o AAFM	5.07M	28.72/0.9142	30.74/0.9495	30.27/0.9382	31.68/0.9258	30.35/0.9319
LF-GAGNet w/o FM	5.06M	28.65/0.9149	30.68/0.9493	30.39/0.9390	31.70/0.9259	30.36/0.9323
LF-GAGNet1	2.74M	28.71/0.9149	30.68/0.9502	30.42/0.9396	31.73/0.9261	30.39/0.9327
LF-GAGNet2	5.12M	28.76/0.9150	30.75/0.9498	30.44/0.9397	31.75/0.9263	30.43/0.9327
LF-GAGNet	5.64M	**28.88/0.9160**	**30.88/0.9503**	**30.47/0.9399**	**31.78/0.9265**	**30.50/0.9332**

results are listed in Table 2. The bicubic interpolation method is used as a baseline in these results.

The Effectiveness of S-AFEM. Extracting the features from SAIs plays an important role in the reconstruction of LF. The previous works [14,16] always adopted a 2D module containing some ResASPP blocks and Resblocks, which ignores the correlations among different angular views. Thus, spatial and angular features of SAIs are extracted by the proposed S-AFEM. To demonstrate the effectiveness of the S-AFEM, we remove the feature extraction module (S-AFEM), replacing the stacked 2D ResASPP blocks and 2D Resblocks. We keep the parameters consistent by adjusting the number of 2D blocks. LF-GAGNet w/o S-AFEM is the variant of feature extraction. In Table 2, the value of the average PSNR and SSIM has decreased by 0.10 dB and 0.0014 for 4× SR, respectively. Our S-AFEM can capture not only the spatial features of each SAIs but also the correlations among different angular views, which is beneficial to improve the performance of LF images.

The Effectiveness of GAGM. The GAGM proposed in our LF-GAGNet is made of an Extractor and an Adjuster, which can adaptively construct the guidance factors based on the global angular information. To verify the effectiveness of the guidance factors produced by the Extractor and the Adjuster in LF-GAGNet, we designed three different variants (LF-GAGNet w/o GAGM_only, LF-GAGNet w/o GAGM (Extractor) and LF-GAGNet w/o GAGM (Adjuster)). As shown in Fig. 1, we keep the top branch with four GAGMs in LF-GAGNet w/o GAGM_only. Specifically, the output of GAGM is directly fed into a 1×1 convolution without the DeforConv and AAFM. For LF-GAGNet w/o GAGM (Extractor) and LF-GAGNet w/o GAGM (Adjuster), we only keep one of the two components (Extractor and Adjuster) compared with LF-GAGNet. All these comparison results on testing datasets for 4× SR are shown in Table 2. It can be observed that the average PSNR and SSIM of LF-GAGNet w/o GAGM_only, LF-GAGNet w/o GAGM (Extractor) and LF-GAGNet w/o GAGM (Adjuster) suffer a decrease of 0.10 dB (0.0008), 0.08 dB (0.0009) and 0.26 dB (0.0040) compared with that of LF-GAGNet, respectively.

The Effectiveness of AAFM. We provide two variants (LF-GAGNet w/o AAFM_only and LF-GAGNet w/o AAFM) to verify the effectiveness of AAFM. The AAFM is a key component to dynamically integrate angular information among all views. For the variant of LF-GAGNet w/o AAFM_only, we remove the GAGM, DeforConv and 1×1 convolution and increase the filters of convolutions for a fair comparison. Meanwhile, we remove the AFAM in LF-GAGNet w/o AAFM. In Table 2, the results of LF-GAGNet w/o AAFM_only and LF-GAGNet w/o AAFM decrease 0.26 dB (0.0040) and 0.15 dB (0.0013) in terms of the average PSNR (SSIM) compared with that of LF-GAGNet, respectively.

The Effectiveness of FM. The FM is the generic module, which is always used in LF image SR. This module can generate attention weights for different channels, blending the concatenated hierarchical features. To demonstrate the effectiveness of our FM, we merely remove the part of channel attention in CAB. This variant is termed as LF-GAGNet w/o FM. It can be observed that the average PSNR (SSIM) value of LF-GAGNet w/o FM suffers a decrease of 0.14 dB (0.0009) compared with that of LF-GAGNet.

The Effectiveness of Sharing Weights in LF-GAGNet. To reduce the number of parameters of our LF-GAGNet, we also conducted two different experiments, which are LF-GAGNet1 and LF-GAGNet2. Specifically, LF-GAGNet1 has the modules of GAGM and AFAM with sharing weights, and only the module of AFAM shares the weights in LF-GAGNet2. For our LF-GAGNet, we do not share the weights for both GAGM and AAFM. The comparative results for $4\times$ LF image SR are shown in Table 2. It can be observed that the performances of LF-GAGNet1 and LF-GAGNet2 are degraded by 0.11 dB and 0.07 dB for the average PSNR with the increase in the number of parameters.

4 Conclusion

In this paper, we propose a new LF-GAGNet network to achieve LF spatial SR in real scenarios. We mainly introduce three components (S-AFEM, GAGM and AAFM) to achieve feature extraction and spatial-angular information fusion. These components are effectively combined by a deformable convolution. Our LF-GAGNet can dynamically supplement global-view information and complementary information from other views. Extensive comparisons with state-of-the-art methods show that our method outperforms them in visual and quantitative results. Moreover, our LF-GAGNet also has a competitive computing efficiency.

In the future, we will continually explore a more lightweight framework based on our LF-GAGNet to reduce computational complexity. Moreover, our group will commit ourselves to an unsupervised learning framework, which may be a novel research trend in the future.

References

1. Zhu, H., Wang, Q., Yu, J.: Occlusion-model guided antiocclusion depth estimation in light field. IEEE J. Sel. Topics Signal Process. **11**(7), 965–978 (2017)
2. Zhang, M., et al.: Memory-oriented decoder for light field salient object detection. In: NeurIPS 2019, pp. 896–906 (2019)
3. Zhang, S., Lin, Y., Sheng, H.: Residual networks for light field image super-resolution. In: Proceedings of the IEEE/CVF Conference on Computer Vision and Pattern Recognition, pp. 11046–11055 (2019)
4. Jin, J., Hou, J., Chen, J., Kwong, S.: Light field spatial super-resolution via deep combinatorial geometry embedding and structural consistency regularization. In: Proceedings of the IEEE/CVF Conference on Computer Vision and Pattern Recognition, pp. 2260–2269 (2020)
5. Y. Wang, L. Wang, J. Yang, W. An, J. Yu, Y. Guo, Spatial-angular interaction for light field image super-resolution, in: European Conference on Computer Vision, Springer, 2020, pp. 290–308
6. Wang, Y., et al.: Disentangling light fields for super-resolution and disparity estimation. In: IEEE Transactions on Pattern Analysis and Machine Intelligence (2022)
7. Zhang, W., Ke, W., Yang, D., et al.: Light field super-resolution using complementary-view feature attention [J]. Comput. Vis. Media **9**(4), 843–858 (2023)
8. Cong, R., Sheng, H., Yang, D., Cui, Z., Chen, R.: Exploiting spatial and angular correlations with deep efficient transformers for light field image super-resolution. IEEE Trans. Multimedia (2023)
9. Wang, S., Zhou, T., Lu, Y., Di, H.: Detail-preserving transformer for light field image super-resolution, arXiv preprint arXiv:2201.00346 (2022)
10. Hu, J., Shen, L., Sun, G.: Squeeze-and-excitation networks. In: Proceedings of the IEEE Conference on Computer Vision and Pattern Recognition, pp. 7132–7141 (2018)
11. Kim, J., Kwon Lee, J., Mu Lee, K.: Accurate image super-resolution using very deep convolutional networks. In: Proceedings of the IEEE Conference on Computer Vision and Pattern Recognition, pp. 1646–1654 (2016)
12. Lim, B., Son, S.,Kim, H., Nah, S., Mu Lee, K.: Enhanced deep residual networks for single image super-resolution. In: Proceedings of the IEEE Conference on Computer Vision and Pattern Recognition Workshops, pp. 136–144 (2017)
13. Yeung, H.W.F., Hou, J., Chen, X., Chen, J., Chen, Z., Chung, Y.Y.: Light field spatial super-resolution using deep efficient spatial-angular separable convolution. IEEE Trans. Image Process. **28**(5), 2319–2330 (2018)
14. Wang, Y., et al.: Light field image super-resolution using deformable convolution. IEEE Trans. Image Process. **30**, 1057–1071 (2020)
15. Zhang, S., Chang, S., Lin, Y.: End-to-end light field spatial super-resolution network using multiple epipolar geometry. IEEE Trans. Image Process. **30**, 5956–5968 (2021)
16. Liu, G., Yue, H., Wu, J., Yang, J.: Intra-inter view interaction network for light field image super-resolution. IEEE Trans. Multimedia **25**, 256–266 (2021)
17. Wang, X., Chan, K.C., Yu, K., Dong, C., Loy, C.C.: EDVR: video restoration with enhanced deformable convolutional networks. In: IEEE/CVF Conference on Computer Vision and Pattern Recognition Workshops (CVPRW) 2019, pp. 1954–1963 (2019). https://doi.org/10.1109/CVPRW.2019.00247

18. Rerabek, M., Ebrahimi, T.: New light field image dataset. In: 8th International Conference on Quality of Multimedia Experience (QoMEX), no. CONF (2016)
19. Le Pendu, M., Jiang, X., Guillemot, C.: Light field inpainting propagation via low rank matrix completion. IEEE Trans. Image Process. **27**(4), 1981–1993 (2018)
20. Vaish, V., Adams, A.: The (new) Stanford light field archive, Computer Graphics Laboratory, Stanford University 6 (7) (2008)
21. Raj, A.S., Lowney, M., Shah, R., Wetzstein, G.: Stanford Lytro light field archive (2016)
22. Zhang, R., Isola, P., Efros, A.A., Shechtman, E., Wang, O.: The unreasonable effectiveness of deep features as a perceptual metric. In: Proceedings of the IEEE Conference on Computer Vision and Pattern Recognition, pp. 586–595 (2018)
23. Liang, Z., Wang, Y., Wang, L., Yang, J., Zhou, S.: Light field image super-resolution with transformers. IEEE Signal Process. Lett. **29**, 563–567 (2022)

Contrastive Learning Augmented Graph Auto-Encoder

Shuaishuai Zu[1], Chuyu Wang[1], Yafei Liu[1], Jun Shen[2], and Li Li[1](✉)

[1] Southwest University, Chongqing, China
{zushuaishuai,wangcy117,stevenlyf,lily}@email.swu.edu.cn
[2] University of Wollongong, Wollongong, NSW, Australia
jshen@uow.edu.au

Abstract. Graph embedding aims to embed the information of graph data into low-dimensional representation space. Prior methods generally suffer from an imbalance of preserving structural information and node features due to their pre-defined inductive biases, leading to unsatisfactory generalization performance. In order to preserve the maximal information, graph contrastive learning (GCL) has become a prominent technique for learning discriminative embeddings. However, in contrast with graph-level embeddings, existing GCL methods generally learn less discriminative node embeddings in a self-supervised way. In this paper, we ascribe above problem to two challenges: 1) graph data augmentations, which are designed for generating contrastive representations, hurt the original semantic information for nodes. 2) the nodes within the same cluster are selected as negative samples. To alleviate these challenges, we propose Contrastive Variational Graph Auto-Encoder (CVGAE). Specifically, we first propose a distribution-dependent regularization to guide the paralleled encoders to generate contrastive representations following similar distributions. Then, we utilize truncated triplet loss, which only selects top-k nodes as negative samples, to avoid over-separate nodes affiliated to the same cluster. Experiments on several real-world datasets show that our model CVGAE advanced performance over all baselines in link prediction, node clustering tasks.

Keywords: graph auto-encoder · contrastive learning · distribution-dependent regularization · truncated triplet loss

1 Introduction

Graph data is commonly applied in many fields to capture relationships among entities, such as citation networks, social networks, and protein interaction networks. Modeling such data is challenging due to the non-Euclidean structure characteristic. Graph embedding that maps the graph data into low-dimensional representation space has emerged as a mainstream for modeling graph structure and node features. Intuitively, the quality of learned node embeddings affects the performance for downstream graph data mining tasks including link prediction [18] and node clustering [1] tasks.

B. Luo et al. (Eds.): ICONIP 2023, CCIS 1965, pp. 280–291, 2024.
https://doi.org/10.1007/978-981-99-8145-8_22

Graph Neural Networks (GNNs) [10], inheriting the power of neural networks and utilizing the neighborhood propagation mechanism, have become powerful tools for learning graph embeddings. Recently, tremendous endeavors have been devoted to graph self-supervised learning because label information is usually difficult to obtain. Generally, these graph embedding methods are designed to exploit the diverse inductive bias exhibited in graphs. Some methods learn the node embeddings by reconstructing the graph topology structure, which inserts an inductive bias that nodes share similarities with their nearby nodes. These methods intend to promote similar representations across nearby nodes, which over-emphasize proximity relations, but ignore the node features [3]. Another group of graph embedding methods attempts to maximize mutual information [17] between each node and the corresponding graph's patch summaries. These methods encourage each node to be mindful of coarser information, increasing the similarities between unconnected nodes and affecting the inherent structural information.

Recently, contrastive learning (CL) [15], a type of self-supervised learning, aims to generate discriminative representations, which can benefit to preserve individual information maximally and bolster the performance on downstream tasks. Therefore, CL has gained great interest and been widely applied for graph embedding learning. Graph contrastive learning (GCL) generally follows the framework of CL in computer vision [7], in which two graph representations are generated for each graph and then maximizes the mutual information between these two representations. In this way, GCL methods can lead to learn discriminative representations distributing on the unit hypersphere [11,23]. To this end, graph data augmentations play a vital role in generating generalized graph representations. However, existing augmentations, including node dropping, edge perturbation, embedding masking, and subgraph replacements, have hurt the information of graph structure and node features. Therefore, existing graph augmentations cannot preserve semantic information of nodes broadly, which significantly limits the generality of learned node embeddings. Besides, existing GCL methods usually involve nodes within the same cluster as negative samples, which refers to as simpling bias [8]. Intuitively, these GCL methods would benefit a coarser-level graph analysis task, e.g., graph classification, and do not benefit a fine-level task, e.g., node clustering, that large.

To address the above problems, we propose to augment the Variational Graph Auto-Encoder (VGAE) with CL framework, namely CVGAE, to improve the generality of learned node embeddings. In this paper, we summarize two challenges waiting for us to solve: 1) existing graph augmentations, that are designed to generate contrastive representations, cannot preserve the original semantic information for nodes. 2) the nodes within the same cluster are usually selected as negative samples. In order to solve the first challenge above, instead of devising more advanced data augmentations for nodes, we propose a novel distribution-dependent regularization for CVGAE to guide its two paralleled encoders. Specifically, we take original graph data as input and Graph Convolutional Network (GCN) [31] models as two paralleled encoders to generate contrastive representations following similar distribution. For the second challenge, we adopt truncated triplet loss by selecting top-k negative samples instead of the widely-adopted InfoNCE loss [20], which takes all negative samples into consideration. VGAE generally endows nearby nodes, that are similar in structure and dissimilar in node features, similar representations. Therefore, most of the top-k negative samples are nearby nodes instead of these nodes affiliated to the same cluster. Owing to the inductive bias of VGAE framework, selecting the top-k negative samples would reduce the sample bias. With the truncated triplet loss as the optimization goal, we avoid to over-separate each node with its unnecessary negative samples affiliated to the same cluster and benefit from the advantage of training efficiency from the reduction of negative samples. At the same time, with a decoder aiming to reconstruct the graph structure, CVGAE can generate discriminative node embeddings by utilizing both the structural information and node features.

2 Related Work

2.1 Graph Embedding

Graph embedding generally learns a low-dimensional representation space according to the information of graph data, and it is essential to facilitate various downstream tasks. Graph Convolutional Network (GCN) [31] is a well-known method, which aggregates information for each node from its neighborhood nodes. Since then, most graph embedding methods utilizes GCNs to extract rich information from graph data [29]. Some reconstruction-based methods insert an inductive bias that nodes share similarities with their nearby nodes for learning graph embeddings. For example, VGAE [16] use GCNs as an encoder to obtain node embeddings and a simple inner product decoder to reconstruct the graph topology structure. VGNAE [3] utilizes L2-normalization to regularize the encoder of GAE for better learning isolated node embedding. There are also some methods centering around the data distribution of node embeddings, which inject an inductive bias that similar nodes share similar feature distribution. Inspired by Deep InfoMax (DIM) [13], Deep Graph Infomax (DGI) [25] maximizes the mutual information between nodes and graph-level summaries obtained by average pooling all node embeddings. Similar to DGI, Graph InfoClust (GIC) [19] maximizes the mutual information between nodes and cluster-level summaries. However, these graph embedding approaches tamper the individual information of nodes themselves [21], which only attain limited performances in downstream tasks.

2.2 Graph Constrative Embedding

Contrastive learning (CL) [15] has been the pinnacle of self-supervised representation learning, in which the embeddings of similar instances are gathered closely, and dissimilar instances are taken apart in the representation space. Up to now, many graph contrastive embedding methods have been proposed for graph embedding learning [30,32]. For GCL, graph data augmentations play an important role in generating positive and negative samples. For example, graph contrastive learning (GraphCL) [30] firstly uses shared two-layer GCNs as an encoder to extract embeddings from two subgraphs, which are dealt with randomly selected graph augmentations. Besides, graph contrastive learning with automated augmentations (JOAO) [32] proposed a unified framework to automatically select the data augmentations for specific datasets. However, these graph data augmentations fail to preserve the information of graph structure and node features. An alternative comes to light when considering that operating the encoders can in turn be utilized to generate contrastive samples. Instead of using these graph data augmentations, Simple Graph Contrastive Learning (SimGRACE) [28] adds perturbs on the encoder respectively to generate corresponding contrastive representations following different distributions. Predominantly, due to prohibitive computations caused by taking all negative nodes into computation, existing graph contrastive embedding methods also suffer from low learning efficiency [27]. Besides, prior work [5] points out that existing GCL methods usually suffer from the problem of sampling bias. Graph Debiased Contrastive Learning (GDCL) [8] utilizes cluster labels as supervisory signals to select negative samples from different clusters. In addition, alleviating the sample bias in a self-supervised way is desired to explore.

3 The Proposed Method

In this paper, we propose CVGAE for guaranteeing the generalization of the learned node embeddings. The overall framework of CVGAE is shown in Fig. 1.

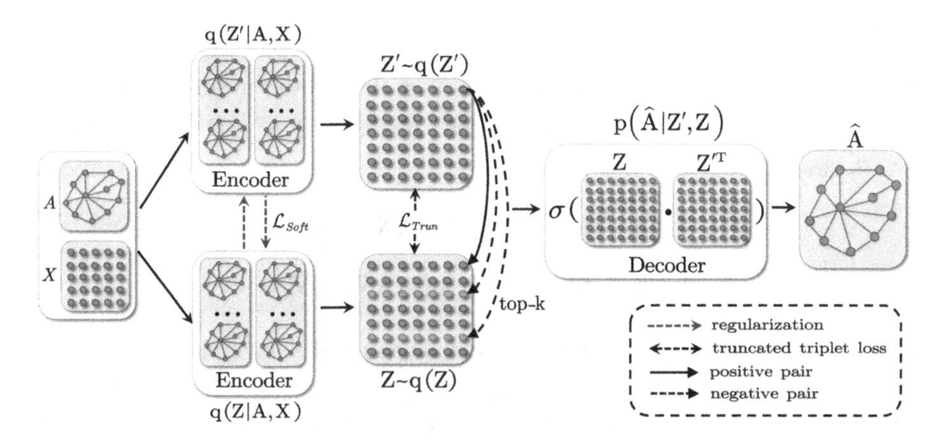

Fig. 1. The framework of CVGAE. The i-th row $z_i, z_i^{'}$ of $\mathbf{Z}, \mathbf{Z}^{'}$ are corresponding positive sample of each other, while the different rows $z_i, z_j^{'}$ are corresponding negative sample of each other.

3.1 Preliminary Work

In this section, we present the preliminary concepts of the graph. A graph G is represented as $G = (V, A, X)$, where $V = \{v_1, v_2, \ldots, v_N\}$ is a node set of G, N is the number of nodes in G. The graph topology structure is captured by an adjacent matrix $A \in \mathbb{R}^{N \times N}$, where $a_{i,j} = 1$ denotes the node v_i connect to v_j mutually, and $a_{i,j} = 0$ otherwise. The embedding matrix $X \in \mathbb{R}^{N \times H}$ preserves the information of node features, where each row $x_i \in \mathbb{R}^H$ denotes the corresponding node feature of v_i, H is the dimension of node feature. The graph embedding methods use $G = (V, A, X)$ as input to learn low-dimensional node embedding matrix $\mathbf{Z} \in \mathbb{R}^{N \times H'}$, which can preserve the information of graph topology structure and node features. Therefore, after obtaining the high-quality node embeddings, we can use them on various downstream tasks, such as link prediction [18] and node clustering [1].

3.2 Variational Graph AutoEncoder

In our model, we use GCNs as encoders to extract information from G. As for CVGAE, the paralleled encoders are designed to obtain the node embedding matrices $\mathbf{Z}, \mathbf{Z}'$ respectively, which follows a multi-variate Gaussian distribution. The generation process is as follows:

$$q(\mathbf{Z} \mid \mathbf{X}, \mathbf{A}) = \prod_{i=1}^{N} q\left(z_i \mid X, A\right) \tag{1}$$

$$q(\mathbf{Z}' \mid \mathbf{X}, \mathbf{A}) = \prod_{i=1}^{N} q\left(z_i' \mid X, A\right) \tag{2}$$

where $\boldsymbol{z}_i, \boldsymbol{z}_i'$ denote the $i-th$ row of $\mathbf{Z}, \mathbf{Z}'$ respectively. Besides, $q\left(z_i \mid X, A\right) = N\left(z_i \mid \boldsymbol{\mu}_i, diag\left(\sigma_i^2\right)\right)$, where $\boldsymbol{\mu}_i \in \boldsymbol{\mu}$, $\boldsymbol{\sigma}_i \in \boldsymbol{\sigma}$ denote the mean and standard deviations of $q\left(z_i \mid X, A\right)$ respectively. The matrices of $\boldsymbol{\mu} \in \mathbb{R}^{N \times H'}$ and $\boldsymbol{\sigma} \in \mathbb{R}^{N \times H'}$ are calculated as follows:

$$\boldsymbol{\mu} = f_\mu(\mathbf{X}, \mathbf{A}) = \tilde{\mathbf{A}}\left(\tilde{\mathbf{A}}\mathbf{X}\mathbf{W}_1\right)\mathbf{W}_2 \tag{3}$$

$$\log \boldsymbol{\sigma} = f_\sigma(\mathbf{X}, \mathbf{A}) = \tilde{\mathbf{A}}\left(\tilde{\mathbf{A}}\mathbf{X}\mathbf{W}_1'\right)\mathbf{W}_2' \tag{4}$$

where f_μ, f_σ denote the mapping functions. For the sake of simplicity, we do not represent the generation process of $\boldsymbol{\mu}_i' \in \boldsymbol{\mu}'$, $\boldsymbol{\sigma}_i' \in \boldsymbol{\sigma}'$, which follows the same process by f_μ', f_σ'. Then, using the reparameterization technique [24] to obtain $\mathbf{Z}, \mathbf{Z}'$. Followed by an inner-product layer to reconstruct the graph structure, and the progress is defined as follows:

$$\mathbf{Z} = \boldsymbol{\mu} + \boldsymbol{\sigma} \odot \boldsymbol{\epsilon} \tag{5}$$

$$\mathbf{Z}^{'} = \boldsymbol{\mu}^{'} + \boldsymbol{\sigma}^{'} \odot \boldsymbol{\epsilon}^{'} \tag{6}$$

$$p(\hat{A} \mid \mathbf{Z}, \mathbf{Z}') = \prod_{i=1}^{N} \prod_{j=1}^{N} \sigma\left(\mathbf{z}_i \mathbf{z}'^{T}_j\right) \tag{7}$$

where $\boldsymbol{\epsilon}, \boldsymbol{\epsilon}^{'} \in \mathbb{R}^{N \times H'}$ are sampled from a multi Gaussian normal distribution with $\boldsymbol{\epsilon}_j \sim \mathcal{N}(0, \mathbf{I}_{H'})$. For CGAE, we only optimize the reconstruction loss $\mathcal{L}_{AE}$. And for CVGAE, we consider the data distribution of node embeddings and hence optimize the variational lower bound $\mathcal{L}_{ELBO}$.

$$\mathcal{L}_{AE} = \mathbb{E}_{q(\mathbf{Z}|\mathbf{X},\mathbf{A})}[\log p(\mathbf{A} \mid \mathbf{Z})] \tag{8}$$

$$\mathcal{L}_{ELBO} = \mathbb{E}_{q(\mathbf{Z}|\mathbf{X},\mathbf{A})}[\log p(\mathbf{A} \mid \mathbf{Z})] - \mathrm{KL}[q(\mathbf{Z} \mid \mathbf{X}, \mathbf{A}) \| p(\mathbf{Z})] - \mathrm{KL}[q(\mathbf{Z}' \mid \mathbf{X}, \mathbf{A}) \| p(\mathbf{Z}')] \tag{9}$$

where $\mathrm{KL}[q(\cdot) \| p(\cdot)]$ measures the Kullback-Leibler divergence [12] between $q(\cdot)$ and $p(\cdot)$. $p(\mathbf{Z}), p(\mathbf{Z}')$ follow Gaussian regularizations, where $p(\mathbf{Z}) = \prod_i \mathcal{N}(\mathbf{z}_i \mid 0, \mathbf{I}_{H'})$ and $p(\mathbf{Z}') = \prod_i \mathcal{N}(\mathbf{z}'_i \mid 0, \mathbf{I}_{H'})$. In this way, the paralleled encoders can lead the $q(\mathbf{Z}' \mid \mathbf{X}, \mathbf{A})$ and $q(\mathbf{Z} \mid \mathbf{X}, \mathbf{A})$ approaching to multi-variate Gaussian distribution.

3.3 Distribution-Dependent Regularization

Different from previous contrastive embedding methods using data augmentations to generate contrastive representations, we propose a unified distribution-dependent regularization leading the encoders to generate contrastive node embeddings, which follow similar distribution. To obtain corresponding positive and negative samples, we must minimize the discrepancy of distribution between embedding matrices $\mathbf{Z}, \mathbf{Z}'$ generated by two paralleled encoders. The well-adopted distribution measure method is Kullback-Leibler divergence, but it is asymmetric and unidirectionally guides the distribution. Hence, we choose Jensen-Shannon divergence [10] to approximate the distribution of $\mathbf{Z}, \mathbf{Z}'$ for CVGAE, which is symmetric and can provide a bidirectional guidance. Here, we follow the independence hypothesis [16] the same as VGAE, and the unidirectional Kullback-Leibler divergence is defined as follows:

$$\begin{aligned} & KL\left(N\left(\mu, \sigma^2\right) | N\left(\mu', \sigma'^2\right)\right) \\ & = \int \frac{e^{-(x-\mu)^2/2\sigma^2}}{\sqrt{2\pi\sigma^2}} \log\left\{\frac{\sqrt{\sigma'^2}}{\sqrt{\sigma^2}} \exp\left\{\frac{1}{2}\left[\frac{(x-\mu')^2}{\sigma'^2} - \frac{(x-\mu)^2}{\sigma^2}\right]\right\}\right\} dx \\ & = \log\sigma' - \log\sigma + \frac{1}{2}\left(\frac{\sigma^2}{\sigma'^2} + \frac{(\mu-\mu')^2}{\sigma'^2} - 1\right) \end{aligned} \tag{10}$$

where $q(\mathbf{Z} \mid \mathbf{X}, \mathbf{A}) = N(\mathbf{Z} \mid \mu, \sigma^{\mathbf{2}})$ and $q(\mathbf{Z}' \mid \mathbf{X}, \mathbf{A}) = N(\mathbf{Z}' \mid \mu', \sigma'^{\mathbf{2}})$. Therefore, the Jensen-Shannon divergence regularization is defined as follows:

$$\begin{aligned} &JS\left(q\left(Z \mid X, A\right) | q\left(Z' \mid X, A\right)\right) \\ &= \frac{1}{2} KL\left(N\left(\mu, \sigma^2\right) | N\left(\mu', \sigma'^2\right)\right) + \frac{1}{2} KL\left(N\left(\mu', \sigma'^2\right) | N\left(\mu, \sigma^2\right)\right) \\ &= \frac{1}{2}\left(\frac{\sigma^2}{\sigma'^2} + \frac{\left(\mu - \mu'\right)^2}{\sigma'^2} + \frac{\sigma'^2}{\sigma^2} + \frac{\left(\mu - \mu'\right)^2}{\sigma^2}\right) - 1 \end{aligned} \tag{11}$$

where $\mathrm{JS}[q(\cdot) \| p(\cdot)]$ is the Jensen-Shannon divergence between $q(\cdot)$, $p(\cdot)$. Hence, we can optimize the lower bound of the Jensen-Shannon divergence between the embedding matrices $\mathbf{Z}, \mathbf{Z}'$. In this paper, we consider a multi-normal Gaussian distribution with an independence hypothesis, which is for simplifying the derivation. And so, only when $\boldsymbol{\mu} = \boldsymbol{\mu}'$ and $\boldsymbol{\sigma} = \boldsymbol{\sigma}'$, the $JS\left(q\left(\mathbf{Z} \mid \mathbf{X}, \mathbf{A}\right) | q\left(\mathbf{Z}' \mid \mathbf{X}, \mathbf{A}\right)\right)$ can achieve the minimum value 0 theoretically, which is the desired distribution situation.

Inspired by previous graph data augmentations [30], we soft the regularization to the node embeddings to some extent, which is beneficial to capture more robust node embeddings [14]. Therefore, we soft the optimization process of maximizing the regularization, which is defined as follows:

$$\mathcal{L}_{Soft} = \mathrm{Re}LU\left(- \triangle + \alpha\right) - ||1 - \alpha||^2 \tag{12}$$

where $\triangle = \mathrm{JS}[q(\cdot) \| p(\cdot)]$, $\boldsymbol{\alpha} \in \mathbb{R}$ is a learnable lower bound of the regularization respectively. $||1 - \alpha||^2$ is a penalty term, which avoids the lower bound approaching to 1. The loss $\mathcal{L}_{Soft}$ aims to constraint $\triangle$ within $[\alpha, 1]$.

3.4 Truncated Triplet Loss

After obtaining the embedding matrices $\mathbf{Z}, \mathbf{Z}'$, we utilize the truncated triplet loss to properly adjust the distance of all node embeddings. For representation simplicity, we only use one node embedding z_i for illustrating the truncated triplet loss. The positive sample of z_i refers to the same index node embeddings $\{z_i, z_i'\}$ of $\mathbf{Z}, \mathbf{Z}'$, while the negative samples refer to the different index node embeddings $\{z_i, z_j'\}(j \neq i, j = 1, 2, 3, \ldots, N)$. As for z_i, its positive node embedding is z_i', and its negative node embeddings are $\{z_j' \mid j \neq i\}$. Then, the initial triplet loss is defined as:

$$\mathcal{L}_{Triplet} = \sum_{i=1}^{N} \sum_{j=1, j \neq i}^{N} \max\left(d\left(z_i, z_i'\right) - d\left(z_i, z_j'\right), \, C\right) \tag{13}$$

where $d(\cdot, \cdot)$ denotes cosine distance, C is a margin value deciding whether or not to drop a negative node embedding. In practice, the loss only focuses on optimizing some negative node embeddings, which are near the margin. In a graph, the scale of one node's nearby neighborhoods is limited, using all negative node embeddings suffers from low learning efficiency [26] and ignoring the cluster information of negative samples. Therefore, we only take part of all negative node embeddings into consideration, which is the meaning of truncation. Firstly, we compute all $d\left(z_i, z_j'\right)$ $(j = 1, 2, \ldots, i-1, i+1, \ldots, N)$. Then, we sort them by ascending. Finally, selecting the **top-k** negative node embeddings into computation, the truncated triplet loss is defined as follows:

$$\mathcal{L}_{Trun} = \sum_{i=1}^{N} \sum_{j \in top-k} \max\left(d\left(\boldsymbol{z}_i, \boldsymbol{z'}_i\right) - d\left(\boldsymbol{z}_i, \boldsymbol{z'}_j\right),\ C\right) \tag{14}$$

4 Experiments

4.1 Datasets

We use three benchmark networks Cora [4], CiteSeer [6] and PubMed [9], in which the nodes are public publications and the connectivities are citation relationships. The features are unique words in each document. The statistical information of each dataset is shown in Table 1.

Table 1. Datasets details.

DataSet	#Nodes	#Edges	#Features	#Clusters	#Avg degree
Cora	2708	5429	1433	7	4.0
CiteSeer	3327	4732	3703	6	2.85
PubMed	19717	44338	500	3	4.5

4.2 Baselines and Implementation Details

In this paper, we compare our proposed models with eight models for link prediction: GAE [16], VGAE [16], ARGA [22], ARVGA [22], GNAE [3], VGNAE [3], DGI [25], GIC [19]. Except for the baselines we compared for link prediction, we include two baselines for node clustering task: K-means [2], GraphCL [30]. Besides, two quantitative metrics are used to evaluate the performances of the link prediction task, which are Area Under Curve (AUC) and Average Precision (AP). Three quantitative metrics are used to evaluate the performances of the node clustering task, which are Average Clustering Accuracy (ACC), Normalized Mutual Information (NMI), and Adjusted Rand Index (ARI). Our proposed model CVGAE is implemented by PyTorch 1.12.0 and trained on a Linux server with GTX 3090. And, we adopt the Adam optimizer with a learning rate 0.001. We use two-layer GCNs as encoders, and the output dimension H' is set to 128. For link prediction, we train our models for a maximum of 500 epochs with an early stopping if the test loss does not decrease in 50 consecutive epochs on datasets Cora and CiteSeer. And, we train our models on the dataset PubMed 1000 epochs for adequate parameters iteration. For all compared baselines, we present their best performances on responding datasets and tasks.

4.3 Link Prediction

Table 2 summarizes the performances of our models and the other eight baseline models. There are some observations from the result: Firstly, DGI is inferior to all other models in most cases, indicating that supplementing graph-level information solely isn't beneficial to the link prediction task. Secondly, GNAE/VGNAE are superior to other baseline models, indicating that using the L2-normalization for node embeddings works well.

Table 2. Link prediction results (in %). The top performances are emphasized in bold.

Models	Cora		CiteSeer		PubMed	
	AUC	AP	AUC	AP	AUC	AP
GAE	91.0	92.0	89.5	89.9	96.4	96.5
VGAE	92.2	93.0	90.8	92.1	94.4	94.7
ARGA	92.4	93.2	91.9	93.0	96.8	97.1
ARVGA	92.4	92.6	92.4	93.0	96.5	96.8
GNAE	95.6	95.7	96.5	97.0	97.5	97.5
VGNAE	95.4	95.8	97.0	97.1	97.6	97.4
DGI	89.8	89.8	95.5	95.7	91.2	92.2
GIC	93.5	93.0	96.8	96.6	93.9	93.5
CVGAE-R	93.5	94.4	95.6	96.2	94.8	94.4
CVGAE-D	94.4	94.4	96.1	95.3	96.0	95.9
CVGAE-A	95.2	95.1	96.2	96.6	96.7	96.4
CVGAE	**96.3**	**96.5**	**97.3**	**97.6**	**97.6**	**97.8**

Specifically, the normalization operation is beneficial to iterations of node embeddings, weakening the inductive bias of GAE. Thirdly, our proposed model CVGAE achieves remarkable improvements to varying degrees on all datasets, which demonstrates that CVGAE can make full use of graph topology structural information.

4.4 Node Clustering

For the node clustering task, we use K-means algorithm [2] on obtained node embeddings from all compared baselines respectively. For our models, the obtained $\mathbf{Z}$ and $\mathbf{Z}'$ are used respectively, and the experiment results show the best performances of one of them. Table 3 summarizes the performances of all models on the task. Compared with ten baseline models, CVGAE achieves the best performances in most cases. Besides, GIC outperforms other baseline models, which is mainly due to the fact that it incorporates cluster-level summary information into node embeddings. The comparison results between GIC and other baseline models demonstrate that node embeddings owning more individual information can bring significant improvements to the node clustering task. Compare with GraphCL, CVGAE performs better on all datasets, which verifies the effectiveness of the triplet truncated loss for alleviating the simpling bias.

Table 3. Node clustering results: Average Clustering Accuracy (ACC), Normalized Mutual Information (NMI) and Adjusted Rand Index (ARI) scores (in %).

	Cora			CiteSeer			PubMed		
	Acc	NMI	ARI	Acc	NMI	ARI	Acc	NMI	ARI
K-means	49.6	32.3	22.6	55.9	31.4	29.4	57.0	22.0	17.8
GAE	61.3	44.4	38.1	48.2	22.7	19.2	64.2	22.5	22.1
VGAE	64.7	43.4	37.5	51.9	24.9	23.8	51.9	24.9	23.8
ARGA	64.0	44.9	35.2	57.3	35.0	34.1	68.1	27.6	29.1
ARVGA	63.8	45.0	62.7	54.4	26.1	24.5	63.5	23.2	22.5
GNAE	72.4	55.6	50.9	67.6	41.8	42.5	68.0	27.8	29.1
VGNAE	72.3	54.7	51.1	67.6	42.1	42.6	69.1	29.1	30.7
DGI	71.3	56.4	51.1	68.8	44.4	45.0	53.3	18.1	16.6
GIC	72.5	53.7	50.8	**69.6**	**45.3**	**46.5**	67.3	29.7	29.1
GraphCL	72.1	55.1	53.4	68.2	42.7	43.9	69.0	28.7	30.4
CVGAE	**74.7**	**57.4**	**54.8**	69.4	44.1	45.5	**69.8**	**31.9**	**31.8**

4.5 Ablation Experiments

In this section, we conduct ablation experiments to verify the effectiveness of different implementations for our proposed model. The details of the three implementations are as follows: (1) CVGAE-R represents that we remove the distribution-dependent regularization. (2) CVGAE-A represents that we replace the truncated triplet loss with InfoNCE loss using all negatives. (3) CVGAE-D represents that we replace the regularization with dropout operation. We pass the same graph to one encoder twice to obtain contrastive representations, and the dropout ratio is set to 0.2. As shown in Table 2, removing the regularization is critically harmful to learning performances. For example, CVGAE outperforms CVGAE-R in all datasets, by at least 1.4% on AUC. CVGAE-D outperforms CVGAE-R in all cases, indicating that the quality of contrastive representations plays a vital role in CL. In addition, the performances of CVGAE are superior to CVGAE-D, which demonstrates that using the distribution-dependent regularization works better than the dropout operation. Meanwhile, considering all negatives is not a wise choice, it slightly hinders performance improvements and affects learning effectiveness.

5 Conclusion

In this paper, we augment the VGAE with CL framework, namely CVGAE, which aims to learn more generalized node embeddings. We argue that existing graph data augmentations fail to preserve original semantic information for nodes. Therefore, different from previous GCL methods using designed graph data augmentations, we proposed a novel distribution-dependent regularization to guide the paralleled encoders to generate contrastive representations. Subsequently, to alleviate the problem of sampling bias in a self-supervised way, the truncated triplet loss is employed to separate the top-k nodes apart in the representation space and improve the learning efficiency. Extensive experiments are evaluated on several real-world datasets and the results demonstrate our proposed model's advanced performance in link prediction, node clustering tasks.

Acknowledgments. This research was partially supported by grants from the National Natural Science Foundation of China (No. 61877051). We acknowledge all the developers and researchers for developing useful tools that enable our experiments.

References

1. Aggarwal, C.C., Wang, H.: A survey of clustering algorithms for graph data. In: Aggarwal, C.C., Wang, H. (eds.) Managing and Mining Graph Data, pp. 275–301. Springer, Boston (2010). https://doi.org/10.1007/978-1-4419-6045-0_9
2. Ahmed, M., Seraj, R., Islam, S.M.S.: The k-means algorithm: a comprehensive survey and performance evaluation. Electronics **9**(8), 1295 (2020)
3. Ahn, S.J., Kim, M.: Variational graph normalized autoencoders. In: Proceedings of the 30th ACM International Conference on Information and Knowledge Management, pp. 2827–2831 (2021)
4. Cabanes, C., et al.: The cora dataset: validation and diagnostics of in-situ ocean temperature and salinity measurements. Ocean Sci. **9**(1), 1–18 (2013)
5. Cai, T.T., Frankle, J., Schwab, D.J., Morcos, A.S.: Are all negatives created equal in contrastive instance discrimination? arXiv preprint arXiv:2010.06682 (2020)
6. Caragea, C., et al.: CiteSeer x: a scholarly big dataset. In: de Rijke, M., et al. (eds.) ECIR 2014. LNCS, vol. 8416, pp. 311–322. Springer, Cham (2014). https://doi.org/10.1007/978-3-319-06028-6_26
7. Chen, T., Kornblith, S., Norouzi, M., Hinton, G.: A simple framework for contrastive learning of visual representations. In: International Conference on Machine Learning, pp. 1597–1607. PMLR (2020)
8. Chuang, C.Y., Robinson, J., Lin, Y.C., Torralba, A., Jegelka, S.: Debiased contrastive learning. Adv. Neural. Inf. Process. Syst. **33**, 8765–8775 (2020)
9. Falagas, M.E., Pitsouni, E.I., Malietzis, G.A., Pappas, G.: Comparison of pubmed, scopus, web of science, and google scholar: strengths and weaknesses. FASEB J. **22**(2), 338–342 (2008)
10. Fuglede, B., Topsoe, F.: Jensen-shannon divergence and hilbert space embedding. In: International Symposium on Information Theory, 2004 (ISIT 2004), Proceedings, p. 31. IEEE (2004)
11. Hassani, K., Khasahmadi, A.H.: Contrastive multi-view representation learning on graphs. In: International Conference on Machine Learning, pp. 4116–4126. PMLR (2020)

12. Hershey, J.R., Olsen, P.A.: Approximating the kullback leibler divergence between gaussian mixture models. In: 2007 IEEE International Conference on Acoustics, Speech and Signal Processing (ICASSP 2007), vol. 4, pp. IV–317. IEEE (2007)
13. Hjelm, R.D., et al.: Learning deep representations by mutual information estimation and maximization. arXiv preprint arXiv:1808.06670 (2018)
14. Hou, Z., et al.: Graphmae: self-supervised masked graph autoencoders. arXiv preprint arXiv:2205.10803 (2022)
15. Jaiswal, A., Babu, A.R., Zadeh, M.Z., Banerjee, D., Makedon, F.: A survey on contrastive self-supervised learning. Technologies **9**(1), 2 (2020)
16. Kipf, T.N., Welling, M.: Variational graph auto-encoders. arXiv preprint arXiv:1611.07308 (2016)
17. Latham, P.E., Roudi, Y.: Mutual information. Scholarpedia **4**(1), 1658 (2009)
18. Liben-Nowell, D., Kleinberg, J.: The link-prediction problem for social networks. J. Am. Soc. Inform. Sci. Technol. **58**(7), 1019–1031 (2007)
19. Mavromatis, C., Karypis, G.: Graph infoclust: leveraging cluster-level node information for unsupervised graph representation learning. arXiv preprint arXiv:2009.06946 (2020)
20. Oord, A.v.d., Li, Y., Vinyals, O.: Representation learning with contrastive predictive coding. arXiv preprint arXiv:1807.03748 (2018)
21. Pan, L., Shi, C., Dokmanić, I.: Neural link prediction with walk pooling. arXiv preprint arXiv:2110.04375 (2021)
22. Pan, S., Hu, R., Long, G., Jiang, J., Yao, L., Zhang, C.: Adversarially regularized graph autoencoder for graph embedding. arXiv preprint arXiv:1802.04407 (2018)
23. Tian, Y., Krishnan, D., Isola, P.: Contrastive multiview coding. In: Vedaldi, A., Bischof, H., Brox, T., Frahm, J.-M. (eds.) ECCV 2020. LNCS, vol. 12356, pp. 776–794. Springer, Cham (2020). https://doi.org/10.1007/978-3-030-58621-8_45
24. Tokui, S., et al.: Reparameterization trick for discrete variables. arXiv preprint arXiv:1611.01239 (2016)
25. Veličković, P., Fedus, W., Hamilton, W.L., Liò, P., Bengio, Y., Hjelm, R.D.: Deep graph infomax. arXiv preprint arXiv:1809.10341 (2018)
26. Wang, G., Wang, K., Wang, G., Torr, P.H., Lin, L.: Solving inefficiency of self-supervised representation learning. In: Proceedings of the IEEE/CVF International Conference on Computer Vision, pp. 9505–9515 (2021)
27. Wang, H., Li, Y., Huang, Z., Dou, Y., Kong, L., Shao, J.: SNCSE: contrastive learning for unsupervised sentence embedding with soft negative samples. arXiv preprint arXiv:2201.05979 (2022)
28. Xia, J., Wu, L., Chen, J., Hu, B., Li, S.Z.: Simgrace: a simple framework for graph contrastive learning without data augmentation. In: Proceedings of the ACM Web Conference 2022, pp. 1070–1079 (2022)
29. Xu, M.: Understanding graph embedding methods and their applications. SIAM Rev. **63**(4), 825–853 (2021)
30. You, Y., Chen, T., Sui, Y., Chen, T., Wang, Z., Shen, Y.: Graph contrastive learning with augmentations. Adv. Neural. Inf. Process. Syst. **33**, 5812–5823 (2020)
31. Zhang, S., Tong, H., Xu, J., Maciejewski, R.: Graph convolutional networks: a comprehensive review. Comput. Soc. Netw. **6**(1), 1–23 (2019). https://doi.org/10.1186/s40649-019-0069-y
32. Zhu, Y., Xu, Y., Yu, F., Liu, Q., Wu, S., Wang, L.: Graph contrastive learning with adaptive augmentation. In: Proceedings of the Web Conference 2021, pp. 2069–2080 (2021)

Enhancing Spatial Consistency and Class-Level Diversity for Segmenting Fine-Grained Objects

Qi Zhao[1], Binghao Liu[1], Shuchang Lyu[1], Chunlei Wang[1], and Yifan Yang[2](✉)

[1] Department of Electrics and Information Engineering, Beihang University, Beijing 100191, China
{zhaoqi,liubinghao,lyushuchang,wcl_buaa}@buaa.edu.cn
[2] Institute of Artificial Intelligence, Beihang University, Beijing 100191, China
stephenyoung@buaa.edu.cn

Abstract. Semantic segmentation is a fundamental computer vision task attracting a lot of attention. However, limited works focus on semantic segmentation on fine-grained class scenario, which has more classes and greater inter-class similarity. Due to the lack of data available for this task, we establish two segmentation benchmarks, CUB-seg and FGSCR42-seg, based on CUB and FGSCR42 datasets. To solve the two major problems in this task, spatial inconsistency and extremely similar classes confusion, we propose the Spatial Consistency and Class-level Diversity enhancement Network. First, we build the Spatial Consistency Enhancement Module to take advantage of the low-frequency information in the feature, enhancing the spatial consistency. Second, Fine-grained Regions Contrastive Loss is designed to make the features of different classes more discriminative, promoting the class-level diversity. Extensive experiments show that our method can significantly improve the performance compared to baseline models. Visualization study also prove the effectiveness of our method for enhancing spatial consistency and class-level diversity.

Keywords: Fine-grained Semantic Segmentation · Spatial Consistency · Contrastive Learning

1 Introduction

Semantic segmentation, as a fundamental computer vision task, aims to assign specific class to each pixel in images. Deep learning methods [1–5] have realized significant achievement in semantic segmentation task. Furthermore, many extended tasks such as few-shot segmentation [6,7], zero-shot segmentation [8] and part segmentation [9] are also proposed to promote the application of semantic segmentation. However, few works [10] focus on semantic segmentation on objects with fine-grained classes.

B. Luo et al. (Eds.): ICONIP 2023, CCIS 1965, pp. 292–304, 2024.
https://doi.org/10.1007/978-981-99-8145-8_23

Fig. 1. Some annotation examples of two datasets. Two rows of (a) are images and their corresponding annotations of CUB-seg, while (b) contains those of FGSCR42-seg.

Fine-grained means there are many objects in the scene that belong to different finer subcategories under a same category. Different from traditional semantic segmentation, objects of different classes in fine-grained semantic segmentation task are extremely similar to each other, accompanying by large intra-class differences. This task can be applied to many practical scenarios, such as biological nature protection, commodity identification and so on.

To improve the fine-grained segmentation ability of models, we first establish two segmentation datasets based on two fine-grained classification datasets, CUB [11] and FGSCR-42 [12]. We use f-BRS model [13] to label the images roughly with pixel-wise annotations and then make some manual adjustments to the mislabeling results. Some final annotations can be seen in Fig. 1. With these two fine-grained semantic segmentation datasets, we can evaluate performance of each model on both natural and remote sensing scenes, which can comprehensively promote the research of this task.

By analyzing the segmentation results of many methods [1,4,14–17] on abovementioned two datasets, we find two major problems which hinder the performances of these methods. First, because of the large intra-class variation and small inter-class difference, the weak class-level feature diversity makes it easy to assign a wrong class to the object, creating confusion among many categories that are very similar. Second, the spatial separation of the different components of an object leads to spatial inconsistency in the object region, driving methods to identify the whole region of the object as a combination of components with different classes. Due to the extreme similarity between components of different classes, the spatial inconsistency becomes increasingly serious. Some failure cases of current methods can be seen in Fig. 2.

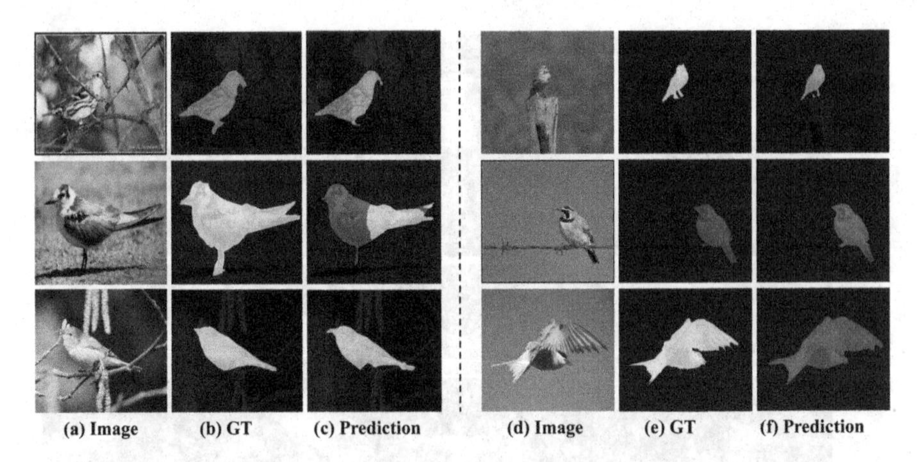

Fig. 2. Some failure cases of current methods on CUB-seg dataset. Columns (a), (b) and (c) show the mistake of segmenting the object into many components with different classes caused by the spatial inconsistency. Columns (d), (e) and (f) show the classification failure caused by the weak class-level feature diversity.

To solve the two problems mentioned above, we propose the Spatial Consistency Enhancement Module (SCEM) and Fine-grained Regions Contrastive Loss (FRC loss) to improve the segmentation performance on fine-grained semantic segmentation task. The SCEM regards the output features of the encoder as high-frequency features, that is, pixel-wise features change dramatically in spatial dimension, resulting in spatial inconsistency. Visualization on the features of current methods also proves this opinion that the hot region of heat map only cover part of the whole object region. So SCEM generates a low-frequency low-resolution feature under the guidance of heat map and then fuse it with original feature to enhance the spatial consistency. Furthermore, to make the features of different classes more discriminative, FRC loss first uses Masked Average Pooling (MAP) to obtain a vector which can represent the object. After that, vectors of different classes in a single batch are generated and dense cosine metrics is utilized to calculate similarities between each pair of vectors, including a vector and itself. Then we use contrastive loss to supervise the similarities, aiming to make representative vectors of different classes far away from each other. With the SCEM and FRC loss, we design the Spatial Consistency and Class-level Diversity enhancement Network (SCCDNet) to address the fine-grained semantic segmentation task specifically.

Extensive experiments prove that our SCCDNet can be built on all encoder-decoder based convolutional neural networks and significantly improve the performance compared to baseline models. We also conduct ablation experiments and visualization study to validate that our SCCDNet is indeed enhance the spatial consistency and class-level diversity of features. Our code, pretrained models and two datasets, CUB-seg and FGSCR42-seg, are available at https://github.com/cv516Buaa/SCCDNet.

The contributions of this paper are listed as follow:

1. We establish two fine-grained semantic segmentation datasets based on CUB and FGSCR-42, providing comprehensive benchmarks for evaluating related methods of this task on natural and remote sensing scenes.
2. We propose Spatial Consistency Enhancement Module (SCEM) to take advantage of the low-frequency information in the feature, enhancing the spatial consistency.
3. Fine-grained Regions Contrastive Loss (FRC loss) is designed to make the features of different classes more discriminative, promoting the class-level diversity.
4. Comprehensive experiments are conducted to prove the effectiveness of our Spatial Consistency and Class-level Diversity enhancement Network (SCCD-Net).

2 Related Work

Semantic segmentation has attracted a lot of attention of researchers and achieves a number of breakthroughs with the emergency of fully convolutional neural network (FCN) based segmentation methods [1,3,5] and large scale segmentation datasets [18–20]. PSPNet [4] introduces pyramid pooling module to extract multi-scale features which contain both local information and global information, solving the problem of insufficient use of context information well. Similarly, Deeplab series models [3,5,15] utilize atrous convolution kernel with different dilation values to get context information.

After transformer [21] has achieved great success in the field of natural language processing, attention based CNN models also obtained greatly developed. SENet [22] squeezes the feature map to a vector by global average pooling and uses several fully connected layers generating channel reweighting vector, which can represent the importance of different channels. DANet [14] adopts position attention module and channel attention module to learn position and channel inter-dependencies. CCNet [16] adopts a criss-cross attention module to capture contextual information from full-image dependencies. Then following ViT [23], Segformer [17] realizes semantic segmentation with a fully transformer architecture.

Fine-grained classification task has also been studied for a long time. Some famous datasets are CUB [11] and Standford dogs [26]. API-Net [24] learns a mutual feature vector to capture semantic differences in the input pair and then compares this mutual vector with individual vectors to generate gates for each input image. Transfg [25] designs a tranformer model to effectively and accurately select discriminative image patches and compute their relations. However, as a similar task, fine-grained semantic segmentation task receives limited attention, so we establish two datasets and study this task in this paper.

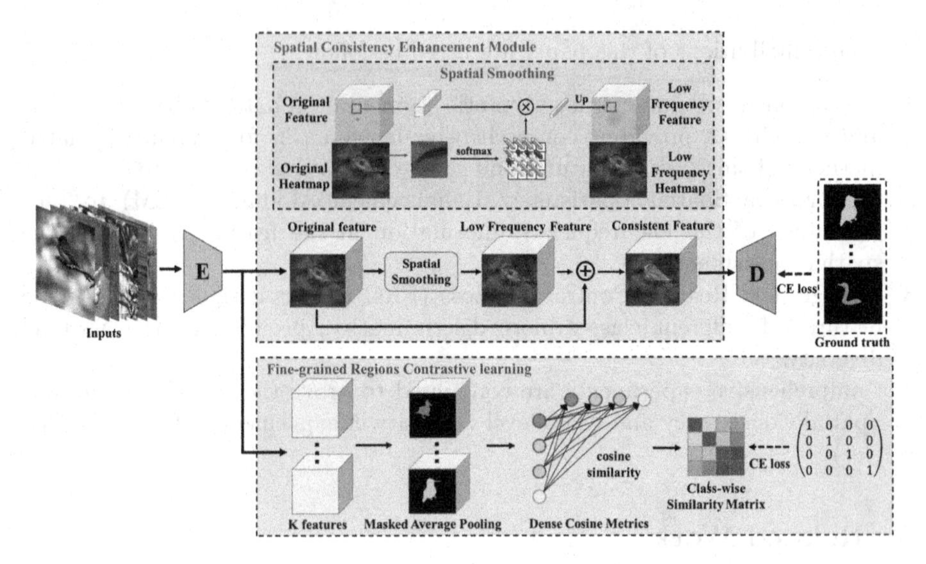

Fig. 3. The overall architecture of our SCCDNet. E and D in figure are encoder and decoder.

3 Method

3.1 Overall Network Architecture

As we discussed above, the Spatial Consistency and Class-level Diversity enhancement Network (SCCDNet) is proposed to solve the two major problems in fine-grained semantic segmentation task. The overall architecture of SCCDNet is shown in Fig. 3, in which the model consists of encoder, decoder, Spatial Consistency Enhancement Module (SCEM) and Fine-grained Regions Contrastive Loss (FRC loss).

Any classification network could be chosen as encoder, such as ResNet [27] and ViT [23]. The decoder is always different in various methods, which is pyramid pooling module (PPM) in PSPNet [4] and atrous spatial pyramid pooling (ASPP) in DeepLabv3 [15]. The SCEM and FRC loss are applied on the output feature of encoder. The output of SCEM will be input to decoder for getting final prediction and FRC loss, which is not used in test phase, is only utilized to promote the feature more discriminative.

3.2 Spatial Consistency Enhancement Module

Due to the large intra-class variation and small inter-class difference, the components of an object could be very similar to components of object with another class, resulting in segmenting an object to multiple parts with different classes. To alleviate this challenge, we propose the Spatial Consistency Enhancement Module (SCEM) to enhance the spatial consistency, making model predict a integral region for a single object.

Define the output feature of encoder as f, so SCEM takes f as input, and output a new feature f_c, whose size is equal to f. Based on the visualization of f, we discover the encoder mainly focuses on part of the whole region of object, inducing decoder to assign different classes to this part and other object regions. As shown in the above dotted box of Fig. 3, the original feature f has a dramatically varying spatial heat distribution. However, we want it to have a lower peak but a smoother heat distribution, in other words, realizing spatial smoothing to change the heat distribution of f from high to low frequency in spatial dimension, while still focusing on the object region. We first calculate the heat map h of f as $h_{(i,j)} = \sum_{n=1}^{c} f_{(i,j)}^{n}/c$, where i, j and n are row index, column index and the number channels of feature respectively.

Consider the certain class fits a particular distribution F, then each feature generated from image with this class is a sample chosen from this distribution. The sampled feature f can be separate to three part along spatial dimension, the background region S_α, the hot region S_β and the other object region S_γ with low value in h. We want get a new feature $\hat{f}$ with low-frequency information compared to f and finally generate $\tilde{f} = \psi\left(f, \hat{f}\right)$, which has both the low frequency spatial consistency and the focus on the whole object region. The $\psi(\cdot,\cdot)$ is a mapping function. Because we mainly aim to increase the heat values of S_γ and maintain those values of S_α and S_β, our objective function of $\hat{h}$ (heat map of $\hat{f}$) in S_γ is shown in Eq. 1.

$$\max \sum\nolimits_{p \in S_\gamma} \left| \frac{\hat{h}_p}{h_p} \right| \tag{1}$$

where $\hat{h}_p$ and h_p denote points in two heat maps. Meanwhile, in S_α and S_β, $\hat{h}_p$ should converge in distribution to ($\xrightarrow{d}$) h as shown in Eq. 2.

$$\hat{h}\left(p \mid p \in (S_\alpha \cup S_\beta)\right) \xrightarrow{d} h\left(p \mid p \in (S_\alpha \cup S_\beta)\right) \tag{2}$$

To satisfy above conditions, we aim to prompt the high heat region of the heat map expand and enlarge. Therefore, as shown in Eq. 3, in each $s \times s$ local region, we get a vector v which has the maximum probability of belonging to the object to represent this region under the guidance of h.

$$v = Softmax(h_1, h_2, \cdots, h_{s\times s}) \cdot (f_1, f_2, \cdots, f_{s\times s})^T \tag{3}$$

where $(h_1, h_2, \cdots, h_{s\times s})$ and $(f_1, f_2, \cdots, f_{s\times s})$ are points in local region of h and f. This method is just similar to the pooling operation, while the output is not average calculation or maximum selection but a weighted average operation based on heat map. By setting the stride of local region as s, the output feature $\hat{f}$ is s times downsampled from f. Two MLPs are applied to $\hat{f}$ and f respectively, aligning the channel distributions of them.

Then we use bilinear interpolation to upsample the $\hat{f}$ to f_l, which has same size with f but lower frequency spatial information because of the information loss caused by the above downsampling. Using $\tilde{f} = (f + f_l)/2$, we can get $\tilde{f}$, which has smoother distribution than f and still focuses on object region with stronger spatial consistency. By using $\tilde{f}$ as input, the decoder will be more inclined to predict the object as a whole in spatial dimension.

3.3 Fine-Grained Regions Contrastive Loss

Although the spatial inconsistency problem can be solved by SCEM to some extent, the small inter-class difference still induces model to confuse different fine-grained classes. To clear the decision boundaries between different classes, we design the Fine-grained Regions Contrastive loss (FRC loss).

The output feature f of encoder contains multiple features extracted from a batch of images. Define the batch size as K, then all features in a single batch can be denoted as $f^1, f^2, \cdots, f^K$. During training phase, we also have mask labels $M^1, M^2, \cdots, M^K$ of samples in the same batch. Then we will use these mask labels to extract representative feature vectors of every objects in this batch, just except the background.

Assume there are m different classes (except background) in images of a single batch and mask M is downsampled to the same size with f, then we can get the representative vector of each class with Masked Average Pooling, as shown in Eq. 4 and Eq. 5.

$$v_t^p = \frac{\sum_{i=1}^{w}\sum_{j=1}^{h} f_{(i,j)}^p \left[M_{(i,j)}^p == t\right]}{\sum_{i=1}^{w}\sum_{j=1}^{h} \left[M_{(i,j)}^p == t\right]} \tag{4}$$

$$v_t = \frac{\sum_{1 \le p \le K, t \in M^p} v_t^p}{\sum_{p=0}^{K} [t \in M^p]} \tag{5}$$

h and w are feature height and width. $[\cdot]$ denotes Iverson bracket, a notation that signifies a number that is 1 if the condition in square brackets is satisfied, and 0 otherwise, i.e. v_t is the vector of class t $(t \le m)$ and v_t^p is the vector generated by the p-th mask M^p.

To enhance the class-level diversity, we introduce contrastive loss on these vectors. At the first, we generate a dense cosine similarity matrix $\mathbf{S}$ by calculating cosine similarity between each pair of vectors (including one and itself). So $\mathbf{S}$ is a $m \times m$ matrix and each point $s_{(a,b)}$ equals to $\langle v_a \cdot v_b \rangle / (|v_a| \cdot |v_b|)$ (a, b are row and column indexes, $1 \le a, b \le m$).

After getting $\mathbf{S}$, we use Softmax operation on it, converting summation of each line to 1 following Eq. 6.

$$s'_{(a,b)} = \frac{exp\left(s_{(a,b)}\right)}{\sum_{i=1}^{m} exp\left(s_{(a,b)}\right)} \tag{6}$$

where $\mathbf{S}'$ is the matrix generated from $\mathbf{S}$ after Softmax. So the FRC loss is obtained by calculating cross entropy loss between $\mathbf{S}'$ and an identity matrix $\mathbf{I}$ as shown in Eq. 7

$$L_{FRC} = -\frac{1}{m}\sum_{a=1}^{m}\left(\frac{1}{m}\sum_{b=1}^{m}\left(I_{(a,b)} \cdot log\ s'_{(a,b)}\right)\right) \tag{7}$$

Table 1. Datasets details and experiment settings of CUB-seg and FGSCR42-seg.

Datasets	Num of Class	Train Set	Test Set	Input Size	Batch Size
CUB-seg	200	5994	5794	416 × 416	8
FGSCR42-seg	42	3924	3854	416 × 416	4

Table 2. Experiment results of six previous models and our SCCDNet on CUB-seg and FGSCR42-seg datasets. PPM and ASPP denote that models are based on PSPNet and DeepLabv3 respectively.

Methods	Backbone	CUB-seg			FGSCR42-seg		
		mIOU	Acc	mAcc	mIOU	Acc	mAcc
FCN [1]	ResNet50	61.91	96.13	73.69	45.31	93.74	55.38
FCN [1]	ResNet101	60.92	96.05	73.04	54.41	94.84	61.60
DANet [14]	ResNet50	63.38	96.04	75.58	68.84	97.35	79.41
DANet [14]	ResNet101	67.85	96.62	79.12	70.28	97.61	80.96
CCNet [16]	ResNet50	68.86	96.79	79.88	63.33	96.47	71.59
CCNet [16]	ResNet101	68.73	96.81	79.58	63.74	96.91	72.31
Segformer [17]	mit-b0	63.37	96.07	73.48	67.43	97.24	77.69
Segformer [17]	mit-b5	69.16	96.88	80.87	70.53	97.56	80.99
PSPNet [4]	ResNet50	69.95	96.93	80.90	75.44	98.10	84.62
PSPNet [4]	ResNet101	70.03	96.97	80.94	73.49	97.82	80.32
DeepLabv3 [15]	ResNet50	69.98	96.98	80.82	72.75	97.69	80.21
DeepLabv3 [15]	ResNet101	70.26	96.97	81.14	70.77	97.89	81.19
SCCDNet (+PPM)	ResNet50	71.12	97.00	81.48	**77.60**	**98.21**	**85.61**
SCCDNet (+PPM)	ResNet101	71.02	97.00	**81.99**	72.24	97.82	82.83
SCCDNet (+ASPP)	ResNet50	**71.29**	**97.04**	81.19	73.57	97.96	83.59
SCCDNet (+ASPP)	ResNet101	70.91	97.01	81.44	71.83	97.71	80.05

By designing the FRC loss, the features of different classes can be more diverse, making it easier to classify.

3.4 Loss

By adding SCEM and FRC loss, the overall loss of our SCCDNet, as shown in Eq. 8, consists of three parts, main cross entropy loss, auxiliary loss and the FRC loss.

$$L = \alpha \cdot L_{ce} + \beta \cdot L_{aux} + \gamma \cdot L_{FRC} \tag{8}$$

L_{aux} is usually based on a FCN decoder and placed after the encoder. In experiments, we set $\alpha = 1$ and $\beta = \gamma = 0.4$.

Table 3. The ablation study of SCEM and FRC loss, the experiments are conducted on CUB-seg datasets and baseline model is PSPNet with ResNet50.

Method	mIOU	Acc	mAcc
PSPNet	69.95	96.93	80.90
PSPNet+SCEM	70.36	96.97	81.02
PSPNet+FRC loss	70.87	96.95	81.33
SCCDNet	71.12	97.00	81.48

Table 4. Ablation study on down-sampling stride s of SCEM, conducted on CUB-seg datasets and baseline model is PSPNet with ResNet50.

Stride	mIOU	Acc	mAcc
2	71.12	97.00	81.48
4	70.94	97.02	81.43
6	69.14	96.90	80.79
8	64.53	96.38	74.62

Table 5. Comparison between "PSPNet+SCEM" and "SCCDNet" with different batch size on CUB-seg dataset using ResNet50 as encoder.

Batch size	2	3	4	5	6	7	8
PSPNet+SCEM (mIOU)	58.94	62.79	65.28	67.94	68.86	70.37	70.36
SCCDNet (mIOU)	56.32	61.43	65.09	67.98	69.50	70.86	71.12

4 Experiments

4.1 Implementation Details

We establish two datasets, CUB-seg and FGSCR43-seg, to evaluate performances of different models, details are shown in Table 1. For CNN based models, ResNet50 and ResNet101 with ImageNet pretrained weights are used as encoders, while mit-b0 and mit-b5 with pretrained weights are used for Segformer. Initial learning rate, momentum and weight decay are set as 0.01, 0.9 and 0.0001, and SGD is chosen. We also utilize "Poly" schedule with 0.9 power and run models for 40k iterations. Methods and experiments are implemented on MMSegmentation framework. Mean IOU (mIOU), pixel accuracy (Acc) and mean pixel accuracy (mAcc) are used as evaluation metrics.

4.2 Results

We conduct extensive experiments on two datasets as shown in Table 2, using six commonly used segmentation models with different backbones and our SCCDNet based on PSPNet and DeepLabv3. It can be seen intuitively that our SCCDNet achieves significant improvement compared to original PSPNet and DeepLabv3. On CUB-seg dataset, there are only delicate difference between same model (such as PSPNet and DeepLabv3) with different encoders. However, on FGSCR42-seg, models based on ResNet101 usually have worse performance than those based on ResNet50. Because FGSCR42-seg datasets has some classes which has few samples, so large backbone has excessive focus on large sample classes and neglect of small sample classes. Overall, our SCCDNet effectively improves performance while introducing very little cost during testing phase.

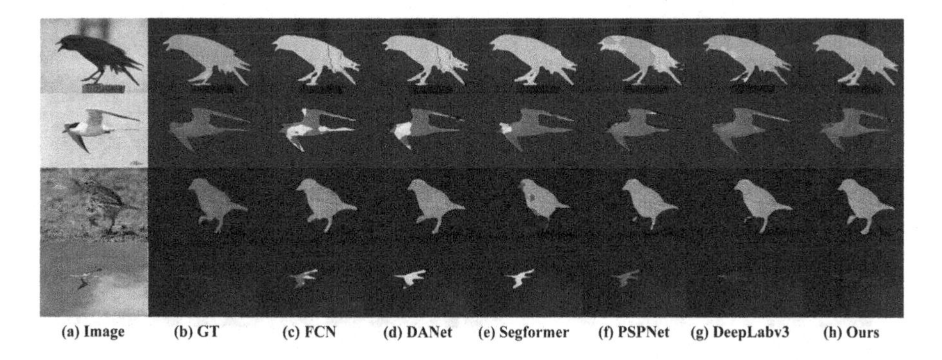

Fig. 4. The prediction visualization of six different models, FCN, DANet, Segformer, PSPNet, DeepLabv3 and our SCCDNet.

4.3 Ablation Study

To further analyze the effect of SCEM and FRC loss, we conduct ablation experiments on CUB-seg dataset as shown in Table 3. The baseline model is PSPNet with ResNet50. We can see that adding SCEM or FRC loss alone to baseline, the performances both can be improved. In addition, the effect of FRC loss is larger than effect of SCEM, which means the small inter-class distinction can be even more detrimental to network performance. Combine SCEM and FRC loss, our SCCDNet achieves the best performance on CUB-seg dataset.

We also validate our SCCDNet on CUB dataset with different stride s of SCEM as shown in Table 4. It is obvious that the segmentation performance going lower when the stride s becomes larger, because as s increases, the negative impact of information loss on the model exceeds the enhancement effect of SCEM on spatial consistency.

As shown in Table 5, as the batch size increases, the performances of two models also increases. However, the larger the batch size, the better the improvement introduced by FRC loss.

4.4 Visualization

To analyze the performance of our SCCDNet intuitively, we conduct several visualization experiments. Figure 4 illustrates the prediction mask of six different models, FCN, DANet, Segformer, PSPNet, DeepLabv3 and our SCCDNet. It can be seen that the prediction results of SCCDNet are more consistency in spatial dimension, and the classification errors are significantly reduced compared with other methods. For example, in the first line, other five models all predict some regions of the black American Crow to another class, but SCCDNet recognize the whole region of it correctly.

We visualize the heat map of features before and after SCEM in Fig. 5. Before SCEM, as shown in Fig. 5 (b), the feature focuses on part of the object region. Figure 5 (c) illustrates the heat map of feature output from the SCEM. We can

Fig. 5. The heat map visualization of features before and after SCEM. (a) original images, (b) heat map of feature before SCEM, (c) heat map of feature after SCEM.

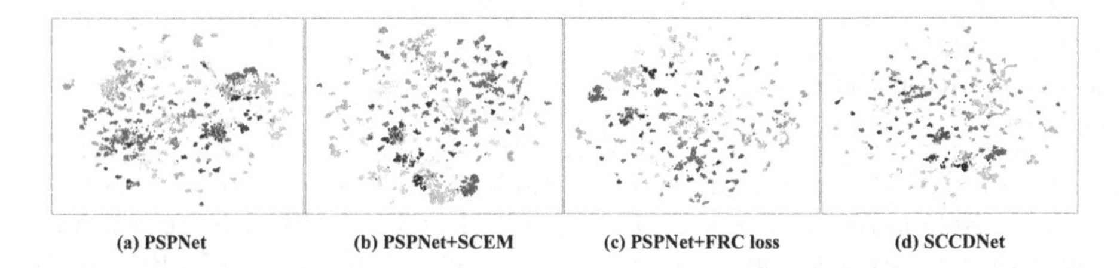

Fig. 6. The t-SNE visualization of vectors generated from MAP on CUB-seg dataset.

intuitively find the hot region where the feature focus is larger than line (b) and cover the whole object region appropriately.

We conduct a t-distributed stochastic neighbor embedding (t-SNE) visualization experiment for the vectors of whole CUB-seg test dataset by masked average pooling. There are 5794 points in Fig. 6 and different colors denote different classes (total 200 classes). We can see that the introduction of FRC loss can shorten distances of same class vector and enlarge them with different classes, significantly enhancing the class-level diversity and promoting the model generating clearer decision boundaries. Otherwise, SCEM has little effect on the distinction between different classes.

5 Conclusion

To explore the fine-grained semantic segmentation which has more classes and greater inter-class similarity, we establish CUB-seg and FGSCR42-seg datasets. To solve the major problems in this task, spatial inconsistency and similar classes confusion, we propose a Spatial Consistency and Class-level Diversity enhancement Network (SCCDNet). We design the Spatial Consistency Enhancement Module (SCEM) to obtain feature with low-frequency and spatial consistent information, which can guide model to segment objects in a more holistic way. Then we introduce Fine-grained Regions Contrastive loss (FRC loss) to prompt features of different classes diverse, realizing clearer decision boundaries and reducing confusion between classes. Extensive experiments have been conducted to prove the effectiveness of our SCCDNet, comprehensive ablation study and visualization work show that SCCDNet indeed enhance the spatial consistency and class-level diversity well. Future work may focus on extraction of key fine-grained discriminative parts and the representation of relationship between part and entirety.

Acknowledgements. This work was supported in part by the National Natural Science Foundation of China under Grant 62072021 and 62002005.

References

1. Long, J., Evan, S., Trevor, D.: Fully convolutional networks for semantic segmentation. In: Proceedings of the IEEE Conference on Computer Vision and Pattern Recognition, pp. 3431–3440 (2015)
2. Badrinarayanan, V., Alex, K., Roberto, C.: SegNet: a deep convolutional encoder-decoder architecture for image segmentation. IEEE Trans. Pattern Anal. Mach. Intell. **39**(12), 2481–2495 (2017)
3. Chen, L., George, P., Iasonas, K., Kevin, M., Alan, L.Y.: Semantic image segmentation with deep convolutional nets and fully connected CRFs. arXiv preprint http://arxiv.org/abs/1412.7062 (2014)
4. Zhao, H., Shi, J., Qi, X., Wang, X., Jia, J.: Pyramid scene parsing network. In: Proceedings of the IEEE Conference on Computer Vision and Pattern Recognition, pp. 2881–2890 (2017)
5. Chen, L., George, P., Iasonas, K., Kevin, M., Alan, L.Y.: DeepLab: semantic image segmentation with deep convolutional nets, atrous convolution, and fully connected CRFs. IEEE Trans. Pattern Anal. Mach. Intell. **40**, 834–848 (2017)
6. Tian, Z., Zhao, H., Shu, M., Yang, Z., Li, R., Jia, J.: Prior guided feature enrichment network for few-shot segmentation. IEEE Trans. Pattern Anal. Mach. Intell. **44**(2), 1050–1065 (2020)
7. Zhao, Q., Liu, B., Lyu, S., Chen, H.: A self-distillation embedded supervised affinity attention model for few-shot segmentation. IEEE Trans. Cogn. Dev. Syst. (2023). https://doi.org/10.1109/TCDS.2023.3251371
8. Bucher, M., Vu, T., Matthieu, C., Patrick, P.: Zero-shot semantic segmentation. In: Advances in Neural Information Processing Systems, vol. 32 (2019)
9. Xia, F., Wang, P., Chen, X., Alan, L.Y.: Joint multi-person pose estimation and semantic part segmentation. In: Proceedings of the IEEE Conference on Computer Vision and Pattern Recognition, pp. 6769–6778 (2017)
10. Zhang, X., Zhao, W., Luo, H., Peng, J., Fan, J.: Class guided channel weighting network for fine-grained semantic segmentation. In: Proceedings of the AAAI Conference on Artificial Intelligence, vol. 36, no. 3, pp. 3344–3352 (2022)
11. Wah, C., Branson, S., Welinder, P., Perona, P., Belongie, S.: The Caltech-UCSD Birds-200-2011 dataset. In: Computation and Neural Systems Technical Report, 2010–001 (2011)
12. Di, Y., Jiang, Z., Zhang, H.: A public dataset for fine-grained ship classification in optical remote sensing images. Remote Sens. **13**(4), 747 (2021). https://doi.org/10.3390/rs13040747

13. Sofiiuk, K., Petrov, I., Barinova, O., Konushin, A.: F-BRS: rethinking backpropagating refinement for interactive segmentation. In: Proceedings of the IEEE Conference on Computer Vision and Pattern Recognition, pp. 8620–8629 (2020). https://doi.org/10.1109/CVPR42600.2020.00865
14. Fu, J., et al.: Dual attention network for scene segmentation. In: Proceedings of the IEEE Conference on Computer Vision and Pattern Recognition, pp. 3146–3154 (2019)
15. Chen, L., George P., Florian S., Hartwig A.: Rethinking atrous convolution for semantic image segmentation. arXiv preprint http://arxiv.org/abs/1706.05587 (2017)
16. Huang, Z., Wang, X., Huang, L., Huang, C., Wei, Y., Liu, W.: CCNet: criss-cross attention for semantic segmentation. In: Proceedings of the IEEE Conference on Computer Vision and Pattern Recognition, pp. 603–612 (2019)
17. Enze, X., Wenhai, W., Zhiding, Y., Anima A., Jose, A., Ping, L.: SegFormer: simple and efficient design for semantic segmentation with transformers. In: Advances in Neural Information Processing Systems, vol. 34, pp. 12077–12090 (2021)
18. Everingham, M., Eslami, S., Van, G., Williams, C., Winn, J., Zisserman, A.: The PASCAL visual object classes challenge: a retrospective. Int. J. Comput. Vis. **111**(1), 98–136 (2015)
19. Lin, T.-Y., et al.: Microsoft COCO: common objects in context. In: Fleet, D., Pajdla, T., Schiele, B., Tuytelaars, T. (eds.) ECCV 2014. LNCS, vol. 8693, pp. 740–755. Springer, Cham (2014). https://doi.org/10.1007/978-3-319-10602-1_48
20. Cordts, M., et al.: The cityscapes dataset for semantic urban scene understanding. In: Proceedings of the IEEE Conference on Computer Vision and Pattern Recognition (2016)
21. Vaswani, A., et al.: Attention is all you need. In: Advances in Neural Information Processing Systems, vol. 30 (2017)
22. Hu, J., Shen, L., Sun, G.: Squeeze-and-excitation networks. In: Proceedings of the IEEE Conference on Computer Vision and Pattern Recognition, pp. 7132–7141 (2018)
23. Dosovitskiy, A., et al.: An image is worth 16x16 words: transformers for image recognition at scale. In: International Conference on Learning Representations (2020)
24. Zhuang, P., Wang, Y., Qiao, Y.: Learning attentive pairwise interaction for fine-grained classification. In: Proceedings of the AAAI Conference on Artificial Intelligence, vol. 34, no. 07, pp. 13130–13137 (2020)
25. He, J., et al.: TransFG: a transformer architecture for fine-grained recognition. In: Proceedings of the AAAI Conference on Artificial Intelligence, vol. 36, no. 1, pp. 852–860 (2022)
26. Khosla, A., Jayadevaprakash, N., Yao, B., Li, F.: Novel dataset for fine-grained image categorization: Stanford dogs. In: Proceedings of the IEEE Conference on Computer Vision and Pattern Recognition Workshops (2011)
27. Kaiming, H., Xiangyu, Z., Shaoqing, R., Jian, S.: Deep residual learning for image recognition. In: Proceedings of the IEEE Conference on Computer Vision and Pattern Recognition, pp. 770–778 (2016). https://doi.org/10.1109/CVPR.2016.90

Diachronic Named Entity Disambiguation for Ancient Chinese Historical Records

Zekun Deng[1], Hao Yang[2,3], and Jun Wang[1,2](✉)

[1] Department of Information Management, Peking University, Beijing, China
[2] Research Center for Digital Humanities, Peking University, Beijing, China
{dzk,yanghao2008,junwang}@pku.edu.cn
[3] Institute of Artificial Intelligence, Peking University, Beijing, China

Abstract. Named entity disambiguation (NED) is a fundamental task in NLP. Although numerous methods have been proposed for NED in recent years, they ignore the fact that a lot of real-world corpora are diachronic by nature, such as historical documents or news articles, which vary greatly in time. As a consequence, most current methods fail to fully exploit the temporal information inside the corpora and knowledge bases. To address the issue, we propose a novel model which integrates temporal feature into pretrained language model to make our model aware of time and a new sample re-weighting scheme for diachronic NED which penalizes highly-frequent mention-entity pairs to improve performance on rare and unseen entities. We present WikiCMAG and WikiSM, two new NED datasets annotated on ancient Chinese historical records. Experiments show that our model outperforms existing methods by large margins, proving the effectiveness of integrating diachronic information and our re-weighting schema. Our model also gains competitive performance on out-of-distribution (OOD) settings. WikiSM is publicly available at https://github.com/PKUDHC/WikiSM.

Keywords: Entity disambiguation · Natural language processing · Diachronic information

1 Introduction

Named entity disambiguation (NED) is the task of mapping mentions in a text to their corresponding entities in a knowledge base (KB). NED is crucial for various downstream natural language processing tasks, such as information retrieval [15], knowledge graph construction [19] and question answering [13]. Despite the numerous methods proposed for this task in recent years [3,14,27], there has been limited focus on NED in diachronic corpora. Many knowledge-intensive texts are diachronic by nature, such as historical documents and news articles. However, most existing methods are designed to be time-agnostic, neglecting the temporal properties embedded in texts and knowledge bases (KBs).

Diachronic NED is faced with three main challenges. First, most pretrained language models are not specifically formulated to capture temporal features,

B. Luo et al. (Eds.): ICONIP 2023, CCIS 1965, pp. 305–319, 2024.
https://doi.org/10.1007/978-981-99-8145-8_24

thus they have a poor perception of the time of input texts and KB entities. Second, NED datasets are often imbalanced, with a very small number of entities with very high frequencies. Third, training resources and benchmarks for diachronic NED are scarce, particularly in non-English languages, posing a significant difficulty for the community to fairly compare and develop new methods.

To address the above challenges, firstly, we propose a new NED model that transforms time signal to natural language expression to jointly encode it with textual inputs. Secondly, we propose to re-weight train samples by their frequency in training set to alleviate the bias caused by data imbalance. Thirdly, we present two new diachronic NED datasets in ancient Chinese history domain.

The contributions of this paper are threefold:

1. We propose a novel diachronic named entity disambiguation method which incorporates temporal information as natural language expressions to improve performance on diachronic corpus. Experiments show that our method outperforms recent baselines by large margins.
2. We propose a novel sample re-weighting scheme for diachronic NED which penalizes highly-frequent mention-entity pairs to improve performance on rare and unseen entities. Experiments show that this scheme effectively enhances model performance on rare or unseen entities.
3. We present WikiCMAG and WikiSM, two new named entity disambiguation datasets annotated on ancient Chinese historical records. The datasets contain a large amount of manually annotated mentions and cover a variety of dynasties spanning nearly 2000 years. We make WikiSM publicly available[1] in hope to boost researches in related areas. To the best of our knowledge, this is the first NED dataset ever in ancient Chinese history domain.

2 Related Works

Previous Methods for NED. Recently, deep learning has achieved enormous success in solving NED task. Gillick et al. [11] propose a multi-layer network to compute the cosine similarity of mention embedding and entity embedding. Chen et al. [7] follow this paradigm by introducing BERT to generate embedding and latent type information to better capture similarity. Logeswaran et al. [18] adopt cross-encoder to improve attention interaction between mention and entity. De Cao et al. [9] propose an autoregressive NED model which generates NED output in a language-model manner under alias trie constraint.

Temporal Methods in NLP. Rijhwani et al. [20] study the effect of temporal drift in evaluation data in named entity recognition (NER) task. They introduce a year-split dataset of English tweets to support their analysis. It is found that temporal information has the potential for improvement in NER and similar NLP tasks. Zaporojets et al. [28] propose TempEL, an entity linking dataset in which entities are time-continual, occurring in different years from 2013 to 2022.

[1] https://github.com/PKUDHC/WikiSM.

Experiments show state-of-the-art models suffer major performance degradation on TempEL. Wang et al. [23] propose TimeBERT, a BERT-based model pre-trained on long-span temporal news collection. It adds a pretraining objective that predicts article timestamp using the representation of [CLS] token. Regarding the NED task, Agarwal et al. [2] propose the first diachronic NED method named diaNED, which introduces time signature from date expressions and is based on static word and entity embeddings and gradient boost trees.

Sample Re-weighting. Sample re-weighting has been shown effective for imbalanced training data [6]. Beigman et al. [5] propose to re-weight samples inversely proportional to class frequency to improve metaphor detection. Su et al. [21] introduce a re-weighting method for zero-shot and rare word sense disambiguation task based on number of senses belonging to the word.

3 Method

Formally, the NED task can be defined as follows. Let $\mathcal{X}$ denote a set of documents and $\mathcal{E}$ denote a knowledge base. An entity $e \in \mathcal{E}$ consists of a title $\tau_t(e)$ and a description $\tau_d(e)$. For a document $\boldsymbol{x} \in \mathcal{X}$, the set of mentions within $\boldsymbol{x}$ is denoted as $M(\boldsymbol{x}) = \{m(\boldsymbol{x})_1, ..., m(\boldsymbol{x})_{|M(\boldsymbol{x})|}\}$. The task of NED requires that for any mention $m \in M(\boldsymbol{x})$ where $\boldsymbol{x} \in \mathcal{X}$, an NED model should assign the correct corresponding entity $e \in \mathcal{E}$ to the mention m.

We propose a novel diachronic NED model based on a two-stage retriever-reranker architecture with time period information as complementary input. The major architecture of our model is illustrated in Fig. 1. Following a common practice in prior works [1,25,29], our model consists of two subsequent modules, namely retriever and reranker. When disambiguating mentions in a chunk of text, the text is first fed into our **retriever** to recall entity candidates that can possibly be linked to the mentions in the chunk, and then the chunk, along with the possible entity candidates collected by the retriever, are provided to the **reranker** for a more scrutinized reranking.

The following parts describe our method formally.

3.1 Retriever

Our retriever is mainly a combination of two encoders, both of which are multi-layer bidirectional transformers initialized from pretrained language models, such as BERT [10] or RoBERTa [17]. Let ϕ_θ^{mnt} and ϕ_θ^{ent} denote the two encoders respectively, where θ denotes the parameters of the corresponding encoder. The parameters of the two encoders are not shared.

We use ϕ_θ^{mnt} to encode the mentions. For a given document $\boldsymbol{x} \in \mathcal{X}$ with mentions $M(\boldsymbol{x})$, we iterate over $M(\boldsymbol{x})$ to obtain a separate representation for each mention (called query mention) individually. The input sequence of the encoder ϕ_θ^{mnt} consists of four parts: the query mention, the left and right context

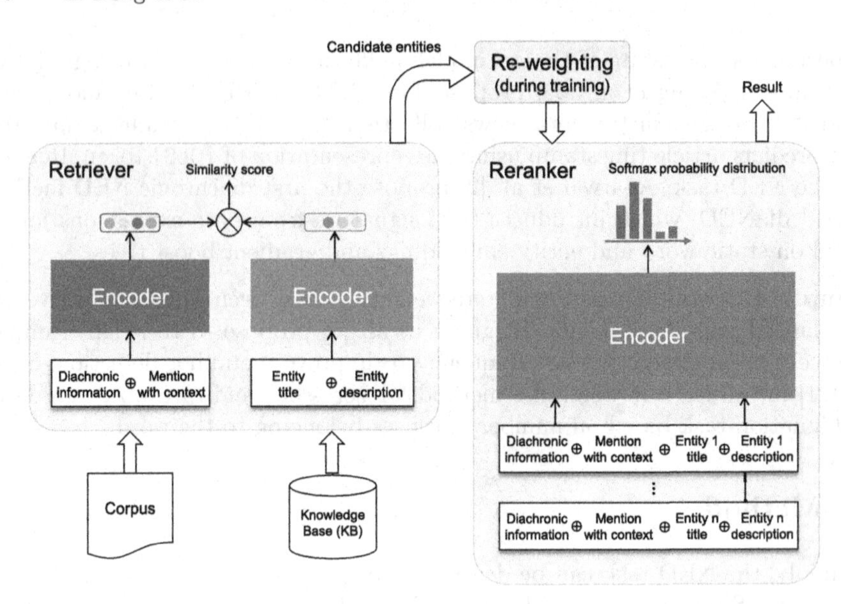

Fig. 1. The overall architecture of our proposed model (⊕ denotes concatenation)

of the mention, and the diachronic information of the text. To be specific, the input token sequence of a mention $m \in M(\boldsymbol{x})$ is defined as:

$$T^{\text{mnt}}(m, \boldsymbol{x}) = [t_{\text{CLS}}, d(\boldsymbol{x}), t_1, c_{\text{left}}(m, \boldsymbol{x}), t_<, m, t_>, c_{\text{right}}(m, \boldsymbol{x}), t_{\text{SEP}}],$$

where $d(\boldsymbol{x})$ is the described time period of the document $\boldsymbol{x}$, $c_{\text{left}}(m, \boldsymbol{x})$ and $c_{\text{right}}(m, \boldsymbol{x})$ are the left and right context of mention m in document $\boldsymbol{x}$ respectively, t_1 is a custom token indicating the concatenation of different segments, $t_<$ and $t_>$ are two custom tokens indicating the position of the query mention among the context, and $t_{\text{CLS}}, t_{\text{SEP}}$ are the special tokens of pretrained language models that indicate the start and the end of a sequence, respectively, such as `[CLS]` and `[SEP]` in BERT. Notably, to inject diachronic information into our model, we prepend time period string to the context to make our retriever aware of the time background.[2]

Entities in knowledge base are encoded by ϕ_θ^{ent}. For a given entity $e \in \mathcal{E}$, the input token sequence is defined as follows:

$$T^{\text{ent}}(e) = [t_{\text{CLS}}, \tau_t(e), t_2, \tau_d(e), t_{\text{SEP}}],$$

where t_2 is a custom token like t_1.

For both encoders, we use the hidden vector from the final layer as the representation of the mention and the entities. For mention encoder ϕ_θ^{mnt}, we use the hidden final vector of corresponding to the $t_<$ token, while for entity encoder

[2] In real scenarios, the time which a piece of text describes is easy to obtain since most texts have known sources and metadata that can help with time identification.

ϕ_θ^{ent}, the t_{CLS} token is used instead. Let $h_{\text{mnt}}(m, \boldsymbol{x}) \in \mathbb{R}^u$ denote the representation of a mention $m \in M(\boldsymbol{x})$ and $h_{\text{ent}}(e) \in \mathbb{R}^u$ denote the representation of an entity $e \in \mathcal{E}$, where u is the hidden size of the pretrained language model. The relevance score between a mention m from document $\boldsymbol{x}$ and an entity e is defined as

$$\text{sim}(m, e; \boldsymbol{x}) = h_{\text{mnt}}(m, \boldsymbol{x})^{\text{T}} h_{\text{ent}}(e). \tag{1}$$

To train the retriever, for each mention m, we construct a batch of entity samples with both a positive sample and negative samples. Exactly 1 positive sample is contained in the batch, which is the entity to which mention m is linked, denoted as e_+. Negative samples are obtained through two criteria: random sampling and hard negative sampling. We obtain k hard negatives per batch by selecting k entities with highest scores predicted by retriever. The resulting negative samples, combining random and hard negatives, are denoted by $Q_-(\boldsymbol{x})$. We use a fixed batch size for training. The loss function of the retriever over m is defined as

$$\mathcal{L}_1 = -\log \frac{e^{\text{sim}(m, e_+; \boldsymbol{x})/\gamma}}{e^{\text{sim}(m, e_+; \boldsymbol{x})/\gamma} + \sum_{e_- \in Q_-(\boldsymbol{x})} e^{\text{sim}(m, e_-; \boldsymbol{x})/\gamma}}, \tag{2}$$

where γ is the temperature hyperparameter.

3.2 Reranker

Our reranker is a cross-encoder that takes a concatenated sequence of mention, context and entities as input. It is also a multi-layer bidirectional transformer initialized from a pretrained language model. Let $Q_r(m, \boldsymbol{x})$ denote the candidate entities obtained by the (fully-trained) retriever. For any $m \in M(\boldsymbol{x})$, $\boldsymbol{x} \in \mathcal{X}$ and $e \in \mathcal{E}$, the input token sequence for the reranker is defined as:

$$T^{\text{rerank}}(m, \boldsymbol{x}, e) = [T^{\text{mnt}}(m, \boldsymbol{x}), \tau_t(e), t_2, \tau_d(e), t_{\text{SEP}}].$$

Let $h_{\text{rer}}(m, \boldsymbol{x})_i$ denote the final layer hidden vector of the reranker corresponding to the "$t_<$" token in $T^{\text{rerank}}(m, \boldsymbol{x}, Q_r(m, \boldsymbol{x})_i)$, where $Q_r(m, \boldsymbol{x})_i$ denotes the i-th entity in the candidate list $Q_r(m, \boldsymbol{x})$. We compute the probability of $Q_r(m, \boldsymbol{x})_i$ being the correct corresponding entity to the query mention by

$$p(m, \boldsymbol{x})_i = \text{softmax}(W^{\text{T}} h_{\text{rer}}(m, \boldsymbol{x})_i), \tag{3}$$

where $W \in \mathbb{R}^u$ is learnable parameter and the softmax is computed over all entities given a certain mention.

The loss function for the reranker is multi-class cross-entropy loss:

$$\mathcal{L}_2 = -\log(p(m, \boldsymbol{x})_y) \tag{4}$$

where y is the index of the gold entity corresponding to the mention m.

We use the candidate entities generated by the fully-trained retriever to train the reranker. To make sure the loss function is valid, we add gold entity into $Q_r(m, \boldsymbol{x})$ if they are missed by the retriever. During inference, we first obtain candidates entities using our fully-trained retriever, and then use the reranker to choose the entity with highest probability.

3.3 Sample Re-Weighting by Frequency

We propose an additional sample re-weighting scheme to reduce the bias induced by imbalanced training data. Since the training data contains mention-entity pairs with very high frequency (e.g. 100x more frequent than average), common training method might misguide the model to memorize high-frequent mentions and entities, thus harming performance of low-frequent or unseen data. Thus, we propose a re-weighting function that assigns a weight monotonically decreasing with respect to frequency. Let $f(m, e)$ denote the number of occurrences of mention m linked to entity e in training set. Then, for every training sample containing mention m linked to entity e, we assign a weight $\alpha(m, e)$ to the sample by the following equation:

$$\alpha(m, e) = \beta_1(f(m, e) + \beta_2)^{-\beta_3}, \tag{5}$$

where $\beta_i > 0 (i = 1, 2, 3)$ are hyperparameters. We provide a plot of the function with specified hyperparameters in Appendix A to illustrate it more clearly. Consequently, the loss function for reranker after re-weighting becomes:

$$\mathcal{L}_2 = -\alpha(m, Q_r(m, \boldsymbol{x})_y) \log(p(m, \boldsymbol{x})_y). \tag{6}$$

3.4 Coreference Resolution Post-processing

It is a prevailing phenomenon in ancient Chinese historical records that a person's surname or the first one or two characters in a person's first name are omitted when the person is mentioned for multiple times. This is problematic because due to complexity constraints, the length of text processed by pretrained language models is limited (e.g. less than 512 in RoBERTa), so the input text must be cut into short chunks. However, sometimes the mentions in the current chunk are incomplete and the full mention is so far away that it can not be seen in the current chunk. This may lead to wrong disambiguation result. To alleviate this issue, we adopt a rule-based coreference resolution algorithm for entity mentions. Concretely, we iterate over all mentions in the text and point the mention to the nearest precedent if the precedent ends with the mention.

4 Dataset

Since there are no publicly available NED dataset on ancient Chinese historical records, we construct two new NED dataset to train and test our method, WikiCMAG and WikiSM. Both datasets are annotated on randomly-selected chapters of classic ancient Chinese history records: WikiCMAG on *Comprehensive Mirror for Aid in Government* (CMAG, 资治通鉴) and WikiSM on *History of Song* (宋史) and *History of Ming* (明史). The datasets are manually annotated by a group of approximately 10 Chinese university students. We instruct the students to link the mentions to the corresponding entity in the KB, if such an entity exists. The KB we use is Chinese Wikipedia. The basic statistics of the

Table 1. Statistics of the train, dev and test split of WikiCMAG and WikiSM

Statistic	WikiCMAG			WikiSM		
	Train	Dev	Test	Train	Dev	Test
# Chapters	29	2	2	7	2	2
# Characters	265K	17.0K	20.4K	66.4K	18.0K	17.0K
# InKB Mentions	23.1K	1.4K	1.6K	2.5K	0.7K	0.8K
# Entities in KB linked to a mention	3.6K	0.3K	0.4K	0.8K	0.4K	0.3K
# Unseen entities	N/A	0.16K	0.27K	N/A	0.28K	0.20K
% Unseen entities	N/A	58.7	67.9	N/A	78.8	62.6
% Unseen dynasties	N/A	100.0	100.0	N/A	0.0	0.0

datasets are shown in Table 1. The scale of WikiCMAG is large, with a total of 26.1K InKB mentions annotated in three splits. As comparison, AIDA-CoNLL [12], one of the most popular English NED benchmarks, contains 27.8K.

Below are some significant characteristics of the two datasets.

Large Number of Unseen Entities. As is shown in Table 1, the portion of unseen entities (those never linked in the training set) in the test set of WikiCMAG and WikiSM are 67.9% and 62.6%, respectively. This indicates that a majority of linked entities in test set are not seen in training set, which brings a big challenge for NED models to link them successfully. The same holds for dynasties in WikiCMAG.

Small Textual Overlap and Large Number of Aliases. The textual overlap between mention surface and entity name is a vital feature in many previous NED methods, such as diaNED [2] and ExtEnD [4]. We calculate the ratio of exact matches between mention surface and entity names in the training set of WikiCMAG, WikiSM and AIDA-CoNLL. It is discovered that AIDA-CoNLL has 43.9% exact matches, while WikiCMAG and WikiSM only have 22.0% and 24.2%. This makes it harder for models to disambiguate according to surface form of mentions. Also, the portion of entities with more than one surface mentions is 46.6% and 24.5% in WikiCMAG and WikiSM, significantly higher than 19.5% in AIDA-CoNLL. These features in overall increases the difficulty of our datasets.

Same Name but Different Time. Our datasets involve many cases where multiple Wikipedia entities in different dynasties share exactly the same name. For instance, the person Jia Kui (贾逵) has 3 entries in Wikipedia, living in Eastern Han, Three Kingdoms and Northern Song (东汉，三国，北宋). The location Ji Zhou (吉州) also has 3 entries, referring to that in Sui, Liao and Jin (隋，辽，金).

Very Long Documents. Many English NED datasets consist of short documents. For instance, the average length of document in AIDA-CoNLL train set is less then 300 (in tokens). However, the average document length of train set in

WikiCMAG and WikiSM are 9.1K and 9.5K (in characters), respectively. This is very different from commonly used NED datasets.

5 Experiments

5.1 Settings

We conduct experiments to show the effectiveness of our method. Three model variants are experimented: (1) full model; (2) model without diachronic information, i.e. temporal information omitted during training and testing; and (3) model without sample re-weighting. All models are evaluated by two ways: with coreference resolution (CR) post-processing and without. We also add an out-of-distribution (OOD) experiment which trains the models on WikiCMAG but evaluates them on WikiSM to assess their OOD generalizability.

We use PyTorch framework for implementation. We use SikuRoBERTa [22] as the pretrained language model for all the encoders in the retriever and the reranker, since it is an ancient Chinese version of RoBERTa that is pretrained on approximately half a billion ancient Chinese characters which is particularly suitable for ancient Chinese understanding. The optimizer is Adam [16]. We use warm-up in training where the learning rate increases linearly from 0 to configured learning rate over the first epoch. See Appendix A for major hyper-parameters and other details in our experiment.

5.2 Baselines and Results

The following baseline methods are compared to our model.

- **BM25** [8] is a classic unsupervised information retrieval algorithm in which is generally considered a strong baseline in information retrieval [24]. We apply it to NED by regarding mention as query and KB entity titles and descriptions as documents.
- **GPT 3.5-ZS**. GPT 3.5, also known as ChatGPT, is a model powered by OpenAI. It is trained by instruction fine-tuning and RLHF, achieving high performance in a wide range of NLP tasks across languages. We test GPT 3.5 in zero-shot (ZS) setting on our datasets by prompting it to output exact Wikipedia entity name.
- **Minimum Edit Distance (MED)**. Edit distance is a common metric for evaluating the similarity of two strings. We take the KB entity with minimum edit distance to a given mention as prediction. [3]
- **diaNED** [2] is the first method for diachronic named entity disambiguation task. We re-implement diaNED and train and evaluate it on our datasets.
- **GENRE** [9] formulates NED as an seq2seq task. It directly generates entity name conditioned on mention and context. We re-implement and train GENRE on our datasets. Note we do not conduct pretraining in Chinese.

[3] If there are more than 1 entity with the same minimum edit distance, a random one is chosen.

- **ExtEnD** [4] formulates NED as a span extraction task by concatenating mention context and candidate entity names (titles) as a sequence, feeding the sequence to a pretrained encoder and extracting answer spans to pick the correct entity. We re-implement and train ExtEnD on our datasets.
- **LUKE**[4] [27] is a global NED method based on BERT. We re-implement and train LUKE on our datasets. Note we do not conduct pretraining in Chinese.

More details of the implementations are described in Appendix A.

Table 2. Performance of our method compared with baselines on WikiCMAG and WikiSM test set (Best method in bold). *w/ CR* and *w/o CR* denote with and without coreference resolution post-processing, respectively. Note that we do **not** conduct pretraining for GENRE and LUKE in our implementations. Also note that diaNED uses pretrained embeddings and mention and entity counts from Wikipedia, meaning that it involves external training resource so that it is **not** directly comparable to other methods which do not include any additional feature.

Type	Method	InKB Micro F1			
		WikiCMAG		WikiSM	
		w/o CR	*w/ CR*	*w/o CR*	*w/ CR*
Non-temporal	BM25	13.54	18.66	4.56	5.32
	GPT 3.5-ZS	25.49	50.35	21.17	25.60
	MED	27.01	55.28	21.29	30.54
	GENRE (2021) [9]	65.53	74.95	52.60	57.67
	LUKE (2022) [27]	40.92	41.68	25.73	24.21
	ExtEnD (2022) [4]	66.80	77.78	55.01	58.43
Temporal	diaNED (2018) [2]	73.12	81.02	60.33	61.60
	Ours	**83.46**	**87.34**	**62.72**	**66.92**
	– *diachronic information*	72.21	81.17	59.44	63.50
	– *sample re-weighting*	80.88	86.43	62.34	66.54

The performance of our method and the baselines are shown in Table 2. Following convention, InKB micro F1 is used to measure the performance of the models. The result clearly indicates that our model outperforms all baselines significantly, with a 6.32 and 5.32 F1 increase compared to the second best on WikiCMAG and WikiSM respectively.

Among non-temporal models, we find that GENRE and ExtEnD have similar performance on both datasets. While the former formulates NED as seq2seq task and the latter as extraction task, both of them utilize *solely* entity title information and ignore entity descriptions, which is also vital for correct disambiguation. LUKE, which relies heavily on entity embeddings, shows poor result without pretraining, which is expensive. All of the above methods ignore

[4] The name comes from that of their public GitHub repository.

diachronic information, which is very helpful when disambiguating entities on a KB that contains entities from a huge time span. GPT 3.5-ZS, without access to external knowledge and without tuning, also performs poorly.

Among temporal methods, diaNED is also outperformed significantly by our model, which does not come as a surprise since diaNED is based on gradient boosting trees and static word and entity embeddings. It is worth noting, however, that diaNED is still competitive with non-temporal neural models. Unlike all above, our model successfully integrates diachronic information and entity description during inference, boosting the performance considerably.

5.3 Analysis

OOD Performance. Table 3 is the result for OOD experiment. It is demonstrated that our model gains a competitive OOD performance when training on WikiCMAG and testing on WikiSM.

Table 3. Out-of-distribution (OOD) performance of our method compared with baselines when training on WikiCMAG and evaluating on WikiSM. *w/ CR* and *w/ CR* denote with and without coreference resolution post-processing, respectively.

Type	Method	InKB Micro F1	
		w/o CR	*w/ CR*
Non-temporal	GENRE [9]	44.36	47.53
	ExtEnD [4]	50.70	55.38
	LUKE [27]	41.83	39.03
Temporal	diaNED [2]	51.33	55.13
	Ours	48.09	52.29

The Effect of Incorporating Diachronic Information. Table 2 shows that reranker's InKB Micro F1 drops sharply by 6.17 and 3.42 on WikiCMAG and WikiSM, indicating that diachronic information is indeed crucial.

The Effect of Sample Re-Weighting. Table 2 shows that removing sample re-weighting causes a performance degradation on both datasets, which is especially significant on WikiCMAG. We furthur compute the performance of models on 0, 1 and 3-shot entities[5] on the two datasets, shown in Fig. 2. It is obvious that across the datasets and different shots, removing sample re-weighting generally leads to a worse performance. This is an evidence that sample re-weighting can help our model perform better on unseen or rare entities. Still, under some settings the performance of model without diachronic information surpasses that of the full model. This might be due to the fact that the number of few-shot samples is too small, causing some randomness in results.

[5] Here k-shot means that the correct entity occurs k times in the training data.

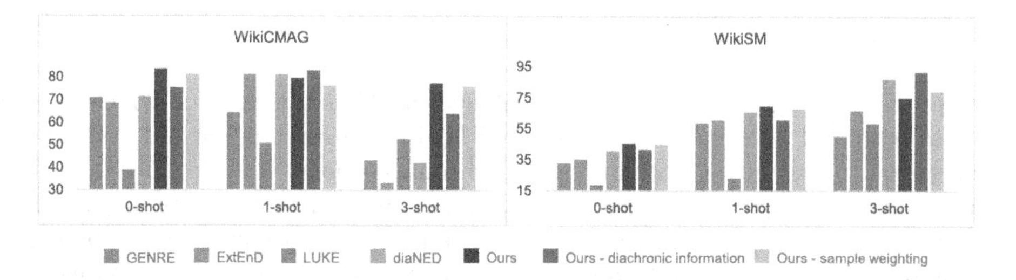

Fig. 2. {0,1,3}-shot performance of our method compared with baselines on WikiCMAG and WikiSM without coreference resolution

Table 4. Cases of predictions of our models and baselines without coreference resolution. Mentions are colored purple. Original text is in traditional Chinese. -DI represents our model variant without diachronic information. *Italicized* entity names are translated directly from Chinese since the entity does not have a corresponding English Wikipedia page.

Dataset	Mention and context	Time	Prediction				
			Ours	Ours (-DI)	diaNED	ExtEnD	GENRE
WikiCMAG	以秦王俊为扬州总管四十四州诸军事	隋	✓扬州(隋朝)	✗扬州(古代)	✓扬州(隋朝)	✗扬州(古代)	✗扬州(古代)
	Appointing Prince Jun of Qin as the Chief Commander of Military Affairs for the forty-four provinces in Yangzhou	Sui	*Yangzhou (Sui)*	Yangzhou (ancient China)	*Yangzhou (Sui)*	Yangzhou (ancient China)	Yangzhou (ancient China)
WikiSM	正统七年十二月奉昭皇后神主庙	明	✓诚孝昭皇后	✗昭怀皇后	✗刘昭妃(明神宗)	✗章献明肃皇后	✗昭慈圣献皇后
	In the twelfth month of the seventh year of the Zhengtong era, sacrifices were offered to the ancestral shrine of the Empress Zhao	Ming	Empress Zhang (Hongxi)	Empress Liu (Zhezong)	*Concubine Zhao (Shenzong)*	Empress Liu (Zhenzong)	Empress Meng

6 Discussion

We conduct case analysis to see how our method improves diachronic NED performance. As is shown in Table 4, for the case from WikiCMAG, despite the fact that all models give the answer Yangzhou (扬州), all non-temporal models point to a wrong entity, Yangzhou (ancient China), which seems correct at the first sight but is actually wrong. In truth, the "Yangzhou" in text refers to the province *created in* 589 AD whose administrative center is Jiangdu (江都, today's Yangzhou city), which is the same as the correct answer. However, Yangzhou (ancient China) mainly represents the administrative zone with the same name *before* 589 AD and during 620–626 AD, which does not have any connection to today's Yangzhou. Nevertheless, ours and diaNED produce right result. For the second case, only our model gives correct entity. Apart from diaNED, all other models that are non-temporal predict persons from Song (宋) dynasty, which conflicts with the time of the document. Although diaNED gives an answer from the correct dynasty, it fails to output the exact correct person due to its limitations in semantic understanding.

7 Conclusion

We propose a novel diachronic named entity disambiguation method which integrates temporal information to pretrained language model as natural language expression to enhance performance on diachronic corpora. We introduce a novel sample re-weighting formula for NED which penalizes mention-entity pairs with high frequency. We present two new human-annotated named entity disambiguation dataset, namely WikiCMAG and WikiSM. Experiments on the two datasets indicate that our method significantly outperforms recent baseline, proving the effectiveness of incorporating diachronic feature. Our method also gains competitive performance in out-of-distribution setting.

This paper also has some limitations. Our study currently only covers Chinese language and may be extended to other languages in the future. The thematic and stylistic diversity of the proposed datasets is also limited. Further research directions might include cross-language diachronic NED and improving out-of-distribution and zero- or few-shot abilities.

Acknowledgements. This research is supported by the NSFC project "the Construction of the Knowledge Graph for the History of Chinese Confucianism" (Grant No. 72010107003).

Appendix A. Implementation Details

Details of our implementations are listed as follows. Unless otherwise specified, we adopt the same hyperparameters as in the original works for baseline methods. Apart from GPT 3.5-ZS, all the baselines below use candidate entities generated by our fully-trained retriever, for it has a better performance than commonly used entity retrieval method in English NED benchmarks.

GPT 3.5-ZS. We use `gpt-3.5-turbo` with following prompt:

```
{{paragraph}}
Which entity does "{{mention}}" in the paragraph above refer to in
Wikipedia? Output the entity title directly.
```

We manually pick out the answer if the response contains irrelevant information. Only exact matches are counted.

diaNED. We use Wikipedia2Vec [26] to obtain pretrained Chinese word and entity embeddings. Temporal vector dimensions are altered to 2500 to represent years 500 BC to 2000 AD. Document creation time is replaced by the approximate year described by the document. We use regular expressions to extract year expressions from Chinese Wikipedia.

GENRE. We use bart-base-chinese[6] to initialize the parameters.

[6] https://huggingface.co/fnlp/bart-base-chinese.

ExtEnD. We use SikuRoBERTa for re-implementation because there is no pre-trained Longformer in ancient Chinese.

LUKE. We use SikuRoBERTa to initialize the parameters.

Ours. See Table 5 for major hyperparameters of our model. We use SikuRoBERTa to initialize the parameters. See Fig. 3 for the plot of our re-weighting function with hyperparameters from Table 5.

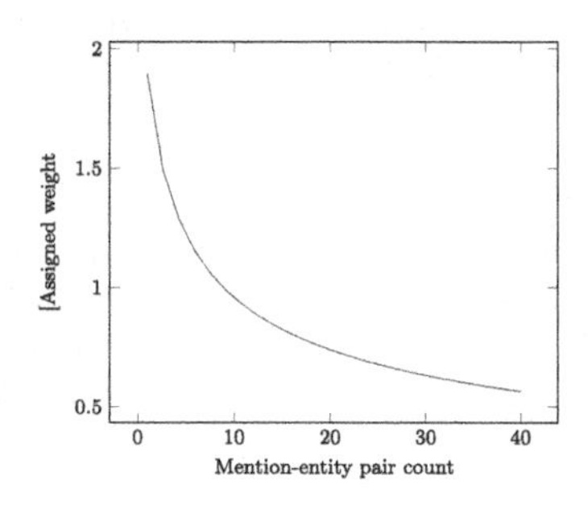

Fig. 3. The plot of re-weighting function with our hyperparameters.

Table 5. Major hyperparameters in the experiment

Hyperparameter	Value
Batch size of retriever	64
# Hard negatives (k)	4
# Candidates for reranker	32
Temperature of loss function (γ)	8
Learning rate (retriever)	2e-6
Learning rate (reranker)	1e-5
Gradient accumulation steps (retriever)	4
Gradient accumulation steps (reranker)	4
Training epochs (retriever)	2
Training epochs (reranker)	2
Length of mention left/right context	64
Re-weighting parameter β_1	2.5
Re-weighting parameter β_2	1.0
Re-weighting parameter β_3	0.4

References

1. Agarwal, D., Angell, R., Monath, N., McCallum, A.: Entity linking via explicit mention-mention coreference modeling. In: Proceedings of NAACL 2022 (2022)
2. Agarwal, P., Strötgen, J., Del Corro, L., Hoffart, J., Weikum, G.: DiaNED: time-aware named entity disambiguation for diachronic corpora. In: Proceedings of the ACL 2018 (Volume 2: Short Papers), pp. 686–693 (2018)
3. Angell, R., Monath, N., Mohan, S., Yadav, N., McCallum, A.: Clustering-based Inference for Biomedical Entity Linking. In: Proceedings of NAACL 2021 (2021)
4. Barba, E., Procopio, L., Navigli, R.: ExtEnD: Extractive Entity Disambiguation. In: Proceedings of ACL 2022 (Volume 1: Long Papers), Dublin, Ireland (2022)
5. Beigman Klebanov, B., Leong, C.W., Flor, M.: Supervised word-level metaphor detection: Experiments with concreteness and reweighting of examples. In: Proceedings of the Third Workshop on Metaphor in NLP, pp. 11–20 (Jun 2015)
6. Byrd, J., Lipton, Z.: What is the effect of importance weighting in deep learning? In: Proceedings of ICML 2019, vol. 97, pp. 872–881. PMLR (09–15 Jun 2019)
7. Chen, S., Wang, J., Jiang, F., Lin, C.Y.: Improving entity linking by modeling latent entity type information. In: Proceedings of the AAAI Conference on Artificial Intelligence, vol. 34, pp. 7529–7537 (2020)
8. Crestani, F., Lalmas, M., Van Rijsbergen, C.J., Campbell, I.: "is this document relevant?...probably": a survey of probabilistic models in information retrieval. ACM Comput. Surv. **30**(4), 528–552 (1998)
9. DeCao, N., Izacard, G., Riedel, S., Petroni, F.: Autoregressive entity retrieval. In: ICLR 2021. OpenReview.net (2021)
10. Devlin, J., Chang, M.W., Lee, K., Toutanova, K.: BERT: pre-training of Deep Bidirectional Transformers for Language Understanding. In: Proceedings of NAACL 2019, pp. 4171–4186 (2019)
11. Gillick, D., Kulkarni, S., Lansing, L., Presta, A., Baldridge, J., Ie, E., Garcia-Olano, D.: Learning dense representations for entity retrieval. In: Proceedings of CoNLL 2019, pp. 528–537 (Nov 2019)
12. Hoffart, J., et al.: Robust disambiguation of named entities in text. In: Proceedings of EMNLP 2011, pp. 782–792 (Jul 2011)
13. Hu, X., Wu, X., Shu, Y., Qu, Y.: Logical form generation via multi-task learning for complex question answering over knowledge bases. In: Proceedings of the 29th International Conference on Computational Linguistics, pp. 1687–1696 (Oct 2022)
14. van Hulst, J.M., Hasibi, F., Dercksen, K., Balog, K., de Vries, A.P.: REL: An Entity Linker Standing on the Shoulders of Giants. In: Proceedings of SIGIR 2020 (2020)
15. Khalid, M.A., Jijkoun, V., de Rijke, M.: The impact of named entity normalization on information retrieval for question answering. In: Advances in Information Retrieval. pp. 705–710. Springer, Berlin Heidelberg, Berlin (2008). https://doi.org/10.1007/978-3-540-78646-7_83
16. Kingma, D.P., Ba, J.: Adam: A method for stochastic optimization. In: Bengio, Y., LeCun, Y. (eds.) ICLR 2015 (2015). http://arxiv.org/abs/1412.6980
17. Liu, Y., et al.: RoBERTa: A Robustly Optimized BERT Pretraining Approach. arXiv:1907.11692 [cs] (Jul 2019)
18. Logeswaran, L., Chang, M.W., Lee, K., Toutanova, K., Devlin, J., Lee, H.: Zero-shot entity linking by reading entity descriptions. In: Proceedings of ACL 2019, pp. 3449–3460. Association for Computational Linguistics, Florence, Italy (Jul 2019)
19. Martinez-Rodriguez, J.L., Lopez-Arevalo, I., Rios-Alvarado, A.B.: Openie-based approach for knowledge graph construction from text. Expert Syst. Appl. **113**, 339–355 (2018)

20. Rijhwani, S., Preotiuc-Pietro, D.: Temporally-informed analysis of named entity recognition. In: Proceedings of ACL 2020, pp. 7605–7617 (Jul 2020)
21. Su, Y., Zhang, H., Song, Y., Zhang, T.: Rare and zero-shot word sense disambiguation using Z-reweighting. In: Proceedings of ACL 2022 (Volume 1: Long Papers), pp. 4713–4723 (May 2022)
22. Wang, D., Liu, C., Zhu, Z., Liu, J., Hu, H., Shen, S., Li, B.: Construction and application of pre-trained models of Siku Quanshu in orientation to digital humanities. Library Tribune **42**(06) (2022)
23. Wang, J., Jatowt, A., Yoshikawa, M.: TimeBERT: extending pre-trained language representations with temporal information. arXiv: 2204.13032 (2022)
24. Wang, S., Zhuang, S., Zuccon, G.: Bert-based dense retrievers require interpolation with BM25 for effective passage retrieval. In: Proceedings of SIGIR 2021 (2021)
25. Wu, L., Petroni, F., Josifoski, M., Riedel, S., Zettlemoyer, L.: Scalable zero-shot entity linking with dense entity retrieval. In: Proceedings of EMNLP 2020 (2020)
26. Yamada, I., et al.: Wikipedia2Vec: An efficient toolkit for learning and visualizing the embeddings of words and entities from Wikipedia. In: Proceedings of EMNLP 2020: System Demonstrations, pp. 23–30. Association for Computational Linguistics (2020)
27. Yamada, I., Washio, K., Shindo, H., Matsumoto, Y.: Global entity disambiguation with BERT. In: Proceedings of NAACL 2022, pp. 3264–3271 (2022)
28. Zaporojets, K., Kaffee, L.A., Deleu, J., Demeester, T., Develder, C., Augenstein, I.: TempEL: linking dynamically evolving and newly emerging entities. In: NeurIPS 2022 Datasets and Benchmarks Track (2022)
29. Zhang, W., Hua, W., Stratos, K.: EntQA: entity linking as question answering. In: ICLR 2022. OpenReview.net (2022)

Construction and Prediction of a Dynamic Multi-relationship Bipartite Network

Hehe Lv[1], Guobing Zou[1(✉)], and Bofeng Zhang[2]

[1] School of Computer Engineering and Science, Shanghai University, Shanghai 200444, China
{hhlv,gbzou}@shu.edu.cn

[2] School of Computer and Information Engineering, Shanghai Polytechnic University, Shanghai 201209, China
bfzhang@sspu.edu.cn

Abstract. Bipartite networks are capable of representing complex systems that involve two distinct types of objects. However, there are limitations to the existing bipartite networks: 1) It is inadequate in characterizing multi-relationships among objects in complex systems, as it is restricted to depict only one type of relationship. 2) It is limited to static representations of complex systems, hampering their ability to describe dynamic changes in the interactions among objects over time. Therefore, the Dynamic Multi-Relationship Bipartite Network (DMBN) model is introduced, which not only models the dynamic multi-relationships between two types of objects in complex systems, but also enables dynamic prediction of the intricate relationships between objects. Extensive experiments were conducted on complex systems, and the results indicate that the DMBN model is significantly better than the baseline methods across multiple evaluation metrics, thereby proving the effectiveness of the DMBN.

Keywords: Multi-relationships aggregation · Feature Representation · Dynamic prediction

1 Introduction

With the development of complex network theory, many complex systems can be described by complex networks [1–5]. A typical complex network comprises nodes and edges, wherein nodes are indicative of objects within a complex system, and edges signify the intricate relationships between these objects [6–8]. A distinctive type of network model exists within complex networks, known as bipartite networks. Bipartite networks consist of two distinct types of nodes, wherein edges are solely present between nodes of different types [9–11]. For instance, bipartite network models can be used to represent purchase relationships between users and items, therapeutic relationships between drugs and diseases, as well as invocation relationships between users and services [12–14].

B. Luo et al. (Eds.): ICONIP 2023, CCIS 1965, pp. 320–331, 2024.
https://doi.org/10.1007/978-981-99-8145-8_25

Traditional bipartite networks are primarily modeled based on a static, single-type relationship between two types of objects. However, in real complex systems, there exist multi-relationships that change over time between two types of objects [15–17]. Each relationship between objects harbors distinct semantic information, making it infeasible to capture the intricate semantic relationship among objects solely based on a singular relationship [18,19]. In addition, traditional bipartite network modeling cannot capture the dynamic interaction information between objects, prompting the need for alternative methodologies in the field of network modeling and analysis.

Therefore, the present paper proposes a novel Dynamic Multi-Relationships Bipartite Network (DMBN) model, designed to effectively model the dynamic multi-relationships that exist between two distinct types of objects in real complex systems. Compared with five existing baseline models, namely DMF, LTSC, TSQP, DLP, and MBN, the DMBN model exhibits superior performance. The main contributions of this paper can be summarized as follows:

(1) A DMBN model is proposed, which can dynamically describe multi-relationships between two types of objects.

(2) Representation methods for multi-relationship aggregation features and preference features are proposed, which can provide features for dynamic prediction.

(3) Experiments on real datasets show that DMBN significantly outperforms baseline methods, proving the effectiveness of the DMBN model.

2 Related Work

A bipartite network is a special type of complex network that consists of two distinct types of nodes and one type of edge, with edges only existing between nodes of different types. The structural characteristics of bipartite networks can be used to describe complex systems consisting of two different types of objects. The characterization and analysis of real complex systems based on bipartite networks can reveal information transmission mechanisms, predict information propagation paths, and explore the relationship between complex system structures and functionalities, providing a scientific basis for decision-making in relevant complex systems. For instance, Fu proposed MVGCN, a robust and effective bipartite network link prediction framework for biomedical applications [20]. The framework is based on bipartite networks in biomedical research and can perform link prediction tasks. Jafari proposed a drug combination strategy method based on bipartite network modeling, which model drug and patient sample (cancer cells) response data as a bipartite network and formulates effective multi-targeted drug combination plans based on community structures [21]. Zhang introduced graph neural on bipartite networks and proposed the BCGNN model for graph classification tasks. This model is able to effectively capture relationships between nodes of the same type in bipartite networks and preserve the structural information of the bipartite graph [22]. Bipartite networks have been proven to be a useful tool for depicting real complex systems and conducting relevant studies and applications. Therefore, this paper implements modeling of multi-relationships between objects in complex systems based on bipartite networks.

3 DMBN Model

3.1 Framework Overview

The overall framework of DMBN is shown in Fig. 1. DMBN models and predicts dynamic multi-relationships between two types of objects in complex systems, which includes two parts: DMBN construction and dynamic prediction. In the DMBN construction part: firstly, the DMBN is proposed to describe the dynamic multi-relationships between objects. Then, different relationships with varying degrees of importance are aggregated based on attention mechanisms to obtain multi-relationships aggregated features. In the dynamic prediction part: firstly, the preference features of nodes are mined based on the similarity between same-type nodes in the network. Then, the temporal features of node pairs, which are represented by the concatenation of the multi-relationships aggregated features and preference features, are fed into the prediction model (GRU) to achieve the prediction of relationships.

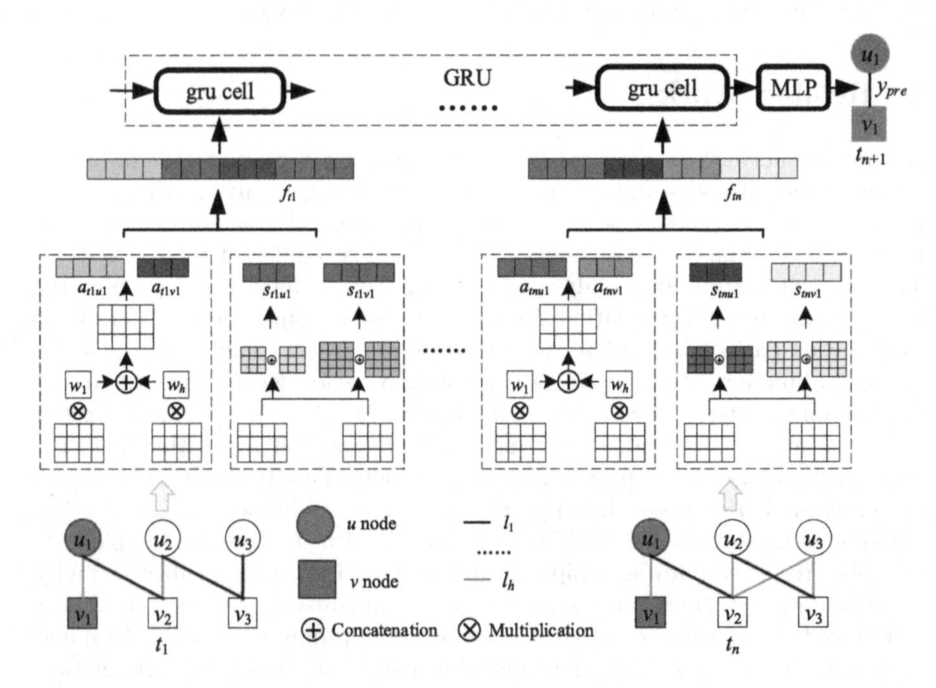

Fig. 1. The overall framework of DMBN.

3.2 DMBN Construction

Definition of DMBN. The DMBN model comprises node sets U and V, type set L, time set T, and edge set E. The DMBN is defined as G_{DMB}=(U, V, L, T, E), where U=$\{u_1, u_2, ...\}$ and V=$\{v_1, v_2, \ldots\}$ respectively represent different types of node sets, L=$\{l_1, l_2, \ldots\}$ represents the set of edge types, T=$\{t_1, t_2, \ldots\}$ represents the set of time instances, and E=$\{e_1, e_2, \ldots\}$ represents the set of edges. The DMBN is represented by the adjacency matrix, A_{DMB}=$\{A^{t1}_{MB}, A^{t2}_{MB}, \ldots\}$, where A^{t1}_{MB}=A^{t1}_{l1} & A^{t1}_{l2} & ... denotes the set of adjacency matrices under different relationship at time t_1.

Multi-relationships Aggregation Based on Attention. There are multi-relationships between two types of nodes in a DMBN, and each relationship contains different semantic information. The key to analyzing and researching DMBNs lies in aggregating the relationships that exist with semantic differences between them. Hence, this paper proposes a multi-relationship aggregation method based on the attention mechanism. The method assigns weights to each relationship based on their relative importance, in order to reflect the significance of different relationships, primarily in terms of weights.

DMBNs can be represented by adjacency matrices, which are based on different time periods and various relationships. For example, A^t_l represents the adjacency matrix at time t and relationship type l. This paper adopts an attention mechanism to aggregate multi-relationships, considering the significance of each relationship, as shown in formula (1).

$$A^t_{MBA} = \sum_{i=1}^{h} W_{l_i} A^t_{l_i} \tag{1}$$

A^t_{MBA} represents the adjacency matrix after multi-relationship aggregation at time t. W_{l_i} denotes the importance of relationship type l_i, $A^t_{l_i}$ represents the adjacency matrix corresponding to relationship type l_i at time t, and h is the number of relationship types.

The multi-relationship aggregation based on attention mechanism can effectively capture the significance of diverse relationships. Therefore, this paper aims to extract the multi-relationship aggregation features for nodes u and v. The corresponding methodology is presented in formulas (2)–(3).

$$a_{tu} = A^t_{MBA}(i:) \tag{2}$$

$$a_{tv} = A^t_{MBA}(:j) \tag{3}$$

In this context, a_{tu} and a_{tv} respectively indicate the multi-relationship aggregation features of node u and node v at time t. $A^t_{MBA}(i{:})$ denotes the i-th row of the adjacency matrix, while $A^t_{MBA}({:}j)$ indicates its j-th column.

3.3 Dynamic Prediction

Feature Representation Based on Similarity. In real complex systems, similarities among objects have the potential to influence the evolution of the complex system, as well as the intricate relationships between objects. Therefore, this paper proposes a preference feature representation method based on the similarity among objects. Based on the initial feature and Pearson correlation coefficient (PCC), the similarity between objects of the same type is calculated. As shown in formula (4). The intuitive feature of complex systems is the interactional relationships among their constituent objects, which can reflect the preferences of these objects to some extent. Therefore, this paper is based on the interaction relationship between objects as the initial feature.

$$sim(x, y) = \frac{\sum_{p=1}^{n}(x_p - \bar{x})(y_p - \bar{y})}{\sqrt{\sum_{p=1}^{n}(x_p - \bar{x})\sum_{p=1}^{n}(y_p - \bar{y})}} \tag{4}$$

Here, x and y represent objects of the same type, p represent the position of the initial feature, and $\bar{x}$ and $\bar{y}$ represent the mean of the initial feature.

Based on the adjacency matrix A_l^t for time t and relationship l, the similarity feature matrices S_{tlu} and S_{tlv} are obtained for u-type nodes and v-type nodes under this time and complex relationship. In this paper, the final similarity feature matrix is obtained by summing the similarity matrices under various complex relationships. The specific formulas are shown in formulas (5)–(6).

$$S_{tu} = \sum_{i=1}^{h} S_{tl_i u} \tag{5}$$

$$S_{tv} = \sum_{i=1}^{h} S_{tl_i v} \tag{6}$$

Here, S_{tu} and S_{tv} represent the similarity feature matrix of u-type nodes and v-type nodes at time t, respectively; $S_{tl_i u}$ represents the similarity feature matrix of u-type nodes under relationship type l_i at time t; h denotes the number of relationship types.

Based on the similarity between nodes of the same type, the preference features of nodes u and v are extracted in this paper, as shown in formulas (7)–(8).

$$s_{tu} = S_{tu}(i :) \tag{7}$$

$$s_{tv} = S_{tv}(j :) \tag{8}$$

Here, s_{tu} and s_{tv} represent the preference features of node u and node v at time t, respectively. $(i{:})$ and $(j{:})$ represent the i-th and j-th rows of the matrix, respectively.

Dynamic Multi-relationships Prediction. The motivation of this paper is to predict the complex relationships among nodes in DMBNs based on the historical features of the nodes. These historical features include multi-relationship aggregation features and preference features. Multi-relationship aggregation features can reflect the importance of different relationships, while preference features can reflect the preferences of nodes. Therefore, this paper is based on the concatenation of multi-relationship aggregation features and preference features to obtain node features, as shown in formulas (9)–(10).

$$g_{tu} = a_{tu} \oplus s_{tu} \tag{9}$$

$$g_{tv} = a_{tv} \oplus s_{tv} \tag{10}$$

In this context, g_{tu} and g_{tv} respectively denote the features of node u and node v at time t, while the symbol $\oplus$ indicates the concatenation of features.

This paper obtains the temporal feature f_t of node pair u-v at time t based on feature concatenation, as shown in formula (11).

$$f_t = g_{tu} \oplus g_{tu} \tag{11}$$

This paper employs the Gate Recurrent Unit (GRU) model to mine implicit information. As a variant of LSTM, this model features a simpler structure, fewer parameters, and faster training speed. By feeding the vector h_k obtained from the final time step output of GRU into a fully connected network, the predicted relationship between nodes u and v can be obtained, as shown in formula (12).

$$\hat{y} = Relu\,(Wh_k + b) \tag{12}$$

Here, W and b are weight matrices for adaptive learning, and $\hat{y}$ represents the predicted value.

In this paper, predicting relationships among objects is regarded as a regression problem, and its loss function is shown in formula (13).

$$L = \alpha^* \frac{1}{M} \sum_{i=1}^{M} (y_i - \hat{y}_i)^2 + (1 - \alpha)^* \sum_j w_j^2 \tag{13}$$

In this equation, y_i represents the true relationship between nodes u and v, $\hat{y}$ represents the predicted relationship, M denotes the number of samples, w_j is a learnable parameter, $\sum_j w_j^2$ represents the regularization term, α is used to balance the importance of the regularization term.

4 Experiment and Results

4.1 Dataset

This paper constructs DMBN based on various complex relationships among complex objects, and performs dynamic prediction of target relationships. The dataset is sourced from various complex relationships between users and items on the Taobao platform [23], including browse (Pv), favorite (Fav), add-to-cart (Cart) and purchase (Buy), among others. The experiments in this paper follow the experimental setup of MBN [28], and three datasets were obtained as experimental data using the same preprocessing method. Moreover, the complex relationships between users and items in the three datasets were rearranged based on their temporal order. In the experimental process, each dataset was divided into five equal parts, where one part was used as a testing set and the remaining parts were used as training sets. This process was repeated five times.

4.2 Evaluation Metrics

This paper constructs a DMBN based on various relationships between users and items, and dynamically predicts the relationships based on the temporal features of DMBN. In the real world, merchants are more concerned with the purchasing relationships. Therefore, this paper regards the purchasing relationship as the ultimate goal. To address the issue of imbalanced data resulting from using negative samples that do not have any relationships between users and items, this paper adopts a regression model-based evaluation metric to measure the performance of DMBN. This approach allows for a more accurate assessment of the predictive performance of the model when dealing with highly imbalanced datasets. The mean absolute error (MAE) and root mean square error (RMSE) were used to evaluate the performance of the predictions in the experiments.

4.3 Baseline Methods

The effectiveness of DMBN is evaluated based on the following five baseline methods. DMF [24]: DMF is a matrix factorization model based on neural network structures. LTSC [25]: LTSC is a feature-enhanced service classification model based on attention mechanisms and convolutional neural networks (CNN). TSQP [26]: TSQP is a QoS prediction method based on deep learning, which aims to perform time-aware service QoS prediction tasks through feature integration. DLP [27]: DLP is a link prediction model that employs the local structures of a bipartite network. MBN [28]: MBN is a network model that is designed to model the various complex relationships between two types of objects in the real world.

4.4 Results

Performance Comparison. The performance comparison results between DMBN and baseline methods based on two evaluation metrics, namely MAE and RMSE, are presented in Table 1. The experimental results demonstrate that DMBN outperforms the baseline methods on all the evaluation metrics. The performance improvements are primarily attributed to the following reasons: 1): Modeling based on the DMBN can characterize the existence of multi-relationships in complex systems. 2): Dynamic modeling can effectively retain both historical and current information among objects. 3): Feature representation based on similarity can to some extent reflect the preference features of objects. Moreover, the following observation results were further summarized:

Table 1. Performance comparison of DMBN and baseline methods.

Method	User Behavior 1		User Behavior 2		User Behavior 3	
	RMSE	MAE	RMSE	MAE	RMSE	MAE
DMF	0.477	0.438	0.488	0.434	0.472	0.424
LTSC	1.503	1.212	1.541	1.502	1.536	1.511
TSQP	1.371	0.784	1.675	0.915	1.583	0.886
DLP	2.718	1.437	2.566	1.051	2.393	1.251
MBN	0.471	0.387	0.473	0.393	0.470	0.388
DMBN	0.319	0.270	0.277	0.197	0.351	0.282

(1) Based on RMSE evaluation metric, the DLP model exhibits the worst performance. DLP predicts the link relationships between nodes based on the local structural information of the bipartite network. This methodology necessitates a significant quantity of edge relationships within the bipartite network to extract more structural features that can distinguish differences in target node. However, the relationships present in real complex systems are typically sparse. When constructing bipartite networks based on complex relationships and extracting local structural features, the limited structural information contained in the features may not be sufficient to provide accurate and effective information for predictions, which consequently may result in suboptimal performance.

(2) Based on the MAE evaluation metric, the LTSC model exhibits poor performance. The LTSC model extracts feature representations based on a word embedding model, and enhances the embedding representation using a label attention mechanism. However, the users and items in the dataset of this paper are mainly represented in the form of IDs, which cannot extract features based on word embedding models. Therefore, utilizing solely the IDs of users and items as features would not be effective in extracting meaningful information.

(3) MBN exhibits superior performance when compared to the other baseline methods. MBN has the ability to model a variety of complex relationships among objects in a complex system and can overcome the limitations of traditional bipartite networks, which can only model a single type of edge relationship. At the same time, an attention mechanism based on the relationship level is designed to fuse multiple relationships and realize the importance distinction of each relationship. Moreover, the superior performance of DMBN relative to MBN indicates that incorporating dynamic factors into modeling can significantly enhance the predictive performance of the model.

Impact of Multi-relationship. In this work, we examine the effects of multi-relationship modeling strategies on performance, with a primary focus on DMBN modeling based on double and multiple relationships.

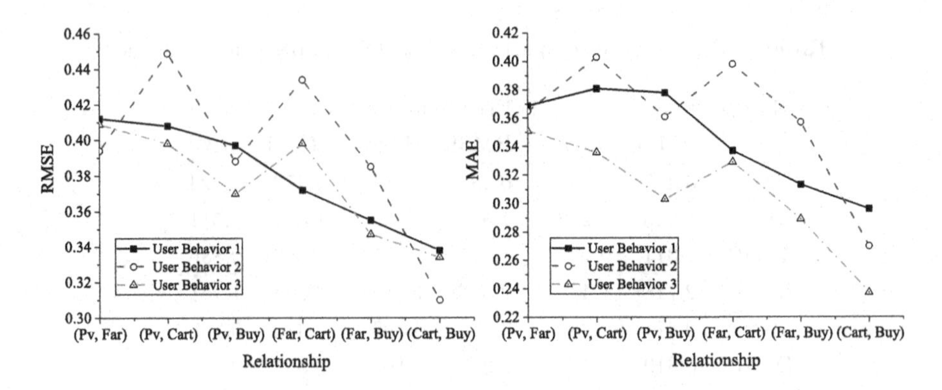

Fig. 2. Modeling based on double relationships.

Initially, we investigate the modeling strategy based on double relationships, where the network is modeled based on two types of relationships. The experimental results demonstrated in Fig. 2 show that the model achieves optimal performance when modeled based on "Cart" and "Buy" relationships. In addition to the "Buy" relationship, adding other relationships still achieves good predictive performance, demonstrating the importance of the "Buy" relationship in enhancing the model performance. Moreover, when modeling is based on a combination of two relationships, that without the "Pv" relationship performs better than that with the "Pv" relationship, indicating the limitations of the "Pv" relationship in improving the model performance.

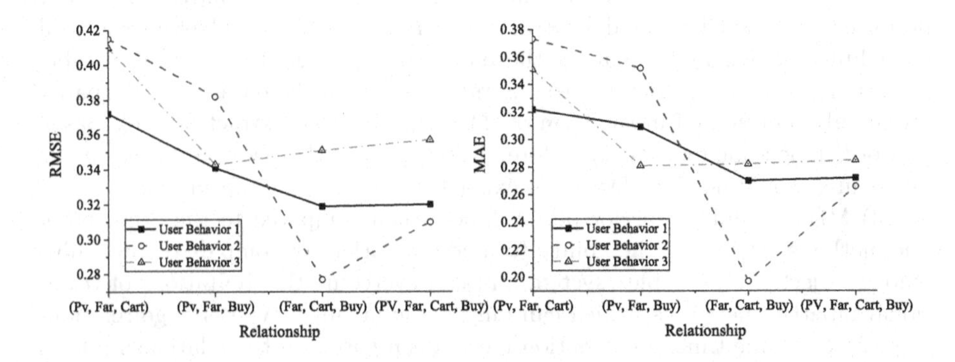

Fig. 3. Modeling based on multiple relationships.

Next, we investigate the modeling strategy based on multi-relationships, where the network is modeled based on a variety of relationships.

The experimental results in Fig. 3 demonstrate that the modeling of the "Far", "Cart", and "Buy" relationships yields the best predictive performance. This indicates the importance of these three relationships for predicting the target relationship. However, it is noteworthy that the predictive performance is not optimal when modeling based on the four relationships. It is proposed that the "Pv" relationship is a ubiquitous factor between users and items. Therefore, including this relationship in the modeling may introduce unnecessary noise, interfere with the model's recognition of the target relationship, and result in a decrease in predictive performance. Furthermore, incorporating the "Buy" relationship in the multi-relationship modeling yields significantly superior performance, thereby underscoring the paramount importance of this relationship in augmenting the predictive capabilities of model.

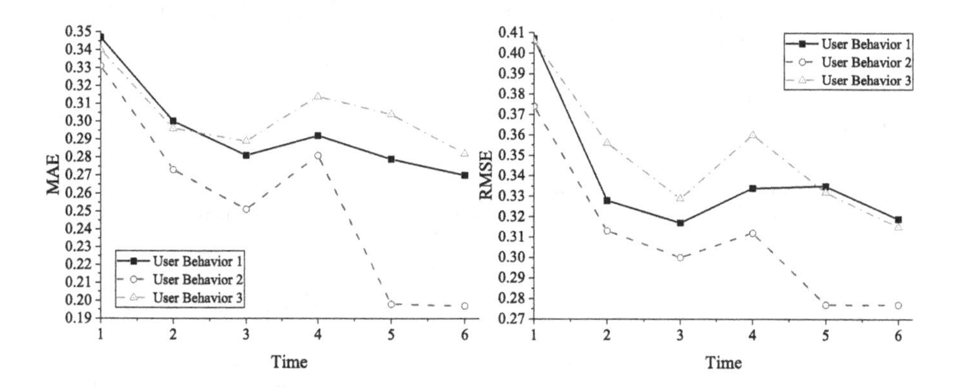

Fig. 4. The impact of time on model performance.

The Impact of Time. This paper performs temporal segmentation of the dataset based on a 24-hour cutoff for partitioning. As illustrated in Fig. 4, the predictive performance of the model improves with the increase of time, which indicates that incorporating historical information can effectively enhance the predictive capability of model. In addition, the introduction of historical information leads to a remarkable improvement in the performance of the model within a short time frame. This could be attributed to the model's incapability of capturing enough historical cues when the temporal window is too brief. At time interval of 4, the model's performance temporarily decreases, possibly due to the redundant interaction information between users and items present in the historical data at that moment. As the time interval increases, the model's performance further improves, highlighting the significant impact of historical information on enhancing model performance.

5 Conclusion

This paper proposes a DMBN model to address the challenge of modeling dynamic multi-relationships between objects in complex systems. The main innovation of DMBN can be summarized as follows: 1) A novel framework, the DMBN model, is proposed to tackle the formidable task of modeling intricate and diverse dynamic relationships that exist between two types of objects in complex systems. 2) A multi-relationship aggregation feature representation method and a preference feature mining method are proposed, and the dynamic prediction of the target relationship is achieved by concatenating these features. 3) Experimental results on a real complex system demonstrate the DMBN's modeling ability and outstanding performance in predicting target relationships. In our future endeavors, we intend to incorporate more intricate objects within the network modeling framework, aiming to enhance the comprehensive characterization of real complex systems.

References

1. Sun, L.H., He, Q., Teng, Y.Y., Zhao, Q., Yan, X., Wang, X.W.: A complex network-based vaccination strategy for infectious diseases. Appl. Soft Comput. **136**, 110081 (2023)
2. Tocino, A., Serrano, D.H., Hernández-Serrano, J., Villarroel, J.: A stochastic simplicial SIS model for complex networks. Commun. Nonlinear Sci. Numer. Simul. **120**, 107161 (2023)
3. Zhou, Y.R., Chen, Z.Q., Liu, Z.X.: Dynamic analysis and community recognition of stock price based on a complex network perspective. Expert Syst. Appl. **213**(1), 118944 (2023)
4. Vatani, N., Rahmani, A.M., Javadi, H.S.: Personality-based and trust-aware products recommendation in social networks. Appl. Intell. **53**(1), 879–903 (2023)
5. Tan, L., Gong, D.F., Xu, J.M., Li, Z.Y., Liu, F.L.: Meta-path fusion based neural recommendation in heterogeneous information networks. Neurocomputing **529**, 236–248 (2023)
6. Wu, H.X., Song, C.Y., Ge, Y., Ge, T.J.: Link prediction on complex networks: an experimental survey. Data Sci. Eng. **7**(3), 253–278 (2022)
7. Zhang, L.L., Zhao, M.H., Zhao, D.Z.: Bipartite graph link prediction method with homogeneous nodes similarity for music recommendation. Multimedia Tools Appli. **79**(19-20), 13197-13215 (2020)
8. Zhou, C.Q., Zhang, J., Gao, K.S., Li, Q.M., Hu, D.M., Sheng, V.S.: Bipartite network embedding with symmetric neighborhood convolution. Expert Syst. Appl. **198**, 116757 (2022)
9. Cimini, G., Carra, A., Didomenicantonio, L., Zaccaria, A.: Meta-validation of bipartite network projections. Commun. Phys. **5**(1), 76 (2022)
10. Valejo, A., Santos, W., Naldi, M.C., Zhao, L.: A review and comparative analysis of coarsening algorithms on bipartite networks. Euro. Phys. J. Special Topics **230**, 2801-2811 (2021)
11. Valejo, A., Faleiros, T., Oliveira, M.C.F., Lopes, A.A.: A coarsening method for bipartite networks via weight-constrained label propagation. Knowl.-Based Syst. **195**, 105678 (2020)

12. Liu, G.G.: An ecommerce recommendation algorithm based on link prediction. Alex. Eng. J. **61**(1), 905–910 (2022)
13. Zhang, G.Z., Li, S.Z., Dou, X.R., Shang, J.L., Ren, Q.Q., Gao, Y.L.: Predicting LncRNA-disease associations based on LncRNA-MiRNA-disease multilayer association network and bipartite network recommendation. In: IEEE International Conference on Bioinformatics and Biomedicine, pp. 2216–2223. IEEE, USA (2022)
14. Wang, G.L., Yu, J., Nguyen, M., Zhang, Y.Q., Yongchareon, S., Han, Y.B.: Motif-based graph attentional neural network for web service recommendation. Knowl.-Based Syst. **269**(7), 110512 (2023)
15. Li, C.Y., et al.: Multiplex network community detection algorithm based on motif awareness. Knowl.-Based Syst. **260**(25), 110136 (2023)
16. Magnani, M., Rossi, L.: Towards effective visual analytics on multiplex and multilayer networks. Chaos, Solitons Fractals **72**, 68–76 (2015)
17. Xia, L.H., Huang, C., Xu, Y., Pei, J.: Multi-behavior sequential recommendation with temporal graph transformer. IEEE Trans. Knowl. Data Eng. **35**(6), 6099–6112 (2023)
18. Singh, D.K., Haraty, R., Debnath, N., Choudhury, P.: an analysis of the dynamic community detection algorithms in complex networks. In: IEEE International Conference on Industrial Technology, pp. 989–994. IEEE, Argentina (2020)
19. Xue, H., Yang, L., Jiang, W., Wei, Y., Hu, Y., Lin, Yu.: Modeling dynamic heterogeneous network for link prediction using hierarchical attention with temporal RNN. In: Hutter, F., Kersting, K., Lijffijt, J., Valera, I. (eds.) ECML PKDD 2020. LNCS (LNAI), vol. 12457, pp. 282–298. Springer, Cham (2021). https://doi.org/10.1007/978-3-030-67658-2_17
20. Fu, H.T., Huang, F., Liu, X., Qiu, Y., Zhang, W.: MVGCN: data integration through multi-view graph convolutional network for predicting links in biomedical bipartite networks. Bioinformatics **38**(2), 426–434 (2022)
21. Jafari, M., et al.: Bipartite network models to design combination therapies in acute myeloid leukaemia. Nat. Commun. **13**(1), 2128 (2022)
22. Zhang, X.H., Wang, H.C., Yu, J.K., Chen, C., Wang, X.Y., Zhang, W.J.: Bipartite graph capsule network. World Wide Web **26**(1), 421–440 (2023)
23. Zhuo, J.W., et al.: Learning optimal tree models under beam search. In: the 37th International Conference on Machine Learning, pp. 11650–11659. PMLR, Virtual Event (2020)
24. Xue, H.J., Dai, X.Y., Zhang, J.B., Huang, S.J., Chen, J.J.: Deep matrix factorization models for recommender systems. In: the Twenty-Sixth International Joint Conference on Artificial Intelligence, pp. 3203–3209. ijcai.org, Australia (2017)
25. Zou, G.B., Yang, S., Duan, S.Y., Zhang, B.F., Gan, Y.L., Chen, Y.X.: DeepLTSC: long-tail service classification via integrating category attentive deep neural network and feature augmentation. IEEE Trans. Netw. Serv. Manage. **19**(2), 922–935 (2022)
26. Zou, G.B., et al.: DeepTSQP: temporal-aware service QoS prediction via deep neural network and feature integration. Knowl.-Based Syst. **241**(6), 108062 (2022)
27. Lv, H.H., Zhang, B.F., Hu, S.X., Xu, Z.K.: Deep link-prediction based on the local structure of bipartite networks. Entropy **24**(5), 610 (2022)
28. Lv, H.H., Zhang, B.F., Li, T.T., Hu, S.X.: Construction and analysis of multi-relationship bipartite network model. Complex Intell. Syst. Early Access (2023)

Category-Wise Fine-Tuning for Image Multi-label Classification with Partial Labels

Chak Fong Chong, Xu Yang(✉), Tenglong Wang, Wei Ke, and Yapeng Wang

Faculty of Applied Sciences, Macao Polytechnic University, Macao SAR, China
{chakfong.chong,xuyang,p1807530,wke,yapengwang}@mpu.edu.mo

Abstract. Image multi-label classification datasets are often partially labeled (for each sample, only the labels on some categories are known). One popular solution for training convolutional neural networks is treating all unknown labels as negative labels, named *Negative* mode. But it produces wrong labels unevenly over categories, decreasing the binary classification performance on different categories to varying degrees. On the other hand, although *Ignore* mode that ignores the contributions of unknown labels may be less effective than *Negative* mode, it ensures the data have no additional wrong labels, which is what *Negative* mode lacks. In this paper, we propose **C**ategory-wise **F**ine-**T**uning (**CFT**), a new post-training method that can be applied to a model trained with *Negative* mode to improve its performance on each category independently. Specifically, CFT uses *Ignore* mode to one-by-one fine-tune the logistic regressions (LRs) in the classification layer. The use of *Ignore* mode reduces the performance decreases caused by the wrong labels of *Negative* mode during training. Particularly, Genetic Algorithm (GA) and binary crossentropy are used in CFT for fine-tuning the LRs. The effectiveness of our methods was evaluated on the CheXpert competition dataset and achieves state-of-the-art results, to our knowledge. A single model submitted to the competition server for the official evaluation achieves mAUC 91.82% on the test set, which is the highest single model score in the leaderboard and literature. Moreover, our ensemble achieves mAUC 93.33% (The competition was recently closed. We evaluate the ensemble on a local machine after the test set is released and can be downloaded.) on the test set, superior to the best in the leaderboard and literature (93.05%). Besides, the effectiveness of our methods is also evaluated on the partially labeled versions of the MS-COCO dataset.

Keywords: Partial Labels · Partial Annotations · Multi-Label Classification · Multi-Label Recognition

1 Introduction

Image multi-label classification (MLC) is a typical computer vision problem that classifies the presence (positive) or absence (negative) of multiple categories in

B. Luo et al. (Eds.): ICONIP 2023, CCIS 1965, pp. 332–345, 2024.
https://doi.org/10.1007/978-981-99-8145-8_26

each image. As an image usually contains multiple objects or concepts, it is more practical than its counterpart single-label classification and hence has a wide range of applications like medical image interpretation [6,7,21].

A crucial challenge of training convolutional neural networks (CNNs) for image MLC is the training data is often partially labeled [17,27]. That is, for each image sample, only the labels on some categories are known, and the rest are unknown. It is because the manual collection of fully labeled data is expensive [13], especially when the numbers of categories and samples are very large.

A popular and effective solution for training CNN with partially labeled data is treating all unknown labels as negative labels [2,3,26,34], named ***Negative* mode** [1]. This mode is based on the prior knowledge of MLC datasets that negative labels are usually much more than positive labels [28]. Nevertheless, this mode produces wrong labels to the training data, as some unknown labels' ground truths are positive labels instead of negative labels. These wrong labels are usually unevenly distributed over different categories [1]. The categories with more wrong labels suffer from more harm. Therefore, different categories suffer from varying degrees of performance decreases.

On the other hand, another solution is ignoring the contributions of unknown labels [1,13], named ***Ignore* mode** [1]. This mode may be less effective than *Negative* mode [26], as it does not utilize the prior knowledge that negative labels are in the majority. Even so, it ensures the training data have no additional wrong labels, which is a vital advantage that *Negative* mode lacks. Therefore, several work utilize this vital advantage of *Ignore* to improve *Negative* mode for training CNNs beginning with initial parameters [1,26].

In this paper, we propose **C**ategory-wise **F**ine-**T**uning (**CFT**), a new post-training method that can be applied to a CNN that has been trained with *Negative* mode to improve its binary classification performance on each category independently. Therefore, CFT is very different from most approaches that train a CNN from initial parameters [1,26]. Specifically, CFT uses *Ignore* mode to one-by-one fine-tune the logistic regressions (LRs) in the classification layer, in which each LR outputs the binary classification result on one category. The use of *Ignore* mode reduces the performance decreases caused by the wrong labels of *Negative* mode during training. The one-by-one fine-tuning can improve the performance on each category independently without affecting the performance on other categories.

While applying CFT to a CNN, the LRs may prefer different fine-tuning configurations (optimization methods, methods for handling untypical labels in particular MLC datasets, etc.) to achieve higher performance. Therefore, we additionally use a greedy selection for CFT to enable choosing the best configuration for each LR from multiple configuration candidates.

During experiments, we found using binary crossentropy (BCE) loss with backpropagation for fine-tuning an LR sometimes unwantedly decreases the performance like AUC (area under the receiver operating characteristic curve). On the other hand, Genetic Algorithm (GA) [29] for fine-tuning can directly improve the performance, avoiding performance drops caused by minimizing BCE.

Sufficient experiments were conducted on the CheXpert [21] competition dataset and the partially labeled versions of the MS-COCO [28] standard MLC dataset to evaluate the effectiveness of our methods. Especially, our methods achieve state-of-the-art on the CheXpert dataset, to the best of our knowledge. We submitted a single CNN to the competition server[1] for the official evaluation on the test set. It achieves mAUC 91.82%, which is the highest single model score in the leaderboard and literature. After that, the competition server was closed and the test set is released. Therefore, our ensemble composed of 5 single CNNs was evaluated on a local machine and achieves mAUC 93.33% on the test set, superior to the best in the leaderboard and literature (mAUC 93.05% [44]).

2 Related Work

Several approaches were proposed to address MLC with partial labels. Binary Relevance [15] converts MLC to multiple binary classification tasks, but it usually fails to model the label dependencies and is less scalable to a large number of categories. [23,41,43] adopted low-rank learning, [39] used a mixed graph to encode a network of label dependency, [3,12] predicted unknown labels by learning label relations, and [8,24,38] predicted unknown labels by posterior inference. However, most of these approaches cannot be well-adapted for training deep models, as they require putting all training data into memory or solving costly optimization problems.

Some approaches train deep models with partial labels by exploiting image and category dependencies. Durand *et al.* [13] proposed predicting unknown labels based on curriculum learning with graph neural networks to model the correlations between categories. IMCL [20] interactively learns a model with a similarity learner which discovers label and image dependencies. SST [5] and HST [4] explore the image-specific occurrence and category-specific feature similarities to complement unknown labels. SARB [32] complements unknown labels by learning and blending category-specific feature representation across different images. However, most of these approaches require particular model architectures or training schemes.

Negative mode and *Ignore* mode are more prevalent in contrast with the complex approaches aforementioned. *Ignore* mode simply ignores the contributions of unknown labels (*e.g.*, partial-BCE loss [13] and partial asymmetric loss [1]) while *Negative* mode [2,3,26,34] treats all unknown labels as negative labels. Several work (including this paper) aim to improve *Negative* mode with *Ignore* mode, as introduced in Sect. 1. Kundu *et al.* [26] proposed a method to soften the signal of the wrong labels of *Negative* mode by exploiting the image and label relationships, but it does not avoid some categories training on too many wrong labels. Ben-Brunch *et al.* [1] proposed *Selective* approach that can adjust the training mode for each category to be either *Negative* or *Ignore*, but it requires the presence frequency of every category which is unavailable in partially labeled datasets.

[1] https://stanfordmlgroup.github.io/competitions/chexpert/.

Unlike most previous approaches that aim to train high performance models beginning with initial parameters, the proposed CFT is a post-training method based on *Ignore* mode that can be applied to models trained with *Negative* mode to further improve the performance. Moreover, CFT can independently improve the classification performance on each category. Hence, CFT may be able to further improve the performance of the models trained with other approaches mentioned above.

3 Methods

This section presents the proposed CFT, the greedy selection for selecting fine-tuning configurations, and GA for fine-tuning, as summarized in Fig. 1.

Notations. Considering a C-category image MLC task with a training set $\mathcal{D} = \{(I, \mathbf{y})_i\}$. Each sample $(I, \mathbf{y})$ consists of an image I and a label vector $\mathbf{y} = [y_1, ..., y_C] \in \{-1, 1, 0\}^C$ where the cth ($c \in \{1, ..., C\}$) element y_c is the label on category c and it is assigned to be either -1 (*negative*), 1 (*positive*), or 0 (*unknown*). A deep neural network (typically CNN) *Baseline* has been trained on the training set $\mathcal{D}$ with *Negative* mode. The architecture of *Baseline* consists of: (1) a backbone $\mathtt{b}$ transforms an input image I to a feature vector $\mathbf{z} = \mathtt{b}(I) \in \mathbb{R}^Z$; and (2) a C-unit fully-connected layer $\mathtt{h}$ with Sigmoid activation transforms a feature vector $\mathbf{z}$ to an output vector $\hat{\mathbf{y}} = \mathtt{h}(\mathbf{z}) = [\hat{y}_1, ..., \hat{y}_C] \in [0, 1]^C$, where the cth element $\hat{y}_c$ is the output representing the binary classification result on category c. To better illustrate CFT, we equivalently regard the fully-connected layer $\mathtt{h}$ as C independent logistic regressions (LRs) $\mathtt{h}_1, ..., \mathtt{h}_C$, as shown in Fig. 1 left. The cth LR $\mathtt{h}_c$ transforms a feature vector $\mathbf{z}$ to an output $\hat{y}_c = \mathtt{h}_c(\mathbf{z})$.

3.1 Category-Wise Fine-Tuning (CFT)

The proposed CFT is a post-training method that can be applied to *Baseline*. CFT uses *Ignore* mode to one-by-one fine-tune the LRs $\mathtt{h}_1, ..., \mathtt{h}_C$ to improve its performance on each category independently. Therefore, the backbone $\mathtt{b}$ is always unchanged.

Specifically, the procedure of CFT has C steps (*i.e.*, determined by the number of categories C). The goal of the cth step ($c = \{1, ..., C\}$) is to independently improve the performance on category c through fine-tuning *Baseline*. That is, the fine-tuning only improves the performance on category c, meanwhile, keeping the performance on other categories unchanged. Hence, each category can be independently improved without any concerns of harming other categories.

To achieve this goal, at the cth step, only the cth LR $\mathtt{h}_c$ is fine-tuned instead of the whole *Baseline*. It is because changing all the parameters of *Baseline* will change the performance on all categories, which does not match the goal. On the other hand, changing the parameters of $\mathtt{h}_c$ only affects the output $\hat{y}_c$ on category c and does not affect the outputs on other categories, which matches the goal.

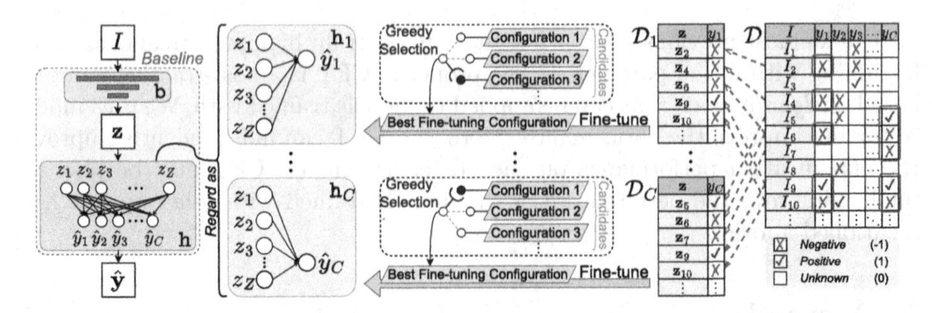

Fig. 1. The overview of CFT and the greedy selection.

At the cth step, the cth LR $\mathbf{h}_c$ is fine-tuned using binary crossentropy (BCE) loss with backpropagation (BP), which is popular for optimizing binary classification models. *Ignore* mode is used to reduce the performance decrease caused by the wrong labels of *Negative* mode during training. Particularly, $\mathbf{h}_c$ is fine-tuned on a new training set $\mathcal{D}_c$ generated from the original training set $\mathcal{D}$ for the use of *Ignore* mode and reducing computation cost, as shown in Fig. 1 right. We first select the samples from $\mathcal{D}$ where the label on category c is known (*i.e.*, $y_c \in \{-1, 1\}$) to be the samples in $\mathcal{D}_c$. This selection ensures $\mathbf{h}_c$ is fine-tuned with *Ignore* mode. Then, as the backbone $\mathtt{b}$ is always the same, we convert the image I of each sample to a feature vector $\mathbf{z} = \mathtt{b}(I)$ in advance to avoid unnecessary computation during fine-tuning. Lastly, the unnecessary labels on other categories are dropped. Formally, the new training set $\mathcal{D}_c = \{(\mathbf{z}, y_c)_i\}$ is generated by: $\mathcal{D}_c = \{\mathtt{T}((I, \mathbf{y})) | (I, \mathbf{y}) \in \mathcal{D}, y_c \in \{-1, 1\}\}$ where $\mathtt{T}((I, \mathbf{y})) = (\mathtt{b}(I), y_c) = (\mathbf{z}, y_c)$.

3.2 Greedy Selection for Fine-Tuning Configuration Selection

While applying CFT to *Baseline*, as the LRs are independent to each other, each LR can be fine-tuned with different configurations to achieve higher performance. The configurations can be different optimization methods (*e.g.*, BCE loss and the below-introduced GA), methods for handling the untypical labels that appear in the CheXpert dataset (see Sect. 4.1), batch sizes, learning rates, etc.

Hence, for each LR, we can additionally compare multiple fine-tuning configuration candidates and select the best one based on the results, referred to as *greedy selection*, as shown in Fig. 1 middle. For example, assume we apply CFT to *Baseline* that has 5 LRs $\mathbf{h}_1, ..., \mathbf{h}_5$ (5 categories). We can additionally compare BCE loss and GA, then choose the best configuration for each LR. A possible result is, $\mathbf{h}_1, \mathbf{h}_4, \mathbf{h}_5$ uses BCE loss, while $\mathbf{h}_2, \mathbf{h}_3$ uses GA.

3.3 Fine-Tuning Logistic Regressions (LRs) Using Genetic Algorithm

During the experiments on the CheXpert dataset (performance metric is AUC, higher is better), we found that fine-tuning an LR using BCE loss sometimes unwantedly decreases AUC. A concrete example is in Fig. 2 which shows the learning curves of fine-tuning the LR of the "Atelectasis" category. In both the training curves and the validation curves, minimizing BCE can cause AUC decreases. It is because minimizing BCE is generally used for optimizing classification accuracy [40], which does not necessarily achieve the best possible AUC [40] or AP (average precision) [33] that are popular metrics for image MLC.

Therefore, we propose using Genetic Algorithm (GA) [29] to fine-tune each LR. GA is a global search algorithm inspired by the principle of the evolution theory. In nature, individuals which are more adapted to the environment have higher chances to survive and produce offspring. This process keeps repeating over generations until the best individual is found.

GA has shown its feasibility for training neural networks [10,18,30] and has several advantages in comparison to BCE loss. (1) GA is a direct search method [37] that can directly improve the performance computed by a metric, which avoids the potential performance decreases caused by minimizing BCE; and (2) BCE loss relies on backpropagation which is easy to trap in local optima and difficult to escape it to find a better solution [18]. GA runs multiple solutions simultaneously, which helps to escape from local optima [37].

4 Experimental Results and Discussion

We conducted sufficient experiments on the CheXpert competition dataset (Sect. 4.1) and the partially labeled versions of the MS-COCO [28] standard MLC dataset (Sect. 4.2) to evaluate the effectiveness of the proposed methods.

4.1 The CheXpert Chest X-Ray Image MLC Competition Dataset

Dataset. CheXpert [21] is a large-scale chest X-ray image 14-category MLC competition dataset. **The training set** has 223,414 image samples. Labels are automatically extracted from the free text reports. Labels are either *positive*, *negative*, *unknown* (the term is *blank* in the original paper), or *uncertain*. Noteworthy, the uncertain labels in this dataset are untypical in partially labeled datasets and have different semantic meanings from unknown labels. An uncertain label captures both the uncertainty in diagnosis and ambiguity in the report, while an unknown label implies no mentions are found in the report. Hence, we do not simply consider the uncertain labels as unknown labels. We handle the uncertain labels in other ways instead, as described in the experimental settings below. **The validation set** has 234 image samples. A label is manually assigned as either *positive* or *negative*. **The test set** has 668 image samples. A label is manually assigned as either *positive* or *negative*. The test set is private and is reserved for the competition. Models must be submitted to the competition server for the official evaluation on the test set. The competition leaderboard is available at https://stanfordmlgroup.github.io/competitions/chexpert/. **The official performance metric** is used, which is computed by the mean AUC (mAUC) on the 5 categories: Atelectasis, Cardiomegaly, Consolidation, Edema, and Pleural Effusion.

Baseline Training. *Baseline* is a DenseNet-121 [19] CNN with an input resolution 224^2. The parameters trained on ImageNet [11] are used as the initial parameters. *Baseline* is trained on the training set for 10 epochs. We follow the previous state-of-the-art [31,44] to treat unknown labels as negative (*Negative* mode) and treat uncertain labels as positive with label smoothing [31]. Images are rescaled to $[0,1]$. We use the same data augmentation as in [6,7]: horizontal flip, rotate $\pm 20°$, and scale $\pm 3\%$. BCE loss with batch size 32 and Adam ($lr = 1 \times 10^{-4}$) [25] is used to update parameters. The checkpoint that achieves the highest validation mAUC is saved. *Baseline* achieves mAUC 89.6% on the validation set (as reported in Table 1) which is already very high for a single CNN. *E.g.*, the single CNN of 2nd place on the competition leaderboard achieves mAUC 89.4% [31].

Ablation Study on CFT. We apply CFT to *Baseline* to improve its performance. The default BCE loss is used to fine-tune each LR, referred to as (CFT-BCE). Besides, we study two variants of CFT-BCE:

1. CFT-BCE-simu: All the LRs are fine-tuned **simu**ltaneously (*i.e.*, fine-tune the whole classification layer), instead of fine-tuning each LR one-by-one. Partial-BCE loss [13] is used to enable *Ignore* mode.
2. CFT-BCE-Nega: Each LR is fine-tuned with ***Nega**tive* mode, instead of *Ignore* mode.

Full-batch gradient descent ($lr = 1 \times 10^{-4}$) is used to update parameters for stability. We treat the uncertain labels as unknown labels, so the uncertain labels are ignored in CFT. The number of epochs is 500.

Table 1 shows the results. CFT-BCE and its variants successfully improve the mAUC of *Baseline*. Particularly, CFT-BCE achieves the highest improvement (mAUC +0.3%). CFT-BCE-simu is less effective (+0.1%), because one-by-one fine-tuning allows individually saving the best checkpoint for each LR, thus achieving better mAUC. CFT-BCE-Negative is also less effective (+0.1%), demonstrating the use *Ignore* Mode can effectively reduce the performance decreases caused by the wrong labels of *Negative* mode during training.

Table 1. Ablation study on CFT, AUC%.

Method	Ate	Car	Con	Ede	P.E	*Mean*
Baseline	85.5	84.2	93.3	92.7	92.3	*89.6*
CFT-BCE-simu	**85.7**	84.0	93.3	92.8	92.4	*89.7*
CFT-BCE-Nega	85.6	84.2	93.4	**92.9**	92.4	*89.7*
CFT-BCE	85.6	**85.0**	**93.5**	**92.9**	**92.5**	***89.9***

Table 2. Ablation study on GA, AUC%.

Config	Ate	Car	Con	Ede	P.E	*Mean*
CFT-BCE	85.6	85.0	93.5	92.9	92.5	*89.9*
CFT-WMW	87.2	87.9	**94.7**	92.9	92.5	*91.0*
CFT-AUCM	**89.1**	87.7	93.8	**93.2**	92.4	*91.2*
CFT-GA	88.8	**88.6**	94.5	93.0	**92.7**	***91.5***

Ablation Study on GA. We study four different optimization methods for fine-tuning LRs to investigate the effectiveness of GA: (1) the default BCE loss used above (CFT-BCE), (2) GA (CFT-GA), (3) the loss proposed in [40], referred to as WMW loss (CFT-WMW), and (4) AUC margin loss (CFT-AUCM) [44]. WMW and AUC margin losses are particularly designed for AUC maximization.

For CFT-GA, we use the GA implementation of PyGAD [14]. The number of generations is 500. An individual represents the parameters of the LR, where one position of the individual represents one parameter. Decoding is the inverse operation of encoding. The number of individuals is 30. All individuals are initialized by encoding the original parameters. The fitness function is set to be the training mAUC. Roulette wheel selection is used to select 14 individuals as parents. 10 of the parents are additionally kept as individuals in the next generation. 2-point crossover is used with a probability of 80%. Mutation probability is set to be 2%. When a mutation occurs, 1% of the positions are mutated by adding random scalars drawn from $[-0.02, 0.02]$. The individual that attains the highest fitness score at every generation is validated instead of all individuals to reduce the risk of overfitting. The individual that achieves the highest validation mAUC is decoded and saved. For CFT-WMW, stochastic gradient descent ($lr = 1 \times 10^{-3}, momentum = 0.9$) with batch size 32768 is used due to memory lack. For CFT-AUCM , we follow the original paper [44] to use PESG ($lr = 1 \times 10^{-2}, margin = 1$) [16]. Full batch size is used.

Table 2 shows all methods successfully improve the AUCs on all 5 categories. Particularly, GA is the most effective (mAUC +1.9%), followed by AUCM loss (+1.6%), WMW loss (+1.4%). BCE loss is the least effective (+0.3%).

Although WMW and AUCM losses are designed for AUC maximization, they are less effective than GA, probably they rely on backpropagation which is easy to trap on local optima. On the other hand, GA can directly optimize AUC and is easier to escape from local optima. BCE loss is the least éffective, as minimizing BCE can lead to AUC drops. *E.g.*, on "Atelectasis" category (Fig. 2).

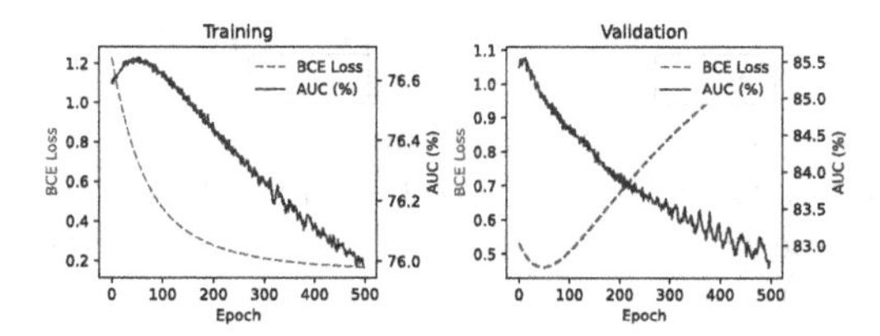

Fig. 2. Learning curves of using BCE loss to fine-tune the LR on Atelectasis. Minimizing BCE loss can decrease AUC.

Table 3. Greedy selection for exploiting uncertain labels, AUC%.

Method	Ate	Car	Con	Ede	P.E	*Mean*
Unknown	**88.8**	**88.6**	94.5	93.0	92.7	*91.5*
Positive	88.6	87.9	93.8	**93.1**	**92.8**	*91.3*
Negative	85.5	88.2	**95.6**	93.0	92.4	*90.9*
Greedy	**88.8**	**88.6**	**95.6**	**93.1**	**92.8**	***91.8***

Greedy Selection for Exploiting Uncertain Labels. In the above ablation studies, treating uncertain labels as unknown may be sub-optimal, as previous studies in this dataset show that treating uncertain labels as positive tends to achieve higher performance [31]. Therefore, we compare three methods for handling uncertain labels with CFT-GA: treat as unknown labels (same as in ablation studies), positive labels [21], and negative labels [21].

Table 3 shows that different categories prefer different methods. Hence, we use the greedy selection to select the best method for each LR, eventually achieving mAUC 91.8%, which is +2.2% higher than *Baseline* mAUC 89.6%. In the following comparison section, we refer to this model as *CFT-GA-Greedy*.

Table 4. Comparison to other state-of-the-art approaches on the **test** set, AUC%.

Model Type	Rank	Approach	Ate	Car	Con	Ede	P.E	*Mean*
Single Model	147	Chong *et al.* [7]	85.67	89.30	82.15	90.92	95.56	*88.72*
	151	Multiview (R-50) [22]	85.60	90.85	81.07	89.45	95.85	*88.60*
	134	Multiview (D-121) [22]	86.49	**90.95**	83.99	89.62	**96.34**	*89.50*
	127	PTRN + Single Model [6]	85.66	89.06	86.89	90.94	95.47	*89.61*
	53	**CFT-GA-Greedy**	**88.58**	90.20	**90.99**	**93.06**	96.26	***91.82***
Ensemble	102	PTRN + Ensemble [6]	85.73	89.90	90.57	91.66	95.04	*90.58*
	98	Stanford Baseline [21]	85.50	89.77	89.76	91.56	96.67	*90.65*
	5	YWW [42]	88.18	**93.96**	**93.43**	92.72	96.15	*92.89*
	2	Hierarchical Learning [31]	90.13	93.18	92.11	92.89	**96.68**	*93.00*
	1	DAM [44]	88.65	93.72	93.21	93.00	96.64	*93.05*
	-	**CFT-GA-Greedy-Ensemble**	**91.52**	93.73	91.57	**93.33**	96.50	***93.33***

Comparison to State-of-the-art Approaches. We compare CFT-GA-Greedy to other state-of-the-art approaches on the **test** set. Most approaches treat unknown labels as negative labels, hence can be considered as strong baselines of *Negative* mode for the comparison. Table 4 shows the comparison.

Single Model. We submitted CFT-GA-Greedy to the competition server for official evaluation. It achieves mAUC 91.82% which is the highest single model AUC in the leaderboard and literature, to the best of our knowledge.

Ensemble. We build an ensemble composed of CFT-GA-Greedy and another 4 CNNs developed by our proposed methods, referred to as *CFT-GA-Greedy-Ensemble*. Similar to 2nd on the competition leaderboard [31], we use test time augmentation [36] for more robust predictions: scale ±5%, rotate ±5°, translate ±5°. Since the competition was suddenly closed, our ensemble cannot be submitted for the official evaluation. After the test set was released and can be downloaded, we evaluate our ensemble on a local machine. Our ensemble achieves mAUC 93.33% which superiors the best in the leaderboard and literature, to the best of our knowledge.

4.2 Partially Labeled Versions of MS-COCO

Dataset. MS-COCO [28] (2014 split) is a standard MLC dataset comprising 80 categories. The training and the validation sets consist of around 80k and 40k image samples, respectively. We follow the work on MS-COCO (*e.g.*, [34]) to use mean AP (mAP) as the performance metric.

As the training data is fully labeled, different schemes of partial labels can be simulated by dropping some labels. Particularly, we study our methods under the proportions of known labels of 10%, 20%, ..., 90%, respectively. To simulate these schemes, we randomly drop 90%, 80%, ..., 10% of labels, respectively.

Table 5. Results on **partially labeled versions** of MS-COCO dataset. In mAP %. "Average" column is the average mAP over label proportions 10% to 90%. (**Bolded** is the best, <u>underlined</u> is the 2nd best)

Method	10%	20%	30%	40%	50%	60%	70%	80%	90%	*Average*
Baseline	54.8	63.1	68.9	72.0	74.1	75.9	77.9	79.2	80.6	71.84
CFT-BCE-simu	54.7	63.1	68.9	73.0	74.9	76.7	78.4	79.6	80.6	72.20
CFT-BCE-Negative	56.6	65.3	70.4	73.4	75.3	77.0	78.8	80.0	81.3	73.11
CFT-BCE	<u>59.3</u>	<u>67.7</u>	**72.6**	<u>75.0</u>	<u>76.6</u>	<u>78.2</u>	<u>79.6</u>	<u>80.7</u>	<u>81.6</u>	<u>74.58</u>
CFT-GA	57.4	65.6	71.0	73.8	75.6	77.4	79.1	80.4	81.4	73.52
CFT-Greedy	**59.3**	**67.7**	**72.6**	**75.0**	**76.6**	**78.2**	**79.7**	**80.8**	**81.7**	**74.61**

Baseline Training. We follow most of the settings of [34] to train *Baseline*, as they achieved state-of-the-art CNN on the original MS-COCO (*i.e.*, fully labeled). *Baseline* is a TResNet-L [35] with an input resolution 448^2 . The parameters trained on ImageNet are used as the initial parameters. *Negative* mode is used to handle the unknown labels. We use batch size 8, asymmetric loss [34], and Adam ($lr = 2 \times 10^{-4}$) to update the parameters. We use AutoAugment [9] with pretrained ImageNet policy as the data augmentation method. Normalization of mean 0 and variance 1 is applied to the input images. The checkpoint that achieves the highest validation mAP is saved. The performance of *Baseline* under different label proportions are reported in Table 5.

Ablation Study on CFT. We apply CFT to *Baseline* to improve its performance. The default BCE loss is used to fine-tune each LR, referred to as *CFT-BCE*. Similar to the experiments on CheXpert, we also study the two variants of CFT-BCE: CFT-BCE-simu and CFT-BCE-Negative. Full-batch gradient descent ($lr = 1 \times 10^{-2}, monentum = 0.9$) is used and the number of epochs is 5000.

CFT-BCE improves the average mAP by 2.74%, CFT-BCE-simu improves 0.36%, and CFT-BCE-Negative improves 1.27%. Both variants are less effective than CFT-BCE, demonstrating the effectiveness of one-by-one fine-tuning and *Ignore* mode.

Noteworthy, CFT-BCE-Negative does not use *Ignore* mode. Although it is less effective than using *Ignore* mode, it still can improve the average mAP. It implies that this improvement is likely to be gained from somewhere else instead of from reducing the performance decreases caused by the wrong labels of *Negative* mode during training. Therefore, CFT may be able to improve models trained with fully labeled data, which requires further investigation.

Ablation Study on GA. We compare GA to the default BCE loss (used in above) for fine-tuning each LR. The number of generations is 2000. The population size is 50. All the individuals of the initial population are encoded from the original parameters. The best individual of the current generation is selected as one individual of the next generation. The parents are selected using roulette wheel selection. During crossover, 20% of the positions of two parents are randomly switched to produce offspring. Each offspring has a 50% chance of being mutated by adding a random scalar between $[-0.001, 0.001]$ to each position.

GA improves the average mAP by 1.68%. However, it is generally less effective than BCE loss (2.74%). The key reasons may be (1) minimizing BCE does not necessarily lead to AP drops, and (2) BCE loss relies on backpropagation which is generally more efficient than GA.

Greedy Selection. We use greedy selection for choosing the best optimization methods between BCE loss and GA for each LR, referred to as *CFT-Greedy*. CFT-Greedy improves the average mAP by 2.77%, which is further higher than CFT-BCE by 0.03%. It implies that the greedy selection has chosen GA for the fine-tuning of a small proportion of LRs.

5 Conclusion

In this paper, we propose a new post-training method called CFT which one-by-one fine-tunes the LRs in a model trained with *Negative* mode to improve its classification performance of each category independently further. Two optimization methods (BCE loss and GA) are tested for fine-tuning LRs. The effectiveness is evaluated on the CheXpert competition dataset and the partially labeled versions of the MS-COCO standard MLC dataset. Especially, CFT achieves state-of-the-art on the CheXpert dataset (single model AUC 91.82% and ensemble AUC 93.33%, on the test set).

Acknowledgements. This work is supported by Macao Polytechnic University under grant number RP/ESCA-01/2021.

References

1. Ben-Baruch, E., et al.: Multi-label classification with partial annotations using class-aware selective loss. In: Proceedings of the IEEE/CVF Conference on Computer Vision and Pattern Recognition, pp. 4764–4772 (2022)

2. Bucak, S.S., Jin, R., Jain, A.K.: Multi-label learning with incomplete class assignments. In: CVPR 2011, pp. 2801–2808. IEEE (2011)
3. Chen, M., Zheng, A., Weinberger, K.: Fast image tagging. In: International Conference on Machine Learning, pp. 1274–1282. PMLR (2013)
4. Chen, T., Pu, T., Liu, L., Shi, Y., Yang, Z., Lin, L.: Heterogeneous semantic transfer for multi-label recognition with partial labels. arXiv preprint arXiv:2205.11131 (2022)
5. Chen, T., Pu, T., Wu, H., Xie, Y., Lin, L.: Structured semantic transfer for multi-label recognition with partial labels. In: Proceedings of the AAAI Conference on Artificial Intelligence, vol. 36, No. 1, pp. 339–346 (2022)
6. Chong, C.F., Wang, Y., Ng, B., Luo, W., Yang, X.: Image projective transformation rectification with synthetic data for smartphone-captured chest X-ray photos classification. Comput. Biol. Med. **164**, 107277 (2023)
7. Chong, C.F., Yang, X., Ke, W., Wang, Y.: GAN-based Spatial transformation adversarial method for disease classification on CXR photographs by smartphones. In: 2021 Digital Image Computing: Techniques and Applications (DICTA), pp. 01–08. IEEE (2021)
8. Chu, H.-M., Yeh, C.-K., Wang, Y.-C.F.: Deep generative models for weakly-supervised multi-label classification. In: Ferrari, V., Hebert, M., Sminchisescu, C., Weiss, Y. (eds.) ECCV 2018. LNCS, vol. 11206, pp. 409–425. Springer, Cham (2018). https://doi.org/10.1007/978-3-030-01216-8_25
9. Cubuk, E.D., Zoph, B., Mane, D., Vasudevan, V., Le, Q.V.: Autoaugment: learning augmentation strategies from data. In: Proceedings of the IEEE/CVF Conference on Computer Vision and Pattern Recognition, pp. 113–123 (2019)
10. David, O.E., Greental, I.: Genetic algorithms for evolving deep neural networks. In: Proceedings of the Companion Publication of the 2014 Annual Conference on Genetic and Evolutionary Computation, pp. 1451–1452 (2014)
11. Deng, J., et al.: ImageNet: a large-scale hierarchical image database. In: 2009 IEEE Conference on Computer Vision and Pattern Recognition, pp. 248–255. IEEE (2009)
12. Deng, J., Russakovsky, O., Krause, J., Bernstein, M.S., Berg, A., Fei-Fei, L.: Scalable multi-label annotation. In: Proceedings of the SIGCHI Conference on Human Factors in Computing Systems, pp. 3099–3102 (2014)
13. Durand, T., Mehrasa, N., Mori, G.: Learning a deep convnet for multi-label classification with partial labels. In: Proceedings of the IEEE/CVF Conference on Computer Vision and Pattern Recognition, pp. 647–657 (2019)
14. Gad, A.F.: PyGAD: an intuitive genetic algorithm python library. arXiv: 2106.06158 (2021)
15. Gong, Y., Jia, Y., Leung, T., Toshev, A., Ioffe, S.: Deep convolutional ranking for multilabel image annotation. arXiv preprint arXiv:1312.4894 (2013)
16. Guo, Z., Yan, Y., Yuan, Z., Yang, T.: Fast objective & duality gap convergence for nonconvex-strongly-concave min-max problems. arXiv preprint arXiv:2006.06889 (2020)
17. Gupta, A., Dollar, P., Girshick, R.: Lvis: A dataset for large vocabulary instance segmentation. In: Proceedings of the IEEE/CVF Conference on Computer Vision and Pattern Recognition, pp. 5356–5364 (2019)
18. Gupta, J.N., Sexton, R.S.: Comparing backpropagation with a genetic algorithm for neural network training. Omega **27**(6), 679–684 (1999)
19. Huang, G., Liu, Z., Van Der Maaten, L., Weinberger, K.Q.: Densely connected convolutional networks. In: Proceedings of the IEEE Conference on Computer Vision and Pattern Recognition, pp. 4700–4708 (2017)

20. Huynh, D., Elhamifar, E.: Interactive multi-label CNN learning with partial labels. In: Proceedings of the IEEE/CVF Conference on Computer Vision and Pattern Recognition, pp. 9423–9432 (2020)
21. Irvin, J., et al.: CheXpert: a large chest radiograph dataset with uncertainty labels and expert comparison. In: Proceedings of the AAAI Conference on Artificial Intelligence, vol. 33, pp. 590–597 (2019)
22. Jansson, P. et al.: Multi-view automated chest radiography interpretation (2021)
23. Jing, L., Yang, L., Yu, J., Ng, M.K.: Semi-supervised low-rank mapping learning for multi-label classification. In: Proceedings of the IEEE Conference on Computer Vision and Pattern Recognition, pp. 1483–1491 (2015)
24. Kapoor, A., Viswanathan, R., Jain, P.: Multilabel classification using bayesian compressed sensing. In: Advances In Neural Information Processing Systems 25 (2012)
25. Kingma, D.P., Ba, J.: Adam: a method for stochastic optimization. arXiv preprint arXiv:1412.6980 (2014)
26. Kundu, K., Tighe, J.: Exploiting weakly supervised visual patterns to learn from partial annotations. Adv. Neural. Inf. Process. Syst. **33**, 561–572 (2020)
27. Kuznetsova, A., et al.: The open images dataset V4. Int. J. Comput. Vis. **128**(7), 1956–1981 (2020). https://doi.org/10.1007/s11263-020-01316-z
28. Lin, T.-Y., et al.: Microsoft COCO: common objects in context. In: Fleet, D., Pajdla, T., Schiele, B., Tuytelaars, T. (eds.) ECCV 2014. LNCS, vol. 8693, pp. 740–755. Springer, Cham (2014). https://doi.org/10.1007/978-3-319-10602-1_48
29. Mitchell, M.: An Introduction to Genetic Algorithms. MIT press (1998)
30. Montana, D.J., et al.: Training feedforward neural networks using genetic algorithms. In: IJCAI, vol. 89, pp. 762–767 (1989)
31. Pham, H.H., Le, T.T., Tran, D.Q., Ngo, D.T., Nguyen, H.Q.: Interpreting chest X-rays via CNNs that exploit hierarchical disease dependencies and uncertainty labels. Neurocomputing **437**, 186–194 (2021)
32. Pu, T., Chen, T., Wu, H., Lin, L.: Semantic-aware representation blending for multi-label image recognition with partial labels. In: Proceedings of the AAAI Conference on Artificial Intelligence, vol. 36, No. 2, pp. 2091–2098 (2022)
33. Qi, Q., Luo, Y., Xu, Z., Ji, S., Yang, T.: Stochastic optimization of areas under precision-recall curves with provable convergence. Adv. Neural. Inf. Process. Syst. **34**, 1752–1765 (2021)
34. Ridnik, T., et al.: Asymmetric loss for multi-label classification. In: Proceedings of the IEEE/CVF International Conference on Computer Vision, pp. 82–91 (2021)
35. Ridnik, T., Lawen, H., Noy, A., Ben Baruch, E., Sharir, G., Friedman, I.: TResNet: high performance GPU-dedicated architecture. In: proceedings of the IEEE/CVF winter conference on applications of computer vision, pp. 1400–1409 (2021)
36. Simonyan, K., Zisserman, A.: Very deep convolutional networks for large-scale image recognition. arXiv preprint arXiv:1409.1556 (2014)
37. Storn, R., Price, K.: Differential evolution-a simple and efficient heuristic for global optimization over continuous spaces. J. Global Optim. **11**(4), 341–359 (1997). https://doi.org/10.1023/A:1008202821328
38. Vasisht, D., Damianou, A., Varma, M., Kapoor, A.: Active learning for sparse bayesian multilabel classification. In: Proceedings of the 20th ACM SIGKDD International Conference on Knowledge Discovery and Data Mining, pp. 472–481 (2014)
39. Wu, B., Lyu, S., Ghanem, B.: ML-MG: multi-label learning with missing labels using a mixed graph. In: Proceedings of the IEEE International Conference on Computer Vision, pp. 4157–4165 (2015)

40. Yan, L., Dodier, R.H., Mozer, M., Wolniewicz, R.H.: Optimizing classifier performance via an approximation to the Wilcoxon-Mann-Whitney statistic. In: Proceedings of the 20th International Conference on Machine Learning (icml-03), pp. 848–855 (2003)
41. Yang, H., Zhou, J.T., Cai, J.: Improving multi-label learning with missing labels by structured semantic correlations. In: Leibe, B., Matas, J., Sebe, N., Welling, M. (eds.) ECCV 2016. LNCS, vol. 9905, pp. 835–851. Springer, Cham (2016). https://doi.org/10.1007/978-3-319-46448-0_50
42. Ye, W., Yao, J., Xue, H., Li, Y.: Weakly supervised lesion localization with probabilistic-cam pooling. arXiv preprint arXiv:2005.14480 (2020)
43. Yu, H.F., Jain, P., Kar, P., Dhillon, I.: Large-scale multi-label learning with missing labels. In: International Conference on Machine Learning, pp. 593–601. PMLR (2014)
44. Yuan, Z., Yan, Y., Sonka, M., Yang, T.: Large-scale robust deep AUC maximization: A new surrogate loss and empirical studies on medical image classification. In: Proceedings of the IEEE/CVF International Conference on Computer Vision, pp. 3040–3049 (2021)

DTSRN: Dynamic Temporal Spatial Relation Network for Stock Ranking Recommendation

Yuhui Zhong, Jiamin Chen, Jianliang Gao(✉), Jiaheng Wang, and Quan Wan

Central South University, Changsha, Hunan, China
{zhongyuhui,chenjiamin,gaojianliang,jhincs,wanquan}@csu.edu.cn

Abstract. The recommendation of stock ranking has always been a challenging task in the financial technology (FinTech) field. Achieving an excellent stock ranking result in stock ranking recommendation (SRR) depends on mining the temporal relations within the stock and the complex spatial relations among the stocks. However, existing studies only consider the temporal relation features of stocks or introduce noise when extracting spatial relation features, which limits the performance of stock ranking recommendation tasks. To address this challenge, we propose the **D**ynamic **T**emporal **S**patial **R**elation **N**etwork (DTSRN), which constructs a spatial relation graph with dynamic stock temporal relation features and extracts dynamic spatial relation features from different views for the stock ranking recommendation. Specifically, we construct learnable global-view and multi-view spatial graphs with stock temporal relation features and then employ efficient graph convolution operations to obtain the final stock representation. We extensively evaluate our method on two real-world datasets and compare it with several state-of-the-art approaches. The experimental results show that our proposed method outperforms the state-of-the-art baseline methods.

Keywords: Temporal spatial relation networks · Graph neural networks · Stock ranking recommendation

1 Introduction

The recent development of financial markets has drawn attention to stock ranking recommendations due to their potentially high returns [18,21]. However, stock ranking recommendations are challenging due to the complex temporal relations within the stock and spatial relations among stocks.

Early research [1,30] treated multiple stock ranking recommendation problems as independent time series forecasting problems with stock temporal feature extraction, ignoring the spatial relation features among stocks, which limits

Y. Zhong and J. Chen—Equal contribution.

B. Luo et al. (Eds.): ICONIP 2023, CCIS 1965, pp. 346–359, 2024.
https://doi.org/10.1007/978-981-99-8145-8_27

the performance of the stock ranking recommendation task. However, in reality, stocks are interrelated. There is a rich signal in the relationship between stocks (or companies) [16]. For example, there is a spatial correlation in stock price trends between companies that are related, as shown in Fig. 1, Kweichow Moutai Company Limited (MOUTAI Inc) and Wuliangye Yibin Company Limited (WULIANGYE Inc), which belong to the same industry, have similar long-term stock price trends. In addition, the stock changes of suppliers interact with the stock price changes of their upstream and downstream companies [8], such as BYD Inc. and Ganfeng Lithium Group Co., Ltd (GLG Inc), Contemporary Amperex Technology Company Limited (CATL Inc), and Guangzhou Tinci Materials Technology Co., Ltd (GTMT Inc). Since there are correlations between companies in the real world, they have a corresponding spatial correlation in the stock price feature space. Mining the spatial relationships between different stock price time series effectively improves the performance of the stock ranking recommendation. To incorporate stock spatial relation features, He et al. [10] proposed a static-dynamic graph neural network (SDGNN) to extract spatial relation features and improve the performance of the stock ranking recommendations. Unfortunately, SDGNN constructs a dense stock static graph using custom stock ID information to extract static spatial relation features without using practical temporal relation features of stock. The process introduces non-negligible noise in spatial feature extraction, causing a performance loss for stock ranking recommendations.

To address this problem, we propose the **D**ynamic **T**emporal **S**patial **R**elation **N**etwork (DTSRN) method. DTSRN utilizes the gate recurrent unit network (GRU) to capture the temporal relation feature matrix of stocks. Next, it constructs a dynamic global-view spatial relation graph using a global similarity matrix with the stock temporal relation feature matrix. DTSRN also creates a dynamic multi-view spatial relation graph by applying a self-attention mechanism to the stock temporal relation feature matrix. Then, the stock dynamic global-view and multi-view spatial relation features are extracted using an efficient graph convolution operation. Finally, DTSRN fuses the dynamic global-view and multi-view spatial relation features for stock ranking recommendations. We validate the performance of our method on multiple real datasets. Our method can outperform all baseline methods, demonstrating the superiority of DTSRN. The main contributions of this work are summarized as follows:

- We intuitively reveal the spatial correlation between stocks in the feature space and analyze the limitations of existing spatial relation feature extraction methods for the stock ranking recommendations.
- We propose the DTSRN, which can simultaneously mine the spatial relation feature of stocks from the global view and local multi-view based on the practical stock relation feature for stock ranking recommendations.
- We have conducted extensive experiments on multiple real-world datasets. The experimental results show that our proposed DTSRN method obtains the best stock ranking recommendation performance compared to the baseline method.

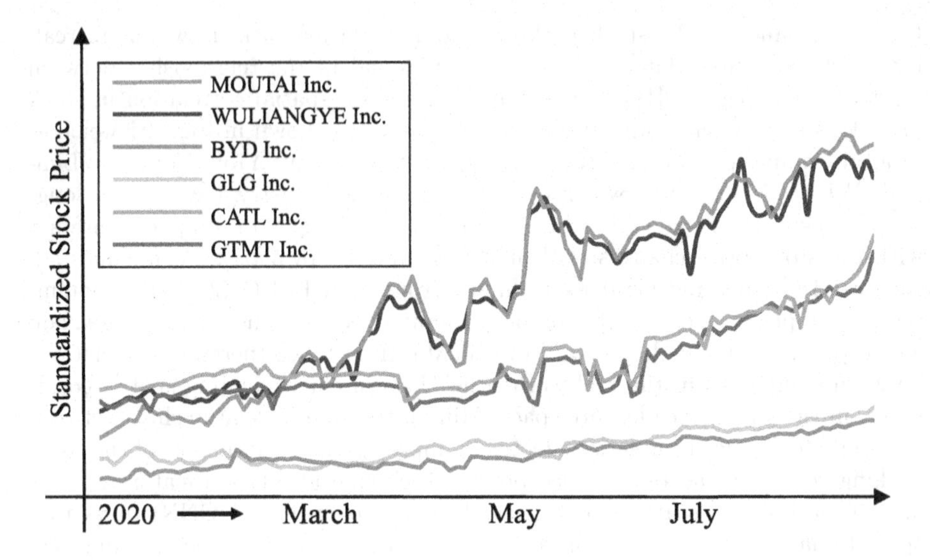

Fig. 1. An example of a correlation of stock price trends between different companies. The Y-axis represents the standardized stock price, and the X-axis represents time. MOUTAI Inc and WULIANGYE Inc belong to the same industry. BYD Inc, GLG Inc, CATL Inc, and GTMT Inc have upstream and downstream industry chain relationships.

2 Related Work

Our work is directly related to recent efforts in utilizing temporal-spatial relations for stock ranking recommendations and dynamic graphs for stock ranking recommendations.

2.1 Stock Ranking Recommendation with Temporal-Spatial Relations

Wang et al. [23] introduced the Deep Co-investment Network Learning (DeepCNL) method. Feng et al. [8] presented a novel deep-learning solution called Relational Stock Ranking (RSR) for the stock ranking recommendation. Existing methods fail to adequately capture temporal trends because the recurrent neural networks (RNNs) or convolutional neural networks (CNNs) employed in these methods cannot capture long-range temporal sequences. Wu et al. [24] propose Graph WaveNet, a novel graph neural network architecture for spatial-temporal graph modeling to address these limitations. To this end, Sawhney et al. [19] introduce STHGCN: Spatio-Temporal Hypergraph Convolution Network, which is the first neural hypergraph model for stock trend forecasting. Zhou et al. [31] propose a generic time series forecasting framework called Dandelion. This framework utilizes the consistency of multiple modalities and explores the relationships among multiple tasks using a deep neural network. Cheong et al. [5]

present the STCNN-RN model, a spatiotemporal convolutional neural network-based relational network that learns complex correlations among multiple financial time-series datasets. They utilize genetic algorithms with a constrained gene to identify outlier time points for companies by fitting the STCNN-RN model and subsequently use these outlier points to detect abnormal situations. Wang et al. [22] propose the Hierarchical Adaptive Temporal-Relational Network (HATR) as a novel method for characterizing and predicting stock evolutions. However, this approach fails to respond effectively to the dynamic changes in relational graphs. Therefore, Xiang et al. [25] propose a temporal and heterogeneous graph neural network-based (THGNN) approach to effectively learn the dynamic relations among price movements in financial time series.

2.2 Stock Ranking Recommendation with Dynamic Graph

Liu et al. [14] conducted a study on anticipating the stock market of renowned companies using a knowledge graph approach. The objective is to develop a model that predicts stock price movements by leveraging a knowledge graph constructed from the financial news of renowned companies. Chen et al. [3] refer to the regression problem involving multiple inter-connected data entities as "dynamic network regression". The primary aim of Matsunaga et al. [15] is to evaluate the effectiveness of this approach across diverse markets and longer time horizons through backtesting using a rolling window analysis. To address this objective, Sawhney et al. [19] propose STHGCN: Spatio-Temporal Hypergraph Convolution Network, which is the first neural hypergraph model for forecasting stock trends. Patil et al. [17] conducted a study on stock market recommendations using an ensemble of graph theory, machine learning, and deep learning models. They proposed a novel approach based on graph theory. Ying et al. [28] propose TRAN, a time-aware graph relational attention network, for a stock recommendation based on return ratio ranking. Yoon et al. [29] propose Anom-Rank, an online algorithm for detecting anomalies in dynamic graphs. Cheng et al. [4] propose AD-GAT, an attribute-driven graph attention network, to tackle the challenges of modeling momentum spillovers. Cai et al. [2] propose StrGNN, an end-to-end structural temporal Graph Neural Network model designed for detecting anomalous edges in dynamic graphs. Gao et al. [9] consider the relationships between stocks (corporations) through a stock relation graph.

3 Preliminary

This section will introduce some definitions of our work and the problem of stock ranking recommendations.

3.1 Stock Series Data

We denote the set of N stocks as $S = \{S_1, S_2, \cdots, S_i, \cdots, S_N\}$. For each stock S_i, the historical series $X_t^i \in R^{L \times U}$ includes features such as opening price, closing price, high price, low price, volume, etc., where L is the length of the series and U is the dimension of the original stock series features. We use $X_t = \{X_t^1, X_t^2, \cdots, X_t^i, \cdots, X_t^N\}$ to represent the set of original stock series features corresponding to the stocks.

3.2 Stock Ranking Recommendation

The stock ranking score y_{t+1}^i for stock S_i on date $t+1$ is defined as the percentage change in the closing price of the stock between day t and $t+1$. The closing price change can be calculated as follows:

$$y_{t+1}^i = \frac{Price_{t+1}^i - Price_t^i}{Price_t^i}, \tag{1}$$

where the stock ranking score y_{t+1}^i represents the label of our recommendations task. $Price_t^i$ is the closing price of stock S_i at date t, and y_{t+1}^i is the rate of change in the closing price of stock S_i at date $t+1$. Our objective is to learn a function f that takes the stock series X_t as inputs and outputs the stock ranking score $\hat{y}_{t+1}^i$ of stock S_i at date $t+1$. We rank all stocks in descending order based on the stock score $\hat{y}_{t+1}^i$ to obtain a ranking set $\hat{Y}_{t+1} = \{\hat{y}_{t+1}^1, \hat{y}_{t+1}^2, \cdots, \hat{y}_{t+1}^i, \cdots, \hat{y}_{t+1}^N\}$ for date $t+1$. Finally, we select the top n ranked stocks as recommended stocks based on the ranking set $\hat{Y}_{t+1}$.

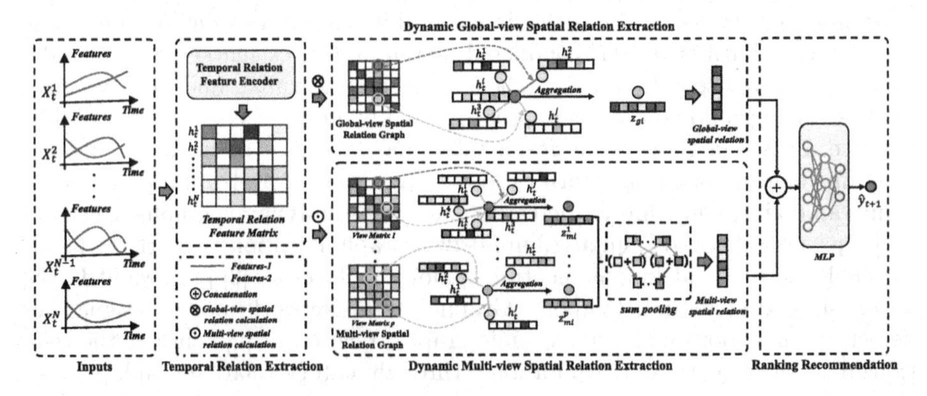

Fig. 2. The overall framework of the proposed DTSRN. The DTSRN contains two paths, the first one is the dynamic global-view spatial relation extraction path, this process utilizes a global similarity matrix with the stock temporal relation feature matrix to construct a dynamic global-view spatial relation graph. The second is a dynamic multi-view spatial relation extraction path, this process utilizes the self-attention mechanism to construct a dynamic multi-view spatial relation graph among stocks based on their temporal features. By employing efficient Graph Convolution Operations (GCO), DTSRN extracts distinct relation features of stock from the dynamic global-view and multi-view spatial relation graph. Finally, DTSRN concatenates global-view and multi-view spatial relation features as perceptron inputs for stock ranking recommendation tasks.

4 Methodology

Figure 2 depicts our proposed model, DTSRN, which comprises various modules, such as temporal feature extraction, dynamic global-view and multi-view spatial relation graph construction, and relation feature fusion. In this section, we elaborate on each module in detail.

4.1 Temporal Relation Extraction

The GRU network is utilized for encoding the stock series and extracting their temporal relation features. To be specific, we set the sliding window size to m and the sliding step size to n in this study. Subsequently, for a sliding window of the time series, $X_t^i = \{x_{t-m}^i, \cdots, x_t^i\}$ of stock S_i, X_t^i is fed as input to the GRU network, and the temporal relation feature representation of stock S_i is obtained at the time t. The calculation formula is as follows.

$$h_t^i = GRU\left(h_{t-1}^i, x_t^i\right), \tag{2}$$

where $h_t^i \in R^{U_h}$ is the hidden state of the GRU network for stock series X_t^i at time t, h_{t-1}^i is the hidden state at time $t-1$, and U_h is the hidden cell dimension of the GRU network. The GRU network outputs a temporal relation feature matrix H_t for N stocks in the sliding window stock series X_t^i, where each row represents the temporal relation feature vector h_t^i of stock S_i at time t.

4.2 Dynamic Global-View Spatial Relation Extraction

We use the temporal relation feature matrix H_t of stocks at different moments to extract dynamic spatial relations. The dynamic global-view spatial relation graph A_g is constructed using cosine similarity to measure the relation between stock series at different moments. A_g is an $N * N$ square matrix, where N denotes the number of stocks and element A_t^{ij} denotes the strength of the dynamic spatial relation between stock S_i and stock S_j at moment t, and element A_t^{ij} is calculated as shown below.

$$A_t^{ij} = Cosine\left(h_t^i, h_t^j\right) = \frac{h_t^i \cdot h_t^j}{\left\|h_t^i\right\| \cdot \left\|h_t^j\right\|}, \tag{3}$$

where $h_t^i \in R^{U_h}$ is the temporal relation features of stock S_i at moment t, $\|\cdot\|$ is the Euclidean distance. We use the row normalization operation to normalize the dynamic global-view spatial relation graph A_g. The normalization process is shown in Eq. 4.

$$\lambda_{gt}^{ij} = \frac{exp\left(A_t^{ij}\right)}{\sum_{k=0}^{N} exp\left(A_t^{ik}\right)}, \tag{4}$$

λ_{gt}^{ij} denotes the element of the i-th row and j-th column of the dynamic global-view spatial relation graph A_g after normalization.

4.3 Dynamic Multi-view Spatial Relation Extraction

We use the multi-view self-attention similarity calculation method to mine dynamic multi-view spatial relations [13]. Firstly, we apply an MLP to transform the temporal relation feature matrix H_t from dimension U_h to U_s, resulting in H_t^{mlp}. Next, we divide each row of H_t^{mlp} equally into v views, where each equally divided feature vector has length U_s/v. We then reconstruct the feature matrix H_t^{mlp} into the temporal matrix H_t^{re} with dimensions $(N, v, U_s/v)$. We transform the dimension of H_t^{re} to $(v, N, U_s/v)$ to obtain the multi-view temporal relation matrix Q_v for v views. Finally, we calculate the spatial relation graph A_{view_p} between N stocks for each view using Eq. 5, resulting in different dynamic view spatial relations.

$$A_{view_p} = dropout\left(\frac{Q_{view_p}Q_{view_p}^T}{\sqrt{d}}\right), \tag{5}$$

where Q_{view_p} is a view of the multi-view temporal relation matrix Q_v. $p \in [1, v]$, d is U_s/v. The A_{view_p} are normalized separately using the softmax function to construct v different dynamic multi-view spatial relation graphs A_m. The normalization process is shown in Eq. 6.

$$\lambda_{mt}^{ij} = \frac{exp\left(A_{view_p}^{ij}\right)}{\sum_{k=0}^{N} exp\left(A_{view_p}^{ik}\right)}, \tag{6}$$

λ_{mt}^{ij} denotes the elements of the i-th row and j-th column in the dynamic multi-view spatial relation graph A_m after normalization. Different dynamic multi-view spatial relation graphs (A_m) can provide a more comprehensive description of the various segmentation patterns between stock series.

4.4 Dynamic Temporal Spatial Relation Aggregation

We utilize efficient graph convolution without learnable parameters to extract the stock spatial relation features from the temporal relation feature matrix H_t based on the dynamic global-view spatial relation graph A_g and different multi-view spatial relation graphs A_m. The aggregation process can be described using the following equation:

$$z_{gi} = \sum_{k=0}^{N} \lambda_{gt}^{ik} h_t^k, \tag{7}$$

z_{gi} is the i-th row feature of the dynamic global-view spatial relation matrix Z_g, λ_{gt}^{ik} denotes the i-th row k-th column element of the dynamic global-view spatial relation graph A_g, and h_t^k is the k-th row stock temporal relation feature vector of the temporal relation feature matrix H_t.

$$z_{mi}^p = \sum_{k=0}^{N} \lambda_{mt}^{ik} h_t^k, \tag{8}$$

z_{mi}^p is the i-th row feature vector of the p-th dynamic multi-view spatial relation feature matrix Z_m^p, $p \in [1, v]$ and λ_{mt}^{ik} denotes the i-th row k-th column element of the multi-view spatial relation graph A_m. To capture the overall relationship between different stocks over time, we fuse v different dynamic multi-view spatial relation feature matrices using the sum pooling method to obtain the dynamic multi-view spatial relation feature matrix Z_m

4.5 Ranking Recommendation

Stock Ranking Score: The stock ranking recommendation is a regression task. To fuse the dynamic global-view spatial relation feature vector $z_{gi} \in \mathbb{R}^{U_h}$ with the dynamic multi-view spatial relation feature vector $z_{mi} \in \mathbb{R}^{U_h}$, DTSRN uses the concatenate operation to obtain the dense representation $h \in \mathbb{R}^{2U_h}$ of the stock in the ranking recommendation module. Then, the multi-layer perception (MLP) as the activation function is used to calculate the recommendations label stock ranking score $\hat{y}_{t+1} \in \mathbb{R}$. The calculation of the recommendations label stock ranking score is shown below.

$$\hat{y}_{t+1} = \left(ReLU\left(\left[z_{gi}; z_{mi}\right] W_1^{'} + b_1^{'}\right)\right) W_2^{'} + b_2^{'}, \tag{9}$$

where $W_1^{'} \in \mathbb{R}^{2U_h \times U_h}$ and $W_2^{'} \in \mathbb{R}^{U_h}$ are the learnable parameters of the fully connected layer; $b_1^{'} \in \mathbb{R}^{U_h}$ and $b_2^{'} \in \mathbb{R}$ are the learnable biases. The [;] is represented as a concatenate operation.

Loss Function: Giving the true value y_{t+1}^i, y_{t+1}^j and predicted ranking score $\hat{y}_{t+1}^i$, $\hat{y}_{t+1}^j$ if $y_{t+1}^i > y_{t+1}^j$, then we expect $\hat{y}_{t+1}^i > \hat{y}_{t+1}^j$, therefore, we combine point-by-point regression loss and pairwise ranking-aware loss as follows.

$$\begin{aligned} L\left(\hat{y}_{t+1}, y_{t+1}\right) = & \left|\left|\hat{y}_{t+1} - y_{t+1}\right|\right|^2 \\ & + \mu \sum_{i=0}^{N} \sum_{j=0}^{N} max\left(0, -\left(\hat{y}_{t+1}^i - \hat{y}_{t+1}^j\right)\left(y_{t+1}^i - y_{t+1}^j\right)\right), \end{aligned} \tag{10}$$

where μ is the hyperparameter that balances the two loss terms.

5 Experiments

In this section, we further conduct extensive experiments on real stock market data to evaluate the validity of our method.

5.1 Datasets

Our experiments use historical stock series data, and the datasets are published on the open-source quantitative investment platform Qlib[1] [27]. Table 1 shows the detailed statistics of these datasets.

[1] https://github.com/microsoft/qlib.

Stock Datasets: We evaluate two popular and representative stock pools: CSI 100 and CSI 300. The CSI 100 and CSI 300 are the largest 100 and 300 stocks in the China A-share Market. The CSI 100 index reflects the performance of the most influential large-cap stock market, and the CSI 300 index reflects the overall performance of the China Ashare market.

Stock Features: We use the stock characteristics of Alpha360 in Qlib. Alpha360 dataset contains six stock data per day, including the opening price, closing price, high price, low price, volume-weighted average price (VWAP), and volume. Alpha360 looks back 60 d to construct 360-dimensional historical stock data as the stock characteristics of the stock. We use the characteristics of CSI 300 and CSI 100 stocks from January 1, 2007, to December 31, 2020, and split them chronologically to obtain the training set (January 2007 to December 1, 2014), the validation set (January 2015 to December 3, 2016) and the test set (January 2017 to December 2020). We used three years of test data, which is sufficient to validate the stability of DTSRN.

Table 1. Statistics of the Datasets

Datasets	CSI100/CSI300
Train(Tr) Period	01/01/2007-12/31/2014
Val(V) Period	01/01/2015-12/31/2016
Test(Te) Period	01/01/2017-12/31/2020

5.2 Experimental Settings

Evaluation Metrics: based on a previous study [26], we assessed the results of the stock ranking recommendation by two widely used metrics: **Information Coefficient (IC)** and **Rank IC**, defined as follows.

$$IC\left(y^{t}, \hat{y}^{t}\right) = corr\left(y^{t}, \hat{y}^{t}\right), \tag{11}$$

$$Rank\,IC\left(y^{t}, \hat{y}^{t}\right) = corr\left(rank_{y^{t}}, rank_{\hat{y}^{t}}\right), \tag{12}$$

where IC is the Pearson correlation coefficient between the labels and recommendations, $Rank\,IC$ is Spearman's rank correlation coefficient. $corr\,(\cdot)$ is the Pearson's correlation coefficient, $rank_{y^t}$ and $rank_{\hat{y}^t}$ are the rankings of labels and recommendations respectively from high to low. We use the average daily IC and $Rank\,IC$ to evaluate the results of stock ranking forecasts.

Moreover, for a stock ranking recommendations model, the accuracy of its first N recommendations is more important for real stock investments. Therefore, we introduce another evaluation metric, **precision@N**, to evaluate the accuracy of the model's top N recommendations. For example, if N is 30 and 15 of the top 30 recommendations have positive labels, the precision@30 is 0.5. We evaluated the model with N values of 3, 5, 10, and 30 for comparison with existing studies, and the detailed evaluation results are presented in Table 2.

Parameter Settings: Our model was implemented using the PyTorch framework and optimized with Adam at a learning rate of 0.0004. We performed a grid search for the parameters to obtain the optimal hyperparameters. Specifically, we searched for the optimal size of the view v in the multi-view spatial relation extraction and the number of hidden units U_h in the GRU by exploring the ranges [2, 3, 4, 5, 6] and [16, 32, 64, 128], respectively. We also adjusted the value of μ in the loss function between 0.1 and 0.5. The model was trained for 100 epochs.

5.3 Baselines Methods

To demonstrate the effectiveness of our method, we compare it with previous baseline methods, including:

- **MLP:** a three-layer multilayer perceptron (MLP) with 512 units on each layer.
- **SFM** [30]: a recurrent neural network (RNN) variant that decomposes hidden states into multiple frequency components to model multiple frequency patterns.
- **GRU** [6]: a stock ranking recommendations method based on gated recursive unit (GRU) networks.
- **LSTM** [11]: a long short-term memory (LSTM) network suitable for dealing with time series forecasting problems.
- **ALSTM** [7]: a variant of LSTM with a time-concerned aggregation layer for aggregating information from all hidden states in previous timestamps.
- **ALSTM+TRA** [12]: an ALSTM extension that uses a temporal routing adapter (TRA) to model multiple transaction patterns.
- **GATs** [20]: a predictive model that uses graphical attention networks (GATs) to embed GRU-encoded stocks into stock graphs. We use stocks as nodes to construct stock graphs, and two stocks are related when they share the same predefined concepts.
- **HIST** [26]: a model based on a bipartite graph of stock concepts to handle spatial correlations between stocks.
- **SDGNN** [10]: a stock ranking recommendation model based on a graph learning module to learn static and dynamic graphs.

6 Results and Analysis

Table 2 presents the experimental results for our model and other baselines, with the best results for each metric marked in bold.

Table 2. Evaluation results on the datasets.

Model	CSI100						CSI300					
	IC	Rank IC	Precision@N(↑)				IC	Rank IC	Precision@N(↑)			
	(↑)	(↑)	3	5	10	30	(↑)	(↑)	3	5	10	30
MLP	0.071	0.067	56.53	56.17	55.49	53.55	0.082	0.079	57.21	57.10	56.75	55.56
SFM	0.081	0.074	57.79	56.96	55.92	53.88	0.102	0.096	59.84	58.28	57.89	56.82
GRU	0.103	0.097	59.97	58.99	58.37	55.09	0.113	0.108	59.95	59.28	58.59	57.43
LSTM	0.097	0.091	60.12	59.49	59.04	54.77	0.104	0.098	59.51	59.27	58.40	56.98
ALSTM	0.102	0.097	60.79	59.76	58.13	55.00	0.115	0.109	59.51	59.33	58.92	57.47
ALSTM+TRA	0.107	0.102	60.27	59.09	57.66	55.16	0.119	0.112	60.45	59.52	59.16	58.24
GATs	0.096	0.090	59.17	58.71	57.48	54.59	0.111	0.105	60.49	59.96	59.02	57.41
HIST	0.120	0.115	61.87	60.82	59.38	56.04	0.131	0.126	61.60	61.08	60.51	58.79
SDGNN	0.126	0.120	62.49	61.41	59.81	56.39	0.137	0.132	62.23	61.76	61.18	59.56
DTSRN	**0.137**	**0.132**	**62.85**	**61.79**	**60.68**	**56.84**	**0.146**	**0.141**	**62.72**	**62.03**	**61.37**	**59.74**

6.1 Overall Performance

Table 2 highlights the importance of IC and Rank IC metrics in the stock ranking recommendation, and our model outperforms all baseline methods on both metrics. For instance, in the CSI 100 and CSI 300 datasets, IC and Rank IC values are 0.137, 0.132, 0.146, and 0.141, respectively. Moreover, our model achieves the highest Precision@N metric score of 62.85% and 62.72%, respectively. Generally, the precision@N value tends to be below 50%. The experimental results validate the effectiveness of DTSRN in leveraging the temporal relation features of stocks to construct dynamic spatial relation graphs from the global view and local multi-view and extract the dynamic spatial relation features of stocks for the stock ranking recommendation.

Table 3. The results of the ablation study. In this table, global-view and multi-view are the dynamic global-view spatial relation extraction pathway and the dynamic multi-view spatial relation extraction pathway, respectively (corresponding to Sects. 4.2 and 4.3). Sum Pooling denotes the last pooling operation module in the multi-view pathway. The symbol of ✓ and ✗ indicate the presence or absence of the component in the variant. The AvgPre is the average of different precision@N values where N are 3,5,10, and 30.

Global-view	Multi-view	Sum Pooling	CSI 100			CSI 300		
			IC(↑)	Rank IC(↑)	AvgPre(↑)	IC(↑)	Rank IC(↑)	AvgPre(↑)
✓	✗	✓	0.131	0.126	60.24	0.140	0.136	61.22
✗	✓	✓	0.135	0.130	60.40	0.144	0.139	61.37
✓	✓	✓	**0.137**	**0.132**	**60.54**	**0.146**	**0.141**	**61.47**

6.2 Ablation Study

Impact of Global-view Spatial Relation: DTSRN-G is a variant of DTSRN that only uses dynamic global spatial relation features for stock ranking recommendation tasks. Compared with the baseline method SDGNN, the DTSRN-G method can obtain performance advantages, which shows that the non-negligible noise introduced by the static relation graph based on the stock ID in SDGNN limits the performance of the stock ranking recommendation (Table 3).

Impact of Multi-view Spatial Relation: DTSRN-M is a variation of DTSRN that preserves the dynamic multi-view spatial relation features while discarding the dynamic global view spatial relation features. By comparing the performance of DTSRN-G and DTSRN-M, we found that the self-attention-based multi-view spatial relation construction method effectively mines segmentation patterns in the original feature space, resulting in better model performance. Our experimental results show that the proposed DTSRN method outperforms other approaches, indicating the effectiveness of the two spatial relation graph construction methods in capturing the spatial relation features of stocks based on temporal relation features, ultimately improving the performance of stock ranking recommendations (Table 3).

7 Conclusion

This paper proposes a novel method called DTSRN for stock ranking recommendation tasks. Two dynamic spatial relation graphs are constructed from the global view and local multi-view to capture the spatial relations among stocks. DTSRN extracts spatial relation features using efficient graph convolution operations from these dynamic spatial relation graphs to improve recommendations performance. Experimental results demonstrate that our model outperforms all baseline models on multiple real-world datasets.

Acknowledgment. This work is supported by the National Natural Science Foundation of China under Grant No.62272487.

References

1. Akita, R., Yoshihara, A., Matsubara, T., Uehara, K.: Deep learning for stock prediction using numerical and textual information. In: Proceedings of the International Conference on Computer and Information Science, pp. 1–6 (2016)
2. Cai, L., Chen, Z., Luo, C., Gui, J., Ni, J., Li, D., Chen, H.: Structural temporal graph neural networks for anomaly detection in dynamic graphs. In: Proceedings of the ACM International Conference on Information & Knowledge Management, pp. 3747–3756 (2021)
3. Chen, Y., Meng, L., Zhang, J.: Graph neural lasso for dynamic network regression. arXiv preprint arXiv:1907.11114 (2019)
4. Cheng, R., Li, Q.: Modeling the momentum spillover effect for stock prediction via attribute-driven graph attention networks. In: Proceedings of the AAAI Conference on Artificial Intelligence, pp. 55–62 (2021)
5. Cheong, M.S., Wu, M.C., Huang, S.H.: Interpretable stock anomaly detection based on spatio-temporal relation networks with genetic algorithm. IEEE Access **9**, 68302–68319 (2021)

6. Chung, J., Gulcehre, C., Cho, K., Bengio, Y.: Empirical evaluation of gated recurrent neural networks on sequence modeling. arXiv preprint arXiv:1412.3555 (2014)
7. Feng, F., Chen, H., He, X., Ding, J., Sun, M., Chua, T.S.: Enhancing stock movement prediction with adversarial training. arXiv preprint arXiv:1810.09936 (2018)
8. Feng, F., He, X., Wang, X., Luo, C., Liu, Y., Chua, T.S.: Temporal relational ranking for stock prediction. ACM Trans. Inform. Syst. **37**, 1–30 (2019)
9. Gao, J., Ying, X., Xu, C., Wang, J., Zhang, S., Li, Z.: Graph-based stock recommendation by time-aware relational attention network. ACM Trans. Knowl. Discov. Data **16**, 1–21 (2021)
10. He, Y., Li, Q., Wu, F., Gao, J.: Static-dynamic graph neural network for stock recommendation. In: Proceedings of the International Conference on Scientific and Statistical Database Management, pp. 1–4 (2022)
11. Hochreiter, S., Schmidhuber, J.: Long short-term memory. Neural Comput. **9**, 1735–1780 (1997)
12. Lin, H., Zhou, D., Liu, W., Bian, J.: Learning multiple stock trading patterns with temporal routing adaptor and optimal transport. In: Proceedings of the ACM SIGKDD Conference on Knowledge Discovery & Data Mining, pp. 1017–1026 (2021)
13. Liu, J., Chen, S., Wang, B., Zhang, J., Li, N., Xu, T.: Attention as relation: learning supervised multi-head self-attention for relation extraction. In: Proceedings of the International Conference on International Joint Conferences on Artificial Intelligence, pp. 3787–3793 (2021)
14. Liu, Y., Zeng, Q., Ordieres Meré, J., Yang, H.: Anticipating stock market of the renowned companies: a knowledge graph approach. Complexity **2019**, 1–15 (2019)
15. Matsunaga, D., Suzumura, T., Takahashi, T.: Exploring graph neural networks for stock market predictions with rolling window analysis. arXiv preprint arXiv:1909.10660 (2019)
16. Nobi, A., Maeng, S.E., Ha, G.G., Lee, J.W.: Effects of global financial crisis on network structure in a local stock market. Phys. A **407**, 135–143 (2014)
17. Patil, P., Wu, C.S.M., Potika, K., Orang, M.: Stock market prediction using ensemble of graph theory, machine learning and deep learning models. In: Proceedings of the International Conference on Software Engineering and Information Management, pp. 85–92 (2020)
18. Sawhney, R., Agarwal, S., Wadhwa, A., Derr, T., Shah, R.R.: Stock selection via spatiotemporal hypergraph attention network: a learning to rank approach. In: Proceedings of the AAAI Conference on Artificial Intelligence, pp. 497–504 (2021)
19. Sawhney, R., Agarwal, S., Wadhwa, A., Shah, R.R.: Spatiotemporal hypergraph convolution network for stock movement forecasting. In: Proceedings of the IEEE International Conference on Data Mining, pp. 482–491 (2020)
20. Veličković, P., Cucurull, G., Casanova, A., Romero, A., Lio, P., Bengio, Y.: Graph attention networks. arXiv preprint arXiv:1710.10903 (2017)
21. Wang, H., Hui, D., Leung, C.S.: Lagrange programming neural networks for sparse portfolio design. In: Proceedings of the Neural Information Processing, pp. 37–48 (2023)
22. Wang, H., Li, S., Wang, T., Zheng, J.: Hierarchical adaptive temporal-relational modeling for stock trend prediction. In: Proceedings of the International Joint Conference on Artificial Intelligence, pp. 3691–3698 (2021)
23. Wang, Y., Zhang, C., Wang, S., Philip, S.Y., Bai, L., Cui, L.: Deep co-investment network learning for financial assets. In: Proceedings of the IEEE International Conference on Big Knowledge, pp. 41–48 (2018)

24. Wu, Z., Pan, S., Long, G., Jiang, J., Zhang, C.: Graph wavenet for deep spatial-temporal graph modeling. arXiv preprint arXiv:1906.00121 (2019)
25. Xiang, S., Cheng, D., Shang, C., Zhang, Y., Liang, Y.: Temporal and heterogeneous graph neural network for financial time series prediction. In: Proceedings of the ACM International Conference on Information & Knowledge Management, pp. 3584–3593 (2022)
26. Xu, W., et al.: Hist: a graph-based framework for stock trend forecasting via mining concept-oriented shared information. arXiv preprint arXiv:2110.13716 (2021)
27. Yang, X., Liu, W., Zhou, D., Bian, J., Liu, T.Y.: Qlib: an ai-oriented quantitative investment platform. arXiv preprint arXiv:2009.11189 (2020)
28. Ying, X., Xu, C., Gao, J., Wang, J., Li, Z.: Time-aware graph relational attention network for stock recommendation. In: Proceedings of the ACM International Conference on Information & Knowledge Management, pp. 2281–2284 (2020)
29. Yoon, M., Hooi, B., Shin, K., Faloutsos, C.: Fast and accurate anomaly detection in dynamic graphs with a two-pronged approach. In: Proceedings of the ACM SIGKDD International Conference on Knowledge Discovery & Data Mining, pp. 647–657 (2019)
30. Zhang, L., Aggarwal, C., Qi, G.J.: Stock price prediction via discovering multi-frequency trading patterns. In: Proceedings of the ACM SIGKDD International Conference on Knowledge Discovery and Data Mining, pp. 2141–2149 (2017)
31. Zhou, D., Zheng, L., Zhu, Y., Li, J., He, J.: Domain adaptive multi-modality neural attention network for financial forecasting. In: Proceedings of The Web Conference, pp. 2230–2240 (2020)

Semantic Segmentation of Multispectral Remote Sensing Images with Class Imbalance Using Contrastive Learning

Zhengyin Liang and Xili Wang(✉)

School of Computer Science, Shaanxi Normal University, Xi'an, China
lzyinxx@foxmail.com, wangxili@snnu.edu.cn

Abstract. Affected by the distribution differences of ground objects, multispectral remote sensing images are characterized by long-tailed distribution, that is, a few classes (head classes) contain many instances, while most classes (tail classes or called rare classes) contain only a few instances. The class imbalanced data brings a great challenge to the semantic segmentation task of multispectral remote sensing images. To conquer this problem, this paper proposes a novel contrastive learning method (CoLM) for semantic segmentation of multispectral remote sensing images with class imbalance. Firstly, we propose a semantic consistency constraint to maximize the similarity of semantic feature embeddings of the same class in the feature space, then a rebalancing sampling strategy is proposed to dynamically select the hard-to-predict samples in each class as anchor samples to impose additional supervision, and use pixel-level supervised contrastive loss to improve the separability of rare classes in the decision space. The experimental results on two long-tailed remote sensing datasets show that our method can be easily integrated into existing segmentation models, effectively improving the segmentation accuracy of rare classes without increasing additional inference costs.

Keywords: Semantic Consistency · Rebalance Sampling Strategy · Contrastive Learning · Multispectral Remote Sensing Images · Semantic Segmentation

1 Introduction

Multispectral remote sensing images have a large imaging range and are usually characterized by a long-tailed distribution due to the distribution differences of ground objects in the real world. Long-tailed remote sensing data or referred to as class imbalance data allows head classes containing the majority of instances to dominate the model training process. Thus, the model can fit head-class instances well during inference, while generalization over rare-class instances is weak, thus compromising the overall recognition performance of the model [1]. Li et al. [2] further explored and found that the model trained by class imbalance data will project rare-class instances closer to or even across classifier decision boundaries, resulting in indistinguishable decision boundaries of rare classes while head classes are virtually unaffected. Traditional methods mainly focus on increasing the training instances of rare classes by data resampling [3] or adjusting

B. Luo et al. (Eds.): ICONIP 2023, CCIS 1965, pp. 360–375, 2024.
https://doi.org/10.1007/978-981-99-8145-8_28

the loss weight ratio of rare classes by designing different loss re-weighting schemes to improve the recognition accuracy of rare classes [4], such as Focal loss [5], WCE loss [6], etc., but they also impair the performance of head classes [7]. There are also two-stage decoupled learning methods that improve the decision boundaries of rare classes by retraining the classifier [8, 9], but the two-stage training approach makes the model very sensitive to the choice of hyperparameters, which also increases the training difficulty.

In recent years, contrastive learning has attracted a lot of attention from researchers. Contrastive learning is a type of metric learning [10], which aims to learn a model that can encode the same class of data similarly and make the encoding results of different classes as different as possible. Contrastive learning regards different augmented views of the same sample as a set of positive sample pairs and other samples as their negative samples, with the core idea of aligning positive sample pairs to increase intra-class compactness and excluding negative sample pairs to increase the inter-class distance. Contrastive learning has made tremendous progress on image classification tasks. Some contrastive learning methods, such as SimCLR [11] and MoCo [12], have achieved state-of-the-art performance on several popular benchmarks, showing great potential for feature representation learning and recognition of long-tailed data [13]. Kang et al. [14] found experimentally that contrastive learning can learn a more linearly separable feature space from long-tailed data and improve robustness to class imbalanced data.

Unlike image classification tasks, semantic segmentation as a pixel-level classification task requires the classification of massive pixel instances. Recently, some works have been investigating the introduction of contrastive learning into semantic segmentation to improve the segmentation performance of the model. For example, Wang et al. [15] proposed a cross-image pixel contrastive method to model the global contextual information of pixels in multiple images, which performs better than semantic segmentation methods that focus only on mining local dependencies between pixels in a single image. Li et al. [16] proposed a global-local matching contrastive learning network to enhance the feature embedding similarity of the same class of remote sensing data at the image level and pixel level, respectively. Zhong et al. [17] proposed pixel consistency contrastive loss to enhance the prediction consistency of different augmented views of the original image, enabling the model can extract common features of the same class of data. Different from image-level contrastive learning, pixel-level contrastive learning in semantic segmentation suffers from the challenge of high computational costs due to a large number of pixels that need to be contrasted. To reduce the potential computational costs, Liu et al. [18] proposed to sample a few samples in each class with confidence less than a preset threshold as the anchor samples of this class to impose contrastive supervision. However, the performance of the model is limited by choice of the threshold value. As mentioned above, current works focus on using contrastive learning to improve the representation ability of semantic segmentation models, and these methods are designed based on balanced data. They are not suitable for remote sensing data that are usually characterized by long-tailed distribution in real situations. To address this problem, this paper proposes a novel contrastive learning method (CoLM) for semantic segmentation of multispectral remote sensing images with class imbalance, which can effectively improve the segmentation accuracy of rare classes while not compromising the overall segmentation performance of the model. To the best of our knowledge, in the field of semantic segmentation, this is the first attempt to use the idea of contrastive learning to solve the long-tailed problem of multispectral remote sensing data. Inspired by the principle of label space consistency [17], we propose the semantic consistency constraint to enhance the similarity of semantic feature embeddings of the same class and reduce the distribution bias in the feature space. The fundamental reason for the degradation of model segmentation performance caused by imbalanced data is that head classes containing most instances dominate the training process, i.e., the model focuses more on the loss convergence of head classes and neglects that of the rare classes. Therefore, a rebalancing sampling strategy is proposed to enhance the attention of the model to rare classes, and the supervised contrastive loss [19] is extended to pixel-level classification tasks to improve the separability of rare classes in the decision space.

The main contributions of this paper are summarized as follows:

1. We propose the semantic consistency constraint to make the semantic feature embeddings of the same class more similar by maximizing the semantic consistency between the high-level feature maps of the original image and its augmented view.
2. We develop a rebalancing sampling strategy to strengthen the attention of the segmentation model to rare classes by dynamically sampling the same number of hard-to-predict samples as possible for each class to impose additional contrastive supervision.
3. This paper provides new insights for the study of semantic segmentation of remote sensing images with class imbalance, and the proposed method can be easily integrated into existing segmentation models, which is very flexible and convenient.

2 Methodology

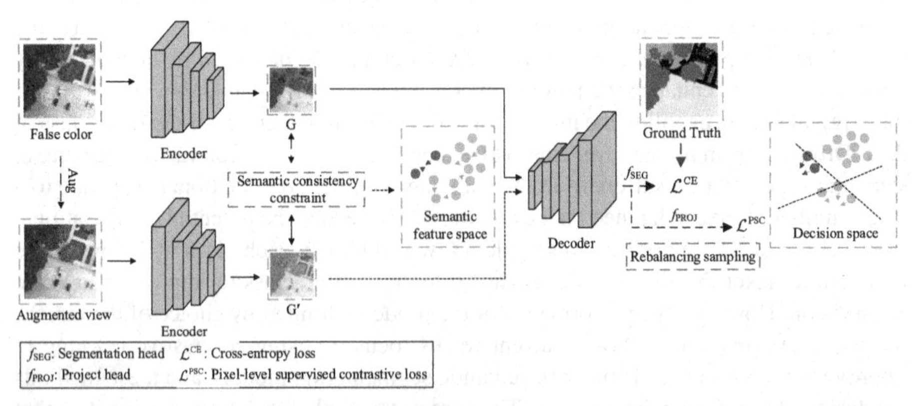

Fig. 1. The framework of the proposed CoLM. The segmentation head f_{SEG} represents the classification layer of the decoder and is used to calculate the cross-entropy loss, and the project head f_{PROJ} consists of a multilayer perceptron for computing the pixel-level supervised contrastive loss.

In this section, we describe the proposed CoLM in detail. Figure 1 illustrates the framework of CoLM, which contains three key components: semantic consistency constraint, rebalancing sampling strategy, and pixel-level supervised contrastive loss ($\mathcal{L}^{\mathrm{PSC}}$). Each part is described separately in the following.

2.1 Semantic Consistency Constraint

The output segmentation masks of different spatially invariant augmented views of the original image should be consistent after passing through the segmentation network. Similarly, the semantic consistency principle should exist in the semantic feature space, i.e., the high-level semantic feature maps extracted from different augmented views after feature extraction by encoder should maintain semantic consistency. Based on semantic consistency theory, we propose a semantic consistency constraint to make the feature embeddings of the same class in the feature space more similar. To improve the generalization ability of the model while preserving the integrity of the original image information as much as possible, we train the model by both the original image and its augmented view instead of learning only the features of different augmented views of the original image as in previous works. The input multispectral remote sensing image is denoted as $\mathrm{X} \in \mathcal{R}^{N \times H \times W}$, where N, H, and W represent the number of spectral bands, height, and width of the image, respectively, and its augmented view is denoted as $\mathrm{X}\prime \in \mathcal{R}^{N \times H \times W}$ (histogram equalization enhancement is used in the experiments and also can be others). The feature maps G and G$\prime$ are obtained separately by the shared parameter encoder (see Fig. 1), after which the semantic consistency constraint is applied to the feature maps. The semantic consistency constraint loss is defined as follows:

$$\mathcal{L}^{\mathrm{SCC}} = -\log\left[\max\left(S\left(g_{(i,j)}, g_{(i,j)\prime}\right)\right)\right] = -\log\left[\max\left(\exp\left(\frac{g_{(i,j)} \cdot g_{(i,j)\prime}}{\left\|g_{(i,j)}\right\| \left\|g_{(i,j)\prime}\right\|}\right)\right)\right] \quad (1)$$

where $S(\cdot)$ denotes the cosine similarity metric function, $g_{(i,j)}$ represents the feature vector with the spatial index (i, j) of the feature map G (from the original image), and $g_{(i,j)\prime}$ represents the feature vector with the spatial index (i, j) of the feature map G$\prime$ (from the augmented view). The similarity of semantic feature embeddings of the same class is improved by minimizing the semantic consistency constraint loss, i.e., maximizing the cosine similarity of the feature vectors $g_{(i,j)}$ and $g_{(i,j)\prime}$, which enables the model to extract semantic invariant features of the same class. To avoid introducing additional computational costs, we only decode the feature maps extracted from the original image.

2.2 Rebalancing Sampling Strategy

The head classes in class imbalance data contain most of the instances and dominate the training process, leading the model to focus more on the loss convergence of head classes and neglect that of rare classes. To address this problem, we propose a rebalancing sampling strategy to sample the same number of difficult samples as possible for each class and impose additional contrastive supervision, thus achieving a class rebalancing paradigm. Generally, a well-trained segmentation model will show good fitting ability for most samples. However, due to the characteristics of remote sensing images, i.e., same objects may have different spectrums and different objects may have the same spectrum, the model may output poor or even wrong predictions for some difficult samples (i.e., hard-to-predict samples), which will affect the overall segmentation performance of the model. Especially, this phenomenon is more obvious in the case of class imbalance. In view of this, we sample a few difficult samples with the lowest prediction probability per class as anchor samples to impose additional contrastive supervision. We define a sparse sampling set $\mathbf{T}$ in each mini-batch, where $\mathbf{T} = \{\boldsymbol{t}_1, \boldsymbol{t}_2, \ldots, \boldsymbol{t}_c\}$ and $\boldsymbol{t}_c$ denotes the sampling queue of class c. The sampling formula is defined as follows:

$$\boldsymbol{t}_c = \text{Rank}_{\text{low}\rightarrow\text{high}} \mathbf{P}_c \big[p_1 : p_s; p_{s+1} : p_n\big]_{\text{row}\downarrow} \tag{2}$$

where $\mathbf{P}_c$ denotes the predicted probability matrix of class c in each mini-batch. We reorder the prediction matrix of each class row by row (row $\downarrow$) according to the magnitude of the prediction probability from low to high (low $\rightarrow$ high), and select the top s samples with the lowest prediction probability in each row as the anchor sample set. In our experiments, s is set to 8.

For class c, if all samples of other classes in each mini-batch are sampled as its negative samples for supervised contrastive learning, it will cause a large amount of space storage and computational costs. Also, randomly sampling negative samples is not an appropriate solution, because some outlier (noisy) samples can also cause misfitting of the model. Therefore, we calculate the prototypes of other classes and find the top k samples that are closest to their prototypes in the feature space as the negative sample set of class c, and use the prototype of class c as the positive sample of this class. Formally, the sampling formula for positive/negative samples is defined as follows:

$$\mathbf{z}_c^+ = \mathbf{e}_c = \frac{1}{|\mathcal{V}_c|} \sum\nolimits_{\mathbf{x} \in \mathcal{V}_c} \mathbf{x} \tag{3}$$

$$\mathbf{z}_c^- \in \text{top}_k \min(\|\mathbf{x}_{\overline{c}} - \mathbf{e}_{\overline{c}}\|) \tag{4}$$

where $\mathbf{z}_c^{+/-}$ denotes the positive/negative samples of class c, $\mathcal{V}_c$ represents the set of class c's feature vectors, $\mathbf{e}_c$ denotes the prototype of class c, $\overline{c}$ represents the other classes (excluding class c), and k is set to 64 in the experiments.

2.3 Pixel-Level Supervised Contrastive Loss

After determining the anchor samples and their corresponding positive/negative samples of each class, additional contrastive supervision is applied to the anchor samples of each class. The pixel-level supervised contrastive loss is defined as follows:

$$\mathcal{L}^{\text{PSC}} = \sum_{c}^{C} \mathcal{L}_c^{\text{PSC}} \tag{5}$$

$$\mathcal{L}_c^{\text{PSC}} = \sum\nolimits_{\mathbf{z}_c \in \mathcal{D}_c} -\log \frac{\exp\big[S\big(\mathbf{z}_c, \mathbf{z}_c^+\big)/\tau\big]}{\exp\big[S\big(\mathbf{z}_c, \mathbf{z}_c^+\big)/\tau\big] + \sum\nolimits_{\mathbf{z}_c^- \in \mathcal{N}_c} \exp\big[S\big(\mathbf{z}_c, \mathbf{z}_c^-\big)/\tau\big]} \tag{6}$$

where $\mathcal{D}_c$ represents the anchor sample set of class c and $\mathcal{N}_c$ denotes its negative sample set. C denotes the total number of classes. τ is a hyperparameter to control the strength of penalties on negative samples with high similarity to the anchor sample, which is set to 0.1 in the experiments. By minimizing the pixel-level supervised contrastive loss, the expression of samples of the same class in the decision space tends to be consistent. For rare classes, additional contrastive supervision can make the decision boundary of rare classes more separable. The total loss of the model is defined as follows:

$$\mathcal{L}^{\mathrm{ALL}} = \mathcal{L}^{\mathrm{CE}} + \mathcal{L}^{\mathrm{SCC}} + \mathcal{L}^{\mathrm{PSC}} \tag{7}$$

where $\mathcal{L}^{\mathrm{CE}}$ denotes the cross-entropy loss function. Through the joint constraint of the three losses, CoLM can significantly improve the segmentation performance of the baseline model for multispectral remote sensing images with class imbalance. Algorithm 1 shows the implementation details of CoLM.

Algorithm 1. Algorithm flow of the proposed CoLM.

Input: Input image X and augmented view X′ ; number of anchor samples s = 8; number of negative samples k = 64; baseline model;
Output: Trained baseline model;
1: Initialize the parameters of model;
2: // Training process
3: **for** t = 1 : max iteration **do**
4: Extract semantic feature maps G and G′ from X and X′ by shared encoder;
5: Calculate the semantic consistency constraint $\mathcal{L}^{\mathrm{SCC}}$ according to Eq. 1;
6: Generate prediction vectors and projection vectors from G by segmentation head f_{SEG} and projection head f_{PROJ};
7: Generate the sample set **T** and the anchor samples **z** for each class according to Eq. 2;
8: Sample positive samples $\mathbf{z}^{+}$ and negative samples $\mathbf{z}^{-}$ according to Eq. 3 and Eq. 4;
9: Calculate the pixel-level supervised contrastive loss $\mathcal{L}^{\mathrm{PSC}}$ according to Eq. 5;
10: Calculate the model loss according to Eq. 7;
11: Update model parameters using mini-batch gradient descent;
12: **end for**
13: // Inference process
14: Conduct pix-level label prediction by trained baseline model;

3 Experiments

In this section, we select several latest semantic segmentation methods that perform well on the benchmark remote sensing dataset (i.e., *ISPRS Potsdam* dataset): Deeplabv3 + [20], MsanlfNet [21], and MANet [22] as baseline models for experiments to verify the effectiveness of the proposed CoLM.

Deeplabv3 +: This method adopts resnet101 [23] as the feature extraction backbone, and contains atrous convolution with three atrous rates of 6, 12, and 18 to expand the effective receptive field of the model.

MsanlfNet: The feature extraction backbone of MsanlfNet is resnet50, which contains a multi-scale attention module and a non-local filtering module to improve the model's ability to capture contextual information.

MANet: MANet uses resnet50 as the feature extraction backbone and introduces a linear attention mechanism in the encoding stage to refine the extracted features.

Four indicators are adopted to evaluate the model comprehensively: overall accuracy (OA), per-class F1 score, mean F1 score (Mean F1), and mean intersection over union (MIoU).

3.1 Datasets

We evaluate the proposed CoLM on two long-tailed remote sensing datasets: RIT-18 [24] and LASA-JILIN.

Table 1. The category distribution rates of datasets.

RIT-18		LASA-JILIN	
Class Name	Percentage (%)	Class Name	Percentage (%)
Road markings	3.93×10^{-1}	Road*	4.26×10^{0}
Tree	1.55×10^{-1}	Farmland*	3.63×10^{0}
Building	4.53×10^{-1}	Snow	7.46×10^{0}
Vehicle	1.12×10^{-1}	Construction land	1.20×10^{1}
Person*	4.20×10^{-3}	Building	5.33×10^{0}
Lifeguard chair*	7.00×10^{-3}	Greenland*	2.54×10^{0}
Picnic table	3.27×10^{-2}	Mountain land	6.25×10^{1}
Orange landing pad*	2.10×10^{-3}	Water*	2.06×10^{0}
Water buoy*	2.30×10^{-3}	-	-
Rocks	1.27×10^{0}	-	-
Other vegetation	1.48×10^{0}	-	-
Grass	3.05×10^{1}	-	-
Sand	2.02×10^{1}	-	-
Lake	1.65×10^{1}	-	-
Pond	2.44×10^{0}	-	-
Asphalt	1.12×10^{1}	-	-
Gini coefficient	0.71	Gini coefficient	0.61

RIT-18: The RIT-18 dataset is a multispectral remote sensing dataset with a severely imbalanced class distribution. It has six spectral bands with a spectral range of 490–900 nm, a ground sampling distance of 4.7 cm, and 16 land-cover classes. The image size of the training set is 9393 × 5642, and the image size of the test set is 8833 × 6918.

We have counted the class distribution rate of the RIT-18 dataset, as shown in Table 1. Classes with a pixel proportion less than 0.01% are defined as rare classes and marked with * for this dataset. The Gini coefficient is used to assess the long-tailed degree of the dataset [25], with values ranging from 0 to 1. A larger Gini coefficient indicates a more imbalanced class distribution of the dataset. The Gini coefficient of the ideal dataset is 0, i.e., an equal number of instances in each class. In contrast, the Gini coefficient of long-tailed datasets is usually above 0.5. As can be seen from Table 1, the Gini coefficient of the RIT-18 dataset even reaches 0.71.

LASA-JILIN: The LASA-JILIN dataset is collected by the JILIN-1 satellite over Lhasa, China. It contains 50 images, each with an average size of 5000×4214, containing four spectral bands, with a ground sampling distance of 0.75m, and containing 8 land-cover classes. We divide 25 of these images into the training set and the rest into the test set. Table 1 shows the class distribution rate of the LASA-JILIN dataset, and the classes with a pixel proportion less than 5% are marked as rare classes with *.

3.2 Experimental Setup

For a fair comparison, all methods are implemented on the Pytorch platform and share the basic experimental settings: the initial learning rate is 0.0003, the batch size is 16, and the maximum numbers of epochs for the RIT-18 and LASA-JILIN datasets are set to 150 and 50, respectively. Additionally, other hyperparameter settings are consistent with the original paper, respectively. All images are cropped to a size of 256×256 for training and testing. The project head f_{PROJ} maps each high-dimensional vector to 32 dimensions for computing pixel-level supervised contrastive loss, and it is applied only during the training phase and removed during inference.

3.3 Experimental Results and Analysis

Experimental results of the RIT-18 dataset: The quantitative experimental results of the RIT-18 dataset are shown in Table 2, and Fig. 2 shows segmentation result maps of different methods. The baseline models have low segmentation accuracy in some classes, especially for rare classes. They do not even detect any instance of some rare classes, resulting in low Mean F1 and MIoU scores. It can be seen that CoLM effectively improves the segmentation accuracy of baseline models for rare classes, and also slightly improve the OA score, showing consistent performance improvement on the three baseline models. Deeplabv3 + adopts multiple atrous convolutions to capture the multi-scale features of the image, and the overall segmentation performance is better than the other two baseline models. However, it still fails to identify the "Water Buoy" class. Deeplabv3 + _CoLM can detect the "Water Buoy" class, and the MIoU score has increased by 5.08%. The MsanlfNet and MANet methods apply multiple attention layers in the encoding stage, which may make the models focus more on the feature extraction of head classes, but neglect rare classes, leading to poor performance in Mean F1 and MIoU scores. Combining CoLM, the segmentation performance of the baseline models is significantly improved, especially for MANet, with 13.00% and 9.88% improvement in Mean F1 and MIoU scores, respectively, proving the effectiveness of the proposed method. For rare classes with very few instances, such as the "Orange landing pad" class,

Table 2. Experimental results obtained by different methods on the RIT-18 dataset, rare classes are marked with *, and their pixel proportion is listed as a percentage.

Method	Road Markings	Tree	Building	Vehicle	Person* (0.0042%)	Lifeguard Chair* (0.0070%)	Picnic Table	Orange Landing Pad* (0.0021%)	Water Buoy* (0.0023%)	Rocks	Other Vegetation	Grass	Sand	Lake	Pond	Asphalt	OA	Mean FI	MIoU
Deeplabv3 +	73.50	85.48	67.64	37.82	0.00	49.77	18.23	54.31	0.00	76.33	48.62	91.86	92.98	91.66	60.37	92.84	88.69	58.84	47.98
Deeplabv3 + _CoLM	80.84	86.97	63.79	50.78	1.88	77.23	17.86	70.26	14.42	86.50	32.03	92.19	92.71	94.28	66.71	93.61	89.72 **(+1.03)**	63.88 **(+5.04)**	53.48 **(+5.50)**
MsanlfNet	65.83	85.69	40.12	30.48	0.00	0.21	1.90	0.00	7.06	85.49	10.97	91.78	89.47	89.48	40.54	87.62	86.67	45.41	37.65
MsanlfNet _CoLM	62.29	85.39	47.08	35.55	3.52	15.92	2.28	63.28	15.29	85.89	9.87	90.77	92.38	93.49	50.93	85.05	87.33 **(+0.66)**	52.44 **(+7.03)**	42.73 **(+5.08)**
MANet	74.76	89.27	59.49	8.17	0.00	0.07	0.00	0.00	0.00	82.35	48.68	93.13	92.24	91.26	46.93	92.03	89.25	48.65	41.36
MANet _CoLM	83.17	87.84	60.78	32.10	3.32	75.58	6.95	52.07	19.69	83.65	49.59	92.42	92.63	93.93	59.14	93.58	89.82 **(+0.57)**	61.65 **(+13.00)**	51.24 **(+9.88)**

the F1 score of MsanlfNet_CoLM increases from 0 to 63.28% compared to the baseline model MsanlfNet. However, for the "Person" class, the improvement is only 3.52%. We further investigate the spectral curves of the RIT-18 dataset (see Fig. 3) and find that the spectral feature distribution of the "Person" class is more similar to that of other classes. In contrast, the "Orange landing pad" class has better spectral differentiation, which can result in a relatively significant improvement.

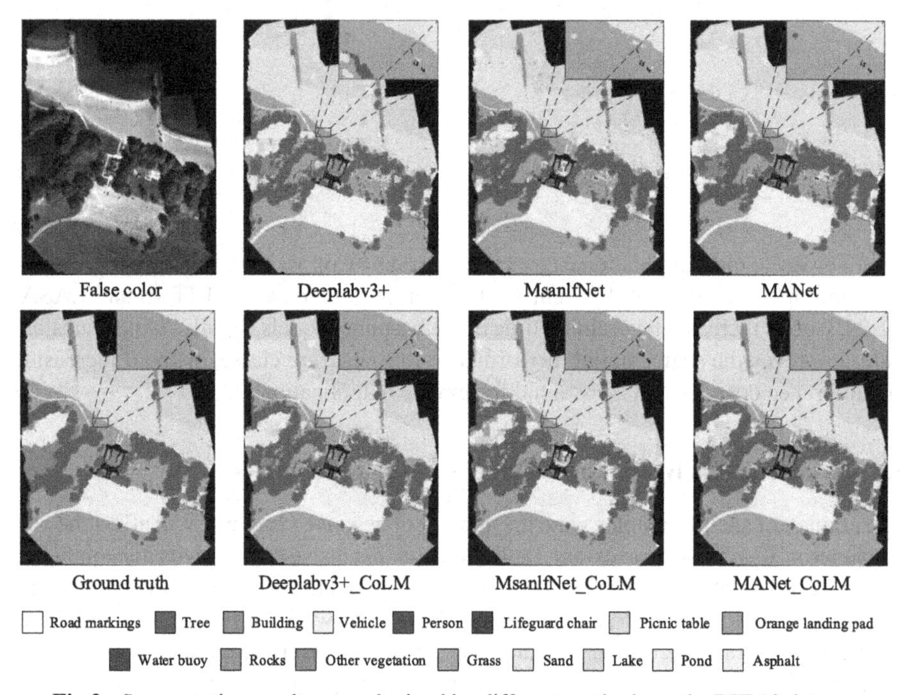

Fig.2. Segmentation result maps obtained by different methods on the RIT-18 dataset.

Experimental results of the LASA-JILIN dataset: Table 3 shows the quantitative results of the LASA-JILIN dataset, and the segmentation result maps obtained by different methods are shown in Fig. 4. Compared with RIT-18, the LASA-JILIN dataset is a large-scale dataset, and its class distribution is also more balanced. Therefore, all baseline models obtain better experimental results on the LASA-JILIN dataset. It is further observed that CoLM is still able to steadily improve the segmentation accuracy of the baseline models on most classes, with 0.75%/3.56%/3.38% improvement in OA/Mean F1/MIoU scores on the MANet baseline model, respectively.

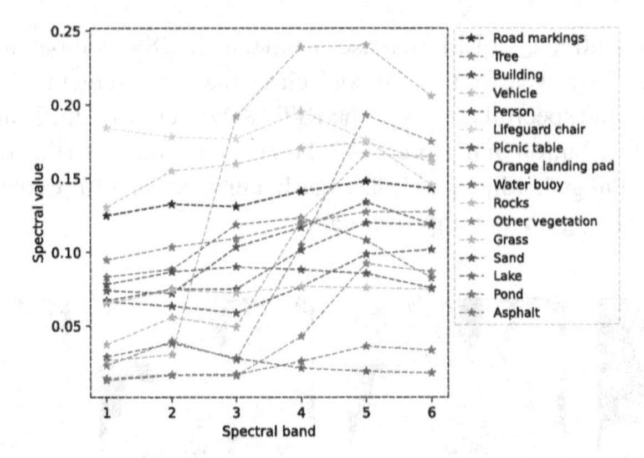

Fig. 3. Spectral curves of the RIT-18 dataset.

To evaluate the computational costs introduced by CoLM, Table 4 shows the training time and inference time of different methods on the two datasets, RIT-18 and LASA-JILIN. CoLM increases the training time of the baseline models slightly. Still, it substantially improves the segmentation performance for most rare classes without increasing any inference costs, which is essential for practical applications.

3.4 Parameter Analysis on the Number of Sampled Samples

The number of anchor samples s and the number of negative samples k will affect the performance of CoLM to some extent. In this section, extensive experiments are conducted on the RIT-18 dataset to explore the optimal combination of these two parameters, and the other experimental settings remain consistent with the previous experiments. Figure 5 shows the experimental results obtained by applying different combinations of parameters on the three baseline models. In general, the model tends to obtain better performance when sampling more anchor and negative samples, but excessively supervised samples also result in more memory usage and computational costs. In addition, we sample negative samples from the center of each class, which may introduce noisy samples when the number of sampled negative samples k is too large. This, in turn, leads to a degradation of the model's performance. The experimental results of three baseline models show that the models can achieve more stable performance when the parameter combination (s, k) is set to $(8, 64)$. Therefore, this parameter combination has also been applied to the LASA-JILIN dataset.

Table 3. Experimental results obtained by different methods on the LASA-JILIN dataset, rare classes are marked with *, and their pixel proportion is listed as a percentage.

Method	Road* (4.26%)	Farmland* (3.63%)	Snow	Construction land	Building	Greenland* (2.54%)	Mountain Land	Water* (2.06%)	OA	Mean F1	MIoU
Deeplabv3 +	40.83	35.73	82.90	62.90	77.59	25.52	91.29	85.77	82.24	62.82	50.14
Deeplabv3 + _CoLM	42.49	40.51	83.38	63.00	79.13	37.92	91.57	87.87	82.38 (**+0.14**)	65.73 (**+2.91**)	52.69 (**+2.55**)
MsanlfNet	34.78	33.85	81.96	63.53	74.72	27.60	91.13	84.24	81.44	61.48	48.69
MsanlfNet_CoLM	37.10	41.07	82.55	64.00	76.03	31.97	90.99	85.65	81.47 (**+0.03**)	63.67 (**+2.19**)	50.59 (**+1.90**)
MANet	30.82	36.38	82.45	63.94	76.41	33.70	90.75	83.95	81.27	62.30	49.39
MANet_CoLM	42.53	39.89	83.38	65.16	78.59	38.45	91.34	87.53	82.02 (**+0.75**)	65.86 (**+3.56**)	52.77 (**+3.38**)

Table 4. The training time in hours (h) and test time in seconds (s) of different methods.

Method	RIT-18		LASA-JILIN	
	Training time (h)	Inference time (s)	Training time (h)	Inference time (s)
Deeplabv3 +	3.17	15.53	2.85	157.61
Deeplabv3 + _CoLM	3.28	15.53	3.69	157.61
MsanlfNet	2.03	14.69	2.15	81.36
MsanlfNet_CoLM	2.70	14.69	3.20	81.36
MANet	3.09	15.12	2.46	134.73
MANet_CoLM	3.19	15.12	3.25	134.73

3.5 Ablation Experiments

As mentioned above, the proposed CoLM consists of three critical components: semantic consistency constraint (SCC), rebalancing sampling strategy (RSS), and pixel-level supervised contrastive loss (PSC).

This section presents some ablation experimental results and analysis to further explore the performance impact of different components on the baseline models. Table 5 shows the results of the ablation experiments on the RIT-18 and LASA-JILIN datasets. SCC and RSS-PSC denote the experimental results obtained by applying semantic consistency constraint and pixel-level supervised contrastive loss (working together with the rebalancing sampling strategy), respectively. Overall, the different components achieve better performance gains on the baseline model, especially on the RIT-18 dataset with the extremely imbalanced class distribution. SCC can prompt the model to extract semantic invariant features of the same class, i.e., common features, thus effectively enhancing the model's feature expression ability. After applying the SCC component on the baseline model MsanlfNet, the MIoU score on the RIT-18 dataset improved by 2.69%. RSS-PSC uses the rebalancing sampling strategy to sample the same number of hard-to-predict samples for each class, and imposes additional contrastive supervision on them through pixel-level supervised contrastive loss, making the decision space more linearly separable and improving the prediction accuracy of difficult samples. Through the collaboration of SCC and RSS-PSC, CoLM significantly enhances the segmentation performance of the baseline models.

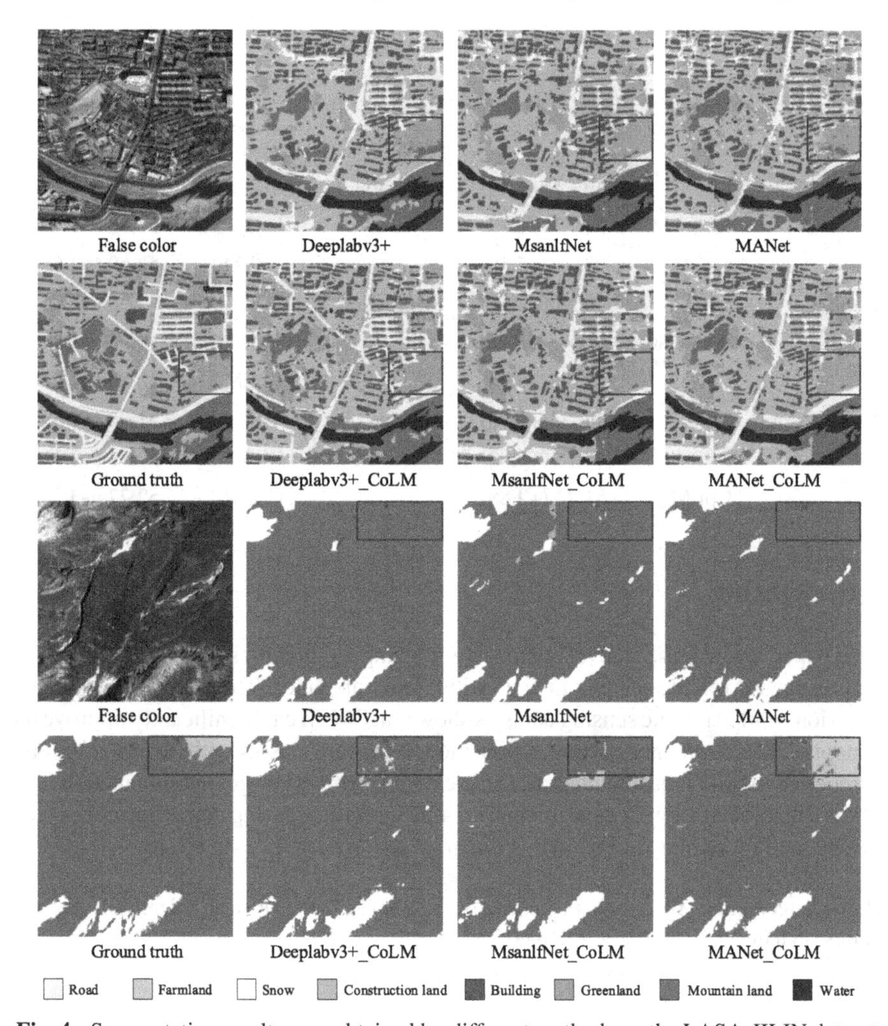

Fig. 4. Segmentation result maps obtained by different methods on the LASA-JILIN dataset.

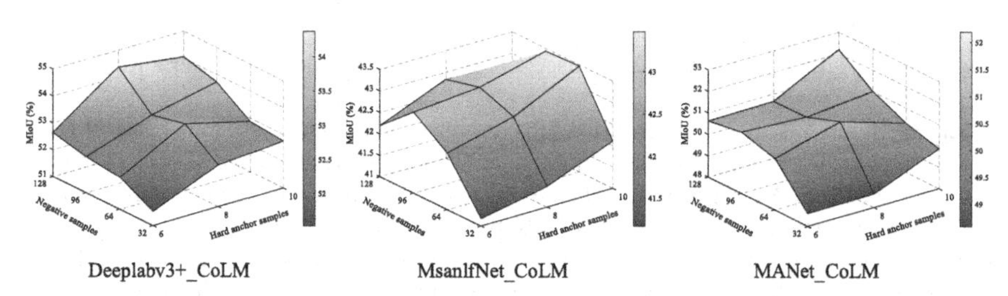

Fig. 5. Parameter sensitivity analysis of the number of anchor samples (s) and the number of negative samples (k) on the RIT-18 dataset.

Table 5. Results of ablation experiments.

RIT-18			LASA-JILIN		
Method		MIoU	Method		MIoU
Deeplabv3 +	SCC	49.42 (+**1.44**)	Deeplabv3 +	SCC	51.43 (+**1.29**)
	RSS-PSC	49.77 (+**1.79**)		RSS-PSC	52.12 (+**1.98**)
	CoLM	53.48 (+**5.50**)		CoLM	52.69 (+**2.55**)
MsanlfNet	SCC	40.34 (+**2.69**)	MsanlfNet	SCC	49.95 (+**1.26**)
	RSS-PSC	41.57 (+**3.92**)		RSS-PSC	49.98 (+**1.29**)
	CoLM	42.73 (+**5.08**)		CoLM	50.59 (+**1.90**)
MANet	SCC	47.26 (+**5.90**)	MANet	SCC	51.08 (+**1.69**)
	RSS-PSC	48.18 (+**6.82**)		RSS-PSC	51.53 (+**2.14**)
	CoLM	51.24 (+**9.88**)		CoLM	52.77 (+**3.38**)

4 Conclusion

In this paper, we propose a novel contrastive learning method for semantic segmentation of multispectral remote sensing images with class imbalance. The experimental results on two long-tailed remote sensing datasets show that CoLM can significantly improve the segmentation performance of the baseline models for rare classes without any additional inference costs, and is easily integrated into existing semantic segmentation models. Our work is a beneficial attempt to segment long-tailed remote sensing data using contrastive learning, and more innovations in this direction are expected in the future.

Acknowledgements. This work is supported by the Second Tibetan Plateau Scientific Expedition and Research under Grant 2019QZKK0405.

References

1. Li, T. et al.: Targeted supervised contrastive learning for long-tailed recognition. In: Proceedings of the IEEE/CVF Conference on Computer Vision and Pattern Recognition, pp. 6918–6928 (2022)
2. Li, Z., Kamnitsas, K., Glocker, B.: Analyzing overfitting under class imbalance in neural networks for image segmentation. IEEE Trans. Med. Imaging **40**(3), 1065–1077 (2020)
3. Buda, M., Maki, A., Mazurowski, M.A.: A systematic study of the class imbalance problem in convolutional neural networks. Neural Netw. **106**, 249–259 (2018)
4. Khan, S. et al.: Striking the right balance with uncertainty. In: Proceedings of the IEEE/CVF Conference on Computer Vision and Pattern Recognition, pp. 103–112 (2019)
5. Lin, T. Y. et al.: Focal loss for dense object detection. In: Proceedings of the IEEE International Conference on Computer Vision, pp. 2980–2988 (2017)
6. Sudre, C.H. et al.: Generalised dice overlap as a deep learning loss function for highly unbalanced segmentations. In: Deep Learning in Medical Image Analysis and Multimodal Learning for Clinical Decision Support, pp. 240–248 (2017)

7. Yang, Y., Xu, Z.: Rethinking the value of labels for improving class-imbalanced learning. In: Advances in Neural Information Processing Systems, pp. 19290–19301 (2020)
8. Kang, B. et al.: Decoupling representation and classifier for long-tailed recognition. In: International Conference on Learning Representations (2020)
9. Zhang, S., Li, Z., Yan, S., He, X., Sun, J.: Distribution alignment: A unified framework for long-tail visual recognition. In: Proceedings of the IEEE/CVF Conference on Computer Vision and Pattern Recognition, pp. 2361–2370 (2021)
10. Ren, Q., Yuan, C., Zhao, Y., Yang, L.: A novel metric learning framework by exploiting global and local information. Neurocomputing **507**, 84–96 (2020)
11. Chen, T., Kornblith, S., Norouzi, M., Hinton, G.: A simple framework for contrastive learning of visual representations. In: International Conference on Machine Learning, pp. 1597–1607 (2020)
12. He, K., Fan, H., Wu, Y., Xie, S., Girshick, R.: Momentum contrast for unsupervised visual representation learning. In: Proceedings of the IEEE/CVF Conference on Computer Vision and Pattern Recognition, pp. 9729–9738 (2020)
13. Wang, P., Han, K., Wei, X.S., Zhang, L., Wang, L.: Contrastive learning based hybrid networks for long-tailed image classification. In: Proceedings of the IEEE/CVF Conference on Computer Vision and Pattern Recognition, pp. 943–952 (2021)
14. Kang, B., Li, Y., Yuan, Z., Feng, J.: Exploring balanced feature spaces for representation learning. In: International Conference on Learning Representations (2021)
15. Wang, W. et al.: Exploring cross-image pixel contrast for semantic segmentation. In: Proceedings of the IEEE/CVF International Conference on Computer Vision, pp. 7303–7313 (2021)
16. Li, H., et al.: Global and local contrastive self-supervised learning for semantic segmentation of HR remote sensing images. IEEE Trans. Geosci. Remote Sens. **60**, 1–14 (2022)
17. Zhong, Y. et al.: Pixel contrastive-consistent semi-supervised semantic segmentation. In: Proceedings of the IEEE/CVF International Conference on Computer Vision, pp. 7273–7282 (2021)
18. Liu, S., Zhi, S., Johns, E., Davison, A.J.: Bootstrapping semantic segmentation with regional contrast. In: International Conference on Learning Representations (2022)
19. Khosla, P. et al.: Supervised contrastive learning. In: Advances in Neural Information Processing Systems, pp. 18661–18673 (2020)
20. Chen, L.C., Zhu, Y., Papandreou, G., Schroff, F., Adam, H.: Encoder-decoder with atrous separable convolution for semantic image segmentation. In: Proceedings of the European Conference on Computer Vision, pp. 801–818 (2018)
21. Bai, L., et al.: MsanlfNet: Semantic segmentation network with multiscale attention and nonlocal filters for high-resolution remote sensing images. IEEE Geosci. Remote Sens. Lett. **19**, 1–5 (2022)
22. Li, R., et al.: Multiattention network for semantic segmentation of fine-resolution remote sensing images. IEEE Trans. Geosci. Remote Sens. **60**, 1–13 (2022)
23. He, K., Zhang, X., Ren, S., Sun, J.: Deep residual learning for image recognition. In: Proceedings of the IEEE Conference on Computer Vision and Pattern Recognition, pp. 770–778 (2016)
24. Kemker, R., Salvaggio, C., Kanan, C.: Algorithms for semantic segmentation of multispectral remote sensing imagery using deep learning. ISPRS J. Photogramm. Remote Sens. **145**, 60–77 (2018)
25. Yang, L., Jiang, H., Song, Q., Guo, J.: A survey on long-tailed visual recognition. Int. J. Comput. Vis. **130**(7), 1837–1872 (2022)

ESTNet: Efficient Spatio-Temporal Network for Industrial Smoke Detection

Shuo Du[1,2], Zheng Lv[1,2(✉)], Linqing Wang[1,2], and Jun Zhao[1,2]

[1] Key Laboratory of Intelligent Control and Optimization for Industrial Equipment, Dalian University of Technology, Ministry of Education, Dalian 116031, China
[2] School of Control Sciences and Engineering, Dalian University of Technology, Dalian 116031, China
lvzheng@dlut.edu.cn

Abstract. Computer vision has emerged as a cost-effective and convenient solution for identifying hazardous smoke emissions in industrial settings. However, in practical scenarios, the performance of existing methods can be affected by complex smoke characteristics and fluctuating environmental factors. To address these challenges, we propose a novel detection model called ESTNet. ESTNet utilizes both smoke texture features and unique motion features to enhance smoke detection. The Shallow Feature Enhancement Module (SFE) specifically enhances the learning of smoke texture features. The Spatio-temporal Feature Learning Module (SFL) effectively differentiates smoke from other interfering factors, enabling the establishment of smoke spatio-temporal feature learning. Notably, this module can be easily integrated into existing 2D CNNs, making it a versatile plug-and-play component. Furthermore, to improve the representation of the video, we employ Multi-Temporal Spans Fusion (MTSF) to incorporate information from multiple frames. This fusion technique allows us to obtain a comprehensive feature representation of the entire video. Extensive experiments and visualizations are conducted, demonstrating the effectiveness of our proposed method with state-of-the-art competitors.

Keywords: Smoke detection · Spatio-temporal learning · Video understanding

1 Introduction

Smoke poses a pervasive threat to various industrial environments, including chemical plants, power plants, and manufacturing facilities, endangering the safety of personnel and causing irreparable harm to the ecosystem. Therefore, smoke detection is critical for accident prevention, personnel safety, and environmental protection. However, conventional smoke detection techniques, such

This work was supported by the Applied Basic Research Program Project of Liaoning Province under Grant 2023JH2/101600043 and the National Natural Science Foundation of China (No.61873048).

B. Luo et al. (Eds.): ICONIP 2023, CCIS 1965, pp. 376–387, 2024.
https://doi.org/10.1007/978-981-99-8145-8_29

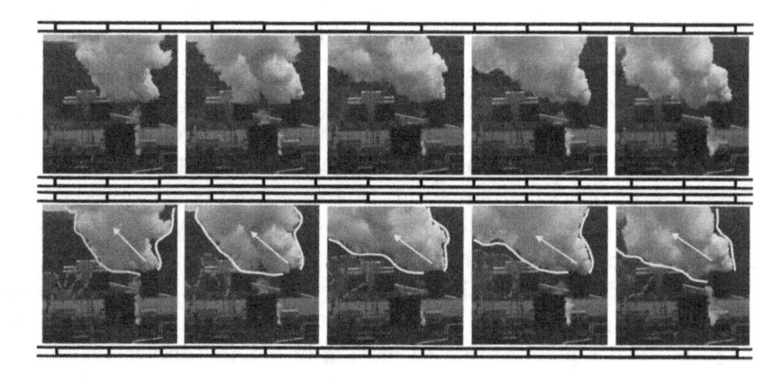

Fig. 1. The elucidation of static and dynamic characteristics of smoke within the video demonstrates that, in comparison to other interference (indicated in green lines), smoke exhibits versatile motion properties and irregular edge features (indicated in red lines). (Color figure online)

as sensors and alarms, exhibit limited efficacy in complex industrial settings, and their accuracy is significantly impacted by environmental factors in the surrounding areas.

To enhance detection precision while ensuring cost-effectiveness, the scientific community has recently explored vision-based methods for detecting smoke. Prior research works relied on physical or manually engineered features to differentiate smoke. For instance, Tian et al. [13] proposed a method for smoke feature extraction based on edge detection and gradient information. Additionally, Yuan et al. [18] extracted smoke characteristics in intricate settings by using a characteristic discriminator and formulated a classifier based on nonlinear dimensionality reduction and Gaussian processes which yielded elevated classification accuracy and robustness. Likewise, Lee et al. [9] employed color and texture as smoke features for image encoding and trained support vector machine (SVM) classifiers using vectorized features. However, those traditional approaches confront limitations in the extraction and comprehension of smoke features, thereby impacting detection's performance.

Compared to conventional techniques, vision-based smoke detection algorithms utilizing deep learning algorithms can automatically extract smoke features, improve algorithm effectiveness and efficiency, and transform the detection and management of smoke hazards in industrial settings. Yin et al. [17] extracted features and applied classification utilizing convolutional neural networks. Similarly, Hsu et al. [7] employed general video understanding techniques for identifying smoke videos, while Cao et al. [1] leveraged two-stream networks for learning smoke motion information. Nevertheless, these approaches did not originate from designing networks based on the fundamental smoke traits. As a result, they may not effectively differentiate smoke from other interfering factors at the primary level.

In smoke detection, a challenge lies in differentiating between smoke and steam. While they are morphologically similar, their physical structures differ: smoke consists of solid particles, while steam is comprised of fine droplets, resulting in distinct textural and color features between the two. As a result, techniques that are solely based on smoke's textural characteristics [9,13] experience reduced accuracy in practical detection. Smoke is commonly observed in blue or discolored hues, owing to pollutants, while steam appears as white. Moreover, smoke and steam display distinguishable motion patterns. Smoke particles disperse outward and downward, influenced by air currents, inducing rotational and turbulent motion, as depicted in Fig. 1. On the contrary, steam typically discharges from fixed pipes, leading to motion in defined directions and having limited boundaries. Smoke's motion display relatively greater randomness due to the lack of a set emission point.

Therefore, we propose an end-to-end network architecture based on spatio-temporal cooperative learning for smoke video detection. The framework utilizes our designed SFL module, which leverages the difference between adjacent frame features to focus on motion information and uses a single-stream network to effectively detect smoke's characteristic spatial and temporal features, allowing for the characterization and differentiation of smoke videos. We design a novel SFE module to learn a multi-level feature representation of smoke based on its spatial characteristics and propose a new MTSF module that takes into account the characteristics of video tasks to help the interaction of temporal dimension information. These modules can operate as plug-and-play modules and can be easily integrated with existing 2D backbone networks. The key contributions of this framework are:

- We propose an end-to-end network architecture incorporating spatio-temporal collaborative learning for smoke video detection. The framework is based on our designed SFL module, by utilizing a Single-Stream Network, it efficiently detects smoke-specific features from both spatial and temporal dimensions for the characterization and distinction of smoke videos.
- We develop a novel SFE module to learn the low-level features of smoke and design MTSF to facilitate better interaction among temporal dimension information. Both modules can function as plug-and-play components and can be easily integrated with existing 2D backbone networks.
- Extensive experiments and visualizations demonstrate the effectiveness of our model, which outperforms other models in both performance and results on the RISE dataset, which is closest to real-world detection scenarios.

2 Our Method

We utilize Xception [3] pretrained on ImageNet [4] as our backbone. As the characteristics of smoke tend to be concentrated in the shallow layers, we employ the SFE structure, illustrated in Fig. 3, to aid in our acquisition of shallow-level texture features. To mitigate any undesired interferences, such as clouds or

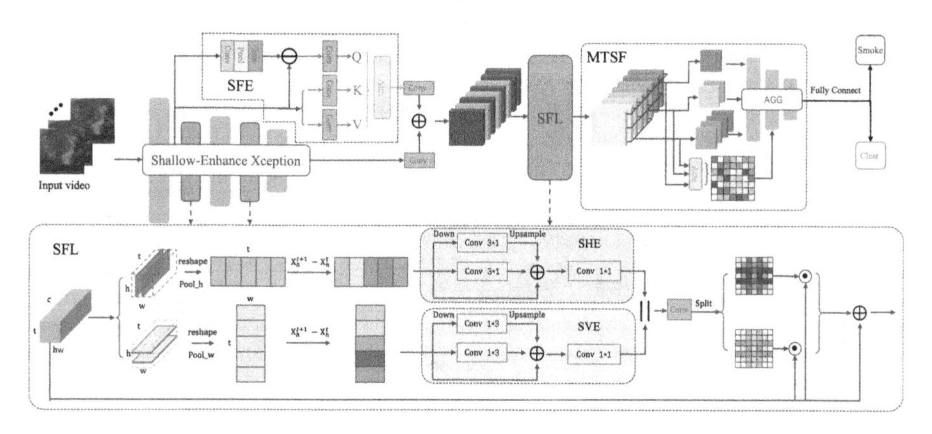

Fig. 2. Overview of the proposed framework. The frame sequence input is enhanced in the 2D CNN backbone network through the utilization of the Shallow Feature Enhancement module (SFE), which facilitates the extraction of texture feature information. The backbone network is integrated with the Spatio-temporal Feature Learning module, which enables the extraction of motion feature information on the temporal difference between the preceding and succeeding frames. Furthermore, the Multi-temporal Spans Fusion module fuses the sequence information to obtain video-level information.

steam, we devise the SFL module to extract the motion features of smoke. The final classification results are obtained by aggregating the standardized video through MTSF, and a diagram of our framework can be seen in Fig. 2.

2.1 Shallow Enhanced Feature Extraction Module

Numerous studies [6,14] on smoke detection have demonstrated that low-level feature information pertaining to color and texture is vital for smoke detection. Consequently, we devise a modified Xception architecture to extract spatial domain features. Employing the last layer features of Xception directly, however, would neglect the shallow texture layer features. Thus, we reconstructe the Xception output using multiscale and designed the SFE module to guide the network with an attention mechanism, compelling the network to learn more elementary information features.

Given the sparse sampled fragment $x \in \mathbb{R}^{T \times C \times H \times W}$, where T, C, H and W are 8, 3, 224, and 224, both shallow output and final output are extracted from different stages of the network flow, yielding shallow feature f_s and final feature f_f of a single-frame image. Generally, smoke features are prominent in the textural information of shallow features. To further enhance the artifacts embedded in the shallow features, SFE module comprising upsampling, downsampling, and attention mechanisms, as illustrated in Fig. 3. The features x_s in the shallow stream of the network are obtained after feature subtraction to acquire the feature x_m reflecting the motion features in the image, whereupon the multi-headed attention mechanism aids in learning the regions that correspond with

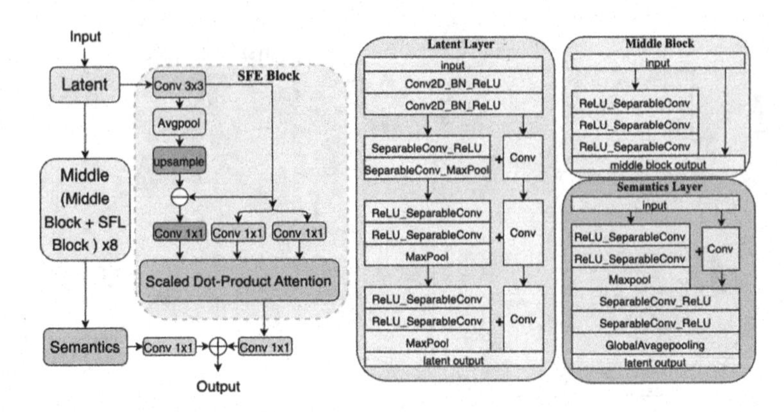

Fig. 3. The architecture of shallow enhance Xception. Our improved Xception enhance latent feature through SFE module.

the motion features of smoke. The shallow texture information f_s is refined by the SFE module:

$$f_s = Conv\left[Atten\left(x_m, x_s, x_s\right)\right], \tag{1}$$

where *Atten* denotes the multi-head attention mechanism, the input x_m serves as the query, while x_s serves as the key and value. *Conv* denotes point-wise convolution. We visualize the output feature maps in Fig. 5 for illustration.

2.2 Smoke Spatio-Temporal Feature Learning

Practical smoke detection is challenging due to the presence of numerous interfering factors, such as steam and clouds. While these factors share spatial texture similarities with smoke, they differ significantly in motion features. To this end, we propose the Spatio-temporal Feature Learning module (SFL), a joint modeling paradigm for smoke spatio-temporal features. SFL captures and utilizes smoke motion within frame sequences without requiring additional inputs or temporal feature pathways.

Specifically, by leveraging SFE-enhanced feature extraction, we acquire features on each frame and construct motion features in both horizontal and vertical directions based on differences in adjacent features to learn the unique spatio-temporal representation of smoke motion, as illustrated in Fig. 2. To achieve optimal performance, we propose a three-path architecture in SFL to provide processing pathways at various resolutions. Given the input feature sequence $X \in \mathbb{R}^{T\times C\times H\times W} = \{x^1, x^2, ..., x^T\}$, we first reduce each frame-level feature channels to $1/r$ for efficiency, then pool them along the vertical and horizontal directions to obtain $x_h \in \mathbb{R}^{C/r\times 1\times H}$ and $x_w \in \mathbb{R}^{C/r\times W\times 1}$. Taking the vertical direction motion feature x_w as an example, the feature-level motion difference at frame t is calculated as follows:

$$m_w^t = Conv_1\left(x_w^{t+1}\right) - x_w^t, t \in \{1, 2, ..., T-1\}, \tag{2}$$

where $Conv_1$ denotes a smooth convolution with a kernel size of 1×3. This smoothing operation mitigates the interference of video jitter on smoke motion extraction. We attain the motion information m_w^T for frame T by averaging the motion features of $T-1$ frames. Subsequently, we obtain the vertical slice motion map M_W by concatenating frame feature differences:

$$M_w = \mathop{\|}_{i=1}^{T} m_w^i, \tag{3}$$

where $M_w \in \mathbb{R}^{T \times C/r \times W \times 1}$ contains the most of vertical direction motion feature of smoke video, $\|$ denotes the concatenation operation along the time dimension, and similarly, the horizontal slice motion features $M_h \in \mathbb{R}^{T \times C/r \times 1 \times H}$. The temporal difference operation can effectively extract motion feature information from the video without introducing any additional parameters.

To capture Smoke-specific motion, we further design a multi-fields-of-views structure for efficient smoke motion extraction, which contains smoke vertical motion enhancement(SVE) and smoke horizontal motion enhancement(SHE). SVE is used to enhance M_w and SHE is used to enhance M_h. We elaborate on the pathway of M_w, as shown in Fig. 2, it takes M_w as input and has three branches to extract multi-level representations: (1) skip connection, (2) downsampling, 1×3 horizontal convolution, and upsampling, (3) 1×3 horizontal convolution. The M_h part has a similar structure except for the 3×1 convolution along the vertical dimension. The features of the three branches are summed up by element-wise addition:

$$M_{svew} = \sum_{i=1}^{3} Branch_i(M_w), \tag{4}$$

where $M_{svew} \in \mathbb{R}^{T \times C/r \times W \times 1}$ denote smoke motion enhanced by SVE, $Branch_i$ represent different branches. Using the same approach, one can obtain M_{sheh} enhanced through SHE. Then, followed by concatenate M_{svew} and M_{sheh} in spatial dimension, the confidence map of smoke motion M_s is calculated:

$$M_s = \sigma \left[Conv_r \left(M_{svew} \,\|\, M_{sheh} \right) \right], \tag{5}$$

where $M_s \in \mathbb{R}^{T \times C \times (H+W) \times 1}$ represents the smoke motion attention map in both directions, σ denotes the sigma function. We can split M_s along the spatial dimension into vertical smoke motion attention map M_{sw} and horizontal smoke motion attention map M_{sh}. By using these two smoke motion attention maps, we can guide the network to learn smoke's specific motion features:

$$f_{SFL} = \frac{1}{2} \times [M_{sw} + M_{sh}] \odot X, \tag{6}$$

where $f_{SFL} \in \mathbb{R}^{T \times C \times H \times W}$ denotes the output spatio-temporal features of SFL, $\odot$ represent the element-wise multiplication.

2.3 Multi-temporal Spans Fusion Module

Smoke is a production accident with no fixed duration from initiation to cessation, nor a designated discharge channel. Thus, smoke motion characteristics not only exist in short-term motion differences between adjacent frames, such as irregular boundary alterations and uncertain movement directions, but also in long-term motion features, including smoke's duration and drifting direction-attributes specific to smoke.

To better obtain a global semantic representation of smoke motion, we propose a global temporal aggregation strategy called MTSF, which adaptively emphasizes the spatio-temporal features of smoke within the video by calculating global information across multiple temporal spans. As shown in Fig. 2, the input of MTSF is the spatio-temporal feature $f \in \mathbb{R}^{T\times C\times (H/8)\times (W/8)} = \{f_1, f_2, \ldots, f_T\}$, obtained from the backbone after being enhanced by SFE and SFL. We employ dilated max-pooling operations on different temporal dimensions and obtain temporal information V_i on various temporal scales by controlling the dilation rate, with the vector length of d.

$$V_i = Maxpool_{(\frac{T}{k},1,k)}\left\{f_1, f_2, ..., f_T\right\}, \tag{7}$$

where $\frac{T}{k}$, 1, and k represent the kernel size, stride, and dilation rate in the maxpooling operation. Following convention, we set k to pyramidal timescale settings, with $k \in \left\{2^0, 2^1, 2^{log_2 T-1}\right\}$, obtaining $V_i \in \mathbb{R}^{(T-1)\times d}$ (for $2^0 + 2^1 + \ldots + 2^{log_2 T-1} = T-1$). Subsequently, the global semantics on different temporal scales are compressed into feature vectors V_{at} reflecting the corresponding temporal scale statistics. The specific operation is as follows:

$$V_{at} = \frac{1}{d}\sum_{i=1}^{d}(V_i), \tag{8}$$

where and the weighted temporal scale aggregated information as $V_{at} \in \mathbb{R}^{(T-1)\times 1}$.

To capture the cross-timescale interdependencies and generalize the dependency on the temporal dimension, we calculate the cosine similarity between time weight features in $V_i = \left\{v_1, v_2, \ldots, v_{(T-1)}\right\}$ at different time steps:

$$S_{i,j} = 1 - \left\langle \frac{v_i}{\|v_i\|_2}, \frac{v_j}{\|v_j\|_2} \right\rangle, \tag{9}$$

where $S_{i,j}$ forms similarity matrix $S \in \mathbb{R}^{(T-1)\times (T-1)}$, We then use a non-linear activation function to perform a weighted perceptive activation on the product of similarity and time features, converting it into an attention map, and the output of MTSF f_{mtsf} is calculated as follows:

$$f_{mtsf} = Agg\left\{V_{at}, Softmax\left[\left(V_{at}W_1 + W_2 S\right)W_3\right]\right\}. \tag{10}$$

where Agg denotes the weighted summation of timescale features V_{at} along temporal dimension, S_a represent the multi-timescale weights, $Softmax$ denotes softmax function, $W_1 \in \mathbb{R}^{1\times(T-1)}$, $W_2 \in \mathbb{R}^{(T-1)\times(T-1)}$ and $W_3 \in \mathbb{R}^{(T-1)\times 1}$ are the learnable parameters of fully-connected layer. At last, we feed f_{mtsf} into the classifier to get the video-level smoke detection result.

3 Experiments

In this section, we first introduce the evaluation datasets and implementation details. Extensive results show the superior performance achieved by our ESTNet compared with baselines and other state-of-the-art methods on the RISE dataset. Then, in ablation studies, we investigate the importance of SFE, SFL, and MTSF for real-time smoke detection. Finally, we visualize these modules for mining spatio-temporal clues of smoke, qualitatively demonstrating ESTNet's superiority in the aspects of smoke spatio-temporal features and motion modeling.

3.1 Datasets

The RISE Smoke Video Dataset [7] consists of 12,567 clips captured from 19 different viewpoints by cameras monitoring three industrial facilities at three locations. The RISE dataset presents a challenging video classification task, as it encompasses a variety of smoke emission features under different weather conditions such as haze, fog, snow, and clouds. According to the different camera positions aCnd shooting times, the RISE dataset has six different partition criteria for training and testing, denoted as S_0 to S_5. We will perform independent training and testing on each partition to simulate real industrial smoke detection tasks.

3.2 Implementation Details

For each video, we employ a global average sampling strategy to extract eight frames from video segments. We apply wildly-used data augmentation, including random resizing, horizontal flipping and cropping, perspective transformation, area erasing, and color jittering. The computational server has an Nvidia A40 GPU and an Intel Xeon Silver 4210R processor. During the training process, we set the batch size to 16 and the initial learning rate to 0.0002, decaying by a factor of 10 after every 10 epochs. We optimize the binary cross-entropy loss function using the Adam optimizer, with a weight decay rate of 0.003.

3.3 Comparisons on RISE

We compared our proposed method with a range of baseline approaches, including Support Vector Machine classifiers and prevalent 2D CNN networks such as Xception [3] and Resnext [16]. As we consider smoke detection a video action classification task, we focused on comparisons with common video understanding networks, such as the Inception-based I3D network [2] (RGB-I3D) and methods incorporating video understanding modules. Specifically, we compared the RGB-I3D-LSTM, which integrates an LSTM [5] at the last layer of the I3D model to establish temporal context relations; the RGB-I3D-NL, which appends Non-Local blocks [15] in the last layer of the I3D model; the RGB-I3D-TSM, which embeds TSM [10] in each layer of the Inception model; and the RGB-I3D-TC, which adds a Timeception [8] layer to the final initial layer. STCNet [1] is a smoke detection-specific two-stream network that utilizes both RGB and frame differences as input. Beyond RGB video input, we also provided optical flow frames based on preprocessed TVL1 optical flow computation. Both Flow-I3D and Flow-SVM adopt the same model architecture as I3D and SVM, respectively, while taking optical flow frames as input.

On the RISE dataset, we tested all models against six different testing set partitions, using F1-score as the evaluation criterion, result as shown in Table 1 and 2 (Fig. 4).

Table 1. F1-Scores for comparing different methods on RISE dataset

Model	S_0	S_1	S_2	S_3	S_4	S_5	Average	Params	Flops
RGB-SVM	.57	.70	.67	.67	.57	.53	.618	–	–
Flow-SVM	.42	.59	.47	.63	.52	.47	.517	–	–
RGB-Xcep [3]	.79	.81	.82	.86	.77	.78	.805	20.83M	36.8G
Plain SE-Resnext [16]	.83	.82	.84	.85	.78	.83	.826	26.6M	34.4G
RGB-I3D [2]	.80	.84	.82	.87	.82	.75	.817	12.3M	62.8G
Flow-I3D [2]	.55	.58	.51	.68	.65	.50	.578	12.3M	62.8G
RGB-I3D-LSTM [5]	.80	.84	.82	.85	.83	.74	.813	38.0M	63.0G
RGB-I3D-TSM [10]	.81	.84	.82	.87	.80	.74	.813	12.3M	62.7G
RGB-I3D-NL [15]	.81	.84	.83	.87	.81	.74	.817	12.3M	62.8G
RGB-I3D-TC [8]	.81	.84	.84	.87	.81	.77	.823	12.5M	62.9G
STCNet [1]	.88	.89	.87	.89	.86	.88	.878	52.1M	68.9G
ESTNet(Ours)	**.90**	.89	**.89**	**.90**	**.87**	.88	**.888**	**23.8M**	75.5G

Table 2. Evaluation of ESTNet(Proposed) on the test set for each split.

Metric	S_0	S_1	S_2	S_3	S_4	S_5	Average
Precision	.95	.94	.95	.94	.93	.92	.938
Recall	.85	.84	.83	.86	.81	.84	.838
F-score	.90	.89	.89	.90	.87	.88	.888

3.4 Ablation Study

This section reports the systematic ablation studies on the RISE dataset for evaluating the efficacy and rationality of each component. To measure the performance, we use the F1-score as the evaluation metric. We investigate the

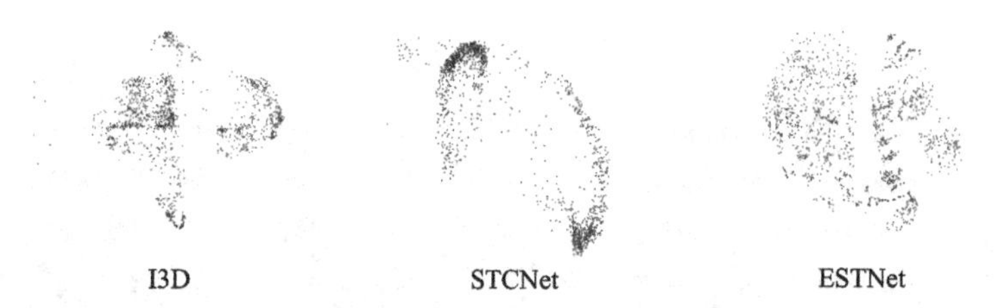

Fig. 4. An illustration of the feature distribution of I3D, STCNet, and our model on the RISE(S_0) dataset. Red points represent video with smoke and blue points represent clear video.

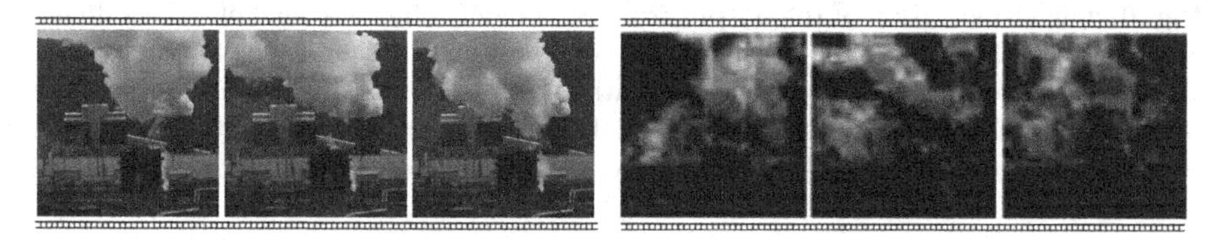

Fig. 5. Visualization of the input and output features of SFE. The characteristics of smoke in the spatio domain can be effectively learned and enhanced through SFE, thus enabling the detection of smoke based on spatio-temporal characteristics via SFL.

individual contribution of every module by designing and separately removing them in the following variants: (1) without any components; (2) SFL only; (3) SFL+MTSF; (4) SFE+SFL; (5) SFE+SFL+MTSF. We present the contributions of each module to the F1-score in Table 3. The results demonstrate that the performance of the baseline model without any additional components is poor, with a notable deficiency in S_4, at 0.77. The experimental outcomes suggest that all three components have a positive influence on the final prediction outcome, with a minimum of 1.4% performance improvement in each category. The most optimal performance is attained when all modules are jointly utilized, achieving a score of 0.9 on S_0. This result demonstrates the effectiveness of our approach (Fig. 6).

Table 3. Study on component effectiveness on the test set for each split.

Model	S_0	S_1	S_2	S_3	S_4	S_5	Average
Baseline(Xception)	.79	.81	.82	.86	.77	.78	.805
SFE	.80	.83	.83	.86	.82	.79	.822
SFL+MTSF	.83	.84	.85	.85	.82	.80	.836
SFE+SFL	.89	.89	.87	.88	.84	.84	.868
SFE+SFL+MTSF	**.90**	.89	**.89**	**.90**	**.87**	**.88**	**.888**

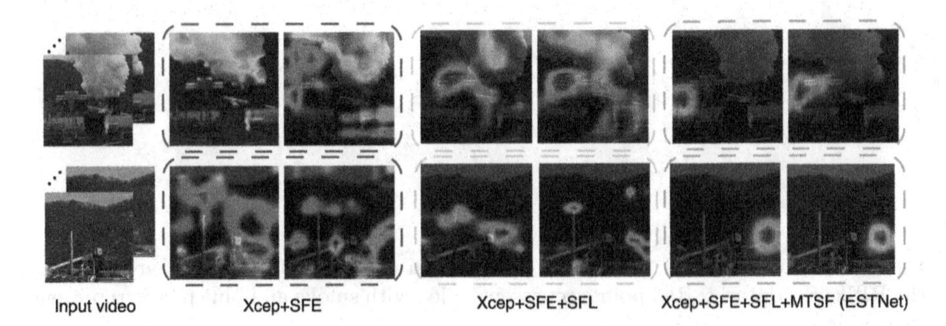

Fig. 6. The visualization of the output from our model. The backbone of our module is Xception. The heatmap region displays how different modules guide the model to focus on smoke's specific motion information rather than just texture information similar to smoke. This approach enables the model to distinguish smoke from other similar textures and identify it based on its specific motion features, which is crucial in smoke detection and tracking.

3.5 Visualization Analysis

In the visualization analysis, we first visualize the features extracted from the backbone network with the smoke texture information-enhanced Shallow Feature Enhancement (SFE) module, see Fig. 5, to confirm that the SFE indeed enhances the smoke texture features. Furthermore, we employ Grad-CAM [12] to investigate how our module progressively guides the network to focus on genuine smoke regions. Lastly, we apply t-SNE [11] to feature visualization of the extracted features from various models to determine if they can distinguish between videos with and without smoke.

4 Conclusion

In this paper, we propose a novel smoke detection algorithm that can detect the presence of smoke in videos. We design our network based on the unique spatio-temporal features of smoke, specifically its texture and motion characteristics. To this end, we employ SFE to force the network to learn texture features, SFL to model smoke motion features, and MTFS to aggregate video representations that reflect smoke features. Leveraging these designs, our end-to-end model only requires RGB video input and has a small model size, making it practical for deployment in industrial settings. With the experiments conducted on the RISE dataset and comparison against the existing state-of-the-art methods, we have validated the effectiveness of the proposed method.

Acknowledgements. This work was supported by the Applied Basic Research Program Project of Liaoning Province under Grant 2023JH2/101600043 and the National Natural Science Foundation of China (No.61873048).

References

1. Cao, Y., Tang, Q., Lu, X.: STCNet: spatiotemporal cross network for industrial smoke detection. Multimedia Tools Appl. **81**(7), 10261–10277 (2022)
2. Carreira, J., Zisserman, A.: Quo vadis, action recognition? A new model and the kinetics dataset. In: Proceedings of the IEEE Conference on Computer Vision and Pattern Recognition, pp. 6299–6308 (2017)
3. Chollet, F.: Xception: deep learning with depthwise separable convolutions. In: Proceedings of the IEEE Conference on Computer Vision and Pattern Recognition, pp. 1251–1258 (2017)
4. Deng, J., Dong, W., Socher, R., Li, L.J., Li, K., Fei-Fei, L.: ImageNet: a large-scale hierarchical image database. In: 2009 IEEE Conference on Computer Vision and Pattern Recognition, pp. 248–255. IEEE (2009)
5. Graves, A., Graves, A.: Long Short-Term Memory. Supervised Sequence Labelling with Recurrent Neural Networks, pp. 37–45 (2012)
6. Gubbi, J., Marusic, S., Palaniswami, M.: Smoke detection in video using wavelets and support vector machines. Fire Saf. J. **44**(8), 1110–1115 (2009)
7. Hsu, Y.C., et al.: Project RISE: recognizing industrial smoke emissions. In: Proceedings of the AAAI Conference on Artificial Intelligence. vol. 35, pp. 14813–14821 (2021)
8. Hussein, N., Gavves, E., Smeulders, A.W.: Timeception for complex action recognition. In: Proceedings of the IEEE/CVF Conference on Computer Vision and Pattern Recognition, pp. 254–263 (2019)
9. Lee, C.Y., Lin, C.T., Hong, C.T., Su, M.T., et al.: Smoke detection using spatial and temporal analyses. Int. J. Innovative Comput. Inf. Control **8**(6), 1–11 (2012)
10. Lin, J., Gan, C., Han, S.: TSM: temporal shift module for efficient video understanding. In: Proceedings of the IEEE/CVF International Conference on Computer Vision, pp. 7083–7093 (2019)
11. Van der Maaten, L., Hinton, G.: Visualizing data using t-SNE. J. Machine Learn. Res. **9**(11), 2579–2605 (2008)
12. Selvaraju, R.R., Cogswell, M., Das, A., Vedantam, R., Parikh, D., Batra, D.: Grad-cam: visual explanations from deep networks via gradient-based localization. In: Proceedings of the IEEE International Conference on Computer Vision, pp. 618–626 (2017)
13. Tian, H., Li, W., Ogunbona, P., Wang, L.: Single image smoke detection. In: Cremers, D., Reid, I., Saito, H., Yang, M.-H. (eds.) ACCV 2014. LNCS, vol. 9004, pp. 87–101. Springer, Cham (2015). https://doi.org/10.1007/978-3-319-16808-1_7
14. Töreyin, B.U., Dedeoğlu, Y., Cetin, A.E.: Wavelet based real-time smoke detection in video. In: 2005 13th European Signal Processing Conference, pp. 1–4. IEEE (2005)
15. Wang, X., Girshick, R., Gupta, A., He, K.: Non-local neural networks. In: Proceedings of the IEEE Conference on Computer Vision and Pattern Recognition, pp. 7794–7803 (2018)
16. Xie, S., Girshick, R., Dollár, P., Tu, Z., He, K.: Aggregated residual transformations for deep neural networks. In: Proceedings of the IEEE Conference on Computer Vision and Pattern Recognition, pp. 1492–1500 (2017)
17. Yin, Z., Wan, B., Yuan, F., Xia, X., Shi, J.: A deep normalization and convolutional neural network for image smoke detection. IEEE Access **5**, 18429–18438 (2017)
18. Yuan, F., Xia, X., Shi, J., Li, H., Li, G.: Non-linear dimensionality reduction and gaussian process based classification method for smoke detection. IEEE Access **5**, 6833–6841 (2017)

Algorithm for Generating Tire Defect Images Based on RS-GAN

Chunhua Li[1,2], Ruizhi Fu[2](✉), and Yukun Liu[2]

[1] School of Humanity and Law, Hebei University Of Science and Technology, Shijiazhuang 050000, China

[2] School of Information Science and Engineering, Hebei University Of Science and Technology, Shijiazhuang 050000, China

1050793649@qq.com

Abstract. Aiming at the problems of poor image quality, unstable training process and slow convergence speed in the data expansion method for generating countermeasures network, this paper proposes a RS-GAN tire defect image generative model. Compared with traditional generative adversarial networks, RS-GAN integrates residual networks and attention mechanisms into an RSNet embedded in the adversarial network structure to improve the model's feature extraction ability; at the same time, the loss function JS divergence of the traditional generation countermeasure network is replaced by Wasserstein distance with gradient penalty term to improve the stability of model training. The experimental results show that the FID value of tire defect images generated by the RS-GAN model can reach 86.75, which is superior to the images generated by DCGAN, WGAN, CGAN, and SAGAN. Moreover, it has achieved more competitive results on SSIM and PSNR. The RS-GAN model can stably generate high-quality tire defect images, providing an effective way to expand the tire defect dataset and alleviating the small sample problem faced by the development of deep learning in the field of defect detection.

Keywords: Generative Adversarial Network · Dataset Expansion · Defect Detection

1 Introduction

At present, many tire manufacturers' detection methods for tire defects are still in the manual observation stage, which makes it difficult to control the errors and easily leads to errors and omissions, greatly reducing the efficiency of tire detection [16].

This work supported by the Key R&D Plan Project of Hebei Province (21351801D) and the Intelligent Operation and Maintenance Platform for Key Equipment of Rail Transit (20310806D).

B. Luo et al. (Eds.): ICONIP 2023, CCIS 1965, pp. 388–399, 2024.
https://doi.org/10.1007/978-981-99-8145-8_30

Deep learning is widely used in defect detection. Data is the core of deep learning [4]. The quality and scope of the data set directly affect the detection effect. If the data set is too small, the model is prone to non convergence and overfitting in the training process [20], resulting in low defect detection accuracy and poor effect. However, in actual industrial production, due to the complex process of collecting defect images, the defect images of tires are severely insufficient, and the relevant dataset is very limited.

Expanding data through generative models is an effective way to solve the small sample problem. Goodfellow et al. [5] proposed the Generative Adversarial Network (GAN) in 2014, which generates new samples by learning feature distributions. However, the generated images have poor quality, unstable training, and are prone to gradient explosions. In recent years, more and more researchers have improved and optimized based on GAN, and many new models have emerged. Mehdi et al. [15] proposed a Conditional Generative Adversarial Network (CGAN) by adding conditional supervised sample generation to the network, which can generate samples in a predetermined direction, but does not solve the problem of training instability. Radforo et al. [13] proposed a deep convolutional generative adversarial network (DCGAN), which uses a convolutional layer with step size to replace the pooling layer and fully connected layer of GAN, and adds a batch normalization layer (BN layer) to increase training stability. The proposal of DCGAN basically determines the basic structure of generating adversarial networks, but this structure still has problems such as limited feature extraction ability and unstable model training.

In order to solve the above problems, this paper proposes a RS-GAN tire defect image generative model based on DCGAN. The model integrates the attention mechanism and residual network into an RSNet and embeds it in DCGAN, and uses Wasserstein distance with gradient penalty term to replace the JS divergence used by the original loss function of DCGAN. The addition of residual networks increases the depth of the network, enabling the model to extract deeper image features. The attention mechanism has the feature of automatically assigning weights, allowing the model to learn more useful texture information and suppress useless noise information.

2 Related Work

2.1 Attention Mechanism

Traditional convolutional neural networks, when used as generators to generate images, are unable to connect two pixels that are far apart due to the fixed and limited size of the convolutional kernel and the limited focus area. During the training process, only local information of the image can be learned, which can easily lead to errors in the generated samples [3]. The attention mechanism SENet [8] first performs the Squeeze operation on the feature map obtained by convolution. The Squeeze operation uses global average pooling to expand the receptive field, encodes the entire spatial feature on a dimension into a global feature, and obtains the weight of each channel. The process can be expressed by formula (1):

$$z = f_{sq}(x) = \frac{1}{H \times W} \sum_{i=1}^{H} \sum_{j=1}^{W} x_{i,j} \tag{1}$$

In the formula, z represents the global feature; f_{sq} represents the Squeeze operation; x represents the input feature map; H represents the height of the feature map; W represents the width of the feature map; $x_{i,j}$ represents the feature vector representing of the i row and j column pixels.

Then perform an Excitation operation on the global features, learn the relationships between each channel, and apply weights to the feature map to obtain the final features. This process can be represented by formula (2):

$$s = f_{ex}(z, W) = \sigma\left(W_2 \delta\left(W_1 z\right)\right) \tag{2}$$

In the formula,s represents the incentive score vector; f_{ex} represents incentive operation; σ represents the sigmoid function; δ represents ReLu activation function.

2.2 Residual Networks and RSNet

For neural networks, ideally, the deeper the layers of the network, the better its performance. However, in the actual training process, as the network gradually deepens, there are often problems such as gradient instability, training difficulties, network degradation, etc. [11]. To solve the above problems, Kaiming He et al. [10] proposed a residual network (ResNet) based on convolutional neural networks (CNN) [1]. The core of ResNet is the introduction of residual blocks, which use skip connections or shortcuts to skip certain convolutional layers, effectively solving problems such as gradient explosion and network degradation.Multiple residual blocks are linearly connected to form a residual network, and the structure of the residual blocks is shown in Fig. 2(d).

Traditional generative adversarial networks have the problem of having fewer convolutional layers and incomplete feature extraction. However, simply deepening the layers of the network can lead to problems such as slow model calculation speed, gradient explosion, and network degradation [18]. Therefore, residual networks with skip connection structures are introduced to suppress gradient explosion and network degradation. However, residual networks do not have a direct advantage in extracting features, but instead achieve the most ideal effect by continuously delving deeper into the representation of more features layer by layer.

To address this issue, it is proposed to embed the attention mechanism SENet into the residual block, and to compensate for the shortcomings of the residual network in feature representation by automatically assigning weights through the attention mechanism. Based on the characteristics of the attention mechanism of "plug and play" and the principle of not damaging the original network structure, SENet is embedded after the residual branch of each residual block before aggregation. In this article, the residual block embedded in SENet is referred to as RSNet. The addition of RSNet only enhances the network's ability to extract features through deeper network layers and attention mechanisms, without changing the number of channels and image size of RSNet input. The structure of RSNet is shown in Fig. 2(c).

3 RS-GAN Tire Defect Image Generative Model

3.1 RS-GAN Model Framework

DCGAN is proposed on the basis of GAN. The idea of the algorithm is derived from the zero sum game in game theory [15]. On the basis of GAN, DCGAN adopts a full convolution network, and uses a convolution layer with step size to replace the pooling layer of generators and discriminators, Use 1×1 convolution replaces the fully connected layer in the structure; Apply the BN layer to every layer except for the output layer of the generator and the input layer of the discriminator; Except for the last layer of the generator, the rest of the layers use ReLu activation function and Leaky ReLu activation function, and the last layer uses Tanh activation function and Sigmoid activation function.

DCGAN consists of a generator and a discriminator. The generator generates as close to real samples as possible, and the discriminator distinguishes between the generated and real samples. The two optimize their respective abilities through continuous iteration, ultimately achieving Nash equilibrium. The objective function of the game between the generator and discriminator can be represented by formula (3):

$$\min_{G} \max_{D} V(D, G) = E_{x \sim P_r(x)}[\log(D(x))] + E_{z \sim P_g(z)}[\log(1 - D(G(z)))] \quad (3)$$

In the formula, x represents the real sample data; $P_r(x)$ represents the probability distribution of x ; E is the mathematical expectation; z is a random variable; $P_g(z)$ is the probability distribution of z; $V(D, G)$ represents the value functions for discriminators and generators.

This article improves on the basis of DCGAN, and the overall process of tire defect image generation is shown in Fig. 1. Firstly, a set of uniformly distributed random noise Z is fed into the generator as input, and the output generates sample G (z); Then, both the real sample and the generated sample G (z) are input into the discriminator to determine the authenticity of the generated sample G (z). The discriminator has a convolutional structure, aiming to determine the probability of true samples as 1. The generator has a transposed convolutional structure, aiming to prevent the discriminator from determining the authenticity of the generated samples. By continuously competing with iterative optimization generators and discriminators, the generated images become closer to real samples to achieve the goal of generating defect images. Although DCGAN is widely used in the field of data augmentation, there are still some problems, such as model collapse, slow convergence speed, and excessively free and uncontrollable sample generation.

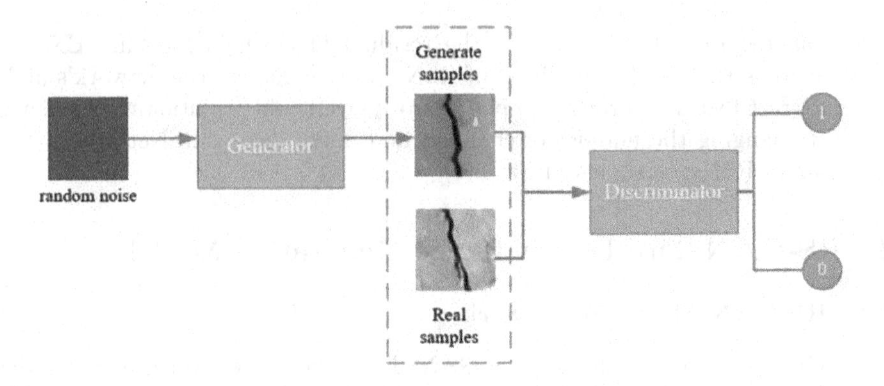

Fig. 1. Overall framework for defect image generation.

3.2 Generators and Discriminators for RS-GAN

As the core of the entire network, the generator aims to generate tire defect images. The structure of the RS-GAN generator is shown in Fig. 2(a). Input a 100 dimensional linear vector into the model, reconstruct it into a 256×8×8 feature map through a linear layer, and introduce RSNet before the feature map is sent to the deconvolution layer. The addition of RSNet only improves the network's feature extraction ability and does not change the size and number of channels of the output feature map. Send the output characteristic maps into the transposed convolution layer, BN layer, and ReLu Activation function layer in turn to get the characteristic maps of 128×16×16. Then, after two deconvolution layers, except that Tanh is used as the Activation function for network output, each other convolution layer uses ReLu as the Activation function. The 256×8×8

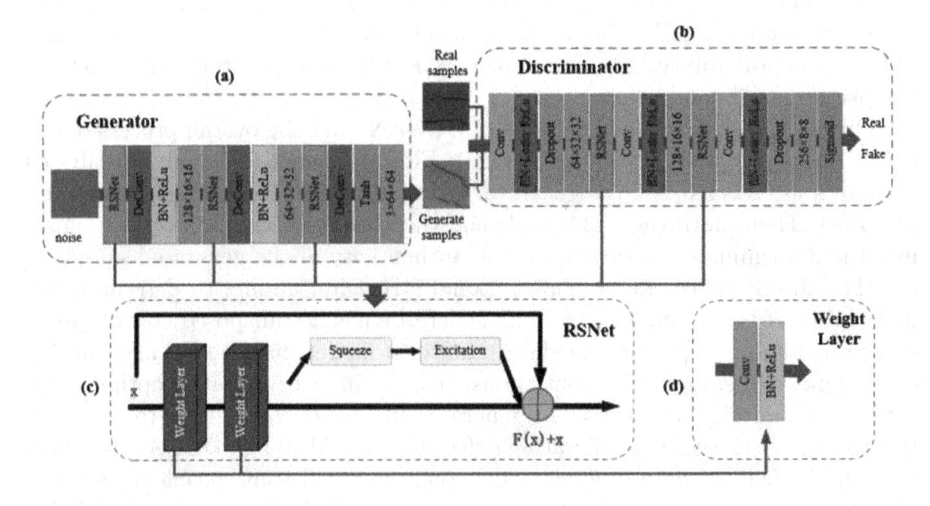

Fig. 2. The framework of the proposed RS-GAN model. (a) Generator structure (b) Discriminator structure (c) RSNet structure (d) Weight Layer structure

feature maps input by the generator are subjected to a series of operations such as 3×64×64 deconvolutions to obtain 3 generated samples as the output of the generator.

The structure of the RS-GAN discriminator is shown in Fig. 2(b). The input of the discriminator is the output image of the generator, and the discriminator consists of three convolutional layers. The feature map of input 3×64×64 is continuously convolved into a feature map of 256×8×8. The activation function of three-layer convolution uses Leaky ReLu. RSNet and BN layers are added before the second and third layer convolutions. The Dropout layer is added after the layering in Volumes 1 and 3 to prevent the discriminator from being too optimized and causing overfitting in training. Finally, the 256×8×8 feature map is passed through a sigmoid activation function to obtain the probability of true or false samples generated by the generator.

3.3 Optimization of Loss Function

For traditional generative adversarial networks, simply improving the structure of the network model cannot fundamentally solve the problems of difficult training and unstable gradients in generative networks. There is a problem with the loss function JS divergence used by the traditional generation countermeasure network. When the real distribution and the generated distribution do not intersect, the JS divergence will become constant, resulting in unstable training of the gradient disappearance model. For this reason, Gulrjani et al. [6] proposed to use Wasserstein distance with gradient penalty term to improve the loss function. Wasserstein distance has the advantage of smoothness, reflecting the distance between the real distribution probability and the generated distribution probability. The smaller Wasserstein distance, the more similar the real distribution and the generated distribution, the smaller the gap between the generated samples and the real samples. The Wasserstein distance can be represented by formula (4):

$$W\left(P_r, P_g\right) = \inf_{\gamma \in \Pi\left(P_r, P_g\right)} E_{(x,y)\sim\gamma}[\|x-y\|] \quad (4)$$

In the formula, P_r represents the true distribution probability; P_g represents the probability of generating a distribution; $\Pi\left(P_r, P_g\right)$ represents the set of joint distribution $\left(P_r, P_g\right)$; γ represents a joint distribution; inf represents the lower bound; x represents a real sample; y represents the generated sample; $\|x-y\|$ represents the distance between x and y.

In order to solve the problem of JS divergence prone to sudden changes, a penalty term Lipschitz function [2] is introduced while using Wasserstein distance to constrain the discriminator weight by limiting the variation amplitude of the objective function. The loss function of generator and discriminator can be expressed by formula (5) and formula (6) respectively:

$$L_G = E_{x\sim P_r}(D(x)) - E_{x\sim P_g}(D(x)) \quad (5)$$

$$L_D = E_{x\sim P_r}(1 - D(G(z))) - E_{x\sim P_g}(D(x)) + \lambda E_{n\sim P_n}\left((\|\nabla_n D(n)\|_P - 1)^2\right) \quad (6)$$

In the formula: n represents the random difference between the real distribution and the generated distribution; P_n represents the difference between the real distribution and the generated distribution; λ represents the coefficient of the regularization term; $\nabla_n D(n)$ represents a gradient constraint.

4 Experimental Results and Analysis

4.1 Evaluation Indicators and Generated Image Display

Reference [14] points out that there is no fixed indicator for evaluating the effectiveness of GAN generated images. In order to more objectively reflect the quality of generated images, FID (Freechet Perception Distance Score) [7], SSIM (Structural Similarity Index) [17], and PSNR (Peak Signal-to-Noise Ratio) [9] were used as the main indicators for evaluating the quality of generated images in this experiment. FID calculates the distance between the feature vectors of the real image and the generated image. The smaller the FID value, the closer the generated image is to the real image [19]. FID extracts the features of both real and generated images through a pre trained Inception v3 network, and the calculation formula can be expressed by formula (7):

$$FID = \|\mu_r - \mu_g\|^2 + \mathrm{Tr}\left(\sigma_r + \sigma_g - 2\left(\sigma_r\sigma_g\right)^{1/2}\right) \tag{7}$$

In the formula,μ_r is the mean of real image features; μ_g is the mean of generated image features; σ_r is the variance of real image features; σ_g is the variance of the generated image features.

SSIM is commonly used to evaluate the similarity between two images. The range of SSIM is 0–1, and a larger SSIM value indicates that the two images are closer. For both real and generated images, SSIM can be represented by formula (8):

$$SSIM(r,g) = \frac{(2\mu_r\mu_g + c_1)(\sigma_{rg} + c_2)}{\left(\mu_r^2 + \mu_g^2 + c_1\right)\left(\sigma_r^2 + \sigma_g^2 + c_2\right)} \tag{8}$$

In the formula, σ_{rg} is the covariance between the real image and the generated image; $c_1 c_2$ is a constant.

PSNR is obtained by calculating the error between the pixels corresponding to the real sample and the generated sample. The higher the PSNR value, the higher the quality of the generated image. PSNR can be represented by formula (9):

$$PSNR = 10 \cdot \log_{10}\left(\frac{MAX_I^2}{MSE}\right) \tag{9}$$

In the formula, MSE is the mean square deviation between the real image and the generated image; MAX is the maximum pixel value in the image.

Figure 3 shows the display of tire defect images generated by RS-GAN. In order to demonstrate in more detail the effects of images generated by different iterations, one generated image is selected as a display after every 200 iterations during the training process, as shown in Fig. 4. As shown in Fig. 4, RS-GAN can

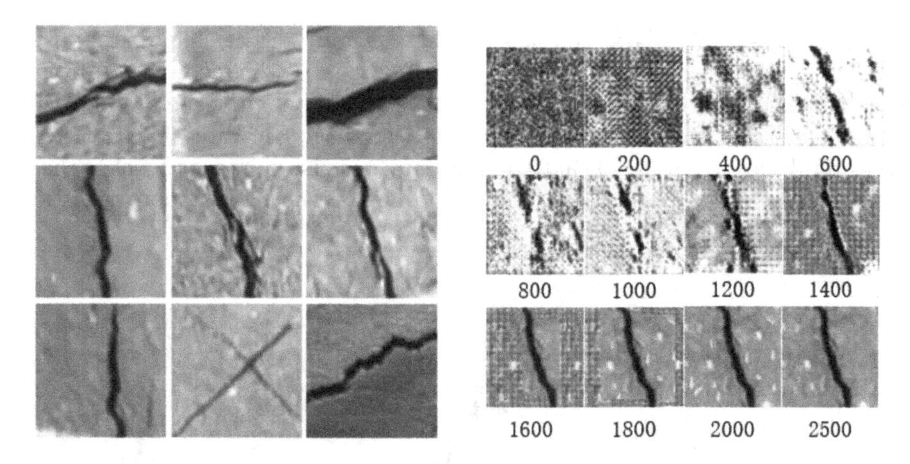

Fig. 3. Generate image display

Fig. 4. Iterative training display

generate defects at 600 iterations, but the generated image is severely distorted at this time; when the iteration reaches 1400 times, the generated image is greatly improved, but the degree of background gridding is very severe; when iterating to 1800 times, the generated image is already close to the real image, except for some grids on the edge of the image; when the iteration reaches 2000–2500 times, the generated image is already very close to the real image

4.2 Results and Analysis of Ablation Experiments

RS-GAN adopts three improvement measures to improve the original DCGAN, namely embedding residual structure and attention mechanism into the DCGAN structure; Wasserstein distance with gradient penalty term is used to replace the loss function of the original DCGAN. In order to verify the gain effect of improvement measures on the generated images of the model, ablation experiments were designed to verify the improvement effects of three improvement measures on the model. Select DCGAN as the baseline model, and add residual network (DCGAN Resnet), attention mechanism (DCGAN SENet), and Wasserstein distance with gradient penalty term (DCGAN Wasserstein) to the baseline model. For these five models, the ablation experiment uses the same Learning rate, iteration number and other indicators to ensure the accuracy of the experiment as much as possible. From both qualitative and quantitative perspectives, it verifies the gains of improvement measures on the model.

Figure 5 qualitatively displays the images generated by five models of ablation experiments. From Fig. 5 (a), (b), (c), and (d), it can be seen that the images generated by the four models already have a rough outline of defects, but the edges of the defects are blurry, easy to fuse with the background, and there are many false textures, resulting in low image quality. Although the images generated by DCGAN have defects, they cannot be fully generated. DCGAN Wasserstein did not experience gradient explosion or other issues during the training process; DCGAN-SENet can roughly generate defect shapes; But the background is severely blurred; DCGAN-ResNet can distinguish defects from the background, and the image background is clearer, but the generated defect shape is not obvious and severely distorted. The image generated by RS-GAN has clearer defect edges, significantly reduced image noise points, and a clear distinction between the overall defect and background, resulting in a more realistic image.

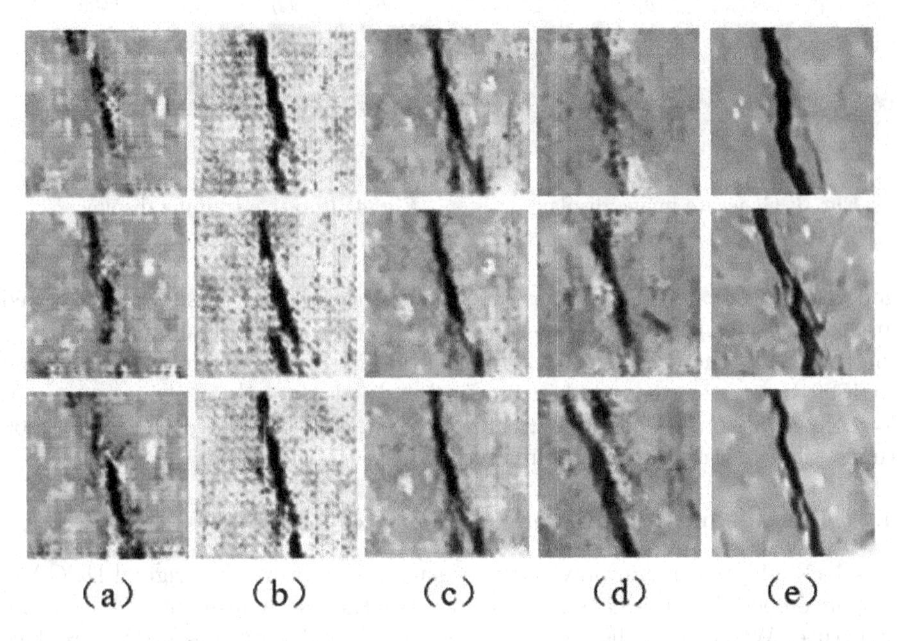

Fig. 5. Results of ablation experiments. (a) Baseline (b) DCGAN-Wasserstein (c) DCGAN-SENet (d) DCGAN-ResNet (e) RS-GAN.

From a quantitative perspective, the FID, SSIM, and PSNR values of the images generated by the five models of ablation experiments were compared, and the results are shown in Table 1. From Table 1, it can be seen that the FID values of the images generated based on the three improvement measures are all lower than Baseline, and the FID value of the images generated by RS-GAN is 86.75, which is much lower than 125.77 of the images generated by Baseline. The SSIM and PSNR values of the generated images based on three improvement measures are higher than those of Baseline. In summary, combining the experimental results from both qualitative and quantitative perspectives, it can be concluded that all three improvement measures for RS-GAN are beneficial for improving the quality of generated images.

Table 1. Quantitative evaluation of ablation experiments.

Models	FID	SSIM	PSNR
Baseline	125.77	0.64	27.45
DCGAN-Wasserstein	120.14	0.65	28.07
DCGAN-SENet	99.98	0.75	31.41
DCGAN-ResNet	102.37	0.73	30.62
RS-GAN	86.75	0.86	35.15

4.3 Results and Analysis of Comparative Experiments

In order to further verify the superiority of RS-GAN compared to existing mainstream models, RS-GAN was compared qualitatively and quantitatively with existing mainstream DCGAN, WGAN, CGAN, and SAGAN. WGAN replaces the loss function of the traditional generation countermeasure network with Wasserstein distance without adding penalty coefficient; CGAN adds conditional constraints to the inputs of the generator and discriminator; SAGAN incorporates self attention mechanism into the model structure.

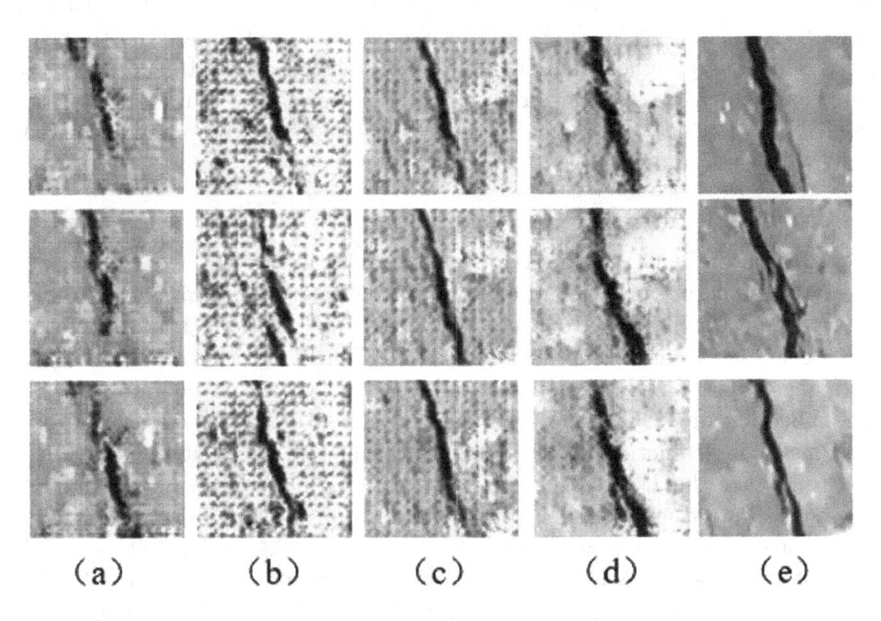

Fig. 6. Results of comparative experiments. (a) DCGAN (b) WGAN (c) CGAN (d) SAGAN (e) RS-GAN.

Figure 6 qualitatively displays the images generated by the five models in the comparative experiment. From Fig. 6, it can be seen that although the images generated by DCGAN generate defects, they cannot be fully generated. WGAN

Table 2. Quantitative evaluation of comparative experiments.

Models	FID	SSIM	PSNR
Baseline	125.77	0.64	27.45
WGAN	113.45	0.72	29.37
CGAN	105.24	0.74	30.86
SAGAN	96.68	0.79	33.88
RS-GAN	86.75	0.86	35.15

only improved the loss function, and there was no gradient explosion and other problems in the training process. The image generated by CGAN has blurry edges and severe background gridding. SAGAN with self attention mechanism can generate clearer defects, but the edges of the defects are blurry and there are many noise points in the background. Compared with the other four models, the image generated by RS-GAN shows clearer defect shapes, more natural fusion of defects and background, significant improvement in background meshing, and no issues such as gradient explosion during training.

From a quantitative perspective, the FID, SSIM, and PSNR values of the images generated by the five models of ablation experiments were compared, and the results are shown in Table 2. From Table 2, it can be seen that the FID value of RS-GAN generated images is 86.75, which is lower than the other four comparative models; The SSIM and PSNR values were 0.86 and 35.15, respectively, which were higher than the four comparative models. In summary, combining experimental results from both qualitative and quantitative perspectives, RS-GAN has better performance in generating tire image compared to DCGAN, WGAN, CGAN, and SAGAN.

5 Conclusion

On the basis of DCGAN, this paper proposes to improve it from two aspects of network structure and loss function, which effectively improves the problems of low image quality, slow convergence speed and unstable training of traditional generation countermeasure network. In terms of network structure, the residual network and attention mechanism (SENet) are fused into an RSNet and embedded into DCGAN to enhance the model's feature extraction ability. In terms of loss function, Wasserstein distance with gradient penalty term is used to replace the JS divergence used by the original DCGAN to improve the convergence speed and stability of the model. Using FID, SSIM, and PSNR as evaluation indicators for image generation, the experimental results show that the RS-GAN model generates images with a FID value of 86.75, and SSIM and PSNR can reach 0.86 and 35.15 respectively, resulting in better image quality than DCGAN, WGAN, CGAN, and SAGAN. This model can generate high-quality tire defect images, providing theoretical reference and methodological basis for expanding the defect sample dataset, and thus promoting the development of deep learning in the field of tire defect detection. However, this model is not ideal in improving the diversity of defect images, and future research will continue with the aim of generating multiple types of defect images.

References

1. Alzubaidi, L., Zhang, J., Humaidi, A.J., et al.: Review of deep learning: concepts, CNN architectures, challenges, applications, future directions. J. Big Data **8**, 1–74 (2021)
2. Arjovsky, M., Chintala, S., Bottou, L.: Wasserstein GAN. In: Proceedings of the 34th International Conference on Machine Learning (ICML), Sydney, pp. 214–223 (2017)
3. Cheng, J., Jiang, N., Tang, J., Deng, X., Yu, W., Zhang, P.: Using a Two-Stage GAN to learn image degradation for image Super-Resolution. In: ICONIP (2021)
4. Dixit, M., Tiwari, A., Pathak, H., Astya, R.: An overview of deep learning architectures, libraries and its applications areas. In: International Conference on Advances in Computing, Communication Control and Networking (ICACCCN), Greater Noida, pp. 293–297 (2018)
5. Goodfellow, I., Pouget-Abadie, J., Mirza, M., et al.: Generative adversarial networks. Commun. ACM, **63**(11), 139–144
6. Gulrajani, I., Ahmed, F., Arjovsky, M., et al.: Improved training of Wasserstein GANs. arXiv preprint arXiv: 1704.00028 (2017)
7. Heusel, M., Ramsauer, H., Unterthiner, T., et al.: GANs trained by a two time-scale update rule converge to a local Nash equilibrium. In: Proceedings of the 31st International Conference on Neural Information Processing Systems (NIPS), Long Beach, pp. 6626–6637 (2017)
8. Hu, J., Shen, L., Sun, G.: Squeeze-and-excitaton networks. In: Proceedings of the IEEE Conference on Computer Vision and Patten Recognition (CVPR), Salt Lake City, pp. 7132–7141 (2018)
9. Huynh-Thu, Q., Ghanbari, M.: Scope of validity of PSNR in image/video quality assessment. Electron. Lett. **44**(13), 800–801 (2008)
10. He, K., Zhang, X., et al.: Deep residual learning for image recognition. In: Proceedings of the IEEE Conference on Computer Vision and Patten Recognition (CVPR), Las Vegas, pp. 770–778 (2016)
11. Krueangsai, A., Supratid, S.: Effects of shortcut-level amount in lightweight resNet of resNet on object recognition with distinct number of categories. In: International Electrical Engineering Congress (iEECON), Khon Kaen, pp. 1–4 (2022)
12. Mirza, M., Osindero, S.: Conditional generative adversarial nets. arXiv preprint arXiv: 1411.1784 (2014)
13. Radird, A., Mez, L., Chintala, S.: Unsupervised representation learning with deep convolutional generative adversarial networks. arXiv preprint arXiv: 1511.06434 (2015)
14. Shmelkov, K., Coroelia, S., Arteek, A.: How good is my GAN?. In: European Conference on Computer Vision (ECCV), Munich, pp. 1–20 (2018)
15. Srivastaa, N., Hinton, G., Krlzhevsky, A., et al.: Dropout a simple way to prevent neural networks from overfitting. J. Mach. Learn. Res. **15**(1), 1929–1958 (2014)
16. Wang, N., Wang, Y.: Review on deep learning techniques for marine object recognition: architectures and algorithms. Control. Eng. Pract. **118**, 104458 (2022)
17. Wang, Z., Bovik, A.C., Sheikh, H.R., et al.: Image quality assessment from error visibility to structural similarity. IEEE Trans. Image Process. **13**(4), 600–612 (2004)
18. Wang, W., Li, Z.: Research progress in generative adversarial networks. J. Commun. **39**(2), 135–148 (2018)
19. Xueyun, C., Xiaoqiao, H., Li, X.: A method for generating adversarial network blood cell image classification and detection based on multi-scale conditions. J. Zhejiang Univ. (Eng. Ed.) **55**(09), 1772–1781 (2021)
20. Ying, X.: An overview of overfitting and its solutions. J. Phys: Conf. Ser. **1168**(2), 022022 (2019)

Novel-Registrable Weights and Region-Level Contrastive Learning for Incremental Few-shot Object Detection

Shiyuan Tang, Hefei Mei, Heqian Qiu, Xinpeng Hao, Taijin Zhao, Benliu Qiu, Haoyang Cheng, Jian Jiao, Chuanyang Gong, and Hongliang Li(✉)

University of Electronic Science and Technology of China, Chengdu, China
{sytang,hfmei,xphao,zhtjww,qbenliu,chenghaoyang,jij2021, cygong}@std.uestc.edu.cn, {hqqiu,hlli}@uestc.edu.cn

Abstract. Few-shot object detection (FSOD) methods learn to detect novel objects from a few data, which also requires reusing base class data if detecting base objects is necessary. However, in some real applications, it is difficult to obtain old class data due to privacy or limited storage capacity, causing catastrophic forgetting when learning new classes. Therefore, incremental few-shot object detection (iFSOD) has attracted the attention of researchers in recent years. The iFSOD methods continuously learn novel classes and not forget learned knowledge without storing old class data. In this paper, we propose a novel method using novel-registrable weights and region-level contrastive learning (NWRC) for iFSOD. First, we use novel-registrable weights for RoI classification, which memorizes class-specific weights to alleviate forgetting old knowledge and registers new weights for novel classes. Then we propose region-level contrastive learning in the base training stage by proposal box augmentation, enhancing the generalizability of the feature representations and plasticity of the detector. We verify the effectiveness of our method on two experimental settings of iFSOD on COCO and VOC datasets. The results show that our method has the ability to learn novel classes with a few-shot dataset and not forget old classes.

Keywords: Few-shot learning · Incremental learning · Object detection · Contrastive learning

1 Introduction

Deep neural networks have achieved excellent results in object detection task [24]. In some daily life applications, object detection systems should update networks to detect objects of new classes continuously. However, deep neural networks often need a large dataset for training and will overfit when training on a few data in a naive way. Researches on few-shot object detection (FSOD) [7,31] train the network on large amounts of data from base classes and then learn on a few

B. Luo et al. (Eds.): ICONIP 2023, CCIS 1965, pp. 400–412, 2024.
https://doi.org/10.1007/978-981-99-8145-8_31

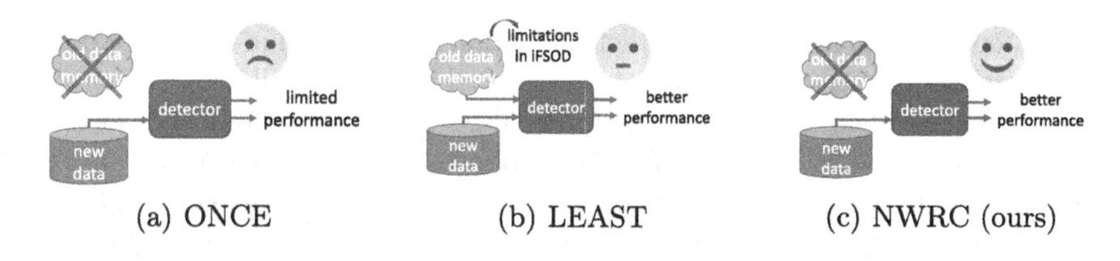

(a) ONCE (b) LEAST (c) NWRC (ours)

Fig. 1. The training data and performance of different methods when learning novel classes. Our method NWRC performs better on iFSOD without old data memory.

data from novel classes, so that the network can detect novel classes objects. However, when fine-tuning on novel classes, it is unworthy to lose the ability to detect the abundant base class objects. Therefore, some researchers propose generalized few-shot object detection (gFSOD) [27], which considers both the performance of base classes and novel classes in the fine-tuning stage. Although the research on gFSOD is very successful [23,28], this training pattern needs memory to store part or all of the base class data and learn all novel classes concurrently. We hope the object detector can continuously learn new classes with only a few training samples after base training on a large dataset and not forget old knowledge without storing old class data. This means the detector needs to have the capability of class-incremental learning. This problem setting is called incremental few-shot object detection (iFSOD) [22].

Pérez-Rúa et al. [22] proposed a meta-learning approach named ONCE after raising the iFSOD problem. ONCE trains a code generator to get weights from the support set of novel classes. As a pioneering work, ONCE has limited performance in solving iFSOD problems, as shown in Fig. 1(a). After that, LEAST [18] introduces knowledge distillation into iFSOD. However, as shown in Fig. 1(b), LEAST still needs to store base-class samples for knowledge distillation, which has limitations in the iFSOD setting.

In this paper, we choose fine-tuning-based TFA [27] as the baseline and consider addressing the iFSOD problem following two aspects: novel-registrable weights and region-level contrastive learning. The advantages of our method compared to other methods are shown in Fig. 1(c).

First, we use novel-registrable weights (NRW) for RoI classification. When using the fully-connected classifier, the outputs of new class neurons will be higher, thereby inhibiting the confidences of old classes, causing catastrophic forgetting. We use the novel-registrable weights that memorize class-specific weights to alleviate forgetting old knowledge and register new weights for new classes. Inspired by [14,27,30], we conduct cosine similarity between classification features and weights in NRW to get balanced outputs. We further remove weights for the background to guarantee separability of potential foreground classes. NRW can avoid forgetting in a fixed feature space and achieve incremental capability.

Second, we propose region-level contrastive learning (RLCL) in the base training stage to enhance the feature extractor. As a self-supervised learning method, contrastive learning can efficiently mine information in images and learn good representations without any labels [2,12]. There are challenges in introducing self-supervised contrastive learning into iFSOD, including plural objects and proposals with semantic overlapping. We propose to view regions as instances for contrast and use proposal box augmentation to realize region-level contrastive learning to the object detection task and apply it in the base training stage. It can enhance the detector to extract more generalized features and have better capability to learn few-shot novel classes.

Our contributions are summarized as follows:

- We use novel-registrable weights for RoI classification, which alleviates forgetting and achieves incremental capability.
- We propose region-level contrastive learning applied in the base training stage, which propels the feature extractor to get more generalized features and enhance the plasticity of the detector.
- Our experimental results on COCO and VOC datasets show that our method has a good ability to learn new classes with a few-shot dataset and does not forget old classes.

2 Related Works

FSOD. Few-shot object detection methods include two categories, meta-learning-based methods [7,15,17] and fine-tuning-based methods [23,26–28]. The former learns a few-shot paradigm on base classes training data and applies the paradigm to novel classes data. The latter usually performs better than the former on the FSOD problem. TFA [27] first proposes a simple FSOD strategy that only needs to fine-tune the RoI classifier and regressor of the detector when learning novel classes. Based on TFA, FSCE [26] carefully adjusts hyperparameters and fine-tunes part of the feature extractor, and introduces supervised contrastive learning into fine-tuning stage to enhance the separability of novel and base classes. DeFRCN [23] decouples features for RPN and RoI, alleviating mismatched goals between different losses. MFDC [28] introduces knowledge distillation into FSOD, which explicitly transfers the knowledge of base classes to novel classes. The above fine-tuning-based approaches freeze all or most layers of the feature extractor in the fine-tuning stage to avoid overfitting.

iFSOD. As same as FSOD, incremental few-shot object detection also has two kinds of methods, meta-learning-based methods [4,22,32] and fine-tuning-based methods [18,19]. ONCE [22] and Sylph [32] train hypernetworks on base training data to generate class codes as part of parameters in the detector. While fine-tuning based approaches have better performance than meta-learning-based ones. LEAST [18] uses knowledge distillation strategy to prevent forgetting old knowledge in the fine-tuning stage. [19] proposes a double-branch framework to decouple the feature representation of base and novel classes. iMTFA [8] uses Mask R-CNN detector, which can also deal with the iFSOD problem while solving incremental few-shot instance segmentation. It implements the incremental mode with a non-trainable cosine similarity classifier based on the TFA strategy. These methods need to store base-classes images or features for knowledge inheritance, which takes up more memory but is inefficient.

Contrastive Learning. Contrastive learning learns representations by contrasting positive and negative pairs [11]. [5] treats each instance as a class and extends contrastive learning as a self-supervised learning paradigm. SimCLR [2] systematically depicts the structure of contrastive learning and expounds on the effects of data augmentation, loss function, and other factors on contrastive learning. MoCo [12] sets up a dynamic queue to expand the number of negative samples in the loss function, and momentum updating that ensures the continuity of the encoder. BYOL [10] proposes a novel strategy without using negative pairs for contrast while the training does not collapse. SimSiam [3] designs a simple Siamese network to learn representations based on the analysis of BYOL. Common contrastive learning methods are self-supervised. SupCon [16] proposes a supervised contrastive learning method. FSCE [26] introduces supervised contrastive learning into the gFSOD fine-tuning to enlarge the difference of inter-class features. Different from FSCE, we introduce self-supervised contrastive learning into iFSOD base training so that the feature extractor can get more generalized features.

3 Method

3.1 Problem Formulation

The fine-tuning based incremental few-shot object detection training process consists of two stages, base training and fine-tuning. The fine-tuning stage is split into different incremental phases, $\{\phi_1, \phi_2, ..., \phi_T\}$. Therefore we can define the base training stage as ϕ_0. Each phase ϕ_t uses only the training dataset $\mathcal{D}_t = \{(x_i^t, y_i^t)\}_{i=1}^{M_t}$, where x_i^t is the input image with N objects and $y_i^t = \{(c_{i,j}^t, b_{i,j}^t)\}_{j=1}^{N_i^t}$ represents the labels and box locations of objects in the image. Note that each label in phase t only belongs to the classes of phase t, namely $c_{i,j}^t \in \mathcal{C}_t$, and the classes of different phases do not overlap, namely $\mathcal{C}_{t_1} \cap \mathcal{C}_{t_2} = \emptyset, t_1 \neq t_2$. Under the setting of K-shot, $\sum_{i=1}^{M_t} N_i^t = K\,|\mathcal{C}_t|$ when $t > 0$, and $\sum_{i=1}^{M_t} N_i^t \gg K\,|\mathcal{C}_t|$ when $t = 0$.

The entire iFSOD training process is as follows: Firstly, we train the detector on abundant base classes training data $\mathcal{D}_0$ to obtain a base detector $\mathcal{F}_0$. Secondly, the detector is fine-tuned on novel classes data. Specifically, the detector $\mathcal{F}_{t-1}$ is pushed to the next new phase for fine-tuning on new classes training data $\mathcal{D}_t$ to obtain a new detector $\mathcal{F}_t$. The detector $\mathcal{F}_t$ can detect objects of old classes $c_{old} \in \bigcup_{\tau=0}^{t-1} \mathcal{C}_\tau$ and new classes $c_{new} \in \mathcal{C}_t$.

3.2 Novel-Registrable Weights

Faster-RCNN uses a fully-connected layer as the RoI classifier [24]. It can linearly map high-dimensional features to scores of classes. We can regard the part before the classifier and regressor as a feature extractor. Few-shot novel class data cannot support the training of complex feature extractor which contains rich knowledge. The knowledge required by novel classes is transferred from the base classes. We freeze the feature extractor in the fine-tuning stage to preserve base-classes knowledge and prevent overfitting a few novel class data.

The feature space is fixed in the fine-tuning stage on account of the frozen feature extractor. To avoid forgetting old classes, the weights of the old class neurons in the classifier should also be retained and fixed [30]. For any input image, the output logits of the old class neurons are constant. During training, the logits will sequentially go through softmax activation layer and cross-entropy loss [9], namely

$$\mathcal{L}_{cls}^{t} = -\frac{1}{N_o}\sum_{i=1}^{N_o} \log \frac{\exp l_{i,gt}}{\sum_{j=1}^{N_c^t} \exp l_{i,j} + \exp l_{i,bg}} \tag{1}$$

where N_o is the number of sampled RoIs, $N_c^t = \sum_{\tau=0}^{t} |\mathcal{C}_\tau|$ is the total number of classes in phase t, and $l_{i,j}$ is the j-th output logit of i-th proposed objects. gt means ground-truth class and bg means background class. Since $gt \in \mathcal{C}_t \cup \{bg\}$ and the distribution of $l_{i,j}(j \notin \mathcal{C}_t)$ is constant, the mean of $l_{i,j}(j \in \mathcal{C}_t)$ distribution will be higher than that of $l_{i,j}(j \notin \mathcal{C}_t)$ as the loss descends. That means the confidences of old classes will be inhibited. The imbalance problem leads to forgetting in the classifier.

To better memorize the old classes and enhance the incremental ability, we use novel-registrable weights (NRW), shown in Fig. 2. Each weight vector in NRW corresponds to a class. When the detector learns new classes, NRW will register a new weight vector for each new class and randomly initialize them. Inspired by [14,27,30], output logits are calculated by cosine similarity of features and weights to avoid inhibiting the confidences of old classes, namely

$$l_{i,j} = \alpha \frac{r_i \cdot w_j}{||r_i|| \, ||w_j||} \tag{2}$$

where r_i is i-th feature, w_j is the weight vector of j-th class, and α is a scaling factor. Compared with the fully-connected layer, the logits calculated by cosine similarity are balanced for different classes, contributing to the registration ability. Note that the NRW can be regarded as a trainable nearest-distance-based classifier with weight vectors as adaptive prototypes. It is undesirable that the features of background boxes are concentrated around one prototype, thereby reducing the separability

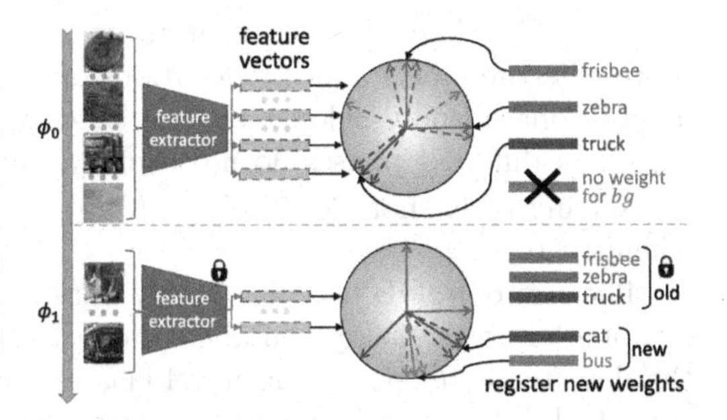

Fig. 2. Novel-registrable weights for RoI classification. Black arrows mean normalization, mapping features and weights to cosine similarity metric space. Locks mean frozen parameters. Weights for the background are removed.

of potential objects. Therefore, we remove the weights for the background class in NRW, and the logit of the background class is calculated indirectly according to the highest logit of foreground classes, namely

$$l_{i,bg} = 1 - \max_{j \in \mathcal{C}_{all}} l_{i,j} \tag{3}$$

where $\mathcal{C}_{all} = \bigcup_{\tau=0}^{t} \mathcal{C}_{\tau}$. In this way, the foreground class outputs of the background instances can be suppressed as the loss decreases.

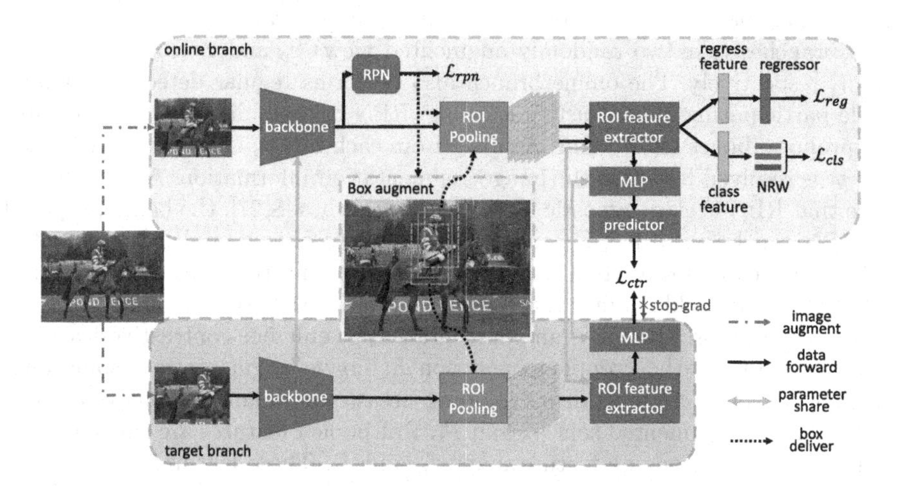

Fig. 3. Illustration of region-level contrastive learning, based on Faster R-CNN detector. The proposal boxes generated by the RPN of the online branch are directly submitted to the RoI head for routine detection training. At the same time, they are augmented into two groups for two branches to perform RoI pooling. Features for classification extracted by the RoI head from these two groups forward propagate through an MLP and a predictor (online-only). The outputs are put into the contrastive loss. Parameters of the entire target branch are shared with the online branch and do not require back-propagation.

3.3 Region-Level Contrastive Learning

We choose to freeze the feature extractor in fine-tuning stage, which limits the plasticity of the network. For few-shot novel classes, the knowledge required for detection is transferred from base-classes knowledge. The potential objects in the background are the key to extracting generalized knowledge. Considering that there is no annotation in the background, we introduce self-supervised contrastive learning, which treats each image instance as a class and guides the network to learn representations by instance discrimination [29].

Applying contrastive learning to object detection has two challenges. Firstly, in the detection task, there could be more than one object in an image. Different locations in an image have noticeable semantic differences. Therefore, instead of viewing each image as an instance, we treat the objects in proposal boxes as instances for contrast. Secondly, we counted the IoU values between proposal boxes proposed by RPN on VOC test set. 59.1% of pairs have overlapping areas, and 20.9% of pairs have more than 0.3 IoU values which cannot be considered negative pairs. We apply contrastive learning without using negative pairs similar to BYOL [10] and SimSiam [3].

To solve the above problems, we propose region-level contrastive learning to enhance the feature extractor, as shown in Fig. 3. Similar to the structure of BYOL, our method has two branches named online branch and target branch. Two branches take two randomly augmented views v_o and v_t from image x as input, respectively. The online branch also performs regular detection training while participating in contrastive learning. RPN of the online branch calculates the proposal boxes set $\mathcal{P}_o$. The mapping from each box $p_o \in \mathcal{P}_o$ to target branch box p_t is resolved based on the image augmentation information. As we hypothesize that RPN is class-agnostic like other approaches [8,27], the boxes proposed by RPN can be regarded as containing base class objects and other potential objects. We cannot directly contrast p_o and p_t, because the convolutional neural network in the backbone has translation invariance [1]. We randomly offset the location of proposal boxes as an augmentation to enhance contrastive learning. Note that due to the cropping operation in image augmentation, some augmented box pairs have small overlapping areas and should be removed. Then we get filtered augmented sets $\hat{\mathcal{P}}_o$ and $\hat{\mathcal{P}}_t$. RoI pooled features are calculated by RoI-Align pooling on the feature maps f_o and f_t. These features are input to RoI feature extractor F. Similar to [10], the outputs are projected by an MLP G, and a predictor P is used in the online branch to predict the outputs of the target branch. The contrastive loss is expressed as follows:

$$\mathcal{L}_{ctr} = 1 - \frac{1}{\left|\hat{\mathcal{P}}_o\right|} \sum_{\hat{p}_o \in \hat{\mathcal{P}}_o, \hat{p}_t \in \hat{\mathcal{P}}_t} \langle P(R(f_o, \hat{p}_o)), R(f_t, \hat{p}_t) \rangle \tag{4}$$

where $\langle \cdot, \cdot \rangle$ is cosine similarity which can be expressed by Eq. 2 without α. $R(\cdot, \cdot)$ is the abbreviation of mapping $G(F(RoIAlign(\cdot, \cdot)))$.

Finally, we define the loss function for base training as

$$\mathcal{L}_{base} = \mathcal{L}_{rpn} + \mathcal{L}_{cls} + \mathcal{L}_{reg} + \lambda \mathcal{L}_{ctr} \tag{5}$$

where the first three items are the regular loss of Faster-RCNN and λ is a scale of contrastive loss.

4 Experiments

4.1 Experimental Setting

We perform experiments on PASCAL VOC [6] and MS COCO [20] benchmarks. For VOC, we adopt the data splits and training examples provided by [27]. 15 of

Table 1. iFSOD performance on COCO val set. AP_n, AP_b, and AP_m are AP of base classes, novel classes, and the mean of all classes. MRCN is Mask R-CNN detector. FRCN is Faster R-CNN. R50 and R101 are ResNet backbones. 10 random sets of novel classes samples are considered in our results. **Bold** and underline indicate the best and the second-best.

Method	Detector	1-shot			5-shot			10-shot		
		AP_n	AP_b	AP_m	AP_n	AP_b	AP_m	AP_n	AP_b	AP_m
ONCE [22]	CentreNet-R50	0.7	17.9	13.6	1.0	17.9	13.7	1.2	17.9	13.7
iMTFA [8]	MRCN-R50-FPN	3.2	27.8	21.7	6.1	24.1	19.6	7.0	23.4	19.3
LEAST [18]	FRCN-R101-C4	**4.4**	24.6	19.6	**9.4**	25.2	21.3	**12.5**	23.1	20.5
Sylph [32]	FCOS-R50-FPN	0.9	29.8	22.6	1.4	35.5	27.0	1.6	35.8	27.3
ours	FRCN-R50-FPN	2.8	**37.3**	**28.7**	7.1	**37.3**	**29.8**	9.1	**37.3**	**30.3**

20 classes are base classes, and others are novel classes. The training data come from trainval sets of VOC 2007 and 2012, and evaluation data come from test set of VOC 2007. We report AP_{50} for VOC results. For COCO, we adopt the data split used by previous few-shot object detection [15,23,26–28] and incremental few-shot object detection works [18,22,32]. COCO has a total of 80 classes, of which 20 classes are included in VOC. These 20 classes are used as novel classes, and the remaining 60 classes are regarded as base classes. Each novel class has $K \in \{1, 5, 10\}$ instances in training data. We evaluate our trained model on the COCO minival set with 5000 images as in [15] and report COCO-style mAP.

We use Faster-RCNN [24] as the basic detection framework, which includes ResNet-50 [13] pre-trained on ImageNet [25] with FPN [21] as the backbone. The difference is that we decouple the classification and regression features in the RoI head. We adopt SGD with a momentum of 0.9 and a weight decay of 0.0001. For VOC, we train the network with a batch size of 2 on 1 TITAN Xp GPU and set the learning rate to 0.0025 for base training and 0.000125 for fine-tuning. For COCO, we train with a batch size of 16 on 8 TITAN X GPUS and set the learning rate to 0.02 for base training and 0.001 for fine-tuning. The hyper-parameter α in Eq. 2 is set to 20 and λ in Eq. 5 is set to 0.2. Both MLP and predictor consist of a fully-connected layer, a ReLU function, and a fully-connected layer.

4.2 Incremental Few-Shot Object Detection

We first conduct experiments under a simple iFSOD setting, similar to the FSOD setting that adds all novel classes simultaneously in the fine-tuning stage. The difference is that no base-class data is used for fine-tuning. We compare our method with existing iFSOD methods: ONCE [22], iMTFA [8], LEAST [18], Sylph [32] in Table 1.

Among those methods, ONCE and Sylph are meta-learning-based methods. The performance of such methods is much worse than other methods based on fine-tuning, though the former do not require fine-tuning that introduces additional training costs [32]. We mainly compare our results with other fine-tuning-based methods. Our method obviously outperforms iMTFA. iMTFA [8] uses Mask-RCNN as the detection framework, which can handle the detection task while aiming at the segmentation task. It uses a cosine similarity classifier which uses the mean of embeddings of all training instances as the weights. In this way, the model can easily overfit the training data on new classes, which results in limited test performance. Note that our method is slightly inferior to iMTFA in the 1-shot setting, because the latter uses a class-agnostic box regressor while we use a class-specific one.

Our approach performs much better than LEAST on base classes. LEAST [18] uses knowledge distillation to alleviate forgetting of old classes and unfreezes the feature extractor to improve the plasticity of new classes. Their experiments show that although naively applying knowledge distillation can perform better for new-classes detection, the ability to alleviate forgetting is limited if old classes data are not stored in the memory. In contrast, our method uses a more straightforward strategy to alleviate the forgetting of old classes to a greater extent. However, our performance on novel classes is slightly lower than LEAST. We think the reason is that the basic detection architectures of the two methods are different, and the unfrozen RoI head has better plasticity.

4.3 Continuous iFSOD

We evaluate the performance of our method in a continuous incremental setting. Only one class is added per phase in the fine-tuning stage. This setting better reflects the incremental few-shot object detection problem compared to Sect. 4.2. For COCO, this means that the fine-tuning stage has a total of 20 phases. The results are shown in Fig. 4.

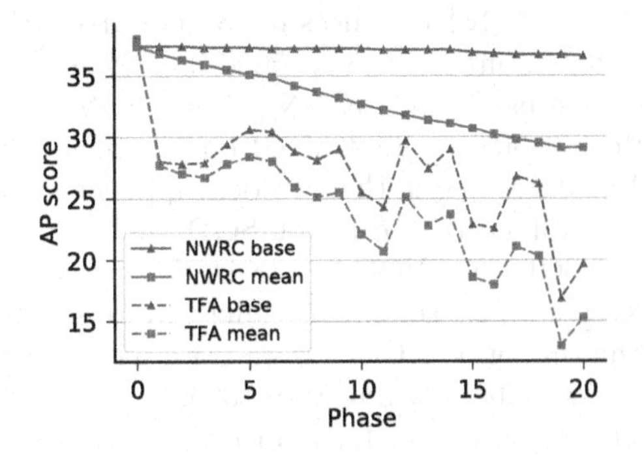

Fig. 4. Continuous iFSOD on COCO dataset.

Compared with the basic method TFA, as the phase increases, the base classes performance of our method has a limited decline. The overall performance is improved significantly. Different from TFA, the AP score of our method is stable over different phases. Our method also outperforms other methods in this setting. The performance of ONCE is stable throughout the incremental process, while the overall AP score is low [22]. LEAST has good performance only with old class samples in memory. However, when the old class data are unavailable, its AP will even decrease to 0 [18].

4.4 Ablation Study

We test the effect of novel-registrable weights (NRW) and region-level contrastive learning (RLCL) on the iFSOD setting, as shown in Table 2. As a method to tackle the FSOD problem, TFA is clearly not equipped to avoid catastrophic forgetting. The forgetting problem is greatly alleviated with NRW. Besides, when RLCL is added to the base training process, the performance of novel classes is further improved. Considering light datasets are closer to scenarios where annotations are not easily attainable, we also conduct ablation studies on VOC dataset, as shown in Table 3. With fewer base classes, training on VOC dataset is more significantly aided by region-level contrastive learning. The results of these experiments show that NRW can greatly alleviate forgetting and improve overall performance. RLCL can promote the generalization of the base model to further improve the performance of novel classes.

Table 2. Ablation study on COCO with 10 shots.

Method	AP_n	AP_b	AP_m
TFA	9.1	19.8	17.1
TFA+NRW	9.1	**37.3**	**30.3**
TFA+NRW+RLCL	**9.4**	37.2	**30.3**

Table 3. Ablation study on VOC split 1 with 3 shots.

Method	AP_n	AP_b	AP_m
TFA	33.8	69.4	60.5
TFA+NRW	51.6	80.5	73.3
TFA+NRW+RLCL	**54.1**	**80.7**	**74.0**

Table 4. Ablation study for different NRW operations.

Method	AP_n	AP_b	AP_m
NRW-fo	31.7	63.1	55.2
NRW-bg	44.2	80.2	71.2
NRW	**51.6**	**80.5**	**73.3**

Table 5. Ablation study for RLCL augmentation.

img	box	AP_n	AP_b	AP_m
	✓	50.9	80.0	72.7
✓		53.5	**80.7**	73.9
✓	✓	**54.1**	**80.7**	**74.0**

We also conduct ablation studies on NRW, as shown in Table 4. NRW-fo means only freezing old class wights, and NRW-bg is NRW that assigns weights to the background class. Experimental results prove the correctness of our theory in Sect. 3.2. Ablation studies on image and proposal box augmentation for region-level contrastive learning are shown in Table 5. The results show the effectiveness of two augmentations.

5 Conclusion

We have proposed NWRC to tackle the incremental few-shot object detection (iFSOD) problem. NWRC uses novel-registrable weights for RoI classification to achieve incremental capability and avoid catastrophic forgetting. We proposed region-level contrastive learning to enhance learning ability for few-shot novel classes. We verify the effectiveness of NWRC by conducting experiments on COCO and VOC datasets. NWRC performs well under two settings of iFSOD.

Acknowledgement. This work was supported in part by National Key R&D Program of China (2021ZD0112001)

References

1. Ajit, A., Acharya, K., Samanta, A.: A review of convolutional neural networks. In: 2020 International Conference on Emerging Trends in Information Technology and Engineering (ic-ETITE), pp. 1–5 (2020). https://doi.org/10.1109/ic-ETITE47903.2020.049
2. Chen, T., Kornblith, S., Norouzi, M., Hinton, G.: A simple framework for contrastive learning of visual representations. In: International Conference on Machine Learning, pp. 1597–1607. PMLR (2020)
3. Chen, X., He, K.: Exploring simple Siamese representation learning. In: Proceedings of the IEEE/CVF Conference on Computer Vision and Pattern Recognition (CVPR), pp. 15750–15758 (2021)
4. Cheng, M., Wang, H., Long, Y.: Meta-learning-based incremental few-shot object detection. IEEE Trans. Circuits Syst. Video Technol. **32**(4), 2158–2169 (2022). https://doi.org/10.1109/TCSVT.2021.3088545
5. Dosovitskiy, A., Springenberg, J.T., Riedmiller, M., Brox, T.: Discriminative unsupervised feature learning with convolutional neural networks. In: Ghahramani, Z., Welling, M., Cortes, C., Lawrence, N., Weinberger, K.Q. (eds.) Advances in Neural Information Processing Systems, vol. 27. Curran Associates, Inc. (2014)
6. Everingham, M., Van Gool, L., Williams, C.K., Winn, J., Zisserman, A.: The pascal visual object classes (VOC) challenge. Int. J. Comput. Vision **88**, 303–338 (2010)
7. Fan, Q., Zhuo, W., Tang, C.K., Tai, Y.W.: Few-shot object detection with attention-RPN and multi-relation detector. In: Proceedings of the IEEE/CVF Conference on Computer Vision and Pattern Recognition (CVPR) (2020)
8. Ganea, D.A., Boom, B., Poppe, R.: Incremental few-shot instance segmentation. In: Proceedings of the IEEE/CVF Conference on Computer Vision and Pattern Recognition (CVPR), pp. 1185–1194 (2021)
9. Girshick, R.: Fast R-CNN. In: Proceedings of the IEEE International Conference on Computer Vision (ICCV) (2015)
10. Grill, J.B., et al.: Bootstrap your own latent - a new approach to self-supervised learning. In: Larochelle, H., et al. (eds.) Advances in Neural Information Processing Systems, vol. 33, pp. 21271–21284. Curran Associates, Inc. (2020)
11. Hadsell, R., Chopra, S., LeCun, Y.: Dimensionality reduction by learning an invariant mapping. In: 2006 IEEE Computer Society Conference on Computer Vision and Pattern Recognition (CVPR 2006), vol. 2, pp. 1735–1742 (2006). https://doi.org/10.1109/CVPR.2006.100

12. He, K., Fan, H., Wu, Y., Xie, S., Girshick, R.: Momentum contrast for unsupervised visual representation learning. In: Proceedings of the IEEE/CVF Conference on Computer Vision and Pattern Recognition (CVPR) (2020)
13. He, K., Zhang, X., Ren, S., Sun, J.: Deep residual learning for image recognition. In: Proceedings of the IEEE Conference on Computer Vision and Pattern Recognition (CVPR) (2016)
14. Hou, S., Pan, X., Loy, C.C., Wang, Z., Lin, D.: Learning a unified classifier incrementally via rebalancing. In: Proceedings of the IEEE/CVF Conference on Computer Vision and Pattern Recognition (CVPR) (2019)
15. Kang, B., Liu, Z., Wang, X., Yu, F., Feng, J., Darrell, T.: Few-shot object detection via feature reweighting. In: Proceedings of the IEEE/CVF International Conference on Computer Vision (ICCV) (2019)
16. Khosla, P., et al.: Supervised contrastive learning. In: Larochelle, H., Ranzato, M., Hadsell, R., Balcan, M., Lin, H. (eds.) Advances in Neural Information Processing Systems, vol. 33, pp. 18661–18673. Curran Associates, Inc. (2020)
17. Li, B., Wang, C., Reddy, P., Kim, S., Scherer, S.: AirDet: few-shot detection without fine-tuning for autonomous exploration. In: Avidan, S., Brostow, G., Cissé, M., Farinella, G.M., Hassner, T. (eds.) Computer Vision – ECCV 2022. ECCV 2022. Lecture Notes in Computer Science, vol. 13699, pp. 427–444. Springer, Cham (2022). https://doi.org/10.1007/978-3-031-19842-7_25
18. Li, P., Li, Y., Cui, H., Wang, D.: Class-incremental few-shot object detection. arXiv preprint arXiv:2105.07637 (2021)
19. Li, Y., et al.: Towards generalized and incremental few-shot object detection. arXiv preprint arXiv:2109.11336 (2021)
20. Lin, T.-Y., et al.: Microsoft COCO: common objects in context. In: Fleet, D., Pajdla, T., Schiele, B., Tuytelaars, T. (eds.) ECCV 2014. LNCS, vol. 8693, pp. 740–755. Springer, Cham (2014). https://doi.org/10.1007/978-3-319-10602-1_48
21. Lin, T.Y., Dollar, P., Girshick, R., He, K., Hariharan, B., Belongie, S.: Feature pyramid networks for object detection. In: Proceedings of the IEEE Conference on Computer Vision and Pattern Recognition (CVPR) (2017)
22. Perez-Rua, J.M., Zhu, X., Hospedales, T.M., Xiang, T.: Incremental few-shot object detection. In: Proceedings of the IEEE/CVF Conference on Computer Vision and Pattern Recognition (CVPR) (2020)
23. Qiao, L., Zhao, Y., Li, Z., Qiu, X., Wu, J., Zhang, C.: DeFRCN: decoupled faster R-CNN for few-shot object detection. In: Proceedings of the IEEE/CVF International Conference on Computer Vision (ICCV), pp. 8681–8690 (2021)
24. Ren, S., He, K., Girshick, R., Sun, J.: Faster R-CNN: towards real-time object detection with region proposal networks. In: Cortes, C., Lawrence, N., Lee, D., Sugiyama, M., Garnett, R. (eds.) Advances in Neural Information Processing Systems, vol. 28. Curran Associates, Inc. (2015)
25. Russakovsky, O., et al.: ImageNet: large scale visual recognition challenge. Int. J. Comput. Vision **115**, 211–252 (2015)
26. Sun, B., Li, B., Cai, S., Yuan, Y., Zhang, C.: FSCE: few-shot object detection via contrastive proposal encoding. In: Proceedings of the IEEE/CVF Conference on Computer Vision and Pattern Recognition (CVPR), pp. 7352–7362 (2021)
27. Wang, X., Huang, T.E., Darrell, T., Gonzalez, J.E., Yu, F.: Frustratingly simple few-shot object detection. arXiv preprint arXiv:2003.06957 (2020)
28. Wu, S., Pei, W., Mei, D., Chen, F., Tian, J., Lu, G.: Multi-faceted distillation of base-novel commonality for few-shot object detection. In: Avidan, S., Brostow, G., Cissé, M., Farinella, G.M., Hassner, T. (eds.) Computer Vision - ECCV 2022, pp. 578–594. Springer Nature Switzerland, Cham (2022)

29. Wu, Z., Xiong, Y., Yu, S.X., Lin, D.: Unsupervised feature learning via non-parametric instance discrimination. In: Proceedings of the IEEE Conference on Computer Vision and Pattern Recognition (CVPR) (2018)
30. Xiang, X., Tan, Y., Wan, Q., Ma, J., Yuille, A., Hager, G.D.: Coarse-to-fine incremental few-shot learning. In: Avidan, S., Brostow, G., Cissé, M., Farinella, G.M., Hassner, T. (eds.) Computer Vision – ECCV 2022. ECCV 2022. Lecture Notes in Computer Science, vol. 13691, pp. 205–222. Springer, Cham (2022). https://doi.org/10.1007/978-3-031-19821-2_12
31. Yan, X., Chen, Z., Xu, A., Wang, X., Liang, X., Lin, L.: Meta R-CNN: towards general solver for instance-level low-shot learning. In: Proceedings of the IEEE/CVF International Conference on Computer Vision (ICCV) (2019)
32. Yin, L., Perez-Rua, J.M., Liang, K.J.: Sylph: a hypernetwork framework for incremental few-shot object detection. In: Proceedings of the IEEE/CVF Conference on Computer Vision and Pattern Recognition (CVPR), pp. 9035–9045 (2022)

Hybrid U-Net: Instrument Semantic Segmentation in RMIS

Yue Wang, Huajian Song(✉), Guangyuan Pan, Qingguo Xiao, Zhiyuan Bai, Ancai Zhang, and Jianlong Qiu

Linyi University, Linyi 276000, China
songhuajian01@sina.cn

Abstract. Accurate semantic segmentation of surgical instruments from images captured by the laparoscopic system plays a crucial role in ensuring the reliability of vision-based Robot-Assisted Minimally Invasive Surgery. Despite numerous notable advancements in semantic segmentation, the achieved segmentation accuracy still falls short of meeting the requirements for surgical safety. To enhance the accuracy further, we propose several modifications to a conventional medical image segmentation network, including a modified Feature Pyramid Module. Within this modified module, Patch-Embedding with varying rates and Self-Attention Blocks are employed to mitigate the loss of feature information while simultaneously expanding the receptive field. As for the network architecture, all feature maps extracted by the encoder are seamlessly integrated into the proposed modified Feature Pyramid Module via element-wise connections. The resulting output from this module is then transmitted to the decoder blocks at each stage. Considering these hybrid properties, the proposed method is called Hybrid U-Net. Subsequently, multiple experiments were conducted on two available medical datasets and the experimental results reveal that our proposed method outperforms the recent methods in terms of accuracy on both medical datasets.

Keywords: Computer vision · Deep learning · Semantic segmentation · Medical image processing · Surgical robotics

1 Introduction

In a class of Robot-Assisted Minimally Invasive Surgery (RMIS) [21], in order for the surgical robot to accurately respond to the surgeon's operation and improve surgical efficiency while ensuring patient safety, it is essential to extract the pixel regions of the surgical instruments in the images captured by the endoscopy system [1,2]. The above process can be performed with the image semantic segmentation with Deep Neural Networks which is a pixel-level classification task and the output image is typically required to have the same size/resolution as the original input image. To provide a more intuitive understanding, we utilize Fig. 1 to directly showcase surgical instruments image segmentation [2,11].

B. Luo et al. (Eds.): ICONIP 2023, CCIS 1965, pp. 413–426, 2024.
https://doi.org/10.1007/978-981-99-8145-8_32

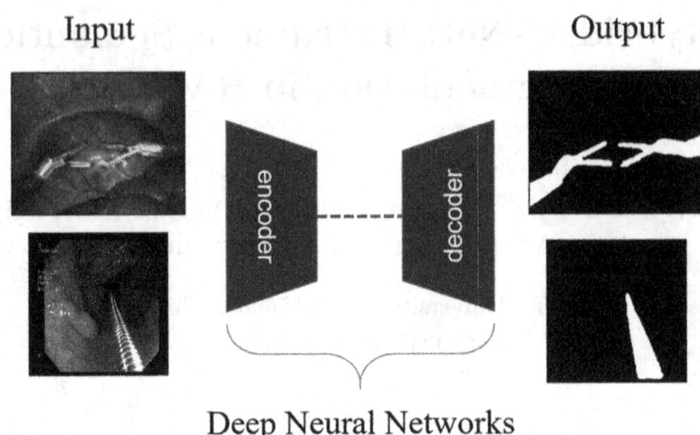

Fig. 1. Example images from the MICCAI EndoVis 2017 Dataset and Kvasir-Instrument, along with the corresponding outputs of the proposed network.

Image segmentation with Deep Neural Networks holds a significant advantage over traditional methods due to its strong generalization capability. There is no need to manually re-select feature spaces or redesign network structures when dealing with different types of surgical instruments. Currently, Vision Transformer (ViT-like) brings a new dynamism to vision tasks. It focuses more on semantic information than Convolutional Neural Networks (CNN) by incorporating Self-Attention mechanisms [6,16,17]. However, the complete ViT-like structure incurs high computational and time costs.

The proposed network is based on an encoder-decoder architecture, with the pre-trained EfficentNetV2-M serving as the encoder, and the decoder is designed to be simple yet effective. An effective module was constructed using the Self-Attention mechanism instead of an entire ViT-like structure. Additionally, a comprehensive feature fusion method is devised to restore detailed feature information at various scales required for network decoding. Experimental results demonstrate that our proposed method achieves the highest MIOU for all three sub-tasks (binary segmentation, instrument parts segmentation, and instrument type segmentation. Figure 2 for an intuitive representation) on the MICCAI EndoVis 2017 test set, and the highest MIOU and Dice (Dice coefficient) were obtained in the binary segmentation task on the test set of Kvasir-Instrument.

We believe that the main contributions of this paper are as follows:

- Hybrid U-net for semantic segmentation of surgical instruments is proposed, which achieves harmony among Skip-Connection, EfficientNet-V2-M, Feature Pyramid Module, and Self-Attention Blocks. Our approach outperforms recent methods on two widely-used medical datasets, demonstrating significant improvements.
- An idea is proposed to use Patch-Embedding with different rates, instead of conventional pooling layers or dilated convolutions, for obtaining receptive fields of different scales. Subsequently, Self-Attention Blocks are concatenated

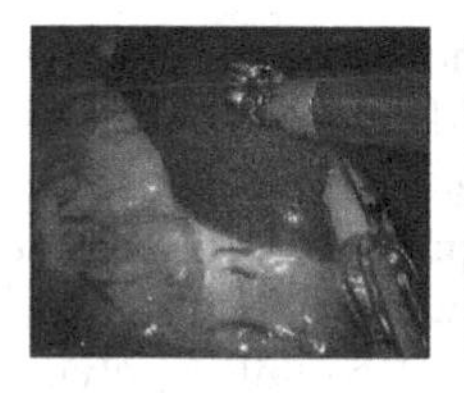  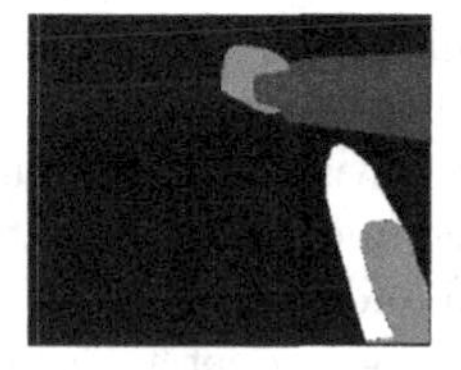 

Fig. 2. Intuitive pictures of the three sub-tasks in the MICCAI EndoVis 2017 Dataset. The leftmost image represents the original image, followed by the labels for binary segmentation, instrument parts segmentation, and instrument type segmentation. Different instruments or instrument parts are marked with distinct colors as required.

to construct a Feature Pyramid Module. This module takes input feature information from all scales, not just the highest level. The purpose of this novel module is to enhance the network's ability to capture short-range details at different scales, thereby achieving improved feature representation and fusion.

Our code and trained model will be published in https://github.com/WY-2022/Hybrid-U-Net soon.

2 Related Work

U-Net [24] introduced a Skip-Connection between each layer of the encoder and decoder to compensate for the loss of image details during feature extraction. By combining this novel feature fusion approach with a step-by-step resolution recovery strategy, UNet-like networks [3,9,10,14,19,22,23] have achieved remarkable results in high-resolution image segmentation. Building upon U-Net, UNet3+ [8] further enhances the architecture by connecting encoder blocks from different stages to all decoder blocks using Skip-Connections. It also applies a similar strategy between decoder blocks. The integration of Skip-Connections and the step-by-step resolution recovery strategy significantly advances high-resolution medical image segmentation. However, their use alone does not greatly enhance the feature representation capability of networks, resulting in relatively limited performance improvement.

For the relatively low-resolution semantic segmentation task, a novel Feature Pyramid Module (FPM) was designed in PSPNet [31] to efficiently utilize the highest-level feature information extracted by the encoder. In the FPM, the highest-level feature information is passed through four down-sampling layers with different rates, resulting in different scales of receptive fields. Subsequently, the feature information is recovered to its initial size and fused as the input of the decoder. A similar FPM (Atrous Spatial Pyramid Pooling) is designed in DeepLabV3+ [5] proposed by Google. The main difference between them lies in how to obtain different scales of receptive fields, the DeepLabV3+ is using an atrous convolution layer with different convolution kernel sizes instead of the pooling layer in PSPNet. UperNet [29] uses a more straightforward idea, which can be seen as replacing the bottom convolutional layer in the U-Net that further processes the highest-level feature information with an FPM, finally the output of the FPM and the feature information of different decoder blocks are fused to an intermediate scale and then the features are more adequate. [13,15] are the combination of FPM and Skip-Connection. An attractive work is DSRD-Net [18] which constructs a sub-encoder to form a dual-stream encoder while applying FPM. Although the semantic information provided by FPM is adequate, relying solely on it may result in a network that is more suitable for object detection rather than segmentation. This is because the latter often demands more precise image restoration and finer-grained levels of detail. Moreover, When dealing with input images of low resolution, the semantic information of the highest-level feature map becomes indistinct, and the incorporation of an FPM may not further enhance the semantic information effectively.

In summary, we aim to combine the advantages of FPM and Skip-Connection while addressing the limitations of FPM in learning short-range information. This integration is intended to achieve better semantic segmentation of surgical instrument images.

3 Method

In this study, we integrate the concepts of Skip-Connections and FPM and incorporate only a few self-attention modules, rather than a complete ViT-like structure, to enhance the segmentation performance. Additionally, we have developed a concise decoder module as part of this study. Figure 3 illustrates the overall architecture of our proposed network.

3.1 Encoder

Inspired by TernausNet [9], this work uses the EfficientNetV2-M [26] to serve the encoder after removing the fully connected layer. The remaining part can be divided into one stem layer and six blocks. Unlike EfficientPS [20], which only selects the feature map with the largest number of channels per size, we use Fusion of Same Size Feature (FSSF) to make full use of the feature maps with the same size and the different number of channels generated by EfficientNetV2-M scaling at the same size. The process of FSSF can be expressed as Fig. 3(a) and (b). In this way, these feature maps can be divided into five stages according to their size, and the decoder will be arranged according to this number.

3.2 Modified Feature Pyramid Module: TSPP

Our modified FPM is internally divided into six branches. The first branch utilizes a 1×1 convolutional layer followed by subsequent batch normalization and nonlinearity layers to enhance salient features in the input feature map. The second, third, and fourth branches perform patch embedding at different rates

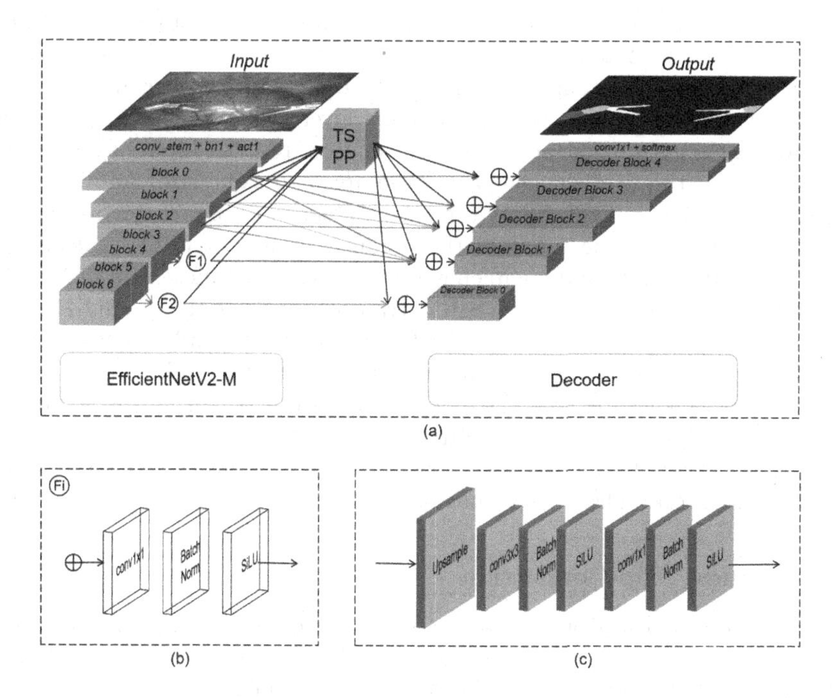

Fig. 3. Network Overview. (a) Overall structure of proposed network. The arrows in this sub-figure represent the Skip-Connection of the feature maps. They are labeled with different colors to easily distinguish which level of feature map they start from, and ⊕ for element-wise tensor concatenation. (b) The structure of Fusion of Same Size Feature (FSSF) operator in (a). (c) Structure of a single decoder block in the proposed network.

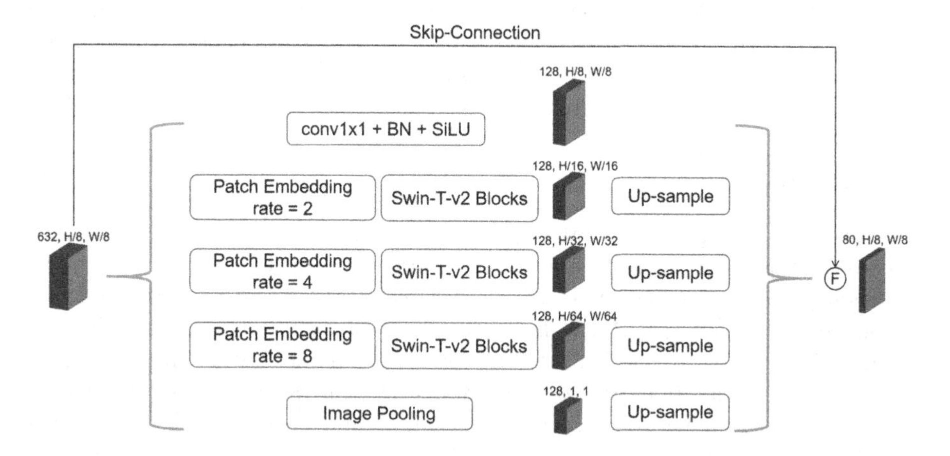

Fig. 4. The TSPP architecture. The shape of feature maps at each step is indicated.

and apply position encoding using linear layers, and these branches are subsequently concatenated and connected to two stacked SwinTransformerV2 blocks to achieve the self-attention mechanism and do not share the weights. The fifth branch incorporates a global average pooling layer, which acts as a spatial-level attention mechanism to extract global information from the input of the module. The outputs of the five branches are up-sampled to same size. Additionally, a Skip-Connection is introduced as the sixth integration branch to fuse the up-sampled results with the input of the module. Finally, a FSSF is utilized to facilitate feature fusion and complementarity across the different branches. The modified FPM is referred to as TSPP (Transformer-Spatial-Pyramid-Pooling), as depicted in Fig. 4. Additionally, the feature maps from all five stages of the encoder, as described in Sect. 3.1, are also input to this module.

This approach ensures that the module retains the idea of capturing different scale receptive fields while avoiding the loss of short-range feature details caused by accepting only the highest-level feature map as input or utilizing pooling layers and high-rate dilated convolutions.

3.3 Seamless Skip-Connection

Inspired by the concepts of information fusion for different stages of the encode-decode process in [8,30], we further enhanced the network structure with the techniques described in Sect. 3.2. The feature maps of each size extracted by the encoder are element-wise concatenated and used as input to the TSPP. And Skip-Connections are set between the TSPP and the decoder, allowing the TSPP's outputs to be introduced before each decoding stage. The arrows of different colors at the bottom of Fig. 3(**a**) provide a helpful visualization to understand the "Seamless" directly.

We believed that the problem of over-segmentation, which may arise from the direct inclusion of background noise information retained in low-level feature maps with the conventional skip-connection strategy, can be avoided by employing a skip-connection strategy similar to UNet3+ between the encoder and decoder, and by feeding feature maps of all scales into the TSPP before they are distributed to their respective decoder blocks.

3.4 Decoder

The decoder block, as illustrated in Fig. 3(**c**), consists of three layers. In each decoding stage, the input is initially up-sampled using bilinear interpolation. Subsequently, a combination layer comprising of a 3×3 convolutional layer, batch normalization, and SiLU is employed to process the input and recover image details. The SiLU activation function is chosen due to its desirable properties, such as lacking upper and lower bounds, smoothness, and non-monotonicity, which have been shown to be effective, especially in deeper networks [7]. Following this, another combination layer involving a 1×1 convolutional operation, batch normalization, and SiLU is utilized. This additional layer serves to further enhance the non-linear mapping capability and adjust the output channel dimension. The resulting output is subsequently forwarded to the next decoder. The decoding process consists of five decoder blocks, aligning with the number of feature map levels extracted by the encoder.

4 Experiments and Results

4.1 Datasets

1. MICCAI EndoVis 2017 Dataset [2]: From the MICCAI Endovis Challenge 2017. It consists of 3000 images with a resolution of 1920*1080 and is divided into 1800 images for training and 1200 images for testing. This dataset provides annotation information for all three sub-tasks. To preprocess the images, we cropped each original frame to remove the black canvas surrounding the valid pixel area and resized the images to 1280*1024 resolution.
2. Kvasir-Instrument [11]: The images and videos were collected using standard endoscopy equipment. For the binary segmentation task, this dataset consists of 472 images for training and 118 images for the test with corresponding labels. According to the experimental details of [11], the resolution of all images in this dataset was resized to 512*512.

4.2 Evaluation Metrics

The evaluation of accuracy follows the two most important metrics consistently used in the field of image segmentation: MIOU and Dice:

$$MIOU(A,B) = \frac{1}{n}\sum_{i=1}^{n}\frac{|A_i \cap B_i|}{|A_i \cup B_i|}, \tag{1}$$

$$Dice(A,B) = \frac{1}{n}\sum_{i=1}^{n}\frac{2|A_i \cap B_i|}{|A_i| + |B_i|}, \tag{2}$$

where n represents the number of image-label pairs in the dataset.

They are used to measure the similarity of set A (images) and set B (lables), both with a value range of 0 to 1, higher is better.

4.3 Experimental Details

The proposed network was implemented using PyTorch 1.10.0, CUDA 11.3, and CUDNN 8.0, leveraging the API provided by timm [28]. For training stability, we employed the AdamW optimizer with a warm-cosine learning rate schedule. Regular augmentation was performed on the training set. The loss function used in all experiments was proposed by TernausNet [9].

To preliminarily evaluate the performance of our approach, we followed a four-fold cross-validation strategy on the MICCAI EndoVis 2017 training set, consistent with previous work. The validation results of the three sub-tasks were

Table 1. Segmentation results per task. Mean Intersection over Union, Dice coefficient, Percentage form. The highest scoring method is shown in bold.

Task:	Binary segmentation		Parts segmentation		Instruments segmentation	
Model	MIOU,%	Dice,%	MIOU,%	Dice,%	MIOU,%	Dice,%
U-Net	75.44	84.37	48.41	60.75	15.80	23.59
TernausNet-11	81.14	88.07	62.23	74.25	34.61	45.86
TernausNet-16	83.60	90.01	65.50	75.97	33.78	44.95
LinkNet-34	82.36	88.87	34.55	41.26	22.47	24.71
A Holistically-Nested U-Net	86.45	92.20				
StreoScenNet [19]	80.82	87.86	66.23	77.57	45.37	56.25
MF-TAPNet [14]	**87.56**	**93.37**	67.92	77.05	36.62	48.01
Ours	84.31	90.17	**68.42**	**78.50**	**51.16**	**60.51**

compared with other outstanding methods, as presented in Table 1. Based on these validation scores, we formulated a training strategy. Specifically, we used a batch size of 8 and an initial learning rate of 1e-5. The learning rate underwent a warm-up phase for the first 3 epochs, linearly increasing to 1e-4, then followed a cosine curve for 51 epochs, gradually decreasing. And then the model was saved for testing. The same training strategy was applied for the Kvasir-Instrument.

4.4 Results on MICCAI EndoVis 2017 Dataset

All three sub-tasks of this dataset were tested, and the comparison results between the proposed method and recent methods in each subset are shown in Table 2 for the binary segmentation task, Table 3 for the instrument parts segmentation task, and Table 4 for the instrument type segmentation task. The calculation of MIOU has taken into account the evaluation rules in the MICCAI Endovis Challenge 2017.

As shown in Table 2, for the binary segmentation task, our approach achieves better performance than the challenge winner MIT in 9 out of 10 test subsets, including the average performance metric. The MIOU of our method is 0.922, which is 3.4 points higher than the team MIT. Although our approach did not yield the best results on Dataset 7, it is still comparable with other methods. Our method also outperforms the recent method ST-MTL [10] on five test subsets, with an improvement of 1.3% points. A visual result example of the binary segmentation task is shown in Fig. 1(a).

In Table 3, quantitative results from the nine participating teams, the recent method, and our method on the instrument part segmentation task are presented (TUM's method is not in plain Deep Learning). As can be seen, our result is 3.8% points better than the challenge winner MIT, and 1.2% points higher than the recent method DSRD-Net [18]. Our method achieves the best results on seven subsets and is also comparable on the other three subsets. A visual result example of the instrument parts segmentation task is shown in Fig. 5(b).

Table 2. The MIOU for the binary segmentation task on ENDOVIS 2017 TEST SET. Our method achieves the best results on 5 subsets. The highest scoring method is shown in bold.

	NCT	UB	BIT	MIT	SIAT	UCL	TUM	Delhi	UA	UW	DSRD [18]	ST-MTL [10]	Ours
Dataset 1	0.784	0.807	0.275	0.854	0.625	0.631	0.760	0.408	0.413	0.337	0.835	0.805	**0.896**
Dataset 2	0.788	0.806	0.282	0.794	0.669	0.645	0.799	0.524	0.463	0.289	0.803	0.826	**0.850**
Dataset 3	0.926	0.914	0.455	0.949	0.897	0.895	0.916	0.743	0.703	0.483	0.959	**0.978**	0.962
Dataset 4	0.934	0.925	0.310	0.949	0.907	0.883	0.915	0.782	0.751	0.678	0.959	**0.968**	0.959
Dataset 5	0.701	0.740	0.220	0.862	0.604	0.719	0.810	0.528	0.375	0.219	**0.933**	0.892	0.902
Dataset 6	0.876	0.890	0.338	0.922	0.843	0.852	0.873	0.292	0.667	0.619	0.933	**0.955**	0.936
Dataset 7	0.846	0.930	0.404	0.856	0.832	0.710	0.844	0.593	0.362	0.325	**0.951**	0.913	0.845
Dataset 8	0.881	0.904	0.366	0.937	0.513	0.517	0.895	0.562	0.797	0.506	0.943	0.953	**0.957**
Dataset 9	0.789	0.855	0.236	0.865	0.839	0.808	0.877	0.626	0.539	0.377	0.817	0.886	**0.925**
Dataset 10	0.899	0.917	0.403	0.905	0.899	0.869	0.909	0.715	0.689	0.603	0.867	0.927	**0.937**
MIOU	0.843	0.875	0.326	0.888	0.803	0.785	0.873	0.612	0.591	0.461	0.878	0.909	**0.922**

Table 3. The MIOU for the parts segmentation task where the metric used is MIOU over all classes. Our method achieves the best results on 7 subsets. The highest scoring method is shown in bold.

	NCT	UB	BIT	MIT	SIAT	UCL	TUM	UA	UW	DSRD [18]	Ours
Dataset 1	0.723	0.715	0.317	0.737	0.591	0.611	0.708	0.485	0.235	0.680	**0.766**
Dataset 2	0.705	0.725	0.294	0.792	0.632	0.606	0.740	0.559	0.244	0.751	**0.825**
Dataset 3	0.809	0.779	0.319	0.825	0.753	0.692	0.787	0.640	0.239	0.846	**0.889**
Dataset 4	0.845	0.737	0.304	0.902	0.792	0.630	0.815	0.692	0.238	0.873	**0.926**
Dataset 5	0.607	0.565	0.280	0.695	0.509	0.541	0.624	0.473	0.240	0.745	**0.824**
Dataset 6	0.731	0.763	0.271	0.802	0.677	0.668	0.756	0.608	0.235	0.833	**0.855**
Dataset 7	0.729	**0.747**	0.359	0.655	0.604	0.523	0.727	0.438	0.207	0.735	0.602
Dataset 8	0.644	0.721	0.300	0.737	0.496	0.441	0.680	0.604	0.236	0.795	**0.869**
Dataset 9	0.561	0.597	0.273	0.650	0.655	0.600	**0.736**	0.551	0.221	0.702	0.687
Dataset 10	0.788	0.767	0.273	0.762	0.751	0.713	**0.807**	0.637	0.241	0.787	0.774
MIOU	0.699	0.700	0.289	0.737	0.667	0.623	0.751	0.578	0.357	0.763	**0.775**

Table 4. The MIOU for the type segmentation task where the metric used is MIOU over all classes. The highest scoring method is shown in bold.

	NCT	UB	MIT	SIAT	UCL	UA	BAR [23]	PAA [22]	ST-MTL [10]	Ours
Dataset 1	0.056	0.111	0.177	0.138	0.073	0.068	0.104	0.106	**0.276**	0.108
Dataset 2	0.499	0.722	0.766	0.013	0.481	0.244	0.801	0.819	**0.830**	0.809
Dataset 3	0.926	0.864	0.611	0.537	0.496	0.765	0.919	0.923	**0.931**	0.924
Dataset 4	0.551	0.680	0.871	0.223	0.204	0.677	0.934	0.945	**0.951**	0.928
Dataset 5	0.442	0.443	0.649	0.017	0.301	0.001	0.830	**0.836**	0.492	0.774
Dataset 6	0.109	0.371	0.593	0.462	0.246	0.400	0.615	0.625	0.501	**0.748**
Dataset 7	0.393	0.416	0.305	0.102	0.071	0.000	**0.534**	0.435	0.480	0.412
Dataset 8	0.441	0.384	0.833	0.028	0.109	0.357	**0.897**	0.869	0.707	0.892
Dataset 9	0.247	0.106	0.357	0.315	0.272	0.040	0.352	0.318	0.409	**0.536**
Dataset 10	0.552	0.709	0.609	0.791	0.583	0.715	0.810	0.858	0.832	**0.875**
MIOU	0.409	0.453	0.542	0.371	0.337	0.346	0.643	0.641	0.633	**0.702**

The test results for the instrument type segmentation task on this dataset are illustrated in Table 4. Although our method achieves the best results on only three subsets, the MIOU is still the highest due to the well done on the two subsets with the most images, 9 and 10, and the scores on other subsets are also comparable. Compared with the recent method BARNet, it is observed that our method can further improve the MIOU from 0.643 to 0.702. Furthermore, a visual result example of the instrument type segmentation task is shown in Fig. 5(c).

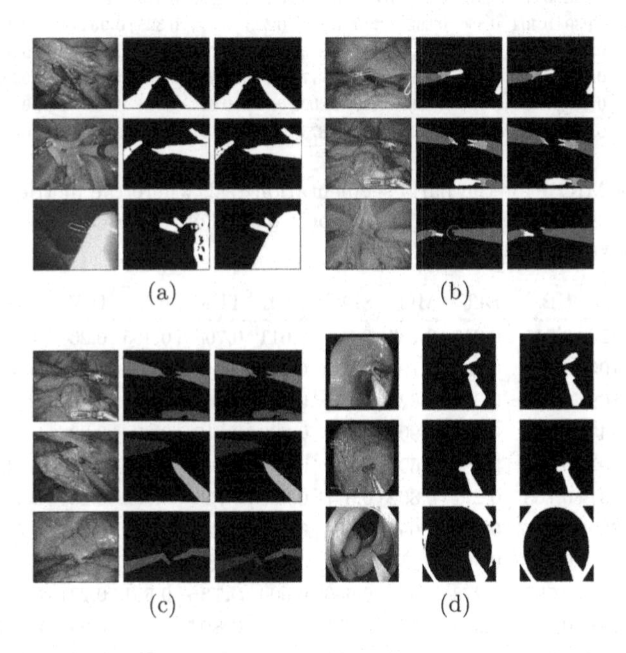

Fig. 5. Examples of three sub-tasks and predicted results on MICCAI EndoVis 2017 Dataset and Kvasir-Instrument: (a) Binary segmentation, (b) Instrument part segmentation, (c) Instrument type segmentation on MICCAI EndoVis 2017 dataset, and (d) Semantic segmentation for Kvasir-Instrument. Each subfigure includes the original image, prediction, and label, and the third row depicts a mismatched case.

4.5 Results on Kvasir-Instrument

To assess the generalization performance of the proposed method, we conducted evaluations on the Kvasir-Instrument dataset. To offer further visualization, Fig. 5(d) presents the segmentation results obtained by our proposed network for the Kvasir-Instrument. A comprehensive list of recent methods tested on this dataset, along with their corresponding results, is presented in Table 5.

Results show that our method achieves 0.9344 MIOU and 0.9646 Dice, exceeding other methods to a not insignificant extent. The second-ranking

Table 5. Comparison of the numerical results of our method with the recent method for the segmentation task on the Kvasir-Instrument. The highest scoring method is shown in bold.

Methods	Backbone Network	MIOU	Dice
U-Net	–	0.8578	0.9158
U-Net with Backbone [27]	Inception-ResNet-v2	0.9167	0.9501
DoubleU-Net [12]	VGG-19	0.8430	0.9038
ResUNet++ [13]	–	0.8635	0.9140
MSDFNet [15]	ResNet50	0.8910	0.9334
SegNet [3]	–	0.7146	0.8086
DeepLabV3+ [5]	–	0.8701	0.9211
DSRD-Net [18]	–	0.9157	0.9513
Ours	EfficientNetV2-M	**0.9344**	**0.9646**

method is U-Net with Inception-ResNet-v2 [25] as the backbone, which achieves 0.9167 MIOU and 0.9501 Dice. Compared with it, our approach has improved by 1.77%, which may have reached a performance bottleneck.

4.6 Ablation Studies

To understand whether the superior performance comes from the EfficientNetV2-M or this Hybrid U-Net design, and to see if each modification works, some ablation experiments were conducted on Kvasir-Instrument.

First, EfficientNetV2-M was used to serve as the encoder in two exceptional networks. As shown in Table 6, with the same training strategy, the proposed network scores higher than both PSPNet and DeepLabV3+, although EfficientNetV2-M brings a significant boost to them.

Table 6. Comparison of the numerical results when using the same backbone.

Methods	Backbone	MIOU	Dice
PSPNet	ResNet50	0.8268	0.8981
	EfficientNetV2-M	0.9002	0.9437
DeepLabV3+ [5]	–	0.8701	0.9211
	EfficientNetV2-M	0.9256	0.9597
Ours	EfficientNetV2-M	**0.9344**	**0.9646**

Second, the decoder blocks of the network were configured as shown in Fig. 3(**c**) to further test various aspects proposed in Sect. 3. As indicated in Table 7, each modification improves the evaluation score compared to the vanilla network (the first row in Table 7), which is roughly equivalent to a U-Net with EfficientNetV2-M as the backbone. Removing any modification results in a decrease in the network's segmentation ability, only if all modifications coexist can the network achieve a state of harmony and attain the highest MIOU and Dice.

Table 7. Experimental results of whether each modification actually works.

FSSF	TSPP	Seamless	MIOU	Dice
			0.9175	0.9535
✓			0.9244	0.9581
	✓		0.9294	0.9583
		✓	0.9299	0.9620
✓	✓		0.9335	0.9638
✓		✓	0.9259	0.9592
	✓	✓	0.9289	0.9610
✓	✓	✓	**0.9344**	**0.9646**

5 Conclusion and Discussion

This work presents the Hybrid U-Net, which achieves harmony among Skip-Connection, EfficientNet-V2-M, Feature Pyramid Module, and Self-Attention Blocks. We introduce Patch-Embedding with varying rates to capture diverse receptive fields and preserve feature information in the Feature Pyramid Module. Additionally, we propose TSPP, a modified Feature Pyramid Module, and utilize feature maps for image detail recovery through TSPP and Seamless Skip-Connection.

Experimental results demonstrate the superior accuracy of the Hybrid U-Net compared to recent methods on MICCAI EndoVis 2017 datasets and Kvasir-Instrument. Ablation studies confirm the effectiveness of the proposed improvements. However, limitations remain, including relatively high training time consumption and memory usage, despite considering these factors during network design. Regarding TSPP, future research may involve utilizing more advanced transformer blocks and exploring Neural Architecture Search techniques [4] to achieve a more optimized internal structure, potentially leading to even more beneficial outcomes.

Acknowledgements. This work was supported in part by the National Natural Science Foundation of Shandong Province under grants No. ZR2022QF058 and Shandong Province Science and Technology-Oriented Minor Enterprise Innovation Capability Enhancement Project under grants No. 2023TSGC0406 and No. 2022TSGC1328. Thanks to Dr. Max Allan for granting us access to the MICCAI EndoVis 2017 dataset, without which our experiments would not have been possible.

References

1. Allan, M., et al.: Image based surgical instrument pose estimation with multi-class labelling and optical flow. In: Navab, N., Hornegger, J., Wells, W.M., Frangi, A.F. (eds.) MICCAI 2015. LNCS, vol. 9349, pp. 331–338. Springer, Cham (2015). https://doi.org/10.1007/978-3-319-24553-9_41
2. Allan, M., et al.: 2017 robotic instrument segmentation challenge. arXiv preprint arXiv:1902.06426 (2019)
3. Badrinarayanan, V., Kendall, A., Cipolla, R.: Segnet: a deep convolutional encoder-decoder architecture for image segmentation. IEEE Trans. Pattern Anal. Mach. Intell. **39**(12), 2481–2495 (2017)
4. Chen, L.C., et al.: Searching for efficient multi-scale architectures for dense image prediction. In: Advances in Neural Information Processing Systems 31 (2018)
5. Chen, L.-C., Zhu, Y., Papandreou, G., Schroff, F., Adam, H.: Encoder-decoder with atrous separable convolution for semantic image segmentation. In: Ferrari, V., Hebert, M., Sminchisescu, C., Weiss, Y. (eds.) ECCV 2018. LNCS, vol. 11211, pp. 833–851. Springer, Cham (2018). https://doi.org/10.1007/978-3-030-01234-2_49
6. Dosovitskiy, A., et al.: An image is worth 16x16 words: transformers for image recognition at scale. In: ICLR (2021)
7. Elfwing, S., Uchibe, E., Doya, K.: Sigmoid-weighted linear units for neural network function approximation in reinforcement learning. Neural Netw. **107**, 3–11 (2018)
8. Huang, H., et al.: Unet 3+: a full-scale connected unet for medical image segmentation. In: ICASSP 2020–2020 IEEE International Conference on Acoustics, Speech and Signal Processing (ICASSP), pp. 1055–1059. IEEE (2020)
9. Iglovikov, V., Shvets, A.: Ternausnet: U-net with vgg11 encoder pre-trained on imagenet for image segmentation. arXiv e-prints, arXiv-1801 (2018)
10. Islam, M., Vibashan, V., Lim, C.M., Ren, H.: St-mtl: spatio-temporal multitask learning model to predict scanpath while tracking instruments in robotic surgery. Med. Image Anal. **67**, 101837 (2021)
11. Jha, D., et al.: Kvasir-instrument: diagnostic and therapeutic tool segmentation dataset in gastrointestinal endoscopy. In: Lokoč, J., et al. (eds.) MMM 2021. LNCS, vol. 12573, pp. 218–229. Springer, Cham (2021). https://doi.org/10.1007/978-3-030-67835-7_19
12. Jha, D., Riegler, M.A., Johansen, D., Halvorsen, P., Johansen, H.D.: Doubleu-net: a deep convolutional neural network for medical image segmentation. In: 2020 IEEE 33rd International symposium on computer-based medical systems (CBMS), pp. 558–564. IEEE (2020)
13. Jha, D., et al.: Resunet++: an advanced architecture for medical image segmentation. In: 2019 IEEE International Symposium on Multimedia (ISM), pp. 225–2255. IEEE (2019)
14. Jin, Y., Cheng, K., Dou, Q., Heng, P.-A.: Incorporating temporal prior from motion flow for instrument segmentation in minimally invasive surgery video. In: Shen, D., et al. (eds.) MICCAI 2019. LNCS, vol. 11768, pp. 440–448. Springer, Cham (2019). https://doi.org/10.1007/978-3-030-32254-0_49
15. Liu, X., et al.: Msdf-net: multi-scale deep fusion network for stroke lesion segmentation. IEEE Access **7**, 178486–178495 (2019)
16. Liu, Z., et al.: Swin transformer v2: Scaling up capacity and resolution. In: International Conference on Computer Vision and Pattern Recognition (CVPR) (2022)
17. Liu, Z., et al.: Swin transformer: hierarchical vision transformer using shifted windows. In: Proceedings of the IEEE/CVF International Conference on Computer Vision, pp. 10012–10022 (2021)

18. Mahmood, T., Cho, S.W., Park, K.R.: Dsrd-net: dual-stream residual dense network for semantic segmentation of instruments in robot-assisted surgery. Expert Syst. Appl. **202**, 117420 (2022)
19. Mohammed, A., Yildirim, S., Farup, I., Pedersen, M., Hovde, Ø.: Streoscennet: surgical stereo robotic scene segmentation. In: Medical Imaging 2019: Image-Guided Procedures, Robotic Interventions, and Modeling, vol. 10951, pp. 174–182. SPIE (2019)
20. Mohan, R., Valada, A.: Efficientps: efficient panoptic segmentation. Int. J. Comput. Vision **129**(5), 1551–1579 (2021)
21. Moustris, G.P., Hiridis, S.C., Deliparaschos, K.M., Konstantinidis, K.M.: Evolution of autonomous and semi-autonomous robotic surgical systems: a review of the literature. Inter. J. Med. Robotics Comput. Assisted Surg. **7**(4), 375–392 (2011)
22. Ni, Z.L., et al.: Pyramid attention aggregation network for semantic segmentation of surgical instruments. In: Proceedings of the AAAI Conference on Artificial Intelligence, vol. 34, pp. 11782–11790 (2020)
23. Ni, Z.L., et al.: Barnet: bilinear attention network with adaptive receptive fields for surgical instrument segmentation. In: Proceedings of the Twenty-Ninth International Conference on International Joint Conferences on Artificial Intelligence, pp. 832–838 (2021)
24. Ronneberger, O., Fischer, P., Brox, T.: U-Net: convolutional networks for biomedical image segmentation. In: Navab, N., Hornegger, J., Wells, W.M., Frangi, A.F. (eds.) MICCAI 2015. LNCS, vol. 9351, pp. 234–241. Springer, Cham (2015). https://doi.org/10.1007/978-3-319-24574-4_28
25. Szegedy, C., Ioffe, S., Vanhoucke, V., Alemi, A.A.: Inception-v4, inception-resnet and the impact of residual connections on learning. In: Thirty-first AAAI Conference on Artificial Intelligence (2017)
26. Tan, M., Le, Q.: Efficientnetv2: smaller models and faster training. In: International Conference on Machine Learning, pp. 10096–10106. PMLR (2021)
27. Watanabe, T., Tanioka, K., Hiwa, S., Hiroyasu, T.: Performance comparison of deep learning architectures for artifact removal in gastrointestinal endoscopic imaging. arXiv e-prints. arXiv-2201 (2021)
28. Wightman, R.: Pytorch image models. https://github.com/rwightman/pytorch-image-models (2019). https://doi.org/10.5281/zenodo.4414861
29. Xiao, T., Liu, Y., Zhou, B., Jiang, Y., Sun, J.: Unified perceptual parsing for scene understanding. In: Ferrari, V., Hebert, M., Sminchisescu, C., Weiss, Y. (eds.) ECCV 2018. LNCS, vol. 11209, pp. 432–448. Springer, Cham (2018). https://doi.org/10.1007/978-3-030-01228-1_26
30. Yu, L., Wang, P., Yu, X., Yan, Y., Xia, Y.: A holistically-nested u-net: Surgical instrument segmentation based on convolutional neural network. J. Digit. Imaging **33**(2), 341–347 (2020)
31. Zhao, H., Shi, J., Qi, X., Wang, X., Jia, J.: Pyramid scene parsing network. In: Proceedings of the IEEE Conference on Computer Vision and Pattern Recognition, pp. 2881–2890 (2017)

Continual Domain Adaption for Neural Machine Translation

Manzhi Yang[1], Huaping Zhang[1], Chenxi Yu[1], and Guotong Geng[2](✉)

[1] Beijing Institute of Technology, Beijing, China
{3120211028,Kevinzhang}@bit.edu.cn
[2] Center for Information Research, Academy of Military Sciences, Beijing, China
ggtong@163.com

Abstract. Domain Neural Machine Translation (NMT) with small data- sets requires continual learning to incorporate new knowledge, as catastrophic forgetting is the main challenge that causes the model to forget old knowledge during fine-tuning. Additionally, most studies ignore the multi-stage domain adaptation of NMT. To address these issues, we propose a multi-stage incremental framework for domain NMT based on knowledge distillation. We also analyze how the supervised signals of the golden label and the teacher model work within a stage. Results show that the teacher model can only benefit the student model in the early epochs, while harms it in the later epochs. To solve this problem, we propose using two training objectives to encourage the early and later training. For early epochs, conventional continual learning is retained to fully leverage the teacher model and integrate old knowledge. For the later epochs, the bidirectional marginal loss is used to get rid of the negative impact of the teacher model. The experiments show that our method outperforms multiple continual learning methods, with an average improvement of 1.11 and 1.06 on two domain translation tasks.

Keywords: Neural machine translation · Continual learning · Knowledge distillation

1 Introduction

Neural Machine Translation (NMT) [19,20] has achieved good performance with large-scale datasets, but it still struggles with domain translation using small datasets. In practical applications, new domain data is usually available in the form of streams through channels such as the Internet. To improve the capabilities of a domain NMT model, it needs to continually incorporate new data. However, fine-tuning the old model with new data directly leads to catastrophic forgetting [7]. Mixing old and new data for retraining solves this problem, but it is inefficient. Additionally, sometimes we cannot access the old data due to data privacy or storage restrictions.

Continual learning, also known as incremental learning, is used to address the problem of catastrophic forgetting. Common methods include knowledge distillation [15,23] and regularization [2,18]. Knowledge distillation uses a teacher-student framework to preserve old knowledge into the new model. Regularization

B. Luo et al. (Eds.): ICONIP 2023, CCIS 1965, pp. 427–439, 2024.
https://doi.org/10.1007/978-981-99-8145-8_33

adds regularization terms to prevent the model from overfitting on new data. In NMT, most studies focus on cross-domain increment [10] and language increment based on multilingual NMT models [3,8]. Little attention has been paid to in-domain increment [4]. Furthermore, most methods only consider single-stage increment, while multi-stage increment [4,11] is ignored.

In order to address the problem of catastrophic forgetting, This paper investigates multi-stage continual learning of domain NMT without accessing translation memory. We propose a multi-stage incremental framework based on knowledge distillation, which uses the last stage model as the teacher model and initialization model for the current stage. The framework fine-tunes the student model using only new data. We analyze the changes of the supervised signal of the golden label and the teacher model within one stage, and find that the teacher model only benefits the student model in the early epochs. However, it prevents the student model from absorbing new knowledge in the later epochs. Based on these findings, we train the student model in two steps using different training objectives. In the first step, we retain the conventional continual learning loss to fully integrate the old knowledge of the teacher model. In the second step, we use the bidirectional marginal loss to ensure that the student model maintains a certain distance from the teacher model. We conduct multi-stage incremental experiments using English-Chinese (En-Zh) and German-English (De-En) domain datasets. The results show that our method provides stable improvement.

Our contributions are as follows:

- We propose a simple multi-stage incremental framework for domain NMT based on knowledge distillation.
- We analyze how knowledge distillation plays a role in continual learning and propose using bidirectional marginal loss to alleviate the negative impact of knowledge distillation.
- We demonstrate the effectiveness of our method through experiments in domain translation tasks.

2 Related Work

2.1 Continual Learning

Continual learning is studied to solve the catastrophic forgetting problem. In NLP, methods are divided into two categories: distillation-based methods and regularization-based methods. The former close the gap between the student model and the teacher model by introducing an additional loss function. The latter make use of regularization terms, such as L2 and EWC [18], to limit the variation of parameters. [11] evaluates various continual learning methods for pretrained language models in multi-stage cross-domain increment and time increment scenarios. Distillation-based methods are found to be the most effective. The translation experiment results of [4] also arrive at similar conclusions.

2.2 Knowledge Distillation

Traditional knowledge distillation involves using a simple-structured student model to simulate the output of a trained teacher model with a complex structure, in order to compress the model [5,6,12]. However, knowledge distillation used for continual learning is to retain the knowledge from the old model to the new model. Knowledge distillation methods in NMT include word-level distillation [14,15] and sequence-level distillation [9,15]. Recent studies are as follows: [22] proposes an online knowledge distillation method to prevent overfitting by generating a teacher model from the checkpoints in real time. [10] limits the updating of parameters to a low forgetting risk regions. These methods are studied in a single-stage increment. [4] proposes a knowledge distillation method for multi-stage increments and conducts multi-stage in-domain incremental experiments. However, these experiments are only carried out on large-scale datasets and lack consideration for small domains. In addition, all the above methods give full trust to the teacher model and use a fixed weight throughout the training process.

3 Method

3.1 Multi-stage Incremental Framework

The multi-stage incremental framework for domain NMT is illustrated in Fig. 1. It consists of sequential teacher-student models, where each stage exclusively employs the new dataset as the training set, last stage model as the teacher model for knowledge distillation, and initialization model for fine-tuning. In particular, the first stage uses either an NMT pretrained model or a general domain model for initialization. To complete domain adaptation quickly, knowledge distillation is not used in $stage_1$. The training objective is shown in Eq. 1, where $\mathcal{L}_{gold}$ and $\mathcal{L}_{KD}$ represent the supervised signal of the golden label and the teacher model, respectively. The weight of knowledge distillation w is set to 0 in $stage_1$.

$$\mathcal{L}_{CL} = \mathcal{L}_{gold} + w\mathcal{L}_{KD} \tag{1}$$

$\mathcal{L}_{gold}$ is the cross-entropy loss function, as in Eq. 2. For a sentence pair (x, y), x and y are the source and target sentences; $\hat{y_s}$ is the student model output; θ_s is the student model parameters.

$$\begin{aligned} \mathcal{L}_{gold}(x, y, \theta_s) &= CE(y, \hat{y_s}) \\ &= -\log P(\hat{y_s} = y|x, \theta_s) \end{aligned} \tag{2}$$

$\mathcal{L}_{KD}$ uses Kullback-Leibler (KL) divergence to measure the difference between the student model output $\hat{y_s}$ and the teacher model output $\hat{y_t}$, as in Eq. 3. Here, θ_t is the teacher model parameters. $\varphi(\cdot)$ is the KL divergence function for the token level. $\hat{y_s}_i$ and $\hat{y_t}_i$ are on behalf of the ith token of the model output. I is

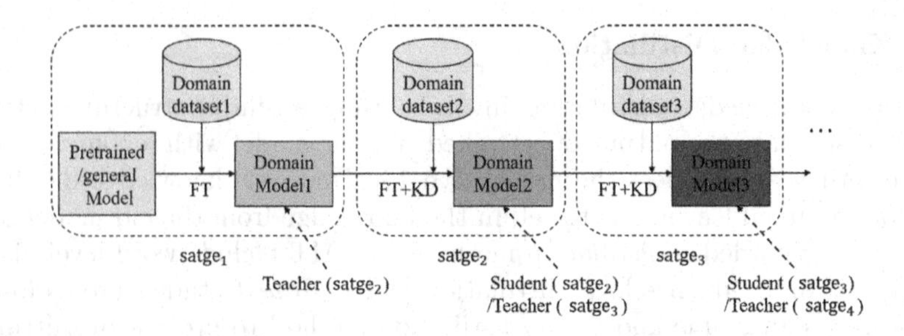

Fig. 1. The multi-stage incremental framework for domain NMT. "FT" represents fine-tuning. "KD" represents knowledge distillation.

the output length. k represents the kth token in the vocabulary V.

$$\begin{aligned}\mathcal{L}_{KD}(x, \theta_s, \theta_t) &= KL\left(\hat{y_s} \| \hat{y_t}\right) \\ &= \sum_{i=1}^{I} \sum_{k=1}^{|V|} \varphi(\hat{y}_{si}, \hat{y}_{ti}, k)\end{aligned} \tag{3}$$

The specific description of $\varphi(\cdot)$ is given in Eq. 4. P and Q are the conditional probability distributions of the output of the student model and the teacher model.

$$\varphi(\hat{y}_{si}, \hat{y}_{ti}, k) = P\left(\hat{y}_{si} = k | \hat{y}_{s1:i-1}, x, \theta_s\right) \log \frac{P\left(\hat{y}_{si} = k | \hat{y}_{s1:i-1}, x, \theta_s\right)}{Q\left(\hat{y}_{ti} = k | \hat{y}_{t1:i-1}, x, \theta_t\right)} \tag{4}$$

3.2 Analysis of KD

To further analyze how $\mathcal{L}_{gold}$ and $\mathcal{L}_{KD}$ work within a stage, we test different knowledge distillation weights w. The BLEU scores are shown in Table 1. The changes of $\mathcal{L}_{gold}$ and $\mathcal{L}_{KD}$ during training is shown in Fig. 2. This is the experimental result of $stage_2$ of En-Zh, and we present the experimental details in Sect. 4.

The best performance is achieved when w is set to 0.5. We find that $\mathcal{L}_{gold}$ keeps decreasing with the training steps, while $\mathcal{L}_{KD}$ has a significant rebound after the decrease. This finding is consistent with the common sense that the teacher model of our proposed framework is a weaker in-domain model that does not always provide beneficial reference for all new samples. We define the rebound starting point as p. Prior to p, the decrease in $\mathcal{L}_{KD}$ indicates that the student model updates in the same direction as the teacher model, providing a beneficial supervised signal. The student model completes the integration of old and new knowledge at these early epochs. After p, the bounce in $\mathcal{L}_{KD}$ indicates that the teacher model can no longer provide favorable guidance to the student model, prevents it from absorbing new knowledge in later epochs.

Table 1. BLEU scores for different knowledge distillation weights.

w	BLEU
0 (fine-tune)	55.75
0.1	55.71
0.3	55.76
0.5	**56.12**
0.7	55.63
0.9	55.92
uncertainty	55.91

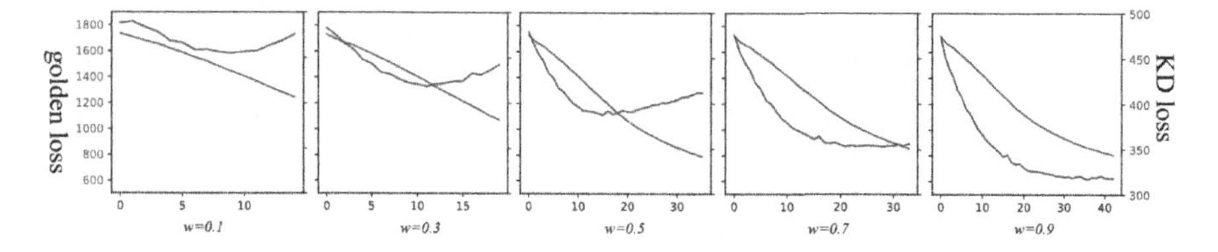

Fig. 2. The changes of $\mathcal{L}_{gold}$ and $\mathcal{L}_{KD}$ with fixed weight. The horizontal coordinates of each subplot are the epoch numbers, and the vertical coordinates arethe average losses of all updates in each epoch.

$\mathcal{L}_{KD}$ also shows a rebound when w is less than 0.5. However, the student model focuses more on fitting the gold label, resulting in premature convergence and insufficient integration of old and new knowledge. When w is larger than 0.5, the BLEU score remains low even though $\mathcal{L}_{KD}$ keeps decreasing. This suggests that placing more emphasis on the teacher model does not serve to improve performance.

In addition to the fixed weight, we also experimented with learnable weights. We construct learnable weights inspired by uncertainty weighting [13] in multi-task learning (MTL), as in Eq. 5. It uses learnable parameters σ_1 and σ_2 to dynamically adjust the weights. The weight of $\mathcal{L}_{gold}$ is $w_1=\frac{1}{2\sigma_1^2}$; The weight of $\mathcal{L}_{KD}$ is $w_2=\frac{1}{2\sigma_2^2}$; the third term on the right-hand side of Eq. 5 is the balance term that prevents the unrestricted decrease of w_1 and w_2. The changes of $\mathcal{L}_{gold}$, $\mathcal{L}_{KD}$, w_1 and w_2 are shown in Fig. 3. We find that w_1 decreases while w_2 increases with the training steps. After the 10th epoch, the increasing trend of w_2 is more obvious because $\mathcal{L}_{KD}$ can no longer decrease. This suggests that uncertainty weighting would give greater weight to emphasize the task with slower loss decline, but it departs from our expectation of reducing the KD weight at later epochs. Therefore, using learnable weight is not appropriate.

$$\mathcal{L}_{MTL} = \frac{1}{2\sigma_1^2}\mathcal{L}_{gold} + \frac{1}{2\sigma_2^2}\mathcal{L}_{KD} + \log \sigma_1 \sigma_2 \tag{5}$$

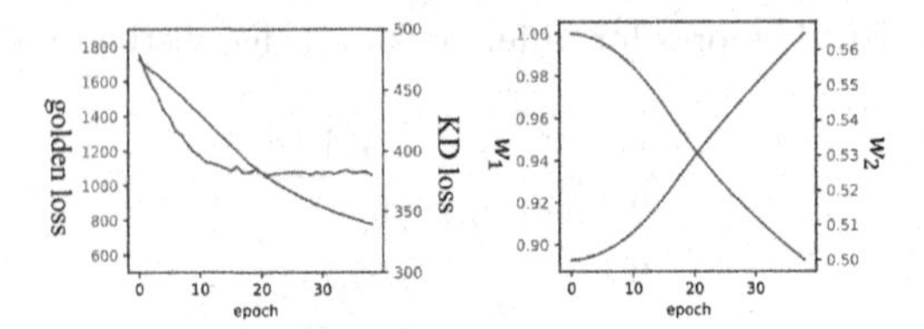

Fig. 3. The changes of losses and weights with uncertainty weighting.

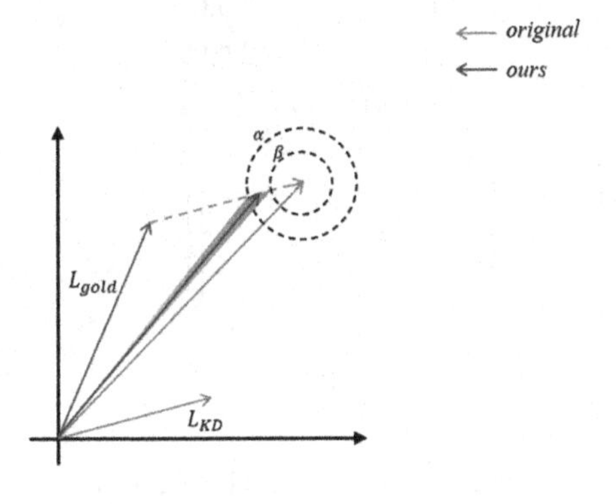

Fig. 4. Diagram of our method.

In conclusion, treating $\mathcal{L}_{gold}$ and $\mathcal{L}_{KD}$ with a fixed weight during a training session is harmful to the student model. Additionally, the learnable weight is not effective in a continual learning scenario, and it is unrealistic to conduct multiple experiments to find the best weight for each stage.

3.3 Alleviate the Negative Impact of KD

To alleviate the negative impact of knowledge distillation in later epochs, we encourage the update direction of the student model using different training objectives in two steps. In the first step, the student model utilizes the old knowledge of the teacher model. In the second step, the model further learns new knowledge from the golden labels. Our method is shown in Eq. 6: before p, we keep the original continual learning loss; After p, we use bidirectional marginal loss (BML) instead of $\mathcal{L}_{KD}$. Here, j represents the jth epoch.

$$\mathcal{L}_{CL} = \begin{cases} \mathcal{L}_{gold} + w\mathcal{L}_{KD}, & j \leq p \\ \mathcal{L}_{gold} + w\mathcal{L}_{BML}, & j > p \end{cases} \tag{6}$$

The bidirectional marginal loss function, as shown in Eq. 7, aims to converge the KL divergence φ to the interval $[\beta, \alpha]$, rather than minimizing it to 0. The lower bound β controls the student model to maintain a certain distance from the teacher model and the upper bound α ensures that the student model is updated in roughly the same direction as the teacher model. When $[\beta, \alpha]$ is $[0, 0]$, Eq. 6 is equivalent to Eq. 1.

Algorithm 1: Multi-stage training process

Data: Initialization model θ_{init}, max stage S, max epoch E, batch set B
Result: Domain model θ_s

```
1  p ← +∞;
2  θ_t ← θ_s ← θ_init;
3  for s ← 1 to S do
4  |   θ_t ← θ_s;
5  |   for j ← 1 to E do
6  |   |   for b ← 1 to |B| do
7  |   |   |   compute the loss l_jb of the bth batch by Eq. 6;
8  |   |   |   θ_s ← θ_s − ∂l_jb/∂θ_s;
9  |   |   end
10 |   |   l_j ← Σ_{b=1}^{|B|} l_jb;
11 |   |   if j>0 and p = +∞ and l_{j−1}<l_j then
12 |   |   |   p ← j;
13 |   |   end
14 |   end
15 end
```

$$\mathcal{L}_{BML} = \sum_{i=1}^{I} \sum_{k=1}^{|V|} Relu(-\varphi + \beta) + Relu(\varphi - \alpha) \tag{7}$$

The diagram of our method is shown in Fig. 4. Compared to the original continual learning loss, our method aligns the model update direction closer to the golden label. The interval $[\beta, \alpha]$ allows the student model to reduce the degree of dependence on the teacher's supervised signal within a certain range. The overall training process is shown in Algorithm 1.

4 Experiments

We conduct experiments on two domain translation tasks: English-Chinese (En-Zh) patent domain and German-English (De-En) IT domain.

4.1 Data Preparation

En-Zh. We use the patent dataset of CCF DBCI[1], which contains 100k parallel sentences. jieba[2] is used as the tool for Chinese text segmentation. The output is evaluated after tokenization using sacremoses[3].

[1] https://www.datafountain.cn/special/BDCI2021/competition.
[2] https://github.com/fxsjy/jieba.
[3] https://github.com/alvations/sacremoses.

De-En. We use the IT dataset clustered by [1], which contains 223k parallel sentences.

To simulate the multi-stage experimental scenario, we divided the training set into 10 parts. Only 10% of the training set is used at each stage. In order to fairly compare the domain translation capability, the valid set and the test set are consistent in all stages. All languages use subword-nmt[4] for BPE.

4.2 Baselines

Our proposed method is compared with 6 baselines:

upper bound All the data from the current and previous stages are mixed and used to fine-tune the pre-trained model. This method makes full use of all knowledge and serves as an upper bound for all methods.

fine-tune Only use the data of the current stage to fine-tune the model from the last stage.

word KD [15] The method uses KL divergence between the output word distribution of the student and teacher model as $\mathcal{L}_{KD}$.

reg KD [14] This method uses the cross entropy between the output word distribution of the student and teacher model as $\mathcal{L}_{KD}$.

selective KD [21] In this method, samples with high word cross entropy are selected for knowledge distillation. Our experiments use the batch-level selection strategy of the method.

CLNMT [4] A method for dynamically adjusting the weight of knowledge distillation at each stage with a hyperparameter setting of 0.5.

4.3 Implementation Details

We use the multilingual pre-trained NMT model mRASP [16] as the initialization model for $stage_1$ and use its language tag and inference method in all subsequent stages. Since large-scale general datasets of En-Zh and De-En are already being used in the mRASP pre-training process, we do not perform additional language transfer training.

The Transformer [20] architecture we use contains 6 layers of encoder, 6 layers of decoder, and 16 attention heads. The hidden size is 1024. The source and target embedding layers share the parameters.

We use the fairseq[5] toolkit to implement the method. For training, we use Adam optimizer; $\mathcal{L}_{gold}$ uses label smoothed cross-entropy with a factor of 0.1, learning rate is 2e-5 and drop-out rate is set to 0.2. For inference, beam is set to 5; SacreBLEU [17] is used to evaluate the accuracy and fluency of the translation. We evaluate the average checkpoint of the five consecutive best checkpoints. For hyperparameters, w for word KD and our method is 0.5, for reg KD is 0.1 and for selective KD is 1. p is the $\mathcal{L}_{KD}$ rebound point by automatic detection. In the main results, $[\beta, \alpha]$ of our method is set to [0, 0.1] for En-Zh and [0.1, 0.2] for De-En.

[4] https://github.com/rsennrich/subword-nmt.

[5] https://github.com/facebookresearch/fairseq.

Table 2. BLEU scores of multiple methods at different stages. The upper and lower subtables show the experimental results of En-Zh and De-En. The first row is the stage number. The last column shows the average comparison results between different methods and fine-tuning.

En-Zh	1	2	3	4	5	6	7	8	9	10	Δavg
upper bound	57.11	59.18	60.00	61.08	61.41	61.94	62.41	62.61	63.18	63.86	+2.21
fine-tune	57.11	58.37	58.69	59.08	59.40	59.80	59.86	59.96	60.34	60.26	-
word KD	57.11	58.70	59.67	59.77	60.22	60.40	60.61	60.63	61.15	61.01	+0.71
reg KD	57.11	58.58	59.54	59.77	60.16	60.19	60.36	60.73	61.02	60.76	+0.59
selective KD	57.11	58.41	58.79	59.11	59.38	59.28	59.50	59.62	60.04	60.02	−0.18
CLNMT	57.11	58.82	59.68	**60.11**	60.47	60.64	60.58	60.87	60.97	60.94	+0.81
ours	57.11	**58.93**	**59.78**	60.07	**60.51**	**60.72**	**60.95**	**61.51**	**61.65**	**61.60**	**+1.11**
De-En	1	2	3	4	5	6	7	8	9	10	Δ avg.
upper bound	36.52	40.68	40.99	41.89	42.50	43.21	43.68	43.82	44.23	44.60	+2.71
fine-tune	36.52	38.38	39.32	39.80	40.19	40.40	40.54	40.94	40.81	40.87	-
word KD	36.52	39.31	39.59	40.38	40.53	41.01	40.97	41.24	41.14	41.40	+0.48
reg KD	36.52	38.94	39.59	40.27	40.43	40.97	41.27	41.48	41.42	41.57	+0.52
selective KD	36.52	38.75	39.44	39.84	40.43	40.62	40.78	40.83	40.85	40.98	+0.14
CLNMT	36.52	39.33	39.69	40.33	40.48	40.89	40.98	40.99	40.98	41.45	+0.43
ours	36.52	**39.76**	**40.11**	**40.69**	**41.02**	**41.34**	**41.93**	**41.67**	**42.03**	**42.25**	**+1.06**

4.4 Main Results

Table 2 shows the main results. Our method outperforms multiple continual learning methods in most stages, but falls short of upper bound. Compared to fine-tune, our method had an average of 1.11 improvement on the En-Zh experiment and 1.06 improvement on the De-En experiment. This indicates that our method is effective in alleviating catastrophic forgetting. The BLEU score of our method on $stage_5$ surpasses that of fine-tuning on $stage_{10}$, indicating that our method has a stronger ability to integrate old and new knowledge. In comparison to three knowledge distillation methods (word KD, reg KD, selective KD), our method exhibits a stable improvement in all stages. This demonstrates the negative impact of the teacher model in later epochs and the benefits of bidirectional marginal loss. When using continual learning, it is important to consider the quality of the teacher model. It is worth noting that selective KD does not perform well in multi-stage incremental scenarios. This suggests that selecting samples based on certain attributes is not suitable for continual learning because the teacher model is not fully utilized. In the first five stages of the En-Zh experiment, the BLEU of CLNMT is similar to that of our method. However, in the latter stages, the BLEU of CLNMT noticeably decreases, indicating that the advantage of dynamically calculating weights does not last for multiple stages. In contrast, our method maintains its advantages even after multiple stages.

Table 3. BLEU scores with different $[\beta, \alpha]$. The upper and lower subtables show the $stage_2$ experimental results of En-Zh and De-En. The first column is the value of β. The first row is the value of α.

En-Zh	0	0.1	0.2	De-En	0	0.1	0.2
0	58.70	**58.93**	58.76	0	39.31	39.64	39.52
0.1	–	58.85	58.81	0.1	–	39.58	**39.76**
0.2	–	–	58.79	0.2	–	–	39.63

4.5 Hyperparameters

We conducted experiments at $stage_2$ with different combinations of β and α, and the results are shown in Table 3. The best BML interval for En-Zh is [0, 0.1] and for De-En is [0.1, 0.2]. We believe that this is related to language and domain features. The diagonal of the table ($\beta = \alpha$) demonstrate an improvement for [0.1, 0.1] and [0.2, 0.2] over [0, 0], suggesting that a certain distance between the student model and the teacher model is beneficial for improving performance. The remaining results ($\beta < \alpha$) suggest that BML provides a more flexible range for updating the parameters of the student model, bringing it closer to the golden label in this range.

4.6 Case Study

This section uses case studies to determine whether our method is effective in alleviating catastrophic forgetting, as shown in Table 4. Fine-tune has obvious catastrophic forgetting, such as *"instant messaging protocols"* and *"a request to send a notification about your reception"* translated correctly in $stage_2$ but mistranslated in $stage_{10}$. Compared to fine-tune, our method alleviates this problem. In addition, fine-tune retains same errors from $stage_2$ to $stage_{10}$, such as *"cause problems with other applications"*, *"send KMail a rejection or a normal replie"*, while our method is correct. This suggests that our method has the ability to integrate old and new knowledge in a way that fine-tune lacks.

Table 4. Samples of the De-En IT domain test set with different methods at different stages. "Red" indicates mistranslation. "Blue" indicates corresponding correct translation.

source	Verwendete Instant-Messeging-Protokolle und installierte -Module (offizielle sowie inoffizielle)
reference	Instant Messaging protocols you use, and plugins you have installed (official and unofficial)
fine-tune ($stage_2$)	Used instant messaging protocols and installed modules (official and unofficial)
fine-tune ($stage_{10}$)	Used instant measurement protocols and installed plugins (official and informal)
ours ($stage_{10}$)	Instant messaging protocols and installed plugins (official and unofficial).
source	Dies kann dazu führen, dass das System sehr langsam reagiert, und es kann zu weiteren Problemen mit anderen Programmen führen. Sind Sie sicher, dass Sie das Bild skalieren möchten?
reference	This can reduce system responsiveness and cause other application resource problems. Are you sure you want to scale the image?
fine-tune ($stage_2$)	This can cause the system to react very slowly, and may cause problems with other applications. Are you sure you want to scale the image?
fine-tune ($stage_{10}$)	This can cause the system to react very slowly, and may cause other applications. Are you sure you want to scale the image?
ours ($stage_{10}$)	This can reduce system responsiveness and cause other application resource problems. Are you sure you want to scale the image?
source	Diese Nachricht enthält die Anforderung einer Empfangsbestätigung, aber die Bestätigung soll an mehr als eine Adresse versendet werden. Sie können die Anforderung ignorieren, KMail eine Ablehnung oder eine normale Antwort senden lassen
reference	This message contains a request to send a notification about your reception of the message, but it is requested to send the notification to more than one address. You can either ignore the request or let KMail send a "denied" or normal response
fine-tune ($stage_2$)	This message contains the request for a receipt confirmation, but the confirmation should be sent to more than one address. You can ignore the request or send KMail a rejection or a normal replie
fine-tune ($stage_{10}$)	This message contains a request for acknowledgement but the confirmation should be sent to more than one address. You can ignore the request or send KMail a refusal or a normal response
ours ($stage_{10}$)	This message contains a request for confirmation of receiving, but the confirmation is to be sent to more than one address. You can ignore the request or let KMail send a denied, or send a normal response

5 Conclusion

In this paper, we propose a multi-stage incremental framework for domain NMT based on knowledge distillation to address catastrophic forgetting and in-domain continual learning. Through extensive experimental analysis, we find that using knowledge distillation with a fixed weight only benefits the student model in the early epochs and harms it in the later epochs. Based on this observation, we propose a two-step training method that uses the bidirectional marginal loss instead of the regular continual learning loss. Our experiments show that our method outperforms others in multiple stages.

References

1. Aharoni, R., Goldberg, Y.: Unsupervised domain clusters in pretrained language models. In: Proceedings of the 58th Annual Meeting of the Association for Computational Linguistics, vol. 1: Long Papers. Association for Computational Linguistics (2020). https://arxiv.org/abs/2004.02105
2. Aljundi, R., Babiloni, F., Elhoseiny, M., Rohrbach, M., Tuytelaars, T.: Memory aware synapses: learning what (not) to forget. In: Proceedings of the European Conference on Computer Vision (ECCV) (2018)
3. Berard, A.: Continual learning in multilingual NMT via language-specific embeddings. In: Proceedings of the Sixth Conference on Machine Translation, pp. 542–565. Association for Computational Linguistics (2021). https://aclanthology.org/2021.wmt-1.62
4. Cao, Y., Wei, H.R., Chen, B., Wan, X.: Continual learning for neural machine translation. In: Proceedings of the 2021 Conference of the North American Chapter of the Association for Computational Linguistics: Human Language Technologies, pp. 3964–3974 (2021)
5. Dabre, R., Fujita, A.: Combining sequence distillation and transfer learning for efficient low-resource neural machine translation models. In: Proceedings of the Fifth Conference on Machine Translation, pp. 492–502. Association for Computational Linguistics (2020). https://aclanthology.org/2020.wmt-1.61
6. Diddee, H., Dandapat, S., Choudhury, M., Ganu, T., Bali, K.: Too brittle to touch: comparing the stability of quantization and distillation towards developing low-resource MT models. In: Proceedings of the Seventh Conference on Machine Translation (WMT), pp. 870–885. Association for Computational Linguistics, Abu Dhabi (Hybrid) (2022). https://aclanthology.org/2022.wmt-1.80
7. French, R.: Catastrophic interference in connectionist networks: can it be predicted, can it be prevented? Adv. Neural Inf. Process. Syst. **6** (1993)
8. Garcia, X., Constant, N., Parikh, A., Firat, O.: Towards continual learning for multilingual machine translation via vocabulary substitution. In: Proceedings of the 2021 Conference of the North American Chapter of the Association for Computational Linguistics: Human Language Technologies, pp. 1184–1192 (2021)
9. Gordon, M.A., Duh, K.: Explaining sequence-level knowledge distillation as data-augmentation for neural machine translation. arXiv preprint arXiv:1912.03334 (2019)
10. Gu, S., Hu, B., Feng, Y.: Continual learning of neural machine translation within low forgetting risk regions. arXiv preprint arXiv:2211.01542 (2022)

11. Jin, X., et al.: Lifelong pretraining: continually adapting language models to emerging corpora. In: Proceedings of the 2022 Conference of the North American Chapter of the Association for Computational Linguistics: Human Language Technologies, pp. 4764–4780 (2022)
12. Jooste, W., Way, A., Haque, R., Superbo, R.: Knowledge distillation for sustainable neural machine translation. In: Proceedings of the 15th Biennial Conference of the Association for Machine Translation in the Americas, vol. 2: Users and Providers Track and Government Track, pp. 221–230. Association for Machine Translation in the Americas, Orlando (2022). https://aclanthology.org/2022.amta-upg.16
13. Kendall, A., Gal, Y., Cipolla, R.: Multi-task learning using uncertainty to weigh losses for scene geometry and semantics. In: Proceedings of the IEEE Conference on Computer Vision and Pattern Recognition, pp. 7482–7491 (2018)
14. Khayrallah, H., Thompson, B., Duh, K., Koehn, P.: Regularized training objective for continued training for domain adaptation in neural machine translation. In: Proceedings of the 2nd Workshop on Neural Machine Translation and Generation, pp. 36–44. Association for Computational Linguistics, Melbourne (2018). https://doi.org/10.18653/v1/W18-2705. https://aclanthology.org/W18-2705
15. Kim, Y., Rush, A.M.: Sequence-level knowledge distillation. In: Proceedings of the 2016 Conference on Empirical Methods in Natural Language Processing, pp. 1317–1327. Association for Computational Linguistics, Austin (2016). https://doi.org/10.18653/v1/D16-1139. https://aclanthology.org/D16-1139
16. Lin, Z., et al.: Pre-training multilingual neural machine translation by leveraging alignment information. In: Proceedings of the 2020 Conference on Empirical Methods in Natural Language Processing (EMNLP), pp. 2649–2663. Association for Computational Linguistics (2020). https://www.aclweb.org/anthology/2020.emnlp-main.210
17. Post, M.: A call for clarity in reporting BLEU scores. In: Proceedings of the Third Conference on Machine Translation: Research Papers, pp. 186–191. Association for Computational Linguistics, Brussels (2018). https://doi.org/10.18653/v1/W18-6319. https://aclanthology.org/W18-6319
18. Saunders, D., Stahlberg, F., de Gispert, A., Byrne, B.: Domain adaptive inference for neural machine translation. In: Proceedings of the 57th Annual Meeting of the Association for Computational Linguistics, pp. 222–228 (2019)
19. Sutskever, I., Vinyals, O., Le, Q.V.: Sequence to sequence learning with neural networks. Adv. Neural Inf. Process. Syst. **27** (2014)
20. Vaswani, A., et al.: Attention is all you need. Adv. Neural Inf. Process. Syst. **30** (2017)
21. Wang, F., Yan, J., Meng, F., Zhou, J.: Selective knowledge distillation for neural machine translation. In: Proceedings of the 59th Annual Meeting of the Association for Computational Linguistics and the 11th International Joint Conference on Natural Language Processing, vol. 1: Long Papers, pp. 6456–6466. Association for Computational Linguistics (2021). https://doi.org/10.18653/v1/2021.acl-long.504. https://aclanthology.org/2021.acl-long.504
22. Wei, H.R., Huang, S., Wang, R., Dai, X.Y., Chen, J.: Online distilling from checkpoints for neural machine translation. In: Proceedings of the 2019 Conference of the North American Chapter of the Association for Computational Linguistics: Human Language Technologies, vol. 1 (Long and Short Papers), pp. 1932–1941. Association for Computational Linguistics, Minneapolis, Minnesota (2019). https://doi.org/10.18653/v1/N19-1192. https://aclanthology.org/N19-1192
23. Wu, Y., et al.: Large scale incremental learning. In: Proceedings of the IEEE/CVF Conference on Computer Vision and Pattern Recognition, pp. 374–382 (2019)

Neural-Symbolic Reasoning with External Knowledge for Machine Reading Comprehension

Yilin Duan[1], Sijia Zhou[1], Xiaoyue Peng[1], Xiaojun Kang[1,2(✉)], and Hong Yao[1,2]

[1] School of Computer Science, China University of Geosciences, Wuhan 430074, China
{duanyl,zsj,pengxiaoyue,kangxj,yaohong}@cug.edu.cn
[2] Hubei Key Laboratory of Intelligent Geo-Information Processing, China University of Geosciences, Wuhan 430074, China

Abstract. Machine reading comprehension is a fundamental in natural language understanding. Existing large-scale pre-trained language models and graph neural network-based models have achieved good gains on logical reasoning of text. However, neither of them can give a complete reasoning chain, while symbolic logic-based reasoning is explicit and explainable. Therefore, we propose a framework LoGEK that integrates symbolic **Lo**gic and **G**raph neural networks for reasoning, while leveraging **E**xternal **K**nowledge to augment the logical graph. The LoGEK model consists of three parts: logic extraction and extension, logical graph reasoning and answer prediction. Specifically, LoGEK extracts and extends logic set from the unstructured text. Then the logical graph reasoning module uses external knowledge to extend the original logical graph. After that, the model uses a path-based relational graph neural network to model the extended logical graph. Finally, the prediction module performs answer prediction based on graph embeddings and text embeddings. We conduct experiments on benchmark datasets for logical reasoning to evaluate the performance of LoGEK. The experimental results show that the accuracy of the method in this paper is better than the baseline models, which verifies the effectiveness of the method.

Keywords: Machine Reading Comprehension · Symbolic Reasoning · External Knowledge · Relational Graph Neural Network

1 Introduction

Machine Reading Comprehension (MRC) is a popular task in natural language processing and understanding. Based on the given context and question, the

This work was supported by the Open Research Project of the Hubei Key Laboratory of Intelligent Geo-Information Processing (No. KLIGIP-2022-B11), and in part by the National Natural Science Foundation of China (NSFC) (No. 61972365, 42071382).

B. Luo et al. (Eds.): ICONIP 2023, CCIS 1965, pp. 440–452, 2024.
https://doi.org/10.1007/978-981-99-8145-8_34

task is to identify the most suitable answer from a set of candidate options. It is expected that the model can give the reasoning process, which is explicit and can be explained. Symbolic logic-based reasoning is usually explainable and transferable while GNN-based reasoning excels in reasoning by modeling paths from the question nodes to the answer nodes, showcasing powerful path modeling capabilities. Another challenge is the lack of reasoning chains due to insufficient context information. Models can only reason from the given context and struggle with scenarios that require external commonsense knowledge. Therefore, external knowledge can enable models to have a better understanding of contextual semantics and background knowledge during the reasoning process.

In recent years, large-scale pre-trained language models (PLMs) have made significant breakthroughs in many NLP tasks. Among them, the most representative are ChatGPT and GPT4, which contribute impressive reasoning capabilities to the community [2,3]. There have been quite a lot of researches on how to make better use of PLMs for reasoning [24,30]. But their effectiveness often deteriorates when specific knowledge is missing or when there is a lack of semantic knowledge in the corpus. Additionally, pre-trained language models are unable to provide explainable predictions since the implicit learning in "black-box" mode makes it challenging to explicitly state the knowledge used in the reasoning process while predicting answers.

This paper proposes the following approach: (1) combining symbolic logic reasoning with graph neural networks to accurately answer questions while outputting explainable reasoning paths; (2) introducing external knowledge to enhance the reasoning chains when background information is insufficient, further improving the accuracy of question answering. The main contributions include:

- We propose a model named LoGEK which combines logic rules with neural networks to infer the logical structure using a relational graph model and enhances the explainability of the model.
- We integrate external knowledge into the existing logical graph. The logical symbols and relations are extracted from the external knowledge and added to the symbolic logical graph to improve the logic chain.
- Comprehensive experiments demonstrate the effectiveness of LoGEK, which outperforms state-of-the-art models on two datasets.

2 Related Work

Symbolic Logic-Based QA. The method based on symbolic logic rules has been widely discussed in the research of question answering and reasoning because of its high accuracy and strong explainability. The mainstream methods of exploring logical rules generally combine either probabilistic logic method or knowledge embedding method. Zhang et al. [28] proposed a probabilistic logical reasoning framework combined with GNN, ExpressGNN. The GQE model [7] embeds a query as a point in a vector space. Ren et al. [19] put forward the Query2Box model, which encodes positive first-order logical queries into rectangular boxes. Their method effectively solves the problem of positive first-order logical query reasoning, and improves the explainability. BetaE model [20] improves the Query2Box model with beta embedding method, and uses probability distributions with bounded support to embed a query or an entity into Beta distribution, thus being able to model uncertainty. Yang et al. [25] proposed an efficient neural network-based inductive learning model (NLIL), transforming the relational path in multi-hop reasoning into a chain-like first-order logic rule to solve the problem of inductive logic programming (ILP), which extends the logical expression effectively. Wang et al. [23] proposed a context extension framework and introduced logic-driven contrastive learning to better capture logical relationships.

GNN-Based QA. The method based on GNN first forms a graph of the question context and the answers, and then deduces the answer on the graph. Lin et al. [12] effectively utilized the external structured commonsense knowledge graph to perform reasoning and proposed KagNet, a knowledge-aware graph network. The multi-hop graph relation network MHGRN [6] proposed by Feng et al. can model multiple relational paths on a large scale demonstrably, and proposes a structured relational attention mechanism for multi-hop path modeling. Asai et al. [1] proposed a graph-based cyclic retrieval method to learn how to retrieve reasoning paths in Wikipedia. This method is divided into a retriever and a reader. Chen et al. [4] proposed HGN which deals with context at both discourse level and word level to provide a more fine-grained relation extraction. Neural-symbolic models have also made some progress. Li et al. [11] proposed AdaLoGN, which applies message passing to logical graphs to achieve mutual iterative reinforcement of neural reasoning and symbolic reasoning. To address the overfitting and poor generalization caused by annotation sparsity, Jiao et al. [8] designed a meta-path guided contrastive learning method to perform self-supervised pre-training on unlabeled text data.

3 Methodology

3.1 Logic Extraction and Extension

Logic Recognition. Logic recognition realizes the recognition and extraction of logical symbols from unstructured text. Specifically, we use a constituency parser [9] to extract the noun phrases and gerund phrases in the text as basic symbols, denoted as $\{\alpha, \beta, \gamma, ...\}$. Then, logical symbols are combined using the set of logical connectors $\{\neg, \rightarrow\}$ to form a set of logical expressions like $\{(\alpha\rightarrow\beta)...\}$, where "$\neg$" represents the negation operation, and "$\rightarrow$" represents the conditional relationship between two logical symbols. If the logical symbol α is negated by a negation word, a new logical symbol $\neg\alpha$ is created by adding a negation connector before α. As shown in the logic recognition module in Fig. 1, there are three logical symbols α, β, γ and two logical expressions: $(\neg\alpha\rightarrow\neg\beta)$ and $(\neg\beta\rightarrow\neg\gamma)$.

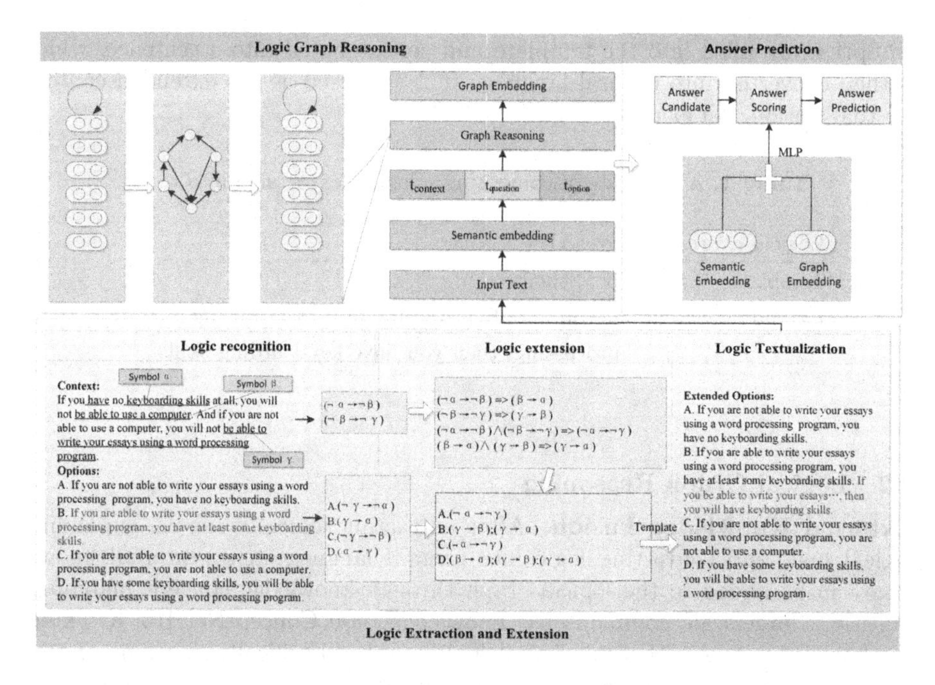

Fig. 1. The overall architecture of LoGEK.

Logic Extension. After the basic logic is identified, there are also some implicit expressions that need to be inferred and extended. First, the logical expressions that have been determined and exist in all sentences in the context are combined into a set of logical expressions S, and the logical equivalence law is used to infer and further expand implicit logical expressions. The logical equivalence law follows the laws of contraposition [21] and transitivity [29]:

$$(\alpha \rightarrow \beta) \Rightarrow (\neg\beta \rightarrow \neg\alpha) \tag{1}$$

$$(\alpha \rightarrow \beta) \wedge (\beta \rightarrow \gamma) \Rightarrow (\alpha \rightarrow \gamma) \tag{2}$$

Therefore, the expanded set of implicit logical expressions forms the extension set S_{E} of the current logical expression set S in Fig. 1.

Logic Textualization. Firstly, relevant expressions for each operation are selected from S_{E}. The identified expressions from the options are compared with the expressions in the expansion set (using text overlap). If an expression in the expansion set contains the same logical symbol as the option expression, the expression from the expansion set is added to the option's expressions to form an expression expansion set for that option. So the expressions related to each option are added to the corresponding option expression, forming the expression expansion set for each option. To convert all logical expressions into natural language text, a conversion template is defined. All logical expressions related to an option are filled into the template and concatenated into a sentence, which is then converted into natural language text and used as the extended context for that option. An example is given in Table 1.

Table 1. A template of converting a logical expression into text.

Logical expression	$(\gamma \rightarrow \alpha)$
Template	If γ, then α.
Extended context	If you are able to write your essays using a word processing, then you have keyboarding skills

3.2 Logical Graph Reasoning

External Knowledge Fusion. After extracting logical symbols, expanding logical text, and converting logic into natural language text, external knowledge can be added to the logical graph through knowledge concept matching. We use a large-scale commonsense knowledge base ConceptNet [13] as external knowledge to enrich the original logic graph. Specifically, n-grams of sentences can be precisely matched with concepts in the knowledge graph. For example, for "sitting too close to watch TV can cause pain", the exact matching results are {sitting, close, watch_tv, pain, ...}. However, the matched concepts are not always needed and can introduce noise. Therefore, this paper uses stemming and stop-word filtering for soft matching to mitigate this effect. We select 17 types of relations in ConceptNet to match with the original logical graph nodes, including "cause", "is_a", "part_of", "related_to", "antonym", "capable_of", "not_capable_of", "derived", "made_of", "desires", "not_desires", "used_for", "has_subevent", "has_context", "has_property", "receives_action" and "at_location". If a node in the original logical graph has one of these relations with a node in ConceptNet, the node from ConceptNet and the relation are added to the logical graph.

Logical Graph Reasoning. We encode the constructed logical graph using Multi-Hop Graph Relation Networks (MHGRN) [6]. First, we use a pre-trained language model to encode the text, obtaining a token sequence and extract node features. Then, node type-specific linear transformations are applied to the input node features to enable the model to perceive node information ϕ:

$$x_i = U_{\phi(i)} h_i + b_{\phi(i)} \tag{3}$$

where U and b are learnable parameters of type specific to node i. For all relational paths, RGCN is used to perform one-hop message passing:

$$\begin{aligned} z_i^k = \sum_{(j,r_1,\ldots,r_k,i)\in\Phi_k} & \alpha\left(j, r_1, \ldots, r_k, i\right)/d_i^k \cdot \\ & W_0^K \cdots W_0^{k+1} W_{r_k}^k \cdots W_{r_1}^1 x_j \quad (1 \le k \le K) \end{aligned} \tag{4}$$

where $(j, r_1, \ldots, r_k, i)$ is the path of length K, Φ_k is the path set, W_r^t is the learnable parameter matrix, $\alpha(j, r_1, \ldots, r_k, i)$ is the attention score and d_i^k is the normalization factor. Then, information from paths of different lengths is aggregated through the attention mechanism:

$$z_i = \sum_{k=1}^{K} softmax\left(bilinear\left(s, z_i^k\right)\right) \cdot z_i^k \tag{5}$$

Finally, the output node embedding is obtained through the nonlinear activation function:

$$h_i' = \sigma\left(Vh_i + V'z_i\right) \tag{6}$$

The structured attention mechanism transforms the problem of computing attention scores into a probability problem based on the semantic vectors of the relation sequence. This allows for an effective parameterization of the attention scores $\alpha(j, r_1, \ldots, r_k, i)$. It is treated as a relation sequence conditioned on the semantic vector s, which can be modeled by a probabilistic graphical model, such as conditional random field:

$$\alpha(j, r_1, \ldots, r_k, i) = p\left(\phi(j), r_1, \ldots, r_k, \phi(i) \mid \mathbf{s}\right) \tag{7}$$

3.3 Answer Prediction

Finally, we calculate the confidence of candidate answer using the embedding of the question q, text s, and graph G. The graph representation is obtained from the node vector using attention pooling. The graph vector and semantic vector are input into an MLP to calculate the confidence score $\rho(q, a) = MLP(s \oplus g)$, and the answer with the highest score is output. The training process aims to maximize the confidence score of the answer $\widehat{a}$ by minimizing the cross-entropy loss:

$$L = E_{q,\widehat{a},C}\left[-\log \frac{\exp(\rho(q, \widehat{a}))}{\sum_{a \in C} \exp(\rho(q, a))}\right] \tag{8}$$

4 Experiments

4.1 Dataset

The Reclor dataset [27], which consists of high-quality practice exam questions from GMAT, LSAT, and other sources. It includes 6138 logical reasoning questions which are divided into 17 categories, such as sufficient assumption and necessary assumption, as shown in Table 2. It contains a training set, a development set, and a test set, with 4638, 500, and 1500 instances, respectively.

The LogiQA dataset [14] is derived from publicly available questions from the National Civil Servants Examination of China. It contains 8678 instances, which are divided in the same way.

Table 2. Statistics of Reclor dataset.

Statistics	Reclor	LogiQA
Context type	Written text	Written text
Number of options	4	4
Number of context	6138	8768
Number of questions	6138	8768
Vocab size	26576	37963
Context length	73.6	76.87
Question length	17.0	12.03
Option length	20.6	15.83

4.2 Experimental Settings

Baseline. The baseline pre-trained models for multiple -choice question answering include GPT [17], GPT-2 [18], BERT [5], XLNet [26], RoBERTa [16], and ALBERT [10]. The logic-driven context extension framework LReasoner [23]. The baseline muli-hop reasoning models include RGCN [22], KagNet [12], and MHGRN [6].

Experimental Environment. The system environment of this paper is the Ubuntu 18.10 Linux operating system, with Python 3.6 as the Python environment. The models used in this study are implemented in Python 3.6 and PyTorch 1.3.1 on GPUs.

Parameter Configuration. This paper uses cross-entropy loss and adopts RAdam [15] as the optimizer. For text semantic vector encoding, RoBERTa-large and ALBERT-xxlarge-v2 are used, both with a learning rate of 1e-5, a hidden layer size of 1024, a batch size of 2, and fine-tuning on ReClor for 20 epochs. The NLTK version used is 3.4.5. The learning rates of the graph encoders RGCN, KagNet, and MHGRN in the compared models are 1e-3, 1e-3, and 1e-4, respectively.

4.3 Experimental Results

Table 3 presents the answer prediction accuracy of our model compared with baseline models on the Reclor dataset. Overall, the prediction accuracy of our model on the validation set of the Reclor dataset is 66.8 ($\text{LoGEK}_{\text{RoBERTa}}$) and 71.8 ($\text{LoGEK}_{\text{ALBERT}}$) respectively. From the results, it can be observed that the answer prediction accuracy of the baseline pre-trained models are generally

Table 3. The accuracy scores on Reclor. Bold indicates the best values. "w/o EK" means remove external knowledge. "w/o RG" means remove relational GNN network.

Models	Dev	Test
GPT	–	45.4
GPT-2	–	47.2
BERT	–	49.8
XLNet	–	56.0
RGCN	68.1	62.5
KagNet	62.5	57.1
MHGRN	66.3	69.4
RoBERTa	62.1	55.6
$\text{Lreasoner}_{\text{RoBERTa}}$	64.7	58.2
$\text{LoGEK}_{\text{RoBERTa}}$(ours)	66.8	60.4
ALBERT	66.5	62.1
$\text{Lreasoner}_{\text{ALBERT}}$	70.8	69.4
$\text{LoGEK}_{\text{ALBERT}}$(ours)	**71.8**	**70.2**
$\text{LoGEK}_{\text{ALBERT}}$ w/o EK	71.6	70.0
$\text{LoGEK}_{\text{ALBERT}}$ w/o RG	70.8	69.4

Table 4. The accuracy scores on LogiQA. Bold indicates the best values.

Model	Dev	Test
RoBERTa	35.8	30.1
$\text{Lreasoner}_{\text{RoBERTa}}$	38.4	33.5
$\text{LoGEK}_{\text{RoBERTa}}$(ours)	41.1	33.9
ALBERT	50.5	44.2
$\text{Lreasoner}_{\text{ALBERT}}$	54.8	47.3
$\text{LoGEK}_{\text{ALBERT}}$(ours)	**55.9**	**48.8**

lower than those of the logical reasoning framework (Lreasoner), which indicates that the logical reasoning framework algorithm is more capable of answering logical questions and demonstrates the effectiveness of the framework for accurately predicting answers. $\text{Lreasoner}_{\text{RoBERTa}}$ and $\text{Lreasoner}_{\text{ALBERT}}$ perform better than their respective baseline models RoBERTa and ALBERT, indicating that the logical reasoning framework is robust and can effectively perform logical reasoning on different pre-trained models. Even though the performance of the baseline model ALBERT is already good, the addition of the logical reasoning framework still achieves higher accuracy.

$\text{Lreasoner}_{\text{ALBERT}}$ achieves the highest accuracy of 71.8, which is 5.5 points higher than MHGRN, indicating that the logical reasoning framework has a significant effect on answering logical questions. The poor performance of the

KagNet and MHGRN is due to their better suitability for answering common sense questions and their insufficient ability to answer logical questions.

Table 4 shows the results of the explainable reasoning models incorporating external knowledge on the LogiQA dataset. It indicates that our method is still effective on different datasets. The model incorporating external knowledge achieves higher accuracy in answer prediction compared to the model without external knowledge, further verifying the effectiveness of incorporating external knowledge in logical reasoning. However, horizontally speaking, the model's predictions on the LogiQA dataset generally have lower accuracy than those on the Reclor dataset. The analysis shows that this may be due to two reasons. Firstly, the LogiQA dataset as a whole is more logical and difficult, making it relatively difficult for the model to understand the problem. Secondly, because most of the questions in the dataset are from the National Civil Servants Examination of China, some obscure words which are difficult to translate into English are expressed in Chinese in the English version of the dataset, causing the model unable to analyze the problem with the original algorithm, resulting in lower accuracy.

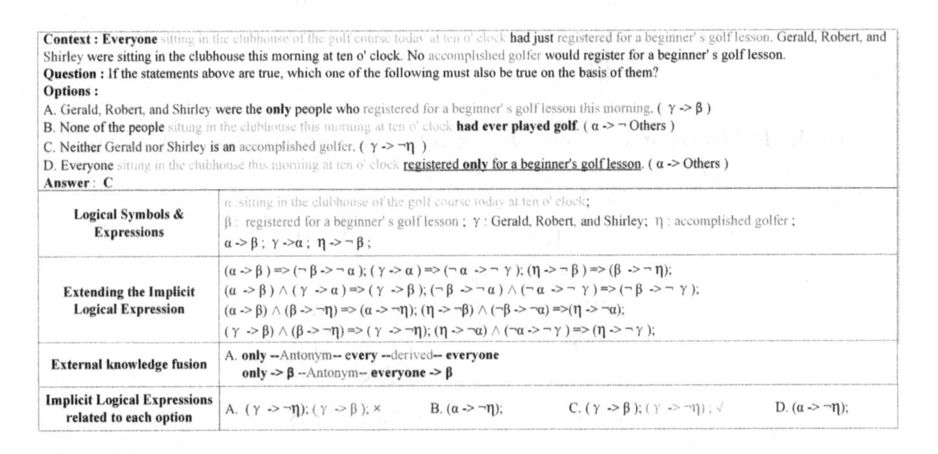

Fig. 2. Case study. It shows a specific case and the reasoning process of the model. Four different colors are used to mark phrases in the context to display different logical symbols. Underlined phrases represent other symbols different from contextual logic symbols, and bold phrases indicate different semantic expressions. The options marked by "× (✓)" are the wrong(predicted) options predicted by the model before(after) adding external knowledge.

4.4 Ablation Study

Table 3 shows the comparison study between ALBERT and RGCN-based baseline models, in order to investigate the effectiveness of different modules. The evaluation metric is Accuracy. From the results of the baseline models, it can be seen that the performance of RGCN is better than ALBERT, indicating that the graph structure with relationships has a more superior and structured text representation compared to PLMs, which makes the semantic expression of the model more sufficient, resulting in more accurate predictions. The advantages of the graph structure are more evident after the logical extension framework is introduced. Furthermore, after incorporating external knowledge, the accuracy of the model's answer prediction is further improved, indicating that external knowledge plays a significant role in enhancing the model's explainability.

4.5 Case Study

Figure 2 shows an example of the reasoning process and reveals the important role of external knowledge fusion in improving model performance. Before combining external knowledge, the correctness of choice A and choice C in the case is indistinguishable, because both options correspond to the expression of logical symbolic reasoning, and the model incorrectly selects A as the correct answer based on the priority of the judgment condition. After combining external knowledge, A special expression with an antisense can be identified in option A. Through the concept matching of external knowledge, "only" in option A matches "every" in ConceptNet as an antonym, and "everyone" in context matches the derivative of "every". Therefore, the expression of choice A does not match the context. The model can select option C as the most reasonable answer, which logically and semantically matches the extended implicit logical expression. It can be seen that the method integrating external knowledge not only improves the reasoning chain, but also makes the model predict the answer more accurately. Meanwhile, the inference path is output, which further improves the explainability of the model.

5 Conclusion

This paper focuses on the complex logical reasoning task. We combine the strengths of neural-based and symbolic logical-based methods to improve the explainability of reasoning. We also integrate external knowledge which can help the model to understand the background knowledge of the context. The model is verified on two logical datasets. The results show that the explainable reasoning algorithm integrated with external knowledge has a significant effect on improving the accuracy and explainability of the model's answer prediction. Although ConceptNet has extensive and diverse knowledge content, each knowledge base has its limitations. Considering the scarcity and varying quality of logical datasets, in future research work, we will look for ways to construct more rigorous and diverse logical datasets to improve the model's robustness and generalization ability.

References

1. Asai, A., Hashimoto, K., Hajishirzi, H., Socher, R., Xiong, C.: Learning to retrieve reasoning paths over wikipedia graph for question answering. In: 8th International Conference on Learning Representations, ICLR 2020, Addis Ababa, Ethiopia, 26–30 April 2020. OpenReview.net (2020)
2. Bang, Y., et al.: A multitask, multilingual, multimodal evaluation of chatgpt on reasoning, hallucination, and interactivity. CoRR abs/ arXiv: 2302.04023 (2023)
3. Bubeck, S., et al.: Sparks of artificial general intelligence: Early experiments with GPT-4. CoRR abs/ arXiv: 2303.12712 (2023)
4. Chen, J., Zhang, Z., Zhao, H.: Modeling hierarchical reasoning chains by linking discourse units and key phrases for reading comprehension. In: Proceedings of the 29th International Conference on Computational Linguistics, COLING 2022, Gyeongju, Republic of Korea, 12–17 October 2022, pp. 1467–1479. International Committee on Computational Linguistics (2022)
5. Devlin, J., Chang, M., Lee, K., Toutanova, K.: BERT: pre-training of deep bidirectional transformers for language understanding. In: Proceedings of NAACL-HLT, pp. 4171–4186. Association for Computational Linguistics (2019)
6. Feng, Y., Chen, X., Lin, B.Y., Wang, P., Yan, J., Ren, X.: Scalable multi-hop relational reasoning for knowledge-aware question answering. In: Proceedings of the 2020 Conference on Empirical Methods in Natural Language Processing, EMNLP 2020, Online, 16–20 November 2020, pp. 1295–1309. Association for Computational Linguistics (2020)
7. Hamilton, W.L., Bajaj, P., Zitnik, M., Jurafsky, D., Leskovec, J.: Embedding logical queries on knowledge graphs. In: Advances in Neural Information Processing Systems 31: Annual Conference on Neural Information Processing Systems 2018, NeurIPS 2018, 3–8 December 2018, Montréal, Canada, pp. 2030–2041 (2018)
8. Jiao, F., Guo, Y., Song, X., Nie, L.: Merit: meta-path guided contrastive learning for logical reasoning. In: Findings of the Association for Computational Linguistics: ACL 2022, Dublin, Ireland, 22–27 May 2022, pp. 3496–3509. Association for Computational Linguistics (2022)
9. Joshi, V., Peters, M.E., Hopkins, M.: Extending a parser to distant domains using a few dozen partially annotated examples. In: Proceedings of the 56th Annual Meeting of the Association for Computational Linguistics (Volume 1: Long Papers), pp. 1190–1199 (2018)
10. Lan, Z., Chen, M., Goodman, S., Gimpel, K., Sharma, P., Soricut, R.: ALBERT: a lite BERT for self-supervised learning of language representations. In: 8th International Conference on Learning Representations, ICLR 2020, Addis Ababa, Ethiopia, 26–30 April 2020. OpenReview.net (2020)
11. Li, X., Cheng, G., Chen, Z., Sun, Y., Qu, Y.: Adalogn: adaptive logic graph network for reasoning-based machine reading comprehension. In: Proceedings of the 60th Annual Meeting of the Association for Computational Linguistics (Volume 1: Long Papers), pp. 7147–7161 (2022)
12. Lin, B.Y., Chen, X., Chen, J., Ren, X.: Kagnet: knowledge-aware graph networks for commonsense reasoning. In: Proceedings of the 2019 Conference on Empirical Methods in Natural Language Processing and the 9th International Joint Conference on Natural Language Processing (EMNLP-IJCNLP), pp. 2829–2839 (2019)
13. Liu, H., Singh, P.: Conceptnet-a practical commonsense reasoning tool-kit. BT Technol. J. **22**(4), 211–226 (2004)

14. Liu, J., Cui, L., Liu, H., Huang, D., Wang, Y., Zhang, Y.: LogiQA: a challenge dataset for machine reading comprehension with logical reasoning. In: Proceedings of the Twenty-Ninth International Joint Conference on Artificial Intelligence, IJCAI 2020, pp. 3622–3628. ijcai.org (2020)
15. Liu, L., et al.: On the variance of the adaptive learning rate and beyond. In: 8th International Conference on Learning Representations, ICLR 2020, Addis Ababa, Ethiopia, 26–30 April 2020. OpenReview.net (2020)
16. Liu, P., Yuan, W., Fu, J., Jiang, Z., Hayashi, H., Neubig, G.: Pre-train, prompt, and predict: a systematic survey of prompting methods in natural language processing. ACM Comput. Surv. **55**(9), 195:1–195:35 (2023)
17. Radford, A., Narasimhan, K., Salimans, T., Sutskever, I., et al.: Improving language understanding by generative pre-training (2018)
18. Radford, A., et al.: Language models are unsupervised multitask learners. OpenAI blog **1**(8), 9 (2019)
19. Ren, H., Hu, W., Leskovec, J.: Query2box: reasoning over knowledge graphs in vector space using box embeddings. In: 8th International Conference on Learning Representations, ICLR 2020, Addis Ababa, Ethiopia, 26–30 April 2020. OpenReview.net (2020)
20. Ren, H., Leskovec, J.: Beta embeddings for multi-hop logical reasoning in knowledge graphs. In: Advances in Neural Information Processing Systems 33: Annual Conference on Neural Information Processing Systems 2020, NeurIPS 2020, 6–12 December 2020, virtual (2020)
21. Russell, S., Norvig, P.: Artificial Intelligence: A Modern Approach, 4th edn. Pearson (2020)
22. Schlichtkrull, M., Kipf, T.N., Bloem, P., van den Berg, R., Titov, I., Welling, M.: Modeling relational data with graph convolutional networks. In: Gangemi, A., et al. (eds.) ESWC 2018. LNCS, vol. 10843, pp. 593–607. Springer, Cham (2018). https://doi.org/10.1007/978-3-319-93417-4_38
23. Wang, S., et al.: Logic-driven context extension and data augmentation for logical reasoning of text. arXiv preprint arXiv:2105.03659 (2021)
24. Wang, X., Wei, J., Schuurmans, D., Le, Q.V., Chi, E.H., Zhou, D.: Self-consistency improves chain of thought reasoning in language models. CoRR abs/ arXiv: 2203.11171 (2022)
25. Yang, Y., Song, L.: Learn to explain efficiently via neural logic inductive learning. In: 8th International Conference on Learning Representations, ICLR 2020, Addis Ababa, Ethiopia, 26–30 April 2020. OpenReview.net (2020)
26. Yang, Z., Dai, Z., Yang, Y., Carbonell, J., Salakhutdinov, R.R., Le, Q.V.: Xlnet: generalized autoregressive pretraining for language understanding. In: Advances in Neural Information Processing Systems 32 (2019)
27. Yu, W., Jiang, Z., Dong, Y., Feng, J.: Reclor: a reading comprehension dataset requiring logical reasoning. In: 8th International Conference on Learning Representations, ICLR 2020, Addis Ababa, Ethiopia, 26–30 April 2020. OpenReview.net (2020)
28. Zhang, Y., et al.: Efficient probabilistic logic reasoning with graph neural networks. In: 8th International Conference on Learning Representations, ICLR 2020, Addis Ababa, Ethiopia, 26–30 April 2020. OpenReview.net (2020)

29. Zhao, J., Rudnick, E.M., Patel, J.H.: Static logic implication with application to redundancy identification. In: 15th IEEE VLSI Test Symposium (VTS'97), Monterey, California, USA, 27 April -1 May 1997, pp. 288–295. IEEE Computer Society (1997)
30. Zhou, D., et al.: Least-to-most prompting enables complex reasoning in large language models. CoRR abs/ arXiv: 2205.10625 (2022)

Partial Multi-label Learning via Constraint Clustering

Sajjad Kamali Siahroudi(✉) and Daniel Kudenko

Leibniz University Hannover, L3S Research Center, Hannover, Germany
{kamali,kudenko}@l3s.de

Abstract. Multi-label learning (*MLL*) refers to a learning task where each instance is associated with a set of labels. However, in most real-world applications, the labeling process is very expensive and time consuming. Partially multi-label learning (*PML*) refers to *MLL* where only a part of the labels are correctly annotated and the rest are false positive labels. The main purpose of *PML* is to learn and predict unseen multi-label data with less annotation cost. To address the ambiguities in the label set, existing popular *PML* research attempts to extract the label confidence for each candidate label. These methods mainly perform disambiguation by considering the correlation among labels or/and features. However, in *PML* because of noisy labels, the true correlation among labels is corrupted. These methods can be easily misled by noisy false-positive labels. In this paper, we propose **P**artial **M**ulti-**L**abel learning method via **C**onstraint **C**lustering (*PML-CC*) to address *PML* based on the underlying structure of data. *PML-CC* gradually extracts high-confidence labels and then uses them to extract the rest labels. To find the high-confidence labels, it solves *PML* as a clustering task while considering extracted information from previous steps as constraints. In each step, *PML-CC* updates the extracted labels and uses them to extract the other labels. Experimental results show that our method successfully tackles *PML* tasks and outperforms the state-of-the-art methods on artificial and real-world datasets.

Keywords: partial multi-label learning · constraint clustering · disambiguation

1 Introduction

Multi-label learning (*MLL*) is a supervised learning task where each sample is associated with multiple labels [20]. *MLL* has been used for many real-world applications such as text, image, audio, video, and gene classification [15] and could successfully address them. However, in some applications such as image annotation, the true label sets of objects are not available. In such a *partially multi-label learning (PML)* setting [7], similar to *MLL*, each instance is labeled with a set of labels. However, for each training instance only a part of the labels are correctly annotated and the rest are false positive labels. *MLL* is a

B. Luo et al. (Eds.): ICONIP 2023, CCIS 1965, pp. 453–469, 2024.
https://doi.org/10.1007/978-981-99-8145-8_35

challenging problem and partially labeled data makes it even more challenging. The main purpose of using partially labeled data is to learn to predict unseen multi-label data with less annotation cost than in the *MLL* case [19]. Since collecting partially labeled training data is less costly and easier, the demand for *PML* solutions in many real-world applications is increasing. For example, many annotation tasks can be done by non-experts with faster and cheaper, but resulting in more label noise [4].

In recent years several methods have been proposed specifically for *PML* tasks. Most of these methods attempt to extract the label confidence for each candidate label, which is then used to assess the probability of a label being the ground-truth. *PML-lc* [14] is proposed to learn the label confidence based on label correlation. *PML-fp* [14] calculates label confidence based on feature correlations. *PML-LMNN* [4] deals with *PML* via Large Margin Nearest Neighbour Embeddings. It attempts to exploit the ground-truth labels by considering feature and label correlations. To deal with asymmetric correlation among labels *PML-SALC* [24] is proposed based on sparse asymmetric label correlations. These methods mainly perform disambiguation by considering the correlation among labels and/or features. However, in *PML* data because of noisy labels the true correlation among labels and/or features are corrupted. Therefore, the methods which consider label/feature correlations lead to the propagation of errors during model updates. To deal with this challenge, *PARTICLE* [22] is proposed to use the structure of the data. By adapting label propagation, *PARTICLE* identifies high-confidence labels based on *KNN* minimum error reconstruction. Utilizing structural information of the feature space can improve the performance of a model. However, in *PML* labels usually correlate only with a subset of the features. Therefore, methods that use the complete feature set are frequently misled by irrelevant features. Nevertheless, due to the noisy labels and imbalanced data, it is normally not possible to determine the relevant features for each label accurately. Thus, a successful solution for *PML* should be able to tackle the following challenges: 1) noisy labels; 2) corrupted correlation; 3) irrelevant features for each label, and 4) imbalanced data.

Existing methods addressing the *MLL* are categorized into three main groups, e.g., first-order, second-order, and higher-order [13]. Methods in the first-order approach consider labels separately. These methods showed promising results especially when there is a weak correlation among labels or the correlation is corrupted [8]. Inspired by the first-order approach, in this paper we propose an adaptive model for the *PML* task (*PML-CC*). *PML-CC* considers each label separately and gradually extracts high-confidence instances and relevant features for each label and uses them to improve its performance. For each label $l \in L$, *PML-CC* keeps three datasets: a positive (S_P^l), negative (S_N^l), and "do-not-know" (S_D^l) dataset. In each iteration, *PML-CC* utilizes fuzzy-c-means clustering to calculate the probability for each instance x and each label l. During clustering *PML-CC* tries to minimize the cost of clustering while considering the member of S_P^l and S_N^l stay in their cluster. To deal with imbalanced data, *PML-CC* adaptively calculates weights for clustering penalty errors based on the imbalance ratio. After clustering, *PML-CC* updates S_P^l and S_N^l and extracts relevant features for each label. *PML-CC* repeats this procedure until there is no change in the datasets.

To demonstrate the success of *PML-CC*, we empirically compare the performance of *PML-CC* with existing state-of-the-art methods on several real-world and synthetic datasets. The empirical results demonstrate that our method successfully tackles *PML* tasks and outperforms the state-of-the-art methods.

2 Related Work

The main goal of the partial multi-label learning framework is to learn a predictor for multi-label data from noisy training data. Partial multi-label learning (*PML*) is a fusion of partial-label learning (*PLL*) and multi-label learning (*MLL*). *PML* and *PLL* both are weakly supervised learning frameworks where trained on partially observed training data. *PML* and *MLL* both deal with multi-label data where each instance is associated with a set of labels.

Multi-Label Learning is a general form of traditional single-label learning, where each instance is associated with a set of labels. *MLL* aims to predict a set of labels for a new unseen instance [13]. Existing methods can be categorized into two main groups: 1) *Algorithm adaption* methods change the existing single label algorithms to tackle the MLL such as *ML-KNN* [23]; 2) *Problem transformation* methods to transform MLL to a traditional learning paradigm [9].

Partial Label Learning is a multi-class weakly supervised learning framework where among a set of candidate labels for each instance of training data only one label is ground-truth and the rest are false positive labels. The main difference between *PML* and *PLL* is that in the *PLL* we know that there is just one true label for each instance where it is unknown in *PML* [6].

Partial Multi-Label Learning is a multi-label weakly learning framework where the only subset of candidate labels of training data are ground-truth and the rest are false positive. Most existing methods perform disambiguation to identify ground-truth labels. *PML-lc* and *PML-fp* [14] are proposed based on label ranking optimization. *PML-lc* uses label correlation and *PML-fp* uses feature correlation to identify the true ranking of labels. *fMLP* [21] is proposed to improve *PML-lc* and *PML-fp* by considering label and feature correlation. *PML-LRS* [11] is another method that is proposed based on label ranking. It uses sparse decomposition and low-rank by considering label and feature interdependencies. The label matrix is decomposed into an irrelevant label matrix and a ground-truth label matrix. During optimization, the ground-truth label matrix and feature mapping matrix are bound to be low rank and the irrelevant label matrix is forced to be sparse. To capture the label confidence *MUSSER* [6] optimizes the label correlation and feature correlation simultaneously. *DRAMA* [12] contains two steps. In the first step, it identifies the label confidence by using the feature manifold. Then a gradient boosting model is utilized to fit the label confidantes. On each boosting round, the feature space is augmented by the elicited labels to explore the label consolations. *PARTICLE* [22] is proposed to decrease the effect of the false positive labels. It extracts reliable labels among candidate labels and then uses these labels to train a multi-label classifier based on pairwise label ranking. *PML-LD* [18] is proposed based on the correlation among labels and topological information of the feature space. To recover the label distribution *PML-LD* uses label enhancement technique [17]. *PML-DM* [16] considered the noisy labels are not random and the noises happen based on ambiguity on contents of the samples. To identify the noisy labels and recover the ground-truth information, *PML-DM* simultaneously optimizes a noisy label identifier and a multi-label classifier.

3 Proposed Method

In the *PML* task the training data with L labels is defined as $D = \{X_i, y_i\}_{i=1}^{N}$, where $X \in N \times R^d$ contains N training instances. The i_{th} instance X_i denotes the d-dimensional feature vector. X_i may associate with more than a label $y_i \in \{0, 1\}^L$. Among these labels, only part of them are ground-truth label(s) and the rest are noisy labels. Each instance is associated at least with one label [14]. Existing methods, mainly perform disambiguating based on the correlation among labels or/and among features. However, due to noisy labels, these relations are corrupted and mislead the algorithms.

We propose **P**artial **M**ulti-**L**abel learning method via **C**onstraint **C**lustering (*PML-CC*). Our proposed method is based on two main characteristics of data. First, the instances of a class are often close to each other rather than instances of different classes (clustering assumption) [1]. Second, each label frequently is associated with some parts of feature space [6,21]. *PML-CC* contains two major components: Clustering and Feature Selection. It uses these components in three steps: 1) *Pre-clustering*: for each label $l \in L$ a fuzzy clustering algorithm [1] clusters the training data. After the high confident instances are identified, it updates the datasets of positive and negative samples (S_P^l and S_N^l); 2) *Feature Selection*: for each label $l \in L$ the irrelevant features are removed from the feature space of the label; 3) *Clustering*: for each label $l \in L$, the clustering algorithm is applied on S_D^l while considering S_P^l and S_N^l as constraints. Then *PML-CC* repeats the second and third steps until there is no change in the datasets.

3.1 Pre-clustering

For each label $l \in L$, *PML-CC* creates three datasets including: 1) S_P^l contains positive samples; 2) S_N^l contains the negative samples, and 3) S_D^l do-not-know samples. *PML-CC* utilizes fuzzy clustering to cluster $S^l = S_P^l \cup S_N^l \cup S_D^l$. Besides the performance, fuzzy clustering does assign a probability to each sample for each label. That probability can be edited in next steps and *PML-CC* can correct its mistake. Then highly confident samples are added to S_P^l. During clustering, instances belonging to S_P^l and S_N^l are used as constraints that the algorithm should keep them in their cluster. Based on the definition of the *PML* problem we know two facts about the labels [4,19,24]. First, each instance is associated with at least one label. Second, if a sample is tagged for a label we are not sure about its real label but if a sample is not tagged for a label we are sure the sample does not belong to the label. So based on these facts we initial the S_P^l and S_N^l as follows. If an instance X_i is not associated with the label $l \in L$ we add it to S_N^l. If an instance X_i is only associated with one label $l \in L$ we add X_i to S_P^l. Then for each label $l \in L$ we utilize a fuzzy clustering algorithm. Equation (1) shows the objective function for the clustering task for each label.

$$J(U,Z) = \sum_{j=1}^{2}\sum_{i=1}^{N} U_{ij} d_{ij}(x_i, z_j) + A_1 \sum_{j=1}^{2}\sum_{i=1}^{N} U_{ij} \log U_{ij} +$$

$$A_2 \sum_{m|X_m \in C^i} \sum_{\substack{n|X_n \in C^i \\ m \neq n}} \sum_{k=1}^{2} \sum_{\substack{p=1 \\ p \neq k}}^{K} U_{mk} U_{np} + A_3 \sum_{m|X_m \in C^i} \sum_{\substack{n|X_n \in C^j \\ i \neq j}} \sum_{k=1}^{2} U_{mk} U_{nk} \quad (1)$$

$$\text{subject to} \sum_{j=1}^{2} U_{ij} = 1, U_{ij} \in (0,1], 1 \leq i \leq N,$$

where Z_j is the center of cluster j. U_{ij} is the probability of instance X_i belongs to cluster j. $d_{ij}(x_i, z_j)$ is square of Euclidean distance of X_i and Z_j. N is the number of instances. A_1, A_2, A_3 are coefficients. In Eq. (1) the first two terms belong to the fuzzy c-means (*FCM*) algorithm. *FCM* attempts to partition instances into 2 different clusters (positive and negative). For reaching this goal *FCM* tries to find the optimal values for centers of cluster $Z = \{Z_0, Z_1\}$ and simultaneously finds the best values for membership for each instance to each cluster. In Eq. (1) the second term is the weighted entropy. This term forces the clusters to contribute to the association of instances. The third penalty term is added for conditions that instances from the same class appear in different clusters. The last term is another penalty term we added for the case that instances of different classes happen in a cluster. The last two-term in Eq. (1) only is applied for the instances of S_D^l and S_N^l.

Equation (1) does not have a closed-form solution. For solving this problem an alternate optimization is used to find optimal values for Z and U. Thus in each step, first Z is fixed and the optimal value of U with respect to the value of the membership matrix is calculated then U is fixed and the optimal value for centers Z with respect to U is calculated. Then the new value of $J(U,Z)$ based on new values of U and Z is calculated. These steps are repeated until there is no difference between the value of $J(U,Z)$ in the last and current steps. To calculate the optimal value for Z_{jp}, in this step we consider fixed value for U and take derivative and set it to zero, then:

$$Z_{jp} = \frac{\sum_{i=1}^{N} U_{ip}x_{ip}}{\sum_{i=1}^{N} U_{ij}} \tag{2}$$

To calculate the optimal value of U_{ij}, the Lagrange multipliers technique is used in order that the sum of membership for each instance will be equal to one. The optimal value for U_{ij} is calculated in the same way the optimal value for Z_{jp} is calculated. By considering Z is fixed and derivative to zero, the optimal value for U_{ij} is obtained as follows:

$$U_{ij} = \frac{\exp\left(\frac{-d_{ij}(x_i, z_j)}{A_1}\right) exp\left(\frac{-A_2\psi_{ij}}{A_1}\right) exp\left(\frac{-A_3\Psi_{ij}}{A_1}\right)}{\sum_{p=1}^{2} \exp\left(\frac{-d_{ip}(x_i, z_p)}{A_1}\right) exp\left(\frac{-A_2\psi_{ip}}{A_1}\right) exp\left(\frac{-A_3\Psi_{ip}}{A_1}\right)} \tag{3}$$

$$\psi_{ij} = \sum_{\substack{n|X_i \in C^m, X_n \in C^m \\ i \neq n}} \sum_{\substack{k=1 \\ k \neq j}}^{K} U_{nk}\,, \; \Psi_{ij} = \sum_{\substack{n|X_i \in C^m, X_n \in C^p \\ m \neq p}} U_{nj} \tag{4}$$

The minimization procedure is given in Algorithm 2. The algorithm returns the optimal values for U and Z. The detailed proof is described in the supplementary material (Appendix .1).

3.2 Feature Selection (FS)

Since it is known that each instance of training data at least is associated with one of the labels, instance $X_i \in S_D^l$ is added to S_P^l where l is the label with the highest membership value $l = \text{argmax}_l(\{U_{i,l}\}_{l=1}^{l=L})$. Then, as a further optimization procedure, *PML-CC* determines the relevant features for each label based on the information that has been extracted from the data. In multi-label data, each label is usually associated with a subset of the feature space and the rest of the features are irrelevant. These irrelevant features hinder classifiers to learn the labels properly [21].

To deal with this challenge, the first norm regularization is used. The L_1 regularization tends to push some of the less important features' corresponding weights to zero. As a result, this effectively removes those features from the model, leading to a simpler and potentially more accurate model by eliminating irrelevant features from the data. For each label $l \in L$ a cost function $L_1(l)$ is defined as follows:

$$L_1(l) = \min_{\omega_l}(\frac{1}{2}\sum_{j=1}^{N}(X_j\omega_i - y_j)^2 + \eta|\omega_l|) = \frac{1}{2}\|y - X\omega\|_2^2 + \eta\|\omega\| \tag{5}$$

where $\omega \in R^m \times R^L$ is matrix and ω_l is weight vector for label $l \in L$. For simplicity, we consider $\omega_l = \omega$ and show how our proposed method solves the problem for label $l \in L$. Then we can generalize the solution for other labels. Equation (5) shows the cost function for L_1 norm. However, there is no closed-form solution for this minimization. To solve this optimization, Proximal Gradient Descent Algorithm (*PGDA*) [2] is used. $L(\omega)$ in Eq. (6) is convex and differentiable. $R(\omega)$ is convex but not differentiable.

$$\underbrace{\frac{1}{2}\|y - X\omega\|_2^2}_{L(\omega)} + \underbrace{\eta\|\omega\|}_{R(\omega)} \tag{6}$$

Then the proximal gradient method iteration will be as follows:

$$\omega^{t+1} = prox_{\alpha R}(\omega^t - \alpha \nabla L(\omega^t)) \tag{7}$$

where

$$\begin{aligned} \nabla L(\omega) &= X^T(X\omega - y) \\ prox_{\alpha}(\hat{\omega}) &= \underset{\omega}{\operatorname{argmin}}\, R(\omega) + \frac{1}{2\alpha}\|\omega - \hat{\omega}\|_2^2 \\ &= \underset{\omega}{\operatorname{argmin}}\, \eta\|\omega\| + \frac{1}{2\alpha}\|\omega - \hat{\omega}\|_2^2 = S_{\eta\alpha}(\hat{\omega}) \end{aligned} \tag{8}$$

$S_{\eta\alpha}(\hat{\omega})$ is the soft-shareholding operator as follows:

$$[S_{\eta\alpha}(\hat{\omega})]_i = \begin{cases} \hat{\omega}_i - \eta & \text{if}\, \hat{\omega}_i > \eta \\ 0 & \text{if } -\eta < \hat{\omega}_i < \eta, i = 1, .., n \\ \hat{\omega}_i + \eta & \text{if}\, \hat{\omega}_i < -\eta \end{cases} \tag{9}$$

Based on Eq. (7) and (8) and (9) proximal gradient update is:

$$\omega^{t+1} = S_{\eta\alpha}(\omega^t + \alpha X^T(y - X\omega^t)) \tag{10}$$

Equation (5) determines relevant features for label $l \in L$ based on the extracted information. Since this information (S_P and S_N) is noisy and incomplete, using only relevant features results in over-fitting. To avoid the over-fitting, inspired by the drop-out algorithm [10], in each step, *PML-CC* randomly removes 50% of detected irrelevant features and keeps the rest.

3.3 Clustering

In the last step, *PML-CC* again uses the constraint fuzzy-c-means algorithm (Eq. 1). In this step, *PML-CC* uses the updated S_P and S_N that were obtained during the first step (3.1) and only considers the relevant features for each label (3.2). Algorithm (1) shows the general workflow of our proposed method.

Algorithm 1 *PML-CC*

Require: Data,Labels,Features,α, η, A_1, A_2, A_3
Ensure: U,Z
1: Initialize S_P, S_N,S_D
2: $Z, U \leftarrow$Minimization($S_P^l, S_N^l, S_D^l, A_1, A_2, A_3$)
3: **for** i in (len(Data)) **do**
4: $k = \text{argmax}_k(\{U_{i,k}\}_{k=1}^{k=L})$
5: Update(S_P^l, S_N^l, S_D^l)
6: **end for**
7: **while** $J(U^{(t)}, Z^{(t)})! = J(U^{(t+1)}, Z^{(t+1)})$ **do**
8: $\hat{S_P^l}, \hat{S_N^l}, \hat{S_D^l} \leftarrow$Feature_SelectionS($S_P^l, S_N^l, S_D^l, \alpha, \eta$)# Based on subsection 3.2
9: $Z, U \leftarrow$Minimization($\hat{S_P^l}, \hat{S_N^l}, \hat{S_D^l}, A_1, A_2, A_3$)
10: Update(S_P^l, S_N^l, S_D^l)
11: $t \leftarrow t + 1$
12: **end while**

4 Experiment

4.1 Dataset

To show the performance of our proposed method, we conduct several experiments on a variety of real-world datasets (*Music-emotion*, *Music-style*, *YeastBP*, and *MIRFlickr* [4]) and synthetic datasets (*Enron*, *CAL500*, *Mediamill*, and *Corel5k*)[1]. The datasets focus on different applications including image annotation (*MIRFlickr*), text categorization (*Enron*), music recognition (*Music-emotion*), biology (*YeastBP*), and video annotation (*Mediamill*). Table 1 shows the characteristics of these datasets [16]. The synthetic datasets are Multi-label but are not *PML*. To construct *PML* data, we add random noise to them. For example, the average number of ground-truth labels (**avg.GLs**) of *Enron* is 3.38, to make *Enron 7*, we added some noises to increase the **avg.GLs** to 7.

Algorithm 2 Minimization.

1: **function** MIZ($S_P^l, S_N^l, S_D^l, A_1, A_2, A_3$)
2: $Z^{(0)}, U^{(0)} \leftarrow RandomNumber$
3: $t \leftarrow 0$
4: **while** *true* **do**
5: calculate $U^{(t+1)}$ by using Eq (3) where $Z^{(t)}$ is fixed.
6: calculate $Z^{(t+1)}$ using Eq (2) where $U^{(t)}$ is fixed .
7: **if** $J(U^{(t)}, Z^{(t)}) == J(U^{(t+1)}, Z^{(t)})$or $J(U^{(t)}, Z^{(t)}) == J(U^{(t)}, Z^{(t+1)})$ **then**
8: Break
9: **end if**
10: $t \leftarrow t + 1$
11: **end while**
12: Return ($Z^{(t)}, U^{(t)}$)
13: **end function**

[1] http://mulan.sourceforge.netdatasets.html/ and https://meka.sourceforge.net/datasets.

Table 1. Characteristics of real-world and Synthetic datasets. **avg. #GLs** is the average number of ground-truth labels and **avg. #CLs** is the average number of candidate labels.

Dataset	#Examples	#Features	#Class	avg.#GLs	avg.#CLs
MIRFlickr	10433	100	7	1.77	3.35
Music-emotion	6833	98	11	2.42	5.29
Music-style	6839	98	10	1.44	6.04
YeastBP	560	5,548	217	21.56	30.43
Enron	1702	1001	53	3.38	7,11
CAL500	502	68	174	26.04	45,65
Corel5k	5000	499	374	3.52	7,9
Mediamill	43907	120	101	4.38	9,13

4.2 Metrics

The evaluation metrics of multi-label learning algorithms are different from traditional single-label learning. Several criteria for multi-label learning have been proposed in the literature: *hamming loss*, *one error*, *average precision*, *coverage*, and *ranking loss* are used to show the performance of our proposed method. These five metrics are commonly used for multi-label and partially multi-label learning. For *hamming loss*, *one error*, *coverage* metrics, and *ranking loss* smaller values show better performance. For *average precision* it is the larger values. More details about multi-label performance metrics can be found in [20]

4.3 Competitors

We compare the performance of *PML-CC* with the following state-of-the-art *PML* algorithms. *PML-LMNNE (LMNNE)* [4], *PARTICLE* [22], *DRAMA* [12], *fPML* [21], *PML-LRS* [11], *MUSER* [6]. The trade-off parameters of all competitors algorithms are set to the values suggested in the respective papers.

4.4 Experimental Results

We report the performance of our method and the Competitor methods on four real-world and eight synthetic datasets. The performance is shown in term of the five performance metrics described in previous section. Table 2 shows the performance of our proposed method in the term of *hamming loss* where the smaller value means better performance. Table 3 shows the performance of our proposed method in the term of *average precision* where the larger value means better performance. Similar results are achieved for other metrics. The result for other metrics are reported in Appendix .2. Based on overall results, the following observation can be made:

- On 83% of datasets (10 out of 12) on each evaluation metric, *PML-CC* outperforms all the competitors.
- Out of 412 (12 datasets × 7 methods × 5 metrics) comparisons, *PML-CC* ranks 1*st* in 83% cases.
- *PML-CC* significantly outperforms *PARTICLE*, *PML-LRS*, *DRAMA*, and *fpml* on all combinations of datasets and metrics.
- *PML-CC* outperforms *MUSER* and *PML-LMNNE* in 86% cases.

Table 2. The result of *PML-CC* and other competitors in term of **hamming loss** on real-world and synthetic datasets (mean±standard deviation). The best method is highlighted by bold text and the runner-up is shown as italic text.

Dataset	PML-CC	LMNNE	PARTICLE	DRAMA	fPML	PML-LRS	MUSER
MIRFlickr	**.143±.021**	*.145±.016*	.193±.017	.219±.014	.223±.022	.237±.012	.193±.056
Music-emotion	**.210±.012**	*.281±.010*	.360±.012	.318±.013	.452±.025	.381±.028	.284±.023
Music-style	**.158±.012**	*.162*±**.011**	.173±.021	.169±.031	.338±.027	.379±.023	.173±.016
YeastBP	**.102±.016**	.161±.014	.236±.012	.227±.021	.214±.015	.182±.009	*.158±.012*
Enron 7	**.067±.013**	*.097±.012*	.286.005	.183±.022	.115±.017	.207±.021	.108±.003
Enron 11	**.069±.017**	*.121±.011*	.303±.005	.209±.013	.128±.018	.209±.014	.123±.014
CAL500 45	**.152±.016**	.260±.013	.271±.024	*.235±.017*	.268±.015	.282±.013	.279±.024
CAL500 65	**.201±.016**	.285±.016	.357±.032	.327±.036	.288±.015	.327±.026	*.283±.014*
Corel5k 7	*.008±.006*	**.007±.002**	.015±.006	.013±.012	.009±.002	.008±.006	.009±.003
Corel5k 11	*.010±.005*	**.011±.004**	.038±.006	.021±.005	.018±.013	.019±.002	.012±.005
Mediamill 9	**.054±.018**	*.059±.013*	.098±.014	.101±.020	.065±.007	.072±.023	.087±.021
Mediamill 13	**.062±.012**	*.122±.010*	.145±.018	.201±.027	.513±.021	.191±.017	.183±.012

The post-hoc Bonferroni-Dunn [3] test is used to further investigate the differences in the results and compute the critical difference. In Fig. 1, for each performance measure, the algorithms not connected with *PML-CC* in the *CD* diagram have a significantly different performance.

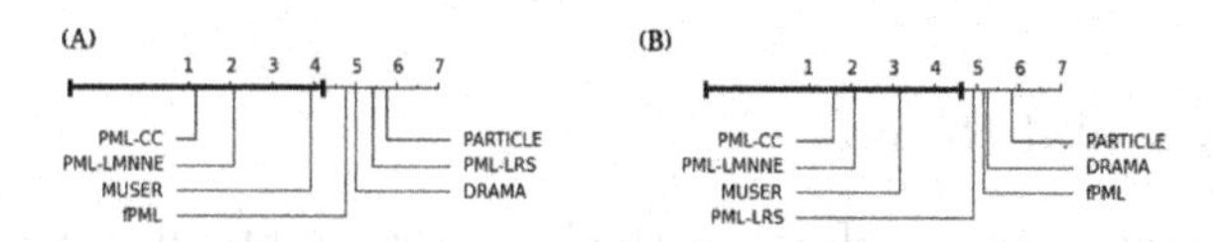

Fig. 1. post-hoc *Bonferroni-Dunn* test, comparing the average rank of *PML-CC* on all datasets, for *hamming loss* (A) *average precision* (B) .

4.5 Parameters Analysis

There are three parameters for the clustering step. These parameters are responsible for the weight of errors (penalty weight). A_1 is the weight for term

Table 3. The result of our proposed method and other competitors in terms of **average precision** on real-world and synthetic datasets (mean±standard deviation). The best method is highlighted by bold text and the runner-up is shown as italic text.

Dataset	PML-CC	LMNNE	PARTICLE	DRAMA	fPML	PML-LRS	MUSER
MIRFlickr	**.861±.015**	*.831±.012*	.685±.017	.707±.014	.731±.015	.796±.012	.801±.016
Music-emotion	**.663±.016**	*.611±.017*	.527±.006	.582±.012	.538±.015	.516±.014	.598±.033
Music-style	**.732±.034**	*.726±.024*	.717±.031	.693±.013	.659±.017	.716±.018	.718±.013
YeastBP	**.206±.024**	.152±.023	.082±.031	.083±.017	.096±.021	.085±.023	**.154±.031**
Enron 7	.680±.018	*.779±.010*	.601±.006	.613±.002	.751±.012	**.782±.011**	.771±.003
Enron 11	.679±.021	**.694±.006**	.587±.006	.556±.012	.670±.006	**.683±.007**	.681±.005
CAL500 45	**622±.021**	.615±.021	.446±.024	.563±.027	.531±.025	.516±.023	**.620±.014**
CAL500 65	**.502±.017**	*.480±.011*	.432±.012	.481±.015	.412±.022	.448±.014	.479±.018
Corel5k 7	**.332±.014**	*.289±.010*	.205±.036	.235±.014	.264±.017	.237±.013	.280±.003
Corel5k 11	**.311±.012**	*.282±.005*	.196±.012	.218±.025	.258±.015	.217±.011	.276±.015
Mediamill 9	**.766±.023**	*.765±.018*	.756±.018	.687±.017	.695±.017	.689±.010	.716±.012
Mediamill 13	**.736±.021**	*.733±.016*	.699±.024	.698±.014	.674±.018	.686±.013	.702±.021

$\sum_{j=1}^{2}\sum_{i=1}^{N} U_{ij}\log U_{ij}$ in Eq. (1). Figure 2A shows that the minimum value of $U_{ij}\log U_{ij}$ occurs when U_{ij} is around 0.4. This term forces the samples to associate with all the clusters and hinders the membership of a sample to a cluster to be very small (zero) or very high (one). Similarly, in Eq. (3), it is shown that the values of the other terms are divided by A_1. That means a big value of A_1, keeps all the membership values close to each other. That makes the optimization process very slow and sometimes leads it to a local minimum. On the other hand, a small value of A_1 magnifies the role of other terms in Eq. (3) (e.g. distance). That leads the optimization process to a local minimum or a loop and never stops.

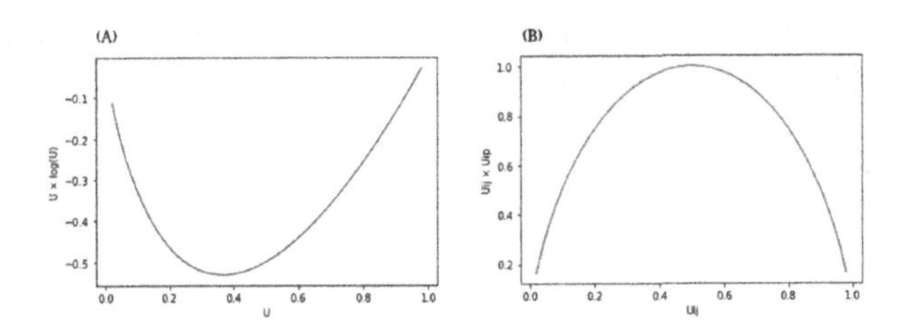

Fig. 2. Shows the behavior of penalty terms in Eq. (1). (A) shows the values for $U \times logU$ for different U. (B) is a diagram that shows the values for $U_{ij} \times U_{ip}$ for different $U_{i,j}$.

A_2 and A_3 are weights for the third and fourth terms in Eq. (1). These penalty terms are applied when two samples of different classes happen in a cluster, or two samples of a class happen in different clusters. As Fig. 2.B shows, the minimum value of $Uij \times Uip$ happens when Uij is close to zero and Uip is close to one, and vice versa. Thus, these terms try to decrease the probability that samples of different classes happen in the same cluster and samples of the same class happen in different clusters. Big values for A_2 and A_3 decrease the role of other terms in Eq. (1) and samples are assigned to the clusters based on theses terms. Small values for theses weights allow mistakes and increase the weight of other terms in Eq. (1). Figure 3 shows the effect of different values of $A1, A2, A3$ on the performance of *PML-CC*. Figure 3A, Fig. 3B, and Fig. 3.C show the performance of *PML-CC* on *Music-emotion* dataset for different values of $A2, A3, A1$ respectively.

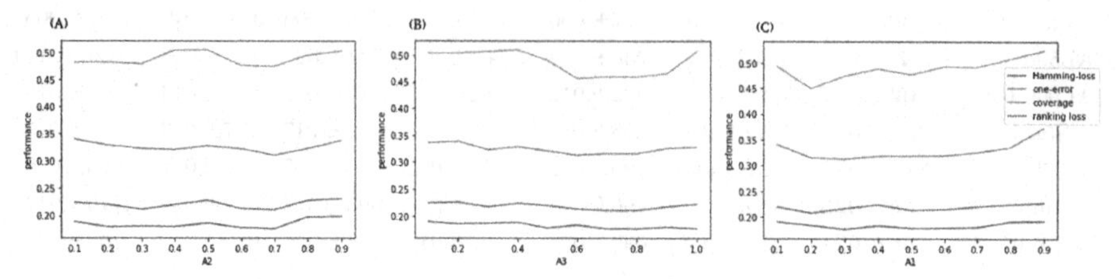

Fig. 3. The effect of different values of $A1, A2, A3$ on the performance of *PML-CC* on *Music-emotion* dataset.

4.6 Time Complexity

The main component of *PML-CC* is *FCM*. The time complexity of *FCM* is $O(N \times C^2 \times d \times i)$ [1]. N in number of samples, C is number of clusters, d is number of attributes and i is number of iteration. In our method, *FCM* is run for each label for two clusters. That means the time complexity of *PML-CC* is $O(N \times d \times i \times L)$ where L is the number of labels. With this time complexity *PML-CC* can handle most of the existing datasets in a reasonable time. However, for a big real-world dataset *FCM* makes *PML-CC* very slow. Several fast versions of *FCM* have been proposed in recent years. For example, *Kolen* [5] proposed a novel method that dramatically decreases the run time of *FCM*. Thus, for the big datasets, *PML-CC* can utilize the fast version of *FCM* to deal with the time complexity.

5 Conclusion

In this paper, we proposed a novel *PML* learning method, *PML-CC*, based on constraint clustering. Unlike existing methods, our method is not based on correlation among labels and/or features. It is a first-order multi-label learning algorithm that considers each label separately. *PML-CC* learns its model in iterations. In each step, it extracts some information from the data and then uses the information to improve its model and performance. Experimental results on a variety of real-world and synthetic datasets using a wide range of performance metrics show that *PML-CC* outperforms existing state-of-the-art algorithms.

Acknowledgment. This work has been partially supported by the Volkswagen foundation.

Appendix

.1 Proof of Formula

In this section, the detail of the optimization of Eq. (1) is given. The goal of this optimization is to find the optimal value for cluster centers ($[Z]_{k\times m\times L}$) and the fuzzy membership ($[U]_{k\times m\times L}$). Where k is the number of classes for each label. Since each label is a binary class we set $k = 2$. m is the size of the feature and L is the number of labels. For making the optimization procedure easier for the reader, the procedure is described for a single label. Thus we consider cluster centers ($[Z]_{2\times m}$) and the fuzzy membership ($[U]_{2\times m}$) for only one label. For the rest of the labels, we repeat the procedure. The Eq. (1) does not have a close form solution. To solve this problem an alternating optimization approach is used. Equation (1) is a constraint non-linear optimization form. By using Lagrange multipliers the following function is obtained.

$$J(U,Z) = \sum_{j=1}^{2}\sum_{i=1}^{N} U_{ij} d_{ij}(x_i,z_j) + A_1 \sum_{j=1}^{2}\sum_{i=1}^{N} U_{ij}\log U_{ij} + A_2 \sum_{m|X_m\in C^i}\ \sum_{\substack{n|X_n\in C^i\\ m\neq n}}\ \sum_{k=1}^{2}\sum_{\substack{p=1\\ p\neq k}}^{K} U_{mk}U_{np}$$
$$+ A_3 \sum_{m|X_m\in C^i}\ \sum_{\substack{n|X_n\in C^j\\ i\neq j}}\ \sum_{k=1}^{2} U_{mk}U_{nk} + \sum_{i=1}^{N}\lambda_i\Big(\sum_{j=1}^{2} U_{ij} - 1\Big) \tag{11}$$

Lemma 1. The optimal value for U_{ij} when Z are fixed is equal to :

$$U_{ij} = \frac{\exp\left(\frac{-d_{ij}(x_i,z_j)-A_2\psi_{ij}-A_3\Psi_{ij}}{A_1}\right)}{\sum_{p=1}^{2}\exp\left(\frac{-d_{ip}(x_i,z_p)-A_2\psi_{ip}-A_3\Psi_{ip}}{A_1}\right)} \tag{12}$$

Proof. To find the optimal value for each U_{ij} we take derivative of Eq. (11) respect to each U_{ij} and set it to zero as follows:

$$\frac{\partial J(U,Z)}{\partial U_{ij}} = 0$$

$$d_{ij}(x_i,z_j) + A_1(1+logU_{ij}) + A_2\Big(\sum_{\substack{n|X_i\in C^m, X_n\in C^p\\ m\neq p}}\ \sum_{\substack{k=1\\ \neq j}}^{2} U_{nk}\Big) + A_3\Big(\sum_{n|X_i,X_n\in C^m} U_{nj}\Big) + \lambda_i = 0 \tag{13}$$

By setting ψ_{ij} and Ψ_{ij} as follows:

$$\psi_{ij} = \sum_{\substack{n|X_i \in C^m, X_n \in C^m \\ i \neq n}} \sum_{\substack{k=1 \\ k \neq j}}^{K} U_{nk}, \quad \Psi_{ij} = \sum_{\substack{n|X_i \in C^m, X_n \in C^p \\ m \neq p}} U_{nj} \tag{14}$$

By solving Eq. (13), U_{ij} will be obtained as follows:

$$U_{ij} = \exp(-1)\exp\left(\frac{-d_{ij}(x_i, z_j) - A_2\psi_{ij} - A_3\Psi_{ij}}{A_1}\right)\exp(\frac{-\lambda_i}{A_1}) \tag{15}$$

Since $\sum_{j=1}^{2} = 1$ the Lagrange multiplier can obtained as follows:

$$\begin{aligned} &\sum_{j=1}^{2}\exp(-1)\exp\left(\frac{-d_{ij}(x_i, z_j) - A_2\psi_{ij} - A_3\Psi_{ij}}{A_1}\right)\exp(\frac{-\lambda_i}{A_1}) = \\ &\exp(-1)\exp(\frac{-\lambda_i}{A_1})\sum_{j=1}^{2}\exp\left(\frac{-d_{ij}(x_i, z_j) - A_2\psi_{ij} - A_3\Psi_{ij}}{A_1}\right) = 1 \\ &=> \exp(\frac{-\lambda_i}{A_1}) = \frac{1}{\exp(-1)\sum_{j=1}^{2}\exp\left(\frac{-d_{ij}(x_i,z_j) - A_2\psi_{ij} - A_3\Psi_{ij}}{A_1}\right)} \end{aligned} \tag{16}$$

By substituting Eq. (16) in Eq. (15) the closed form solution for uij (Eq. (12)) will be obtained and completes the proof of lemma.

Lemma 2. *If the U (fuzzy memberships) are fixed, the optimal value for Z (cluster centers) are equal to equation (17).*

$$Z_{jp} = \frac{\sum_{i=1}^{N} U_{ij}x_{ip}}{\sum_{i=1}^{N} U_{ij}} \tag{17}$$

Proof. Again the alternative approach is used. First, The U are fixed then the optimal values for Z is obtained by taking derivative of Eq. (11) respect to each cluster center and set it to zero.

$$\frac{\partial J(U,Z)}{\partial Z_{jp}} = 0 \quad => \sum_{i=1}^{N} 2U_{ij}(Z_{jp} - x_{ip}) = 0 \quad => Z_{jp} = \frac{\sum_{i=1}^{N} U_{ij}x_{ip}}{\sum_{i=1}^{N} U_{ij}} \tag{18}$$

Lemma 3. *U and Z are local optimum of $J(U,Z)$ if Z_{ij} and U_{ij} are calculated using Eq. (17) and (12) and $A_1, A_2, A_3 > 0$*

Proof. Let $J(U)$ be $J(U,Z)$ when Z are fixed, $J(Z)$ be $J(U,Z)$ when U are fixed and $A_1, A_2, A_3 > 0$. Then, the Hessian $H(J(Z))$ and $H(J(U))$ matrices are calculated as follows:

$$h_{fg,ij}(J(U)) = \frac{\partial}{\partial_{fg}}\frac{\partial J(U)}{\partial U_{ij}} = \begin{cases} \frac{A_1}{U_{ij}}, & iff{=}i,g{=}j \\ 0, & otherwise \end{cases} \tag{19}$$

$$h_{fg,il}(J(Z)) = \frac{\partial}{\partial_{fg}}\frac{\partial J(Z)}{\partial Z_{ip}} = \begin{cases} \sum_{j=1}^{2} 2U_{ij}, & iff{=}i,g{=}p \\ 0, & otherwise \end{cases} \tag{20}$$

Equation (19) and (20) shows $H(J(Z))$ and $H(J(U))$ are diagonal matrices. Since $A_1 > 0$ and $0 < U_{ij} \leq 1$, the Hessian matrices are positive definite. Thus Eq. (12) and (17) are sufficient conditions to minimize $J(U)$ and $J(Z)$.

.2 Additional Excremental Result

Tables 4,5,6 show the performance of our proposed method in the term of *ranking loss* and *coverage* respectively.

Table 4. The result of our proposed method and other competitors in term of **ranking loss** on real-world and synthetic datasets (mean±standard deviation)

Dataset	PML-CC	LMNNE	PARTICLE	DRAMA	fPML	PML-LRS	MUSER
MIRFlickr	**.070±.007**	*.072±.004*	.203±.007	.189±.010	.146±.008	.107±.002	.093±.006
Music-emotion	**.161±.023**	*.182±.019*	.260±.005	.218±.013	.254±.006	.281±.005	.189±.021
Music-style	**.163±.014**	*.169±.012*	.182±.021	.178±.013	.267±.013	.179±.007	.171±.006
YeastBP	**.240±.029**	.347±.038	.436±.032	.407±.021	.382±.036	.418±.031	*.341±.015*
Enron 7	**.040±.019**	*.110±.012*	.297±.007	.194±.012	.338±.004	.207±.021	.114±.003
Enron 11	**.050±.019**	*.119±.014*	.312±.006	.210±.015	.341±.002	.215±.017	.123±.004
CAL500 45	**.087±.011**	.183±.013	.353±.014	.235±.007	.316±.008	.281±.013	*.179±.024*
CAL500 65	**.129±.011**	.256±.017	.471±.012	.317±.016	.365±.014	.347±.016	*.213±.021*
Corel5k 7	.039±.013	**.013±.007**	.345±.070	.193±.052	.161±.013	.193±.016	*.015±.004*
Corel5k 11	.045±.011	**.017±.003**	.383±.056	.201±.065	.171±.014	.202±.021	*.017±.005*
Mediamill 9	**.046±.029**	*.121±.010*	.135±.008	.219±.012	.226±.009	.193±.107	.173±.002
Mediamill 13	**.053±.026**	*.153±017*	.198±.004	.301±.007	.316±.015	.212±.023	.187±.021

Table 5. The result of our proposed method and other competitors in term of **one error** on real-world and synthetic datasets (mean±standard deviation).

Dataset	PML-CC	LMNNE	PARTICLE	DRAMA	fPML	PML-LRS	MUSER
MIRFlickr	**.127±.012**	*.187±.010*	.263±.013	.289±.016	.346±.021	.497±.031	.223±.026
Music-emotion	**.450±.017**	*.521±.016*	.560±.015	.568±.023	.554±.026	.581±.025	.539±.022
Music-style	**.401±.016**	*.417±.014*	.460±.025	.458±.014	.454±.025	.481±.028	.439±.021
YeastBP	**.660±.027**	*.911±.021*	.936±.030	.941±.025	982±.016	.918±.021	.907±.027
Enron 7	**.164±.020**	*.207±.014*	.297±.027	.294±.022	.338±.004	.215±.027	.214±.024
Enron 11	**.124±.021**	*.218±.012*	.312±.016	.310±.015	.341±.022	.307±.021	.223±.013
CAL500 45	**.110±.013**	*.169±.011*	.171±.014	.235±.007	.265±.008	.181±.013	.176±.024
CAL500 65	**.161±.014**	*.236±.017*	.363±.012	.337±.016	.356±.014	.247±.016	.233±.021
Corel5k 7	.252±.023	**.213±.011**	.345±.025	.281±.052	.261±.013	.292±.016	*.215±.014*
Corel5k 11	.280±.027	**.225±014**	.383±.056	.293±.035	.297±.024	.293±.011	*.227±.019*
Mediamill 9	**.211±.009**	*.212±.011*	.235±.021	.297±.016	.311±.022	.293±.017	.237±.012
Mediamill 13	**.196±.012**	*.231±.016*	.298±.015	.301±.012	.316±.016	.312±.023	.277±.021

Table 6. The result of *PML-CC* and other competitors in term of **coverage** on real-world and synthetic datasets (mean±standard deviation).

Dataset	PML-CC	LMNNE	PARTICLE	DRAMA	fPML	PML-LRS	MUSER
MIRFlickr	**.190±.031**	*.210±.007*	.263±.007	.289±.010	.248±.008	.287±.002	.233±.016
Music-emotion	**.300±.012**	*.381±.010*	.461±.005	.428±.013	.434±.006	.483±.005	.387±.021
Music-style	**.195±.025**	*.203±.017*	.208±.021	.217±.032	.367±.013	.279±.007	.216±.006
YeastBP	**.321±.021**	*.426±.022*	.731±.032	.417±.025	.489±.036	.668±.031	.431±.015
Enron 7	**.195±.003**	*.306±.005*	.397±.007	.394±.012	.431±.004	.415±.021	.314±.033
Enron 11	**.270±.011**	*.311±.019*	.416±.026	.415±.015	.446±.022	.417±.017	.323±.024
CAL500 45	*.704±.035*	.715±.031	.872±.024	.838±.027	.936±.038	.865±.033	**.679±.027**
CAL500 65	*.718±041*	.736±.026	.953±.012	.917±.016	.955±.014	.878±.016	**.712±.012**
Corel5k 7	**.178±.015**	*.227±.013*	.415±.070	.439±.052	.361±.011	.372±.015	.238±.014
Corel5k 11	**.186±.019**	*.255±.021*	.463±.026	.451±.025	.373±.014	.393±.021	.278±.005
Mediamill 9	**.105 ±.014**	*.185±.016*	.211±.018	.317±.012	.266±.013	.291±.107	.212±.022
Mediamill 13	**.125 ±.026**	*.193±.022*	.298±.024	.321±.027	.362±.015	.322±.021	.215±.010

References

1. Bezdek, J.C., Ehrlich, R., Full, W.: FCM: the fuzzy c-means clustering algorithm. Comput. Geosci. **10**(2–3), 191–203 (1984)
2. Chen, A.I.A.: Fast distributed first-order methods. Ph.D. thesis, Massachusetts Institute of Technology (2012)
3. Demšar, J.: Statistical comparisons of classifiers over multiple data sets. J. Mach. Learn. Res. **7**, 1–30 (2006)
4. Gong, X., Yuan, D., Bao, W.: Partial multi-label learning via large margin nearest neighbour embeddings (2022)
5. Kolen, J.F., Hutcheson, T.: Reducing the time complexity of the fuzzy c-means algorithm. IEEE Trans. Fuzzy Syst. **10**(2), 263–267 (2002)
6. Li, Z., Lyu, G., Feng, S.: Partial multi-label learning via multi-subspace representation. In: IJCAI, pp. 2612–2618 (2020)
7. Lyu, G., Feng, S., Li, Y.: Partial multi-label learning via probabilistic graph matching mechanism. In: Proceedings of the 26th ACM SIGKDD International Conference on Knowledge Discovery & Data Mining, pp. 105–113 (2020)
8. Read, J., Pfahringer, B., Holmes, G., Frank, E.: Classifier chains for multi-label classification. Mach. Learn. **85**(3), 333–359 (2011)
9. Siahroudi, S.K., Kudenko, D.: An effective single-model learning for multi-label data. Expert Syst. Appl. **232**, 120887 (2023)
10. Srivastava, N., Hinton, G., Krizhevsky, A., Sutskever, I., Salakhutdinov, R.: Dropout: a simple way to prevent neural networks from overfitting. J. Mach. Learn. Res. **15**(1), 1929–1958 (2014)
11. Sun, L., Feng, S., Wang, T., Lang, C., Jin, Y.: Partial multi-label learning by low-rank and sparse decomposition. In: Proceedings of the AAAI Conference on Artificial Intelligence, vol. 33, pp. 5016–5023 (2019)
12. Wang, H., Liu, W., Zhao, Y., Zhang, C., Hu, T., Chen, G.: Discriminative and correlative partial multi-label learning. In: IJCAI, pp. 3691–3697 (2019)
13. Wang, R., Kwong, S., Wang, X., Jia, Y.: Active k-labelsets ensemble for multi-label classification. Pattern Recogn. **109**, 107583 (2021)

14. Xie, M.K., Huang, S.J.: Partial multi-label learning. In: Proceedings of the AAAI Conference on Artificial Intelligence, vol. 32 (2018)
15. Xie, M.K., Huang, S.J.: Partial multi-label learning with noisy label identification. IEEE Trans. Pattern Anal. Mach. Intell. **44**, 3676–3687 (2021)
16. Xie, M.K., Sun, F., Huang, S.J.: Partial multi-label learning with meta disambiguation. In: Proceedings of the 27th ACM SIGKDD Conference on Knowledge Discovery & Data Mining, pp. 1904–1912 (2021)
17. Xu, N., Liu, Y.P., Geng, X.: Label enhancement for label distribution learning. IEEE Trans. Knowl. Data Eng. **33**(4), 1632–1643 (2019)
18. Xu, N., Liu, Y.P., Geng, X.: Partial multi-label learning with label distribution. In: Proceedings of the AAAI Conference on Artificial Intelligence, vol. 34, pp. 6510–6517 (2020)
19. Yan, Y., Guo, Y.: Adversarial partial multi-label learning with label disambiguation. In: Proceedings of the AAAI Conference on Artificial Intelligence, vol. 35, pp. 10568–10576 (2021)
20. Yan, Y., Li, S., Feng, L.: Partial multi-label learning with mutual teaching. Knowl.-Based Syst. **212**, 106624 (2021)
21. Yu, G., et al.: Feature-induced partial multi-label learning. In: 2018 IEEE International Conference on Data Mining (ICDM), pp. 1398–1403. IEEE (2018)
22. Zhang, M., Fang, J.: Partial multi-label learning via credible label elicitation. IEEE Trans. Pattern Anal. Mach. Intell. **43**(10), 3587–3599 (2021)
23. Zhang, M.L., Zhou, Z.H.: ML-KNN: a lazy learning approach to multi-label learning. Pattern Recogn. **40**(7), 2038–2048 (2007)
24. Zhao, P., Zhao, S., Zhao, X., Liu, H., Ji, X.: Partial multi-label learning based on sparse asymmetric label correlations. Knowl.-Based Syst. **245**, 108601 (2022)

Abstractive Multi-document Summarization with Cross-Documents Discourse Relations

Mengling Han, Zhongqing Wang, Hongling Wang[(✉)], Xiaoyi Bao, and Kaixin Niu

School of Computer Science and Technology, Soochow University, Suzhou, China
hlwang@suda.edu.com

Abstract. Generating a summary from a set of documents remains a challenging task. Abstractive multi-document summarization (MDS) methods have shown remarkable advantages when compared with extractive MDS. They can express the original document information in new sentences with higher continuity and readability. However, mainstream abstractive models, which are pre-trained on sentence pairs rather than entire documents, often fail to effectively capture long-range dependencies throughout the document. To address these issues, we propose a novel abstractive MDS model that aims to succinctly inject semantic and structural information of elementary discourse units into the model to improve its generative ability. In particular, we first extract semantic features by splitting the single document into discourses and building the discourse tree. Then, we design discourse Patterns to convert the raw document text and trees into a linearized format while guaranteeing corresponding relationships. Finally, we employ an abstractive model to generate target summaries with the processed input sequence and to learn the discourse semantic information. Extensive experiments show that our model outperforms current mainstream MDS methods in the ROUGE evaluation. This indicates the superiority of our proposed model and the capacity of the abstractive model with the hybrid pattern.

Keywords: Discourse Rhetorical Structure · Abstractive Summarization · Multi-document · LongT5

1 Introduction

Multi-document summarization (MDS) is a technique that compresses multiple topic-related documents into a concise summary without losing important information. Previously, most MDS tasks are approached using extractive summarization [18], which involved scoring and ranking sentences in the documents to extract critical sentences while ensuring diversity. However, extractive summarization faced challenges in handling redundancy and contradictions across multiple documents [22]. In recent studies, abstractive summarization has emerged as

B. Luo et al. (Eds.): ICONIP 2023, CCIS 1965, pp. 470–481, 2024.
https://doi.org/10.1007/978-981-99-8145-8_36

a preferred approach, generating summaries with new words or sentences based on semantic analysis. This method aims to achieve better diversity and reduce redundancy, resulting in summaries that resemble human-written summaries. However, these methods primarily concentrate on optimizing the summary generation process without adequately addressing the challenge of capturing the long-term dependencies present throughout the document. Documents are not simply a stack of text sequences. Instead, they are compositions of Elementary Discourse Units (EDUs) linked to each other.

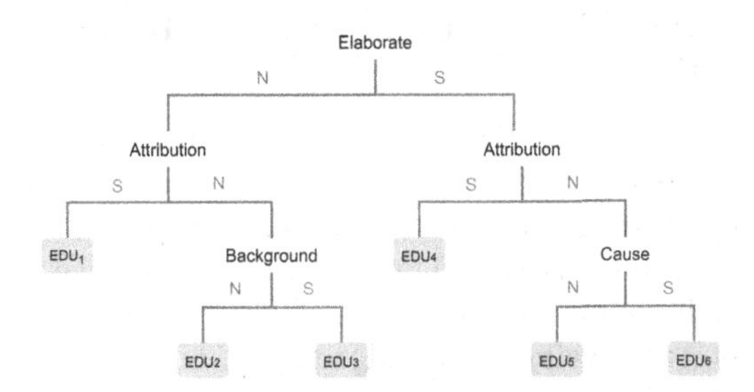

Fig. 1. An example RST-style discourse tree.

Rhetorical Structure Theory (RST), recognized as the primary linguistic framework for Discourse Rhetorical Structure (DRS) [14], offers a comprehensive representation of the document in the form of a discourse tree. The discourse tree, depicted in Fig. 1, exhibits a hierarchical structure where each leaf node corresponds to an Elementary Discourse Unit (EDU). These EDUs are interconnected by rhetorical and nuclearity relations, resulting in the formation of higher-level discourse units (DUs). Rhetorical relations serve the purpose of describing the functional and semantic relationships between EDUs, while nuclearity is characterized by nucleus (N) and satellite (S) tags, attributing greater importance to the nucleus. By incorporating structural elements, rhetorical analysis, and the notion of nucleus, discourse trees effectively capture the text's organization, rendering them valuable for text summarization purposes.

Previous studies have shown that discourse trees can capture structural and semantic information between EDUs, which is effective for single document summarization (SDS). Therefore, we attempt to apply the discourse rhetorical structure information in the MDS task. In accordance with the characteristics of MDS task, we propose to construct a discourse tree for each single document and design discourse patterns to merge them. This method not only reduces computational complexity, but also better preserves the distinct characteristics of each single document while capturing the relationships between them, as opposed to constructing the discourse tree for the entire set of documents.

The main contribution is threefold: (i) We propose a discourse-aware abstractive summarization model for MDS task, which operates on an elementary discourse unit level to capture rich semantic and structural information and generate overall coherent summaries. (ii) We design new patterns to form the documents and discourse trees in a flat sequence as the input of the abstractive model. These patterns are well-designed to avoid explicit modeling of structure with the guarantee of corresponding relationships. (iii) Our model achieves a new state of the art on the Multi-Document Summarization Dataset, outperforming other abstractive models.

2 Related Work

2.1 Discourse Rhetorical Structure Based Multi-document Summarization

Rhetorical Structure Theory (RST) is a comprehensive theory of text organization [14]. It has gained increasing attention and has been applied to various high-level NLP applications, including text summarization, following Marcu's earlier works on RST parsing [17]. The authors of RST have hypothesized that the nucleus in the discourse tree (DT) can serve as an adequate summarization of the text, a notion first validated by Marcu [16]. Louis et al. [13] have demonstrated that the structure features, such as the position in the global structure of the whole text, of the DT are the most useful for computing the salience of text spans. For MDS, Zahri et al. [23] address the redundancy issue by utilizing DT for cluster-based MDS. They leverage rhetorical relations between sentences to group similar sentences into multiple clusters, thereby identifying themes of common information from which candidate summary sentences are extracted.

2.2 Abstractive Multi-document Summarization

With the development of representation learning for NLP [2] and large-scale datasets [4], some studies have achieved promising results on abstractive MDS [7]. See at all [20] propose Pointer Generator Network (PGN) to overcome the problems of factual errors and high redundancy in the MDS. Liu et al. [11] introduce Transformer to MDS tasks, aiming to generate a Wikipedia article from a given topic and set of references. Their model selects a series of top-K tokens and feeds them into a Transformer based decoder-only sequence transduction model to generate Wikipedia articles. Raffel et al. [19]. propose T5, which is a transformer based text-to-text pre-trained language model that is gaining popularity for its unified framework that converts all text-based language problems into a text-to-text format. More recently, Guo et al. [5]. extend the original T5 encoder with global-local attention sparsity patterns to handle long inputs. In this work, we propose an effective method to combine pre-trained LMs with our discourse tree and make them able to process much longer inputs effectively.

3 Discourse Patterns Construction

In this section, we first introduce the process of generating discourse trees. And then explain how we construct discourse patterns used in the abstractive model.

3.1 Tree Generation Model

The process of the discourse tree construction can be separated into two stages: Elementary Discourse Unit segmentation (EDU segmentation) and Discourse Rhetorical Structure parsing (DRS parsing).EDU segmentation approach builds upon the work of Zhang et al. [25], employing a sequence-to-sequence model for EDU segmentation. The segmentation model learns to capture the intra-document features and cast the EDU segmentation problem as a sequence labeling task. Given the token sequence $Doc_i = \{x_1, x_2, ..., x_n\}$ as input. Firstly, they employ bi-directional GRUs and intra-sentence dependency structures to obtain the refined vectors $\{h'_0, h'_1, ..., h'_n\}$. After that, they take the refined vectors as input to a BiGRU encoder to generate candidate EDU boundaries $\{e_0, e_1, ..., e_n\}$. Finally, taking the concatenation of the last hidden states e_n in both directions as the input of the decoder. The decoder's outputs $\{d_0, d_i, d_j, ...\}$ are employed to determine the segmentation boundaries $\{e_{i-1}, e_{j-1}, ..., e_n\}$ for the EDUs.

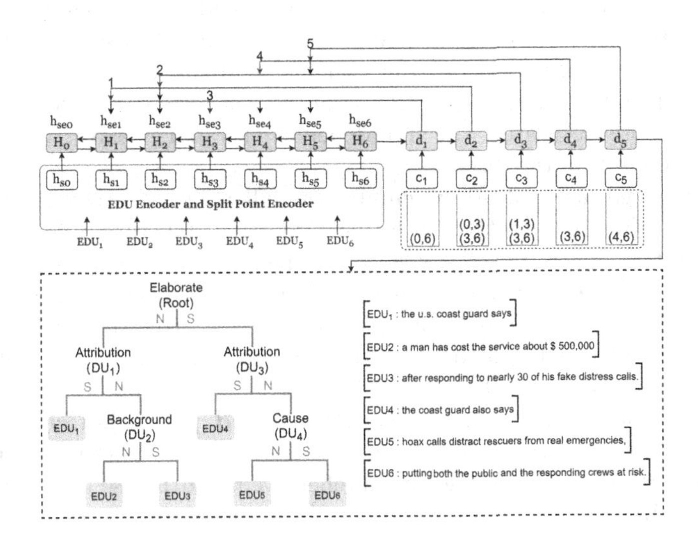

Fig. 2. A parsing example of the attention-based encoder-decoder.

After obtaining EDUs, we follow the setting of Zhang et al. [26], employing the pre-trained DRS parser to construct the discourse tree. The parsing process is illustrated in Fig. 2. Given a text containing six $EDUs = \{EDU_1, EDU_2, ..., EDU_6\}$, the encoder encodes them to obtain contextual representations $h_e = \{h_{e1}, h_{e2}, ..., h_{e6}\}$. We name the split position between any two EDUs as a split point. The split point encoder is responsible for encoding each split point. In particular, we feed the sequence of encoded EDUs h_e into a bi-directional GRU network to get the final sequence of encoded EDUs $h'_e = \{h'_{e0}, h'_{e1}, ..., h'_{e7}\}$. We use a CNN net with a window of 2 and a stride size of 1 to compute the final split point representation $h_s = \{h_{s0}, h_{s1}, ..., h_{s6}\}$.

The decoding process is essentially the process of iteratively searching for the segmentation points of the subregions within a given sequence (1, 2, 3, 4, 5, 6). The initial stack contains the segmentation point between the beginning and end of the entire discourse (0, 6). And then, the arrows in red indicate the selected split points at each time step. The discourse tree is built after 5 iterations with the split points (3, 1, 2, 4, 5) detected in turn. When generating the discourse tree, for each edge containing an N-R (Nuclearity and Rhetoric) label, it is pruned from the original tree and labeled as an unknown label, and then a classifier is used to classify it, resulting in the final N-R label.

3.2 Discourse Patterns

Upon obtaining the discourse tree, the sentence's structure, logical meaning, and functional implications become apparent. To augment the abstractive model's ability to comprehend tree information, we introduce eight types of discourse patterns. In particular, the discourse tree is encoded in a parenthetical format that preserves its hierarchical structure and then transformed into a flat sequence using a predefined ordering scheme. Subsequently, we inject rhetorical and nuclear information into the parenthetical format, allowing the model to capture more accurately the interdependencies and relative significance of various EDUs within the discourse.

As illustrated in Fig. 3 (a), we employ the following two strategies to encode the discourse tree in parenthetical format, injecting structural information into the original text, where the EDUs represent the content of each discourse. **DFS-based parenthetical format** follows the natural traversal order of the tree, i.e., the root node is processed first, followed by the left subtree, and finally the right subtree. Since the EDUs in the discourse tree are arranged in a specific order, during the DFS process, each EDU will be visited before its parent node, ensuring the preservation of its original order. **BFS-based parenthetical format** is employed to systematically investigate the discourse tree structure by traversing it level by level, thereby offering a comprehensive breadth-wise perspective. This strategy enables the analysis of hierarchical relationships by prioritizing nodes within the same level before proceeding to the subsequent level.

Rhetorical relations showcase the semantic relationships and information flow between EDUs. For example, in Fig. 2, EDU2 and EDU3 are linked together through a "background" relation, forming a higher-level discourse unit (DU2), which is subsequently integrated with EDU1 through an "attribution" relation. By recognizing rhetorical relations, the abstractive summarization model can accurately capture important arguments, resulting in more logical summaries. We utilize the label embedding technique to incorporate rhetorical relations into the parenthetical format, resulting in novel discourse patterns. This discourse pattern seamlessly integrates the structural and rhetorical information from the original text, serving as the input text for the abstractive summarization model. By leveraging the embedded labels, the abstractive model is able to better understand the structure and relationships within the document, guiding the genera-

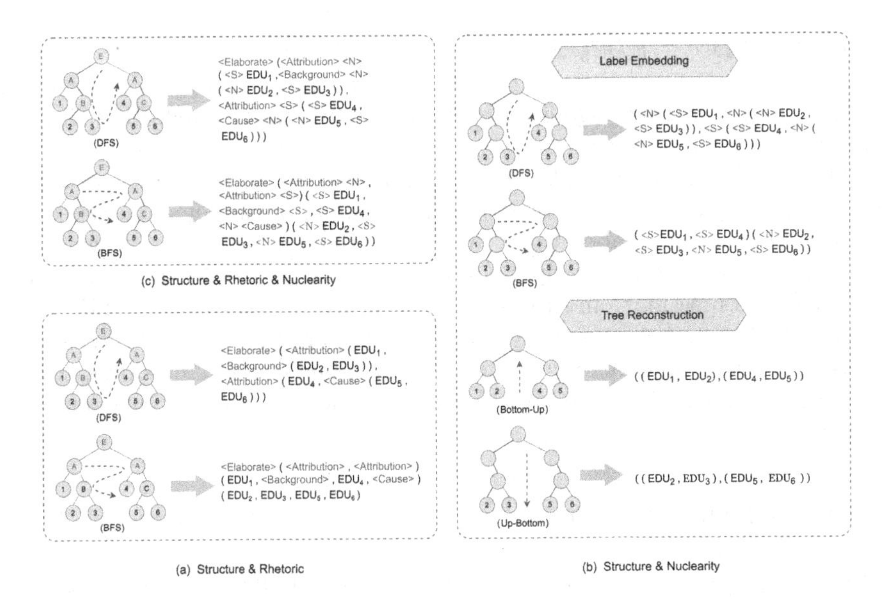

Fig. 3. Example of discourse patterns.

tion of summaries that maintain consistent logical and semantic relations with the original text.

The nodes of the discourse tree are connected by nuclear relations. Nuclearity includes nucleus (N) and satellite (S) tags, where the nucleus is considered more important than the satellite. As illustrated in Fig. 3 (b), label embedding and tree reconstruction are employed to integrate the nuclear relations into the parenthetical format to form novel discourse patterns. Similar to Fig. 2 (a), We use $< N >$ and $< S >$ labels help the abstractive model recognize the importance of the text adjacent to the labels. On the other hand, the reconstruction of the discourse tree is specifically focused on re-establishing the hierarchical organization of discourse based on nuclear information. In the discourse tree, the lower level conventionally pertains to sentence or phrase structures, whereas the higher level encompasses paragraph or document structures. Employing a **Bottom-Up** selection process, we preserve the nucleus and satellite relations found within the initial three levels. Subsequently, at other levels, we adopt a bottom-up approach to amalgamate the EDUs associated with the nucleus relations, thereby forming DUs. For instance, in the case of the discourse tree depicted in Fig. 2, we convert it into a serialized representation as $((EDU_1, EDU_2), (EDU_4, EDU_5))$. This merging process purposefully retains and accentuates the nuclear relations that are identified as pivotal and pertinent, effectively manifesting them within higher-level structures. Furthermore, through the **Up-Bottom** selection process, we conserve the nucleus and satellite relations at the initial and final layers. For the intermediate layers, we employ an up-bottom approach to amalgamate the EDUs linked to the nuclear relations into DUs. For example, for the discourse tree in Fig. 2, we transform it into a serialized representation as $((EDU_2, EDU_3), (EDU_5, EDU_6))$.This merging process allows for selective transmission and expansion of the information related to upper-level nuclear relations based on their significance and relevance, thereby reflecting them in lower-level structures.

The cohesive relations in RST combine nuclear relations with rhetorical relations, offering a more comprehensive understanding of the semantic aspects of the text. Specifically, Elaboration : the nucleus that expresses a state, while the satellite provides further description of that state. Attribution : the nucleus consists of a discourse unit, and the satellite indicates the speaker's identity. The example in Fig. 2 also serves as evidence to support this claim. Therefore, we integrate nucleus and rhetorical information through label embedding to add new semantic information for the abstractive model, forming discourse patterns as depicted in Fig. 3 (c).

4 Model Description

This section describes our model, which is an abstractive model based on discourse tree. The overall architecture is presented in Fig. 4. Given a set of documents $\{Doc_1, Doc_2, ...Doc_m\}$, the goal of our model is to generate a word sequence $S = \{y_1, y_2, ..., y_n\}$ as the summary. Our model consists of four major components: EDU Segmentation, DRS Parsing, Discourse Patterns and Summarization Generation.

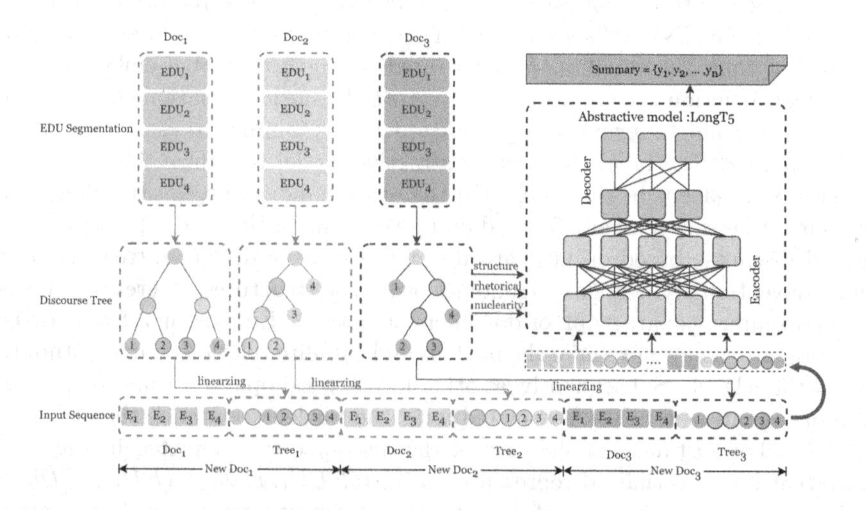

Fig. 4. Overview of proposed model

We utilize the LongT5 model [5] to generate summaries. After obtaining the discourse patterns for each single document, we concatenate them to form the new input sequence X for the LongT5 model. And then, we provide the input sequence X to the encoder and the target sequence Y to the decoder. The long sequence X is turned into digital by the encoder. It uses Transient Global Attention (TGlobal), which focuses on the words in each encoder layer. The decoder gets the output sequence $\hat{Y}$ by using the encoding information. The output sequence $\hat{Y}$ with the target sequence Y using the CrossEntropy Loss function. And the model performance is improved by continuously reducing the loss value:

$$\text{loss} = -\frac{1}{N}\sum_{i=1}^{N}\sum_{j=1}^{K} Y_i \log \widehat{Y}_i + \frac{\lambda}{2}\left|\theta_y\right|^2 \tag{1}$$

where i is the index of the data samples, j is the index of the word list, Y_i is the target word, $\hat{Y}$ is the predicted word, N is the total number of samples, K is the word list size, θ_y is the model parameter, and λ is the $L2's$ regular parameter.

The LongT5 model has undergone extensive pre-training, resulting in a substantial enhancement of its language knowledge and semantic comprehension. Notably, it combines the traditional encoder self-attention layer with local attention or Transient Global Attention (TGlobal), thereby augmenting its capacity to process long text input. To further bolster its performance, the model adopts the GSG strategy in PEGASUS [24], a prominent approach that successfully guides the generation of summaries. Through meticulous fine-tuning, the LongT5 model demonstrates remarkable adaptability to the demanding conditions of MDS tasks [3].

5 Experiments

This section starts with describing the dataset, evaluation metrics and training details. We then analyze model performance and finally conduct ablation studies.

5.1 Dataset and Evaluation Metrics

In this study, we use Multi-News [4] dataset for our experiments. This dataset includes 56,216 article-summary pairs, with each example consisting of 2–10 source documents and a summary. Each article is collected from real-life scenarios and the golden summaries are written by multiple experts, which ensures the data quality. Following the setting from Fabbri et al. [4], we divide the original dataset into a training set (80%, 44,972), a validation set (10%, 5,622), and a testing set (10%, 5,622). Like most previous works, we use the F1 of the ROUGE-N and ROUGE-L for performance evaluation [9].

5.2 Training Details

The segmentation system is implemented with PyTorch framework. We employed the 300D word embeddings provided by GloVe and used the Stanford CoreNLP toolkit [15] to obtain POS tags and intra-sentence dependency structures. For DRS parsing model learning, The learning rate is $1e-6$, the batch size is 5 and the training epochs are 20. Additionally, we choose LongT5 large (770M) as the generative model. All models were trained on a V100 (GPU). Due to the limitation of GPU memory, the input document is truncated to the first 1500 words. The output length is set to 300 words at maximum and 200 words at minimum. The learning rate is $1e-6$. AdaFactor [21] is the optimizer. Beam Search is used in the decoder with beam size set to 9.

5.3 Main Result

We compare the proposed model with various strong baselines in Table 1, where,

- **LEAD-3** [6]: LEAD-3 concatenates the first three sentences of each article on the same topic as a summary.
- **TextRank** [10]: It is a graph-based ranking model to extract salient sentences from documents.
- **MMR** [1]: It calculates the relevance between sentences and raw documents to score candidate sentences for the summary generation.
- **BERTSUM** [12]: It applies BERT to label each sentence and sentences labeled 1 are selected as summary.
- **PGN-MMR** [8]: PG-MMR is based on PGN and incorporates the MMR algorithm to reweight the importance distribution of sentences used for summary extraction.
- **LongT5** [5]: It is an extension of the T5 model. LongT5 model can handle longer input sequences based on GSG and TGlobal.

Table 1. Evaluation Results

Model	ROUGE-1	ROUGE-2	ROUGE-L
LEAD-3	39.78	11.92	18.18
TextRank	41.42	13.37	19.44
MMR	41.89	13.34	19.04
BERTSUM	42.93	13.98	19.74
PGN-MMR	42.22	13.65	19.06
LongT5(1.5k input)	42.86	13.28	20.39
Our(1.5k input)	**45.23**	**14.59**	**21.35**

Table 1 summarizes the evaluation results on the Multi-News dataset. The first block shows four popular extractive baselines, and the second block shows two strong abstractive baselines. The last block shows the results of our models. Our model adopts the DFS-based parenthetical format strategy illustrated in Fig. 3 (c). This approach incorporates three types of information: structural, rhetorical, and nuclear. The results demonstrate that our model outperforms other models. Compared to the LongT5 baseline model, our model achieves 2.37/1.31/0.96 improvements on ROUGE-1, ROUGE-2, and ROUGE-L. The evaluation results of the Multi-News dataset demonstrates the effectiveness of our model in capturing semantic relationships between discourse segments, leading to significant improvements in MDS tasks.

5.4 Ablation Study

As shown in Fig. 5, we employ ablation experiments to analyze the influence of different information on the generated summaries. Compared to using LongT5 alone, when only the structure information of the discourse tree (ST) is added, the performance is improved by 0.35/0.01 on ROUGE-1 and ROUGE-L; When structural and rhetorical information (ST+RH) are added, the performance is improved by 1.78/0.32; When structural and nuclear information (ST+NU) are added, the performance is improved by 1.73/0.60. The results show that the discourse tree's structural, rhetorical, and nuclear information all contribute to the MDS tasks. By analyzing ROUGE metrics and summary content, it can be deduced that the structural information of discourse trees can aid models in identifying which portions are crucial and which ones can be omitted. Rhetorical information can assist models in recognizing key concepts and details within the text, thereby enhancing the capture of the essential content of the original text. Nuclear information can aid generative models in accurately identifying and extracting pivotal textual information. Therefore, by integrating these three pieces of information, our proposed model (ST+NU+RH) achieves the best performance, the ROUGE value is improved by 2.37/0.96. Compared to the DFS-based parenthetical format, it is evident that the BFS-based parenthetical format places a stronger emphasis on the correlation between elements at the same level. However, it also sacrifices continuity. Consequently, its performance on the Rouge-L is lower than that of the DFS-based parenthetical format.

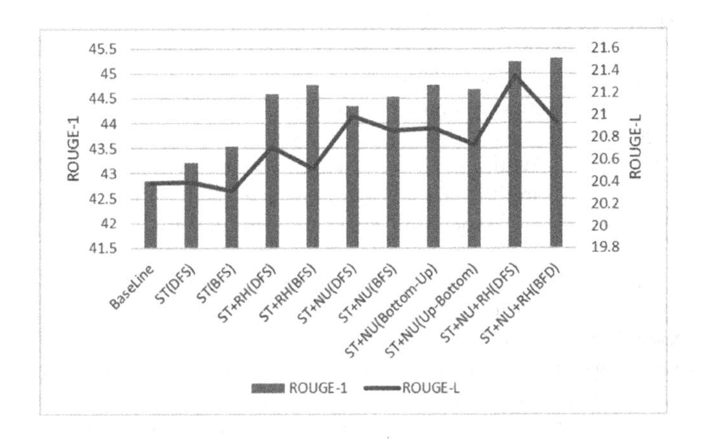

Fig. 5. Ablation study on the DRS

Nevertheless, its performance on the Rouge-1 surpasses that of the DFS-based parenthetical format.

6 Conclusion

In this paper, we propose a novel abstractive MDS model that integrates a joint DRS to capture semantic information. Experimental results demonstrate that our model achieves the-state-of-the art results on summarization. Our study is still a primary effort toward abstractive MDS. Future work we can do includes alleviating the requirement of a good pre-trained abstractive summarization model, designing better methods to help the abstractive model understand the discourse tree, and investigating our approach based on other model architectures [18].

References

1. Carbonell, J., Goldstein, J.: The use of MMR, diversity-based reranking for reordering documents and producing summaries. In: Proceedings of the 21st Annual International ACM SIGIR Conference on Research and Development in Information Retrieval, pp. 335–336 (1998)
2. Devlin, J., Chang, M., Lee, K., Toutanova, K.: BERT: pre-training of deep bidirectional transformers for language understanding, pp. 4171–4186. Association for Computational Linguistics (2019)
3. Dong, L., et al.: Unified language model pre-training for natural language understanding and generation. In: Advances in Neural Information Processing Systems, vol. 32 (2019)
4. Fabbri, A., Li, I., She, T., Li, S., Radev, D.: Multi-News: A Large-Scale Multi-Document Summarization Dataset and Abstractive Hierarchical Model. Association for Computational Linguistics, Florence, Italy (Jul (2019)
5. Guo, M., et al.: LongT5: Efficient text-to-text transformer for long sequences. In: Findings of the Association for Computational Linguistics: NAACL 2022. Seattle, United States (2022)
6. Ishikawa, K.: A hybrid text summarization method based on the TF method and the lead method. In: Proceedings of the Second NTCIR Workshop Meeting on Evaluation of Chinese & Japanese Text Retrieval and Text Summarization, pp. 325–330 (2001)
7. Jin, H., Wang, T., Wan, X.: Multi-granularity interaction network for extractive and abstractive multi-document summarization. In: Proceedings of the 58th Annual Meeting of the Association for Computational Linguistics, pp. 6244–6254 (2020)
8. Lebanoff, L., Song, K., Liu, F.: Adapting the neural encoder-decoder framework from single to multi-document summarization. In: Conference on Empirical Methods in Natural Language Processing (2018)
9. Lin, C.Y.: Rouge: A package for automatic evaluation of summaries. In: Text Summarization Branches Out, pp. 74–81 (2004)
10. Lin, D., Wu, D.: Proceedings of the 2004 conference on empirical methods in natural language processing. In: Proceedings of the 2004 Conference on Empirical Methods in Natural Language Processing (2004)

11. Liu, P.J., Saleh, M., Pot, E., Goodrich, B., Sepassi, R., Kaiser, L., Shazeer, N.: Generating Wikipedia by summarizing long sequences. OpenReview.net (2018)
12. Liu, Y.: Fine-tune BERT for extractive summarization. arXiv preprint arXiv:1903.10318 (2019)
13. Louis, A., Joshi, A.K., Nenkova, A.: Discourse indicators for content selection in summaization (2010)
14. Mann, W.C., Thompson, S.A.: Rhetorical structure theory: toward a functional theory of text organization. Text-Interdisc. J. Study Discourse **8**(3), 243–281 (1988)
15. Manning, C.D., Surdeanu, M., Bauer, J., Finkel, J.R., Bethard, S., McClosky, D.: The stanford corenlp natural language processing toolkit. In: Proceedings of 52nd Annual Meeting of the Association for Computational Linguistics: System Demonstrations, pp. 55–60 (2014)
16. Marcu, D.: From discourse structures to text summaries. In: Intelligent Scalable Text Summarization (1997)
17. Marcu, D.: The theory and practice of discourse parsing and summarization. MIT Press (2000)
18. Radev, D.: A common theory of information fusion from multiple text sources step one: cross-document structure. In: 1st SIGdial Workshop on Discourse and Dialogue, pp. 74–83 (2000)
19. Raffel, C., et al.: Exploring the limits of transfer learning with a unified text-to-text transformer. J. Mach. Learn. Res. **21**(140), 1–67 (2020)
20. See, A., Liu, P.J., Manning, C.D.: Get to the point: Summarization with pointer-generator networks, pp. 1073–1083. Association for Computational Linguistics (2017)
21. Shazeer, N., Stern, M.: Adafactor: adaptive learning rates with sublinear memory cost. In: International Conference on Machine Learning, pp. 4596–4604. PMLR (2018)
22. Sun, S., Shi, H., Wu, Y.: A survey of multi-source domain adaptation. Inf. Fusion **24**, 84–92 (2015)
23. Zahri, N.A.H., Fukumoto, F., Suguru, M., Lynn, O.B.: Exploiting rhetorical relations to multiple documents text summarization. Int. J. Netw. Secur. Appl. **7**(2), 1 (2015)
24. Zhang, J., Zhao, Y., Saleh, M., Liu, P.: PEGASUS: pre-training with extracted gap-sentences for abstractive summarization. In: International Conference on Machine Learning, pp. 11328–11339. PMLR (2020)
25. Zhang, L., Kong, F., Zhou, G.: Syntax-guided sequence to sequence modeling for discourse segmentation. In: Zhu, X., Zhang, M., Hong, Yu., He, R. (eds.) NLPCC 2020. LNCS (LNAI), vol. 12431, pp. 95–107. Springer, Cham (2020). https://doi.org/10.1007/978-3-030-60457-8_8
26. Zhang, L., Kong, F., Zhou, G.: Adversarial learning for discourse rhetorical structure parsing. In: Proceedings of the 59th Annual Meeting of the Association for Computational Linguistics and the 11th International Joint Conference on Natural Language Processing (Volume 1: Long Papers), pp. 3946–3957 (2021)

MelMAE-VC: Extending Masked Autoencoders to Voice Conversion

Yuhao Wang(✉) and Yuantao Gu

Tsinghua University, Beijing, China
yuhaowan20@mails.tsinghua.edu.cn, gyt@tsinghua.edu.cn

Abstract. Voice conversion is a technique that generates speeches with text contents identical to source speeches and timbre features similar to reference speeches. This paper proposes MelMAE-VC, a neural network for non-parallel many-to-many voice conversion that utilizes pre-trained Masked Autoencoders (MAEs) for representation learning. Our neural network mainly consists of transformer layers and no recurrent units, aiming to achieve better scalability and parallel computing capability. We follow a similar scheme of image-based MAE in the pre-training phase that conceals a portion of the input spectrogram; then we set up a vanilla autoencoding task for training. The encoder yields latent representation from the visible subset of the full spectrogram; then the decoder reconstructs the full spectrogram from the representation of only visible patches. To achieve voice conversion, we adopt the pre-trained encoder to extract preliminary features, and then use a speaker embedder to control timbre information of synthesized spectrograms. The style transfer decoder could be either a simple autoencoder or a conditional variational autoencoder (CVAE) that mixes timbre and text information from different utterances. The optimization goal of voice conversion model training is a hybrid loss function that combines reconstruction loss, style loss, and stochastic similarity. Results show that our model speeds up and simplifies the training process, and has better modularity and scalability while achieving similar performance compared with other models.

Keywords: Voice conversion · Masked autoencoders · Audio processing

1 Introduction

In recent years, speech synthesis has benefited from the development of computing hardware and the introduction of advancing deep-learning models. Depending on specific scenarios, there are two major approaches to speech synthesis: voice conversion (VC) and text-to-speech (TTS). Voice conversion transfers the timbre of speech segments to another timbre while preserving its semantic contents, and TTS generates speech waveform from given text contents. Both techniques have found their social or industrial applications *e.g.* identity obscuring

B. Luo et al. (Eds.): ICONIP 2023, CCIS 1965, pp. 482–494, 2024.
https://doi.org/10.1007/978-981-99-8145-8_37

and encryption, speech enhancing and augmentation [22], and media content creation [12,24] *etc.*

Compared to TTS, voice conversion models are capable of generating speech with richness and vividness in terms of expression, because speech input contains higher-level information in comparison with either plain or labeled text. As a practical result, voice conversion models would be preferable to TTS in applications that require precise control over emotion and expression. Commonly, voice conversion is performed on time-frequency spectrograms (*e.g.* Mel spectrograms) followed by a vocoder to convert spectrograms back to waveforms [14,18,19]. Few end-to-end approaches that directly manipulate raw audio waveforms were proposed [17].

Voice conversion technique presents three main challenges: the lack of parallel data, the ability to generalize, and the balance between complexity and performance. Parallel data *i.e.* voice data with the same text contents but spoken by different people, are useful in disentangling text from other information in latent space, but can be difficult to collect or collate. In contrast, non-parallel data are almost unlimited and easier to obtain. The generalization ability of voice conversion models is mainly reflected by their source and target domain of transformation and their performance on unseen data. Models with many-to-many conversion functionality and few- or zero-shot conversion capability are more challenging. Finally, although scaling up a network and processing more categories of information are common and effective approaches to achieve better performance, the improvement does not always justify the higher computational cost required.

Many voice conversion networks rely on independent modules to process specific information for better conversion results. For example, HiFi-VC utilizes an automatic speech recognition (ASR) module to explicitly match output text [11]; and AutoVC utilizes an embedder that maps the input spectrogram into a latent space where embeddings are clustered by speakers' identity [20]. The approaches with independent modules usually lead to additional training or even incompatibility issues because they may use different configurations. Being similar to a multi-functional taschenmesser that can handle various needs, a universal framework for audio tasks would be ideal and suitable in practice.

In this paper, we propose MelMAE-VC, a voice conversion network capable of many-to-many conversion with non-parallel training. The network consists of three components: MelMAE encoder, speaker embedder, and decoder for voice style transfer. MelMAE encoder is based on a variant of the Masked Autoencoder (MAE) [4] that is designed for spectrogram learning as a universal framework and pre-train method. With proper pre-training, the encoder gains the capability of extracting latent representation from input spectrogram. The embedder maps the encoder latent representation into speaker embedding that describes speakers' timbre. The style transfer decoder disentangles voice features from text contents and recomposites them to yield converted spectrogram, enabling voice conversion from an arbitrary speaker to another. The loss function of voice conversion network is a hybrid loss that combines style, content, and reconstruction error. The MAE framework variant shows great flexibility and scalability in MelMAE-VC and significantly accelerates pre-training and fine-tuning for voice conversion tasks. As a byproduct, we also verify the acceleration effect of MAE framework when training and fine-tuning for speaker verification tasks. Moreover, our network mainly utilizes vision transformer layers and is free of recurrent units, which can further reduce the computational cost of training and inference.

In summary, our main contributions are in three aspects:

- We propose a voice conversion network that performs voice conversion between arbitrary source and target speaker, without parallel data required in training.
- We introduce the Masked Autoencoder to audio processing, not only for classification tasks, but also style transfer tasks which are more challenging.
- The network we proposed has lower complexity, better scalability, and parallel computing capability of training and inference.

The rest of this paper is organized as follows. Section 2 summarizes studies that directly relate to or inspire our work. Section 3 shows an overview of the voice conversion network as well as detailed designs. Section 4 demonstrates the experiments and analyzes the results. Section 5 concludes this paper.

2 Related Work

Models with Masking Technique. The idea of concealing a portion of information has been commonly utilized in neural networks. By masking part of the input and training model to reconstruct masked contents, the model gains better generalization ability to downstream tasks. This idea is implemented in various pre-training models in natural language processing (NLP), *e.g.* GPT [21] and BERT [3] that large-scale downstream models usually derive from. Masked autoencoder [4] introduces a similar pre-training approach to computer vision (CV) and further extends itself to audio signal processing [5]. Besides, pre-training approach, reconstruction from masked information can also be the main goal in certain applications [10]. In these contexts, masking and reconstruction is an approach of adding noise and denoising that serves as an auxiliary task to encourage neural networks to generate content based on visible information, similar to image inpainting.

Voice Conversion Networks. VC networks transfer acoustic characteristics of a given speech while preserving its semantic contents. The conversion could be performed on raw waveforms [17] or various types of time-frequency spectrograms. GANs and (variational) autoencoders are commonly used, while transformers for voice conversion still remain under-explored. CycleGAN-VC and its incremental iterations [7–10] are based on GAN architecture that transfers styles via traditional residual networks and variations of GAN loss. AutoVC [20] is based on autoencoder architecture that uses a bottleneck to filter and separate text and identity information to achieve voice conversion.

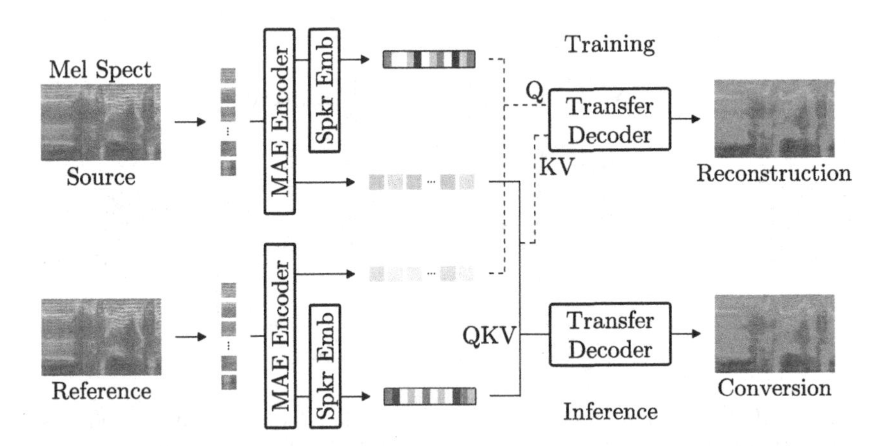

Fig. 1. Workflow of MelMAE-VC network. The details of implementation are omitted. We cross-assemble speaker embedding and latent representation, then train the voice conversion network with an autoencoding task. As for inference, the latent representation of source and the speaker embedding of reference are combined to yield a converted spectrogram. Converted spectrogram can also be generated by combining latent representation and speaker from source and target separately, then feeding them into the transfer decoder as query and key-value input respectively.

3 MelMAE-VC

In Sects. 3 and 4, the notation scheme is described as follows. Let $\boldsymbol{x}$ denote Mel spectrograms, $\boldsymbol{y}$ denote encoder latent representation, and $\boldsymbol{z}$ denote speaker embedding. Superscript denotes speaker's identity or convert sequence *e.g.* $\boldsymbol{x}^A$ and $\boldsymbol{x}^{A \rightarrow B}$ for Mel spectrograms of the original utterance from speaker A and its conversion with speaker B's timbre respectively. Prime mark ($'$) in superscript suggests the value originates from another utterance of the same speaker *e.g.* $\boldsymbol{x}^{A \rightarrow A'}$ for converted spectrogram with source and reference being different utterances from the same speaker.

In this section, we present the architecture of a voice style transfer network based on MAE. We name it MelMAE-VC, as it is an adaptation of MAE that performs voice conversion on Mel spectrograms. Figure 1 demonstrates the basic training and inference workflow of MelMAE-VC. As shown in the figure, our MelMAE-VC has three major components: the MAE encoder, the speaker embedder, and the style transfer decoder. The output of the decoder is converted Mel spectrograms with the text contents of source and timbre features of reference. The converted spectrograms eventually get transformed into waveforms by a HiFi-GAN vocoder [14] .

3.1 Pre-train Network MelMAE

Our voice conversion model MelMAE-VC derives from a pre-train network which we name MelMAE. The pre-train network shares a similar architecture with its

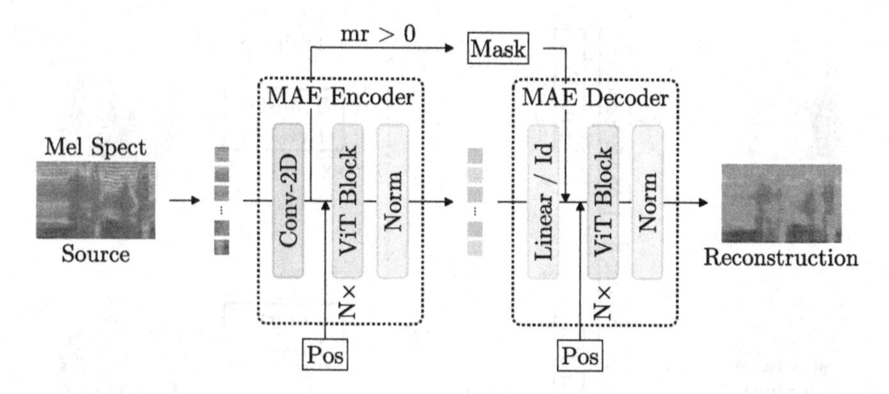

Fig. 2. MelMAE pre-training network. Implementation details of MAE encoder (same as in Fig. 1) and MAE decoder are demonstrated. Mask tokens are randomly generated by the encoder with mask ratio configuration, and later fed into the decoder to indicate whether a patch is visible or not. Fixed sine-cosine position embeddings for 2D patch grids are added to MAE encoder and decoder respectively.

visual prototype MAE as Fig. 2 shows. Minor changes have been made, mainly to enable MelMAE learning from Mel spectrograms with flexible duration. MelMAE accepts logarithm amplitude Mel spectrograms as input of the encoder and generates latent representation of visible patches; then the decoder reconstructs the full spectrogram from the latent representation and mask tokens. The decoder could be a relatively simple and shallow network, as long as the reconstructed audio waveforms (synthesized by HiFi-GAN vocoder [14] from reconstructed Mel spectrograms) have good perceptual quality and similarity compared with original audio.

The objective of MelMAE pre-training phase is patch-wise averaged MSE between source and reconstructed spectrograms. Although the original MAE for visual tasks prefers MSE on masked patches over MSE on all patches, the difference is minor in reconstruction and other downstream tasks [4]. We assume such characteristic is similar in spectrogram tasks and choose patch-wise averaged MSE on all patches because it is relatively simple and indeed shows negligible difference in performance of spectrogram learning tasks.

The patch-wise weighted reconstruction loss $\mathcal{L}_{Rec}$ is described as Eq. 1a, where $\boldsymbol{x}$ and $\hat{\boldsymbol{x}}$ denotes source and reconstructed spectrogram respectively, $\boldsymbol{p}$ denotes patch-wise weight and $\otimes$ denotes patch-wise multiplication. In our case, we mainly used a simplified version that evaluates error on all patches (*i.e.* $\boldsymbol{p} = \mathbf{1}$ and $\sum \boldsymbol{p}$ is the number of patches n_{Pat}) as Eq. 1b shows.

$$\mathcal{L}_{Rec}(\boldsymbol{x}, \hat{\boldsymbol{x}}, \boldsymbol{p}) = \mathbb{E}\left[\frac{1}{\sum \boldsymbol{p}} \|(\boldsymbol{x} - \hat{\boldsymbol{x}}) \otimes \boldsymbol{p}^{1/2}\|_2^2\right] \tag{1a}$$

$$\mathcal{L}_{Rec}(\boldsymbol{x}, \hat{\boldsymbol{x}}) = \mathbb{E}\left[\frac{1}{n_{Pat}} \|(\boldsymbol{x} - \hat{\boldsymbol{x}})\|_2^2\right] \tag{1b}$$

3.2 Speaker Embedder

The speaker embedder is an extension of MelMAE that maps patched latent representation tensors to D-vectors for speaker verification. As Fig. 3 shows, it has several layers of ViT blocks, followed by a linear projection layer, and finally an L2-normalization layer.

GE2E Softmax Loss [23] is used when training the speaker embedder. Although the original decoder is unnecessary in the inference phase of speaker verification, we keep the decoder and yield latent representation either directly or with an autoencoding route before feeding into the speaker embedder. As a result, the speaker embedding of raw and reconstructed spectrogram would be nearly the same. This technique ensures forward consistency, thus enhancing robustness of the network and avoiding overfitting.

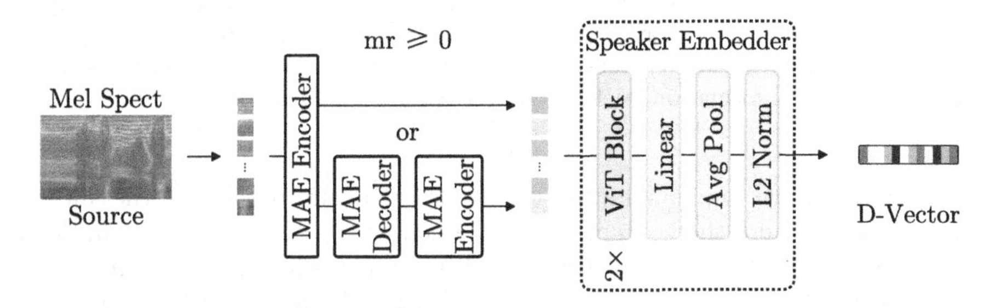

Fig. 3. Speaker embedder. In the training phase, an alternative autoencoding workflow can enhance robustness and avoid overfitting.

3.3 Style Transfer Decoder

The style transfer decoder is demonstrated in Fig. 4. Latent representation and speaker embedding from source and reference spectrogram are cross-assembled and fed into the decoder. An intuitive explanation is that, by setting up an autoencoding task, the decoder learns to decouple content and style information in latent representation.

The loss function is a linear combination of reconstruction loss, content loss, and style loss. The reconstruction loss is similar to Eq. 1a. Given the source spectrogram $\boldsymbol{x}^A$ and an intra-class conversion $\boldsymbol{x}^{A\to A'}$ or $\boldsymbol{x}^{A\to B\to A}$, the patch-wise weighted reconstruction loss is defined as follows.

$$\mathcal{L}_{Rec1}(\boldsymbol{x}^A, \boldsymbol{x}^{A\to A'}, \boldsymbol{p}) = \mathbb{E}\left[\frac{1}{\sum \boldsymbol{p}} \|(\boldsymbol{x}^A - \boldsymbol{x}^{A\to A'}) \otimes \boldsymbol{p}^{1/2}\|_2^2\right] \tag{2a}$$

$$\mathcal{L}_{Rec2}(\boldsymbol{x}^A, \boldsymbol{x}^{A\to B\to A}, \boldsymbol{p}) = \mathbb{E}\left[\frac{1}{\sum \boldsymbol{p}} \|(\boldsymbol{x}^A - \boldsymbol{x}^{A\to B\to A}) \otimes \boldsymbol{p}^{1/2}\|_2^2\right] \tag{2b}$$

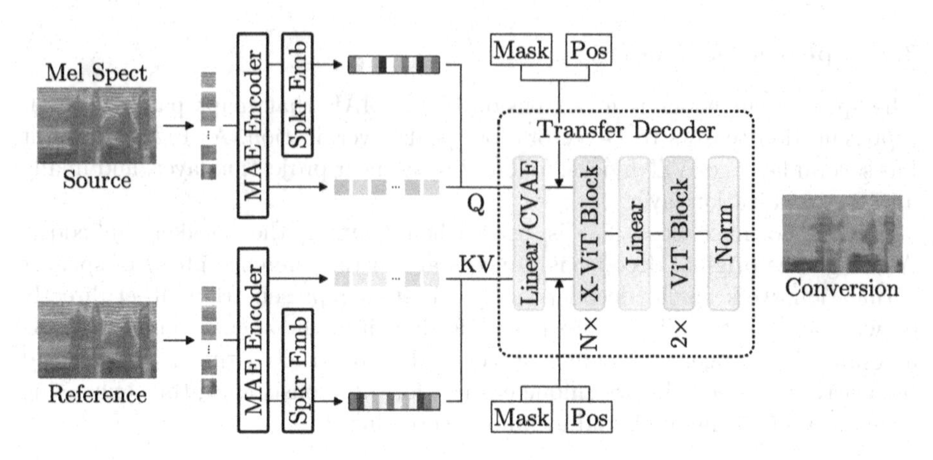

Fig. 4. MelMAE-VC style transfer decoder. In inference, by combining speaker embedding and latent representation of source and reference then feeding them into query and key-value input of cross-attention block, the network decouples and reassembles content and style information, yielding converted spectrogram.

A relatively simple implementation of the style transfer decoder is the autoencoder variant, where the content loss is defined as the MSE of source and converted encoder latent representation. In CVAE implementation, the content loss is normalized Kullback-Leibler divergence (KLD) of two Gaussian distribution that describes source and converted spectrogram.

$$\mathcal{L}_{CntAE}(\boldsymbol{y}^A, \boldsymbol{y}^{A\to B}) = \mathbb{E}\left[\|(\boldsymbol{y}^A - \boldsymbol{y}^{A\to B})\|_2^2\right] \tag{3a}$$

$$\mathcal{L}_{CntCVAE}(\boldsymbol{\mu}^A, \boldsymbol{\mu}^{A\to B}, \boldsymbol{\sigma}^A, \boldsymbol{\sigma}^{A\to B}) = KLD(\boldsymbol{\mu}^A, \boldsymbol{\mu}^{A\to B}, \boldsymbol{\sigma}^A, \boldsymbol{\sigma}^{A\to B}) \tag{3b}$$

The style loss is the MSE between speaker embedding of reference and converted spectrogram.

$$\mathcal{L}_{Sty}(\boldsymbol{z}^B, \boldsymbol{z}^{A\to B}) = \mathbb{E}\left[\|(\boldsymbol{z}^B - \boldsymbol{z}^{A\to B})\|_2^2\right] \tag{4}$$

The hybrid loss function is the linear combination of reconstruction loss, content loss, and style loss.

$$\mathcal{L} = \lambda_{Rec}\mathcal{L}_{Rec} + \lambda_{Cnt}\mathcal{L}_{Cnt} + \lambda_{Sty}\mathcal{L}_{Sty} \tag{5}$$

4 Experiments

4.1 Conditions of Experiments

Dataset. The dataset we used in this paper is VoxCeleb1 [16] and VoxCeleb2 [2]. VoxCeleb1 contains $\sim$ 150K utterances from 1211 speakers in the dev partition, and $\sim$ 5K utterances from 40 speakers in the test partition. VoxCeleb2 contains $\sim$ 1M utterances from 5994 speakers in the dev partition, and $\sim$ 36K utterances from 112 speakers in the test partition. The utterances are spoken in English and extracted from online videos of celebrities, without fully eliminating background noise. Speakers in dev sets and test sets have no overlap. To balance the number of data used in training, up to 100 utterances of each speaker are randomly selected from the dev partition of both datasets, summing up to $\sim$ 620K utterances.

The original sample rate of VoxCeleb1 and VoxCeleb2 data is 16000 Hz. We up-sampled the waveforms to 22050 Hz and normalized audio waveforms by loudness before computing Mel spectrograms. We used the following configuration to generate Mel spectrograms from raw waveforms: number of Mel bands $n_{Mel} = 80$, FFT window length of STFT $n_{FFT} = 1024$, hop length of STFT $d_{hop} = 256$, $f_{min} = 0$, $f_{max} = 8000$, and Mel spectrogram magnitude $p = 1$. Finally, we took the logarithm of the amplitude of Mel spectrograms.

Network Configurations. The pre-trained model is MelMAE. The encoder consists of 8 layers of 16-head ViT blocks that encode each 80×4 Mel spectrogram patch into a 256-dimension latent representation. The decoder uses an identity mapping instead of linear projection at the very beginning and consists of 8 layers of 16-head ViT blocks, eventually reconstructing the spectrogram from the encoder latent representation.

The speaker embedder is a shallow network that consists of only 2 layers of 8-head ViT block and a linear projection layer. It digests masked or fully-visible encoder latent representation to yield a 64-dimension D-vector.

We implement a CVAE-based style transfer decoder in MelMAE-VC. The encoder latent representation is projected into a 16-dimension vector per patch. The D-vector is duplicated to match the number of patches and then concatenated with the projected encoder latent representation to form a joint tensor. The CVAE encodes the joint tensor into parameters of 16-dimension Gaussian distributions for every patch, eventually yielding a 256-dimension decoder latent representation. The rest of the style transfer decoder mainly contains 8 layers of 16-head cross-attention ViT blocks and 2 layers of 16-head regular ViT blocks.

Training Process. We pre-train the MAE encoder and decoder on the dev set of VoxCeleb1. The batch size is set to 64. In the first 800K iterations, we use patch-wise normalized MSE loss on unmasked patches and AdamW as optimizer [13,15], with momentum $\beta_1 = 0.9, \beta_2 = 0.95$, learning rate initialized as 1e-3 and gradually reduced to 1e-5, mask ratio initialized with 10% infimum and supremum and gradually increased supremum from 10% to 75%. Then we switched to normalized MSE loss on all patches and trained for 1.6M iterations, with the mask ratio randomly chosen between 0 and 75% every batch.

The speaker embedder is trained on a randomly selected subset of VoxCeleb1 and VoxCeleb2 that has relatively balanced speaker classes. Each batch contains 64 speaker classes and 8 utterances each class, summing up to a total batch size of 256. We freeze the encoder and decoder and train only the embedder using GE2E Softmax loss [23] and SGD optimizer with the learning rate initialized as 1e-2 and gradually reduced to 1e-5. The scalar weight and bias in GE2E Softmax loss are set to $w = 10, b = 5$ respectively, and would not vary in the training process. We apply a gradient scaling technique described in [23] on the linear projection layer of the embedder, the scale factor is 1e-3. We have trained the speaker embedder for 1M iterations and achieved a 2.05% equal error rate (EER).

The style transfer decoder is trained on the same balanced subset of VoxCeleb1 and VoxCeleb2. Each batch contains 16 speaker classes and 8 utterances in each class. The centroid speaker embedding is calculated from 32 utterances of a speaker. The linear combination multipliers of each term in hybrid loss (Eq. 5) are simply set to 1. We freeze the rest of MelMAE-VC network and train only the decoder for nearly 2M iterations using AdamW optimizer, with momentum $\beta_1 = 0.9, \beta_2 = 0.95$, learning rate initialized as 1e-3 and gradually reduced to 1e-5, mask ratio randomly chosen in the range of $0 \sim 50\%$ each batch.

Patch size is a key hyperparameter that would affect voice conversion yield. Our configuration set patch size to 80×4, which means that each patch contains 4 frames of Mel-scale STFT spectrum and all 80 coefficients on Mel-frequency domain. We have attempted to train with other configurations that only differ in patch size which does not include the entire frequency scale within a single patch. Although such configurations work well on speaker verification tasks, they usually do not yield audible voice conversion results. This result contradicts the behavior of the network on visual style transfer tasks. Empirically, visual styles are mostly local pixel characteristics, while voice styles are contrarily defined by harmonic characteristics among the entire of frequency scale. This causes diverted practical behaviors of a similar network on image-based tasks and spectrogram-based tasks.

4.2 Objective Evaluation

The purpose of the objective evaluation is to investigate if text contents of reconstructed and converted utterances are preserved. We evaluated this based on two criteria: character error rate (CER) and word error rate (WER). The ASR model we chose is implemented with Wav2vec2 [1] and Flashlight's CTC beam search decoder [6]. Ground truth transcripts are generated with this ASR inference model on re-sampled speech waveforms of VoxCeleb1 and VoxCeleb2 which are trimmed to match the duration of Mel spectrograms.[1] Both original data and trimming operation would result in incomplete words at the beginning and the end of the audio segment, resulting in measurement errors brought by the context-based ASR method.

Results are shown in Table 1. We select AutoVC and MaskCycleGAN-VC (one-to-one conversion) as the benchmark. We also list the result of HiFi-GAN vocoder reconstruction and vanilla MelMAE reconstruction to demonstrate measurement error brought by the ASR method. The AutoVC network has been

[1] In our configuration, this duration is approximately 5.94 s. It is neither too short for the context-based ASR method to process nor too long to exceed the average duration of audio files.

trained for 2.5M iterations on a single RTX 3090 for 170 h, while our network achieved better results with less than 2M iterations trained on a single RTX 3090 for 140 h.

Table 1. Objective evaluation: comparison of character error rate (CER) and word error rate (WER). Our method can achieve better CER and WER than AutoVC (many-to-many conversion). The performance lies between AutoVC and MaskCycleGAN-VC (one-to-one conversion).

Model	CER (%)	WER (%)	#Parameters
MelMAE-VC	25.33	36.51	13M
AutoVC	29.04	40.02	20M
MaskCycleGAN-VC	22.71	30.75	17M
HiFi-GAN Vocoder	9.44	14.77	–
MelMAE Reconstruction	11.43	17.55	–

4.3 Subjective Evaluation

The final output is segments of audio waveform which are perceived by the human auditory system, hence listening tests for subjective evaluation are essential for any voice conversion tasks. Besides, audio generated by three conversion models (MelMAE-VC, AutoVC, MaskCycleGAN-VC), raw audio, and MAE-reconstructed audio are also tested, which sets a rough reference of data degradation brought by the vocoder. Note that speech segments from the original VoxCeleb1 and VoxCeleb2 dataset are collected without noise cancellation. Quality and similarity are demonstrated in Fig. 5.

Quality and Naturalness. We measure the quality by the mean opinion score (MOS) of reconstructed or converted samples. Participants are presented with shuffled original and synthesized utterances (without parallel data) and asked to give a score between 0 and 4 with an interval of 1. Clarity of utterance, background noise details, and stuttering or synthesized artifacts were considered. Each aspect is worth 1 point, meaning that if an aspect delivers noticeably degraded audio perception, then 1 point is subtracted. And if any of the aspects significantly worsen the audio experience, then up to 1 extra point is subtracted. Naturalness is defined by whether synthesized waveforms are statistically distinguishable from original ones.

Similarity. We measure the similarity also by MOS. Participants are presented with multiple sets of waveforms generated from three models without knowing which exact model generated the waveform. They are asked to give a score between 0 and 4 with an interval of 1. 4 means very similar to the reference speaker; 0 means very similar to the source speaker; 2 means similar to neither of the speakers.

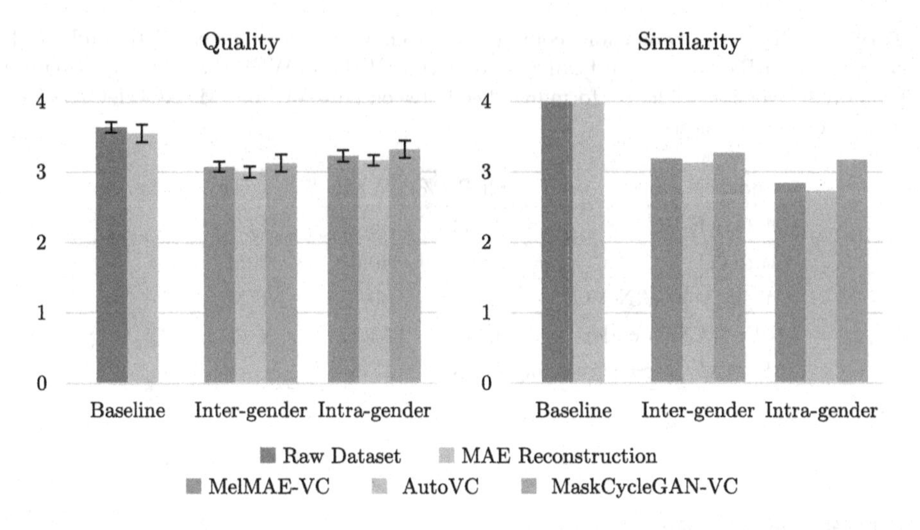

Fig. 5. Subjective evaluation: MOS of quality and similarity. Inter-gender conversion has relatively higher MOS similarity because gender difference makes the audio more distinguishable.

5 Conclusions

In this paper, we propose MelMAE-VC, a voice conversion network that performs few- or zero-shot many-to-many style transfer on Mel spectrograms. We pre-trained and fine-tuned masked autoencoder framework on non-parallel open datasets, then made modifications and added sub-modules to the framework to perform style transfer. The results show that our network is capable of yielding converted speech segments while maintaining the simplicity of network architecture and training process. We also experimented with various network designs under the MAE framework. As a result, we confirmed that MAE architecture is highly scalable and flexible to adapt to a variety of objectives.

While our study has demonstrated that MelMAE network could be applied in complex audio tasks, it has potential limitations that we would focus on and conduct further incremental studies. One major limitation is that hyperparameters related to network architecture could be sensitive to conversion results. To reduce such sensitivity, we would like to investigate how patch size affects the non-local transform of features in the frequency domain, especially when certain frequencies are shifted across patch borders. In addition, we could further improve the fidelity and quality of converted audio. We believe that this improvement can mainly come from various enhancements, which includes applying better denoising and augmentation technique to training data, performing in-depth analysis and modeling of the information structure in speech audio, and introducing loss functions defined by spectrum distortion. Moreover, we hope to extend the applications of audio-based MAE framework for more tasks with complexity, *e.g.* ASR and TTS.

References

1. Baevski, A., Zhou, Y., Mohamed, A., Auli, M.: wav2vec 2.0: a framework for self-supervised learning of speech representations. In: Advances in Neural Information Processing Systems, vol. 33, pp. 12449–12460 (2020)
2. Chung, J.S., Nagrani, A., Zisserman, A.: Voxceleb2: deep speaker recognition. arXiv preprint arXiv:1806.05622 (2018)
3. Devlin, J., Chang, M.W., Lee, K., Toutanova, K.: Bert: pre-training of deep bidirectional transformers for language understanding. arXiv preprint arXiv:1810.04805 (2018)
4. He, K., Chen, X., Xie, S., Li, Y., Dollár, P., Girshick, R.: Masked autoencoders are scalable vision learners. In: Proceedings of the IEEE/CVF Conference on Computer Vision and Pattern Recognition, pp. 16000–16009 (2022)
5. Huang, P.Y., et al.: Masked autoencoders that listen. Adv. Neural. Inf. Process. Syst. **35**, 28708–28720 (2022)
6. Kahn, J.D., et al.: Flashlight: enabling innovation in tools for machine learning. In: International Conference on Machine Learning, pp. 10557–10574. PMLR (2022)
7. Kaneko, T., Kameoka, H.: CycleGAN-VC: non-parallel voice conversion using cycle-consistent adversarial networks. In: 2018 26th European Signal Processing Conference (EUSIPCO), pp. 2100–2104. IEEE (2018)
8. Kaneko, T., Kameoka, H., Tanaka, K., Hojo, N.: CycleGAN-VC2: improved cycleGAN-based non-parallel voice conversion. In: ICASSP 2019–2019 IEEE International Conference on Acoustics, Speech and Signal Processing (ICASSP), pp. 6820–6824. IEEE (2019)
9. Kaneko, T., Kameoka, H., Tanaka, K., Hojo, N.: CycleGAN-VC3: examining and improving CycleGAN-VCs for mel-spectrogram conversion. arXiv preprint arXiv:2010.11672 (2020)
10. Kaneko, T., Kameoka, H., Tanaka, K., Hojo, N.: MaskcycleGAN-VC: learning non-parallel voice conversion with filling in frames. In: ICASSP 2021–2021 IEEE International Conference on Acoustics, Speech and Signal Processing (ICASSP), pp. 5919–5923. IEEE (2021)
11. Kashkin, A., Karpukhin, I., Shishkin, S.: Hifi-VC: high quality ASR-based voice conversion. arXiv preprint arXiv:2203.16937 (2022)
12. Kim, J., Kong, J., Son, J.: Conditional variational autoencoder with adversarial learning for end-to-end text-to-speech. In: International Conference on Machine Learning, pp. 5530–5540. PMLR (2021)
13. Kingma, D.P., Ba, J.: Adam: a method for stochastic optimization. arXiv preprint arXiv:1412.6980 (2014)
14. Kong, J., Kim, J., Bae, J.: Hifi-GAN: generative adversarial networks for efficient and high fidelity speech synthesis. Adv. Neural. Inf. Process. Syst. **33**, 17022–17033 (2020)

15. Loshchilov, I., Hutter, F.: Decoupled weight decay regularization. arXiv preprint arXiv:1711.05101 (2017)
16. Nagrani, A., Chung, J.S., Zisserman, A.: VoxCeleb: a large-scale speaker identification dataset. arXiv preprint arXiv:1706.08612 (2017)
17. Nguyen, B., Cardinaux, F.: NVC-Net: End-to-end adversarial voice conversion. In: ICASSP 2022–2022 IEEE International Conference on Acoustics, Speech and Signal Processing (ICASSP), pp. 7012–7016. IEEE (2022)
18. Oord, A.v.d., et al.: WaveNet: a generative model for raw audio. arXiv preprint arXiv:1609.03499 (2016)
19. Prenger, R., Valle, R., Catanzaro, B.: WaveGlow: a flow-based generative network for speech synthesis. In: ICASSP 2019–2019 IEEE International Conference on Acoustics, Speech and Signal Processing (ICASSP), pp. 3617–3621. IEEE (2019)
20. Qian, K., Zhang, Y., Chang, S., Yang, X., Hasegawa-Johnson, M.: AutoVC: zero-shot voice style transfer with only autoencoder loss. In: International Conference on Machine Learning, pp. 5210–5219. PMLR (2019)
21. Radford, A., Narasimhan, K., Salimans, T., Sutskever, I., et al.: Improving language understanding by generative pre-training (2018)
22. Toda, T., Nakagiri, M., Shikano, K.: Statistical voice conversion techniques for body-conducted unvoiced speech enhancement. IEEE Trans. Audio Speech Lang. Process. **20**(9), 2505–2517 (2012)
23. Wan, L., Wang, Q., Papir, A., Moreno, I.L.: Generalized end-to-end loss for speaker verification. In: 2018 IEEE International Conference on Acoustics, Speech and Signal Processing (ICASSP), pp. 4879–4883. IEEE (2018)
24. Zhang, Y., Cong, J., Xue, H., Xie, L., Zhu, P., Bi, M.: Visinger: variational inference with adversarial learning for end-to-end singing voice synthesis. In: ICASSP 2022–2022 IEEE International Conference on Acoustics, Speech and Signal Processing (ICASSP), pp. 7237–7241. IEEE (2022)

Aspect-Based Sentiment Analysis Using Dual Probability Graph Convolutional Networks (DP-GCN) Integrating Multi-scale Information

Yunhui Pan, Dongyao Li, Zhouhao Dai, and Peng Cui(✉)

Key Laboratory of Blockchain and Fintech of Department of Education of Guizhou, School of Information, Guizhou University of Finance and Economics, Guiyang, China
pcui@mail.gufe.edu.cn

Abstract. Aspect-based sentiment analysis (ABSA) is a fine-grained entity-level sentiment analysis task that aims to identify the emotions associated with specific aspects or details within text. ABSA has been widely applied to various areas such as analyzing product reviews and monitoring public opinion on social media. In recent years, methods based on graph neural networks combined with syntactic information have achieved promising results in the task of ABSA. However, existing methods using syntactic dependency trees contain redundant information, and the relationships with identical weights do not reflect the importance of the aspect words and opinion words' dependencies. Moreover, ABSA is limited by issues such as short sentence length and informal expression. Therefore, this paper proposes a Double Probabilistic Graph Convolutional Network (DP-GCN) integrating multi-scale information to address the aforementioned issues. Firstly, the original dependency tree is reshaped through pruning, creating aspect-based syntactic dependency tree corresponding syntactic dependency weights. Next, two probability attention matrixes are constructed based on both semantic and syntactic information. The semantic probability attention matrix represents the weighted directed graph of semantic correlations between words. Compared with the discrete adjacency matrix directly constructed by the syntax dependency tree, the probability matrix representing the dependency relationship between words based on syntax information contains rich syntactic information. Based on this, semantic information and syntactic dependency information are separately extracted via graph convolutional networks. Interactive attention is used to guide mutual learning between semantic information and syntactic dependency information, enabling full interaction and fusion of both types of information before finally carrying out sentiment polarity classification. Our model was tested on four public datasets, Restaurant, Laptop, Twitter and MAMS. The accuracy (ACC) and F1 score improved by 0.14% to 1.26% and 0.4% to 2.19%, respectively, indicating its outstanding performance.

Keywords: Aspect-based sentiment analysis · Graph neural network · Attention mechanism · Syntactic dependency tree

B. Luo et al. (Eds.): ICONIP 2023, CCIS 1965, pp. 495–512, 2024.
https://doi.org/10.1007/978-981-99-8145-8_38

1 Introduction

Sentiment analysis is an important research direction in the field of natural language processing, aimed at identifying the emotional bias in text. Compared to article-level sentiment analysis and sentence-level sentiment analysis, ABSA is a more fine-grained entity-level sentiment analysis task that aims to analyze and differentiate the emotional polarity expressed by different aspects in the same sentence. For example, in the sentence "The material of this clothes is very good but the price is expensive", "material" and "price" are aspect words of two aspects of the clothes. However, the emotional polarity of "material" is positive, and the emotional polarity of "price" is negative.

The key to the ABSA task is to establish a dependency relationship between all aspect words and their corresponding opinion words in the sentence, distinguishing each aspect word and its associated contextual information. In earlier research, Wang [1], Tang [2], Ma [3], Chen [4], and Fan [5] proposed various attention mechanisms to generate sentence representations specific to aspect words and model the relationship between aspect words and context words, achieving good results. For example, Wang [1] proposed an attention-based long short-term memory network for ABSA tasks, where the attention mechanism can focus on different parts of the sentence when different aspects are inputted. Tang [2] proposed a neural attention model that adds external memory to deep memory networks to capture the importance of each context word for inferring the emotional polarity of aspect words. Fan [5] proposed a multi-granularity attention network model (MGAN) to capture word-level interactions between aspect words and context words. However, models based on attention mechanisms are prone to mistakenly focusing on context information unrelated to aspect words, hence the attention mechanism is easily affected by additional information. Recently, with the development of graph neural networks (GNNs), using dependency parsers to parse the syntactic structure of sentences and generate syntactic dependency trees has gradually become a trend in solving ABSA tasks. Some researchers, such as Zhang [6], Liang [7], Wang [8], Li [9], have constructed different graph convolutional networks (GCNs) and graph attention networks (GATs), using the syntactic structure of sentences on the dependency tree to model the syntactic relationship between aspect words and context words. However, existing dependency trees not only contain a lot of redundant information but also assign the same weight to the dependency relationships of each edge in the sentence, resulting in a tree structure that neglects the importance of the dependency relationship between aspect words and their corresponding opinion words. In addition, some sentences with short lengths and informal expressions can cause models to perform poorly on data that is not sensitive to syntactic information.

In this paper, we propose a dual-probability graph convolutional network (DP-GCN) that combines multi-scale information to address the above two problems. For the first problem, we first obtain the original syntactic dependency tree of the sentence through the StanfordNLP parser, then reshape and prune the original tree to construct a syntactic dependency tree with aspect words as root nodes and with attached syntactic dependency weights. The syntactic dependency tree reshaped in this way can not only clarify the syntactic dependency relationship between aspect words and their corresponding opinion words but also reveal the importance of the syntactic dependency information of individual words in the sentence with respect to aspect words. For the second problem, we extract and combine both linear structural semantic information and tree structural syntactic dependency information, respectively constructing probability attention matrices based on semantic and dependency information. We use graph convolutional networks to extract both semantic and syntactic dependency information, and then use an interactive attention mechanism to guide mutual learning between the two types of information.

The main contributions of this paper are as follows:

(1) We propose a dual-probability graph convolutional network (DP-GCN) that combines multi-scale information. We construct two probability attention matrices for semantic and syntactic dependency information, respectively, and send them into two graph convolutional networks. We utilize an interactive attention module to interactively learn semantic information and syntactic dependency information.
(2) We propose a syntactic dependency tree based on aspect words with attached dependency weights. The syntactic dependency weight reflects the importance of the syntactic dependency information of individual words in the sentence with respect to aspect words, making the syntactic dependency tree more suitable for ABSA.
(3) We conducted extensive experiments on the Restaurant dataset and Laptop dataset of SemEval2014 [24], Twitter dataset [25], and MAMS dataset [26] to evaluate our model, and the experimental results demonstrate the effectiveness of the DP-GCN model.

2 Related Work

The key to the ABSA task is to establish the relationship between aspect words and their corresponding opinion words to distinguish the emotional tendencies corresponding to different aspect words in the same sentence. In earlier methods, feature vectors were usually designed manually and combined with machine learning algorithms to capture opinion words related to aspect words [10–13]. However, this approach cannot model the dependency relationship between aspect words and their context. Subsequently, various attention-based models [14–17] emerged, which implicitly model the semantic relationship between aspect words and context words to obtain opinion words corresponding to sentences and aspect words, and achieved good performance. Huang et al. proposed an attention over-attention (AOA) network, which models both aspects and sentences jointly to capture interactions between aspects and contextual sentences. The AOA network learns representations of aspects and sentences together and automatically focuses on important parts of the sentences. Wang et al. combined a multi-level interactive bidirectional gated recurrent unit (MI-bi-GRU), attention mechanism, and position features to allow their model to focus on target and contextual words that are important for sentiment analysis. Li et al. proposed a hierarchical attention position-aware network (HAPN), which introduces positional embeddings to learn position-aware representations of sentences and further generates target-specific representations of contextual words. Tan et al. argued that expressing conflicting emotions towards an aspect (i.e., expressing both positive and negative emotions towards it simultaneously) is a common phenomenon. They suggested that excluding conflicting opinions is problematic and proposed a multi-label classification model with dual attention mechanism to address the issue of identifying conflicting opinions in existing models.

In addition, the pre-trained language model BERT [18] has achieved significant performance in natural language processing (NLP) tasks. Currently, many researchers [19–21] apply BERT pre-trained models to ABSA tasks, improving the performance of models in modeling semantic information of sentences, and better preparing for semantic interaction information between context and aspect words. For example, Sun et al. [19] construct an auxiliary sentence from the aspect and transform ABSA into a "sentence pair" classification task, and use fine-tuning BERT pre-trained models for ABSA tasks. Liang [21] proposed a bilingual syntax-aware graph attention network (BiSyn-GAT+), which fully utilizes the compositional tree information of a sentence's syntax (e.g., phrase segmentation and hierarchical structure) to simulate sentiment contexts of each aspect (intra-contexts) and cross-aspect sentiment relations (inter-contexts) for learning.

Currently, ABSA research mainly focuses on graph neural networks (GNNs) based on dependency trees. These methods explicitly utilize the syntactic structure information of sentences by extending graph convolutional network (GCN) and graph attention network (GAT) models through syntactic dependency trees, better handling the semantic and syntactic dependency relationships between aspect words and context, and proposing some outstanding models. For example, Zhang et al. [6] first applied GCN to ABSA, proposing a graph convolutional network on sentence dependency tree to solve the sentiment classification problem by utilizing dependency relationships in syntax information. Liang et al. [7] proposed an interactive graph convolutional network, identifying important aspect words and context words by constructing a heterogeneous graph for each sentence. Tang et al. [22] proposed a dependency graph enhanced dual Transformer network (DGEDT), which simultaneously considers both plane representation learned from Transformers and graph-based representation learned from corresponding dependency graph to iteratively model in an interactive manner. Specifically, DGEDT utilizes rich structural information by constructing a text sequence graph and an enhanced dependency graph, and designs a dual Transformer to model the structural information of the two graphs and learn sentence representations from two different perspectives. Wang et al. [8] created a unified aspect-oriented dependency tree structure, where the target aspect is the root node, by adjusting and refining a regular dependency parse tree. They proposed a relation graph attention network (R-GAT) to encode the new tree structure for sentiment prediction. Tian et al. [23] explicitly employed dependency types and used an attention mechanism to identify different types of dependencies. Li et al. [9] proposed a dual graph convolutional network model that simultaneously considered the complementarity of syntactic structures and the relationship of semantics.

3 Reshaped Syntactic Dependency Trees and Multi-scale Information

3.1 Aspect-Based Syntactic Dependency Tree Corresponding Syntactic Dependency Weights

The syntactic dependency tree obtained by a regular syntactic parser contains the dependency relationships of all the words in the sentence, and all dependency relationships have the same weight. As shown in Fig. 1, where there are many redundant dependency relationship types that are irrelevant to the ABSA task. However, the key to ABSA is to establish the relationship between aspect words and their opinion words. Therefore, reshaping and pruning the obtained syntactic dependency tree is necessary to obtain a syntactic dependency tree that is tailored to aspect words.

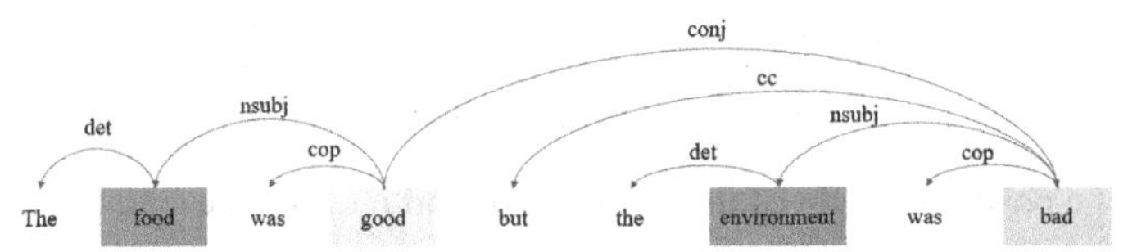

Fig. 1. Syntactic dependency tree including two aspect items, "food" and "environment", and two corresponding opinion words, "good" and "bad," in their context. The arrows in the figure indicate the dependency relationships between the two words, and the labels on the arrows represent the type of dependency relationship.

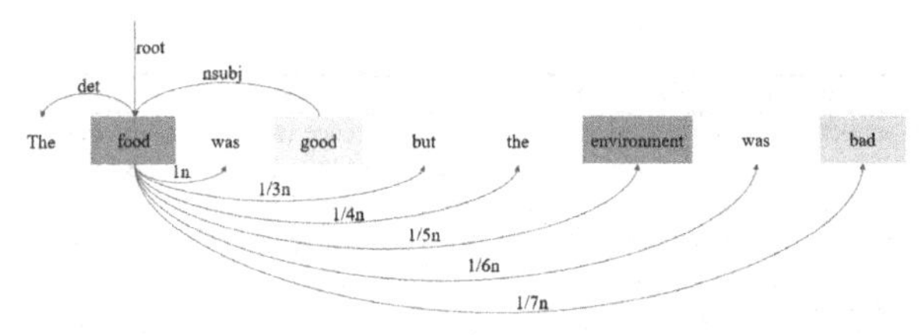

Fig. 2. In the figure, "food" is the root node, and all other dependency relationships are direct connections with "food".

Here are the steps to reshape and prune a regular syntactic dependency tree into a syntactic dependency tree based on aspect words and corresponding syntactic dependency weights: Firstly, we use a regular parser (StanfordNLP) to obtain the dependency tree of the input sentence. Then, we set the aspect word as the root node and generate a dependency tree based on the aspect word. If a sentence contains multiple aspects, then a tree based on each aspect will be constructed. Finally, the dependency tree is pruned so that words directly dependent on the aspect word have a dependency weight of 1 on their edge, the dependency weight of a word that does not have a direct dependency relationship with an aspect word is set to the reciprocal of its relative position to the aspect word. Figure 2 shows the aspect-based syntactic dependency tree obtained after reshaping and pruning.

3.2 Multi-scale Information

To address the lack of contextual information, this paper simultaneously uses linear-structured semantic information and tree-structured syntactic dependency information to reveal hidden information in the sentence.

Positional Distance. In linear-structured sentences, the position and relative distance of each word in the sentence hold important information. By extracting the relative positional distances between each word and aspect words in the sentence, we can emphasize information from words closer to the aspect words and weaken information from words farther away from the aspect words. We can then use the positional distance to calculate the weights of each word in the sentence based on the aspect word. The calculation formula is as follows:

$$p_i = \begin{cases} 1 - (j_s - i)/n, & 0 \le i \le j_s \\ 0, & j_s \le i \le j_{s+m} \\ 1 - (i - j_{s+m})/n, & j_{s+m} \le i \le n \end{cases} \quad (1)$$

Here, p_i is the position weight of the i-th word, j_s and j_{s+m} are the start and end indexes of the aspect word.

Dependency Distance. In the syntactic dependency information of a tree structure, dependency distance is the shortest distance between a word in a sentence and the aspect word in the syntactic dependency tree. Based on the dependency tree we constructed, the formula for constructing the dependency distance is shown below.

Algorithm 1 Dependency distance algorithm based on aspect-based syntactic dependency tree

Input: index of aspect words (aspect_idx), length of this sentence (n_words), adjacency matrix (adj);
Output: distances: dependent distance sequence of each word based on the aspect word;
1: Use StanfordNLP for syntax analysis to get dependency tree and POS tags;
2: Reshape the dependency tree and transform it into a syntactic dependency tree based on aspect words.
3: distances = create an array of size n_words, the initial value is -1
4: Set distances[aspect_idx] to 0
5: Create an empty queue
6: Add aspect_idx to queue
7: **When** the queue is not empty, execute the following steps in a loop:
8: Take a node from the left side of the queue
9: Traversing all neighbor nodes and corresponding weights of nodes in the adjacency matrix
10: **If** weight ! = 0 and distances[neighbor]= -1
11: distances[neighbor] = distances[node] + weight
12: Add the neighbor to the queue
13: **Return** distances

Dependency Relationship. Dependency relationships can represent the syntactic relationships between words in the sentence's tree structure. If a word has a dependency relationship with the aspect word, then the corresponding edge in A_{rel} is set to the weight of the dependency for that word. If there is no dependency relationship, then the edge is set to 0. Thus, A_{rel} is constructed for the sentence, as shown in Fig. 3.

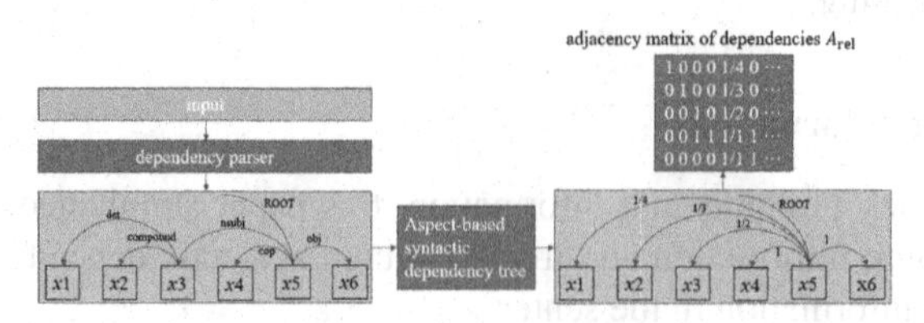

Fig. 3. Adjacency matrix $\boldsymbol{A}_{\mathbf{rel}}$ of the dependency relationship.

Dependency Type. The type of dependency relationship is a special and important piece of information. This paper first counts all dependency types in the dataset and generates a dependency types dictionary. Then, a randomly initialized vector for the initial dependency type corresponding to the text sequence S is generated, and a BiLSTM is used to obtain a feature vector $h_{\text{type}} \in \mathbb{R}^{n \times d}$, where n represents the length of the dependency type dictionary and d is the word vector dimension of the dependency type. Dependency types are embedded, as shown in Fig. 4.

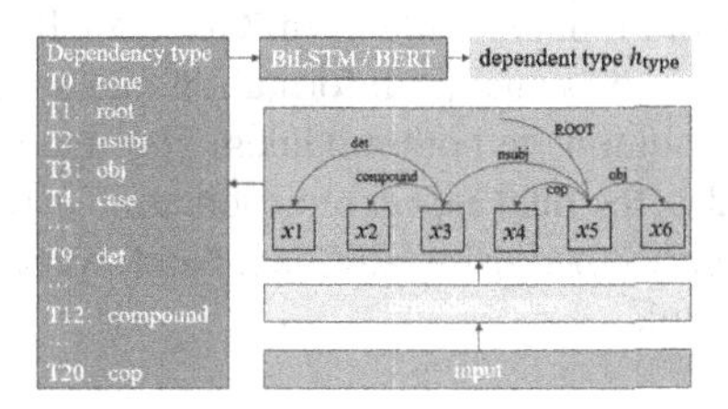

Fig. 4. Dependency type dictionary and dependency relationship types of the sentence.

4 Proposed DP-GCN Model

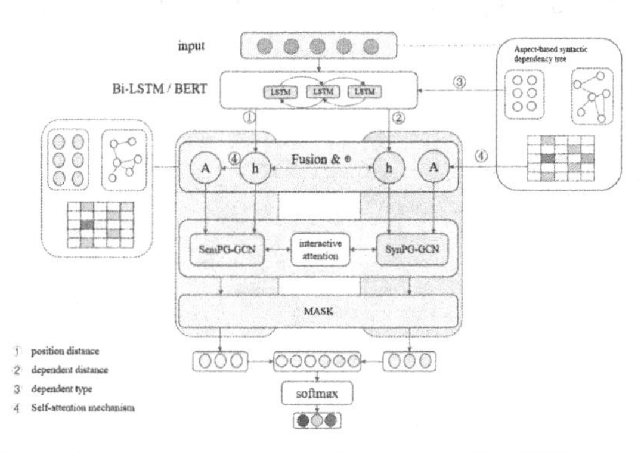

Fig. 5. Double probabilistic graph convolutional network (DP-GCN) integrating multi-scale information

Our proposed DP-GCN model is shown in Fig. 5. In the ABSA task, given a sentence $W^c = \{w_1^c, w_2^c, \cdots, w_{r+1}^a, \cdots, w_{r+m,}^a \cdots, w_n^c\}$ containing n words, where $W^a = \{w_{r+1}^a, \cdots, w_{r+m}^a\}$ is the aspect word sequence. Firstly, the words in the sentence are embedded into a low-dimensional vector space using an embedding matrix $E \in \mathbb{R}^{|v| \times d_e}$, where $|v|$ is the vocabulary size and d represents the dimension of the word embedding. We use the StanfordNLP syntactic parser to parse the sentence and obtain its syntactic dependency information. Next, the obtained dependency type information is embedded into the low-dimensional vector space $E \in \mathbb{R}^{|v| \times d_e}$, where $|v|$ is the size of the dependency type vocabulary and d is the dimension of the dependency type word embedding. Then, BiLSTM or BERT is used as the sentence encoder to extract the hidden contextual semantic representation h_{sem} and the dependency type representation h_{type}.

The hidden contextual semantic representation h_{sem} and the dependency type representation h_{type} of the sentence are fused with the multi-scale information. The fused representation h_{input} with multi-scale information is obtained by interacting the information through an interactive attention mechanism. Then, h_{input} is separately fed into the semantic probability graph convolutional module (SemPG-GCN) and the syntactic probability graph convolutional module (SynPG-GCN). Interacting attention is used to guide the communication of semantic information and syntactic dependency information during graph convolutions in both modules. Through masking, connection, and aggregation of aspect nodes, the final aspect representation is formed. Finally, sentiment polarity classification is performed using softmax. Next, we will describe the details of the DP-GCN model in detail.

4.1 Interactive Attention

The implementation of the interactive attention layer is mainly based on self-attention mechanism, which enables the model to simultaneously calculate the attention of contextual semantic features and dependency type features. Through the interactive attention mechanism, the dependency type features guide the learning of contextual features, while the contextual features guide the learning of dependency type features, as shown in Fig. 6.

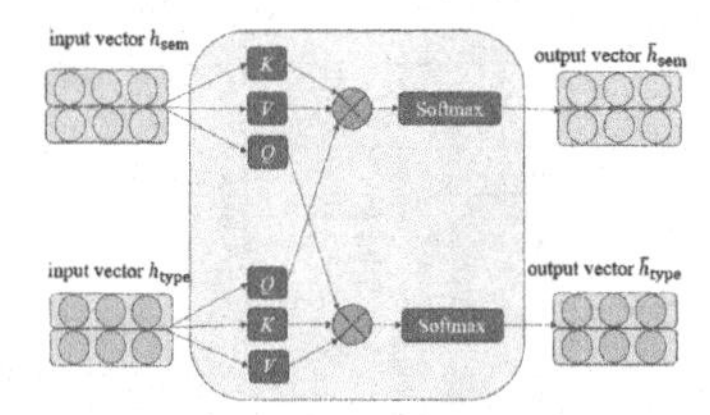

Fig. 6. Structure diagram of Interactive Attention.

4.2 Fusion of Multi-scale Information

This paper utilizes and integrates the multi-scale semantic information and multi-scale syntactic dependency information mentioned above as inputs to the model.

Fusing Contextual Semantic Information. The positional distance is incorporated into the contextual representation as a weight parameter of the linear structure. This context semantic information fused with the position distance can reflect the semantic association between different words and the aspect word in terms of distance. The fusion formula is as follows:

$$h_{\text{sem}} = \mathrm{F}(h_{\text{sem}}) = p_i \cdot h_{\text{sem}} \tag{2}$$

Here, F is the positional weight function, and p_i is the positional weight of the i-th word. Thus, the closer the distance between words and the aspect word, the greater their relevance in the sentence, and the more significant their contribution to the judgment of sentiment polarity, since they have a higher weight value.

Fusing Syntactic Dependency Information. Integrating dependency type and dependency distance information. The dependency distance reflects the importance of the syntactic dependency between each word in the sentence and the aspect word, which strengthens the words that have a direct syntactic dependency relationship with the aspect word and weakens those that do not have a direct relationship with the aspect word. The formula for fusing dependency type h_{type} with dependency distance is as follows:

$$h_{\text{type}} = \mathrm{F}\left(h_{\text{type}}\right) = T * h_{\text{type}} \tag{3}$$

Here, F is the function for element-wise matrix multiplication, T is the dependency weight matrix composed of all dependency distances t_i. The multiplication is performed between the dependency type hidden vector and the corresponding element in the dependency weight matrix.

Fusing Semantic Information and Syntactic Dependency Information. In this paper, interactive attention is used to fuse semantic information and syntactic dependency information, and the result of the fused information is used as the input to the model. Figure 6 shows the process of the interactive attention between the multi-scale semantic information h_{sem} and the multi-scale syntactic dependency information h_{type} to guide each other's learning.

Here, the multi-scale semantic information h_{sem} and the multi-scale syntactic dependency information h_{type} are used as inputs to the interactive attention. According to the Transformer model, h_{sem} and h_{type} are mapped to query (Q_{sem} and Q_{type}), key (K_{sem} and K_{type}), and value (V_{sem} and V_{type}) matrices through linear layers. The formula for calculating h_{sem} using h_{type} is as follows:

$$C_{\text{sem}} = \mathrm{softmax}\left(\frac{Q_{\text{type}} K_{\text{sem}}^{\mathrm{T}}}{\sqrt{d}}\right) V_{\text{sem}} \tag{4}$$

$$\overline{h}_{\text{sem}} = \mathrm{LN}(h_{\text{sem}} + C_{\text{sem}}) \tag{5}$$

Here, LN is a standardization function. Similarly, h_{sem} is used to guide h_{type}, as given in the following equation:

$$C_{\text{type}} = \mathrm{softmax}\left(\frac{Q_{\text{sem}} K_{\text{type}}^{\mathrm{T}}}{\sqrt{d}}\right) V_{\text{type}} \tag{6}$$

$$\overline{h}_{\text{type}} = \mathrm{LN}\left(h_{\text{type}} + C_{\text{type}}\right) \tag{7}$$

Here, $\overline{h}_{\text{sem}} \in \mathbb{R}^{n \times d}$ and $\overline{h}_{\text{type}} \in \mathbb{R}^{n \times d}$ are both outputs of the interactive attention, and they use each other's feature information to enhance their own hidden representation abilities. Finally, the concatenated representation of the interactive semantic and syntactic dependency information is used as the input to the model, as shown in the following equation:

$$h_{\text{input}} = \overline{h}_{\text{sem}} \oplus \overline{h}_{\text{type}} \tag{8}$$

4.3 Semantic Probabilistic Graph Convolution Module

Semantic Probabilistic Graph Convolution (SemPG-GCN). In order to fully focus the DP-GCN model on the aspect words and the corresponding opinion words, we use the self-attention mechanism to construct a probabilistic attention matrix A_{sem} about the multi-scale contextual semantic hidden representation h_{sem}, which is used as the input to the graph convolution. The specific formula is as follows:

$$A_{\text{sem}} = \text{softmax}\left(\frac{QW^{Q} \times \left(KW^{K}\right)^{\text{T}}}{\sqrt{d}}\right) \tag{9}$$

Here, Q and K are both the multi-scale contextual semantic hidden representation h_{sem}, while W^{Q} and W^{K} are learnable weight matrices, and d is the dimension of the multi-scale contextual semantic hidden representation h_{sem}.

Then, learn semantic information through graph convolutional networks. The specific formula of graph convolution is as follows:

$$h_i^l = \sigma(\textstyle\sum_{j=1}^{n} A_{ij} W^l h_j^{l-1} + b^l) \tag{10}$$

where h_i^l represents the hidden representation of node i in layer l, the initial value of the first layer is h_{input}. A_{ij} represents the element value in the i-th row and j-th column of matrix A_{sem}. W^l is a learnable parameter matrix, h_j^{l-1} is the hidden representation of neighboring nodes of h^l in layer $l-1$, and b^l is the bias term of the graph convolution.

4.4 Syntactic Probabilistic Graph Convolutional Module

Syntactic Probabilistic Graph Convolutional Networks (SynPG-GCN). Nodes that have no dependency relationship with the aspect word are assigned 0, resulting in many zero elements in the generated adjacency matrix, which leads to the problem of missing information. The self-attention mechanism is applied to the matrix to obtain a continuous 0–1 probability matrix, which makes the model more robust and advanced.

$$A_{\text{rela}} = \text{softmax}\left(A_{\text{rel}} * W * U^{\text{T}}\right) \tag{11}$$

where W and U are learnable weight matrices, A_{rela} is the probabilistic attention matrix of syntactic information, and A_{rel} is the adjacency matrix of the dependency relationship.

Then, learn syntactic dependency information through graph convolutional networks. The specific formula of graph convolution is as follows:

$$h_i^l = \sigma(\textstyle\sum_{j=1}^{n} A_{ij} W^l h_j^{l-1} + b^l) \tag{12}$$

where h_i^l represents the hidden representation of node i in layer l, the initial value of the first layer is h_{input}, A_{ij} represents the element value in the i-th row and j-th column of matrix A_{rela}, W^l is a learnable parameter matrix, h_j^{l-1} is the hidden representation of neighboring nodes of h^l in layer $l-1$, and b^l is the bias term of the graph convolution.

4.5 Sentiment Classification

h_{input} Obtains h_{semPG} and h_{synPG} through SemPG-GCN and SynPG-GCN. Then, it multiplies with the aspect word masking matrix to extract the corresponding parts of the aspect word. The mask operation obtains $\overline{h}_{\text{semPG}}$ and $\overline{h}_{\text{synPG}}$. They are concatenated and sent to the softmax layer to calculate the probability distribution of the input text in positive, negative, and neutral sentiment. The specific operation is as follows:

$$M_{i,j} = \begin{cases} 1, & i = j = p \\ 0, & \text{otherwise} \end{cases} \tag{13}$$

where $M_{i,j}$ represents the element value in the i-th row and j-th column of the mask matrix. If $i = j = p$, which means the current position is the corresponding position of the aspect word, then the corresponding element value is set to 1. Otherwise, the corresponding element value is set to 0.

$$\overline{h}_{\text{semPG}} = h_{\text{semPG}} * M \tag{14}$$

$$\overline{h}_{\text{synPG}} = h_{\text{synPG}} * M \tag{15}$$

$$h_{\text{out}} = \left[\overline{h}_{\text{semPG}}; \overline{h}_{\text{synPG}}\right] \tag{16}$$

The probability of h_{out} after softmax is:

$$P(a) = \text{softmax}(Wh_{\text{out}} + b) \tag{17}$$

where W and b are both learnable parameters, $p(a)$ is the emotion probability distribution of the aspect word. In the model training process, cross-entropy is used as the loss function, and its formula is:

$$J = -\frac{1}{N}\sum\nolimits_{i=1}^{N}\sum\nolimits_{k=1}^{K} y_{i,k}\log\left(\hat{y}_{i,k}\right) \tag{18}$$

where N denotes the number of samples, K denotes the number of classes, $y_{i,k}$ is the true label of sample i belonging to class k, and $\hat{y}_{i,k}$ is the predicted probability of the model that the sample i belongs to class k.

5 Experiments

5.1 Dataset and Evaluation Criteria

This paper verifies the effectiveness of the DP-GCN model by conducting experiments on four publicly available datasets, which are Laptop and Restaurant datasets from SemEval2014 [24], Twitter dataset [25], and MAMS dataset [26]. Each sample in these four datasets is annotated with a sentiment label of one or more aspect words in a sentence, and the sentiment labels have three classifications: Positive, Negative, and Neutral. The statistical data for the number of samples in each category of the dataset is shown in Table 1.

Table 1. Experiment Data Statistics

Dataset	Positive		Neutral		Negative	
	Train	Test	Train	Test	Train	Test
Laptop	994	341	464	169	870	128
Restaurant	2164	728	637	196	807	182
Twitter	1561	173	3127	346	1560	173
MAMS	3380	400	5042	607	2764	329

This experiment uses two evaluation metrics, accuracy (Acc) and macro-average F1 score (MF1), to evaluate the effectiveness of the DP-GCN model.

5.2 Parameter Setting

In this experiment, the experimental parameters of Glove and Bert are set as follows for the four datasets. The specific experimental parameters are shown in Table 2.

Table 2. Experimental Hyperparameter Settings

Experimental parameters	Set value
Num-epoch	50
Batch-size	16
Number of GCN layers	2
Number of LSTM layers	1
Number of interaction attention layers	2
Dependency type embedding dimension	40
BiLSTM hidden layer dimension	50
GCN hidden layer dimension	50
Max-length	85
L2 regularization	10^{-5}
Adam learning rate	0.002
Input/BiLSTM/GCN dropout	0.7/0.1/0.1
Early-stopping	500

5.3 Baseline Methods

To comprehensively evaluate the performance of our model (DP-GCN), we compared it with the following baseline models on the four datasets:

1. ATAE-LSTM: Weighted the output of LSTM based on the attention mechanism to extract emotional words and features.
2. IAN: Simultaneously considered information at both the word and sentence levels in the text, and calculated the text representation using an interactive way so that the model can better capture the relationship between words and sentences.
3. RAM: When calculating the sentiment polarity of each aspect, not only the information of that aspect is considered, but also the information of other aspects is taken into account. The memory vector of each aspect is matched with the current input word vector sequence to obtain the attention vector of that aspect.
4. CDT: Used convolution on the dependency tree model to learn sentence features representation.
5. R-GAT: Used bidirectional GAT as the basic model, and employed relation-aware graph attention mechanism to capture the relationship between words and better capture information in the text sequence.
6. DGEDT: This model is based on a dual-channel LSTM, combined with a dynamic graph augmentation mechanism, which enables the utilization of both sentiment-embedded information and semantic information present in the text.
7. DualGCN: Built two graph convolution modules to process semantic information and syntactic dependency information.
8. T-GCN+BERT: Proposed a method that utilized a type-aware graph convolutional network (T-GCN) to explicitly depend on the ABSA type. Attention was used in T-GCN to distinguish different edges in the graph.
9. R-GAT+BERT: Used the pre-trained model BERT as the encoder instead of BiLSTM.

5.4 Experimental Results and Analysis

We conducted a three-class ABSA experiment on the four datasets from Sect. 4.1. The experimental results are shown in Table 4. The results in Table 4 indicate that our model (DP-GCN) has achieved a certain degree of improvement in both ACC and F1-score on the four public datasets.

From the experimental results of our model and the baseline model, it can be found that the performance of the DP-GCN model is better than models that solely use attention mechanism to capture aspect words and contextual words for modeling, such as ATAE-LSTM, IAN, etc. This suggests that the attention mechanism may only consider the semantic information of the sentence and cannot effectively capture the syntactic dependency information corresponding to the opinion words related to the aspect words. When dealing with longer sentences where aspects words and opinion words have distant dependencies, it is difficult to effectively identify the relationship between them. Models that consider the multiple aspect features of a sentence, such as RAM and CDT, introduce additional syntactic dependency information on the basis of the attention mechanism. However, the attention mechanism is easily affected by additional noise, making it difficult for the model to handle both semantic information and syntactic dependency information effectively. Models that use graph neural networks (such as R-GAT, DGEDT, DualGCN) can capture words with long-distance dependencies in the context, which can better establish the relationship between aspect words and their opinion words. However, when dealing with informal datasets such as Twitter, these models have some limitations and do not consider the role of semantic information and syntactic dependency information in identifying relationships.

The DP-GCN model achieved good results in terms of ACC and F1-score on the four public datasets, indicating that the fusion of multi-scale information in the model input has integrated more semantic and syntactic dependency information of the sentences. The probability graph convolution module combined with an interactive attention mechanism enables the model to fully consider the semantic and syntactic information of the sentences. The enhancement in the model's performance indicates that to some extent, the syntactic dependency tree constructed by our model can mitigate the issue of the attention mechanism being susceptible to disruptions from noise.

In addition, the overall performance of the DP-GCN+BERT model proposed in this paper is also better than R-GAT+BERT and T-GCN+BERT, further demonstrating that the probability attention matrix with weighted syntactic dependency tree, semantic information, and syntactic dependency information has good effects on downstream tasks. Compared with the Glove-based DP-GCN model, DP-GCN+BERT improved the ACC by 1.27%–2.85% and F1-score by −0.46%–2.61%, and achieved better results than the non-BERT models in Table 3.

Table 3. Experimental results of different models on four public datasets

Models	Restaurant		Laptop		Twitter		MAMS	
	Acc	F1	Acc	F1	Acc	F1	Acc	F1
ATAE-LSTM	77.20	–	68.70	–	–	–	–	–
IAN	78.60	–	72.10	–	–	–	–	–
RAM	80.23	70.80	74.49	71.35	69.36	67.30	–	–
CDT	82.30	74.02	77.19	72.99	74.66	73.66	–	–
R-GAT	83.30	76.08	77.42	73.76	75.57	73.82		
DGEDT	83.90	75.10	76.80	72.30	74.80	73.40	–	–
DualGCN	84.27	78.08	78.48	74.74	75.92	74.29		
Our DP-GCN	84.76	78.48	79.74	76.20	76.06	76.48	81.96	81.15
R-GAT+BERT	86.60	81.35	78.21	74.07	76.15	74.88	–	–
T-GCN+BERT	86.16	79.95	80.88	77.03	76.45	75.25	83.68	83.07
Our DP-GCN+BERT	87.31	81.09	81.01	77.96	76.80	76.02	84.85	83.49

5.5 Ablation Experiment

In order to further study the role of a certain module in DP-GCN model, we conducted extensive ablation experiments. The results are shown in Table 4, and the specific experiments are as follows:

(1) **w/o location distance.** Remove the location distance information of the model, that is, reduce the dependency degree of the position distance in the semantic information. As shown in Table 4, on the Restaurant, Laptop and MAMS datasets, the ACC and F1-score have decreased to some extent after removing the position information, while on the Twitter dataset, there is little change in ACC and F1-score. This suggests that the position information of the words in the sentence has little effect on the model's performance in datasets containing a large number of informal expressions.
(2) **w/o dependent type.** Remove the dependence type information, and the input of the model have only semantic information without the syntactic dependency information. The ACC and F1-score on all four datasets have decreased after removing the dependent type information, indicating that the dependency type information in the sentences can supplement the semantic information to some extent, allowing the model to learn more effective information.
(3) **w/o dependent tree.** Remove the tree based on the aspect word corresponding syntactic dependency weight, use StanfordNLP to generate the syntactic dependency tree, and also remove the dependency distance but retain the dependency type. The ACC and F1-score have shown a significant decrease on all four datasets after removing the dependent tree, indicating that reshaping the syntactic dependency tree is effective for ABSA tasks, and also suggesting that the original syntactic dependency tree contains redundant information.
(4) **w/o SemPG-GCN.** Remove the semantic information graph convolution module, and the ACC and F1-score have significantly decreased on all four datasets, indicating that the graph convolution module of the semantic information is the core module of this model, and suggesting that semantic information is essential for ABSA tasks.
(5) **w/o SynPG-GCN.** Remove the syntactic information graph convolution module, and the ACC and F1-score have decreased on all four datasets. From the experimental results, it can be seen that the syntactic information graph convolution module can complement the semantic information graph convolution module to some extent, and jointly improve the performance of the model.

In summary, deleting distance information and dependency distance information will decrease the accuracy of our DP-GCN model, which illustrates the importance of the semantic information of the hidden linear structure and the syntactic information of the tree structure for the input information of the model. It can solve the problem of short sentences and informal expressions to some extent. Deleting the probability attention matrix constructed by the self-attention mechanism of the SynPG-GCN module also leads to a decrease in accuracy, indicating that constructing a probability matrix about syntactic information through attention mechanisms can alleviate the influence of dependency parsing errors. Compared to comments from Restaurant and Laptop datasets, comments from Twitter are largely informal and insensitive to grammar information. Finally, the dependency tree and probability graph convolutional network that are based

Table 4. Experimental results of ablation experiments

Models	Restaurant		Laptop		Twitter		MAMS	
	Acc	F1	Acc	F1	Acc	F1	Acc	F1
w/o location distance	83.41	76.25	78.63	75.21	75.95	76.31	81.06	80.26
w/o dependent type	82.53	73.14	76.48	73.21	74.63	73.69	80.22	80.32
w/o dependent tree	82.22	74.29	76.30	72.99	73.11	71.52	80.84	79.87
w/o SemPG-GCN	81.59	73.75	75.79	72.77	74.35	74.28	80.26	80.73
w/o SynPG-GCN	83.57	73.46	76.30	72.68	75.22	75.12	81.34	80.64

on aspect words and weighted dependencies are better suited for the MAMS dataset with multiple aspect words, as the relationship modeling between aspect words and corresponding opinion words becomes increasingly reliant on syntactic information as sentence complexity rises.

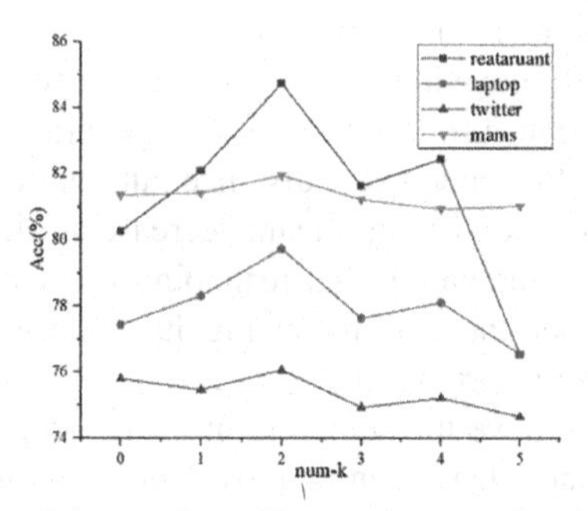

Fig. 7. The impact of the number of interactive attention layers on the model

Interactive attention is a critical module for the exchange of semantic information and syntactic dependency information. To explore the impact of the number of interactive attention layers on model performance, we investigated the number of interactive attention layers by setting the number of layers num-k = {0, 1, 2, 3, 4, 5}, respectively, and obtained the accuracy (ACC) of the four datasets, as shown in Fig. 7.

As shown in Fig. 7, the impact of the number of interactive attention layers on the model is nonlinear, and too few or too many layers can affect the performance of the model. In this experiment, when the number of interactive attention layers was 2, the highest accuracy was achieved in all four datasets. This may be because the interactive attention introduces different levels of interaction information while maintaining the consistency of the input feature space, which has a positive effect on improving the model performance. However, too many layers of interaction may introduce too much noise, leading to a decrease in model performance. Therefore, to obtain better performance in practical applications, it is necessary to adjust the number of interactive attention layers according to the specific dataset and task.

6 Conclusion

In this paper, we aimed to address the issue of redundant information in the current syntactic dependency trees for ABSA tasks. We proposed a tree structure based on aspect words corresponding syntactic dependency weights to systematically process ABSA tasks. We also proposed a dual probability graph convolutional network (DP-GCN) that combines multiscale information, which constructs two probability attention matrices to accommodate unclear or insignificant syntax and context semantic information. We used the interactive attention mechanism to guide the mutual learning of semantic and syntactic dependency information, thereby enhancing model expressiveness. Experimental results on datasets indicate that our DP-GCN model outperforms baseline models. However, our model still has limitations when processing datasets with many informal and biased expressions, such as the Twitter dataset. In future work, we will consider extracting other useful information related to semantic and syntactic information and optimizing the fusion of these two types of information. Additionally, we will improve the graph convolutional network model to enhance its generalization performance for ABSA tasks.

Acknowledgment. This research was supported by the Top Talent Project of Guizhou Provincial Department of Education (QJJ-[2022]-080) and the Research Project for Students at Guizhou University of Finance and Economics (2022ZXSY163).

References

1. Wang, Y., Huang, M., Zhu, X., Zhao, L.: Attention-based LSTM for aspect-level sentiment classification. In: Conference on Empirical Methods in Natural Language Processing (2016)
2. Tang, D., Qin, B., Liu, T.: Aspect level sentiment classification with deep memory network. arXiv preprint arXiv:1605.08900 (2016)
3. Ma, D., Li, S., Zhang, X., Wang, H.: Interactive attention networks for aspect-level sentiment classification. arXiv preprint arXiv:1709.00893 (2017)
4. Chen, P., Sun, Z., Bing, L., Yang, W.: Recurrent attention network on memory for aspect sentiment analysis. In: Conference on Empirical Methods in Natural Language Processing (2017)
5. Fan, F., Feng, Y., Zhao, D.: Multi-grained attention network for aspect-level sentiment classification. In: Conference on Empirical Methods in Natural Language Processing (2018)
6. Zhang, C., Li, Q., Song, D.: Aspect-based sentiment classification with aspect-specific graph convolutional networks. arXiv preprint arXiv:1909.03477 (2019)
7. Liang, B., Yin, R., Gui, L., Du, J., Xu, R.: Jointly learning aspect-focused and inter-aspect relations with graph convolutional networks for aspect sentiment analysis. In: International Conference on Computational Linguistics (2020)
8. Wang, K., Shen, W., Yang, Y., Quan, X., Wang, R.: Relational graph attention network for aspect-based sentiment analysis. In: Annual Meeting of the Association for Computational Linguistics (2020)
9. Li, R., Chen, H., Feng, F., Ma, Z., Wang, X., Hovy, E.H.: Dual graph convolutional networks for aspect-based sentiment analysis. In: Annual Meeting of the Association for Computational Linguistics (2021)

10. Titov, I., McDonald, R.T.: Modeling online reviews with multi-grain topic models. arXiv preprint arXiv:0801.1063 (2008)
11. Ding, X., Zhang, Y., Liu, T., Duan, J.: Deep learning for event-driven stock prediction. In: International Joint Conference on Artificial Intelligence (2015)
12. Kiritchenko, S., Zhu, X.-D., Cherry, C., Mohammad, S.M.: NRC-Canada-2014: detecting aspects and sentiment in customer reviews. In: International Workshop on Semantic Evaluation (2014)
13. Huang, B., Ou, Y., Carley, K.M.: Aspect level sentiment classification with attention-over-attention neural networks. arXiv preprint arXiv:1804.06536 (2018)
14. Wang, X., Chen, X., Tang, M., Yang, T., Wang, Z.: Aspect-level sentiment analysis based on position features using multilevel interactive bidirectional GRU and attention mechanism. Discrete Dyn. Nat. Soc. **2020**, 1–13 (2020)
15. Li, L., Liu, Y., Zhou, A.: Hierarchical attention based position-aware network for aspect-level sentiment analysis. In: Conference on Computational Natural Language Learning (2018)
16. Tan, X., Cai, Y., Zhu, C.: Recognizing conflict opinions in aspect-level sentiment classification with dual attention networks. In: Conference on Empirical Methods in Natural Language Processing (2019)
17. Devlin, J., Chang, M.-W., Lee, K., Toutanova, K.: BERT: pre-training of deep bidirectional transformers for language understanding. arXiv preprint arXiv:1810.04805 (2019)
18. Sun, C., Huang, L., Qiu, X.: Utilizing BERT for aspect-based sentiment analysis via constructing auxiliary sentence. In: North American Chapter of the Association for Computational Linguistics (2019)
19. Xu, H., Liu, B., Shu, L., Yu, P.S.: BERT post-training for review reading comprehension and aspect-based sentiment analysis. arXiv preprint arXiv:1904.02232 (2019)
20. Liang, S., Wei, W., Mao, X.-L., Wang, F., He, Z.: BiSyn-GAT+: Bi-syntax aware graph attention network for aspect-based sentiment analysis. In: Findings (2022)
21. Tang, H., Ji, D.-H., Li, C., Zhou, Q.: Dependency graph enhanced dual-transformer structure for aspect-based sentiment classification. In: Annual Meeting of the Association for Computational Linguistics (2020)
22. Tian, Y., Chen, G., Song, Y.: Aspect-based sentiment analysis with type-aware graph convolutional networks and layer ensemble. In: North American Chapter of the Association for Computational Linguistics (2021)
23. Pontiki, M., Galanis, D., Pavlopoulos, J., Papageorgiou, H., Androutsopoulos, I., Manandhar, S.: SemEval-2014 task 4: aspect based sentiment analysis. In: International Workshop on Semantic Evaluation (2014)
24. Dong, L., Wei, F., Tan, C., Tang, D., Zhou, M., Xu, K.: adaptive recursive neural network for target-dependent Twitter sentiment classification. In: Annual Meeting of the Association for Computational Linguistics (2014)
25. Jiang, Q., Chen, L.-T., Xu, R., Ao, X., Yang, M.: A challenge dataset and effective models for aspect-based sentiment analysis. In: Conference on Empirical Methods in Natural Language Processing (2019)
26. Sun, K., Zhang, R., Mensah, S., Mao, Y., Liu, X.: Aspect-level sentiment analysis via convolution over dependency tree. In: Conference on Empirical Methods in Natural Language Processing (2019)

Privacy-Preserving Image Classification and Retrieval Scheme over Encrypted Images

Yingzhu Wang, Xiehua Li(✉), Wanting Lei, and Sijie Li

College of Computer Science and Electronic Engineering, Hunan University, Changsha, Hunan, China
beverly@hnu.edu.cn

Abstract. Image retrieval is a crucial function in several emerging computer vision applications, including online medical diagnosis and image recognition systems. With an increasing number of images being generated and outsourced to public clouds, there is a growing concern about privacy leaks of image contents. To address this issue, we propose an efficient privacy-preserving image classification and retrieval scheme (PICR) that employs low-dimensional vectors to represent image categories and feature vectors, thereby improving retrieval efficiency and reducing index storage costs. First, we design a feature extraction model based on convolutional neural network (CNN) to generate segmented hash codes that represent both image categories and features. Next, the cryptographic hierarchical index structure based on the category hash code is designed to improve retrieval accuracy and efficiency. Then, we employ random vectors and the Learning With Errors (LWE)-based secure k-Nearest Neighbour (kNN) algorithm to preserve the privacy of segmented hash codes and file-access patterns. Finally, we provide the security analysis that verifies our PICR scheme can protect image privacy as well as indexing and query privacy. Experimental evaluation demonstrates that our proposed scheme outperforms the existing state-of-the-art schemes in terms of retrieval accuracy, search efficiency and storage costs.

Keywords: Image privacy · encrypted image retrieval · convolutional neural network · learning with errors

The popularity of mobile devices and various applications makes the multimedia data such as personal pictures and videos grow exponentially. Individuals and business users are more inclined to outsource these multimedia data to the cloud to reduce the pressure on local storage and improve data reliability. However, storing plaintext images on an untrusted third party may cause privacy leakage and make them vulnerable to various attacks [1]. The recent released

---

This work was supported by the National Natural Science Foundation of China under Project 92067104, 61872134, and is also supported by the Hunan Provincial Natural Science Foundation under Project 2021JJ30140.

B. Luo et al. (Eds.): ICONIP 2023, CCIS 1965, pp. 513–525, 2024.
https://doi.org/10.1007/978-981-99-8145-8_39

IBM *Cost of a Data Breach Report* shows that the average global cost of data breach reached USD 4.35 million in 2022, which is a 12.7% increase from the last year, and the highest ever noted across the history of IBM reports [2]. Encryption helps ensure the confidentiality of outsourced data but makes subsequent processing difficult. To solve this problem, some privacy-preserving content-based image retrieval(CBIR) schemes have been proposed, which can protect the image content and meanwhile support the searching of similar images. The existing privacy-preserving CBIR schemes mainly aim to solve the retrieval precision, efficiency or security issues. For example, visual word frequencies are used as the image features and then be encrypted to enable retrieval over encrypted images [3]. Local sensitive hash is employed to improve the retrieval accuracy of encrypted images [4]. Homomorphic encryption is applied to encrypt image features to support similarity calculation [5,6]. With the rapid development of deep learning, Convolutional Neural Network (CNN) models have emerged as a prevalent tool for image feature extraction [7] [8]. The pre-trained CNN models facilitates more precise depiction of image content, consequently enhancing search accuracy [9] [10]. Although these schemes can perform similarity searches on encrypted images, the search accuracy and efficiency are not satisfactory due to the computational complexity and the adopted high-dimensional feature vectors. In this paper, we propose an encrypted image retrieval scheme with segmented hash features, which allows users to retrieve images quickly and accurately, while having less index construction time and low storage costs. Our contributions are set out as follows:

- A CNN-based feature extraction and image classification model is designed to generate a low-dimensional two-level segmented hash code which can present the image category and feature simultaneously. The fine-designed segmented hash code can provide a more accurate representation of image with less feature vectors, thereby guaranteeing the accuracy of image searching and reducing feature vectors storage overhead.
- A hierarchical index structure is build based on the segmented hash codes. The top level of the index is generated by category hash code clustering, which avoid traversing the entire image database when searching. The underlying low-dimensional image features speed up the retrieval process and improve search efficiency.
- Different encryption algorithms are applied to protect the privacy of images, index and features, while enabling accurate search and similarity ranking. In addition, we proposed a secure trapdoor generation algorithm that can not only protect the security of queries but also helps breaking linkability between queries and search results.

The rest of this paper is organized as follows. In Sect. 1, we briefly discuss related work. Section 2 describes the system and security assumptions. In Sect. 3, we describe the main idea of our proposed image classification model and the detailed search and retrieval procedure of PICR. Section 4 provides the security analysis and proof of PICR. We report our experimental and comparison results in Sect. 5. Finally, we conclude this paper in Sect. 6.

1 Related Work

The main contributions of existing privacy-preserving CBIR technology are to improve retrieval accuracy, efficiency and guarantee information security.

Improve Retrieval Accuracy. In 2009, Lu *et al.* proposed the first CBIR-based encrypted image search scheme which used visual word frequencies as image features [3]. There are many features that can also be used for image retrieval such as text description, semantic matrix, image related GPS and multi-feature fusion of contextual information [11,12]. CNN are applied to extract more accurate image features to improve retrieval accuracy [13]. Learning with errors (LWE)-based secure kNN encryption methods have been proposed to improve ranking accuracy in ciphertext domain [8].

Improve Retrieval Efficiency . Zhu *et al.* proposed a clustering algorithm CAK (Combination of AP and K-means clustering) to solve the problem of K-means clustering and achieved more accurate classification [14]. Li *et al.* adopted Zhu's CAK-means algorithm to construct a tree-based index, and applied LWE encryption method to achieve higher retrieval accuracy and efficiency [10]. Other solutions like secure kNN and secret sharing [15] have been used to accomplish search and retrieval tasks in encrypted image domain. Guo *et al.* [16] proposed a scheme to find the exact nearest neighbor over encrypted medical images to remove candidates with low similarity.

Protect Image and Feature Privacy. To support security computation of ciphtertext, homomorphic encryption is commonly used in secure image retrieval schemes. Hsu *et al.* adopted homomorphic encryption to encrypt images, which allowed cloud servers to extract the Scale Invariant Feature Transform(SIFT) features for retrieval [17]. Xia *et al.* proposed a scheme based on the Bag-of-Encrypted-Words(BOEW) model that used the frequency of visual word occurrences as image features [18,19]. Li *et al.* proposed a CBIR scheme that combines the homomorphic encryption and scalar-product-preserving encryption (ASPE) to keep the key confidentiality [20]. There are some other schemes that employ secure Local Binary Pattern(LBP) to calculate Euclidean or Manhattan distance on encrypted images to improve searching security and accuracy [19].

Although many encrypted CBIR schemes have been proposed, they still have some limitations in ensuring accuracy, efficiency, and security simultaneously. Aiming to break all these limitations, we propose a privacy-preserving image classification and retrieval(PICR) scheme that can satisfy all these goals.

2 The System Model and Security Assumption

2.1 System Model

In the proposed PICR system, there are three principals: image owner, user, and cloud server.

Image owner generates secret keys, extracts image features, constructs index and encrypts both images and index. Then it submits the encrypted index and images to the cloud server and distributes secret keys to authorized users.

User builds up query trapdoor and submits it to cloud server. After receiving the search results, search user decrypted results with the authorized keys.

Cloud Server stores encrypted index and images. When receiving queries, it runs search algorithm and returns top-k most similar images to search user.

2.2 Security Assumption

We assume that the cloud server is "honest but curios", the cloud server honestly executes the protocol, but it will analyze the data and search the index for more information related to the plaintext image. Based on the available information to the cloud server, the following threat model is defined for our scheme.

Known Ciphertext Model. In this model, the cloud server only knows encrypted information, such as encrypted images, encrypted index, and query trapdoor.

3 Image Classification and Privacy Protection

In order to improve both retrieval efficiency and accuracy, we propose the segmented hash codes to represent image category and feature simultaneously. Meanwhile, PICR can also protect the privacy of images and indexes.

3.1 System Structure of PICR

The system infrastructure of PICR is a tuple of several algorithms, namely: **GenKey, ExtractFea, EncImage, GenIndex, GenQuery, Search, DecImage**, which are depicted in Fig. 1. In step ① image owner generates keys and assigns them to the authorized search users. ②- ④, image owner generates the segmented hash codes with the proposed training model. Then, it encrypts images and builds the encrypted hierarchical index, finally it sends encrypted images and index to the cloud server. In the step ⑤, search user generates trapdoor from the query images and submits them to the cloud server. Step ⑥ shows the cloud server implements searching algorithm and return top-k similar ciphertext images to the querying user. In the step ⑦, the search user decrypts returned results to obtain images similar to the query images.

3.2 Image Feature Extraction Model of PICR

In this paper, we design a feature extraction model to get segmented hash codes as image features. The segmented hash code consists of two parts, the first part is category hash code representing the image classification information. The second part is feature vectors representing image features. To obtain the category hash codes, we use a hash code generation algorithm to assign the generated hash codes as target hash codes to each image category.

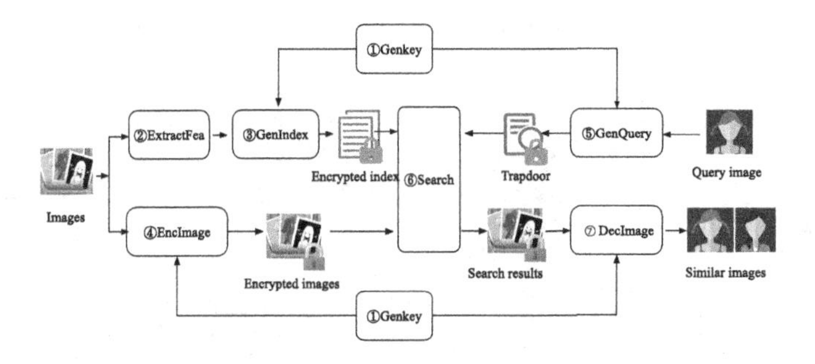

Fig. 1. System infrastructure of proposed PICR

Category Hash Code Generation. The design of category hash code needs to satisfy certain criteria. First, the Hamming distance between any two categories needs to be large enough to reduce the query error. Second, the dimensionality of the category hash code should be as low as possible to improve the retrieval efficiency and reduce index storage costs. $H^{(1)}$ is the image category hash code. The number of image categories is K, the length of category hash code is d_1. The minimum Hamming distance between any two category hash codes should be larger than a threshold t. We design a random hash code generation algorithm to find the hash code that can satisfy Eq. 1.

$$d(H_i^{(1)}, H_j^{(1)}) > t, i \neq j \tag{1}$$

where $d()$ is the Hamming distance between category hash codes, $H_i^{(1)}, H_j^{(1)}$ are the hash codes representing two different image categories. The hash code generation process repeats until there are K hash codes have been obtained.

Image Feature Extraction. In order to reduce the dimensionality of the feature vector, we proposed a CNN-based feature extraction model to generate the low-dimensional segmented hash codes as the representation vector of each image. We add two layers after the basic CNN model. **Hash Layer 1** is the classification layer used to train the category hash codes, the output is $H_i^{(1)} = \{h_{i,1}^{(1)}, h_{i,2}^{(1)}, \ldots, h_{i,d_1}^{(1)}\}$. **Hash Layer 2** is used to train the image features, the output is a d_2-dimensional vector $H_i^{(2)} = \{h_{i,1}^{(2)}, h_{i,2}^{(2)}, \ldots, h_{i,d_2}^{(2)}\}$ containing the feature of image i. The proposed segmented hash codes training model is shown in Fig. 2.

PICR scheme uses DenseNet-169 as the basic CNN model. The final representation vector of each image is described as:

$$f_i = (H_i^{(1)} || H_i^{(2)}) = \{h_{i,1}^{(1)}, h_{i,2}^{(1)}, \ldots, h_{i,d_1}^{(1)}, h_{i,1}^{(2)}, h_{i,2}^{(2)}, \ldots, h_{i,d_2}^{(2)}\} \tag{2}$$

Where "$||$" is the concatenation character. Based on our experimental observation, we use a 12-dimensional category hash code to represent the set of 50-category 4000 images in Caltech-256, and the 32-dimensional hash code to represent image features.

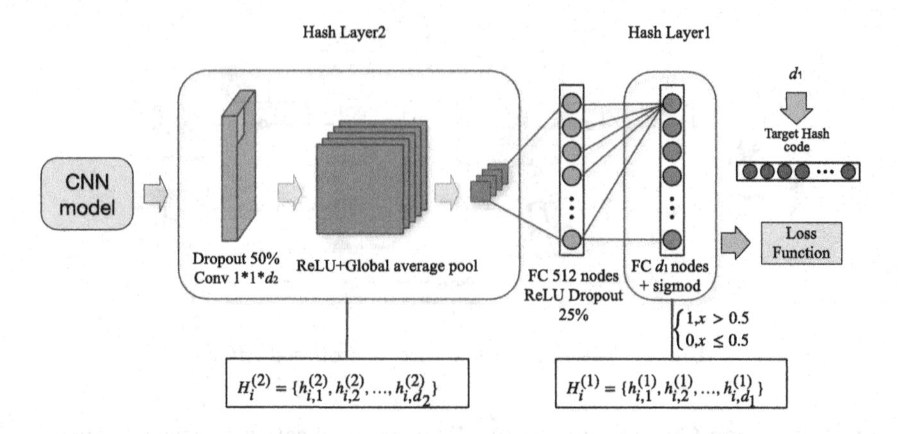

Fig. 2. Segmented hash codes training model of PICR

3.3 Privacy Protection Algorithms

PICR uses different encryption methods to ensure the security of images, category hash codes and image features. The main functions for privacy protection are **GenKey**, **EncImage** and **GenIndex**.

GenKey(SK). The image owner generates different encryption keys and distributes them to the cloud servers and users for later usage.

- First, image owner generates two d_1-dimensional vectors $V = \{v_1, v_2, \ldots, v_{d_1}\}$, $R = \{r_1, r_2, \ldots, r_{d_1}\}$, and a permutation π, where $v_j \in [-p_1, p_1]$, $r_j \in [-p_2, p_2]$, $p_1 >> p_2$ are randomly selected, $sk_1 = \{V, R, \pi\}$.
- Second, image owner randomly choses $\gamma \in \mathbb{Z}_{p_1}$, and generates matrix M and its inverse form M^{-1}. $M_{IO}, M'_{IO} \in \mathbb{Z}^{2d_2 \times 2d_2}$ are randomly divided from M. $M_U, M'_U \in \mathbb{Z}^{2d_2 \times 2d_2}$ are randomly divided from M^{-1}, $sk_2 = M_{IO}$.
- Then, image owner generates AES key sk_3. Finally, it issues the $SK_U = (sk_1, M_U, \gamma, sk_3)$ to users, M'_{IO}, M'_U to cloud server via secure channel.

EncImage$(\{m_i\}_{i=1}^N, sk_3) \rightarrow \{m'_i\}_{i=1}^N$. Image owner encrypts each image in $\{m_i\}_{i=1}^N$ with key sk_3, then sends the ciphertext $\{m'_i\}_{i=1}^N$ to the cloud server.

GenIndex$(\{m_i\}_{i=1}^N, SK, M_{IO}, M'_{IO}) \rightarrow \tilde{I}$. After obtaining f_i of each image, image owner needs to cluster the images with their category hash codes $H_i^{(1)}$ and build up the secure hierarchical index.

- Hierarchical index building. K-means algorithm is applied to cluster the images using $H_i^{(1)}$ as input, the output is K class centers $\{C_l\}_{l=1}^K$. The index is constructed with $C_l = \{c_{l,1}, c_{l,2}, \ldots, c_{l,d_1}\}$ as the top category level. In each category, there are several images organized with their feature vectors $H_i^{(2)}$.
- Index encryption. Image owner first replaces all 0 in C_l with -1 to get $\hat{C}_l$, and then encrypts it with Eq. 3.

$$\tilde{C}_l = \pi(V \cdot \hat{C}_l \cdot R) = \pi\{(v_j \cdot \hat{c}_{l,j} \cdot r_j)\}_{j=1}^{d_1} \quad (3)$$

When encrypts feature vectors $H_i^{(2)}$, image owner first extends $H_i^{(2)}$ to $\hat{H}_i^{(2)}$.

$$\hat{H}_i^{(2)} = \{h_{i,1}^{(2)}, h_{i,2}^{(2)}, \ldots, h_{i,d_2}^{(2)}, -\frac{1}{2}\sum_{j=1}^{d_2} h_{i,j}^{(2)}, \alpha_1, \alpha_2, \ldots, \alpha_{d_2-1}\} \tag{4}$$

where $\alpha_1, \alpha_2, \ldots, \alpha_{d_2} \in \mathbb{A}_{p_2}$ are random numbers chosen by the image owner. Then, image owner encrypts $\hat{H}_i^{(2)}$ and gets the ciphertext $\tilde{H}_i^{(2)}$.

$$\tilde{H}_i^{(2)} = (\gamma \cdot \hat{H}_i^{(2)} + \zeta_i) \cdot M_{IO} \tag{5}$$

$\gamma \in \mathbb{Z}_{p_1}, p_1 \gg p_2, \gamma \gg 2|max(\zeta_i)|$. $\zeta_i \in \mathbb{Z}_{p_2}^{2d}$ is an integer error vector randomly chosen from probability distributions.

The final encrypted index is represented as $\tilde{I} = \{\tilde{C}_l, \tilde{H}_i^{(2)}\}$, $l \in [1, K], i \in [1, N]$. Image owner uploads the encrypted images and index to the cloud server.

3.4 Privacy-Preserving Image Retrieval

Search users extract feature vectors with the same pre-trained PICR model. Then, they generate retrieval trapdoors using **GenQuery** and send them to the cloud server. Finally, the cloud server implements the **Search** function and returns the top-k ranked results to search users.

GenQuery$(m_q, SK) \rightarrow Q = \{\tilde{f}_q, W_q\}$. Search user extracts features $f_q = (H_q^{(1)}||H_q^{(2)})$ from the query image m_q, and then generates the trapdoor $Q = \{\tilde{f}_q, W_q\}$ with the key SK_U. The process is described with the followed steps.

- First, search user replaces all 0 in $H_q^{(1)}$ with -1 and get $\hat{H}_q^{(1)}$.
- Second, search user generates a d_1-dimensional random vector U_q and encrypts $\hat{H}_q^{(1)}$ with the permutation π and random vector $R = \{r_1, r_2, \ldots, r_{d_1}\}$.

$$\tilde{H}_q^{(1)} = \pi(U_q \cdot \hat{H}_q^{(1)} \cdot R) = \pi\{(u_{q,j} \cdot \hat{h}_{q,j}^{(1)} \cdot r_j)\}_{j=1}^{d_1} \tag{6}$$

In different queries, search user generates different U_q.
- Third, search user extends and encrypts the feature vector $H_q^{(2)}$.

$$\hat{H}_q^{(2)} = \{\rho_q h_{q,1}^{(2)}, \ldots, \rho_q h_{q,d_2}^{(2)}, \rho_q, \beta_1, \ldots, \beta_{d_2-1}\} \tag{7}$$

$$\tilde{H}_q^{(2)} = M_U \cdot (\gamma \cdot \hat{H}_q^{(2)\top} + \zeta_j^\top) \tag{8}$$

where $\rho_q, \beta_1, \beta_2, \ldots, \beta_{d_2-1} \in \mathbb{Z}_{p_2}$, $\rho_q > 0$, are random numbers selected by search user for each query. The query feature $\tilde{f}_q$ is described in Eq. 9.

$$\tilde{f}_q = \tilde{H}_q^{(1)}||\tilde{H}_q^{(2)} \tag{9}$$

- Finally, search user uses SK_U to generate $W_q = \pi \cdot \{(v_j + u_{q,j})\}_{j=1}^{d_1}$ and sends $Q = \{\tilde{f}_q, W_q\}$ to cloud server for image search and retrieval.

Search$(\tilde{I}, Q) \rightarrow R'$. Cloud server runs the followed steps and returns top-k most similar results.

- First, the secure Hamming distance between $\tilde{H}_q^{(1)}$ and $\tilde{C}_l$ is calculated.

$$EnD(\tilde{H}_q^{(1)}, \tilde{C}_l) == \sum_{j=1}^{d_1} \{[(u_{q,j}\hat{h}_{q,j}^{(1)} r_j + v_j \hat{c}_{l,j} r_j) \text{ mod}]\ (u_{q,j} + v_j)] \oplus 0\} \quad (10)$$

  $EnD()$ is the secure Hamming distance calculation algorithm. The cloud server stops the calculation when $EnD(\tilde{H}_q^{(1)}, \tilde{C}_l) < t$, then it sets the current C_l as the category which the query image belongs to, and proceeds to retrieve the specific images in the C_l category using $\tilde{H}_q^{(2)}$.
- Second, within the category C_l, cloud server calculates the similarity between $\tilde{H}_q^{(2)}$ and $\tilde{H}_i^{(2)}$ with the followed formula.

$$\begin{aligned} EnS(\tilde{H}_i^{(2)}, \tilde{H}_q^{(2)}) &= \left(\frac{\tilde{H}_i^{(2)} \cdot M'_{IO} \cdot M'_U \cdot \tilde{H}_q^{(2)}}{\gamma^2} \right) \bmod p_1 \\ &= -\frac{\rho_q}{2} \left(||H_i^{(2)} - H_q^{(2)}||^2 - ||H_i^{(2)}||^2 \right) + \sum_{j=1}^{d_2-1} \alpha_j \beta_j \end{aligned} \quad (11)$$

$Ens()$ is the secure Euclidean distances. The inner product of the encrypted feature vectors between the query and stored index approximates to the Euclidean distance of their plaintexts. This allows the cloud server to rank the results. The cloud server sends the top-k most similar images back to search user.

DecImage$(m'_i, sk_3) \rightarrow m_i$. User uses the pre-assigned decryption key to get the original plaintext image.

4 Security Analysis of PICR

In this section, we prove that our PICR scheme meets the privacy preserving requirements of index and query privacy. The image privacy is guaranteed by the security of traditional symmetric encryption.

4.1 Security of Category Hash Code Encyption

Theorem 1. *PICR can guarantee the privacy of category hash code in both index and queries.*

Proof. Security of category hash code in the index. The category hash code C is first transformed to $\hat{C}$ and then encrypted with $sk_1 = \{V, R, \pi\}$. Since V, R and π are randomly selected by the image owner. C is protected by the random number and the difficulty of factoring very large numbers.

Security of category hash label in queries. The category hash code in queries are encrypted in the same way as $\hat{C}$. However, search user choses a different random vector U_q for each query encryption. Therefore, the same feature vector $H_q^{(1)}$ have different values in each query and it not feasible to obtain the plaintext.

4.2 Privacy Guarantee on Feature Vectors

The feature vectors of images are encrypted with LWE-based encryption.

Theorem 2. *PICR can guarantee the security of feature vectors in index and queries under the known ciphertext attack model.*

Proof. In PICR, the feature vectors are first extended to $2d_2$ dimensions and then encrypted to $\tilde{H}^{(2)}$.

$$\tilde{H}_i^{(2)} = (\gamma \cdot \hat{H}_i^{(2)} + \zeta_i) \cdot M_{IO} = \gamma \cdot \hat{H}_i^{(2)} \cdot M_{IO} + \zeta_i \cdot M_{IO} \qquad (12)$$

Since feature vectors of images are encrypted with LWE-based encryption, recovering $\hat{H}_i^{(2)}$ from $\tilde{H}_i^{(2)}$ is the LWE problem [10], so it is considered a hard problem and computational infeasible. In summary, the privacy of feature vectors in index and queries can be protected under the *known ciphertext attack model.* 2 is prooved.

5 Experimental Evaluation

We evaluate the performance of PICR and compare it with the most relevant schemes CASHEIRS [7], SEI [21], SVMIS [10]. The experiment is carried out on a PC with Windows 10 OS with 3.6GHz Intel Core i7 CPU and 8GB memory. We choose 50 categories in Caltech256, there are 80 images in each category. The performance of PICR is evaluated in terms of the search accuracy, efficiency and storage cost.

5.1 Accuracy Comparison

We use Precision at top-k(P@k) to measure the retrieval accuracy. P@k = $correct\_num/k$, where *correct_num* represents the number of *correct* images in the top-k returned ones. The retrieval accuracy comparison among schemes PICR, CASHEIRS [7], SEI [10] and SVIMS [21] with different k is shown in Table 1, [10,21] both use the LWE-based kNN to encrypt the 128-dimensional feature vectors.

As k increases, the retrieval accuracy of other schemes decreases. However, the retrieval accuracy maintains stable in PICR. The reason is that the design of category hash code and pre-trained model of PICR enables the cloud server to pin the categories of queried images with high probability and stability.

Table 1. Accuracy Comparison with 50 Categories

Scheme	k=5	k=10	k=20	k=30	k=40
CASHEIRS [7]	0.545	0.561	0.558	0.563	0.546
SEI [10]	0.618	0.619	0.618	0.611	0.608
SVIMS [21]	0.814	0.806	0.783	0.763	0.741
PICR	**0.879**	**0.879**	**0.879**	**0.879**	**0.879**

5.2 Performance Evaluation and Comparison

The performance comparison among PIRC, CASHEIRS [7], SEI [10] and SVIMS [21] includes index building time, storage costs and search time. The number of category chosen from Caltech256 are 10, 20, 30, 40 and 50, the number of images various from 800 to 4000. There are 80 images per category, the returned image number k = 40.

The index generation time for different schemes is shown in Fig. 3 a). The index building time is directly proportional to the dimension of the feature vector and the size of the data set, and is related to the complexity of the retrieval structure. CASHEIRS [7] only uses K-means to build its index tree. SEI [10] applies CAK algorithm to construct the deeper index tree. SVIMS [21] establishes the linear index, its index building time mainly depends on the feature vector dimensions and the total number of images. These three schemes use 128-dimensional features to construct the index, but in PICR we use the short segmented hash code (12-dimensional category hash code + 32-dimensional features) as the image feature, and the index construction time mainly relies on K-means clustering time.

The storage overhead is mainly composed of the index structure and encrypted features, which increases with the dimensionality of the feature vector and the total number of images. Compared with PICR scheme, the other three schemes use a 128-dimensional vector as the image features, SVIMS and SEI extend the feature vector to 256 dimensions after using the LWE-based encryption. In contrast, our PICR scheme only needs to extend the 32-dimensional feature vector to 64 dimensions, and the dimensionality of the category hash labels remains the same, so the overall number of encrypted features in the PICR scheme is 12 + 64 dimensions. As shown in Fig. 3 b), our PICR scheme requires less storage space on the cloud server.

As shown in Fig. 3 c), the tree-based index structure in CASHEIRS and SEI are more efficient in searching, but has low retrieval accuracy. The linear index structure in SVIMS is less efficient, but has more accuracy. In PICR, we balance the retrieval efficiency and accuracy by introducing the hierarchical index structure. The category-feature index structure can first find the image category, and then find similar images. This avoids the cloud server traversing the entire database, thereby improving query efficiency while ensuring accuracy.

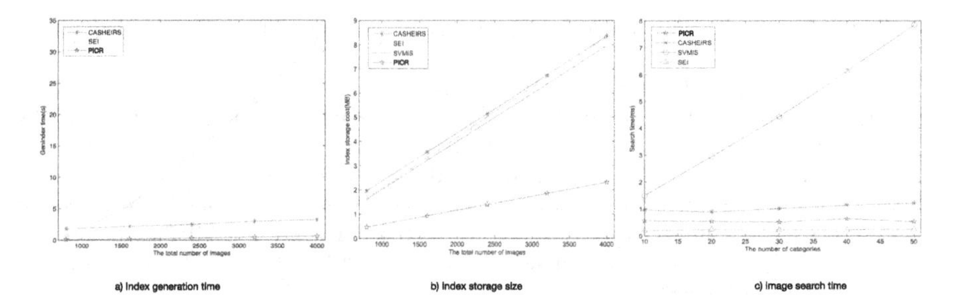

Fig. 3. Performance evaluation and comparison

6 Conclusion

In this paper, we propose a privacy-preserving image retrieval scheme PICR that can provide accurate and efficient search and retrieval of encrypted images. Specifically, we propose a CNN-based training model to get the segmented hash codes as the representation of image features. The proposed training model can provide a precise image classification and feature extraction, and also reduce the dimensionality of the feature vectors, thus improving both the retrieval accuracy and reduce the index storage costs. In addition, by constructing a hierarchical retrieval structure, we improve the retrieval efficiency while guaranteeing accuracy. The employment of random vectors and the LWE-based secure kNN algorithm can preserve the privacy of segmented hash codes and file-access patterns. We also give a rigorous security analysis and conduct experiments on the real-world dataset. The results show that compared to other similar schemes, PICR can guarantee image retrieval privacy, accuracy and efficiency simultaneously. In the future, we intend to extend this work in two directions. One is to make our scheme applicable to computation constrained IoT devices by introducing edge servers to perform feature extraction. The other is the application of access control and watermarking techniques for retrieval control and copyright protection of images.

References

1. Ahmad, L.: A Survey on DDoS Attacks in Edge Servers, Master's thesis, University of Taxes at Arlington, Arlington (2020)
2. Cost of a data breach report, Tech. rep., IBM Security Community (2022)
3. Lu, W., Swaminathan, A., Varna, A., Wu, M.: Enabling search over encrypted multimedia databases. In: SPIE/IS&T Media Forensics and Security, pp. 7254–18 (2009)
4. Song, F., Qin, Z., Zhang, J., Liu, D., Liang, J., Shen, X.S.: Efficient and privacy-preserving outsourced image retrieval in public clouds. In: GLOBECOM 2020–2020 IEEE Global Communications Conference, pp. 1–6 (2020). https://doi.org/10.1109/GLOBECOM42002.2020.9322134

5. Zhang, L., et al.: Pic: enable large-scale privacy preserving content-based image search on cloud. IEEE Trans. Parallel Distrib. Syst. **28**(11), 3258–3271 (2017). https://doi.org/10.1109/TPDS.2017.2712148
6. Engelsma, J.J., Jain, A.K., Boddeti, V.N.: Hers: homomorphically encrypted representation search. IEEE Trans. Biometrics Behav. Identity Sci. **4**(3), 349–360 (2022). https://doi.org/10.1109/TBIOM.2021.3139866
7. Li, X., Xue, Q., Chuah, M.C.: Casheirs: cloud assisted scalable hierarchical encrypted based image retrieval system. In: IEEE INFOCOM 2017 - IEEE Conference on Computer Communications, pp. 1–9 (2017). https://doi.org/10.1109/INFOCOM.2017.8056953
8. Li, Y., Ma, J., Miao, Y., Liu, L., Liu, X., Choo, K.K.R.: Secure and verifiable multikey image search in cloud-assisted edge computing. IEEE Trans. Industr. Inf. **17**(8), 5348–5359 (2021). https://doi.org/10.1109/TII.2020.3032147
9. Wang, X., Ma, J., Liu, X., Miao, Y.: Search in my way: practical outsourced image retrieval framework supporting unshared key. In: IEEE INFOCOM 2019 - IEEE Conference on Computer Communications, pp. 2485–2493 (2019). https://doi.org/10.1109/INFOCOM.2019.8737619
10. Li, Y., Ma, J., Miao, Y., Wang, Y., Liu, X., Choo, K.K.R.: Similarity search for encrypted images in secure cloud computing. In: IEEE Transactions on Cloud Computing, pp. 1–1 (2020). https://doi.org/10.1109/TCC.2020.2989923
11. Zhu, L., Huang, Z., Chang, X., Song, J., Shen, H.T.: Exploring consistent preferences: discrete hashing with pair-exemplar for scalable landmark search. In: Proceedings of the 25th ACM International Conference on Multimedia, pp. 726–734 (2017). https://doi.org/10.1145/3123266.3123301
12. Zhu, L., Huang, Z., Li, Z., Xie, L., Shen, H.T.: Exploring auxiliary context: discrete semantic transfer hashing for scalable image retrieval. IEEE Trans. Neural Networks Learn. Syst. **29**(11), 5264–5276 (2018). https://doi.org/10.1109/TNNLS.2018.2797248
13. Li, Y., et al.: DVREI: dynamic verifiable retrieval over encrypted images. IEEE Trans. Comput. **71**(8), 1755–1769 (2022). https://doi.org/10.1109/TC.2021.3106482
14. Zhu, Y., Yu, J., Jia, C.: Initializing k-means clustering using affinity propagation. In: 2009 Ninth International Conference on Hybrid Intelligent Systems, vol. 1, pp. 338–343 (2009). https://doi.org/10.1109/HIS.2009.73
15. Aminuddin Mohd Kamal, A.A., Iwamura, K., Kang, H.: Searchable encryption of image based on secret sharing scheme. In: 2017 Asia-Pacific Signal and Information Processing Association Annual Summit and Conference (APSIPA ASC), pp. 1495–1503 (2017). https://doi.org/10.1109/APSIPA.2017.8282269
16. Guo, C., Su, S., Choo, K.K.R., Tang, X.: A fast nearest neighbor search scheme over outsourced encrypted medical images. IEEE Trans. Industr. Inf. **17**(1), 514–523 (2021). https://doi.org/10.1109/TII.2018.2883680
17. Hsu, C.Y., Lu, C.S., Pei, S.C.: Image feature extraction in encrypted domain with privacy-preserving sift. IEEE Trans. Image Process. **21**(11), 4593–4607 (2012). https://doi.org/10.1109/TIP.2012.2204272
18. Xia, Z., Jiang, L., Liu, D., Lu, L., Jeon, B.: Boew: a content-based image retrieval scheme using bag-of-encrypted-words in cloud computing. IEEE Trans. Serv. Comput. **15**(1), 202–214 (2022). https://doi.org/10.1109/TSC.2019.2927215
19. Xia, Z., Wang, L., Tang, J., Xiong, N.N., Weng, J.: A privacy-preserving image retrieval scheme using secure local binary pattern in cloud computing. IEEE Trans. Network Sci. Eng. **8**(1), 318–330 (2021). https://doi.org/10.1109/TNSE.2020.3038218

20. Li, J.S., Liu, I.H., Tsai, C.J., Su, Z.Y., Li, C.F., Liu, C.G.: Secure content-based image retrieval in the cloud with key confidentiality. IEEE Access **8**, 114940–114952 (2020). https://doi.org/10.1109/ACCESS.2020.3003928
21. Li, Y., Ma, J., Miao, Y., Liu, L., Liu, X., Choo, K.K.R.: Secure and verifiable multi-key image search in cloud-assisted edge computing. IEEE Trans. Industr. Inf. **17**(8), 5348–5359 (2020). https://doi.org/10.1109/TII.2020.3032147

An End-to-End Structure with Novel Position Mechanism and Improved EMD for Stock Forecasting

Chufeng Li and Jianyong Chen(✉)

School of Computer Science and Software Engineering, Shenzhen University, Shenzhen, China
jychen@szu.edu.cn

Abstract. As a branch of time series forecasting, stock movement forecasting is one of the challenging problems for investors and researchers. Since Transformer was introduced to analyze financial data, many researchers have dedicated themselves to forecasting stock movement using Transformer or attention mechanisms. However, existing research mostly focuses on individual stock information but ignores stock market information and high noise in stock data. In this paper, we propose a novel method using the attention mechanism in which both stock market information and individual stock information are considered. Meanwhile, we propose a novel EMD-based algorithm for reducing short-term noise in stock data. Two randomly selected exchange-traded funds (ETFs) spanning over ten years from US stock markets are used to demonstrate the superior performance of the proposed attention-based method. The experimental analysis demonstrates that the proposed attention-based method significantly outperforms other state-of-the-art baselines. Code is available at https://github.com/DurandalLee/ACEFormer.

Keywords: Financial Time Series · Empirical Mode Decomposition · Attention · Stock Forecast

1 Introduction

Stock trend prediction is an important research hotspot in the field of financial quantification. Currently, many denoising algorithms and deep learning are applied to predict stock trends [1]. The Fractal Market Hypothesis [2] points out that stock prices are nonlinear, highly volatile, and noisy, and the dissemination of market information is not uniform. What's more, if future stock trends can be accurately predicted, investors can buy (or sell) before the price rises (or falls) to maximize profits. Therefore, accurate prediction of stock trends is a challenging and profitable task [3,4].

As early as the end of the last century, Ref. [5] exploited time delay, recurrent, and probabilistic neural networks (TDNN, RNN, and PNN, respectively) to forecast stock trends, and showed that all the networks are feasible. With the

B. Luo et al. (Eds.): ICONIP 2023, CCIS 1965, pp. 526–537, 2024.
https://doi.org/10.1007/978-981-99-8145-8_40

rapid development of deep learning, many deep learning methods, especially RNN and Long Short-Term Memory (LSTM) [6], have been widely used in the field of financial quantification. Nabipour et al. [7] proved that RNN and LSTM outperform nine machine learning models in predicting the trends of stock data. With the proposal of the attention mechanism [8], such as Transformer [9] which is based on the attention mechanism and has achieved unprecedented results in the field of natural language processing, the focus of time series research has also shifted to the attention mechanism. Zhang et al. [10] proved that, in the field of stock forecast, LSTM combined with Attention and Transformer only had subtle differences, but both better than LSTM. Wang et al. [11] combine graph attention network with LSTM in forecasting stock and get a better result. Ji et al. [12] build a stock price prediction model based on attention-based LSTM (ALSTM) network. But, unlike time series data such as traffic, due to the trading rules of the stock market, the time interval of stock data is not regular. The self-attention mechanism has the ability to focus on the overall relevance of the data, but it is not only weak to capture short-term and long-term features in multi-dimensional time series data [13] but also weak to extract and retain positional information [14]. However, positional information is very important for time series.

Quantitative trading seeks to find long-term trends in stock volatility. However, short-term high-frequency trading can conceal the actual trend of the stock. This means that short-term high-frequency trading is a kind of noise that prevents the right judgment of long-term trends. Highly volatile stock data greatly affects the effectiveness of deep learning models. In the current stock market, there are many indicators used to smooth stock data and eliminate noise, such as moving averages. However, these indicators are usually based on a large amount of historical stock data. They are lagging indicators [15] and cannot timely reflect the actual fluctuations of the long-term trend. In the field of signal analysis, algorithms such as Fourier Transform (FT), Wavelet Transform (WT), and Empirical Mode Decomposition (EMD) can effectively eliminate signal noise and avoid lag. Compared with FT [16] and WT [17], EMD has been proven to be more suitable for time series analysis [16,17]. It is a completely data-driven adaptive method that can better handle non-linear high-noise data and eliminate data noise. However, EMD also has disadvantages [18] such as endpoint effects and modal aliasing.

Since short-term high-frequency trading has great impact on the long-term trend of stocks, removing short-term trading noise can effectively increase the likelihood of the model finding the correct rules for long-term trends. To solve this problem, we introduced a denoising algorithm called Alias Complete Ensemble Empirical Mode Decomposition with Adaptive Noise (ACEEMD). The noise is eliminated by removing the first intrinsic mode function (IMF) [19]. ACEEMD not only solves the endpoint effect problem but also avoids over-denoising and effectively keeps key turning points in stock trends. In this paper, we propose a stock trend prediction solution, **ACE**EMD Attention **Former** (ACEFormer). It mainly consists of ACEEMD, time-aware mechanism, and attention mechanism. The time-aware mechanism can overcome the weak ability of the attention mechanism to extract positional information and the irregularity of stock data intervals. The main contributions of this paper are summarized as follows:

- We propose a stock trend prediction solution called ACEFormer. It consists of a pretreatment module, a distillation module, an attention module, and a fully connected module.
- We propose a noise reduction algorithm called ACEEMD. It is an improvement on EMD which can not only address the endpoint effect but also preserve the critical turning points in the stock data.
- We propose a time-aware mechanism that can extract temporal features and enhance the temporal information of input data.
- We conduct extensive experiments on two public benchmarks, NASDAQ and SPY. The proposed solution significantly outperforms several state-of-the-art baselines such as Informer [20], TimesNet [21], DLinear [22], and Non-stationary Transformer [23].

2 Methodology

In this section, we first present the proposed ACEFormer. Next, we introduce the noise reduction algorithm ACEEMD. Finally, the time-aware mechanism is designed.

2.1 ACEFormer

The architecture of our proposed model is shown in Fig. 1 which includes pretreatment module, distillation module, attention module and fully connected module.

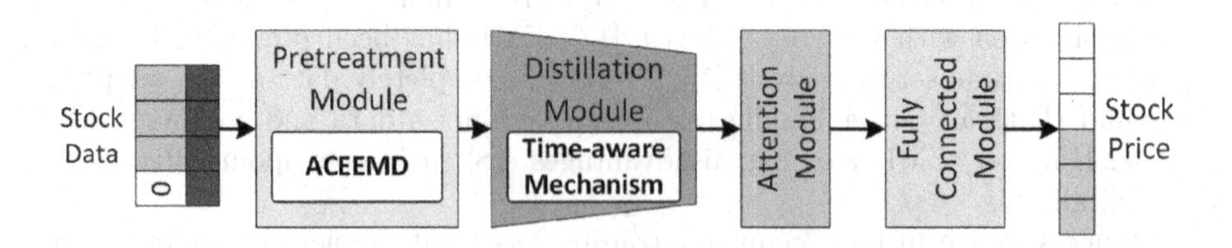

Fig. 1. The architecture of ACEFormer.

The pretreatment module preprocesses the input data, which is conducive to the model for better extracting the trend rules of stock data. Among them, our proposed ACEEMD algorithm is also added to the pretreatment module which is shown in Fig. 2.

Let $S = \{s_1, s_2, ..., s_n\}$ represent the stock data input for model training, where s_i includes the price and trading volume of the stock itself and two overall stock market indices on the i-th day. Since the data to be predicted is unknown, it is replaced with all zeros. Let $[0:p]$ represent a sequence of p consecutive zeros, where p is the number of days to be predicted. The input data for the model is $D = S||[0:p]$. Let f and g denote functions, $*$ denotes the convolution operation, and $PE(\cdot)$ denotes position encoding. We define the output of pretreatment module denoted by D which is given by:

$$X_{pre} = ACEEMD((f * g)(D)) + PE(D) \tag{1}$$

The distillation module extracts the main features using the probability self-attention mechanism, and reduces the dimension of the feature data using convolution and pooling. In addition, the time-aware mechanism in it is used to extract position features to increase the global position weight.

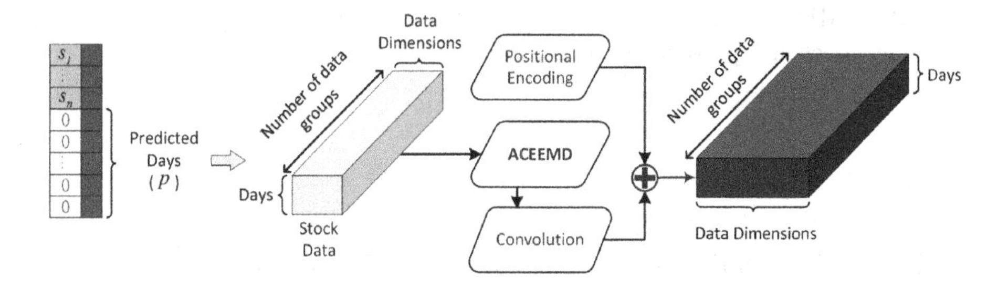

Fig. 2. The pretreatment module of ACEFormer.

The distillation module, as shown in Fig. 3, includes the probability attention [20], the convolution, the max pooling, and the time-aware mechanism which is described in detail in the Sect. 3.3. The output features of probability attention contain features of different levels of importance, so the convolution and pooling allow the main features to be retained and the dimensionality of the data to be reduced. It can reduce the number of parameters afterward.

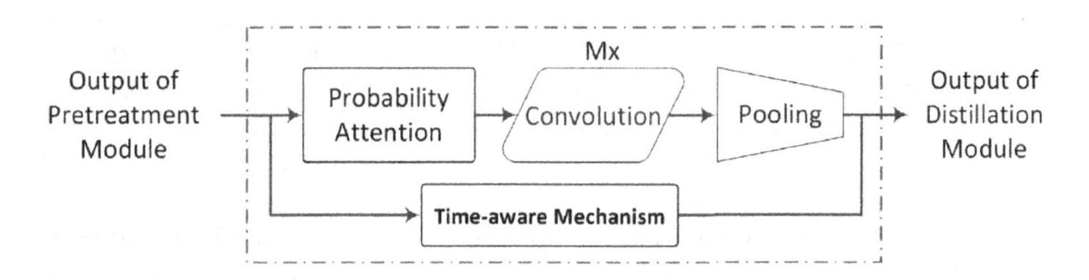

Fig. 3. The distillation module of ACEFormer.

The attention module is used to further extract the main features. It can focus on the feature data from the distillation module and extract more critical features. The fully connected module is a linear regression which produces the final predicted values.

Because dimension expansion can generate redundant features, the use of probability attention can increase the weight of effective features based on dispersion. Meanwhile, the convolution and the pooling can eliminate redundant features. In addition, the position mechanism can retain valid features which may be unintentionally eliminated. Because the whole process progressively extracts important features and reduces dimensions of stock data, the self-attention only gets features from the distillation. In the case, it can focus on the compressed data and extract more critical features. Meanwhile, it can effectively eliminate out irrelevant features.

2.2 ACEEMD

The ACEEMD can improve the fitting of the original curve by mitigating the endpoint effect and preserving outliers in stock data, which can have significant impacts on trading.

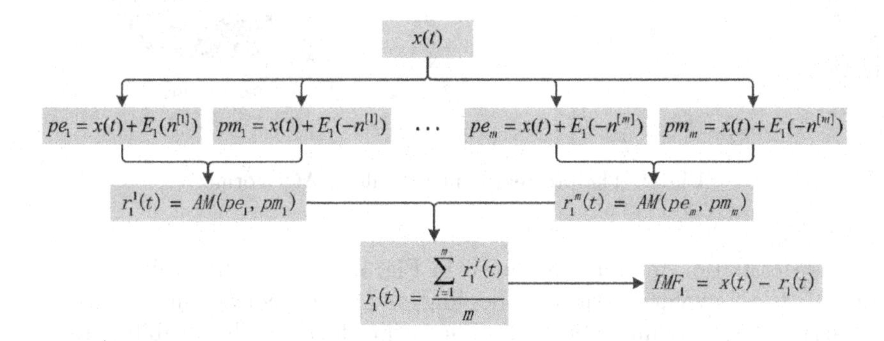

Fig. 4. The flowchart of the ACEEMD algorithm architecture.

Figure 4 shows the ACEEMD algorithm. $x(t)$ refers to the input data, i.e. the stock data. $n^{[i]}$ represents the i-th Gaussian noise, where the total number of noises m is an adjustable parameter and the default value is 5. $E(\cdot)$ denotes the first-order IMF component of the signal in parentheses. pe_i and pm_i both represent the result of the input data and a Gaussian noise, but the difference between them is that the Gaussian noise they add is opposite in sign to each other. The generating function $AM(pe_i, pm_i)$ is used to denoise the data and is also the core of ACEEMD. IMF_1 represents the first-order IMF component of ACEEMD, which is the eliminable noise component in the input data. $r_1^i(t)$ represents the denoised data of the input data with the i-th group of added Gaussian noise, and $r_1(t)$ represents the denoised data obtained by processing the input data with ACEEMD.

ACEEMD algorithm has two improvements. First, to avoid the endpoint effect, the input data points for cubic interpolation sampling are constructed

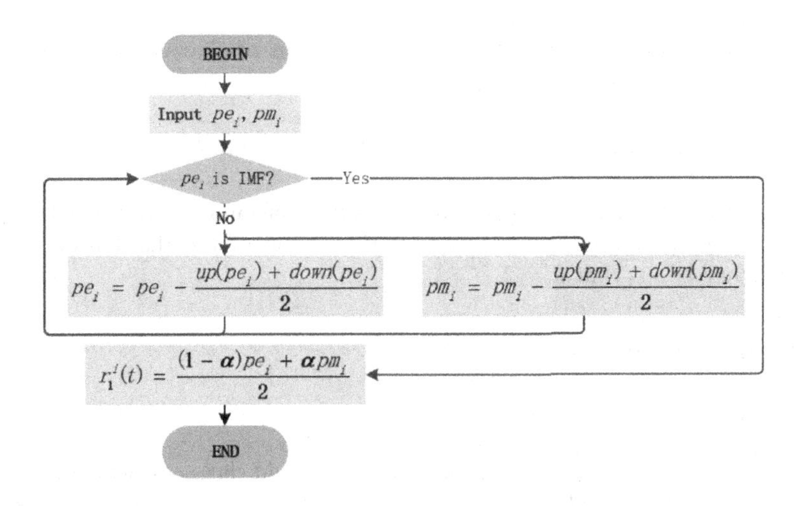

Fig. 5. The flowchart of the core function $AM(pe_i, pm_i)$ of ACEEMD.

using the endpoints and extreme points of the input data. Second, the middle point of a sequence is defined as the data corresponding to the midpoint of the abscissa between the peaks and troughs. The paired data sets with opposite Gaussian noise are pe_i and pm_i shown in Fig. 4. To further preserve the short-term stock trend, the input data points for cubic interpolation sampling of pm_i include not only the extreme points and the endpoints, but also the middle points.

The core of ACCEMD is $r_1^i(t)$ shown in the Fig. 5, which is referred to as the aliased complete algorithm. It applies cubic interpolation sampling to pe_i and pm_i, and the termination condition of the loop in it is that the intermediate signal of pe_i is IMF. The i-th first-order IMF component is obtained by taking a weighted average of the intermediate signals of pe_i and pm_i, with a default weight $\alpha = 0.5$.

2.3 Time-Aware Mechanism

The time-aware mechanism is constructed by linear regression. It is the bias of the distillation module and generates a matrix of the same size as the max pooling result. As part of the distillation module output, it can increase the feature content and minimize information loss of the output features.

Let W_t denote the weight, which is used to multiply the input matrix, b_t denotes the bias matrix, T denotes the matrix of the time-aware mechanism, and X_{pre} is defined as (1). So $T = X_{pre} \times W_t + b_t$. Because the input data of the time-aware mechanism is the same as the data received by probability attention, it can effectively extract features from input data with complete features.

We can express the i-th row and j-th column of the output D from the distillation module, represented by (2),

$$D_{ij} = \max_{0 \leq m < k, 0 \leq n < k} (f * g)(\bar{A}(X_{pre}))_{i \times k+m, j \times k+n} + T_{ij} \tag{2}$$

where, $\bar{A}$ denotes the probability attention operator, and k denotes the window length of the max pooling. The feature dimension of the distillation module output is halved, so that $k = 2$.

3 Experiments

3.1 Datasets

We evaluate the proposed method on two real-world datasets, which are NASDAQ100 and SPY500 [24], from US stock markets spanning over ten years. The NASDAQ100 is a stock market index made up of 102 equity stocks of non-financial companies from the NASDAQ. The SPY500 is Standard and Poor's 500, which is a stock market index tracking the stock performance of 500 large companies listed on stock exchanges in the United States.

We selected historical data[1] ranging from Jan-03-2012 to Jan-28-2022 for our experiments. First, we aligned the trading days in the history by removing the lack of data during weekends and public holidays. Then, we split the historical data into training set (Jan-03-2012 to Jun-25-2021), validation set (Jun-28-2021 to Sept-07-2021), and testing set (Sept-07-2021 to Jan-28-2022). In our experiments, we also include mainstream indices (DJIA and NASDAQ) as secondary data when using the two datasets.

3.2 Model Setting

In order to avoid the impact of randomly initialized parameters on the prediction results during the training process and obtain stable experimental results we train the model with multiple times. In the experiment of each model, we first train five model results independently using the training set, then we select the model result with the best experimental index performance in the validation set, and finally we use the selected model result to predict the test set.

3.3 Evaluation Metrics

Trend. We evaluate the performance of forecast trends with two metrics, Accuracy (Acc) and Matthews Correlation Coefficient (MCC) [6] of which the ranges are in $[0, 100]$ and $[-1, 1]$. Note that better performance can be get by higher value of the metrics.

Return. Sawhney [24] points out that classification task evaluation metrics can not prove the actual performance of the solution in terms of profit. Therefore,

[1] https://www.investing.com/.

as Sawhney did, we also introduced investment return ratio **(IRR)** [24] and the Sharpe Ratio **(SR)** as metric for solution return. The IRR is defined as $IRR = \sum_{i=1}^{n} R_i + 1$ where R_i denotes the ratio of profit on day i with the range $[-100\%, 100\%]$. The SR is a measure of the return of a portfolio compared to per unit of risk. It is defined as $SR = \frac{E[R_a - R_f]}{std[R_a - R_f]}$ where R_a is the earned return and R_f is risk-free of US[2].

3.4 Competing Methods

We will select the top three models on the authoritative time series prediction leaderboard[3], TimesNet [21], Non-stationary Transformer [23], and DLinear [22], as well as the Informer [20] which using the probability self-attention mechanism as comparison models.

4 Result

4.1 Trend Evaluation

We use five models to conduct trend prediction experiments on two datasets. The prediction curve of the test set is shown in the Fig. 6.

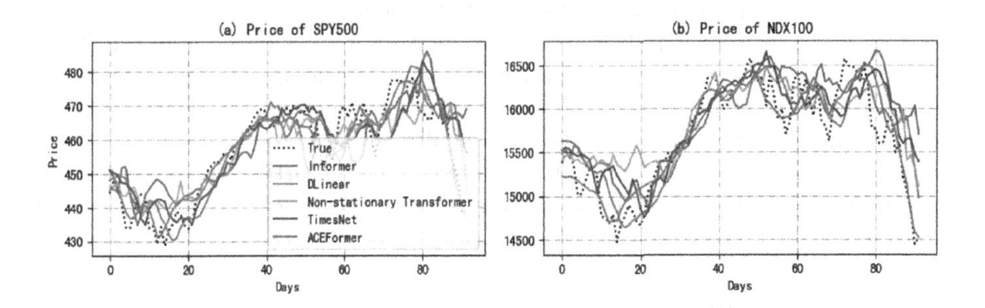

Fig. 6. The result of stock trends forecasting by five different models.

In order to quantitatively evaluate the prediction effect of each model, this article uses four indicators for evaluation, and the results are shown in the Table 1. The best results are shown in **bold**.

According to the experimental results in Table 1, it can be clearly seen that among all experimental models, ACEFormer performs the best. In terms of trend evaluation metrics, the ACC and MCC results of ACEFormer on the SPY500 (NASDAQ100) dataset are 69.23% and 0.379 (69.23% and 0.382), respectively. In contrast, only the ACC of the Non-stationary Transformer is slightly larger

[2] https://home.treasury.gov/.

[3] https://github.com/thuml/Time-Series-Library.

Table 1. Standard

Model	SPY500				NASDAQ100			
	ACC	MCC	IRR	SR	ACC	MCC	IRR	SR
Benchmark			−0.97%	−0.27			−5.68%	−0.80
Informer [20]	43.96%	−0.145	−8.38%	−2.05	45.05%	−0.110	−10.67%	−1.93
DLinear [22]	48.35%	−0.041	−2.65%	−0.90	49.45%	−0.021	−5.76%	−1.22
TimesNet [21]	48.35%	−0.028	−6.86%	−1.97	45.05%	−0.105	−8.52%	−1.59
Non-stationary Transformer [23]	56.04%	0.116	2.13%	0.48	60.44%	0.195	−0.86%	−0.23
ACEFormer (Ours)	**69.23%**	**0.379**	**16.62%**	**5.71**	**69.23%**	**0.382**	**22.31%**	**6.43**

than 60%, and only the MCC of the Non-stationary Transformer is larger than 0. The ACC intuitively indicates that the fitting degree of the ACEFormer prediction curve is better than other models. At the same time, the MCC proves that ACEFormer can better predict the rise and fall. Based on the ACEFormer prediction results, IRR and SR on the SPY500 (NASDAQ100) dataset are 16.62% and 5.71 (22.31% and 6.43), respectively. In contrast, only the performance of Non-stationary Transformer is better than Benchmark. The IRR shows that ACEFormer can achieve a return of 16.62% (22.31%) on SPY500 (NASDAQ100) within one hundred trading days, and at the same time it can achieve an excess return of 5.71 (6.43) when undertaking a unit of risk. This means that ACEFormer can predict the rise and fall more timely, provide better buying and selling opportunities for trading, obtain greater benefits, and avoid greater losses.

The above statement indicates that in the field of stock prediction, the ACEFormer model performs better than other state-of-the-art models. There are three reasons. First, the ACEEMD algorithm can eliminate as much noise as possible in stock data and reduce the difficulty of predicting long-term trends. Second, the cross-use of multiple attention mechanisms further optimizes feature extraction capabilities. Third, the time-aware mechanism can retain more stock position features and strengthen the temporal coherence of overall features.

4.2 ACEEMD Effect

To elaborate on the impact of ACEEMD, we have presented evidence of its effectiveness on stock data of various lengths. Since the unit length of stock data in our solution is 30, we use a 30-day segment of the NASDAQ100 closing price as an example to illustrate the effect of ACEEMD, as shown in Fig. 7. To facilitate the description of ablation experiments, we name ACEEMD without middle points as ECEEMD.

From the endpoints of Fig. 7(a), it is evident that the denoised data obtained by EMD has a significant deviation from the original curve, which is the endpoint effect. On the other hand, the other two denoising algorithms can effectively avoid this issue. Moreover, the curve from day 4 to day 19 is shown in Fig. 7(b). It is observed that the denoised data obtained by ACEEMD can retain the trend of the stock data. In contrast, the other denoised data appear excessively smooth and fail to capture some of the fluctuations presenting in the stock

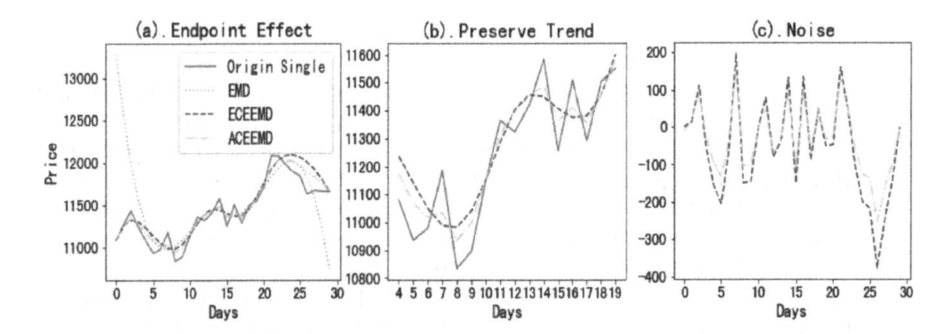

Fig. 7. Results of processing stock data using multiple noise reduction algorithms. (a) The effect of the endpoint effect of the EMD algorithm on the noise reduction results. (b) The core function of ACEEMD retaining more stock trends. (c) The noise removed by ECEEMD and ACEEMD respectively.

data. Figure 7(c) displays the noise, which is the first-order IMF component, extracted by two different algorithms. It can be observed that the fluctuation trend of the two noises is completely consistent, and at some positions, the fluctuation degree of the noise extracted by ACEEMD is relatively small. This indicates that the positions of noise identified by the two methods are the same. However, ACEEMD is capable of retaining more useful features, and resulting in the preservation of more trends in the stock data.

Thus, we can demonstrate that ACEEMD can not only avoid endpoint effects but also further preserve short-term stock trends.

5 Conclusion

In this paper, we address the challenge of predicting nonlinear and highly volatile stock movements and propose a stock trend prediction solution, ACEFormer, that achieves more accurate predictions. In the structure, a denoising algorithm, ACEEMD, is proposed which outperforms existing methods in removing noise from stock data. By using the distillation module and the time-aware mechanism, ACEFormer extracts the key features of denoised stock data and generates more precise predictions. In experimental evaluations, ACEFormer demonstrates improved performance in forecasting stock trends.

Acknowledgement. This work is supported in part by the National Nature Science Foundation of China under Grant U2013201 and in part by the Pearl River Talent Plan of Guangdong Province under Grant 2019ZT08X603.

References

1. Cavalcante, R.C., Brasileiro, R.C., Souza, V.L.F., Nobrega, J.P., Oliveira, A.L.I.: Computational intelligence and financial markets: a survey and future directions. Exp. Syst. Appl. **55**(C), 194–211 (2016)

2. Peters, E.E.: A chaotic attractor for the S&P 500. Financ. Anal. J. **47**(2), 55–62 (1991)
3. Guo, Y., Han, S., Shen, C., Li, Y., Yin, X., Bai, Yu.: An adaptive SVR for high-frequency stock price forecasting. IEEE Access **6**, 11397–11404 (2018)
4. Liu, Y., Cao, C., Huang, W., Hao, S.: A deep neural network based model for stock market prediction. In: 2021 IEEE 2nd ICBAIE, pp. 320–323 (2021)
5. Saad, E.W., Prokhorov, D.V., Wunsch, D.C.: Comparative study of stock trend prediction using time delay, recurrent and probabilistic neural networks. IEEE Trans. Neural Netw. **9**(6), 1456–1470 (1998)
6. Feng, F., Chen, H., He, X., Ding, J., Chua, T.S.: Enhancing stock movement prediction with adversarial training. In: Twenty-Eighth International Joint Conference on Artificial Intelligence, IJCAI-19, pp. 1–8 (2019)
7. Nabipour, M., Nayyeri, P., Jabani, H., Shahab, S., Mosavi, A.: Predicting stock market trends using machine learning and deep learning algorithms via continuous and binary data; a comparative analysis. IEEE Access **8**, 150199–150212 (2020)
8. Bahdanau, D., Cho, K., Bengio, Y.: Neural machine translation by jointly learning to align and translate. CoRR, abs/1409.0473 (2015)
9. Vaswani, A., et al.: Attention is all you need. In: Advances in Neural Information Processing Systems, vol. 30, pp. 5998–6008 (2017)
10. Zhang, S., Zhang, H.: Prediction of stock closing prices based on attention mechanism. IEEE Access **9**, 36591–36600 (2021)
11. Wang, C., Ren, J., Liang, H.: MSGraph: modeling multi-scale k-line sequences with graph attention network for profitable indices recommendation. Electron. Res. Arch. **31**(5), 2626–2650 (2023)
12. Ji, Z., Wu, P., Ling, C., Zhu, P.: Exploring the impact of investor's sentiment tendency in varying input window length for stock price prediction. Multimedia Tools Appl. **82**, 27415–27449 (2023)
13. Feng, S., Feng, Y.: A dual-staged attention based conversion-gated long short term memory for multivariable time series prediction. IEEE Access **10**, 368–379 (2022)
14. Ding, Q., Wu, S., Sun, H., Guo, J., Guo, J.: Hierarchical multi-scale gaussian transformer for stock movement prediction. In: IJCAI, pp. 4640–4646 (2020)
15. Dinesh, S., Rao, N.R., Anusha, S.P., Samhitha, R.: Prediction of trends in stock market using moving averages and machine learning. In: 2021 International Conference on Advances in Computing, Communication Control and Networking (ICACCCN), pp. 1–5 (2021)
16. Mandic, D.P., Rehman, N.U., Wu, Z., Huang, N.E.: Empirical mode decomposition-based time-frequency analysis of multivariate signals: the power of adaptive data analysis. IEEE Sig. Process. Mag. **30**(6), 74–86 (2013)
17. Hu, J., Wang, X., Qin, H.: Novel and efficient computation of Hilbert-Huang transform on surfaces. Comput. Aided Geom. Des. **43**, 95–108 (2016)
18. Ge, S., Rum, S.N.B.M., Ibrahim, H., Marsilah, E., Perumal, T.: An effective source number enumeration approach based on SEMD. IEEE Access **10**, 96066–96078 (2022)
19. Zhang, X., Zhang, X., Li, Y.: Coal thickness prediction method based on VMD and LSTM. J. Sens. **2021**, 1–10 (2021)
20. Zhou, H., et al.: Informer: Beyond efficient transformer for long sequence time-series forecasting. Proc. AAAI Conf. Artif. Intell. **35**(12), 11106–11115 (2021)
21. Wu, H., Hu, T., Liu, Y., Zhou, H., Wang, J., Long, M.: TimesNet: temporal 2D-variation modeling for general time series analysis. In: International Conference on Learning Representations (2023)

22. Zeng, A., Chen, M., Zhang, L., Xu, Q.: Are transformers effective for time series forecasting? (2023)
23. Liu, Y., Wu, H., Wang, J., Long, M.: Non-stationary transformers: exploring the stationarity in time series forecasting (2022)
24. Sawhney, R., Agarwal, S., Wadhwa, A., Derr, T., Shah, R.R.: Stock selection via spatiotemporal hypergraph attention network: a learning to rank approach. In: AAAI Conference on Artificial Intelligence, vol. 35, pp. 497–504 (2021)

Multiscale Network with Equivalent Large Kernel Attention for Crowd Counting

Zhiwei Wu, Wenhui Gong, Yan Chen, Xiaofeng Xia, and Jun Sang(✉)

School of Big Data and Software Engineering, Chongqing University, Chongqing 401331, China
jsang@cqu.edu.cn

Abstract. Most of the existing crowd counting methods are based on convolutional neural networks (CNN) to solve the crowd scale and background noise problems. These methods can effectively extract local features, but their convolutional kernel sizes are limited so that it is hard to obtain global information which is also crucial for scale awareness and noise discrimination. In this paper, we propose a Multiscale Network with Equivalent Large Kernel Attention for Crowd Counting (MELANet), which can extract both global and local information based on CNN. MELANet is composed of three parts: feature extraction module (FEM) for original feature extraction, multiscale equivalent attention module (MEAM) for global and local information combination, and fusion module (FM) for multiscale feature fusion. In MEAM, by decomposing large convolution kernels into equivalent combinations of small convolution kernels, the model obtains receptive fields equivalent to the large convolutional kernels with lower complexity and less parameters. It enables local and global correlation in the attention mechanism based on CNN, which makes the model focus more on the crowd head region to resist the background noise. Besides, we use a multiscale structure and different convolution kernel sizes to encode contextual information at different scales into the feature maps to deal with head scale transformations. Furthermore, we add gate channel attention units in MEAM to enhance the channel adaptivity of the model. Extensive experiments demonstrate that MELANet can achieve excellent counting performance on three popular crowd counting datasets.

Keywords: Crowd counting · Multiscale · Equivalent kernel · Attention

1 Introduction

As one of the research topics in computer vision, crowd counting has developed rapidly in the last few years. It aims to analyze crowd scene in a given image or video, and then quickly count the number of people in the scene, which has a wide range of applications in areas such as behavior analysis, crowd control, and smart city planning. Although a variety of crowd counting networks have been designed to improve the accuracy of crowd counting tasks, scale variation and background noise remain challenging issues that hinder the improvement of crowd counting accuracy. Therefore, crowd counting is still a computer vision topic worthy of further study.

B. Luo et al. (Eds.): ICONIP 2023, CCIS 1965, pp. 538–550, 2024.
https://doi.org/10.1007/978-981-99-8145-8_41

With the proposal of multi-column convolutional neural network (MCNN) [1], the methods combining CNN with density map estimation are gradually becoming the mainstream crowd counting methods. For the crowd counting methods based on density map estimation, they aim to estimate the crowd density maps corresponding to the input images by using CNN and then obtain the total counts by integration. These methods not only effectively improve the accuracy of crowd counting, but also make full use of the spatial distribution information of the crowd with the ground truth. To solve the problem of scale variation in crowd images, some researchers have used convolutional kernels of different sizes in the feature extraction stage so that the models acquire different sizes of receptive fields to deal with head scale variations [1–6]. For the background noise, many methods use visual attention mechanism to generate attention maps with weights, which are combined with feature maps to highlight the targets to be counted [7–11]. The models constructed by these methods are a great improvement on the CNN and can effectively obtain local feature correlation. However, they use limited convolutional kernel sizes, which make the models lack of global information.

To capture the global information of images, some recent approaches [12–14] build crowd counting models based on Transformer, which fully utilize the self-attention and global information modeling capability of Transformer. However, Transformer-based counting models usually need to stack multiple encoder blocks to obtain multiscale feature map information. This makes the model structure complex and the parameters are much higher than the general CNN-based crowd counting models. It leads to higher requirements for training equipment as well.

To overcome the above issues, we propose a Multiscale Network with Equivalent Large Kernel Attention for Crowd Counting (MELANet) to extract global attention and local attention simultaneously. It uses a simple structure with parallel multiscale large kernel attention units to effectively adapt to crowd image head scale transformations and reduce background noise. Excellent counting performance is achieved with a small number of parameters and costs.

In summary, the main contributions of our work are as follows:

- We propose a crowd counting network MELANet. It combines global and local correlation in the attention mechanism through multiscale and different convolutional kernel sizes to deal with crowd head scale transformations and reduce background noise.
- To reduce the parameters and the costs of the network, we design a multiscale equivalent attention module by decomposing large convolutional kernels into equivalent combinations of small convolutional kernels.
- We design gate channel attention units to further enhance the channel adaptability of the network.
- The experimental results show that the proposed MELANet achieves excellent counting performance on the three existing crowd counting benchmark datasets.

2 Related Works

2.1 Multiscale Crowd Counting Methods

Due to the different distances of the crowd from the camera, there are usually differences in head sizes in different regions of the scene, resulting in the inability to accommodate all head size variations using a single size convolution kernel. With regards to this, many methods use a multi-column convolutional structure to capture different sizes of receptive fields to accommodate variations in head scales. Zhang et al. [1] designed a multi-column convolutional neural network (MCNN) for crowd counting. They first introduced the concept of mapping images to their crowd density maps, laying the foundation for density map counting. Sam et al. [2] proposed Switching-CNN, it divided the image into image patches and dynamically selected the appropriate network branches to extract multiscale features based on the size of the crowd heads within the image patches. Following the above work, Sindagi et al. [3] proposed a novel contextual pyramid convolutional neural network (CP-CNN). It employed two additional branches to incorporate local and global contextual information into the estimated density map to generate a high-quality estimated density map. Cao et al. [4] designed a simple and effective end-to-end crowd counting network (SANet). In order to reduce the effect of scale differences on the counts, they performed multiple iterations of the input images by using four different sizes of convolution kernels.

Recent approaches have attempted to utilize new strategies to learn crowd head scale variations. Liu et al. [5] took advantage of right contextual information at each image location, thus incorporating multiscale contextual information into the estimated density map to deal with changes in crowd size. Song et al. [6] proposed a novel multilevel-based scale adaptive selection network (SASNet). It adaptively selected a learning strategy based on the heads of people of different sizes in the image, so that the network adapted to image scale changes.

2.2 Attention-Based Methods

Attention mechanisms are widely used in the field of computer vision [15, 16] with the aim of enabling models to assign higher attention to task-specific targets like the human visual system. Attention mechanisms have also been used with great success when applied to the field of crowd counting. Liu et al. [7] designed a framework decision network (DecideNet). Through an attention mechanism, it adaptively assigned weights to detection network branches and regression network branches to achieve optimal counting. To obtain long-range contextual information, Gao et al. [8] came up with SCAR for crowd counting, which combined Spatial-wise Attention Model and Channel-wise Attention Model. Zhang et al. [9] proposed a new network structure MRA-CNN. It guided the network to focus on the crowd head region by generating the attention map corresponding to the feature map separately. Guo et al. [10] proposed DADNet and they used a scale-aware attention fusion module to capture the visual representation of multiscale features, which not only effectively resisted scale variations but also reduced the interference of background noise. Sindagi et al. [11] designed a hierarchical attention-based crowd counting network (HA-CNN). It adopted spatial attention module and global attention module to generate attention maps of multi-level features separately, and fused them to generate higher quality density maps.

In fact, some researchers usually adopt a combination of multiscale and attention for network design when using attention mechanism for crowd counting [17–20]. Hossain et al. [17] proposed an end-to-end network SAAN, which contained multiscale features and scale attention information. Wang et al. [20] came up with a hybrid attention network (HANet), and it used a progressive learning strategy to embed multiscale contextual information into the estimated density map. Chen et al. [21] first applied the variational attention to the field of crowd counting, extracting information about the differences within and between crowd images through attention to improve counting performance.

3 Proposed Approach

In this section, we first introduce the architecture of the proposed Multiscale Network with Equivalent Large Kernel Attention for Crowd Counting (MELANet), and then describe the specific modules in MELANet.

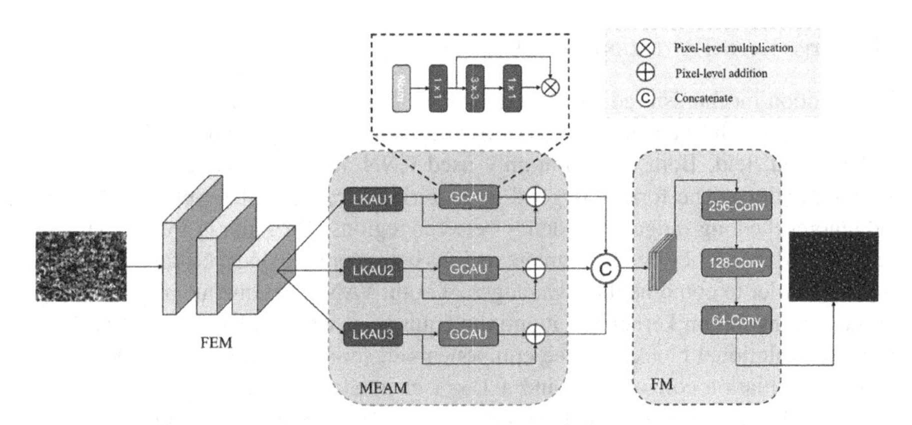

Fig. 1. The architecture of MELANet.

As shown in Fig. 1, MELANet contains three modules: feature extraction module (FEM), multiscale equivalent attention module (MEAM), and fusion module (FM). VGG [22] is widely used as a feature extraction module in computer vision because of its simple network structure and excellent feature extraction ability. Hence, similar to [11, 20, 23], we use the pretrained VGG16-BN as FEM, which is commonly used in crowd counting. In FEM, the input images are output with richer original feature maps as inputs of the MEAM. The MEAM consists of three branches of large kernel attention units (LKAUs) with different convolution kernel sizes and gate channel attention units (GCAUs) with the same structure. LKAU can obtain receptive fields equivalent to large convolutional kernels, while capturing local attention through small convolutional kernels. The combination of both global and local information helps the network to pay more attention to the head regions for distinguishing background noise. The purpose of using three scale branches is to obtain the feature maps of contextual information at different scales to deal with the crowd head scale transformations. Since LKAU pays less attention to channel features, we add adaptive GCAU after each LKAU to enhance the adaptability of network in the channels. In FM, the enhanced feature maps output from MEAM which contain rich contextual information with different scales are concatenated in the channel dimension. Then they pass through a series of convolutional layers in FM to generate the final crowd estimated density map, as illustrated in Fig. 1. 256-Conv, 128-Conv, and 64-Conv represent 3×3 convolutional layers with output channels of 256, 128, and 64, respectively. We use Euclidean Loss (L_{MSE}) and Optimal Transport Loss (L_{OT}) proposed in [24] as loss function for training. The final loss function is shown in Eq. 1.

$$L = \lambda_1 L_{MSE} + \lambda_2 L_{OT} \tag{1}$$

where λ_1, λ_2 represent the weight factors of L_{MSE} and L_{OT}, respectively. λ_2 uses the same value 0.1 as in [24], and λ_1 is set to 1.

Specific implementations of the proposed LKAU and GCAU are introduced in the following sections.

3.1 Large Kernel Attention Unit (LKAU)

The attention mechanism adaptively assigns weight to different regions by generating attention maps. This concept can effectively resist the interference of background noise in the vision field. Both the commonly used CNN attention and self-attention have some limitations. The former is good at acquiring local receptive fields in the image and ignores the long-range dependency between regions. The latter is able to extract the long-range dependency between image regions but ignores the local correlation between pixels. In order to combine the advantages of both, VAN [25] equivalently decomposes the large convolution kernel for attention acquisition, where a large kernel is divided into a convolutional block of three components in series: a depth-wise convolution, a depth-wise dilation convolution, and a 1×1 channel convolution. Inspired by it, we first apply the idea of large kernel attention to crowd counting study, that is, large kernel attention unit (LKAU) in each branch of MEAM. LKAU can focus more on head areas to distinguish background noise. The structure of LKAU is shown in Fig. 2. This process can be expressed in Eq. 2.

$$F_{LKAU} = \sigma[G_1(f_{DWD}(f_{DW}(X)))] \otimes X \tag{2}$$

where X is the input feature map and $X \in R^{C \times H \times W}$. After X goes through a depth-wise convolution f_{DW}, a depth-wise dilation convolution f_{DWD} and a 1×1 channel convolution G_1, Sigmoid activation function σ is applied to generate the attention map. Finally, the attention map and the input feature map X are multiplied at the pixel level to obtain the enhanced feature map F_{LKAU}.

In VAN, it simply uses a single scale of large kernel attention, ignoring the continuity of scale information as well as smoothing, making the global and local information obtained by the network limited. To this end, our approach improves it by using large kernel attention units with multiple scales and different kernel sizes to incorporate rich global and local information into the attention map. It makes the network adapt to the smooth transformation of crowd head scale while distinguishing background noise. In addition, compared with using large convolution kernel directly, this combination method can greatly reduce the number of parameters in the network. Specifically, the parameters of our designed multiscale LKAU are shown in Table 1.

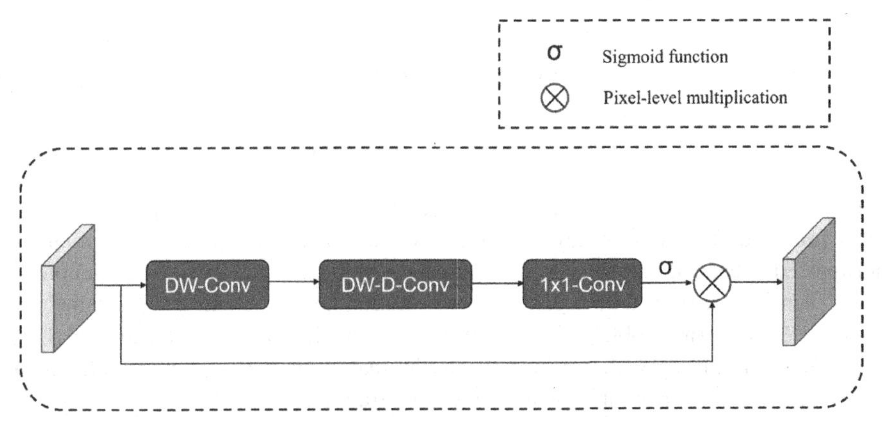

Fig. 2. The specific architecture of LKAU. The convolution kernel size of DW-Conv is (2d-1) and the convolution kernel size of DW-D-Conv is (K/d), where d denotes the dilation rate of DW-D-Conv and K is the equivalent large convolution kernel.

Table 1. The parameters of multiscale LKAU in MEAM.

Unit	DW-Conv	DW-D-Conv	EQ-Conv
LKAU1	7	9	35
LKAU2	5	7	21
LKAU3	3	5	9

In Table 1, the equivalent decomposition of the large kernel convolution is referred to the rules in VAN. DW-Conv represents a depth-wise convolution with a convolution kernel of 2d-1 and DW-D-Conv represents a depth-wise dilation convolution with a convolution kernel of K/d, where d denotes the dilation rate of DW-D-Conv and K is the equivalent large convolution kernel (EQ-Conv). The reason for MEAM to use three scale branches will be explained in detail later in the ablation study.

3.2 Gate Channel Attention Unit (GCAU)

The LKAU achieves spatial adaptivity of the network well by depth-wise convolution and dilation convolution, but it simply uses 1×1 convolution to obtain the channel information of the features. This is not enough. To further enhance the channel adaptivity of the network, a gate channel attention unit (GCAU) is connected behind each

LKAU in the multiscale equivalent attention module. In addition, due to the use of dilated convolution in the equivalent representation of the large convolution kernel, the enhanced feature maps will have features that do not exist in the original images. This phenomenon is called gridding artifact. The addition of GCAU mitigates this gridding artifact phenomenon. As shown in Fig. 1, GCAU consists of a norm layer, two 1×1 channel convolution layers, and a 3×3 convolution layer. The specific process is as follows.

$$Z_{NC} = G_1(f_N(F_{LKAU})) \tag{3}$$

$$F_{GCAU} = \alpha[Z_{NC} \otimes G_1(G_3(Z_{NC}))] \tag{4}$$

where F_{LKAU} represents the enhanced feature map output by LKAU. F_{LKAU} is sequentially processed by the norm layer f_N and 1×1 channel convolution G_1 to obtain the normalized enhanced feature map representation in the channel dimension, denoted as Z_{NC}. Then, Z_{NC} is passed through 3×3 convolution layers G_3 and 1×1 channel convolution G_1 in turn to obtain the attention map. Finally, the attention map and Z_{NC} are multiplied at the pixel level to generate the feature map F_{GCAU} with rich channel information. α is the learnable parameter during training.

4 Implementation Details

4.1 Ground Truth Generation

Our proposed method is based on density maps for crowd counting estimation. However, the ground truth provided by existing crowd counting datasets are discrete crowd head coordinates. Therefore, we need to convert the discrete crowd head coordinates into a continuous crowd density map, namely the ground truth density map. Assuming that x_i is the head coordinate of the crowd in the ground truth, we use the Gaussian function G to convolve it and obtain the ground truth density map D_{GT} as shown in Eq. 5.

$$D_{GT} = \sum_{i=1}^{N} \delta(x - x_i) \times G_\sigma \tag{5}$$

where N is the total number of people in an original label dot map, δ is a delta function and σ is the gaussian kernel of G.

4.2 Datasets

UCF_CC_50 [26]: This is a dataset consisting of 50 crowd images containing various scenes, each crowd image contains between 94 and 4543 head center annotations. Although the number of images in this dataset is small, the images in it are all dense crowd images. Therefore, it has certain reference value to the evaluation of crowd counting models.

ShanghaiTech [1]: ShanghaiTech dataset consists of two parts, ShanghaiTech Part A and ShanghaiTech Part B. It contains a total of 1198 crowd images. There are 300 training images and 182 testing images in ShanghaiTech Part A, while ShanghaiTech Part B has 400 training images and 316 testing images. This dataset is recognized as the authoritative dataset in crowd counting.

UCF-QNRF [27]: UCF-QNRF is a large dataset that includes a total of 1535 crowd images. In the dataset, all images are divided into 1201 training images and 334 testing images. Due to the drastic scale and size transformation between different crowd images in this dataset, it is considered to be a dataset that more closely to the real crowd scene than UCF_CC_50 and ShanghaiTech. It is also a challenging dataset in crowd counting.

4.3 Training Details

The feature extraction network of the proposed MELANet employs the first ten layers of the pre-trained VGG16-BN. The other convolutional layers in the network structure are randomly initialized by a Gaussian distribution with a mean of 0 and a standard deviation of 0.01. We adopt AdamW optimizer with a learning rate of 1e−4 to train the model, and set the weight decay parameter to 1e−4. As for batch size and epoch, they are 16 and 2000 respectively.

In order to enhance the training data with limited images in the dataset, we improve the diversity of input images in various ways in the training phase. In each epoch, we set different crop sizes to crop the images randomly according to the characteristics of different crowd datasets. In addition, we also randomly flip the images with 0.5 probability after cropping to make the model more generalizable and robust.

5 Experiments

5.1 Evaluation Metrics

To evaluate the performance of the proposed model, we adopt the mean absolute error (MAE) which represents the accuracy of the model to calculate the error between the estimated and the ground truth number. Further, the deviation degree between the estimated number and the ground truth number is indicated by the root mean square error (RMSE). The MAE and RMSE are expressed as shown in Eqs. 6 and 7, respectively.

$$MAE = \frac{1}{N}\sum_{i=1}^{N}\left|y_i^{ES} - y_i^{GT}\right| \tag{6}$$

$$RMSE = \sqrt{\frac{1}{N}\sum_{i=1}^{N}(y_i^{ES} - y_i^{GT})^2} \tag{7}$$

where N is the number of images in the dataset, y_i^{ES} represents the estimated crowd counts and y_i^{GT} is the corresponding ground truth number.

Table 2. The results of MAE and RMSE on ShanghaiTech dataset.

Method	Part A		Part B	
	MAE	RMSE	MAE	RMSE
MCNN [1]	110.2	173.2	26.4	41.3
Switch-CNN [2]	90.4	135.0	21.6	33.4
CSRNet [23]	68.2	115.0	10.6	16.0
SANet [4]	67.0	104.5	8.4	13.6
SCAR [8]	66.3	114.1	9.5	15.2
TEDNet [28]	64.2	109.1	8.2	12.8
KDMG [29]	63.8	99.2	7.8	12.7
ADCrowdNet [30]	63.2	98.9	7.7	12.9
HA-CNN [11]	62.9	94.9	8.1	13.4
RANet [31]	59.4	102.0	7.9	12.9
DMCNet [32]	58.5	**84.6**	8.6	13.7
CRNet [33]	56.4	90.4	7.4	11.9
MELANet (Ours)	**55.5**	89.0	**6.8**	**11.3**

5.2 Comparisons with State-of-the-Art

To demonstrate the efficiency of the proposed MELANet, comparative experiments are conducted on three popular crowd counting datasets.

For ShanghaiTech dataset, our proposed MELANet achieves the lowest MAE in both ShanghaiTech Part A and ShanghaiTech Part B datasets compared to 12 crowd counting methods, as shown in Table 2. In ShanghaiTech Part A, the MAE of MELANet is 55.5 and the RMSE is 89.0. The MAE of MELANet is reduced by 0.9 compared to CRNet. In ShanghaiTech Part B, MELANet achieves the best counting results, where MAE is 6.8 and RMSE is 11.3.

As shown in Table 3, the proposed MELANet reaches the lowest MAE (84.7), but the RMSE (147.8) is slightly higher than that of SASNet in the UCF-QNRF dataset. Although MELANet does not have the lowest RMSE on UCF-QNRF, it has a lower MAE of 0.5 compared to SASNet. Because the UCF_CC_50 dataset is small and the crowd images inside are badly occluded, it leads to a higher MAE for all methods. Nevertheless, our proposed method is able to achieve a lowest MAE with a value of 201.9, as shown in Table 4.

Table 3. The results of MAE and RMSE on UCF-QNRF dataset.

Method	UCF-QNRF	
	MAE	RMSE
MCNN [1]	227.0	426.0
Switch-CNN [2]	228.0	445.0
TEDNet [28]	113.0	188.0
DMCNet [32]	96.5	164.0
DM-Count [24]	85.6	148.3
SASNet [6]	85.2	**147.3**
MELANet (Ours)	**84.7**	147.8

Table 4. The results of MAE and RMSE on UCF_CC_50 dataset.

Method	UCF_CC_50	
	MAE	RMSE
MCNN [1]	377.6	509.1
Switch-CNN [2]	318.1	439.2
TEDNet [28]	249.4	354.5
RANet [31]	239.8	319.4
DM-Count [24]	211.0	291.5
CRNet [33]	203.3	**263.4**
MELANet (Ours)	**201.9**	266.0

Above all, our proposed MELANet achieves lower mean absolute errors on all three of the popular crowd counting benchmark datasets. This indicates that our proposed method can reach more accurate counting results. Whether in dense or sparse crowd scenes, the accuracy of MELANet is improved compared to some existing crowd counting methods. This shows that MELANet can effectively adapt to scale transformations and background occlusion by multiscale large kernel attention units. In addition, we select some samples from the testing data of ShanghaiTech Part A for the visualization of MELANet, as shown in Fig. 3.

5.3 Ablation Study

To verify the effectiveness of the proposed MELANet and confirm the optimal scale branch number in MEAM, we perform ablation experiments on the ShanghaiTech Part A dataset, as shown in Table 5. The Baseline uses FEC and FM in MELANet directly connected, without adding MEAM. Based on Baseline, we add MEAM with two scale

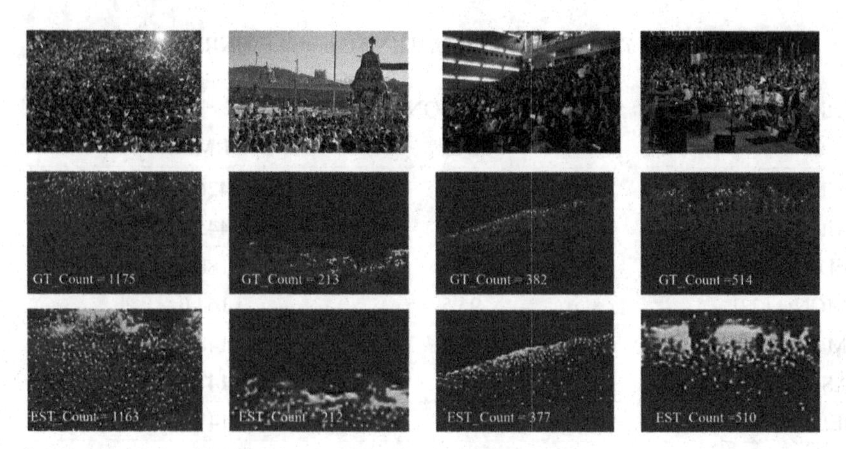

Fig. 3. Visualization of MELANet on ShanghaiTech Part A. The three rows from top to bottom represent the input images, the ground truth density maps, and the estimated density maps.

Table 5. Different combinations of our proposed method on ShanghaiTech Part A.

Index	Module	MAE	RMSE
A	Baseline (without MEAM)	65.4	102.2
B	Baseline + two scale branches in MEAM	60.4	92.3
C	Baseline + three scale branches in MEAM	**55.5**	**89.0**
D	Baseline + four scale branches in MEAM	56.0	92.3

branches, MEAM with three scale branches, and MEAM with four scale branches respectively for comparison experiments. According to row A and other rows, we find that the counting accuracy is significantly improved after adding MEAM on the basis of Baseline. This indicates that our proposed MEAM can effectively improve the performance of crowd counting. According to the comparison of rows B, C and D in Table 5, it can be seen that MAE and RMSE are the lowest when three scale branches are adopted in MEAM. At this point, our method is able to achieve MAE of 55.5 and RMSE of 89.0 on the ShanghaiTech Part A dataset. Therefore, we use MEAM with three scale branches in the proposed MELANet.

6 Conclusion

In this paper, a multiscale network based on equivalent large kernel attention is proposed for crowd counting. It combines the advantages of global attention and local attention by fusing long-range dependence and local relevance, which enables the network to adapt to crowd head scale transformations and resist background noise. On three popular crowd counting datasets, our method achieves better counting performance with smaller number of parameters and costs compared to existing crowd counting methods.

Acknowledgments. This work was supported by National Natural Science Foundation of China (No. 61971073).

References

1. Zhang, Y., Zhou, D., Chen, S., Gao, S., Ma, Y.: Single-image crowd counting via multi-column convolutional neural network. In: CVPR, pp. 589–597 (2016)
2. Sam, D.B., Surya, S., Babu, R.V.: Switching convolutional neural network for crowd counting. In: CVPR, pp. 4031–4039 (2017)
3. Sindagi, V.A., Patel, V.M.: Generating high-quality crowd density maps using contextual pyramid CNNs. In: ICCV, pp. 1879–1888 (2017)
4. Cao, X., Wang, Z., Zhao, Y., Su, F.: Scale aggregation network for accurate and efficient crowd counting. In: Ferrari, V., Hebert, M., Sminchisescu, C., Weiss, Y. (eds.) ECCV 2018. LNCS, vol. 11209, pp. 757–773. Springer, Cham (2018). https://doi.org/10.1007/978-3-030-01228-1_45
5. Liu, W., Salzmann, M., Fua, P.: Context-aware crowd counting. In: CVPR, pp. 5094–5103 (2019)
6. Song, Q., et al.: To choose or to fuse? Scale selection for crowd counting. In: AAAI, pp. 2576–2583 (2021)
7. Liu, J., Gao, C., Meng, D., Hauptmann, A.G.: DecideNet: counting varying density crowds through attention guided detection and density estimation. In: CVPR, pp. 5197–5206 (2018)
8. Gao, J., Wang, Q., Yuan, Y.: SCAR: spatial-/channel-wise attention regression networks for crowd counting. Neurocomputing **363**, 1–8 (2019)
9. Zhang, Y., Zhou, C., Chang, F., Kot, A.C.: Multi-resolution attention convolutional neural network for crowd counting. Neurocomputing **329**, 144–152 (2019)
10. Guo, D., Li, K., Zha, Z., Wang, M.: DADNet: dilated-attention-deformable ConvNet for crowd counting. In: ACM MM, pp. 1823–1832 (2019)
11. Sindagi, V.A., Patel, V.M.: HA-CCN: hierarchical attention-based crowd counting network. TIP **29**, 323–335 (2020)
12. Bakhtiarnia, A., Zhang, Q., Iosifidis, A.: Single-layer vision transformers for more accurate early exits with less overhead? Neural Netw. **153**, 461–473 (2022)
13. Liang, D., Chen, X., Xu, W., Zhou, Y., Bai, X.: TransCrowd: weakly-supervised crowd counting with transformers. Inf. Sci. **65**(6) (2022)
14. Liang, D., Xu, W., Bai, X.: An end-to-end transformer model for crowd localization. In: Avidan, S., Brostow, G., Cissé, M., Farinella, G.M., Hassner, T. (eds.) ECCV 2022. LNCS, vol. 13661, pp. 38–54. Springer, Cham (2022). https://doi.org/10.1007/978-3-031-19769-7_3
15. Cao, C., et al.: Look and think twice: capturing top-down visual attention with feedback convolutional neural networks. In: ICCV, pp. 2956–2964 (2015)
16. Yoo, D., Park, S., Lee, J.Y., Paek, A.S., Kweon, I.S.: AttentionNet: aggregating weak directions for accurate object detection. In: ICCV, pp. 2659–2667 (2015)
17. Hossain, M.A., Hosseinzadeh, M., Chanda, O., Wang, Y.: Crowd counting using scale-aware attention networks. In: WACV, pp. 1280–1288 (2019)
18. Tian, Y., Lei, Y., Zhang, J., Wang, J.: PaDNet: pan-density crowd counting. TIP **29**, 2714–2727 (2020)
19. Jiang, X., et al.: Density-aware multi-task learning for crowd counting. IEEE Trans. Multimedia **23**, 443–453 (2021)
20. Wang, F., Sang, J., Wu, Z., Liu, Q., Sang, N.: Hybrid attention network based on progressive embedding scale-context for crowd counting. Inf. Sci. **591**, 306–318 (2022)

21. Chen, B., Yan, Z., Li, K., Li, P., Wang, B., Zuo, W., Zhang, L.: Variational attention: propagating domain-specific knowledge for multi-domain learning in crowd counting. In: ICCV, pp. 16045–16055 (2021)
22. Simonyan, K., Zisserman, A.: Very deep convolutional networks for large-scale image recognition. Comput. Sci. (2014)
23. Li, Y., Zhang, X., Chen, D.: CSRNet: dilated convolutional neural networks for understanding the highly congested scenes. In: CVPR, pp. 1091–1100 (2018)
24. Wang, B., Liu, H., Samaras, D., Hoai, M.: Distribution matching for crowd counting. In: Proceedings of the 34th International Conference on Neural Information Processing Systems, Article no. 135. Curran Associates Inc., Vancouver, BC, Canada (2020)
25. Guo, M., Lu, C., Liu, Z., Cheng, M., Hu, S.: Visual attention network arXiv preprint arXiv:2202.09741 (2022)
26. Idrees, H., Saleemi, I., Seibert, C., Shah, M.: Multi-source multi-scale counting in extremely dense crowd images. In: CVPR, pp. 2547–2554 (2013)
27. Idrees, H., et al.: Composition loss for counting, density map estimation and localization in dense crowds. In: Ferrari, V., Hebert, M., Sminchisescu, C., Weiss, Y. (eds.) ECCV 2018. LNCS, vol. 11206, pp. 544–559. Springer, Cham (2018). https://doi.org/10.1007/978-3-030-01216-8_33
28. Jiang, X., et al.: Crowd counting and density estimation by trellis encoder-decoder networks. In: CVPR, pp. 6126–6135 (2019)
29. Wan, J., Wang, Q., Chan, A.: Kernel-based density map generation for dense object counting. IEEE Trans. Pattern Anal. Mach. Intell. **44**(3), 1357–1370 (2022)
30. Liu, N., Long, Y., Zou, C., Niu, Q., Pan, L., Wu, H.: ADCrowdNet: an attention-injective deformable convolutional network for crowd understanding. In: CVPR, pp. 3220–3229 (2019)
31. Zhang, A., et al.: Relational attention network for crowd counting. In: ICCV, pp. 6787–6796 (2019)
32. Wang, M., Cai, H., Dai, Y., Gong, M.: Dynamic mixture of counter network for location-agnostic crowd counting. In: Proceedings of the IEEE/CVF Winter Conference on Applications of Computer Vision, pp. 167–177 (2023)
33. Liu, Y., et al.: Crowd counting via cross-stage refinement networks. TIP **29**, 6800–6812 (2020)

M^3FGM: A Node Masking and Multi-granularity Message Passing-Based Federated Graph Model for Spatial-Temporal Data Prediction

Yuxing Tian[1], Jiachi Luo[1], Zheng Liu[2], Song Li[3], and Yanwen Qu[1(✉)]

[1] Jiangxi Normal University, Nanchang, China
qu_yw@jxnu.edu.cn
[2] Nanjing University of Posts and Telecommunications, Nanjing, China
[3] Shanghai Enflame Technology Co. Ltd., Shanghai, China

Abstract. Researchers are solving the challenges of spatial-temporal prediction by combining Federated Learning (FL) and graph models with respect to the constrain of privacy and security. In order to make better use of the power of graph model, some researchs also combine split learning (SL). However, there are still several issues left unattended: 1) Clients might not be able to access the server during inference phase; 2) The graph of clients designed manually in the server model may not reveal the proper relationship between clients. This paper proposes a new GNN-oriented split federated learning method, named node **M**asking and **M**ulti-granularity **M**essage passing-based Federated Graph Model (M^3FGM) for the above issues. For the first issue, the server model of M^3FGM employs a MaskNode layer to simulate the case of clients being offline. We also redesign the decoder of the client model using a dual-sub-decoders structure so that each client model can use its local data to predict independently when offline. As for the second issue, a new GNN layer named Multi-Granularity Message Passing (MGMP) layer enables each client node to perceive global and local information. We conducted extensive experiments in two different scenarios on two real traffic datasets. Results show that M^3FGM outperforms the baselines and variant models, achieves the best results in both datasets and scenarios.

Keywords: Federated learning · split learning · spatial-temporal data prediction · graph neural network · data privacy

1 Introduction

Utilizing graph structure to model spatial-temporal data in the prediction task has been popular in recent years [6,8,13,23,25,29]. It is critical for various applications including traffic flow prediction, forecasting, and user activity detection.

B. Luo et al. (Eds.): ICONIP 2023, CCIS 1965, pp. 551–566, 2024.
https://doi.org/10.1007/978-981-99-8145-8_42

Most of these works train models under the assumption that a massive amount of real-world spatial-temporal data can be centralized. However, with increasing concerns about data privacy and access restrictions due to existing licensing agreements and commercial competition, there are numerous real-world cases in which spatial-temporal data is decentralized. For instance, in traffic flow prediction, different organizations or companies collect traffic data by their private deployed road sensors and these data cannot exchange it due to privacy preservation or commercial reasons.

As an effective solution to data privacy protection, Federated Learning (FL) [18] has attracted significant research efforts recently. FL is a learning paradigm for model training that collaborates with clients (i.e., local data owners) without exposing their original data. By integrating all client model weights or gradients, the FL-trained model demonstrates superior generalization capabilities.

Recent research has introduced a series of FL-based models for spatial-temporal data prediction while preserving privacy [16,32]. However, these models do not consider the inherent spatial dependencies of the data. Current works focus on integrating FL with graph neural networks (GNNs), which can be divided into two categories: *1) Client-side GNN model training for local model updates*: A common characteristic of these approaches [2,15] is their emphasis on training with well-established graph-structured data. In practice, not all clients possess built-in graph structure datasets, which raises the question of how to process node-level data using GNNs in such contexts. *2) Server-based GNN model training for enhanced FL aggregation*: Techniques such as PFL [4] employ GCN to perform model parameters aggregation according to the clients' relational graph structure, introducing a supervised loss function with graph smoothness regularization for training both local and server models. BiG-Fed [30] devises bi-level optimization schemes for training local models and GNN models with dual objective functions and proposes an unsupervised contrastive learning loss function. Despite these methods consider the structural relationships among clients and offer GNN-based model parameter aggregation techniques, they do not fully exploit the true capabilities of GNNs as they are unable to directly model the dependency relationships within spatial-temporal data. Consequently, their performance is significantly distant from that of centralized GNN approaches.

In recent years, there has been an architectural approach called Split Federated Learning (SFL) that divides a complete model into several parts, placing them on the client and server sides respectively, such as [26] and [7]. This approach is primarily adopted due to the limited computational resources of the devices participating in federated learning. However, recently, Meng et al. have successfully employed this framework to enable GNNs to directly participate in spatial-temporal data processing, proposed CNFGNN [19]. Specifically, CNFGNN partitions the complete model into two components: employing identical encoder-decoder models on all clients, with the encoder used to extract local temporal embeddings, and the decoder utilized to generate predictions. Graph Network (GN) [1] is employed on the server side to obtain spatial embeddings by aggregating the local temporal embeddings uploaded from the clients. CNFGNN can be regarded as a GNN-oriented SFL method.

Nonetheless, two significant issues remain. (1) For CNFGNN, when employing trained model for inference, some clients might be unable to connect to the server due to network disconnection. While it is feasible to replace missing embeddings with all-zero data, this approach significantly diminishes predictive performance. Moreover, these offline clients cannot generate predictions without the server model. (2) The performance of GNN training relies heavily on the accuracy of the graph structure. However, the graph structure of clients in existing methods [19,31,33] is constructed manually in a heuristic way, which might not represent client relations properly, leading to deteriorated performance.

In this paper, we propose a new GNN-oriented split federated learning method for spatial-temporal data prediction, named node **M**asking and **M**ulti-granularity **M**essage passing-based Federated Graph Model (M^{3}FGM) to overcome the above issues. To address the concern of offline clients, we propose a MaskNode layer to the server model to simulate that clients are offline during the training phase. Additionally, we devise a dual-sub-decoders structure for the client model's decoder, permitting offline clients to make predictions during the inference phase. For the issue of graph structures, a new GNN layer, named Multi-Granularity Message Passing (MGMP) layer, is proposed. We construct a comprehensive coarse-grained graph, referred to as the cluster graph, by applying spectral clustering on the client graph. The MGMP layer empowers each client node to aggregate fine-grained local information from neighbors in the client graph and global coarse-grained information from the cluster graph.

The contribution of this paper is summarized as follows:

(1) As far as we know, this paper is the first to consider the non-ideal scenario when designing a GNN-oriented SFL method. We propose MaskNode to enhance the model robustness and design a dual-sub-decoders structure, enabling offline clients to make independent predictions.
(2) We propose a novel GNN Layer, MGMP Layer, which enables client nodes to perceive local and global information through multi-granularity message passing.
(3) We propose M^{3}FGM for spatial-temporal data prediction under privacy protection. The extensive experiments demonstrate the effectiveness of our model on two real-world traffic datasets.

2 Related Work

Our method combines elements from graph neural networks, split federated learning. We now review related works in these areas and discuss their relevance to our work.

2.1 Graph Neural Networks

Graph Neural Networks (GNNs) have demonstrated outstanding efficacy across a diverse range of learning tasks involving graph-structured data, such as node classification [12,27,28], link prediction [3,17], spatial-temporal data modeling [20,22]. Although GNNs exploit a powerful inductive bias to extract meaningful information from graph-structured data, there are challenges that need to be addressed to fully exploit their potential. One critical aspect of GNN performance is the accurate representation of graph structure. In real-world scenarios, graph structures can be highly complex, making it difficult to manually construct them using only prior knowledge. Another challenge faced by GNNs is the difficulty in capturing long-range dependencies within a limited number of message passing steps. This limitation can hinder the learning capabilities of GNNs, especially in scenarios where long-range interactions play a significant role. Therefore, in this paper, we propose a novel GNN Layer to address above issues through multi-granularity message passing.

Furthermore, most studies necessitate centralized data during training and inference processes. This reliance on centralized data leads to privacy concerns, especially when dealing with sensitive information in domains like healthcare, finance, or social networks. Consequently, there is a burgeoning interest in developing privacy-preserving GNNs that facilitate distributed learning across multiple entities, ensuring data confidentiality and compliance with data protection regulations.

2.2 Split Federated Learning

Federated learning (FL) [18] is a machine learning paradigm that enables multiple entities, such as mobile devices, edge nodes, or data centers, to collaboratively train a model while maintaining the privacy and decentralization of their local data. In a typical federated learning setting, multiple clients and a central server participate in training a global model. The global model is copied in multiple copies and deployed on each client. Each participating client trains the model locally using its data and sends only the updated model parameters to the central server for aggregation.

Split learning (SL) [9] is a technique that divides a complete model into several components to enable efficient utilization of computational resources across a network of devices, leveraging their individual strengths while minimizing the overall computational burden. SL can also achieve increased scalability in large-scale distributed systems.

Recent research has integrated SL with FL to address high training latency for clients with limited resources [7,26]. This combination, referred to as Split Federated Learning (SFL), typically divides the global model into two components: client-side and server-side components. Clients access only the client-side component, while the server exclusively accesses the server-side component. In SFL, clients send their processed data (outputs from the client-side model) to the server, where the server-side model continues training. After calculating the loss and updating the gradient, the server adjusts the server-side model, and the gradients of the processed data are sent back to the clients. Clients then update the client-side model based on the gradient. By training part of the model on the server, SFL significantly reduces the computational burden for resource-constrained devices. Collaborative training between clients and the server ensures that the original data remains stored locally on the client, preventing sensitive information disclosure. However, most studies primarily focus on standard deep learning models such as CNN and RNN, with GNN-oriented SFL being rarely studied. CNFGNN [19] can be regarded as an example of a GNN-oriented SFL method. The GNN-oriented SFL method can truly unleash the potential of graph models in modeling graph-structured data, as the input for the GNN portion of the model is processed data rather than model parameters.

3 Problem Formulation

We introduce notions and definitions in this section, followed by a brief introduction to the GNN-oriented split federated learning. Let us denote the client graph constructed by the server as $G = \{V, E\}$, where V is the set of client nodes, and E represents the edge set. $v_i \in V$ denotes the i-th client node in the G. $N = |V|$ is the number of client nodes (the number of clients). c_i represents the client corresponding to the client node v_i. Let $x_i^{t_1:t_2}$ denotes the local graph signals recorded between the timestamp t_1 and t_2 at client i . $X^{t_1:t_2}$ denotes the graph signals observed at all clients between the timestamp t_1 and t_2.

The GNN-oriented split federated learning method aims to learn a client model f_i for each client c_i, and a GNN modelf_{ser} for the server. At each time step t, each client model c_i uses an encoder f_i^{enc} to extract local temporal embedding h_i^t according to $x_i^{(t-S:t)}$. Server model f_{ser} computes the spatial embeddings $\{s_i^t\}_{i=1}^N$ according to G and the local temporal embeddings $\{h_i^t\}_{i=1}^N$ collected from all clients. Each client model c_i then uses a decoder f_i^{dec} to output prediction according to h_i^t and s_i^t. Thus, the mapping from S historical graph signals $X^{(t-S):t}$ to future T graph signals $X^{(t+1):(t+T)}$ can be achieved.

$$[X^{(t-S):t}, G] \xrightarrow{\{f_i=\{f_i^{enc}, f_i^{dec}\}\}_{i=1}^N, f_{ser}} X^{(t+1):(t+T)} \tag{1}$$

4 Methodology

Figure 1 shows the overall architecture of M^{3}FGM, and we will cover the details of the model in terms of the server model, the client model, and the training and inference process, respectively.

4.1 Server Model

The MaskNode Layer: The MaskNode (MN) layer is employed exclusively during the training phase. Prior to model training, we select a mask rate mr. Upon feeding data into the MN layer, a certain number of client nodes, $mr \times N$, are randomly sampled. When $mr \times N$ is a noninteger, we round it down. The uploaded local temporal embeddings of these sampled client nodes are replaced

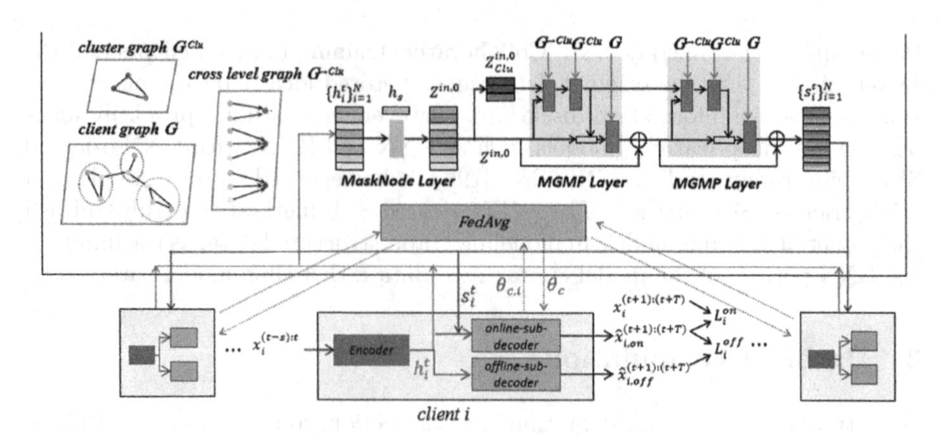

Fig. 1. The overall architecture of M^{3}FGM

with a shared trainable tensor h_s. The set of sampled nodes is denoted as V_{off}, while the set of remaining nodes is denoted as V_{on}. The operation of the MaskNode layer can be expressed by Eq. (2).

$$\left|\begin{cases} z_i^{in,0} = h_i^t, if\ v_i \in V_{on} \\ z_i^{in,0} = h_s, if\ v_i \in V_{off} \end{cases}\right| \tag{2}$$

$Z^{in,0} = \{z_i^{in,0}\}_{i=1}^N$ is the output of the MN layer and will be fed into the first MGMP layer. When the model training is completed and deployed, if client c_i is offline, the server model will utilize the trained tensor h_s as h_i^t to conduct inference. Next, we briefly describe the differences between the MaskNode operation and two related techniques. Unlike the DropEdge operation [21], the MaskNode operation does not perturb the graph structure. In contrast to the masked self-supervised task [11], we only replace the masked node embeddings with shared trainable tensors. We do not attempt to reconstruct or forecast the node embeddings.

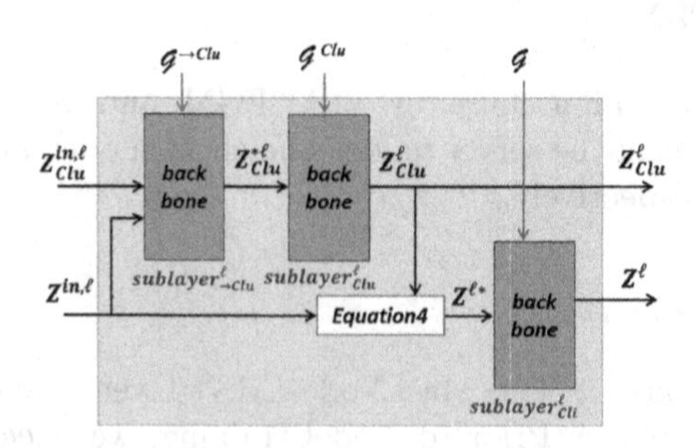

Fig. 2. The structure of MGMP Layer

The MGMP Layer: The MGMP layer employs the following three graph structures for message passing: 1) client graph $G = \{V, E\}$, 2) cluster graph $G^{Clu} = \{V^{Clu}, E^{Clu}\}$, and 3) cross-level graph $G^{\to Clu} = \{V^{\to Clu}, E^{\to Clu}\}$.

The client graph G is constructed manually in a heuristic way. To obtain the cluster graph G^{Clu}, we apply spectral clustering on the Laplacian matrix of the client graph G and get M clusters. Each cluster is regarded as a coarse node of the cluster graph G^{Clu}. Denote the set of client nodes in the m-th cluster as $V_m^{Clu} \subset V$. The edge of the cluster graph G^{Clu} is constructed based on the client graph G, for example, if $V_i \in V_m^{Clu}$ and $V_j \in V_n^{Clu}$, and the V_i connects to V_j in client graph G, then the V_m^{Clu} connects to V_n^{Clu} in cluster graph G^{Clu}.

$G^{\to Clu}$ is a bipartite graph for accelerating the message transfers. $V^{\to Clu} = V \cup V^{Clu}$, $E^{\to Clu}$ is the edge set contains directed edges which are from client nodes in V to the cluster nodes in V^{Clu} corresponding to the cluster of starting node. The diagram of these three graphs is given on the left side of Fig. 1.

Figure 2 shows the internal structure of the l-th MGMP layer, which contains three sub-layers with same backbone. The backbone can be any graph model. Let us assume that the inputs of the l-th layer, which include the input embeddings $Z^{in,l} = \{z_i^{in,l}\}_{i=1}^N$ of client nodes and the input embeddings $Z_{Clu}^{in,l} = \{z_{m,c}^{in,l}\}_{m=1}^M$ of cluster nodes are known. In particular, $Z_{Clu}^{in,0}$ are calculated by Eq. (3):

$$z_{m,c}^{in,0} = \sum_{v_i \in V_m^{Clu}} z_i^{in,0} / |V_m^{Clu}|, \ m = 1, ..., M \tag{3}$$

We will describe how client nodes can perceive the local and global information with the help of MGMP. First, information is propagated on $G^{\to^{Clu}}$ ($sublayer^l_{\to Clu}$) so that the cluster nodes can perceive cluster-level information from the client graph. Then, information is propagated on G^{Clu}($sublayer^l_{Clu}$) to obtain $Z^l_{Clu} = \{z^l_{m,c}\}_{m=1}^M$ which represents the embeddings of cluster nodes. Afterwards, the information on the cluster graph is passed back to the client graph according to Eq. (4), where $Z^{l*} = \{z_i^{l*}\}_{i=1}^N$ represents the embeddings of the client nodes after perceiving the coarse-grained global information, W^l is a trainable matrix. Finally, fine-grained local information is propagated on G($sublayer^l_{Cli}$) to obtain the $Z^l = \{z_i^l\}_{i=1}^N$, which represents the embeddings of client nodes.

$$z_i^{l*} = z_i^{in,l} || W^l z^l_{m,c}, \ v_i \in V_m^{Clu} \tag{4}$$

Computational Flow of Server Model: The computational flow first passes the MaskNode layer and subsequently passes two MGMP layers with residual connection [10]. Note that we only add the residual connection to the input Z^l. The outputs of the last MGMP layer are sent back to clients.

4.2 Client Model

We employ the encoder-decoder architecture on each client for the modeling of local temporal embeddings. Given an input sequence $x_i^{(t-S):t} \in R^{S \times D}$ on the

i-th client, the encoder sequentially reads the whole sequence and outputs the hidden state h_i^t as the temporal embedding of the input sequence.

$$h_i^t = f_i^{enc}(x_i^{(t-S):t}) \tag{5}$$

The Dual-Sub-decoders Structure. Unlike the usual Encoder-Decoder architecture, to enable offline clients to make independent predictions, we propose the dual-sub-decoders structure. As shown in Fig. 1, the dual-sub-decoders structure includes an online-sub-decoder and an offline-sub-decoder. The online-sub-decoder $f_{i,on}^{dec}$ employs local temporal embedding h_i^t and spatial embedding s_i^t to generate the predictions $\hat{x}_{i,on}^{(t+1):(t+T)}$. The offline-sub-decoder $f_{i,off}^{dec}$ only employs local temporal embedding to output predictions $\hat{x}_{i,off}^{((t+1):(t+T)}$.

$$\hat{x}_{i,on}^{(t+1):(t+T)} = f_{i,on}^{dec}(h_i^t, s_i^t) \tag{6}$$

$$\hat{x}_{i,off}^{((t+1):(t+T)} = f_{i,off}^{dec}(h_i^t) \tag{7}$$

The backbone of Encoder and dual-subdecoders can be any model. In experiments, for fair, we use GRU as the backbone.

Loss Function. To train the two sub-decoders alternately, we designed two loss functions: L_i^on and L_i^{off}, By taking a single training sample $(x_i^{(t-S):t}, x_i^{(t+1):(t+T)})$ owned by client c_i as an example, the two loss functions are as follows:

$$L_i^{on} = \sum_{k=1}^{T}(x_i^{t+k} - \hat{x}_{i,off}^{t+k})^2/T \tag{8}$$

$$L_i^{off} = \sum_{k=1}^{T}(\hat{x}_{i,off}^{t+k} - \hat{x}_{i,off}^{t+k})^2/T \tag{9}$$

It can be observed that L_i^{off} is MSE function between the outputs of the two sub-decoders. This design aims to bring the prediction of offline-sub-decoder as close as feasible to online-sub-decoder.

4.3 Training and Inference Process

Training Step. We use the alternating training method proposed in [19] to train our model to reduce communication consumption. The training and inference process of M^{3}FGM is slightly different from [19] because of the dual-subdecoders architecture. Here we briefly describe the training process:

Step 1: Initially, the clients' models are trained for R_c round with the server model and spatial embeddings fixed. Taking client i for example, in each round, the offline-sub-decoder$f_{i,off}^{dec}$ is fixed, and the encoder f_i^{enc} and online-sub-decoder $f_{i,on}^{dec}$ are trained by minimizing L_i^{on}. Then the f_i^{enc} and $f_{i,off}^{dec}$ are fixed and the $f_{i,off}^{dec}$ is trained by minimizing L_i^{off}.

Step 2: After completing R_c round, all clients' model parameters $\{\theta_{c,i}\}_{i=1}^{N}$ and local temporal embeddings $\{h_i^t\}_{i=1}^{N}$ are uploaded to the server, and then the

Algorithm 1. Training pipeline for M^3FGM with one training sample

Require: Client graph G and data $(X^{(t-S):t}, X^{(t+1):(t+T)})$. Initial each client model weights as $\theta_c = \{\theta^{enc}, \theta^{dec}_{on}, \theta^{dec}_{off}\}$, initial server model weights θ_{server}. Initial spatial embeddings $\{s_i^t\}_{i=1}^N = s_0$, s_0 is a zero-valued vector. Masknode rate mr.
Ensure: Trained client model weights θ_c, trained server model weights θ_{server}
1: **for** global training round $r_g = 1, 2, ...R_g$ **do**
2: *Step 1:*
3: **for** client $i (i = 1, ..., N)$ in parallel **do**
4: **for** local training round $r_c = 1, 2, ...R_c$ **do**
5: $f_i^{enc}(x_i^{(t-S):t}) \rightarrow h_i^t$
6: $f_{i,on}^{dec}(h_i^t, s_i^t) \rightarrow \hat{x}_{i,on}^{(t+1):(t+T)}$
7: $f_{i,off}^{dec}(h_i^t) \rightarrow \hat{x}_{i,off}^{(t+1):(t+T)}$
8: Calculate L_i^{on} and L_i^{off} according to Equations (8) and (9)
9: update $\{\theta_i^{enc}, \theta_{i,on}^{dec}\}$ according to L_i^{on}
10: update $\theta_{i,off}^{dec}$ according to L_i^{off}
11: **end for**
12: **end for**
13: *Step 2:*
14: Send latest embedding $\{h_i^t\}_{i=1}^N$ and the client model weights $\{\theta_{c,i}\}_{i=1}^N$ to the server
15: Fix all client models' weights.
16: **for** server training round $r_s = 1, 2, ...R_s$ **do**
17: construct cluster graph G^{clu} and cross-level graph $G^{\rightarrow Clu}$ according to client graph G
18: $f_{server}(\{h_i^t\}_{i=1}^N, G, G^{clu}, G^{\rightarrow Clu}, mr) \rightarrow \{s_i^t\}_{i=1}^N$
19: send $\{s_i^t\}_{i=1}^N$ to corresponding clients
20: $f_{i,on}^{dec}(h_i^t, s_i^t) \rightarrow \hat{x}_{i,on}^{(t+1):(t+T)}, i = 1, ..., N$
21: Calculate $\sum_{i=1}^N L_i^{on}$ and update θ_{server}
22: **end for**
23: *Step 3:*
24: Update latest graph embedding $\{s_i^t\}_{i=1}^N$
25: Use FedAvg to aggregate $\{\theta_{c,i}\}_{i=1}^N \rightarrow \theta_c$
26: $\{s_i^t\}_{i=1}^N$ is send to corresponding clients respectively and θ_c is send to all clients as the new model weights for next global training round.
27: **end for**

training of the server model begins for R_s rounds with clients' model fixed. $\sum_{i=1}^N L_i^{on}$ is used to update the server model.

Step 3: Once the server model is trained, the FedAvg algorithm [18] is employed by the server to aggregate $\{\theta_{c,i}\}_{i=1}^N$ to obtain θ_c. The server subsequently sends θ_c back to all clients and spatial embeddings $\{s_i^t\}_{i=1}^N$ are returned to their corresponding clients.

The above process is repeated R_g times.

Inference Step. If a client can connect to the server, the client feeds its local data to the encoder to obtain local temporal embedding, then upload embedding to the server to compute spatial embedding. After that, the client receives the spatial embedding transmitted back by the server and makes predictions using the online-sub-decoder. Conversely, when a client is unable to establish a connection to the server, it utilizes the encoder and offline-sub-decoder independently to make predictions.

5 Experiments

5.1 Datasets

Traffic data are commonly in the format of spatial-temporal graphs. We verify M^{3}FGM on two real-world traffic datasets: METR-LA and PEMS-BAY, which are released by Li et al. [14]. (1) METR-LA: which records traffic speed information collected from 207 loop detectors in the highway of Los Angeles County over 4 months. (2) PEMS-BAY: which contains 6 months of traffic speed information ranging on 325 sensors in the Bay Area.

For both two datasets, the readings of sensors are aggregated into 5 min windows. We standardize the data by removing the mean and scaling to unit variance. And then we split 70% into training set, 20% into testing set and 10% into validation set, in chronological order. And We adopt the same data pre-processing method as [19].

5.2 Compared Models and Settings

Since our primary focus is on the architecture of the federated graph, rather than the specific models, we have not made comparisons with some SOTA centralized spatial-temporal graph methods. We follow the setup of [19], comparing M^{3}FGM with four baselines. We compare M^{3}FGM with 4 baselines: GRU, GRU+ FedAvg, GRU+FMTL [24] and CNFGNN. These baselines all use the GRU-based encoder-decoder model [5] as the client-side model. For each baseline, there are 2 variants of the GRU model to show the effect of on-device model complexity: one with 63K parameters and the other with 727K parameters. For CNFGNN, the encoder-decoder model on each client has 64K parameters and the GN model has 1M parameters. The experimental results of the baseline models, as reported in [19], are utilized in the subsequent analysis. Additionally, to facilitate an objective ablative analysis, two variant models have been constructed: *CNFGNN+MN*: Add the MaskNode layer to the server model of CNFGNN. *M^3FGM w/o MN*: M^{3}FGM without MaskNode layer.

We conduct experiments under two scenarios to verify the effectiveness of our model: an ideal scenario in which all nodes are online during the inference phase and a non-ideal scenario in which some nodes are offline during the inference phase. To ensure fair evaluation and comparison, GRU is used as the backbone of the encoder and sub-decoder in the client model when implementing M^{3}FGM and M^{3}FGM w/o MN. To optimize the model, the Adam optimizer is employed with a learning rate set at 1e−3. The root mean squared error (RMSE) metric is utilized to evaluate the predictive performance.

5.3 Performance Comparison Under the Ideal Scenario

Table 1 reveals that M^{3}FGM achieves the lowest prediction error on both datasets. Specifically, M^{3}FGM and CNFGNN demonstrate superior performance

Table 1. Comparison of performance with the Rooted Mean Squared Error (RMSE) as the evaluation metrics.

Method	PEMS-BAY	METR-LA
GRU (central,63k)	4.124	11.730
GRU (central,727k)	4.128	11.787
GRU+GN (central,64k+1M)	3.816	11.471
GRU (local,63k)	4.010	11.801
GRU (local,727k)	4.152	12.224
GRU (63k)+FedAvg	4.512	12.132
GRU (727k)+FedAvg	4.432	12.058
GRU (63k)+FedMTL	3.9561	11.548
GRU (727k)+FedMTL	3.955	11.570
CNFGNN (64k+1M)	3.822	11.487
CNFGNN (64k+1M) +MN	3.831	11.504
M^3FGM w/o MN	3.697	11.371
M^3FGM ($mr = 25\%$)	**3.684**	**11.352**

compared to GRU+FedAvg and GRU+ FedMTL by taking into account the spatial correlation of client nodes.

Ablation Analysis: In Table 1: 1) M^3FGM outperforms M^3FGM w/o MN, indicating that the MaskNode layer contributes to enhanced prediction performance under the ideal scenario. 2) M^3FGM w/o MN surpasses CNFGNN. Given that the client models of the two methods share the same structure under the ideal scenario, this result suggests that the MGMP Layer is instrumental in improving prediction performance. 3) CNFGNN+MN exhibits slightly inferior performance compared to CNFGNN on both datasets. We hypothesize that this is because the server model of CNFGNN+MN struggles to aggregate valuable information within a few message-passing steps when certain node embeddings are masked. In contrast, M^3FGM with the MGMP layer addresses this issue by passing neighbor and global information.

5.4 Performance Comparison Under the Non-Ideal Scenario

To simulate non-ideal situations, we set two different client offline rates: 25% and 35%. We calculate the RMSE of the models separately for online and offline nodes. Given that each client of GRU (local) makes predictions independently during the inference phase, we also compare M^3FGM with GRU (local). We find that the prediction performance of M^3FGM on offline clients surpasses that of GRU (local) on all clients. The results show that the training method adopted by M^3FGM enables the offline-sub-decoder, which is used for local independent prediction, to outperform GRU (local). Moreover, M^3FGM exhibits better robustness than CNFGNN as the offline rate increases. Comparing the predic-

Table 2. Performance comparison under the non-ideal scenario

	online| *offline*	75%|25%	65%|35%
270PEMS-BAY	GRU (local)	4.010	4.010
	CNFGNN	3.972|*	4.232|*
	CNFGNN+MN ($mr = 10\%$)	3.904|*	4.163|*
	M^{3}FGM w/o MN	3.837|3.967	4.021|3.969
	M^{3}FGM ($mr = 25\%$)	3.741|3.934	3.836|3.938
270METR-LA	GRU (local)	11.801	11.801
	CNFGNN	11.637|*	11.809|*
	CNFGNN+MN ($mr = 10\%$)	11.563|*	11.704|*
	M^{3}FGM w/o MN	11.516|11.787	11.633|11.788
	M^{3}FGM ($mr = 25\%$)	11.423|11.782	11.513|11.782

tion error of M^{3}FGM with CNFGNN on online nodes, the increase rate of RMSE is 2.5% vs. 6.5% on PEMS-BAY and 0.8% vs. 1.5% on METR-LA.
Ablation Analysis: Upon analyzing the results in Table 2, we observe that 1) CNFGNN+MN outperforms CNFGNN on online clients, and M^{3}FGM surpasses M^{3}FGM w/o MN on both online and offline clients. These results demonstrate that the MaskNode layer improves the model's robustness. 2) Comparing the experimental results of M^{3}FGM and CNFGNN+MN on online clients, we deduce that employing the MGMP layer enhances the prediction performance on online clients under the non-ideal scenario. This finding highlights the importance of incorporating the MGMP layer in non-ideal scenarios to achieve improved prediction accuracy and model robustness.

5.5 Effect of Mask Node Rate and Discussion

In order to investigate the effect of mask rate on model prediction performance, we selected five mask rates to train M^{3}FGM: 10%, 20%, 25%, 30%, and 40%, and conducted inference on two datasets under different offline rates:0%, 25%, 35%. Figure 4 displays the performance of the model on online nodes. From these results, it can be observed that: (1) On the two datasets, it is not the case that the lower or higher the mask rate, the better. When the offline rate is fixed, compared to other mask rates, selecting a mask rate closer to the offline rate leads to better performance of the model. When the offline rate is 0%, which is the ideal scenario, choosing a mask rate within the range of 10% to 25% would be better. (2) When the mask rate is fixed, as the offline rate increases, the performance of model decreases. (3) The prediction error of the model on the PEMS-BAY dataset is significantly lower than that on the METR-LA dataset. However, the model's performance on the PEMS-BA dataset exhibits greater fluctuations with mask rate variation compared to its performance on the METR-LA dataset.

To understand the underlying principles of these trends, we analyzed the data used in the experiments. We selected six nodes from the first 100 nodes ranked by

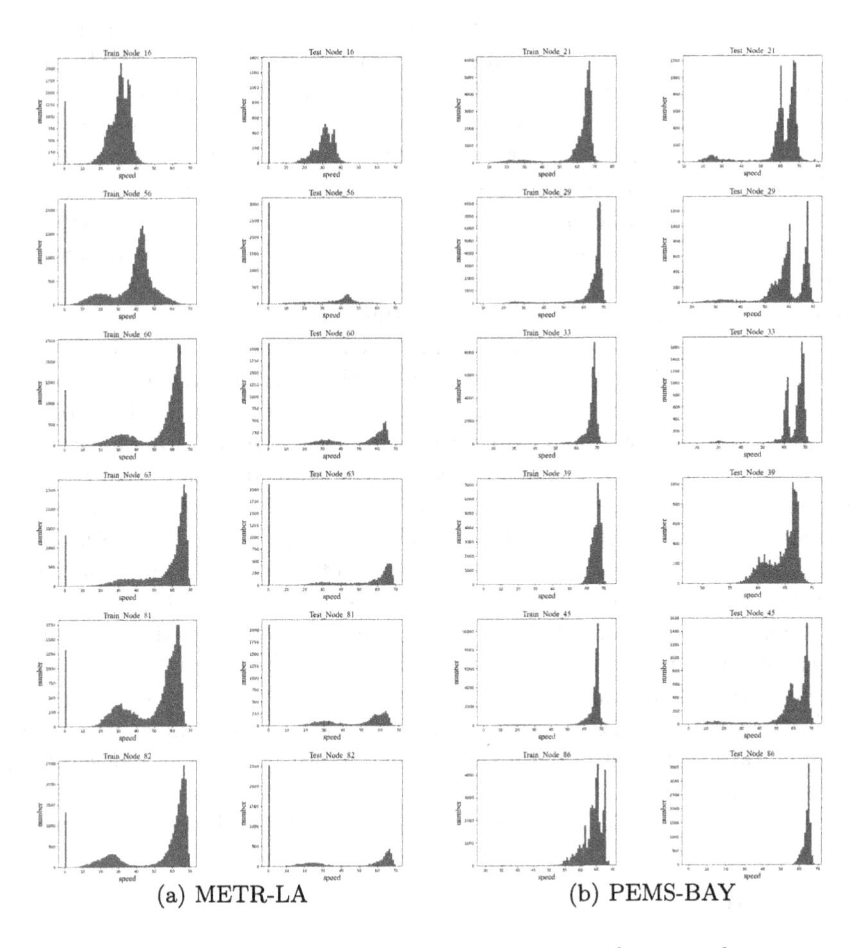

(a) METR-LA (b) PEMS-BAY

Fig. 3. Data distribution of training set data and test set data

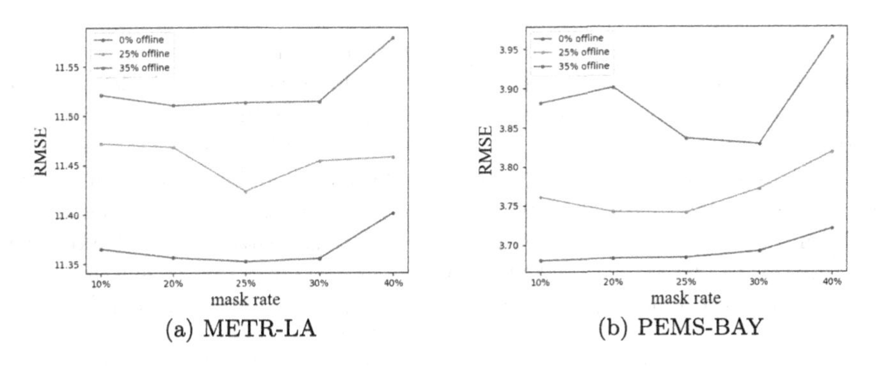

(a) METR-LA (b) PEMS-BAY

Fig. 4. Comparison of performance under different mask rate and offline rate with RMSE.

their IDs in METR-LA and PEMS-BAY and illustrated the statistical histogram of traffic speed of training data and test data of different nodes of METR-LA and PEMS-BAY in Fig. 3(a) and (b), respectively. The analysis revealed the following key insights: (1) On the METR-LA dataset, the histograms show that the data distribution varies with nodes, and most importantly, their training and test data distributions exhibit considerable discrepancies. (2) On the PEMS-BAY dataset, however, the differences between the training and test data distributions are much smaller. Additionally, the data distributions among different nodes are more similar to each other.

Based on this analysis, we can conclude that on the METR-LA dataset, existing a strong shift in data distribution. The occurrence of data distribution shift can result in a significant decline in the predictive performance of a model. For instance, when employing traffic forecasting model trained on the data collected in sunny days for rainy or foggy environments, inevitable performance drop can often be observed in such scenarios. Because the trained models tend to overfit the training data and show vulnerability to the statistic changes at testing time, substantially limiting the generalization ability of the learned representations. Thus, selecting an appropriate mask rate can effectively prevent model overfitting and reduce prediction errors. In contrast, The data distribution among various nodes in the PEMS-BAY dataset is relatively similar, and the differences between the training data distribution and the testing data distribution within each node are not substantial. This suggests that the correlation among nodes in the PEMS-BAY dataset is stronger, resulting in a more significant impact of the offline rate and mask rate on the model's performance. These observations above emphasize the importance of selecting an appropriate mask rate based on the specific characteristics of the dataset to achieve optimal model performance.

6 Conclusion

In this paper, we propose a new GNN-oriented split federated learning method, named node Masking and Multi-granularity Message passing-based Federated Graph Model (M^{3}FGM) specifically developed for spatial-temporal data prediction in scenarios where data decentralization is imperative due to privacy concerns. We improve robustness of model by introducing the MaskNode layer and the proposed dual-sub-decoders structure enables independent offline prediction. In addition, a new GNN layer named Multi-Granularity Message Passing (MGMP) layer enables each client node to perceive global and local information in a short message passing steps. We conducted evaluations under both ideal and non-ideal scenarios, the comprehensive experimental results demonstrate the superiority of the proposed M^{3}FGM model in comparison to existing methods in terms of prediction accuracy and robustness under various conditions.

Acknowledgments. This work was supported by the National Science Foundation of China 61562041.

References

1. Battaglia, P.W., Hamrick, J.B., Bapst, V., Sanchez-Gonzalez, A., Zambaldi, V., et al.: Relational inductive biases, deep learning, and graph networks. arxiv arXiv:1806.01261 (2018)
2. Caldarola, D., Mancini, M., Galasso, F., Ciccone, M., Rodolà, E., Caputo, B.: Cluster-driven graph federated learning over multiple domains. CoRR abs/2104.14628. arXiv arxiv:2104.14628 (2021)
3. Chang, X., et al.: Continuous-time dynamic graph learning via neural interaction processes. In: Proceedings of the 29th ACM International Conference on Information & Knowledge Management, pp. 145–154 (2020)
4. Chen, F., Long, G., Wu, Z., Zhou, T., Jiang, J.: Personalized federated learning with a graph. In: IJCAI (2022)
5. Cho, K., Merrienboer, B., Gulcehre, C., Bougares, F., Schwenk, H., Bengio, Y.: Learning phrase representations using RNN encoder-decoder for statistical machine translation. In: EMNLP (2014)
6. Fang, Z., Long, Q., Song, G., Xie, K.: Spatial-temporal graph ODE networks for traffic flow forecasting. In: KDD (2021)
7. Gao, Y., et al.: End-to-end evaluation of federated learning and split learning for internet of things. CoRR abs/2003.13376. arXiv arXiv:2003.13376 (2020)
8. Guo, S., Lin, Y., Feng, N., Song, C., Wan, H.: Attention based spatial-temporal graph convolutional networks for traffic flow forecasting. In: AAAI (2019)
9. Gupta, O., Raskar, R.: Distributed learning of deep neural network over multiple agents. CoRR abs/1810.06060. arXiv arXiv:1810.06060 (2018)
10. He, K., Zhang, X., Ren, S., Sun, J.: Deep residual learning for image recognition. In: CVPR (2015)
11. Hou, Z., et al.: GraphMAE: self-supervised masked graph autoencoders. In: KDD (2022)
12. Kipf, T.N., Welling, M.: Semi-supervised classification with graph convolutional networks. arXiv preprint arXiv:1609.02907 (2016)
13. Lan, S., Ma, Y., Huang, W., Wang, W., Yang, H., Li, P.: DSTAGNN: dynamic spatial-temporal aware graph neural network for traffic flow forecasting. In: ICML (2022)
14. Li, Y., Yu, R., Shahabi, C., Liu, Y.: Graph convolutional recurrent neural network: data-driven traffic forecasting. In: ICLR (2018)
15. Litany, O., Maron, H., Acuna, D., Kautz, J., Chechik, G., Fidler, S.: Federated learning with heterogeneous architectures using graph hypernetworks. CoRR abs/2201.08459. arXiv arXiv:2201.08459 (2022)
16. Liu, Y., Yu, J.J., Kang, J., Niyato, D., Zhang, S.: Privacy-preserving traffic flow prediction: a federated learning approach. IEEE IoT J. **7**, 7751–7763 (2020)
17. Luo, L., Haffari, R., Pan, S.: Graph sequential neural ode process for link prediction on dynamic and sparse graphs, November 2022. https://doi.org/10.48550/arXiv.2211.08568
18. McMahan, H.B., Moore, E., Ramage, D., Hampson, S., y Arcas, B.A.: Communication-efficient learning of deep networks from decentralized data. In: AISTATS (2016)
19. Meng, C., Rambhatla, S., Liu, Y.: Cross-node federated graph neural network for spatio-temporal data modeling. In: KDD (2021)
20. Peng, H., et al.: Dynamic graph convolutional network for long-term traffic flow prediction with reinforcement learning. Inf. Sci. **578**, 401–416 (2021)

21. Rong, Y., Huang, W., Xu, T., Huang, J.: DropEdge: towards deep graph convolutional networks on node classification. In: ICLR (2020)
22. Shao, W., et al.: Long-term spatio-temporal forecasting via dynamic multiple-graph attention. arXiv preprint arXiv:2204.11008 (2022)
23. Shao, Z., et al.: Decoupled dynamic spatial-temporal graph neural network for traffic forecasting. In: VLDB (2022)
24. Smith, V., Chiang, C.K., Sanjabi, M., Talwalkar, A.: Federated multi-task learning. In: NIPS (2017)
25. Song, C., Lin, Y., Guo, S., Wan, H.: Spatial-temporal synchronous graph convolutional networks: a new framework for spatial-temporal network data forecasting. In: AAAI (2020)
26. Thapa, C., Chamikara, M.A.P., Camtepe, S.: SplitFed: when federated learning meets split learning. CoRR abs/2004.12088. arXiv arXiv:2004.12088 (2020)
27. Wang, Y., Wang, W., Liang, Y., Cai, Y., Liu, J., Hooi, B.: NodeAug: semi-supervised node classification with data augmentation, pp. 207–217, August 2020. https://doi.org/10.1145/3394486.3403063
28. Wu, L., Lin, H., Gao, Z., Tan, C., Li, S.Z.: GraphMixup: improving class-imbalanced node classification on graphs by self-supervised context prediction. CoRR abs/2106.11133. arXiv arXiv:2106.11133 (2021)
29. Wu, Z., Pan, S., Long, G., Jiang, J., Zhang, C.: Graph WaveNet for deep spatial-temporal graph modeling. In: IJCAI (2019)
30. Xing, P., Lu, S., Wu, L., Yu, H.: BiG-Fed: bilevel optimization enhanced graph-aided federated learning. IEEE Trans. Big Data, 1–12 (2022)
31. Yu, B., Yin, H., Zhu, Z.: Spatio-temporal graph convolutional networks: a deep learning framework for traffic forecasting. In: IJCAI (2018)
32. Zhang, C., Cui, L., Yu, S., Yu, J.J.: A communication-efficient federated learning scheme for IoT-based traffic forecasting. IEEE IoT J. **9**, 11918–11931 (2021). https://doi.org/10.1109/JIOT.2021.3132363
33. Zhang, C., Zhang, S., Yu, J.J., Yu, S.: FASTGNN: a topological information protected federated learning approach for traffic speed forecasting. IEEE Trans. Ind. Inf. **17**, 8464–8474 (2021)

LenANet: A Length-Controllable Attention Network for Source Code Summarization

Peng Chen, Shaojuan Wu, Ziqiang Chen, Jiarui Zhang, Xiaowang Zhang(✉), and Zhiyong Feng

College of Intelligence and Computing, Tianjin University, Tianjin, China
{net07,shaojuanwu,zqchen99,jiaruizhang,xiaowangzhang,zyfeng}@tju.edu.cn

Abstract. Source code summarization aims at generating brief description of a source code. Existing approaches have made great breakthroughs through encoder-decoder models. They focus on learning common features contained in translation from source code to natural language summaries. As a result, they tend to generate generic summaries independent of the context and lack of details. However, specific summaries which characterize specific features of code snippets are widely present in real-world scenarios. Such summaries are rarely studied as capturing specific features of source code would be difficult. What's more, only the common features learned would result in only the generic short summaries generated. In this paper, we present LenANet to generate specific summaries by considering the desired length information and extracting the specific code sentence. Firstly, we introduce length offset vector to force the generation of summaries which could contain specific amount of information, laying the groundwork for generating specific summaries. Further, forcing the model to generate summaries with a certain length would bring in invalid or generic descriptions, a context-aware code sentence extractor is proposed to extract specific features corresponding to specific information. Besides, we present a innovative sentence-level code tree to capture the structural semantics and learn the representation of code sentence by graph attention network, which is crucial for specific features extraction. The experiments on CodeXGLUE datasets with six programming language demonstrate that LenANet significantly outperforms the baselines and has the potential to generate specific summaries. In particular, the overall BLEU-4 is improved by 0.53 on the basis of CodeT5 with length control.

Keywords: code summarization · length control · code splitting · graph attention network

1 Introduction

Source Code Summarization aims at automatically generating summaries in natural language for the function and purpose of identifiers from structured code

B. Luo et al. (Eds.): ICONIP 2023, CCIS 1965, pp. 567–578, 2024.
https://doi.org/10.1007/978-981-99-8145-8_43

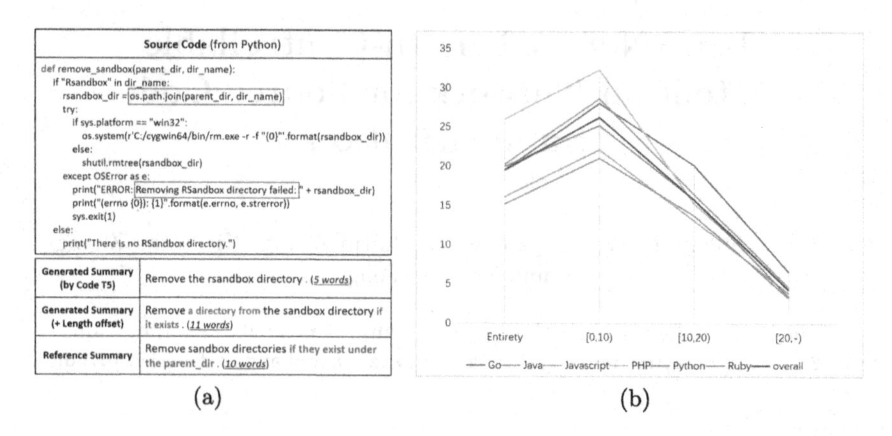

Fig. 1. The motivation of our approach. (a) An example comes from the CodeXGLUE dataset for code summarization. (b) BLEU-4 of CodeT5 for reference summaries with different length intervals

snippets [10,15,20], which is crucial for maintaining software development efficiency and reducing developers' tedious workload.

Recently, code summarization is treated as a task of Machine Translation, which usually adopts encoder-decoder framework and Seq2seq models [1,3,8,22]. These endeavors depend on sequential models, facing challenges in capturing the structural semantics of source code. Thus some recent works [18,22] attempted to encode code structure (e.g. abstract syntax tree) to incorporate structural semantics, but it is still difficult to capture the sentence-level semantics. In addition, models pre-trained on programming language [5,24] have achieved impressive improvements on code summarization and many other code-related tasks. Even so, these models for code summarization focus on learning common and united translation from source code to natural language summaries, but ignore discrepancies of source codes and corresponding summaries (e.g. only the file-related code needs to determine whether the certain file exists). As a result, they tend to generate **generic summaries** independent of the context, yet do not perform well generating **specific summaries** which match the unique corresponding code snippet.

As illustrated in Fig. 1(a), the generated summaries of Codet5 with and without length offset is different. They are same at "Remove the sandbox directory", which is the function of source code and applies to any code snippets that include sandbox directory, regardless of the condition. But "if they exist under the parent_dir" is the specific description only matching the above source code. After introducing length offset, we note that "Remove a directory from the sandbox directory if it exists" is almost as long as the reference summary, but it also results in generating some invalid information as noise, which means existing methods still have difficult in extracting accurate and specific features in codes to fill in the summaries, even if given the desired length.

In addition, we have compared BLEU-4 scores of generated summary for three different length intervals (of reference summaries): [0, 10), [10, 20), [20, -) on six programming language of CodeXGLUE dataset. As shown in Fig. 1(b), the performance of the SOTA significantly experiences a notable decline as the length of reference summary is longer. Specifically, overall results are 26.23 and 4.30 for length intervals [0, 10) and [20, -), which differ by 21.93. It's obvious that the result of [0,10) is overall 6.68 higher than the entirety result. But for summaries whose length exceeding 20, the result is overall 15.25 lower than entirety. This situation suggests that existing models have the ability to generate shorter summaries, but they fail to generate summaries whose length exceeding 20. As a result, the desired length of code summaries plays a key role to specific summaries, which tend to be longer than generic summaries.

Based on the above findings, we present LenANet, a length-controllable attention network that adaptively generates specific summaries via introducing length offset. As shown in Fig. 2, we first introduce the length offset to force the model to generate summaries with specific length corresponding to different amounts of information, laying the groundwork for generating specific summaries. Then, a novel sentence-level code tree is constructed to capture the structural semantics. Lastly, a graph attention network is employed to reason the relationship among code sentences and further select key code sentences representing specific features of source codes. In addition, experiments conducted on the CodeXGLUE dataset have shown that our model is significantly superior to the baselines in all six programming language subdataset. In particular, the overall BLEU-4 is improved by 0.53 on the basis of CodeT5 with length control. Further, it is verified that the performance of our model is improved to a certain extent in all three length intervals.

In a nutshell, the contributions of our work are listed as follows:

- We propose a framework called LenANet to generate code summaries through adaptively length control, which introduces a length offset for generating specific summaries.
- We create an innovative sentence-level code tree, which is applied to capture the sentence-level structural semantics of code snippets.
- We propose a context-aware code sentence extractor that employs a GAT to reason the relationship among code sentences and extract key sentences to obtain the specific features.

2 Method

In this section, we present a comprehensive overview of our methodology. It mainly contains four components: Encoder-Decoder Framework, Length Offset Vector, Sentence-level Code Tree Construction and Attention-based Code Sentence Extractor.

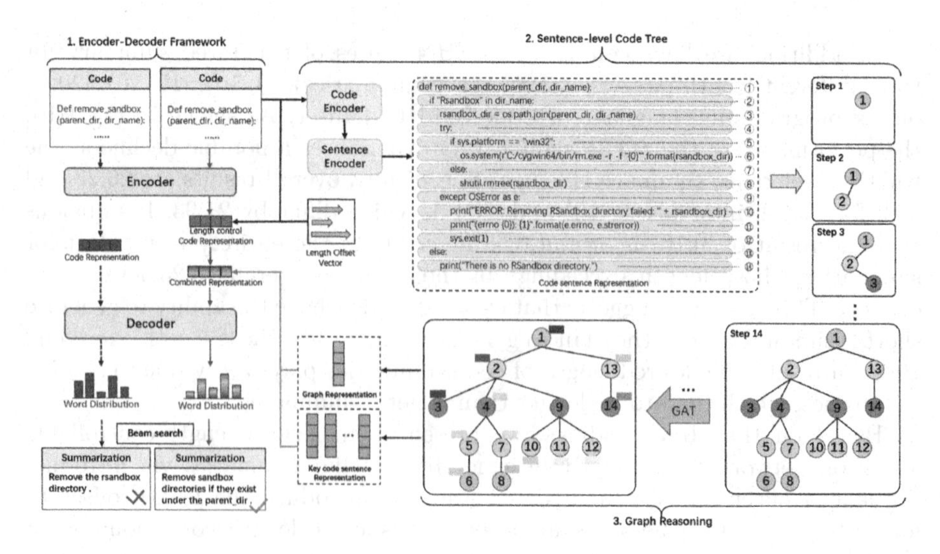

Fig. 2. The overall of LenANet.

2.1 Encoder with Length Offset Vector

Given a code snippet C containing a set of tokens $C = \{token_1, token_2, ..., token_n\}$, code summarization aims at generating a summary S in natural language to describe C. S is consisted of a sequence of tokens $S = \{s_1, s_2, ..., s_L\}$, where L denotes the length of S. Following prior works [1,5], our approach adopt encoder-decoder framework and encode code tokens using Transformer-based models that leverage the multi-head self-attention module. Fusing the length information makes it possible for the encoder to decide the importance of the tokens based on the specific desired length of the summary and encode better. In this paper, we directly add the length offset vector to the input sequence to modify the desired summary length of encoded sequence. Ambiguous length offset vector denotes the desired length interval (short for 0$\sim$10 tokens, median for 10$\sim$20 tokens and long for 20^+ tokens), while specific length offset vector denotes the specific length, which is suitable for more accurate length control. We have tried both two types of length offset vectors and chose ambiguous length offset vector which is more flexible.

2.2 Sentence-Level Code Tree

In this section, we devise an innovative sentence-level code tree to capture the sentence-level structural information of code snippets.

A Sentence-level Code Tree (SCT), denoted as $T = (V, E)$, is a undirected graph in which each node V corresponds to a statement along with its respective feature and each edge E signifies the hierarchical relationship between statements. As shown in Algorithm 1, initially, we split source code into n code sentences as tree nodes, $V = \{v_1, v_2, ..., v_n\}$. Then, the source code is traversed

Algorithm 1: Sentence-Level Code Tree Construction.

Input: Source code snippets: C; Code sentences: S_i.
Output: The sentence-level code tree: T.

```
1  Initialize v_0 := the first node in V; c_0 := the first code line in C; T := an
     empty tree;
2  for v_i in V and c_i in C, i ∈ {1, 2, ..., n} do
3      I_i := the indentation of c_i;
4      I_{i-1} := the indentation of c_{i-1};
5      if I_i > I_{i-1} then
6          The parent node of v_i is v_{i-1};
7          AddNode(v_i, T);
8      else if I_i = I_{i-1} then
9          The parent node of v_i is the parent node of v_{i-1};
10         AddNode(v_i, T);
11     else
12         j := i;
13         repeat
14             I_j := the indentation of c_{i-1};
15             if I_j = I_i then
16                 The parent node of v_i is the parent node of v_j;
17                 AddNode(v_i, T);
18                 Break
19         until j = 0;
20 return T
```

to delineate the relationship between nodes. The indentation of adjacent code sentences (i.e. nodes) is applied to establish the presence of edges between nodes. Specifically, if the indentation of current code sentences is shorter than the previous code sentences, then the current code sentences is the child of previous. The two code sentences are sibling nodes on condition that the same indentation. And while the indentation of current code sentences is longer, preceding nodes are traversed to find the code sentences with the same indentation as sibling node. The SCT T will be applied in graph reasoning to extract key statements.

2.3 Context-Aware Code Sentence Extractor

As mentioned earlier, specific summaries not only describes the function of the code snippet, but also focus on the specific information (e.g. scope of a variable), which is corresponding to the specific sentences in source code. As a result, we present a Context-aware Code Sentence Extractor (CCSE), which is applied to reason the relationship among code sentences and extract key sentences as key feature.

Specifically, a graph attention networks (GAT [23]) is applied to reason the relationship among code sentences. The input comprises a collection of node features from the Sentence-level Code Tree. The node features are sentence-level features aggregated from encoded code tokens $\mathbf{H} = \{\boldsymbol{h}_1, \boldsymbol{h}_2, \ldots, \boldsymbol{h}_n\}, \boldsymbol{h}_i \in \mathbb{R}^F$, where N represents the number of code tokens and F represents the dimension of features of each token. The output is $\mathbf{H}' = \{\boldsymbol{h}'_1, \boldsymbol{h}'_2, \ldots, \boldsymbol{h}'_n\}, \boldsymbol{h}'_i \in \mathbb{R}^{F'}$, where F' is potentially different cardinality. Then, CCSE takes in the graph-encoded presentation and employ a full connected layer to score the centrality of each statement. Next, key statement with the highest score will be fused into the hidden states $\mathbf{H}$, which is fed into decoder and generate code summaries. When training time, we fuse $\mathbf{H}$, the output of GAT $\mathbf{H}'$ and the embedding of the Ground-truth sentences as follows:

$$e^{fuse} = [\mathbf{H}, \mathbf{H}', E_{Ground-truth}] \tag{1}$$

when test time, we fuse $\mathbf{H}$, $\mathbf{H}'$ and the embbedding of the predicted sentences as follows:

$$e^{fuse} = [\mathbf{H}, \mathbf{H}', E_{predicted}] \tag{2}$$

To make CCSE more effective, Ground-truth labels of sentences are produced heuristically to train the extractor. Denoting the labels of sentences as $L = [l_1, l_2, ..., l_m]$ where $l_i \in [0, 1]$, we set the label of the most prominent statement to 1 and others to 0. To obtain the heuristic labels, we measure the importance and informativity of sentences by computing the co-occurrence frequency of the sentences and the reference summaries.

3 Experiments

3.1 Experimental Setup

Datasets. Our experiments are conducted on CodeXGLUE dataset [14], including 14 sub-datasets that span 10 diversified code intelligence tasks. For our experimentation, we focus on the code-text dataset, specifically targeting code summarization. It encompasses six programming languages (i.e., Python, Java, JavaScript, PHP, Ruby and Go). Following CodeT5 [24], we adopt smoothed BLEU-4 as automatic metrics.

Hyper-parameter. During the training phase, we configure the learning rate to 5e−5, the batch size to 8, and limit the maximum epochs to 5. Our method are based on codet5[1] with AdamW optimizer and dropout of 0.2. Moreover, we establish the maximum source length at 256 and the maximum target length at 128. Patience is used to early stop training which is set to 2 in our approach.

Pre-processing. The dataset we used is originally from CodeSearchNet [13]. Based on CodeSearchNet, CodeXGLUE refine it by four methods. (e.g. Removed the examples which can be parsed into Abstract Syntax Tree.) What's more, we split the source code by line for constructing a sentence-level code tree. Two extra tokens, <extra_id_0> and <extra_id_1>, are inserted into the start and end of each code sentence.

[1] https://github.com/salesforce/CodeT5.

Table 1. The BLEU-4 of our approach compared with other baselines

Models	Go	Java	JavaScript	PHP	Python	Ruby	Overall
Seq2Seq	13.98	15.09	10.21	21.08	15.93	9.64	14.32
Transformer	16.38	16.26	11.59	22.12	15.81	11.18	15.56
RoBERTa	17.72	16.47	11.90	24.02	18.14	11.17	16.57
CodeBERT	18.07	17.65	14.90	25.16	19.06	12.16	17.83
DOBF	-	19.05	-	-	18.24	-	-
PLBART	18.91	18.45	15.56	23.58	19.30	14.11	18.32
CodeT5	19.56	20.31	16.16	26.03	20.01	15.24	19.55
CodeT5(+length)	21.94	23.14	18.71	28.15	**21.29**	17.56	21.71
LenANet (ours)	**22.41**	**24.08**	**18.91**	**28.50**	21.14	**18.37**	**22.24**

3.2 Baselines

We consider two categories of state-of-the-art models as our baselines for comparison. In addition to basic methods in the field of text generation such as Seq2Seq [21], Transformer and RoBERTa [13], We compare with the SOTA models pre-trained on code and natural languages, including CodeBERT [6], DOBF [17], PLBART [2] and CodeT5 [24].

CodeT5 [24] is a novel pre-trained encoder-decoder model for programming languages, which is pre-trained using a corpus of 8.35M functions in 8 programming languages. It achieves state-of-the-art performance on multiple code-related downstream tasks including code-text generation tasks.

3.3 Main Results

As shown in Table 1, the overall BLEU-4 of our approach is 22.24, which achieves the best performance. We can observe a significant improvement from 20.31 to 24.08 in Java. In addition, the improvement on the Ruby dataset is more pronounced than the others, which means that our approach is also adaptive for smaller datasets. On the basis of only training with length offset, the model with SCT and CCSE is more effective. By acquiring an understanding of the structural semantics within source code, it is capable of generating specific summaries with extracting specific features.

4 Discussion

4.1 Specific Feature Selection Without Length Offset

To demonstrate the efficiency of Sentence-level Code Tree and Context-aware Code Sentence Extractor, we conduct further experiment without introducing length offset.

Table 2. The BLEU-4 of CodeT5 and our approach for reference summaries of different length ranges. '↑' denotes the result is better than the entirety and '↓' denotes the result is worse than the entirety.

model	Length	Go	Java	JavaScript	PHP	Python	Ruby	Overall
CodeT5	Entirety	19.56	20.31	16.16	26.03	20.01	15.24	19.55
	[0, 10)	28.02(↑ 8.46)	28.61(↑ 8.30)	22.13(↑ 5.97)	32.34(↑ 6.31)	25.21(↑ 5.20)	21.05(↑ 5.81)	**26.23(↑ 6.68)**
	[10, 20)	20.16(↑ 0.60)	16.48(↓ 3.83)	13.14(↓ 3.02)	15.30(↓ 10.73)	15.81(↓ 4.20)	13.78(↓ 1.46)	15.78(↓ 3.77)
	[20, −)	6.59(↓ 12.97)	4.64(↓ 15.67)	3.86(↓ 12.30)	4.51(↓ 21.52)	3.44(↓ 16.57)	3.27(↓ 11.97)	**4.30(↓ 15.25)**
LenANet	Entirety	22.41	24.08	18.91	28.50	21.14	18.37	22.24
	[0, 10)	29.94(↑ 7.53)	31.86(↑ 7.78)	24.11(↑ 5.20)	33.50(↑ 5.00)	26.17(↑ 5.03)	23.86(↑ 5.49)	**28.24(↑ 6.00)**
	[10, 20)	21.41(↓ 1.00)	18.17(↓ 5.91)	15.69(↓ 3.22)	19.19(↓ 9.31)	17.10(↓ 4.04)	15.20(↓ 3.17)	17.79(↓ 4.45)
	[20, −)	11.95(↓ 10.46)	10.66(↓ 13.42)	9.15(↓ 9.76)	12.03(↓ 16.47)	9.61(↓ 11.53)	9.00(↓ 9.37)	**10.40(↓ 11.84)**

Table 3. The BLEU-4 scores of LenANet without length offset

	Go	Java	Javascript	PHP	Python	Ruby	Overall
Codet5	19.56	20.31	16.16	26.03	20.01	15.24	19.55
Ours	19.44	20.61	16.06	26.26	20.59	15.50	19.74

As shown in Table 3, our overall results achieve better BLEU-4 than CodeT5. Comparable outcomes are evident in previous experiment (Table 1). Experimental results indicate that our approach is more effective on the basis of length offset. Without length offset, our approach improve by only 0.19, while applying our framework can improve by 0.53 with length offset. What's more, specific feature selection is important whatever the desired length of the summary. But it's easy to see that CCSE plays a more critical role in generating long summaries from Table 2.

4.2 Visualization Analysis

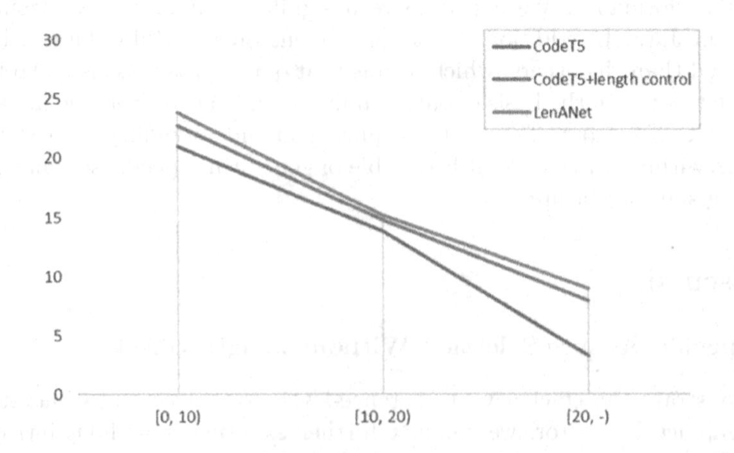

Fig. 3. Test result for different length offsets in the test set of Ruby.

Experimental Result for Each Length Offset. In this subsection, we visualize the experimental result of Ruby dataset for each length offset. As shown in Fig. 3, adding length offset performs significantly better compared with the baseline, especially for longer summaries, which demonstrates that length information is important for planning the content of the summaries. What's more, our method further promote the quality of the summaries.

Sentence-Level Graph Encoding. To visualize the sentence-level graph encoding, we draw the attention scores among code sentences by heat map. As shown in Fig. 4, Sentence-level Code Tree with Context-aware Code Sentence Extractor encodes the sentence-level structural semantics well. For example, the high attention scores among sentences 4˜6 demonstrate the structural information of the loop statements. And the high attention scores between sentences 0 and 7 demonstrate that return function represents exit of the whole program.

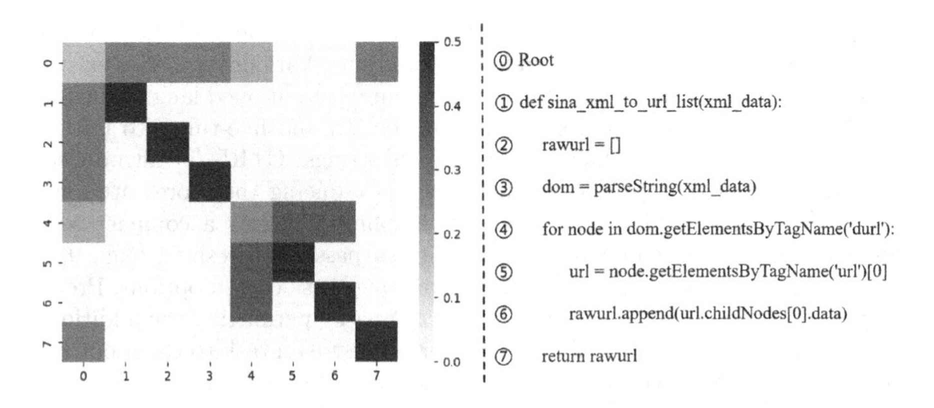

Fig. 4. An example of sentence-level graph encoding.

5 Related Work

5.1 Code Summarization

As a pivotal research topic in software engineering, code summarization has garnered substantial attention in recent years. Most methods on code summarization can be primarily categorized as rule based approaches [19] and deep learning based approaches [1,3,8]. For example, Moreno et al. [16] select key information for the code heuristically and combine the information based on rules. However, rule-based approaches are constrained by the quality and diversity of the rules and have difficult in generating high-quality summaries. Recently, most works concentrate on deep-learning-based approaches. To generate better summaries, there exists numerous models include RNN [8,26] and Transformers

[1,22]. What's more, Pre-trained models for programming language [5,24] have achieved impressive improvements on code summarization.

Recent works mainly focus on the structural features of the code. For example, Abstract Syntax Tree (AST), Data Flow Graph (DFG) and other parsed forms are encoded to obtain the structure information to help with summarization [18,22]. In addition, since code duplication is common in "big code", some studies search for similar codes and summaries based on information retrieval for code summarization [12,25]. Our approach pays attention to the length of the summaries for the first time to help with information selection and generated content planning. More importantly, noticed that some summaries may focus on key statements, our work encode the relation of the statements instead of the relations of the tokens.

5.2 Controllable Text Generation

Controllable text generation seeks to enhance the manageability of the generated text, ensuring alignment with intended expectations. Various strategies exist for enabling controlled content generation within an unconditioned language model, including decoding strategies, smart prompt design and fine-tune. To mitigate the recurrent issue of producing duplicate substrings, CTRL [9] introduced a sampling method that penalizes repetitions by reducing the scores attributed to previously generated tokens. Nucleus sampling [7] elects a compact set of top candidates whose cumulative probability surpasses a threshold (e.g., 0.95), subsequently recalibrating the distribution among these chosen options. Prompt Tuning [11] simplifies the concept of prefix tuning by permitting the addition of only a certain number of tunable tokens per downstream task to the input text, a tactic conducive to controllable text generation. Recently, many researchers adopt fine-tuning to control text generation, commonly by training on supervised datasets or by reinforcement learning. PPLM [4] updates the gradient according to the target text through the small differentiable attribute model, then the gradient is transmitted back to the language model to control the gradient to control over the pre-trained language model.

6 Conclusions

In this paper, we present LenANet to generate specific summaries by considering the desired length information and extracting the specific feature contained in specific code sentences. Firstly, We introduce length offset vector to determine the amount of specific information as well as the length of summary. Further, to extract corresponding specific feature, a context-aware code sentence extractor is proposed to extract specific information of the source code. Additionally, we present a novel sentence-level code tree to capture the structural information and acquire the representation of code sentence by GAT. Experimental results demonstrate the superiority of LenANet on CodeXGLUE datasets with six programming language. What's more, further experiments and discussions verified that our approach has the potential of length control and specific summaries generation.

References

1. Ahmad, W.U., Chakraborty, S., Ray, B., Chang, K.: A transformer-based approach for source code summarization. In: Proceedings of the 58th Annual Meeting of the Association for Computational Linguistics, ACL 2020, Online, July 5–10, 2020, pp. 4998–5007. Association for Computational Linguistics (2020)
2. Ahmad, W.U., Chakraborty, S., Ray, B., Chang, K.: Unified pre-training for program understanding and generation. CoRR abs/2103.06333
3. Allamanis, M., Peng, H., Sutton, C.: A convolutional attention network for extreme summarization of source code. In: Proceedings of the 33nd International Conference on Machine Learning, ICML 2016, New York City, NY, USA, June 19–24, 2016. JMLR Workshop and Conference Proceedings, vol. 48, pp. 2091–2100. JMLR.org
4. Dathathri, S., et al.: Plug and play language models: a simple approach to controlled text generation. In: 8th International Conference on Learning Representations, ICLR 2020, Addis Ababa, Ethiopia, April 26–30, 2020. OpenReview.net
5. Feng, Z., et al.: Codebert: a pre-trained model for programming and natural languages. In: Findings of the Association for Computational Linguistics: EMNLP 2020, Online Event, 16–20 November 2020. Findings of ACL, vol. EMNLP 2020, pp. 1536–1547. Association for Computational Linguistics (2020)
6. Feng, Z., et al.: Codebert: a pre-trained model for programming and natural languages. CoRR abs/2002.08155
7. Holtzman, A., Buys, J., Forbes, M., Choi, Y.: The curious case of neural text degeneration. CoRR abs/1904.09751
8. Iyer, S., Konstas, I., Cheung, A., Zettlemoyer, L.: Summarizing source code using a neural attention model. In: Proceedings of the 54th Annual Meeting of the Association for Computational Linguistics, ACL 2016, August 7–12, 2016, Berlin, Germany, Volume 1: Long Papers. The Association for Computer Linguistics (2016)
9. Keskar, N.S., McCann, B., Varshney, L.R., Xiong, C., Socher, R.: CTRL: a conditional transformer language model for controllable generation. CoRR abs/1909.05858
10. LeClair, A., Haque, S., Wu, L., McMillan, C.: Improved code summarization via a graph neural network. In: ICPC '20: 28th International Conference on Program Comprehension, Seoul, Republic of Korea, July 13–15, 2020, pp. 184–195. ACM (2020)
11. Lester, B., Al-Rfou, R., Constant, N.: The power of scale for parameter-efficient prompt tuning. CoRR abs/2104.08691
12. Liu, S., Chen, Y., Xie, X., Siow, J.K., Liu, Y.: Retrieval-augmented generation for code summarization via hybrid GNN. In: 9th International Conference on Learning Representations, ICLR 2021, Virtual Event, Austria, May 3–7, 2021. OpenReview.net
13. Liu, Y., et al.: Roberta: a robustly optimized BERT pretraining approach. CoRR abs/1907.11692
14. Lu, S., et al.: Codexglue: a machine learning benchmark dataset for code understanding and generation. CoRR abs/2102.04664
15. McBurney, P.W., McMillan, C.: Automatic source code summarization of context for Java methods. IEEE Trans. Softw. Eng. **42**(2), 103–119 (2016)

16. Moreno, L., Aponte, J., Sridhara, G., Marcus, A., Pollock, L.L., Vijay-Shanker, K.: Automatic generation of natural language summaries for java classes. In: IEEE 21st International Conference on Program Comprehension, ICPC 2013, San Francisco, CA, USA, 20–21 May, 2013, pp. 23–32. IEEE Computer Society (2013)
17. Rozière, B., Lachaux, M.A., Szafraniec, M., Lample, G.: Dobf: A deobfuscation pre-training objective for programming languages. In: NeurIPS
18. Shido, Y., Kobayashi, Y., Yamamoto, A., Miyamoto, A., Matsumura, T.: Automatic source code summarization with extended tree-lstm. In: International Joint Conference on Neural Networks, IJCNN 2019 Budapest, Hungary, July 14–19, 2019, pp. 1–8. IEEE (2019)
19. Sridhara, G., Hill, E., Muppaneni, D., Pollock, L.L., Vijay-Shanker, K.: Towards automatically generating summary comments for java methods. In: ASE 2010, 25th IEEE/ACM International Conference on Automated Software Engineering, Antwerp, Belgium, September 20–24, 2010, pp. 43–52. ACM (2010)
20. Sun, W., et al.: An extractive-and-abstractive framework for source code summarization. CoRR abs/2206.07245 (2022)
21. Sutskever, I., Vinyals, O., Le, Q.V.: Sequence to sequence learning with neural networks. CoRR abs/1409.3215
22. Tang, Z., Li, C., Ge, J., Shen, X., Zhu, Z., Luo, B.: Ast-transformer: encoding abstract syntax trees efficiently for code summarization. In: 36th IEEE/ACM International Conference on Automated Software Engineering, ASE 2021, Melbourne, Australia, November 15–19, 2021, pp. 1193–1195. IEEE (2021)
23. Veličković, P., Cucurull, G., Casanova, A., Romero, A., Liò, P., Bengio, Y.: Graph attention networks. In: International Conference on Learning Representations
24. Wang, Y., Wang, W., Joty, S., Hoi., S.C.: Codet 5: Identifier-aware unified pre-trained encoder-decoder models for code understanding and generation. In: Proceedings of the 2021 Conference on Empirical Methods in Natural Language Processing, EMNLP 2021
25. Zhang, J., Wang, X., Zhang, H., Sun, H., Liu, X.: Retrieval-based neural source code summarization. In: ICSE '20: 42nd International Conference on Software Engineering, Seoul, South Korea, 27 June - 19 July, 2020, pp. 1385–1397. ACM (2020)
26. Zhao, Y., Shen, X., Bi, W., Aizawa, A.: Unsupervised rewriter for multi-sentence compression. In: Proceedings of the 57th Conference of the Association for Computational Linguistics, ACL 2019, Florence, Italy, July 28- August 2, 2019, Volume 1: Long Papers, pp. 2235–2240. Association for Computational Linguistics (2019)

Self-supervised Multimodal Representation Learning for Product Identification and Retrieval

Yiquan Jiang[1](✉), Kengte Liao[2](✉), Shoude Lin[2](✉), Hongming Qiao[1], Kefeng Yu[1], Chengwei Yang[1], and Yinqi Chen[3]

[1] Research Institute of China Telecom, Guangzhou 510630, China
{jiangyq6,qiaohm,yukef,yangcw2}@chinatelecom.cn
[2] Department of Computer Science and Information Engineering, National Taiwan University, Taipei 106319, Taiwan
{d05922001,sdlin}@csie.ntu.edu.tw
[3] Department of Information Science, University of Arkansas at Little Rock, Little Rock, AR 72204, USA

Abstract. Solving object similarity remains a persistent challenge in the field of data science. In the context of e-commerce retail, the identification of substitutable and similar products involves similarity measures. Leveraging the multimodal learning derived from real-world experiences, humans can recognize similar products based solely on their titles, even in cases where significant literal differences exist. Motivated by this intuition, we propose a self-supervised mechanism that extracts strong prior knowledge from product image-title pairs. This mechanism serves to enhance the encoder's capacity for learning product representations in a multimodal framework. The similarity between products can be reflected by the distance between their respective representations. Additionally, we introduce a novel attention regularization to effectively direct attention toward product category-related signals. The proposed model exhibits wide applicability as it can be effectively employed in unimodal tasks where only free-text inputs are available. To validate our approach, we evaluate our model on two key tasks: product similarity matching and retrieval. These evaluations are conducted on a real-world dataset consisting of thousands of diverse products. Experimental results demonstrate that multimodal learning significantly enhances the language understanding capabilities within the e-commerce domain. Moreover, our approach outperforms strong unimodal baselines and recently proposed multimodal methods, further validating its superiority.

Keywords: Multimodal Learning · Self-Supervised Learning · Product Similarity

1 Introduction

In recent years, the proliferation of products on e-commerce platforms has led to the emergence of applications for product similarity. The identification of similar products involves diverse use cases that yield substantial benefits. For instance, it enables retailers

B. Luo et al. (Eds.): ICONIP 2023, CCIS 1965, pp. 579–594, 2024.
https://doi.org/10.1007/978-981-99-8145-8_44

to monitor how competitors adjust prices for competing products over time, facilitating timely optimization of their product pricing strategies. Additionally, a high-quality similarity system must address the ranking problem. By quantifying the similarity between different products, the e-commerce platform can effectively classify and aggregate products based on varying granularities, forming cohesive product categories. Therefore, identifying substitutable products and measuring the similarity between products play pivotal roles in driving important applications in the e-commerce ecosystem.

Product similarity matching poses a formidable challenge, primarily attributed to the pronounced degree of product heterogeneity, scarcity of labeled data, and variable data quality. Prior studies [1–3] have often approached product matching as a binary classification problem, yielding promising outcomes in fine-grained matching tasks. However, these approaches commonly necessitate a substantial number of labeled structured data, resulting in significant resource allocation for data annotation. It is noteworthy that while certain sellers furnish comprehensive and structured descriptions, the majority solely provide images and titles. In this study, our primary focus lies in extracting low-dimensional representations for products from such unstructured data in an unsupervised learning paradigm. Leveraging these discriminative representations, we can apply similarity measures to establish product similarities, facilitating streamlined and scalable solutions for product matching and retrieval.

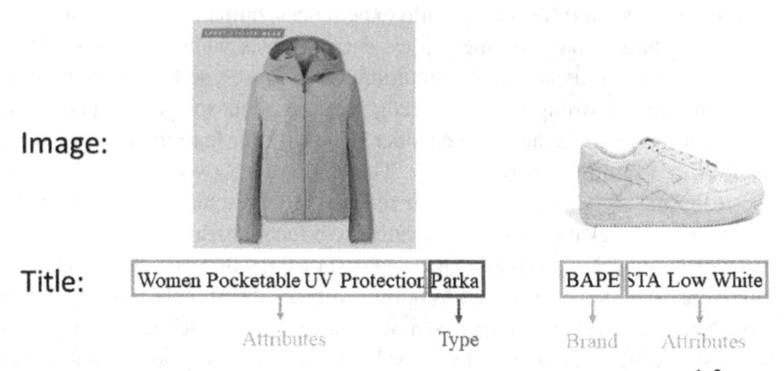

Fig. 1. Examples of images and titles of e-commerce products[1,2]

A product title serves as a sequence of free text, comprised of tokens that convey information about the product's type, brand, color, and other attributes (refer to Fig. 1). These tokens exhibit varying degrees of influence when measuring product similarity. While tokens indicating the product type often facilitate product identification, certain product titles may lack such specific tokens. For instance, consider the product title "BAPE STA Low White", which poses challenges to identify the product literally. Moreover, diverse sellers within the e-commerce domain employ varying terminology. For instance, both "sneakers" and "running shoes" refer to the same products. In the absence of additional

[1] UV Protection Pocketable Parka. [online]. Available from: https://image.uniqlo.com/UQ/ST3/WesternCommon/imagesgoods/419912/item/goods_03_419912.jpg.

[2] BAPE STA Low White Leather. [online]. Available from: https://cdn.modesens.com/product/24628313_69.

supervision signals, models face challenges in quantifying product similarity. However, images depicting products within the same category frequently exhibit noticeable visual similarities, enabling them to serve as valuable cues for assessing the similarity between products. Consequently, incorporating images as a source of supervision holds the potential to enhance the model's understanding of specialized terminology prevalent in the e-commerce domain.

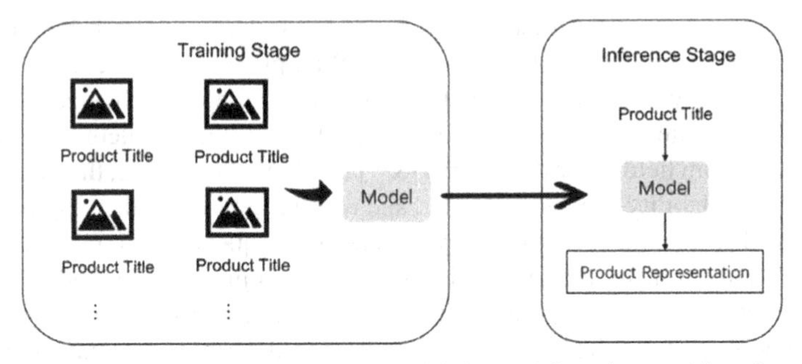

Fig. 2. Working flow of our approach. Our model is learned from image-title pairs, while only titles are required during inference.

With the recent advancements in multimodal models, researchers [4, 5, 13] have explored the utilization of images and texts to learn representations for measuring product similarity. However, most existing methods focus on building joint multimodal representations, rendering them unsuitable for text-only unimodal scenarios. In cross-platform application scenarios, product matching and retrieval predominantly rely on textual information. To address this, our study proposes a novel multimodal approach for learning product representations, allowing the same model to handle scenarios with solely textual data, thus expanding its applicability to a wider range of tasks. The working flow is depicted in Fig. 2. The similarity between two products is quantified using the cosine distance between their respective representations. Our experimental results demonstrate that incorporating image signals enhances the model's language understanding capabilities, thereby leading to improved performance on unimodal tasks.

The main contributions of this paper can be summarized as follows:

1. We propose a self-supervised model that learns a discriminative representation for product similarity from image-title pairs. Importantly, the same model can be applied to unimodal downstream tasks where only free-text inputs are available.
2. We introduce a self-supervision module that leverages hierarchical clustering algorithms to extract prior knowledge related to item categories from multimodal data. During encoder training, the acquired prior knowledge is utilized to enhance the model's ability to embed key information about product categories.
3. We propose a novel attention regularization term that utilizes the prior knowledge generated by the self-supervision module to further reinforce the attention mechanism, leading to a significant improvement in the model's performance.

2 Related Work

This section presents previous related work on product similarity and multimodal representation learning.

2.1 Product Similarity

A substantial portion of the literature in product similarity research has focused on text-based data, particularly in the context of online platforms where product descriptions are commonly presented as free text. This has prompted investigations into the extraction of attribute-value pairs for product matching. Although methods based on the conditional random field algorithm (CRF) [8, 9] have shown promise, their effectiveness relies on the availability of well-annotated and diverse datasets. Other approaches have explored the direct extraction of features from textual sequences. For instance, previous works [2, 3] have utilized bidirectional long short-term memory networks (Bi-LSTM) [35] to extract features from product titles for matching purposes. However, many of these methodologies require additional classifiers to determine the matching status of two products. Moreover, the generation of numerous candidate matching pairs poses significant computational challenges when dealing with large datasets.

In recent years, pre-trained language models [10, 11] have made significant strides across a wide array of tasks. For entity matching, [1, 12] utilize pre-trained Transformer-based language models to enhance language understanding capabilities. Moreover, [1] developed an advanced blocking technique to reduce the number of candidate matching pairs. While these approaches have shown success, they often rely on large amounts of labeled training data. More recently, several studies have sought to overcome this challenge through the utilization of multimodal architectures and weakly supervised learning. [13] exploited images and attributes as inputs for their model, leveraging attribute-aware embedding spaces and attention mechanisms capable of localizing relevant regions of interest to derive fine-grained fashion similarities. Building upon this foundation, [5] extended the scope of their work to incorporate other product catalog data, employing raw catalog data for weakly supervised product representation learning.

Our proposed model distinguishes itself from previous approaches in the following ways: (1) Product representations are learned from multimodal data in an unsupervised setting, (2) product representations can be integrated with vector similarity search for efficient retrieval of similar products, and (3) the model can be applied to unimodal tasks with only unstructured text data as input, thereby enabling its extension to a broader range of application scenarios.

2.2 Multimodal Representation

Multimodal data is considered to be more informative as compared to modality-specific data. Multimodal representation learning has become a popular research problem in the field of natural language processing (NLP). In the context of product similarity, our research objective is to enhance language understanding on downstream tasks by utilizing image-text pairs as a source of supervision for product representation learning.

Studies [14] have demonstrated that incorporating information from linguistic, visual, and auditory modalities can enhance the performance of NLP models.

Fusion-based multimodal approaches incorporate linguistic and visual features through concatenation, weighting, and cross-modal attention to capture complementary information about word semantics. Recurrent fusion networks for multi-view sequential learning, such as audio, video, and language, have been proposed to tackle a range of NLP problems [15, 16]. Another class of methods focuses on integrating multimodal information indirectly using modified loss functions that consider different modalities. [17, 18] utilize the multimodal extended Skip-gram [36] objective function to enhance word representations, achieving better results than unimodal baselines.

In recent years, multimodal transformers have emerged as a general model for learning representations. Inspired by unimodal pre-training schemes such as masked language modeling and masked image modeling, self-supervised multimodal pre-training has become an effective way to train these models [19, 20]. These pre-training objectives typically include unimodal masking prediction, cross-modal masking prediction, and multimodal alignment. Similar to previous fusion-based studies, fusion encoders use a single transformer for multimodal signal fusion [20, 21] or a joint attention transformer layer for cross-attention and cross-modal encoding [19]. In contrast, the indirect fusion method encodes different modalities separately and connects different modalities through the shallow interaction layer of downstream tasks [22, 23]. Although more suitable for unimodal retrieval tasks, such methods lack deeper modality fusion compared to fusion encoders. Recently, more general models have emerged that combine these two types of approaches into an overall model. VATT [24] jointly trains shared models on video, audio, and text data, benefiting a variety of downstream tasks. FLAVA [25] is pre-trained with unpaired unimodal data and image-text pairs, exhibiting strong performance on both unimodal and cross-modal retrieval tasks.

3 Methodology

This study focuses on enhancing the language understanding capabilities of text encoders through self-supervised multimodal learning, with the objective of obtaining discriminative product representations. We present an indirect fusion method employing multi-task learning to integrate multimodal signals. Our framework, depicted in Fig. 3, comprises a text encoder and a self-supervision module. The text encoder consists of a Gated Recurrent Unit (GRU) [37] layer and a self-attention layer, which generate the title embedding, the target multimodal product representation we aim to learn. The title embedding is then passed through two parallel fully connected layer (FC) branches, connected to image feature prediction (L_{Image}) and token ranking prediction (L_{Token}), respectively. The self-supervised module extracts prior knowledge from image-title pairs, which subsequently supervises the training of the text encoder via the proposed attention regularization term ($L_{Regul.}$) and L_{Token}.

We adopt a two-stage training strategy: First, the self-supervision module clusters titles based on image features and subsequently assigns weights to tokens to acquire prior knowledge. Second, this prior knowledge, along with image features, is utilized as supervision signals for self-supervised learning of the text encoder through the aforementioned three losses: $L_{Regul.}$, L_{Image}, and L_{Token}. For the image features required during the

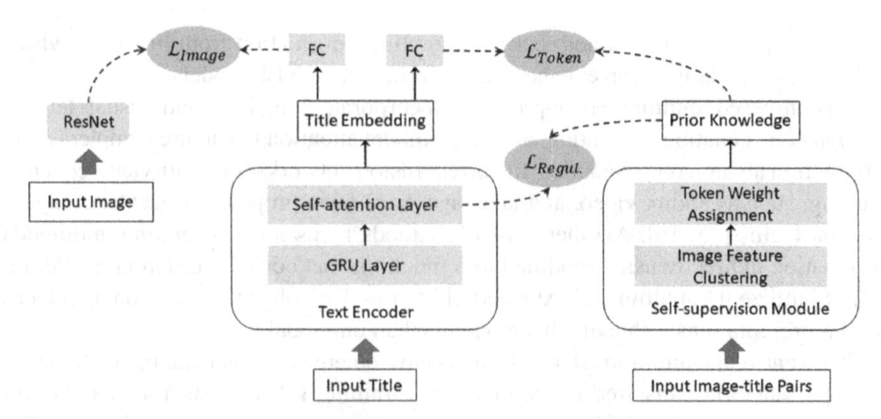

Fig. 3. Overview of our model. The self-supervision module learns prior knowledge from image-title pairs, which in turn supervises the training process of the text encoder. The text encoder transforms product titles into title embeddings, which serve as product representations.

training phase, we utilize the ResNet-50 [26] pre-trained on ImageNet [27], extracting the hidden representation from its last layer (i.e., the layer preceding the Softmax layer) as the image feature vector. During the inference stage, this lightweight text encoder can be applied independently to a wide range of unimodal scenarios. In this section, we will delve into three main aspects of our method: 1) self-attention mechanism, 2) self-supervision module, and 3) multimodal training objectives.

3.1 Self-attention Mechanism

Given that the product title needs to be transformed into a single representation, where each word contributes to varying degrees in identifying the product (e.g., the product's type being a more critical attribute compared to color or size), self-attention [28] can be employed to guide the model's attention towards the token associated with the product category.

The embedding module comprises a GRU layer and a self-attention layer. The tokenized input title is processed using a bidirectional GRU, resulting in the hidden state h_t. This hidden state h_t is obtained by concatenating the hidden states of the GRU in both forward and backward directions. Subsequently, the title can be represented as a matrix H.

$$h_t = \overrightarrow{GRU}(w_t, \overrightarrow{h_{t-1}}) \oplus \overleftarrow{GRU}(w_t, \overleftarrow{h_{t-1}}) \tag{1}$$

$$H = (h_1, h_2, \cdots, h_n) \tag{2}$$

The attention mechanism takes the entire matrix of hidden states H from the GRU as input and performs a nonlinear transformation to produce the weight vector, denoted as the attention vector a, as shown in Eq. 3. This transformation involves the weight matrix W_{s1} and the parameter vector w_{s2}.

$$a = softmax\left(w_{s2} tanh\left(W_{s1} H^T\right)\right) \tag{3}$$

the GRU hidden states H are summed up based on the weights provided by the attention vector a, resulting in a vector representation m of the input title.

$$m = a^T H \tag{4}$$

3.2 Self-supervision Module

The objective of solving the product metric problem is to ensure that representations of similar products are in close proximity, while representations of dissimilar items are farther apart. In the absence of labeled training data, capturing the association between products solely through learning multimodal features from each image-title pair becomes challenging. Although the image features generated by the pre-trained ResNet can offer strong supervision, our dataset encompasses a substantial portion of the product catalogs on the e-commerce platform, leading to noise and variability in the image features that can disrupt the encoder training process.

Nonetheless, products within the same category often exhibit similar images. Leveraging this characteristic, we propose a self-supervision module to capture the associations between products. This module initially clusters the image-title pairs based on the image features generated by the pre-trained ResNet. Subsequently, the tokens within the titles are assigned weights based on their distribution across the clusters. These weights, referred to as prior knowledge, serve as a stronger supervision signal during the training phase of the text encoder. Consequently, the model is empowered to achieve enhanced performance in product recognition and retrieval tasks.

The self-supervision module comprises two steps: image feature clustering and token weight assignment. In the first step, agglomerative clustering is performed on image features to roughly categorize titles according to product category. In the second step, the following two assumptions were made:

Hypothesis 1: Tokens frequently observed within the same cluster are more likely to describe product types and hold greater significance for similarity measures.

Hypothesis 2: Tokens frequently observed in different clusters are more likely to describe common attributes of products and hold less significance for similarity measures.

To assign higher weights to tokens describing product types, as stated in Hypothesis 1, each token is assigned a weight using an equation similar to the concept of TF-IDF. Specifically, for a token t_i observed in cluster c_j, the left term of Eq. 5 calculates the frequency of t_i in c_j, where $n_{i,j}$ denotes the number of times t_i occurs in c_j. The right term corresponds to the logarithmically scaled inverse fraction of clusters that contain t_i. This is computed by dividing the total number of clusters (referred to as $|C|$) by the number of clusters that contain t_i, and then taking the logarithm of this quotient.

$$s_{i,j} = \frac{n_{i,j}}{\sum_k n_{k,j}} \times lg \frac{|C|}{|\{j : t_i \in c_j\}|} \tag{5}$$

In this manner, the self-supervision module assigns weights to tokens within each cluster based on these hypotheses. The weights of tokens within the same cluster remain unaffected by the diversity of titles to which they belong. As illustrated in Fig. 4, a selection of clusters is randomly chosen, and the top ten tokens with the highest weights in descending order are listed. It is evident that these three clusters correspond to the product categories of jeans, mouse, and facial cleansers, respectively. The tokens with the highest weights in these clusters typically reflect the product category, while subsequent tokens with higher weights, such as "Lee" and "Logitech," often denote the brand or other important attributes specific to the product category. The prior knowledge conveyed by these token weights serves as an additional source of supervision.

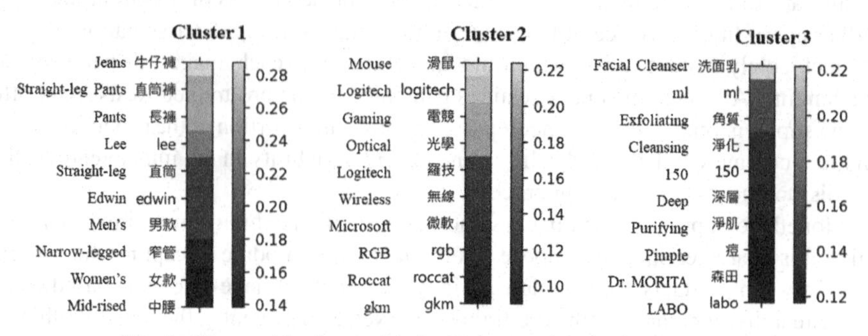

Fig. 4. The top 10 tokens with the highest weight in three clusters.

3.3 Multimodal Training Objectives

During the training phase of the text encoder, the image features and prior knowledge serve as sources of supervision to enhance the encoder's language understanding capability. Drawing from these supervision signals, we employ multimodal signal fusion on text encoders through multitask learning. In particular, we introduce a novel attention regularization term that encourages the model to attend to product category-related signals within titles, exploiting the prior knowledge. The training process of the text encoder encompasses the following multimodal training objectives.

Image Feature Prediction. Previous work [29] has demonstrated the efficacy of learning language-to-vision mappings as a straightforward approach to constructing multimodal representations. In our approach, we facilitate this process by mapping the title embedding, generated by the text encoder, onto the same space as the image feature using a fully connected layer. Subsequently, we compute the loss between the predicted image feature $\hat{\mathbf{y}}$ and the target image feature $\mathbf{y}$ via the mean squared error (MSE).

$$L_{Image}(\mathbf{y}, \hat{\mathbf{y}}) = MSE(\mathbf{y}, \hat{\mathbf{y}}) \tag{6}$$

Attention Regularization. As mentioned in Sect. 3.1, the weights in the attention vector reflect the significance of respective tokens in constructing product representations. It is desirable for the ranking of attention weights to align with the ranking of token weights in the prior knowledge. To accomplish this, we employ a listwise loss function, derived from the learning to rank [30], as a means to regularize attention based on prior knowledge. Specifically, for a given input title **t**, the attention vector, the attention vector $\boldsymbol{a}$ is obtained from the self-attention mechanism (Eq. 7). The top one probability P_{a_i}, which indicates the likelihood of the attention weight a_i being ranked first in the attention vector $\boldsymbol{a}$, is computed using Eq. 8. Similarly, for the corresponding token t_i in the prior knowledge, the top one probability P_{s_i} of its prior weight s_i in the prior weight vector s_i can be calculated (Eq. 9).

$$\boldsymbol{a} = (a_1, a_2, \cdots, a_n) \tag{7}$$

$$P_{a_i} = \frac{exp(a_i)}{\sum_{k=1}^{n} exp(a_k)} \tag{8}$$

$$P_{s_i} = \frac{exp(s_i)}{\sum_{k=1}^{n} exp(s_k)} \tag{9}$$

The attention regularization term is formulated as the cross-entropy between the top one probability distribution of the attention vector and the top one probability distribution of the prior weights.

$$L_{Regul.}(\boldsymbol{s}, \boldsymbol{a}) = -\sum\nolimits_{k=1}^{n} P_{s_k} log\left(P_{a_k}\right) \tag{10}$$

Token Ranking Prediction. Previous studies [31] have demonstrated that autoencoders are effective in capturing intra-sentence information by reconstructing input sentences. In the token ranking prediction task, our goal is not only to capture intra-sentence information but also to capture the association information between products under the guidance of prior knowledge. In this task's downstream branch, the title embedding is transformed into a vector of token scores via a fully connected layer, with the size of this vector determined by the vocabulary size. Similarly, the vector of corresponding token weights in prior knowledge is padded with zeros to match the dimensions of the vocabulary space. As depicted in Eq. 11, the loss term, utilizing the padded vector of prior weights as the target, employs the listwise loss function to quantify the listwise loss between this target $\boldsymbol{c}$ and the output token score vector $\boldsymbol{u}$.

$$L_{Token}(\boldsymbol{c}, \boldsymbol{u}) = -\sum\nolimits_{k=1}^{|V|} P_{c_k} log\left(P_{u_k}\right) \tag{11}$$

The final multimodal training objective of the text encoder consists of the three aforementioned loss functions:

$$L = \lambda_1 L_{Image} + \lambda_2 L_{Regul.} + \lambda_3 L_{Token}, \tag{12}$$

where we assign weights λ_1, λ_2, and λ_3 to the respective loss functions as 0.5, 1, and 0.5.

4 Experiments

4.1 Datasets

The dataset employed for model training comprises 0.55M products sourced from the five most prominent online retail platforms in Taiwan. These products exhibit an extensive spectrum of categories, encompassing the prevalent product catalogs found across diverse e-commerce platforms. Each product item within the dataset is accompanied by a corresponding image and a title in the Traditional Chinese language. Additionally, we apply a filtering process to eliminate items with duplicate titles, thereby ensuring that all titles within the dataset are unique.

4.2 Evaluation Data and Metrics

We evaluate the representation of products in two distinct applications: 1) Identical product classification and 2) Similar product retrieval.

Identical Product Classification (Scenario 1). We collaborated with experienced personnel from e-commerce platforms to collect product pairs and manually assign labels to them. Our evaluation dataset encompasses 10k labeled product pairs, with an equal distribution of positive and negative labels. A positive label indicates that the product pair is identical, while a negative label suggests otherwise. To ascertain the degree of similarity, we leverage cosine similarity measurements between the two product representations and utilize a predefined threshold to determine whether the products are identical. The evaluation of this task incorporates four fundamental metrics: accuracy, precision, recall, and F1.

Similar Product Retrieval (Scenario 2). For product retrieval evaluation, the dataset comprises 163 product groups, each consisting of identical products. We establish a hierarchical, tree-like category structure where higher-level nodes represent coarse-grained categories, and the leaves correspond to product groups. Manual labeling is performed for each group, based on the defined category structure. To assess the quality of retrieval, we employ the Normalized Discounted Cumulative Gain (nDCG) as an evaluation metric. The Discounted Cumulative Gain (DCG) quantifies the ranking quality of a list by accumulating the relevance of the top K search items while considering penalties according to their respective ranks.

$$DCG_K = \sum_{i=1}^{K} \frac{rel_i}{\log_2(i+1)} \tag{13}$$

$$nDCG_K = \frac{DCG_K}{IDCG_K} \tag{14}$$

Here, rel_i represents the graded relevance of the item ranked at position i. The graded relevance between any two products within the dataset is determined by evaluating the distance between the nodes to which the products belong within the category structure. The nDCG metric is calculated as the normalization of DCG by the ideal DCG value (IDCG). The calculation method for IDCG is similar to that of DCG, as it represents the DCG under ideal conditions. Therefore, for each item in the testing set, we can employ the k-nearest neighbors algorithm to search for its top K nearest neighbors and calculate the nDCG.

4.3 Evaluation Results

We evaluate our model and the following representation learning models on two scenarios: identical product classification and similar product retrieval.

Unimodal Baselines. Regarding image representations, we adopt a pre-trained Resnet to extract image features from the evaluation dataset, following the same image feature extraction process utilized during the training phase. These extracted image features serve as the product representations. On the linguistic side, we employ two simple yet effective language models, namely TF-IDF and FastText [32]. Furthermore, we compare the performance of our model against the widely-utilized pre-trained language model BERT [11]. The 12-layer BERT weights are initialized by a Chinese pre-trained model and subsequently fine-tuned on our product title corpus using the Masked Language Modeling. The vector corresponding to the [CLS] token in the model's sequence output is employed as the product representation.

Multimodal Baselines. We conduct a comparison with two recently proposed models: Chinese CLIP [33] and FLAVA [25]. Chinese CLIP is an implementation of CLIP [22] trained on a large-scale dataset comprising approximately 200 million Chinese image-text pairs. Both Chinese CLIP and FLAVA encode images and text separately, making them suitable for unimodal scenarios involving only text. In our experimental setup, we fine-tune the pre-trained models of Chinese CLIP and FLAVA using product image-title pairs. The fine-tuning objectives align with the pre-training objectives described in their respective original papers. Since the pre-trained FLAVA model is trained on an English corpus, we replace its text encoder with a pre-trained Chinese RoBERTa model [34]. During the evaluation, we extract the hidden state vector corresponding to the [CLS] token in the language sequences from the text encoders of both models, utilizing them as product representations.

Table 1. The performance of baseline models and our model (Ours) on the identical product classification (Scenario 1) and similar product retrieval (Scenario 2).

Model	Scenario 1				Scenario 2
	Accuracy	Precision	Recall	F1	nDCG
ResNet	0.675	0.697	0.62	0.657	-
TF-IDF	0.727	0.739	0.702	0.72	0.739
FastText	0.7	0.679	0.757	0.716	0.673
BERT	0.825	0.839	0.805	0.822	0.752
ChineseCLIP	0.762	0.806	0.691	0.744	0.797
FLAVA	0.816	0.809	0.828	0.818	0.789
Ours	**0.85**	**0.855**	**0.843**	**0.849**	**0.835**

Table 1 presents the evaluation results of the aforementioned comparison methods, alongside the performance of our model, across two distinct scenarios about product similarity problems. The first four rows of the table correspond to unimodal models, whereas the remaining rows represent multimodal approaches. The results for ResNet in Scenario 2 are currently unavailable due to the absence of image data in the testing data.

Overall, our model outperforms the aforementioned methods in both scenarios. It is worth noting that in Scenario 2, the multimodal methods exhibit significantly better performance than the unimodal models. This disparity suggests that incorporating image signals during training can enhance the performance of multimodal models for the task of product title retrieval.

4.4 Ablation Studies

Table 2. Results of ablation study on two scenarios. L_{Image}: Image feature prediction loss, $L_{Regul.}$: Attention regularization, Self-super: Self-supervision module, Hypo. 1: Hypothesis 1, Hypo. 2: Hypothesis 2, w/o: without.

Model	Scenario 1				Scenario 2
	Accuracy	Precision	Recall	F1	nDCG
w/o L_{Image}	0.817	0.824	0.805	0.815	0.696
w/o $L_{Regul.}$	0.825	0.824	0.828	0.826	0.807
w/o Self-super.	0.763	0.767	0.755	0.761	0.798
w/o Hypo. 1	0.812	**0.858**	0.748	0.799	0.719
w/o Hypo. 2	0.829	0.849	0.8	0.824	0.812
Ours	**0.85**	0.855	**0.843**	**0.849**	**0.835**

We conducted ablation experiments to assess the importance of different components within our proposed framework. The experimental outcomes are presented in Table 2. To validate the necessity of a self-supervision module, we performed the third experiment where prior knowledge was not utilized as a supervision signal during encoder training. Specifically, we removed the attention regularization loss term and replaced the downstream task of token ranking prediction with token probability prediction. This task can be seen as a multi-label classification problem, with the loss computed using binary cross-entropy. By comparing the results of the first three experiments with the complete model (row 6), it is evident that the removal or modification of the loss term leads to a significant decrease in performance.

Moreover, ablation experiments were performed to examine two hypotheses in the self-supervision module, namely Hypothesis 1 and Hypothesis 2. As indicated in Sect. 3.2, these hypotheses govern the computation of token prior weights, representing the within-cluster token frequency term and the inverse cluster frequency term, respectively. To verify each hypothesis, the corresponding frequency term in the equation for prior weight calculation was set to a constant value of 1. The results of these experiments are presented in the fourth and fifth rows of Table 2. Notably, the models that exclude Hypothesis 1 demonstrate a more pronounced decrease in performance across both scenarios. This observation highlights the substantial impact of Hypothesis 1 compared to Hypothesis 2.

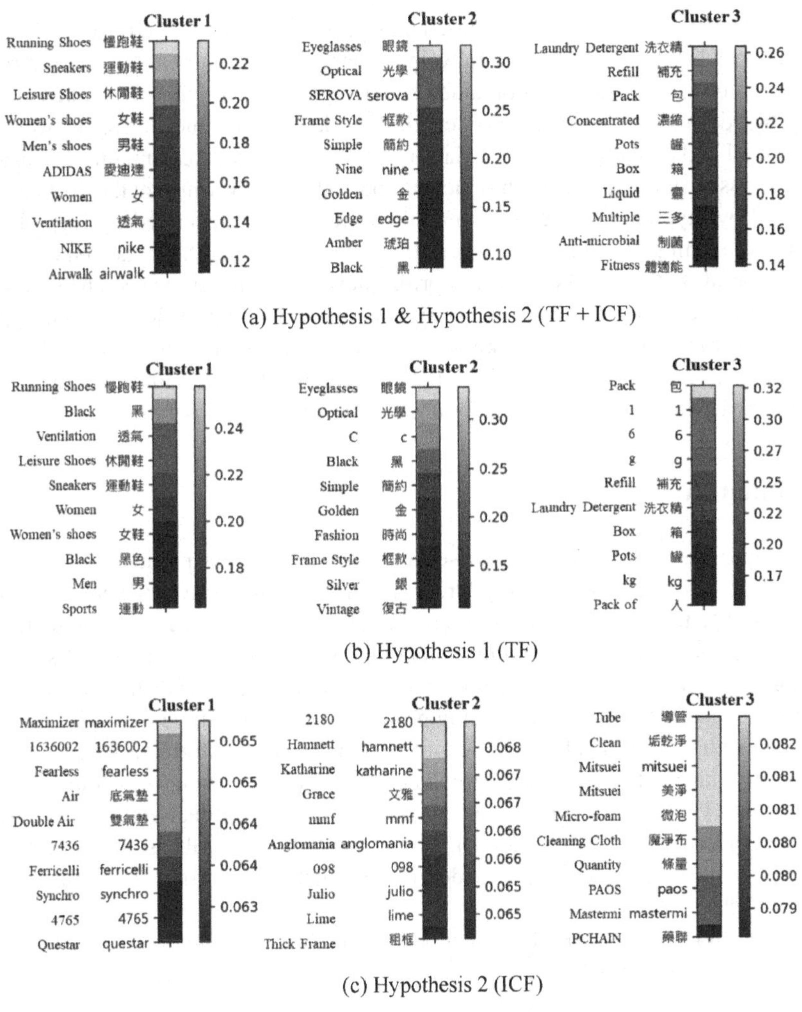

Fig. 5. The top 10 tokens with the highest prior weights in the three clusters under various hypotheses.

To further investigate the validity of these two hypotheses, we randomly selected three clusters from the output of the self-supervision module and visualized the token rankings based on their weights within each cluster. Figure 5 presents the top 10 tokens with the highest prior weights in each cluster for the complete model (TF + ICF), as well as the models utilizing only Hypothesis 1 (TF) or Hypothesis 2 (ICF). Notably, the complete model demonstrated token rankings that aligned with our expectations. This alignment can be inferred from the highest weighted tokens, which corresponded to running shoes, eyeglasses, and laundry detergent, indicating that a significant portion of the products within these clusters belonged to these respective categories. Additionally, other highly ranked tokens often represented brand names or other significant attributes associated with specific product categories.

Regarding tokens ranked solely based on Hypothesis 1, we noticed that while tokens implying product types received high weights, certain less significant product attributes such as color and volume also obtained relatively higher weights. For tokens ranked solely based on Hypothesis 2, we observed that the weight values of tokens were generally small and exhibited limited variation. The top-ranked tokens mainly consisted of meaningless attribute values like product numbers. It is noteworthy that the token weights produced by Hypothesis 2 serve to modulate the weights generated by Hypothesis 1 thereby weakening the weights of tokens that refer to less significant attributes.

In general, both image signals and prior knowledge can serve as sources of supervision during encoder training. Image signals primarily impact downstream applications about ranking, while prior knowledge influences both recognition and ranking applications. The token weights within the prior knowledge are primarily influenced by hypothesis 1, with hypothesis 2 predominantly serving a regulatory role.

5 Conclusion

In this paper, we present a self-supervised approach for learning product representations from image-title pairs, aiming to address the product similarity problem. Our model leverages only text as input during inference, enabling more efficient computations in practical applications and extending its applicability to a broader range of scenarios. Through a thorough analysis of e-commerce product data, we observe the complementary nature of images and titles in describing products. We demonstrate that self-supervised learning mechanisms can effectively capture the association information between products from image-title pairs, thereby enhancing the model's language comprehension capabilities. The experiments conducted on two real-world datasets demonstrate the superior performance of our model compared to strong unimodal baselines and recently proposed multimodal models in both identical product classification and similar product retrieval tasks.

The training dataset used for our model comprises real online data acquired from multiple shopping websites, presenting an imbalanced category distribution. This inherent imbalance among product categories can lead to varying performance across categories. Addressing the challenge of imbalanced product categories represents a realistic problem for future investigation. Moreover, possible future directions of the current work include parameterizing the self-supervision module and integrating it with the text encoder to form a joint optimization framework, which is anticipated to enhance the overall performance.

References

1. Li, Y., Li, J., Suhara, Y., et al.: Deep entity matching with pre-trained language models. Proc. VLDB Endowment **14**(1), 50–60 (2020)
2. Shah, K., Kopru, S., Ruvini, J.D.: Neural network based extreme classification and similarity models for product matching. In: Proceedings of the 2018 Conference of the North American Chapter of the Association for Computational Linguistics: Human Language Technologies, vol. 3 (Industry Papers), pp. 8–15 (2018)

3. Li, J., Dou, Z., Zhu, Y., et al.: Deep cross-platform product matching in e-commerce. Inf. Retrieval J. **23**, 136–158 (2020)
4. Ristoski, P., Petrovski, P., Mika, P., et al.: A machine learning approach for product matching and categorization. Seman. Web **9**(5), 707–728 (2018)
5. Das, N., Joshi, A., Yenigalla, P., Agrwal, G.: MAPS: multimodal attention for product similarity. In: Proceedings of the IEEE/CVF Winter Conference on Applications of Computer Vision, pp. 3338–3346 (2022)
6. Petrovski, P., Bryl, V., Bizer, C.: Integrating product data from websites offering microdata markup. In: Proceedings of the 23rd International Conference on World Wide Web, pp. 1299–1304 (2014)
7. Köpcke, H., Thor, A., Thomas, S., Rahm, E.: Tailoring entity resolution for matching product offers. In: Proceedings of the 15th International Conference on Extending Database Technology, pp. 545–550 (2012)
8. Melli, G.: Shallow semantic parsing of product offering titles (for better automatic hyperlink insertion) In: Proceedings of the 20th ACM SIGKDD International Conference on Knowledge Discovery and Data Mining, pp. 1670–1678 (2014)
9. Ristoski, P., Mika, P.: Enriching product ads with metadata from HTML annotations. In: Sack, H., Blomqvist, E., d'Aquin, M., Ghidini, C., Ponzetto, S., Lange, C. (eds.) The Semantic Web. Latest Advances and New Domains. ESWC 2016. LSCS, vol. 9678, pp. 151-167. Springer, Cham (2016). https://doi.org/10.1007/978-3-319-34129-3_10
10. Radford, A., Wu, J., Child, R., et al.: Language models are unsupervised multitask learners. OpenAI blog **1**(8), 9 (2019)
11. Devlin, J., Chang, M.W., Lee, K.: BERT: pre-training of deep bidirectional transformers for language understanding. arXiv preprint arXiv:1810.04805 (2018)
12. Brunner, U., Stockinger, K.: Entity matching with transformer architectures-a step forward in data integration. In: 23rd International Conference on Extending Database Technology, Copenhagen, 30 March-2 April 2020. OpenProceedings (2020)
13. Dong, J., Ma, Z., Mao, X., et al.: Fine-grained fashion similarity prediction by attribute-specific embedding learning. IEEE Trans. Image Process. **30**, 8410–8425 (2021)
14. Zadeh, A., Liang, P.P., Morency, L.P.: Foundations of multimodal co-learning. Inf. Fusion **64**, 188–193 (2020)
15. Zadeh, A., Liang, P.P., Mazumder, N., Poria, S., Cambria, E., Morency, L.P.: Memory fusion network for multi-view sequential learning. In: Proceedings of the AAAI Conference on Artificial Intelligence, vol. 32, no. 1 (2018)
16. Liang, P.P., Liu, Z., Zadeh, A., Morency, L.P.: Multimodal language analysis with recurrent multistage fusion. arXiv preprint arXiv:1808.03920 (2018)
17. Lazaridou, A., Pham, N.T., Baroni, M.: Combining language and vision with a multimodal skip-gram model. arXiv preprint arXiv:1501.02598 (2015)
18. Zablocki, E., Piwowarski. B., Soulier. L., Gallinari, P.: Learning multi-modal word representation grounded in visual context. In: Proceedings of the AAAI Conference on Artificial Intelligence, vol. 32, no. 1 (2018)
19. Lu, J., Batra, D., Parikh, D.: ViLBERT: pretraining task-agnostic visiolinguistic representations for vision-and-language tasks. In: Advances in Neural Information Processing Systems, vol. 32 (2019)
20. Su, W., et al.: Vl-BERT: pre-training of generic visual-linguistic representations. arXiv preprint arXiv:1908.08530 (2019)
21. Chen, Y.-C., et al.: UNITER: UNiversal Image-TExt representation learning. In: Vedaldi, A., Bischof, H., Brox, T., Frahm, J.-M. (eds.) ECCV 2020. LNCS, vol. 12375, pp. 104–120. Springer, Cham (2020). https://doi.org/10.1007/978-3-030-58577-8_7
22. Radford, A., et al.: Learning transferable visual models from natural language supervision. In: International Conference on Machine Learning, pp. 8748–8763. PMLR (2021)

23. Jia, C., et al.: Scaling up visual and vision-language representation learning with noisy text supervision. In: International Conference on Machine Learning, pp. 4904–4916. PMLR (2021)
24. Akbari, H., Yuan, L., Qian, R., et al.: VATT: transformers for multimodal self-supervised learning from raw video, audio and text. Adv. Neural Inf. Process. Syst. **34**, 24206–24221 (2021)
25. Singh, A., et al.: FLAVA: a foundational language and vision alignment model. In: Proceedings of the IEEE/CVF Conference on Computer Vision and Pattern Recognition, pp. 15638–15650 (2022)
26. He, K., Zhang, X., Ren, S., Sun, J.: Deep residual learning for image recognition. In: Proceedings of the IEEE Conference on Computer Vision and Pattern Recognition, pp. 770–778 (2016)
27. Deng, J., Dong, W., Socher, R., Li, L.J., Li, K., Fei-Fei, L.: ImageNet: a large-scale hierarchical image database. In: 2009 IEEE Conference on Computer Vision and Pattern Recognition, pp. 248–255. IEEE (2009)
28. Lin, Z., et al.: A structured self-attentive sentence embedding. arXiv preprint arXiv:1703.03130 (2017)
29. Collell, G., Zhang, T., Moens, M.F.: Imagined visual representations as multimodal embeddings. In: Proceedings of the AAAI Conference on Artificial Intelligence, vol. 31, no. 1 (2017)
30. Cao, Z., Qin, T., Liu, T.Y., Tsai, M.F., Li, H.: Learning to rank: from pairwise approach to listwise approach. In: Proceedings of the 24th International Conference on Machine Learning, pp. 129–136 (2007)
31. Gan, Z., Pu, Y., Henao, R., Li, C., He, X., Carin, L.: Learning generic sentence representations using convolutional neural networks. In: Proceedings of the 2017 Conference on Empirical Methods in Natural Language Processing, pp. 2390-2400 (2017)
32. Mikolov, T., Grave, E., Bojanowski, P., Puhrsch, C., Joulin, A.: Advances in pre-training distributed word representations. In: Proceedings of the Eleventh International Conference on Language Resources and Evaluation (LREC 2018) (2018)
33. Yang, A., et al.: Chinese CLIP: contrastive vision-language pretraining in Chinese. arXiv preprint arXiv:2211.01335 (2022)
34. Cui, Y., Che, W., Liu, T., Qin, B., Wang, S., Hu, G.: Revisiting pre-trained models for Chinese natural language processing. In: Findings of the Association for Computational Linguistics: EMNLP 2020, pp. 657–668 (2020)
35. Schuster, M., Paliwal, K.K.: Bidirectional recurrent neural networks. IEEE Trans. Sig. Process. **45**(11), 2673–2681 (1997)
36. Mikolov, T., Chen, K., Corrado, G., Dean, J.: Efficient estimation of word representations in vector space. arXiv preprint arXiv:1301.3781 (2013)
37. Chung, J., Gulcehre, C., Cho, K., Bengio, Y.: Empirical evaluation of gated recurrent neural networks on sequence modeling. arXiv preprint arXiv:1412.3555 (2014)

Author Index

B. Luo et al. (Eds.): ICONIP 2023, CCIS 1965, pp. 595–597, 2024.
https://doi.org/10.1007/978-981-99-8145-8

Printed in the United States
by Baker & Taylor Publisher Services